30

Handbook of GROWTH FACTORS

Volume II: Peptide Growth Factors

Handbook of Growth Factors
Enrique Pimentel, M.D.

Volume I
General Basic Aspects

Regulation of Cell Functions
Growth Factor Receptors
Postreceptor Mechanisms of Growth Factor Action
Cyclic Nucleotides
Guanosine Triphosphate-Binding Proteins
The Calcium-Calmodulin System
Phosphoinositide Metabolism
Protein Phosphorylation
Proto-Oncogene and Onco-Suppressor Gene Expression
Role of Growth Factors in Neoplastic Processes

Volume II
Peptide Growth Factors

Insulin
Insulin-Like Growth Factors
Epidermal Growth Factor
Fibroblast Growth Factors
Neurotrophic Growth Factors
Organ-Specific Growth Factors
Cell-Specific Growth Factors
Transforming Growth Factors
Regulatory Peptides with Growth Factor-Like Properties

Volume III
Hematopoietic Growth Factors and Cytokines

Hematopoietic Growth Factors
Interleukins and Cytokines
Colony-Stimulating Factors
Interferons
Tumor Necrosis Factors
Erythropoietic Growth Factors
Platelet-Derived Growth Factor
Transferrins

Handbook of
GROWTH FACTORS
Volume II: Peptide Growth Factors

Enrique Pimentel, M.D.
National Center of Genetics
Institute of Experimental Medicine
Central University of Venezuela
Caracas, Venezuela

CRC Press
Boca Raton Ann Arbor London Tokyo

Library of Congress Cataloging-in-Publication Data

Pimentel, Enrique.
Handbook of growth factors.
Includes bibliographical references and index.
Contents: v. 1. General basic aspects. —
v. 2. Peptide growth factors — v. 3. Hematopoietic
growth factors and cytokines.
1. Growth factors. I. Title. [DNLM: 1. Growth
Substances. QU 100 P644h 1994]
QP552.G76P56 1994 612.6 93-40108

International Standard Book Number 0-8493-2505-6 (Volume I)
International Standard Book Number 0-8493-2506-4 (Volume II)
International Standard Book Number 0-8493-2507-2 (Volume III)
Library of Congress Card Number 93-40108

Printed in the United States of America 2 3 4 5 6 7 8 9 0
Printed on acid-free paper

CONTENTS

Chapter 1
Insulin

Chapter 2

Insulin-like Growth Factors

Chapter 3

Epidermal Growth Factor

Chapter 4

Fibroblast Growth Factors

Chapter 6
Organ-Specific Growth Factors

Chapter 7
Cell-Specific Growth Factors

Chapter 8
Transforming Growth Factors

Chapter 9

Regulatory Peptides with Growth Factor-Like Properties

THE AUTHOR

Enrique Pimentel, M.D., is Professor of General Pathology and Pathophysiology at the School of Medicine, Central University of Venezuela, Caracas. He was formerly Director of the Institute of Experimental Medicine at the same university and founded and directed the National Center of Genetics in Venezuela.

Born on April 7, 1928, in Caracas, Venezuela, he obtained an M.D. degree from the Universities of Madrid, Spain, and Caracas, Venezuela, in 1953. He is President of the National Academy of Medicine in Venezuela and an honorary, corresponding, or active member of 32 national and international scientific academies and societies. He is an active member of the New York Academy of Sciences and Vice President of the International Academy of Tumor Marker Oncology (IATMO). He has received several decorations in his own country and the Grosse Verdienstkreuz (Great Cross to the Merit) of the Federal Republic of Germany. In 1982 he received the National Award of Science in Venezuela. On many occasions Dr. Pimentel has been invited to give lectures and seminars at universities and other scientific institutions in North America and Europe.

Dr. Pimentel is the author of more than 100 papers and co-author of seven books on topics related to endocrinology, genetics, and oncology. In addition, he is the author of *Hormones, Growth Factors, and Oncogenes* and *Oncogenes* (first edition in one volume, second edition in two volumes) published by CRC Press. He is editor of the bimonthly journal *Critical Reviews in Oncogenesis.*

Chapter 1

Insulin

I. INTRODUCTION

The islets of Langerhans in the pancreas are composed mainly of three types of cells, termed α or A, β or B, and δ or D. The different cell types synthesize and secrete different hormones. Insulin and glucagon are produced by the β cells and α cells, respectively. The structure and expression of the insulin and glucagon genes is well characterized.[1] The δ cells produce somatostatin, a peptide of 14 amino acids that was initially identified in the hypothalamus but which is also synthesized in the stomach, the intestine, and the pancreas.

Diabetes mellitus is a common human disease that is associated with alterations in the secretion of insulin and other hormones. Cellular and molecular defects occurring at both central (pancreatic) and peripheral (extrapancreatic) sites are implicated in the origin and development of diabetes.[2] Two main types of diabetes exist: type I diabetes (insulin-dependent diabetes) and type II diabetes (insulin-independent or essential diabetes). Type I diabetes is a chronic autoimmune disease characterized by absolute deficiency of insulin due to a specific destruction of the pancreatic β cells.[3] Other endocrine cells of the islets, such as those secreting glucagon and somatostatin, are preserved in type I diabetes. Both viruses and cytokines are implicated in the pathogenesis of type I human diabetes in genetically susceptible individuals.[4] The presence of an amino acid residue other than aspartic acid at position 57 of the human leukocyte antigen (HLA)-DQ β chain is associated with susceptibility to type I diabetes at the worldwide level.[5] Viral infection of the β cell may cause enhanced expression of major histocompatibility complex (MHC) proteins, and the presence of these alterations in genetically susceptible individuals may predispose to the development of anti-β-cell immunity with the infiltration of the islets by activated lymphocytes and macrophages, resulting in the characteristic lesion called insulitis. Locally produced cytokines such as interleukin 1 (IL-1), IL-6, interferon-γ (IFN-γ), and tumor necrosis factor-α (TNF-α) may augment MHC protein expression by the β cells, which may result in their destruction by activated T and NK lymphocytes. The progression of these lesions may eventually reduce the total population of β cells to a point where insulin secretion becomes insufficient to cover the normal requirements of the body, leading to the clinical manifestation of diabetes. In contrast to type I diabetes, type II diabetes usually affects presenile or senile persons and follows a more benign course than type I diabetes. Whereas insulin treatment is usually required for the therapeutic control of type I diabetes, most patients with type II diabetes can be controlled by diet alone. Type II diabetes is attributed to complex etiopathogenetic factors, including both alterations in islet function and peripheral resistance to insulin action. Growth factors may play a role in the derangements of cell proliferation that occur frequently in diabetes, such as mesangial cell hyperplasia and atherosclerosis.[6]

Insulin is an anabolic hormone with a mechanism of action which is closely related to that of the insulin-like growth factor I (IGF-I). It has a wide range of effects on various essential metabolic functions, including stimulation of the synthesis of DNA, RNA, proteins, glycogen, and lipids. In addition, insulin is involved in the control mechanisms of cell proliferation and differentiation both *in vivo* and *in vitro*.[7-9] Insulin and the IGFs have important roles in developmental processes.[10] The structure of insulin in the human and other species is well characterized. Recently, there have been important advances in the structural and functional characterization of insulin receptors as well as in the knowledge of postreceptor mechanisms of insulin action in different types of cells.[11-14]

II. INSULIN STRUCTURE AND BIOSYNTHESIS

Human insulin is a polypeptide composed of 51 amino acids arranged in two chains (A and B), linked by two disulfide bonds. The mature hormone is synthesized in the form of a large precursor, called preproinsulin, which is processed via an intermediate precursor, proinsulin, to the mature insulin molecule. Human proinsulin is composed of 86 amino acid residues, from which residues 1 to 32 correspond to the B chain and residues 64 to 86 to the A chain of the mature insulin molecule. The insulin A chain contains an internal disulfide bridge between residues 71 and 76. The residue 33 to 64 segment in the proinsulin molecule, which joins the A and B chains of the hormone, is referred to as the connecting peptide (C peptide). The C peptide is separated within the β cell from the rest of the molecule by proteolytic cleavage during the processes associated with biosynthesis of the mature hormone.

Insulin is synthesized and secreted only by the β cells (B cells) of the islets of Langerhans in the pancreas. The physiological stimuli related to the development of the endocrine pancreas are little known. The growth hormone may act as a growth factor for differentiated β cells, promoting the proliferation and insulin production of these cells.[15] Prolactin and/or placental lactogen could have a similar role during periods of somatic growth. Many endogenous and exogenous factors are involved in the regulation of islet cell functions. The levels of glycemia are involved in the regulation of insulin secretion by the β cells, and glucose is involved in the regulation of insulin gene expression.[16] A high level of glucose in the blood stimulates insulin release from the β cells. The rate of glucose usage by β cells plays a crucial role in the secretory response of β cells to glucose, but the metabolic intermediates responsible for this effect are unknown.[17] Cyclic adenosine-3′,5′-monophosphate (cAMP) and Ca^{2+}, as well as phospholipid metabolites and protein kinase C, may act as mediators of extracellular factors involved in the regulation of insulin secretion.[18-20] Arachidonic acid may have a function as second messenger in glucose-induced insulin secretion from the β cells.[21] Insulin release from the β cells is stimulated by phospholipase D, suggesting a potential role for endogenous phosphatidic acid in pancreatic islet function.[22]

Functional interactions within the islets include locally produced hormones and growth factors.[23] Somatostatin synthesized within the islets acts in a paracrine fashion as a negative regulator of the secretion of insulin and glucagon. Interleukins also participate in the regulation of insulin and glucagon secretion. IL-1 produced in the islets may influence the rate of insulin secretion.[24] IL-6 produced in the islets can affect the secretion of insulin and the metabolism of glucose in rat pancreatic islets *in vitro*.[25,26] Inhibition of insulin secretion in the rat islets by IL-6 is an effect distinct from that of IL-1.[27] TGF-β may be involved in the local control of insulin secretion.[28] The autonomic nervous system may contribute to regulate the secretion of insulin through the release of epinephrine, acetylcholine, and other hormones and neurotransmitters. Insulin itself is involved in feedback inhibition of insulin gene expression.[29]

There is evidence that insulin is synthesized in small amounts at extrapancreatic sites including the lung, the intestine, and the central nervous system, although these sites would lack the capacity to secrete the hormone.[30-32] However, these results are a subject of controversy. An appropriate ensemble of tests should be applied in order to confirm or reject the possibility of an extrapancreatic synthesis of insulin.[33]

The structural gene for insulin *(INS)* is located on the short arm of human chromosome 11, at region 11p14.1, which corresponds to the same region where the human c-H-*ras*-1 proto-oncogene (*RAS*^{H}) has been mapped.[34-36] Human chromosome 11p also contains the genes for parathormone *(PTH)* and β-globin *(HBG)*. The linear order of genes in human chromosome 11p is cen-*PTH-HBG-RAS*^{H}-*INS*)-pter.[36,37] The gene coding for the β subunit of the follicle-stimulating hormone (FSH) is also located close to this chromosome region.[38] Deletions of human chromosome region 11p13-p14 have been detected in the tumor cells of children with the aniridia-Wilms' tumor association.[39,40] The deletion may result in either hemizygosity or homozygosity of the corresponding allele, the latter being produced by duplication of the respective segment remaining in the unaffected chromosome.[41-44] In some Wilms' tumors one allele of the insulin gene and one allele of the c-H-*ras* proto-oncogene are deleted on the tumor cells.[43,45] Deletion of the same chromosome region (11p13-p14) was also detected in the tumor cells of a human hepatocellular carcinoma in relation to the integration of a hepatitis B virus genome copy.[46] A similar deletion may occur in bladder cancer.[47] The precise biological significance of such partial deletions of chromosome 11 is unknown but some of them may be associated with the loss of tumor suppressor genes.[48]

The complete nucleotide sequence of the human insulin gene has been determined.[49-54] The gene is composed of two exons and two introns and is associated with elements of the *Alu* family of short (<400 bp) interspersed repeated DNA sequences. Tandemly repeating DNA sequences have been identified in the 5′-flanking region of the human insulin gene, with specific allelic sequences showing different frequencies among the human population.[55,56] It has been suggested that particular polymorphisms in the 5′-flanking region of the human insulin gene may be associated with increased risk for non-insulin-dependent diabetes (type II diabetes),[57-60] but this association was not confirmed.[61,62] Population associations between DNA polymorphisms at the 5′-flanking region of the human insulin gene and susceptibility genes for different diseases would be, in any case, too weak to be of practical benefit.[63]

Point mutations of the human insulin gene coding sequences may result in alterations in the primary structure of the hormone and may be associated with rare cases of diabetes mellitus.[64,65] Such cases are described under the generic name of insulinopathies. The diabetes-associated insulin mutations are concentrated at sites that are important for the function of the insulin molecule. A mutation consisting in substitution for phenylalanine at position 25 of the insulin B chain resulted in conformational changes affecting ligand-receptor interaction.[66]

In addition to the insulin gene on chromosome region 11p14.1, two insulin-related DNA sequences — termed hIr1 and hIr2 — have been localized to human chromosomes 2 and 11, respectively.[67] It could not be established whether these two sequences are functional or are nonfunctional remnants of insulin gene family progenitors. While the human genome contains only one functional insulin gene, two structural genes coding for insulin are present in the genome of rat, mouse, and three fish species.[68] The rat preproinsulin gene I contains a single intron and is a functional gene generated by reinsertion into the genome (retroposition) of a cDNA copy of preproinsulin gene II transcripts.[69] In other words, the rat preproinsulin I gene is a functional retroposon. Insulin I and insulin II genes are asyntenic in the mouse but are syntenic in the rat, both being located on chromosome 1.[70] Point mutations close to the AUG initiator codon may affect the efficiency of translation of rat preproinsulin gene II mRNA *in vivo*.[71] The two nonallelic rat insulin genes are regulated coordinately *in vivo*.[72] The transcriptional activity of rat insulin genes depends on the tissue-specific binding of proteins to the promoter region of the genes.[73] A promoter element located –53 to –46 relative to the transcription start site of the rat insulin II gene is necessary for the expression of the gene, and a mutation within this sequence drastically decreases promoter activity *in vivo* and *in vitro*.[74] This DNA sequence binds a protein which is the same transcription factor that binds to the promoter of the chicken ovalbumin gene. However, the binding sequences of the chicken ovalbumin and rat insulin promoters have only limited structural similarities. The insulin I locus of the rat is polymorphic due to the presence or absence of a 2.7-kb repeated element in a region located upstream of the coding sequence.[75] The possible biological significance of DNA polymorphisms due to the insertion of short or long repeated nucleotide sequences is unknown, but alleles of the rat insulin I gene associated with restriction fragment length polymorphisms (RFLPs) in the 5′-flanking region of the gene are not associated with the appearance of spontaneous diabetes in the animal.[76] A negative regulatory element that suppresses enhancer-dependent transcriptional activity was identified in the 5′ region of the rat preproinsulin I gene.[77] This element, called silencer, is a member of a family of long interspersed repetitive sequences (LINES) of the rat genome.

The essential biological importance of insulin is indicated by its presence in all vertebrates tested so far. Insulin-like proteins, or factors with insulin-like activity, are present also in a diversity of nonvertebrate species. Insulin-related proteins have been detected in insects such as *Drosophila melanogaster* and annelids such as the earthworm *Annelida oligocheta*.[78,79] The amino-terminal portion of the prothoracicotropic hormone of the silkworm *Bombyx mori* exhibits significant homology with insulin and IGFs.[80] The curious phenomenon of somatic growth stimulation associated with infections of mice with the plerocercoid stage of the tapeworm, *Spirometra mansonoides,* is due to a factor, termed plerocercoid growth factor (PCF), which exhibits insulin-like activity in rats.[81] Analysis of a cDNA encoding an insulin-like peptide from a primitive chordate species, the amphioxus *(Branchiostoma californiensis)*, shows that the product is a hybrid molecule containing features characteristic of both insulin and IGF.[82] This finding suggests that the IGFs emerged at a very early stage in vertebrate evolution from an ancestral insulin-type gene. Cells of a wall-less strain of the fungus *Neurospora crassa* possess high-affinity insulin binding sites on the surface, and binding of porcine insulin to these sites elicits tyrosine kinase activity.[83] A molecule closely related to insulin is produced by *Escherichia coli* and *Tetrahymena*.[84] The possible role of insulin-like compounds produced by microbes is unknown. In any case, an insulin-related gene has not been detected as yet in microbes, and the results of studies performed with immunologic methods should be checked out by means of techniques working at the nucleic acid level.[33]

Inbred mice strains constitutively expressing in the β cells an endogenous retrovirus, the intracisternal A particle (IAP), may develop severe diabetes mellitus.[85] Glucose enhances the transcription and translation of IAP genomes in the β cells. Mice strains unable to express IAP retroviruses, as assessed by the p73 antigen equivalent to unprocessed viral *gag* gene polypeptides, are resistant to diabetes; whereas strains expressing p73 in either β cells or thymocytes are very susceptible to the diabetogenic action of the autosomal recessive obesity-diabetes mutation, "diabetes" *(db)*. These results suggest a mechanism of the glucose-stressed β cells that may lead to the development of diabetes in genetically susceptible animals.

Little is known about the possible role of proto-oncogenes in β-cell function and proliferation. Studies with the rat islet cell line RIN established from a transplantable, radiation-induced islet cell tumor showed that incubation of the cells with glucose results in a transient increase in the level of c-*myc* mRNA, which is followed by DNA synthesis.[86] However, the possible role of the c-Myc protein in the glucose-induced activation of β cells is not understood. Regulation of insulin gene expression by glucose appears to be exerted through multiple cAMP-response elements (CREs) and c-*jun* gene expression.[87] The c-Jun protein represses the basal and cAMP-induced activities of the human insulin gene promoter, and c-Jun may be important in the decrease of insulin gene expression at low glucose concentration. The possible influence of other proto-oncogenes in insulin gene expression is unknown. Introduction into islets cells of cloned oncogenes linked to the rat insulin II gene promoter resulted in stimulation of DNA synthesis.[88] The effect was most marked after transfection of the v-*src* oncogene and less pronounced with a combination of a c-*myc* gene and a mutated c-H-*ras* gene.

Transgenic mice expressing recombinant genes composed of the upstream region of the rat preproinsulin gene II linked to sequences encoding simian virus 40 (SV40) large T antigen develop hyperplasia of the islets, followed by the formation of β-cell tumors.[89] Significant levels of the endogenous antigen p53 were produced in β cells expressing SV40 large T antigen, while p53 was undetectable in normal β cells.[90] In the β-cell tumors the viral T antigen and the cellular p53 antigen were associated in the form of a complex. However, only a small fraction of the islets developed into tumors, indicating that factors in addition to the expression of large T and p53 antigens are required for tumorigenesis.

The human insulin gene may be appropriately expressed in the transgenic mice.[91,92] The gene was expressed in a tissue-specific manner in the islets of such mice; and the levels of human insulin in the serum of these animals were properly regulated by glucose, amino acids, and the oral hypoglycemic agent tolbutamide.[93] Thus, serum glucose homeostasis was normal in the transgenic mice expressing the human insulin gene. These findings indicate that the human DNA fragments microinjected into fertilized mouse eggs carry the sequences necessary for tissue-specific expression of the insulin gene and that the human regulatory DNA sequences respond to homologous signals in the mouse. Unexpectedly, a high incidence of serum antibodies to SV40 large T antigen was observed in three of four of the transgenic mice.[94] Both nontolerance and autoimmune response would result from delayed onset of large T-antigen expression during pancreatic β-cell development in the transgenic animals.

III. THE INSULIN RECEPTOR

The insulin receptor is a protein kinase located on the cell surface.[95-100] The kinase activity of the insulin receptor is absolutely specific for tyrosine residues in its acceptor substrate.[101] The kinetic properties of the insulin receptor kinase demonstrated with whole cells or membrane preparations are preserved after purification to homogeneity of the receptor molecule, indicating that these properties are intrinsic to the receptor.[102] The kinase activity of the insulin receptor from rat liver is greater than that of the insulin receptor from human placenta.[103] This difference in enzymatic activity could reflect the physiological role of each insulin receptor in tissue-specific responses to insulin. The kinase activity of the insulin receptor depends on the presence of divalent cations (Mg^{2+} and Mn^{2+}) and may be regulated or modulated by other proteins, including intracellular basic proteins.[104,105]

A. STRUCTURE AND EXPRESSION OF THE INSULIN RECEPTOR GENE

The insulin receptor gene, *INSR,* is located on the short arm of human chromosome 19, at bands 19p13.2-p13.3.[106] The general structure and nucleotide sequence of the human *INSR* gene have been determined.[107-111] The gene spans a region of over 120 kb and comprises 22 exons and 21 introns. The exons vary in size from 36 to >2500 bp. The organization of the human *INSR* gene partially reflects the organization of its protein product, because many of the exons code for structural or functional modules of the protein.[112] Comparison of the exon structure of the tyrosine kinase domain of the human *INSR* gene with the corresponding regions of the human proto-oncogenes c-*src,* c-*ros,* and c-*erb*-B-2 shows that although the exon-intron organization of this region has not been well conserved in evolution, there are similarities in the positions of some introns between pairs of the compared genes. Sequentiation of a 1.5-kbp fragment of the 5′-flanking region of the human *INSR* gene indicates the existence of a transcriptional start site located 203 bp upstream from the site of translation start.[109] The promoter region of the human insulin receptor gene is similar to that of constitutively expressed genes in containing neither a TATA box nor a CAAT box. The promoter is characterized by an extremely high G+C content.

cDNA clones for the human insulin receptor have been expressed in mammalian cells.[113] The human receptor expressed in Chinese hamster ovary (CHO) cells appeared as functionally normal in the heterologous cell system, specifically binding insulin but not IGF-I, displaying insulin-stimulated autophosphorylation, and mediating insulin-stimulated 2-deoxyglucose uptake. A full-length human insulin receptor cDNA inserted in a vector was also transfected into mouse National Institute of Health (NIH)/3T3 cells, which expressed thereafter high levels of functionally normal insulin receptor protein.[114]

Overexpression of the human insulin receptor in rodent cells may serve as a suitable system for the investigation of insulin receptor function by site-directed mutagenesis. A vector containing only the sequences coding for the extracellular domain of the human insulin receptor protein was constructed by mutagenic deletion of the sequences corresponding to the transmembrane and cytoplasmic domains of the insulin receptor molecule.[115] Expression of this construct in NIH/3T3 cells resulted in the synthesis and secretion of a protein with insulin-binding properties. The cytoplasmic, tyrosine kinase domain of the human insulin receptor was expressed in an insect cell line transfected with a baculovirus expression vector.[116] The efficient expression of biologically active insulin receptor-related molecules in heterologous cell systems may provide a method for producing these proteins in sufficient quantity for structural and functional studies. Expression of the insulin receptor gene in intact cells is regulated by complex factors which can either enhance or inhibit the expression. Insulin itself downregulates the steady-state level of its receptor mRNA.[117] Other hormones and growth factors may have important influences on the levels of expression of the insulin receptor in different types of cells.

1. Insulin Receptor Heterogeneity

The mature human insulin receptor is a heterotetramer composed of two α subunits of 719 or 731 amino acids and two β subunits of 620 amino acids. The insulin-binding α subunit and the membrane-spanning β subunit of the receptor are generated by proteolytic processing of a common single-chain precursor of 1370 or 1382 amino acids, which includes a signal peptide of 27 amino acids in addition to the 1343 or 1355 proreceptor molecule. The differences in size of the α subunit and its precursors are due to tissue-specific, possibly developmentally regulated, alternative RNA splicing of exon 11, which codes for a 12-amino acid segment at the carboxyl-terminal region of the α subunit.[118] Brain and spleen express a 719-amino acid α subunit almost exclusively, whereas other tissues including placenta, liver, kidney, and adipose tissue express α subunits of both 719 and 731 amino acids. The biological significance of this heterogeneity is unknown but the two alternative forms of the human insulin receptor — termed HIR-A and HIR-B — when expressed in Rat 1 fibroblasts, exhibit distinct insulin-binding characteristics; this suggests a genetic regulation of cell type-specific responsiveness to stimulation by the same hormone using a single receptor gene.[119]

2. Phylogenetic Aspects of the Insulin Receptor

Insulin receptor homologous DNA, RNA, and protein sequences are present in the fruit fly, *Drosophila melanogaster*.[120-122] The receptor of *Drosophila* is similar in its structure and function to the human insulin receptor, and binding of the physiological ligand results in activation of an intrinsic protein kinase. The insulin receptor gene of *Drosophila* produces two types of transcripts, one of 8.6 kb which is most abundant in early embryos and the other of 11.0 kb which is predominantly expressed during midembryogenesis. High levels of insulin receptor mRNA are expressed in the nervous tissue of the insect during development. In addition to the high-affinity insulin receptor, a 100-kDa protein that has dual binding specificity for both insulin and EGF is present in *Drosophila*.[123] This protein may be derived from an evolutionary precursor of the mammalian insulin and EGF receptors. These results suggest a role for insulin in *Drosophila* development and physiology.

Insulin at relatively high concentrations mimics the effect of progesterone by inducing cell division in *Xenopus laevis* oocytes. These cells possess few, if any, insulin receptors but express IGF-I and IGF-II receptors; and the effects of insulin on *Xenopus laevis* oocytes are mediated by the IGF-I receptor.[124] Both insulin and IGF-I are able to activate β subunit phosphorylation of the amphibian oocyte IGF-I receptor.

B. STRUCTURE OF INSULIN RECEPTORS

The mature human insulin receptors expressed on the cell surface are heterotetramers of 350 kDa and are composed of two α subunits of 135 kDa and two β subunits of 95 kDa, linked by sulfhydryl groups as a (β-s-s-α)-s-s-(α-s-s-β) complex. The α subunit (719 or 731 amino acids), which is external to the cell membrane, comprises the amino-terminal portion of the precursor and contains the insulin-binding domain. The β subunit (620 amino acids), corresponding to the carboxyl-terminal portion of the precursor, includes an extracellular domain of 194 amino acids, a transmembrane domain of 23 amino acids, and a cytoplasmic domain of 403 amino acids with the structural elements required for tyrosine kinase activity. Results from electron microscopic analyses of the soluble insulin receptor ectodomain (IR921 protein) show that it has a structure similar to that of a "Y".[125] Each arm of the Y contains two domains (L1 and L2) involved in binding of the hormone, which results in loss of segmental flexibility of the ectodomain. Mutational analysis suggests that the transmembrane domain of the receptor may play a functional role in regulation of the receptor tyrosine kinase activity.[126] Selective proteolysis at the carboxyl-terminal region of the β subunit results in inactivation of the specific protein kinase activity.[127] The molecular weight of the predicted polypeptide coded by a cDNA of the human insulin receptor is 153,917, whereas the estimated molecular weight of the receptor solubilized from human placenta is 220,000.[107] Thus a substantial fraction of the mass of the insulin receptor molecule must be carbohydrate or other compounds.

In addition to the typical $\alpha_2\beta_2$ form, the insulin receptor from rat hepatoma cells exists under other different forms, including free α and β subunits and combinations of disulfide-linked oligomers.[128] Two hydrodynamic forms of the receptor correspond to larger and smaller oligomeric forms.[129] The possible physiological role of the insulin receptor isoforms in hepatoma cells is unknown but the α and β subunits may be in close physical association in the membrane when they are not linked by disulfide bonds. The association of two αβ subunits is of critical importance for the formation of an insulin receptor with high affinity for ligand and for the insulin-dependent activation of the receptor tyrosine kinase activity.[130-132] Although some tetrameric molecular forms of the receptor have a proteolytically derived fragment of the β subunit, only the intact $\alpha_2\beta_2$ holoreceptor is enzymatically active.[133] The two intact β subunits are required for the insulin receptor to respond to the hormone with an increase in autophosphorylation rate and extent. The insulin receptor tyrosine kinase activity depends on αβ heterodimeric interactions, but it does not necessarily require covalent disulfide bond formation between the individual αβ heterodimeric species.[134]

Insulin receptors isolated from different cells or tissues may exhibit minor structural differences. Subtle differences in molecular weight have been found between the α subunits of insulin receptors isolated from circulating monocytes and erythrocytes.[135] The precise structural basis for such differences remains to be characterized.

Since only the α chains of the receptor bind insulin and each receptor molecule has two α chains, it has been proposed that the receptor is bivalent for the physiological ligand. Bivalent insulin receptors have been purified by affinity chromatography and photoaffinity labeling.[136,137] However, there is evidence that the insulin receptor purified from human placenta binds only one molecule of insulin with high affinity.[138] Only one of the two α subunits of the receptor would bind insulin.[139] Insulin binding occurs in the cysteine-rich region of the receptor α subunit which contains amino acid residues 205 to 316.[140] However, the study of chimeric insulin receptors containing the cysteine-rich domain of the IGF-I receptor indicates that the high affinity of binding of insulin to its receptor can occur in the absence of insulin receptor-specific residues encoded by exon 3 (the cysteine-rich region) and that insulin binding to the receptor involves interactions outside the cysteine-rich region.[141] The crystal structure of normal and mutant human insulin molecules suggests that detachment of the carboxyl-terminal region of the insulin β chain occurs in native insulin on binding to its receptor.[142] This unfolding exposes hydrophobic side chains in the molecule that would be accessible for direct contact with the receptor.

Disulfide bonds and sulfhydryl exchange reactions are involved in interconversion of different molecular forms of insulin receptors as well as in the formation of groups of insulin receptors on the cell surface, which may be important for the biological actions of insulin.[143-146] The insulin receptor subunits are held together by two sets of disulfide bonds designated as class I and class II disulfides. Class I disulfides link the αβ halves of the receptor and are relatively easily reduced, whereas class II disulfides join the α subunits to the β subunits and require stronger physicochemical conditions to obtain a complete reduction. Treatment of the receptor with dithiothreitol, which produces a reduction of disulfide bonds, results in reduction of insulin binding by 40 to 50% and inhibition of the associated tyrosine kinase activity.[147] The receptor is a glycoprotein with high mannose content, but the precise arrangement of the carbohydrate residues in the receptor molecule is unknown.[148,149]

There is a specific association between MHC class I antigens and insulin receptors in mouse liver plasma membranes.[150,151] The biological significance of this association is not understood but it is well known that human insulin-dependent, type I diabetes mellitus is closely associated with particular types of MHC antigens, especially to HLA-DQ antigens.[152]

C. INSULIN RECEPTOR BIOSYNTHESIS

Insulin receptor biosynthesis is initiated by the production of a precursor polypeptide.[153] The α and β subunits of the receptor are synthesized from a single-chain precursor of 210 kDa which is posttranslationally modified by glycosylation and proteolytic cleavage.[154,155] The proreceptor acquires insulin-binding activity through a subtle structural change probably involving disulfide bond isomerization.[156] Defective processing of the receptor due to enzyme defects leading to uncleaving of the proreceptor may lead to the accumulation of nonfunctional proreceptors in the plasma membrane and extreme insulin resistance.[157] Studies performed on a patient with insulin-resistant diabetes due to a point mutation that prevents insulin proreceptor processing suggest that partial proteolysis of the proreceptor is necessary for the acquisition of normal full insulin-binding sensitivity and signal-transducing activity.[158] A cellular protease with more stringent specificity than trypsin would be involved in processing of the insulin receptor precursor. The addition of *N*-linked oligosaccharides to the receptor precursor is also essential for acquisition of binding activity.

Results from studies of IM-9 human lymphoblastoid cells and HEP-G2 hepatoma cells indicate the following steps in the biosynthesis of the human insulin receptor.[159] The insulin receptor mRNA is translated into a 170-kDa protein that is cotranslationally *N*-glycosylated to form a 190-kDa high-mannose precursor. Within 30 min this precursor is activated so that it can bind insulin. Subsequently, the receptor is transported out to the endoplasmic reticulum by a vesicular mechanism, and on fusing with the Golgi apparatus the precursor is cleaved proteolytically into two subunits of 120 and 90 kDa. Then, in the Golgi complex, their oligosaccharide chains are further glycosylated and terminally capped with galactose and sialic acid to form the mature 130-kDa α subunit and 95-kDa β subunit that are transported to the cell surface. Structural differences detected in the insulin receptors derived from the two main target organs of insulin, skeletal muscle and liver, may be due to differences in glycosylation, in particular in the sialic acid content of β subunits.[160] Posttranslational modification of the insulin and IGF-I receptors have been detected in the chick embryo, with higher molecular weight forms of the receptor being present in the liver, intermediate forms in heart and skeletal muscle, and lower molecular weight forms in the brain.[161] The functional implications of insulin receptor heterogeneity remain speculative.

In addition to glycosylation, the insulin receptor in cultured human IM-9 lymphocytes undergoes incorporation of fatty acids.[162] Both the α and β subunits of the receptor incorporate myristic and palmitic acids in a covalent form. The insulin receptor represents the first example of a eukaryotic protein with both types of bound fatty acids and is the first membrane receptor known to contain myristic acid. The physiological role of this modification is still unclear but it may provide hydrophobic domains to the proteins and may thus serve the function of static anchorage points to the plasma membrane.

D. FORMATION AND PHOSPHORYLATION OF THE INSULIN RECEPTOR COMPLEX

The cellular actions of insulin are initiated by its binding to the specific receptors on the cell surface. The binding produces a conformational change of the receptor which may reflect the insulin-induced autophosphorylation of the receptor molecule.[163] Binding of insulin to its receptor is subjected to negative cooperativity, i.e., unlabeled insulin accelerates the dissociation of labeled insulin prebound to the receptor. Insulin binding is pH- and temperature-dependent but the reaction does not conform to a simple reversible bimolecular model; this is due to the existence of negative cooperativity in the hormone-receptor interaction as well as to internalization and degradation of the insulin receptor complex, which results in downregulation of receptor expression on the cell surface. However, negative cooperativity and downregulation are not interrelated phenomena.[164] Furthermore, occupancy of binding sites is not stoichiometrically related to the biological effects of insulin; and in most tissues there is an excess of insulin receptors, called "spare receptors".[165]

Insulin receptors may apparently exist in either a lower or a higher affinity state, and hormone binding alters the equilibrium between the two states, favoring the formation of that of higher affinity. Moreover, insulin binding produces conformational changes in the insulin receptor; and the first, rapid change, exposes parts of the receptor to tryptic degradation.[166] Insulin binding initiates two processes that occur with similar courses: an increase of receptor affinity for hormone and degradation of the 135-kDa α subunit of the insulin receptor to a fragment of 120 kDa.[167] Stimulation of insulin binding to the α subunit of the receptor occurs when adenosine 5′-triphosphate (ATP) binds to the β subunit.[168] The conversion between different states of affinity may play a role in the regulation of receptor proteolysis and physiological activity.

Trypsin exerts insulin-like effects in intact cells as well as in purified preparations of insulin receptors. Trypsin-catalyzed cleavage within the α subunit of the insulin receptor may result in truncation of the receptor molecule and separation of the $\alpha_2\beta_2$ heterotetrameric holoreceptor into heterodimers. The insulin-binding site would be lost during truncation and the β subunit kinase is released from α subunit control.[169] However, activation of the insulin receptor kinase by trypsin may not involve physical loss of the α subunit.[170] Trypsin-induced cleavage of the receptor α subunit, possibly within the insulin-binding site, would lead to the same conformational change as that induced by insulin binding and may result in receptor-specific signal transduction. The effects of insulin stimulation and trypsin-induced activation of the insulin receptor on its autophosphorylation may be indistinguishable. The α subunit of the receptor would thus inhibit the constitutively activated β subunit, and the inhibitory effects of the α subunit could be overcome by insulin binding to or removal of the α subunit. It appears that in the absence of the physiological ligand the extracellular α subunit inhibits the constitutively activated β subunit kinase, whereas in the presence of insulin the β subunit is released from inhibitory control.

Receptor cross-linking or aggregation would be necessary and sufficient for activation of the insulin receptor kinase.[171] Therefore, this aggregation may be an important step in the insulin-induced transmembrane signaling process. Basic proteins such as protamine sulfate, histone Hf2b, and polylysine can activate the insulin receptor protein kinase.[104] Polylysine induces, in the presence of divalent cations (Mg^{2+} and Mn^{2+}), significant changes in subunit interactions within the native insulin holoreceptor complex; the result is a marked potentiation of insulin-stimulated protein kinase activity. A unique endogenous basic protein may be involved in the regulation of $\alpha_2\beta_2$ heterotetrameric holoreceptor kinase activity *in vivo.*

1. Insulin Receptor Phosphorylation

Insulin binds to the α subunit of the receptor, and after this binding it stimulates the phosphorylation of the β subunit of the receptor.[172-174] In intact normal cells (cultured rat hepatocytes) insulin initially stimulates phosphorylation of its receptor predominantly on tyrosine residues.[175] Of the 13 tyrosine residues located on the β subunit intracellular extension of the insulin receptor, 6 are phosphorylated by influence of insulin in intact rat hepatoma cells.[176] The β subunit of the receptor undergoes autophosphorylation on tyrosine as early as 10 s after insulin binding.[177] Autophosphorylation alters the conformation of the β subunit of the receptor.[178] The autophosphorylated insulin receptor acquires protein kinase activity with specificity for tyrosine residues of different cellular proteins.[179-186]

Phosphorylation of the insulin receptor on tyrosine residues is crucially associated with the functional properties of the insulin receptor molecule, including insulin-directed glucose uptake and glycogen synthesis.[187-191] At least five sites of tyrosine phosphorylation have been identified in the β subunit of the human insulin receptor, three of them (Tyr-1146, Tyr-1150, and Tyr-1151) on the regulatory region and the other two (Tyr-1316 and Tyr-1322) in the carboxyl-terminus of the subunit. Autophosphorylation of the receptor begins in the regulatory region immediately after insulin binding. Phosphorylation of all three tyrosines in the regulatory region of the receptor appears to be required for the complete activation of its tyrosine kinase activity, whereas phosphorylation of the carboxyl-terminus is probably not required. The twin tyrosine residues Tyr-1150 and Tyr-1151 (equivalent to Tyr-1162 and Tyr-1163 in another nomenclature[107]) are involved in the control of aggregation of insulin receptor molecules, may be a sufficient signal to trigger the stimulation of glycogen synthesis, and may also play a critical role in insulin-induced hormone and receptor internalization.[192] Phosphorylation of Tyr-1162/1163 but not Tyr-1158 is critical for the activation of the insulin receptor kinase.[193] The mechanism of activation of the receptor kinase by autophosphorylation is more complex than just an introduction of a cluster of negative charges in this region of the receptor.

In addition to tyrosine, other sites of phosphorylation of the insulin receptor exist on serine and threonine residues.[194] While in cell-free systems insulin-dependent receptor phosphorylation occurs only on tyrosine, in intact cells the phosphorylation occurs on serine and, to a lesser extent, on tyrosine.[195] The Thr-1348 residue is the major, if not the only, insulin-stimulated threonine phosphorylation site in simian COS cells.[196] Ser-1305 and Ser-1306 are also phosphorylated by insulin in these cells; however, substitution of these two residues, as well as substitution of Thr-1348, with neutral or negatively charged amino acids has no effect on insulin-stimulated autophosphorylation of the receptor. Tyrosine phosphorylation is one of the earliest events that occurs in intact cells after insulin binding, preceding serine phosphorylation of the β subunit.[197] Tyrosine autophosphorylation occurs in insulin receptors containing little or no phosphoserine and phosphothreonine, suggesting that receptor activity may be regulated intracellularly by phosphorylation processes.[198] Insulin-mimetic agents such as trypsin may cause phosphorylation of the receptor β subunit on tyrosine in intact adipocytes, and activate the insulin receptor-associated kinase activity.[199]

The possible physiological role of the phosphorylation of the insulin receptor on serine and threonine residues is not known. Protein kinase C is able to phosphorylate directly the insulin receptor *in vitro,* which results in reduction of the tyrosine kinase activity of the receptor.[200] The insulin receptor is a substitute for protein kinase C in intact hepatoma cells (Fao cells), and the increase in serine phosphorylation of the β subunit of the receptor produced by phorbol ester treatment inhibits the protein-tyrosine kinase activity of the receptor.[201] These results suggest that protein kinase C may exert a negative regulation on the functional activity of the receptor through phosphorylation of the receptor on serine residues. However, phosphorylation of the insulin receptor on serine residues *in vivo* may not only depend on protein kinase C activity.[202]

The enzymes responsible for serine phosphorylation of the insulin receptor in intact cells remain incompletely characterized. In extracts prepared from the lymphocyte cell line IM-9, insulin-responsive serine-specific protein kinase activity can be separated from the tyrosine kinase activity carried out by the β subunit of the receptor by using antireceptor antibodies.[203] The serine kinase activity is distinct from the insulin receptor but could be physically associated with the receptor. Binding of insulin to the receptor would result in separation of the serine kinase. Phosphorylation of the serine kinase by the receptor protein-tyrosine kinase may not be required for its activation. However, the nature and the role of this serine kinase are unknown. The purified human insulin receptor is a substrate for cAMP-dependent protein kinase.[204] This phosphorylation does not affect the insulin-binding activity of the receptor but partially inhibits its intrinsic kinase activity. The possible physiological role of cAMP-dependent protein kinase-induced insulin receptor phosphorylation is not understood.

The insulin receptor and other cell surface receptors with tyrosine kinase activity may utilize closely related or identical mechanisms for signal transduction across the cell membrane. The EGF receptor kinase domain of a constructed chimeric receptor molecule comprising the extracellular portion of the insulin receptor joined to the transmembrane and intracellular domains of the EGF receptor is activated by insulin binding.[205] However, the specific signals transmitted by hybrid receptor molecules may be different.

2. Physiological Role of Insulin Receptor Phosphorylation

It is generally believed that phosphorylation of the insulin receptor is crucially required for its activation. However, the particular roles of phosphorylations occurring on specific tyrosine and nontyrosine residues of the receptor by the action of insulin or insulin-mimetic agents are not understood in molecular terms. Insulin stimulation of the receptor tyrosine kinase activity can occur in the complete absence of β subunit autophosphorylation.[206] Prephosphorylation of the receptor in the absence of insulin was found to have no significant effect on the amount of insulin-stimulated exogenous protein kinase activity, whereas phosphorylation in the presence of insulin resulted in a dramatic increase in the insulin-stimulated kinase activity of the receptor. These results suggest that insulin binding stimulates the exogenous substrate protein kinase of the receptor as well as the autophosphorylation of its β subunit *in vivo* and that this autophosphorylation allows maximal activation of the receptor tyrosine kinase activity by insulin.

Evidence that the kinase activity of the insulin receptor is essential for at least the biological properties of the receptor molecule has been obtained by transfection of expression vectors carrying DNA sequences coding for either a normal human insulin receptor or an insulin receptor produced by site-directed mutagenesis with substitution of alanine for lysine at position 1018, in the ATP-binding domain of the β subunit.[207] The mutated receptor bound insulin but lacked tyrosine kinase activity. In contrast to the normal receptor, the mutated receptor was unable to stimulate the following functions in CHO cells expressing the vector: insulin-dependent deoxyglucose uptake, S6 kinase activation, endogenous substrate phosphorylation, glycogen synthesis, and thymidine incorporation into DNA. It was concluded that the kinase activity of the insulin receptor is crucially associated with apparently all its biological function. However, this conclusion may not be supported by other studies. CHO cell transfectants expressing insulin receptors in which Tyr-1162 and Tyr-1163 were replaced by phenylalanine exhibited a total inhibition of insulin-mediated tyrosine kinase activity toward exogenous substrates, and the mutation abolished the effect of the hormone on glycogen synthesis. In an independent study it was found that insulin receptors mutated at Tyr-1162 and Tyr-1163, when expressed in CHO cells, can retain normal signaling of the stimulatory effect of insulin on glucose transport activity and GLUT-1 expression, but not on glycogenesis and overall protein synthesis.[208] Thus, the long-term stimulatory effects of insulin may be controlled in a differential manner by tyrosine 1162 and 1163 of the major insulin receptor autophosphorylation domain. CHO cell transfectants expressing the insulin receptors mutated at Tyr-1162 and Tyr-1163 may show insulin-stimulated mitogenic effects with the same insulin concentration-response curve for DNA synthesis as those cells expressing intact insulin receptors.[209] Studies using a mutant insulin receptor in which Tyr-1158 was replaced with phenylalanine indicated that autophosphorylation of this residue is a critical step in the autophosphorylation cascade activating the insulin receptor kinase, and that it plays an essential role in the internalization of the receptor and the regulation of growth but may not be required for some metabolic signals.[210] These results suggest the existence of at least two signal transduction pathways branching from the insulin receptor, one leading to changes such as those related to glycogen synthase activation and another leading to DNA synthesis. It may be concluded that site-directed mutational studies on the human insulin receptor strongly suggest that different signaling pathways can be triggered by distinct domains of the activated insulin receptor β subunit.

The physical state and the chemical structure of the insulin receptor contribute to determine its functional properties. Treatment of the insulin receptor with 1m*M* dithiothreitol completely reduces disulfide linkages between the receptor subunits, and it has been found that the monomeric $\alpha\beta$ form of the receptor exhibits much higher insulin-dependent protein-tyrosine kinase activity than the intact receptor in the $\alpha_2\beta_2$ form.[211] Both insulin and antireceptor antibodies induce cooperative interactions between the two linked α subunits of the receptor $\alpha_2\beta_2$ dimer leading to a decrease in the insulin binding of this receptor form.[212] The possible relationship between the increased tyrosine kinase activity of the $\alpha\beta$ insulin receptor monomer and its decreased activity for the ligand remains to be elucidated. Removal of sialic acid from the purified insulin receptor results in enhanced binding and tyrosine kinase activities.[213] The production of an antipeptide antibody that specifically inhibits insulin receptor autophosphorylation and protein kinase activity has been reported.[214]

Studies with polyclonal and monoclonal antibodies may contribute to elucidation of the structure and function of the insulin receptor kinase. Antibodies to different epitopes of the human insulin receptor may produce either insulin-like and insulin-inhibitory effects.[215] Both the acute and the long-term effects of insulin would depend on the activation of the specific insulin receptor-associated kinase. Insulin-mimetic anti-insulin receptor-specific monoclonal antibodies stimulate biological responses in intact cells via their ability to stimulate the specific protein kinase activity of the receptor.[216] Injection of a monoclonal antibody that inhibits tyrosine kinase activity results in a decreased ability of insulin to stimulate the uptake of 2-deoxyglucose in CHO cells and rat adipocytes, the phosphorylation of S6 ribosomal protein in CHO cells, and the synthesis of glucose in the human hepatoma cell line HepG2.[217] The ability of insulin and IGFs to stimulate glucose uptake in TA1 mouse adipocytes is also inhibited by the monoclonal antibody. Microinjection of a monoclonal antibody with specific inhibitory effects on the insulin receptor tyrosine-specific kinase into *Xenopus* oocytes blocks the ability of insulin to stimulate oocyte maturation.[218] In general, multisite phosphorylation of the insulin receptor is important for the regulation of its specific kinase activity.[219]

The physiological role of insulin receptor phosphorylation at specific sites is not totally clear. After binding to its receptor, insulin can stimulate *in vitro* phosphorylation not only of its own receptor but also of exogenously added substrates such as casein, histone H2b, and a synthetic peptide.[220] Insulin receptor phosphorylation may not be a prerequisite for acute insulin action,[221] and the effects of insulin on its cell surface receptor are not absolutely specific because treatment of rat adipocytes or human placenta with trypsin also stimulates phosphorylation of the insulin receptor on tyrosine residues.[222] Studies with rat and hamster cells transfected with expression vectors containing genes for either normal human insulin receptors or receptors with inactivated tyrosine kinase domains indicate that the insulin receptor signaling pathway for activation of pyruvate dehydrogenase bypasses the receptor associated tyrosine kinase activity.[223] Thus, at least in certain cases, the specific kinase activity of the insulin receptor may not play an obligatory role in the insulin-signaling pathway.

Studies using monoclonal antibodies directed against the α subunit of the insulin receptor suggested that some of the physiological actions of insulin may not depend on phosphorylation of the receptor on tyrosine residues and activation of its specific kinase activity. Three monoclonal antibodies that react with the α subunit of the insulin receptor in IM-9 lymphocytes failed to stimulate receptor autophosphorylation and receptor-mediated phosphorylation of exogenous substrates, but two of these antibodies stimulated glucose transport in isolated human adipocytes.[224] These studies suggested that monoclonal antibodies to the insulin receptor can mimic a major function of insulin (stimulation of glucose transport) without activating receptor kinase. However, the interpretation of these data is obscured by evidence indicating that insulin receptor-binding antibodies and insulin may act through separate pathways.[225] A polyclonal antihuman placental insulin receptor antibody enhances insulin binding to the receptor probably by causing conformational perturbation of the receptor molecule.[226] Another polyclonal anti-insulin receptor antibody binds to both wild-type and mutant insulin receptors, but it induces effects characteristic of insulin action (stimulation of hexose transport and thymidine incorporation into DNA) only in cells expressing the wild-type receptor.[227] Moreover, these effects were associated with activation of the insulin receptor kinase. In general, the results obtained in most studies, especially those performed with sensitive methods, support the concept that the specific kinase activity of the ligand-activated insulin receptor is essential for insulin action in intact cells.[216] In any case, it is clear that various exogenous factors may be capable of modifying the enzymatic and functional activities of the insulin receptor.[228,229]

In conclusion, ligand-binding-induced activation of the specific kinase activity of the insulin receptor is an integral part of the cellular mechanisms of insulin action. The mitogenic effects of insulin crucially depend on the phosphorylation cascade of cellular proteins initiated by the binding of insulin to its receptor on the cell surface. However, some of the physiological effects of insulin, especially the immediate responses such as glucose transport and glycogen synthesis may directly depend on the ligand binding and would be elicited through mechanisms not necessarily involving activation of the insulin receptor-associated tyrosine kinase. The precise nature of such mechanisms and their physiological importance remains to be elucidated.

3. Insulin Receptor Dephosphorylation

Insulin regulates not only the phosphorylation but also the dephosphorylation of its own receptor in a dynamic fashion.[230] Insulin-induced dephosphorylation of the β subunit of the insulin receptor may be mediated by membrane-bound phosphatases and is specifically enhanced by guanosine triphosphate (GTP). A number of protein-tyrosine phosphatases are involved in insulin receptor dephosphorylation. Preparations of rat liver plasma membrane contain a protein-tyrosine phosphatase which is distinct from alkaline phosphatase; the enzyme catalyzes the dephosphorylation of the insulin receptor β subunit.[231] Three protein-tyrosine phosphatases are expressed in insulin-sensitive liver and muscle tissue; and one of them, termed leukocyte common antigen-related phosphatase (LAR), is highly active in dephosphorylating tyrosine residues 1146, 1150, and 1151, which form a domain crucially involved in the activation of the insulin receptor kinase.[232]

E. MUTATIONS OF THE INSULIN RECEPTOR

Different types of alterations in the mechanism of action of insulin at either the receptor or postreceptor level may result in various degrees of metabolic abnormalities. The importance of insulin receptor phosphorylation for the action of insulin is demonstrated by the observation that insulin resistance in a diabetic patient was associated with normal insulin binding but defective phosphorylation of the β subunit of the receptor.[233] However, a selective defect in the phosphorylation of this subunit is not responsible for most cases of resistance to insulin in humans.[234] Insulin receptor-associated tyrosine kinase activity may be defective in insulin-resistant obese mice.[235] This defect in the enzyme function of the receptor molecule was observed for both receptor autophosphorylation and the ability of the receptor to catalyze phosphorylation of a synthetic peptide substrate.

Mutations of the insulin receptor gene, consisting of either point mutations or deletions, have been found in a number of patients with insulin resistance.[236-238] Impairment of insulin action in such cases may be due to decreased numbers of insulin receptors expressed on the surface of the target cells, decreased affinity of the receptor to bind insulin, or decreased protein kinase activity of the receptor molecule. Several clinical syndromes associated with insulin resistance and hyperinsulinemia have been described, some of them characterized by the presence of acanthosis nigricans, which consists of a hyperkeratotic and hyperpigmented skin lesion localized predominantly in skin folds such as the axillae and antecubital fossae. Type A insulin resistance is observed in young female patients with acanthosis nigricans who may exhibit virilization and polycystic ovaries and may have hyperinsulinemia and severe resistance to insulin due to structural defects in the insulin receptor molecule.

Deletion of insulin receptor gene sequences have been found in a number of patients with insulin resistance. The deleted gene (deletion type A Chiba) found in one of these patients lacked almost the entire kinase domain (exons 17 to 22) of the receptor molecule.[239] In another patient, the deletion (type A Yamanishi) affected exon 14 of the gene.[240] In this case the truncated receptor lacked 488 amino acid residues in the transmembrane and cytoplasmic domain of the β subunit. Homozygous or heterozygous point mutations of the insulin receptor gene have been found in patients with insulin resistance. In one of these patients the receptor contained a mutation in which valine was substituted for glycine at position 966, the third glycine in a conserved motif in the putative ATP-binding site.[241] Expression of this receptor by transfection into CHO cells confirmed that the mutation impaired the tyrosine kinase activity of the protein.

Leprechaunism is a heritable (autosomal recessive) form of insulin resistance that is characterized by intrauterine and postnatal growth restriction, loss of glucose homeostasis, and very high concentrations of circulating insulin. This syndrome is associated with mutations in the gene encoding the insulin receptor subunits, which frequently result in decreased insulin-stimulated protein kinase activity of the receptor molecule. The insulin receptors from a patient with this rare disease were markedly defective in insulin-stimulated autophosphorylation, and insulin binding did not activate the tyrosine kinase of the receptor β subunit.[242] The insulin receptor from this patient had a mutation at position 233 of the α subunit, which is outside the domains involved in insulin binding and kinase activity of the receptor molecule. The region of the insulin receptor containing the mutation may be required for transmitting the insulin binding signal to the kinase region. The cells from another patient with leprechaunism showed increased sugar transport, even though insulin binding was markedly reduced; and they exhibited constitutively increased insulin receptor autophosphorylation and kinase activity.[243] In an exceptional leprechaun patient who expressed normal amounts of insulin receptor RNA and protein in his/her cultured fibroblasts, glucose transport and insulin receptor kinase activity were constitutively elevated; and the patient was found to be homozygous for a point mutation in the insulin receptor gene, converting the arginine-86 to proline in the α subunit of the receptor.[244] The domain of the insulin receptor affected by the mutation impaired receptor phosphorylation and kinase activity and membrane glucose transport but not insulin-sensitive cellular growth.

Insulin resistance may be associated with postreceptor defects in insulin action. A postbinding defect in the action of insulin was found in a family with an autosomal dominant form of insulin resistance, although the defect could not be characterized at the molecular level.[245] Impaired action of insulin on RNA synthesis has been observed in the Alström syndrome, a disorder associated with a genetically defective insulin receptor.[246] In the cultured fibroblasts of patients with the Alström syndrome, insulin receptor binding and insulin-stimulated glucose uptake are in the normal range. The molecular basis of the insulin resistance in patients with the Alström syndrome is unknown; however, the results obtained with Alström patients suggest that, after binding of insulin to its receptor, the events involved in the early cellular action of insulin, such as glucose uptake, may be mediated by mechanisms which are different from those involved in the late effects, such as RNA synthesis. The possible role of insulin receptor phosphorylation at different sites in relation to early and late insulin effects is not understood. Another familial syndrome, observed in three siblings, was characterized by a combined defect of the action of insulin, IGF-I, and EGF.[247] Studies with fibroblasts cultured from these patients showed that receptor binding occurred with normal capacity and affinity; nevertheless, the growth factors had a markedly reduced capacity to stimulate RNA synthesis and 2-deoxyglucose uptake, as compared with normal controls, suggesting some defect at the postreceptor level.

F. RELATIONSHIPS BETWEEN THE INSULIN RECEPTOR AND ONCOPROTEINS

A homology between the insulin receptor and oncoproteins was initially suggested by the fact that the insulin receptors present in cultured human lymphocytes (IM-9 cells) are specifically immunoprecipitated by antibodies to the v-Src oncoprotein.[248] The precipitation was found to be competitively inhibited by purified v-Src protein but not by the v-Raf protein. Src proteins do not represent truncated forms of the insulin receptor because other antibodies to the v-Src protein, including a monoclonal antibody, are unable to induce precipitation of the receptor.[249] It was concluded that there is a similarity between certain epitopes of the insulin receptor and the v-Src oncoprotein.

The solubilized insulin receptor can be phosphorylated and activated by the v-Src oncoprotein kinase.[250,251] The biological significance of this modification is not clear since the v-Src kinase also phosphorylates exogenous proteins such as angiotensin and calmodulin.[252,253] The insulin receptor is not a substrate for phosphorylation by two other tyrosine kinases, the EGF receptor and the v-Abl oncoprotein.[251] On the other hand, the activated insulin receptor kinase is apparently unable to phosphorylate the Src kinase.

The insulin receptor is closely related to the Src/tyrosine kinase family of oncoproteins.[254] In particular, there is a high degree of sequence homology between the insulin receptor and the v-Ros oncoprotein.[108,255] A human insulin receptor/v-Ros hybrid protein is capable of activating in an insulin-dependent manner the S6 and microtubule-associated protein 2 (MAP-2) kinases.[256] The precise function of the normal c-Ros protein is unknown, but its structural similarity with the receptors for insulin EGF suggests that it may be a receptor for an unidentified growth factor. The c-*ros* gene is the vertebrate homolog of the *sevenless* tyrosine kinase receptor of *Drosophila melanogaster*.[257] Expression of c-*ros* mRNA in the chicken is restricted mainly to the kidney, and the highest levels of this mRNA are present in 7- to 14-d-old chickens. In the mouse embryo, c-*ros* transcripts are expressed in kidney, intestine, and lung.[258] A constructed hybrid receptor molecule composed of the extracellular domain of the human insulin receptor and the transmembrane and cytoplasmic (tyrosine kinase) domains of the v-*ros* oncoprotein was expressed in CHO cells.[259] The molecule was expressed at high levels on the cell surface, bound insulin; was phosphorylated on tyrosine residues upon ligand binding; and was capable of expressing transmembrane signaling. However, the hybrid molecule was unable to elicit insulin-specific responses such as activation of glucose uptake and stimulation of DNA synthesis. The responses mediated by ligand-activated tyrosine kinases may utilize distinct intracellular mechanisms for postreceptor signaling.

A constructed hybrid molecule contained the 5′ portion of the UR2 *gag* gene fused to 46 amino acid residues of the extracellular domain and the entire transmembrane and cytoplasmic domains of the β subunit of the human insulin receptor.[260] The constructed retrovirus, termed UIR, coded for a hybrid polypeptide with capability for inducing transformation of CEF cells and promoting formation of colonies in soft agar, but the infected cells did not form tumors *in vivo*. However, a variant that arose from the parental UIR was capable of efficiently inducing sarcomas *in vivo*. UIR-transformed cells exhibited higher rates of growth and glucose uptake than normal cells. The UIR genome codes for a membrane-associated, glycosylated Gag-human insulin receptor fusion protein of 75 kDa that possesses tyrosine kinase activity and is capable of autophosphorylation. This protein is phosphorylated on serine and threonine residues *in vivo* and is capable of phosphorylating foreign substrates *in vitro*. Activation of the oncogenic potential of the human insulin receptor in the constructed molecule is due to the deletion of a sequence located immediately upstream from the transmembrane protein domain that imposes a negative effect on the transforming potential of the fusion protein, which results in a constitutively activated kinase.[261]

G. INSULIN RECEPTOR EXPRESSION AND FUNCTION

A wide diversity of endogenous and exogenous factors are capable of regulating the expression of insulin receptors, including age, menstrual cycle, pregnancy, diet, physical exercise, and various pathological conditions. In general, the expression of insulin receptors on the cell surface is inversely correlated with the concentrations of circulating insulin. Obese individuals (who are frequently hyperinsulinemic) are characterized by a decreased number of insulin receptors and are resistant to the physiological effects of insulin.

The number of growth factor receptors expressed on the surface of different types of cells is regulated by the humoral milieu, including the environmental concentration of the extracellular signaling agents. Both autologous and heterologous regulation of insulin receptor expression may occur. Chronic exposure to high concentrations of insulin induce a loss of insulin receptor expression on the cell surface, which is known as receptor downregulation.[262-265] The mechanisms involved in downregulation of insulin receptors are not totally clear. Unidentified factors present in serum may play a role in such regulatory phenomena.[266] Chronic hyperinsulinism may also result in an uncoupling of insulin binding from activation of the intrinsic kinase activity of the receptor.[267] Insulin-induced desensitization of the insulin receptor kinase does not correlate with the extent of β subunit serine/threonine phosphorylation. Homologous downregulation of the receptor is also associated with increased postreceptor biosynthesis and processing into mature units, which may represent a mechanism compensating for insulin-induced receptor loss.[268]

Heterologous regulation of insulin receptor expression is related to the action of multiple hormones, growth factors, and other extracellular signaling agents. Glucocorticoids regulate insulin receptor expression at the transcriptional level and can induce a rapid increase in insulin receptor mRNA in a diversity of cells, including cultured lymphocytes, the human lymphoblastoid cell line IM-9, and rat pancreatic acinar cells.[269-271] Glucocorticoids have no significant effect on the degradation rate of insulin receptor gene transcripts. Thyroid hormone may be involved in regulating autophosphorylation of the insulin receptor β subunit.[272] Thyroidectomy increases insulin receptor autophosphorylation without changing the number or affinity of the receptor.

External factors including quantitative and qualitative aspects of the diet may contribute to the regulation of insulin receptor expression. Exposure of animals to cold results in a decreased tyrosine kinase activity of the insulin receptor, which is probably induced by endogenous catecholamines that activate cAMP-dependent protein kinase.[273] However, the mechanism of this effect is unknown and there is no evidence that it involves changes in insulin receptor phosphorylation on serine residues.

Phorbol esters stimulate phosphorylation of the insulin receptor in intact hepatoma cells and other types of cells at serine and threonine residues.[274] The phosphorylation occurs at nine sites or more in the receptor β subunit. Phorbol esters may increase the equilibrium constant on a molality basis (K_m) of the ATP-binding site of the insulin receptor kinase from rat adipocytes.[275] Phorbol ester action on the phosphorylation of the insulin receptor could be mediated by protein kinase C activity, but there is evidence that this phosphorylation depends on a different enzyme not characterized as yet.[202]

1. Functional Role of the Insulin Receptor

The role of surface receptors in the cellular responses to insulin is not totally clear. The presence of insulin receptors alone does not ensure insulin responsiveness because insulin receptors are present in many types of cells where a specific action of insulin has not been found. Madin-Darby canine kidney (MDCK) cells lack insulin receptors and do not respond to insulin by increasing the incorporation of radiolabeled glucose into glycogen or the uptake of α-aminobutyrate. However, MDCK cells display some responses considered as characteristic of insulin action, including stimulation of glycogen synthesis, when they are exposed to insulin mimickers such as insulin-ricin B hybrid molecules; this suggests that other membrane receptors may act as functional alternates for insulin receptors.[276] Moreover, lectins and hydrogen peroxide can elicit insulin-like responses in MDCK cells, which indicates that it is possible to bypass plasma membrane insulin-binding sites but still elicit cellular responses characteristically mediated by insulin.[277] Microinjection of insulin into *Xenopus laevis* oocytes obviates the interaction of the hormone with its receptor on the cell surface; however, the insulin microinjected into the oocytes stimulates transcription and translation, thus suggesting that the nucleus may be a site for direct insulin action.[278] Intact insulin, but not the insulin receptor, accumulates in nuclei of insulin-treated rat and mouse cells, suggesting that insulin dissociates from its receptor prior to entering the nucleus.[279] A nuclear insulin receptor would be contained in a chromatin fraction which is bound to a specific DNA restriction fragment.[280] Thus, at least some of the effects of insulin on nuclear function may be caused by the translocation of the intact and biologically active hormone to the nucleus and its binding to nuclear components in the heterochromatin, which may directly affect gene transcription and functions related to cell growth.

2. Developmental Changes in Insulin Receptor

Insulin may have an important influence on the processes of cell proliferation and differentiation that are associated with fetal development, and this influence may be modulated by structural and functional changes occurring in the insulin receptor during development. The α subunits of insulin and IGF-I receptor of adult rat brain are about 10 kDa lower in molecular size than the corresponding α subunits from rat liver.[281] During fetal development of the rat brain, there is a progressive decrease in the molecular mass of the α subunit of the insulin receptor, which may be due to changes in glycosylation, with a decrease in the content of sialic acid.[282] These structural changes are associated with functional changes reflected in the binding capacity and the insulin-stimulated autophosphorylation of the receptor. The peak of insulin receptor expression in the rat brain occurs at a time of tremendous growth and differentiation of the organ.

3. Insulin Receptor Abnormalities in Neoplastic Cells

Malignant cells may exhibit various structural and/or functional abnormalities of the insulin receptor molecule including changes in the expression and phosphorylation, as well as in the ligand binding properties, or the receptor. For example, an abnormality in insulin binding and receptor phosphorylation is present in the insulin-resistant mouse melanoma cell line Cloudman S91.[283] In contrast to the almost universal proliferation-inducing action of insulin in cultured cells, insulin acts as a potent and reversible inhibitor of proliferation in Cloudman S91 cells.[284] The insulin receptor present in these cells is defective both in its affinity for insulin and its autophosphorylation properties, one site of autophosphorylation being absent in the altered receptor, which would make it unable to stimulate cell growth. An abnormally large but functionally normal insulin receptor is present in the human monocyte cell line U-937, which was derived from a patient with generalized histiocytic lymphoma.[285] The altered size of the insulin receptor in U-937 cells could be due to either differences in the core protein or to abnormalities in the patterns of insulin receptor glycosylation.

Decreased insulin binding and receptor autophosphorylation have been observed in hepatocellular carcinomas induced in rats with carcinogens such as acetylaminofluorene (AAF) and diethylnitrosamine (DENA).[286] Similar changes were detected in the epidermal growth factor (EGF) receptors from the same tumors. These alterations are associated with decreased transcription of the mRNAs for insulin and EGF receptors.[287] Although these changes seem to represent an initial chemical effect on the great majority of liver cells, only a small minority of the cells become truly initiated and retain the altered characteristics up to the tumor stage. The fact that similar changes occur in both insulin and EGF receptors suggests the possibility of a common underlying mechanism. Treatment of minimal deviation H4 hepatoma cells with the lectin, wheat germ agglutinin, or mild proteolytic treatment with trypsin results in a rapid and marked increase in the affinity of the insulin receptor molecule.[288]

Translocations involving the insulin receptor gene, which is located on human chromosome 19 at region 19p13.2-p13.3, may alter the expression of the insulin receptor. Childhood pre-B ALL is frequently associated with translocation t(1;19) near the insulin locus, and the leukemic cells bearing this translocation may constitutively express high levels of insulin receptor at the cell surface.[289] The human pre-B ALL cell line 697, which contains a t(1;19), expresses 10- to 20-fold more insulin receptor mRNA than leukemic cells with structurally normal chromosome 19. However, the structure of the insulin receptor is apparently normal in ALL cells with the t(1;19), and the significance of the quantitative alteration of the insulin receptor in relation to the leukemic process is unknown.

Insulin receptor expression may be altered during the process of differentiation induced in neoplastic cell lines by chemical agents. The HL-60 human promyelocytic leukemia cell line can be induced to undergo differentiation *in vitro* toward either a mature monocyte or granulocyte. Induction of monocytic differentiation in HL-60 cells by specific agents is accompanied by a significant increase in expression of the insulin receptor, whereas induction of granulocytic differentiation by other types of agents is accompanied by decreased insulin receptor expression.[290] This difference correlates with the insulin binding characteristics in normal human peripheral monocytes and granulocytes.

4. The pp63 Insulin Receptor Tyrosine Kinase Inhibitor

A natural inhibitor of the insulin receptor tyrosine kinase, termed pp63, is produced by adult rat hepatocytes and has been characterized as a glycoprotein of 63 kDa.[291,292] The pp63 inhibitor is secreted in the serum and is biologically active in the phosphorylated form. pp63 is capable of inhibiting both the tyrosine kinase activity and the autophosphorylation of the insulin receptor. pp63 antagonizes the growth-promoting effect of insulin in the insulin-sensitive rat hepatoma cell line Fao but does not affect insulin-mediated amino acid transport capacity or tyrosine aminotransferase induction in these cells. pp63 may participate in the maintenance of a quiescent state of normal adult hepatocytes, acting as a growth-inhibitory factor. However, the levels of pp63 mRNA are increased in spontaneously occurring liver tumors. The mechanism of action and the precise physiological role of the pp63 insulin receptor inhibitor are unknown.

H. INTERNALIZATION AND DEGRADATION OF THE INSULIN RECEPTOR COMPLEX

The insulin receptor complex is internalized and processed intracellularly, which may result in a downregulation of insulin receptor expression on the cell surface.[262,265,293,294] Insulin induces internalization of the receptor into the cell, and either tyrosine kinase activity of the receptor or its phosphorylation state would be essential for ligand-mediated receptor downregulation.[295,296] The intrinsic tyrosine kinase activity of the receptor is required for insulin to initiate the signal to stimulate receptor internalization but would not be required for the constitutive internalization of the receptor.[297] However, studies on the well-differentiated rat hepatoma cell line Fao, depleted of ATP by treatment with 2,4-dinitrophenol, indicate that insulin-stimulated internalization of the insulin receptor may be independent of receptor β subunit tyrosyl autophosphorylation.[298] Calcium and protein kinase C may have an important role in the process of endocytosis of insulin receptor complexes.[299]

Internalized insulin receptors may be recycled back to the cell surface.[300] Studies with photoaffinity-labeled insulin receptors in rat adipocytes show that dissociation of insulin receptor complexes is not required for receptor recycling.[301] After internalization, insulin receptor complexes may be degraded by lysosomes.[263] However, there is cell-specific heterogeneity among different cell types in the processing of insulin receptor complexes.[302] In hepatocytes most of the internalized insulin receptors are recycled to the cell surface.[303]

Insulin molecules may have different fates after their binding to the receptor at the level of the plasma membrane. After binding, insulin can be released from the cell intact, be partially degraded on the membrane, or be internalized into the cell via membrane-bounded vesicles or endosomes.[304,305] Internalization of insulin into endosomes occurs subsequent to the localization of receptor-bounded insulin in coated pits on the cell membrane, invagination of these pits into the cytoplasm, and then pinching off of the invagination to form the endosome. Tyrosine kinase-defective receptors undergo insulin-induced microaggregation but do not concentrate in coated pits.[306] Endosomes may fuse with lysosomes, resulting in the degradation of insulin and some of its receptors; however, trafficking of internalized insulin is complex, and it is not clear whether the lysosome is the primary mechanism for insulin degradation within the cell. The initial limited hydrolysis of insulin, and possibly also subsequent steps, may occur via a nonlysosomal pathway. It is not known whether insulin fragments produced during the intracellular degradation of the hormone have biological activity. Intermediary degradation products of insulin may be found not only within the cell, but also, after incubation of isolated rat hepatocytes with labeled insulin, peptide products smaller than insulin can be detected predominantly in the incubation medium.[307]

The primary insulin degrading enzyme is insulin protease, a neutral thiol metalloproteinase of 110 kDa located in the cytoplasm, which is involved in degrading the insulin molecule in isolated hepatocytes and possibly also in other types of cells.[308] Seven peptide bonds on the amino-terminal side of the insulin β chain are susceptible to cleavage by insulin protease. A cDNA coding for insulin protease was isolated and sequenced.[309] The full-length cDNA contains an open reading frame (ORF) capable of encoding a polypeptide of 1019 amino acids with a molecular weight of 117,863. Interestingly, the amino acid sequence of insulin protease is homologous to that of *Escherichia coli* protease III. The two proteins may be members of a family of proteases that are involved in intercellular peptide signaling. The genes encoding insulin protease have been assigned to human chromosome 10 and mouse chromosome 19.[310]

Different types of cells may show differences in the pathways related to the internalization and processing of insulin-receptor complexes.[311] In certain types of cells the intact hormone may be shuttled across the cell in order to be delivered on the other side of an epithelial cell barrier. This process, called transcytosis, has been demonstrated to occur in the receptor-mediated delivery of insulin across endothelial cells to the target tissues.[312] In other types of cells such as hepatocytes, insulin dissociates from the receptor a few minutes after internalization and may not be degraded but may partially enter a nondegradative pathway, termed retroendocytosis, by which the hormone is released in an intact form.[303] The retroendocytosis pathway is proportional to the amount of insulin bound and internalized by the hepatocyte and is distinct from the pathway of insulin degradation. These observations suggest that insulin binding and processing may be controlled by mechanisms that are, at least in part, cell specific.

Intracellular insulin "receptors" have been described, and some of them may be localized in the nucleus.[313-315] In different types of cell lines insulin, apparently bound to its receptor, would become tightly associated with chromatin.[316] It has been suggested that the internalized insulin receptor complex may retain a weak, but significant, capacity to stimulate both glucose transport and phosphodiesterase activities.[317] Nuclear translocation of the insulin receptor complex would mediate the long-term effects of insulin.[318] However, the precise biological significance of these observations is unknown. Endocytosis of insulin receptor complexes is apparently not required, or not sufficient, to mediate the effects of insulin.[187] The insulin molecule may be dissociated from its receptor before entering the nucleus and would become associated with chromatin.[279] However, it has been suggested that the results of insulin entering the nucleus would be due to the presence of a heptapeptide sequence in the B chain of the insulin molecule which is homologous to a sequence present in the first zinc finger of the DNA-binding domain of the glucocorticoid receptor.[319] It may be concluded that the possible action of insulin at the nuclear level remains uncharacterized.

Loss of insulin receptors from the cell surface is observed in cells treated with various polyclonal and monoclonal antibodies. The mechanism related to antibody-induced downregulation of the insulin receptor is apparently different from that of insulin exposure since it can occur in the absence of receptor-associated kinase activity, receptor phosphorylation, and even in the absence of an intact receptor cytoplasmic domain.[320] Replacement of the human insulin receptor and cytoplasmic domains by corresponding domains of the v-Ros oncoprotein leads to an accelerated internalization, degradation, and downregulation of the receptor molecule expressed in CHO cells.[321] The hybrid receptor molecule exhibits insulin-stimulated transmembrane signaling, including autophosphorylation and activation of kinase function, but does not mediate insulin-stimulated uptake of 2-deoxyglucose or incorporation of thymidine into DNA. Further studies are required in order elucidate the relationships existing between insulin receptor activation, internalization, and degradation.

IV. TRANSDUCTIONAL MECHANISMS OF INSULIN ACTION

The postreceptor mechanisms of action of insulin, which comprise transductional and posttransductional events, are still little understood in spite of numerous studies performed on this subject during the last few decades. It is generally accepted that insulin receptor-mediated phosphorylation of specific cellular proteins is of critical importance for transduction of the insulin signal. It may be true that most or all insulin effects at the cellular level, including an increased rate of glucose transport,[322] depend directly on the kinase activity of the insulin receptor complex. The possible role of classical mediators of hormone action, such as cyclic nucleotides and calcium ions, in the mechanisms of action of insulin is a subject of controversy.[323-330] However, there is evidence that nontyrosine kinase-dependent pathways may be important for some of the physiological actions of insulin.[331] Two main lines of evidence in favor of this possibility are based on the facts that monoclonal and polyclonal antibodies to the receptor can mediate many of the actions of insulin with little or no stimulation of the receptor protein kinase and the demonstration that insulin receptor mutants with reduced or no tyrosine kinase activity are able to mediate several of the known actions of insulin.

A. THE ADENYLYL CYCLASE SYSTEM

The precise role of the adenylyl cyclase system in the mechanism of action of insulin is poorly understood. Both cAMP-dependent and cAMP-independent pathways are involved in the mechanisms of action of insulin in target cells such as adipocytes.[332] In these cells, lipolysis is promoted by agents that increase intracellular cAMP concentrations and antagonized by agents that decrease synthesis and/or increase degradation of cAMP. Insulin is a potent inhibitor of lipolysis, and this action may be mediated by the activation of serine protein kinases that catalyze the phosphorylation and activation of cAMP phosphodiesterase.[333] In addition, insulin may act at the level of cAMP-dependent protein kinases by decreasing their affinities for cAMP.[334] Studies with intact rat hepatocytes suggest that insulin stimulation of cAMP phosphodiesterase in these cells may be independent of G protein pathways involved in adenylyl cyclase regulation.[335] The precise relationship between changes in intracellular cAMP levels and activation of the insulin receptor kinase by ligand binding remains to be elucidated.

B. THE CALCIUM/CALMODULIN SYSTEM

A possible role for divalent cations, in particular Ca^{2+}, in the cellular action of insulin was suggested by the results from early studies.[323-330] This possibility is supported by the fact that the insulin receptor kinase contains a calmodulin-binding domain.[336] Calmodulin enhances insulin-mediated receptor kinase activity, and insulin stimulates phosphorylation of calmodulin in rat adipocyte and liver cell preparations.[337-339] In rat hepatocytes, this phosphorylation exhibits an absolute requirement for insulin receptors, divalent cations, and certain basic proteins. Insulin-induced calmodulin phosphorylation takes place only on tyrosine residues.[340] In rat adipocytes, calmodulin phosphorylation by the activated insulin receptor kinase can occur at both the Tyr-99 and Tyr-138 residues, which are located on the third and fourth Ca^{2+}-binding pockets of the calmodulin molecule, respectively.[341] The insulin receptor purified from human placenta induces tyrosine phosphorylation of calmodulin.[342]

C. THE NA^+/H^+ ANTIPORT

The possible role of the Na^+/H^+ antiport in the transduction of the insulin-elicited signal is little known. Stimulation of hexose transport observed in 3T3-L1 undifferentiated mouse fibroblasts acutely exposed to insulin does not depend on activation of Na^+/H^+ exchange across the membrane and cytoplasmic alkalinization.[343] The specific messenger signals that transfer information from the activated insulin receptor to the glucose transporter remain to be elucidated.

D. POLYAMINE METABOLISM

The possible role of polyamines in the mechanism of insulin action is unknown. Ornithine decarboxylase (ODC) activity is essential for the generation of polyamines, and insulin induces ODC expression in cells such as Reuber H35 rat hepatoma cells.[344] The phorbol ester phenylmercuric acetate (PMA) has a similar effect on these cells, but it acts through a mechanism which is different from that of insulin. The effects of insulin and PMA on the increase of ODC activity are additive.

E. GUANINE NUCLEOTIDE-BINDING REGULATORY PROTEINS

G proteins are involved in the transductional mechanisms of insulin action, as suggested by the observation that insulin proliferative pathways are sensitive to both pertussis and cholera toxins.[345] Adenosine can modulate the insulin antagonistic effect of β adrenergic stimulation on rat adipocytes via G_i protein through both cAMP-dependent and cAMP-independent mechanisms.[332] A noncovalent interaction occurs between the insulin receptor kinase and a regulatory G protein system during the process of insulin signaling in murine myocytes, indicating that the G protein mediates some aspects of the insulin action in the myocytes.[346] The kinase activity of the insulin receptor may not play a significant role in the communications established between the receptor and the G protein system. These observations suggest the existence of two parallel pathways by which insulin transmits its signal to the target cell: on one hand, the tyrosine kinase activity of the receptor initiates a cascade of events in which phosphorylation of specific proteins plays a central role; on the other hand, direct interactions of the receptor with G proteins may lead to activation of specific effectors one of which is most likely a specific phospholipase C involved in the generation of insulin mediators. Both pathways should probably be activated for the production of the complex biochemical and functional effects of insulin in intact cells and tissues.

F. LOW MOLECULAR WEIGHT POLYPEPTIDE INSULIN MEDIATORS

It has been proposed that after binding of insulin to its receptor several insulin-specific intracellular mediators may be formed, possibly by limited proteolytic processes.[331,347,348] These mediators would be peptides of mol wt 1000 to 3000 acting at a number of intracellular sites and involved in the control of different enzyme activities. A putative insulin mediator of mol wt 2500 was purified from livers of insulin-treated rats with streptozotocin-induced diabetes.[349] The mediator would act as an inhibitor of the catalytic subunit of adenylyl cyclase and would also inhibit cAMP-dependent protein kinase. However, the efforts oriented to the complete purification and characterization of specific insulin mediators have so far encountered little success. If such mediators exist, they could be responsible for the insulin-like effects produced by treatment of cells with trypsin.

G. PHOSPHOINOSITIDES AND GLYCOLIPIDS

Insulin, as other hormones, induces rapid changes in phospholipid metabolism.[350,351] However, the precise role of phospholipids in the cellular mechanism of insulin action remains a subject of high controversy. Insulin enhances incorporation of orthophosphate onto different subcellular fractions from rat adipose tissue incubated *in vitro*.[352,353] It increases the incorporation of [^{32}P]-labeled orthophosphate into phosphatidate, phosphatidylinositol, and diacylglycerol in the same tissue.[354] Treatment of rats bearing R3230AC mammary carcinoma with insulin results in inhibition of tumor growth, and this effect is associated with a reduction in phosphatidylinositol kinase activity.[355] However, activation of phosphoinositide metabolism may not represent a universal pathway of insulin action and may show variations among different tissues.

In rat liver plasma membrane preparations or isolated hepatocytes insulin does not stimulate the synthesis of phosphatidylinositol, phosphatidylinositol 4-phosphate, or phosphatidylinositol 4,5-bisphosphate, suggesting that the insulin receptor kinase does not act on phosphoinositide pathways in the liver.[356,357] Insulin and IGFs may not be members of the family of hormones that generate inositol trisphosphate as a second messenger in the liver.[358] EGF alone or in combination with insulin can stimulate G_0-arrested hamster fibroblastic cells to undergo DNA replication and to proliferate without activating phosphatidylinositol turnover.[359] Insulin does not stimulate phosphoinositide degradation in BALB/c 3T3 mouse fibroblasts at concentrations at which its actions are determined primarily by an interaction with its own receptor, or at higher concentrations at which it may cross-react with IGF-I receptors.[360] In isolated fat cell preparations, insulin does not stimulate phosphoinositide breakdown but only increases the synthesis of phosphatidylinositol and phosphatidylinositol 4,5-bisphosphate.[361] The insulin receptor tyrosine kinase, purified from human placenta, is devoid of phosphatidylinositol kinase activity.[362]

1. Glycolipid Hydrolysis

Glycophospholipid anchors of membrane proteins may be potential precursors of insulin mediators.[363] In cultured myocytes, insulin causes a rapid hydrolysis of a defined membrane glycolipid, which results in the production of two related complex carbohydrates as well as in generation of 1,2-diacylglycerol.[364-366] Such results suggest that insulin stimulates an endogenous phospholipase C activity that hydrolyzes a particular glycolipid, glycosyl-phosphatidylinositol, generating an oligosaccharide (inositol-phosphate glycan) which would mediate some of the physiological actions of the hormone. The inositol-phosphate glycan generated in these metabolic processes would regulate cAMP phosphodiesterase, and perhaps other insulin-sensitive enzymes. The purification of an enzyme from rat liver membranes that is highly specific for molecules containing phosphatidylinositol has been reported.[367] The activity of this enzyme, phosphatidylinositol-glycan-specific phospholipase C, is stimulated by insulin and it may represent an effector for some of the metabolic actions of insulin. The proof of a role for inositol-phosphate glycan as a messenger of insulin action depends on the determination of its structure.[368]

2. Enzymatic Methylation of Phospholipids

Enzymatic methylation of phospholipids plays a role in the transduction of receptor-mediated signals through the membranes of a variety of cells, and it has been suggested that this process may also play an important role in the transductional mechanisms of insulin action.[369] Plasma membranes prepared from rat adipocytes contain a phosphatidylethanolamine methyltransferase system, and insulin stimulates this enzyme in a concentration-dependent manner with the effects being observed as early as 15 s after the addition of the hormone.

3. Generation of Diacylglycerol

Insulin-induced changes in plasma membrane glycerolipid metabolism may result in an increased generation of 1,2-diacylglycerol, which would function as a signal capable of exerting regulatory changes in insulin-responsive cellular components, including responses not only at the plasma membrane level but also at the genomic level.[370] However, the evidence in favor or diacylglycerol as a mediator of insulin action on gene expression is circumstantial and is based on the fact that phorbol esters such as 12-*O*-tetradecanoylphorbol-13-acetate (TPA) can mimic insulin-induced changes in gene expression.

Insulin-induced stimulation of glycerolipid synthesis and phospholipid hydrolysis, resulting in the generation of membrane diacylglycerol, may play a role in mediating an acute effect of insulin on the stimulation of glucose transport in cells such as BC3H-1 myocytes.[371] In these cells, insulin stimulates the generation of diacyalglycerol and inositol glycan.[372] Stimulation of 2-deoxyglucose uptake in CHO cells expressing a transfected human insulin gene occurs by a mechanism involving myristoyl-diacylglycerol production and protein kinase C activation.[373] The Tyr-1162 and Tyr-1163 residues of the insulin receptor may control these effects of the hormone. Insulin causes a rapid increase in 1,2-diacylglycerol synthesis in mouse 3T3 fibroblasts, which could be explained by a mechanism involving a G protein-mediated activation of phospholipase C.[374] However, insulin does not appear to activate a phosphoinositide-specific phospholipase C in cells such as adipocytes.[375]

Insulin could stimulate the synthesis of 1,2-diacylglycerol not only by an increased hydrolysis of phospholipids but also by inducing a burst of phosphatidic acid synthesis. Administration of insulin *in vivo* may provoke a rapid increase in the concentration of phosphatidic acid, phosphatidylinositol, and phosphoinositides in rat adipose tissue and also increases phosphatidylinositol levels *in vitro*.[376] The results suggest that stimulation of phosphatidic acid synthesis may serve as an effector mechanism for insulin, at least in adipose tissue. Insulin rapidly and massively increases the synthesis of phosphatidic acid in cultured BC3H-1 myocytes by increasing the activity of glycerol-3′-phosphate acyltransferase, and the newly synthesized phosphatidic acid is preferentially converted directly into diacylglycerol.[377] This effect may provide a mechanism for the activation of protein kinase C without a concurrent increase of inositol trisphosphate production and Ca^{2+} mobilization.

4. Activation of Protein Kinase C

Protein kinase C may play a role in mediating some aspects of insulin action.[378,379] Insulin may induce changes in the activity of protein kinase C and its association with cell membranes in defined systems, for example, in BC3H-1 myocytes and macrophage chemotactic factor 7 (MCF-7) mammary tumor cells.[380,381] Similar increases in the activity of protein kinase C are induced by insulin preparations of rat skeletal muscle and adipocyte plasma membranes.[382,383] There is direct evidence that protein kinase C activity is required for insulin-stimulated hexose uptake in rat adipocytes.[384] Particular isoforms of protein kinase C are activated and translocated to the plasma membrane by the action of insulin on rat adipocytes.[385,386] The mechanism of insulin-induced activation of protein kinase C is unknown but is probably not identical to that of phorbol esters.

Phorbol esters can mimic the short-term effects of insulin on hepatic fatty acid metabolism.[387] Phorbol esters stimulate phosphorylation of the insulin receptor on serine and threonine, altering insulin receptor autophosphorylation on tyrosine and insulin activation of cellular enzymes (glycogen synthetase and tyrosine aminotransferase).[274] Since the cellular actions of phorbol esters are associated with activation of protein kinase C, these results suggest a physiological relation between insulin action and protein kinase C. However, protein kinase C is not essential for the insulin-stimulated activation of hexose transport across the cell membrane.[343] Phorbol ester-inducible protein kinase C is apparently not necessary for the effect of insulin on DNA synthesis in rat H4 hepatoma cells.[388] Both phorbol ester and 1,2-diacylglycerol produce immediate and reversible inhibition of insulin binding to its receptor by altering the affinity of the insulin receptor.[389,390] Internalization of the insulin receptor complex is also affected by treatment of cells with either phorbol ester or diacylglycerol. Protein kinase C is able to phosphorylate the insulin receptor *in vitro,* which may result in reduction of the tyrosine kinase activity of the receptor.[200] However, the physiological significance of this activity of protein kinase C *in vivo* is unclear. An enzyme other than protein kinase C may be involved in phosphorylation of the insulin receptor on serine and threonine in living cells.[202]

5. Activation of Phosphatidylinositol 3-Kinase

Phosphatidylinositol 3-kinase (PI 3-kinase) phosphorylates the inositol ring of phosphatidylinositol at the D3 position to produce phosphatidylinositol 3-monophosphate, as well as 3,4-bisphosphate and 3,4,5-trisphosphate phosphorylated at the D3 position, which may function as second messenger molecules distinct from the classical phosphoinositide pathway. Purification of PI 3-kinase has demonstrated that the enzyme is a heterodimer consisting of 85- and 110-kDa subunits and that the 85-kDa subunit (p85) contains two SH2 domains. Other proteins containing two SH2 domains include phospholipase C-γ and GTPase-activating protein (GAP). The SH2 domains bind specifically to phosphotyrosine-containing amino acid sequences in growth factor receptors and other proteins, and are involved in the regulation of protein-tyrosine phosphorylation. The insulin receptor and its 160- to 185-kDa protein substrate, the insulin receptor substrate 1 (IRS-1), present in 3T3-L1 adipocytes and other cell types associate with SH2 domains contained in the PI 3-kinase, but not those of GAP or phospholipase C-γ.[391] The association between the PI 3-kinase and the insulin receptor may result in PI 3-kinase activation.[392,393] The possible biological significance of this activation is not clear. The PI 3-kinase would be a nontyrosine phosphorylated member of the insulin receptor signaling complex.[394] In CHO cells overexpressing the human insulin receptor, insulin induces the association of p85 with tyrosine-phosphorylated IRS-1 but not with the activated insulin receptor.[395] However, there are two types of p85, α and β; and there is evidence that treatment of cells with insulin may result in the stimulation of tyrosine phosphorylation of α-type p85.[396] Specifically, the phosphorylation of the tyrosine residues 368, 580, and 607 of α-type p85 may be stimulated by the activated insulin receptor.[397] In addition to the IRS-1 protein, treatment of adipocytes with insulin results in the phosphorylation of a 60-kDa protein (pp60), which is also associated with the PI 3-kinase.[398] The possible functional consequence of this association is unknown. Further studies are required for a better knowledge of the possible role of the PI 3-kinase in the mechanism of insulin action.

In conclusion, the role of phosphoinositide metabolism in the mechanisms of insulin action is not clear and may show variations from one tissue to another. Studies using a monoclonal antibody with specificity to phosphatidylinositol 4,5-bisphosphate clearly indicate that phosphatidylinositol breakdown is not crucially required for the mitogenic action of insulin in NIH/3T3 cells.[399] In the final analysis, it may be true that the cellular actions of insulin are exerted directly by the specific kinase activity of its receptor, which may be responsible for the phosphorylation of different cellular substrates, including other kinases that become activated by this modification. One of such activated kinases may be the PI 3-kinase.

V. POSTTRANSDUCTIONAL MECHANISMS OF INSULIN ACTION

The formation of an insulin receptor complex in insulin responsive cells induces an array of functional changes at the level of the plasma membrane, including stimulation of ion transport and enhanced uptake of essential nutrients, especially glucose and amino acids.[400,401]

The metabolic actions of insulin do not depend solely on the transport of glucose into the cells,[402] but an important target of the activated insulin receptor tyrosine kinase is the glucose transporter.[403] In L6 muscle cells, insulin causes a rapid translocation of the transporter from internal membranes to the plasma membrane.[404] In 3T3-L1 adipocytes, insulin regulates the expression of two types of glucose transporters, termed GLUT-1 and GLUT-4, by causing their translocation from an intracellular location to the plasma membrane.[405,406] There are about 950,000 and 280,000 copies of GLUT-1 and GLUT-4, respectively, per cell. The main insulin-responsive glucose transporter regulated by insulin in adipose and muscle tissues is the 509-amino acid integral membrane protein GLUT-4. The steady-state levels of GLUT-4 mRNA are markedly decreased in adipose tissue from fasted rats or rats made insulin deficient by treatment with the diabetogenic compound, streptozotocin.[407] Refeeding of the fasted rats or insulin treatment of the diabetic rats causes a rapid recovery of the GLUT-4 mRNA to levels above those observed in untreated control animals. However, the role of insulin in GLUT-4 expression may vary according to the experimental system. The rates of synthesis and degradation of GLUT-1 and GLUT-4 proteins in mouse 3T3-L1 adipocytes are regulated independently.[408] In these cells, insulin induces downregulation of GLUT-4 mRNA and protein by a dual mechanism involving changes in both transcription and mRNA turnover rates.[409] In addition to insulin, other hormones, growth factors, and regulatory peptides are involved in regulating the expression of glucose transporters. IL-1 stimulates hexose transport in fibroblasts by increasing the expression of glucose transporters.[410] Noradrenaline and insulin independently stimulate the translocation of glucose transporters from intracellular stores to the plasma membrane in mouse brown adipocytes.[411]

A cDNA encoding a glucose transporter which is expressed in insulin-responsive tissues such as skeletal muscle, heart, and adipose tissue was cloned and sequenced recently.[412] The β cells of the pancreas express glucose transporters, and there is evidence that a marked downregulation of GLUT-2 in these cells may play a causal role in the development of hyperglycemia in ZDF diabetic rats.[413] An insulin-regulatable glucose transporter is apparently not present in endothelial cells.[414] The Src protein may have a role in the differential regulation of glucose transporter isoform expression at both the transcriptional and posttranscriptional level.[415] This regulation may be exerted through a phosphorylation of the transporter protein on specific amino acid residues.

The activity of a number of enzymes is modified by insulin in different tissues. Enzyme activation by insulin action can occur by phosphorylation processes depending on the tyrosine-specific kinase activity of the insulin receptor complex as well as by interaction of this complex with serine/threonine-specific kinases.[416] This interaction may result in a cascade of protein phosphorylations which may contribute to determining the cellular response to the hormone. Furthermore, the total amount of phosphotyrosine in cellular proteins depends not only on the activity of tyrosine kinases such as the insulin receptor, but also on the dephosphorylating activity of protein phosphatases with specificity for tyrosine or serine/threonine residues.[417] Microinjection of a 35-kDa tyrosine-specific protein phosphatase (PTPase 1B) into *Xenopus laevis* oocytes inhibits insulin-stimulated phosphorylation of tyrosyl residues on endogenous proteins, including a protein having a molecular mass in the same range of the β subunit of the insulin and IGF-I receptor.[418] The protein tyrosine phosphatase (PTPase) also blocked the activation of an S6 peptide kinase. The molecular mechanism by which insulin stimulates glycogen synthesis in mammalian skeletal muscle is associated with phosphorylation of the regulatory subunit of type I protein phosphatase at a specific serine residue.[419]

Addition of insulin to specific types of sensitive cells in culture may produce rapid morphological effects. Insulin, as well as IGFs and EGF, cause rapid membrane ruffling in KB human epidermoid carcinoma cells and other types of cells, including nontransformed cells.[420] The morphological change can be observed within 1 min and is accompanied by microfilament reorganization. The physiological role of these changes is not clear at present. In T84 human colonic epithelial cells, insulin modulates the permeability of the occluding junction (paracellular permeability) through a receptor-mediated process which involves alterations in protein synthesis and cytoskeletal structure.[421]

The synthesis of specific proteins and/or other cellular components may be increased or decreased by the action of insulin, depending on the type of cell, its stage of differentiation, and other physiological conditions. It is difficult, however, to make a distinction between the direct and indirect actions of insulin on particular metabolic reactions occurring within insulin-responsive cells. For example, it is well known that insulin promotes the synthesis of glycogen in the liver, but this effect cannot be ascribed to a specific hormonal action on hepatic glucose utilization mediated by direct stimulation of insulin on particular enzyme activities. Apparently, this action of insulin is only associated with the insulin-induced increased intrahepatic glucose concentration reflecting changes in hepatocyte glucose uptake.[422] Insulin may also influence a number of mitochondrial functions. There is evidence that the mitochondrial Krebs cycle may be a selective intracellular site of insulin action.[423] However, the molecular mechanisms involved in the action of insulin at the mitochondrial level are unknown.

In certain types of cells and tissues, prolonged insulin action may result in stimulation of DNA synthesis and mitogenic effects. In density-inhibited chick embryo cell cultures the following sequence of events has been observed after stimulation with microgram quantities of insulin: early increase in sugar uptake and decrease in leucine uptake, increase in cell volume, stimulation of RNA and protein synthesis, increase in thymidine uptake, DNA synthesis, mitosis, and cell division.[424] However, very little effect of insulin is observed in this system in the absence of serum, suggesting that some other factors present in serum are required for the stimulation. The precise sequence of the complex insulin-induced cellular changes, the relationships existing between them, and the interactions of insulin with other hormones and growth factors in relation to the development of such changes are not understood. Studies with defined cellular systems *in vitro* may contribute to a better knowledge of these changes. Protein accumulation is observed in cell cultures maintained in serum-free media stimulated by either insulin or IGF-I, which is a consequence of both an increase in the synthesis of protein and a decrease in protein breakdown.[425] In the whole animal, insulin usually behaves as a potent anabolic agent. Treatment with insulin can reverse cachexia in tumor-bearing rats, with suppression of anorexia and preservation of host weight and without stimulation of tumor growth.[426] In spite of such beneficial effects, long-term insulin treatment slightly, but significantly, shortens survival in tumor-bearing animals.

A. INSULIN ACTION ON PROTEIN PHOSPHORYLATION

Protein phosphorylation/dephosphorylation processes are of crucial importance in the action of insulin on its target cells, but the exact role of these processes in the multiple physiological actions of insulin remains unknown.[427,428] A number of substrates of the insulin receptor kinase have been identified, but the function of any of them in the cells is little understood. It is also unknown whether the tyrosine phosphorylation of any of the known substrates is required to elicit a particular biological response.

The insulin receptor is an insulin-activatable tyrosine kinase capable of autophosphorylation and is able to catalyze the phosphorylation of exogenous and endogenous protein substrates. The receptor may be considered as an integrated system for transmembrane hormone signaling, and this function is intimately associated with protein phosphorylation. The insulin receptor complex is associated with two types of protein kinase activities, namely, tyrosine specific and serine specific. Whereas the tyrosine-specific kinase is a constituent of the insulin receptor itself being contained in the β subunit of the receptor, the serine-specific kinase would be associated with the insulin receptor on the cell membrane, but would not form an integral part of the receptor. The intrinsic kinase activity of the insulin receptor is absolutely specific for tyrosine. Insulin-induced phosphorylation of proteins on serine and threonine residues depends on the secondary activation of protein kinases other than the insulin receptor. Insulin enhances, for example, casein kinase II activity in 3T3-L1 mouse adipocytes and H4-IIE rat hepatoma cells.[429] This effect appears to be mediated by direct phosphorylation of the enzyme and may result in the secondary phosphorylation of various cellular proteins.

In addition to its effect on protein phosphorylation through protein kinase activation, insulin may induce alterations in the phosphorylation of cellular proteins through activation of the dephosphorylation process mediated by specific types of protein phosphatases.[430,431] Other growth factors (EGF and platelet-derived growth factor [PDGF]) have similar effects on phosphatase activation.

1. Membrane Proteins

The activated insulin receptor kinase can phosphorylate a number of plasma membrane-associated substrates in isolated membranes or intact cells, but the precise physiological significance of these modifications is little understood. Homodimers of G proteins are phosphorylated *in vitro* by the activated insulin receptor protein-tyrosine kinase.[432] Tyrosine phosphorylation of a 46-kDa membrane protein is rapidly stimulated by insulin in intact cells.[433] In human and rat liver plasma membranes, insulin promotes the phosphorylation of proteins with molecular weight of 43, 50, and 120 kDa.[434-436] The 120-kDa membrane protein, pp120, is also a substrate for the IGF-I receptor kinase. The pp120 protein is identical to HA4, a membrane glycoprotein localized to the apical (bile canalicular) side of the hepatocyte plasma membrane.[437] The function of the pp120/HA4 protein at the bile front of hepatocytes is unknown.

2. Ribosomal Protein S6

Several cellular proteins are phosphorylated on serine, threonine, and/or tyrosine as a consequence of the interaction of insulin with its receptor. One of them is the 40S basic ribosomal protein S6,[438] which may be involved in the biochemical processes leading to initiation of protein synthesis. Protein S6 is phosphorylated on serine by the action of serum, EGF, prostaglandins, and phorbol esters.[439-443] S6 is also phosphorylated by oncogene products with protein kinase activity, including the v-Src oncoprotein.[444] The insulin-modulated kinases responsible for S6 phosphorylation remain little characterized.[445] It has been suggested that the enzyme mediating insulin-stimulated phosphorylation of protein S6 is a soluble S6 kinase, S6 kinase II (S6 KII), which is neither cyclic AMP-dependent nor phospholipid- and Ca^{2+}-dependent.[446] The S6 KII kinase is rapidly activated by exposure of differentiated 3T3-L1 adipocytes to nanomolar concentrations of insulin or phorbol ester. Similar S6 kinase activity is found in insulin-treated 3T3-L1 cells and chick embryo fibroblasts transformed by Rous sarcoma virus (RSV).[447,448] Activation of the S6 KII enzyme may depend on the MAP-2 protein kinase, which is itself a phosphoprotein that can be deactivated by protein phosphatase 2A.[449] Thus, a step in insulin signaling would involve sequential activation by phosphorylation of at least two serine/threonine protein kinases. In addition, insulin regulates S6 phosphorylation through modulation of phosphatase activities, in particular the activity of type I phosphatase.[450]

Insulin-mediated phosphorylation of the ribosomal protein S6 may be related to induction of cell proliferation,[451] but a direct regulation of cellular DNA synthesis through S6 phosphorylation seems unlikely. Under conditions in which EGF maximally stimulates S6 phosphorylation and DNA synthesis, insulin further stimulates DNA synthesis but not S6 phosphorylation.[452] Probably, insulin and EGF use different pathways for stimulating DNA synthesis.

3. Proteins p15, p46, and p62

The precise physiological significance of the phosphorylation of specific cellular proteins of tyrosine in cells stimulated by insulin is not clear. One exception could be a 15-kDa protein (p15) in which phosphorylation on a tyrosine residue is stimulated by insulin on 3T3-L1 adipocytes.[453] The p15 protein is apparently involved in the insulin receptor-initiated signal transduction to the glucose transport system.

A cytosolic protein of 46 kDa (p46) is phosphorylated in the rat liver by insulin as well as by growth hormone and glucagon.[454] The hormone effects on p46 phosphorylation are rapid and comparable in magnitude, maximally increasing phosphorylation about threefold. Although the identity of p46 has not been determined, the fact that it is modified by three different hormones suggests that it may have an important physiological role.

Incubation of rat adipocytes with insulin results in a very rapid and marked (approximately 80-fold) increase in the phosphorylation of a soluble protein of 62 kDa (p62) which is recognized by a monoclonal antibody against the type II regulatory subunit (RII) of cAMP-dependent protein kinase.[455] However, peptide mapping indicates that p62 and RII are distinct polypeptides. The physiological role of p62 is unknown.

4. High Molecular Weight Proteins

A number of high molecular weight cellular proteins are phosphorylated by the insulin receptor kinase. A monomeric membrane protein of 180 kDa found in human placenta — as well as in rat liver, skeletal muscle, heart, and brain — is phosphorylated in a direct manner by the insulin receptor kinase.[456] Another substrate for insulin-stimulated phosphorylation is a 96-kDa cytosolic protein which may be involved in the initiation of DNA synthesis.[457] Cellular proteins of 110, 120, 185, and 240 kDa are substrates for the insulin receptor kinase,[177,458,459] but the functions of these proteins are not known. A 180-kDa substrate of the insulin receptor kinase appears to be represented by a protein-tyrosine phosphatase that is elevated in diabetic plasma membranes.[460] Both insulin and EGF stimulate serine phosphorylation of a 170-kDa protein, related to lipocortin 1, in rat hepatocytes.[461]

A protein of 160 to 185 kDa (p160 or p185) is prominent among the cellular substrates of the insulin receptor kinase and is a putative participant in insulin signaling.[462,463] The p160/p185 protein is expressed in a variety of insulin-responsive tissues and is rapidly phosphorylated on tyrosine in the intact animal.[464] The activated type I IGF receptor kinase can also phosphorylate p160/p185, suggesting that it represents a common link in the signal transmission of insulin and IGF-I.[465] Cloning of a genomic sequence, *IRS*-1, which encodes a component of p160/p185, indicates that the product of the *IRS*-1 gene contains ten potential tyrosine phosphorylation sites.[466] The IRS-1 protein binds PI 3-kinase, suggesting that the IRS-1 component of p160/p185 represents a link between the insulin kinase and enzymes of the phosphoinositide pathway involved in the regulation of cell proliferation and metabolism.[467]

A 90-kDa phosphoprotein (pp90), an associated protein kinase, and a specific phosphatase are involved in the regulation of the Cloudman melanoma cell line proliferation by insulin.[468] Growth of the wild-type Cloudman melanoma cell line is inhibited by insulin, and pp90 phosphorylation/dephosphorylation may be one step in the insulin-mediated control of growth of this cell line.

5. Calmodulin

The insulin receptor kinase may catalyze the phosphorylation of calmodulin and calmodulin-dependent protein kinase.[337,416,469] Insulin-induced phosphorylation of calmodulin may occur in intact cells such as rat adipocytes and affect two tyrosine residues contained in the calmodulin molecule (Tyr-99 and Tyr-138).[338-341] The phosphorylation may alter the activity of calmodulin, which could lead to changes in the activities of Ca^{2+}/calmodulin-dependent enzymes, including protein kinases. As a consequence, cellular proteins may be modified by phosphorylation and dephosphorylation processes. Phosphorylation of a 40-kDa protein is stimulated by insulin in a Ca^{2+}-dependent manner.[470] Elevated calmodulin levels have been found in the heart and kidney of two types of diabetic mice, and in muscle and abdominal fat of streptozotocin-induced diabetes in mice.[471]

6. Microtubule Components

Cytoskeletal proteins are important substrates for tyrosine kinases, including the insulin and EGF receptor kinases as well as the v-Src kinase.[472] The cytoskeletal substrates of the insulin receptor kinase may be different from those of other tyrosine kinases. Microtubules are involved in the mechanisms of insulin action, and tubulin and microtubule-associated proteins are phosphorylated by the action of the insulin receptor-associated tyrosine kinase activity.[458,473-475] A microtubule-associated kinase (MAP kinase) is phosphorylated on tyrosine, and to a lesser extent on threonine, *in vivo* when 3T3-L1 cells are treated with insulin.[476] As is true for many other cellular proteins, it is not known which cellular proteins are primary targets of the insulin-receptor complex kinase activity and which others are phosphorylated by other, secondarily activated kinases.

7. Oncoproteins

The human c-K-Ras oncoprotein, which contains an extremely basic domain (residues 172 to 182), can be phosphorylated *in vitro* by the insulin receptor kinase.[477] The c-K-Ras protein would play a role in the mechanism of insulin action by stimulating in an insulin-independent way autophosphorylation of the insulin receptor.[478] In contrast, the c-H-Ras protein, which lacks the basic domain, is phosphorylated by insulin *in vitro,* but only in the presence of poly(L-lysine), which is required for establishing an interaction between the insulin receptor and its substrate. Phosphorylation of calmodulin by the insulin receptor *in vitro* also requires cofactors such as protamine and poly(L-lysine). The c-Src protein is a substrate for phosphorylation in BC3H-1 murine myocytes stimulated by insulin or phorbol ester, although this effect is mediated in each case by distinct mechanisms.[479] The possible biological significance of oncoprotein phosphorylation by the activated insulin receptor kinase is not known.

Recent evidence indicates that Ras proteins play an important role in signal transmission from cell surface receptors with tyrosine kinase activity — including the EGF, nerve growth factor (NGF), and PDGF receptors — and possibly also the insulin receptor.[480] The protein Grb2, which contains in its amino-terminal end one SH2 domain flanked by two SH3 domains, couples the receptor tyrosine kinases to a protein, Sos, which activates the Ras signaling pathway.[481] In mammalian CHO cells expressing the insulin receptor, insulin stimulation induces the formation of a stable complex between Grb2 and two tyrosine-phosphorylated proteins, IRS-1 and Shc.[482] IRS-1, a widely expressed 165- to 185-kDa phosphoprotein, is a major target for the insulin and IGF-I receptor tyrosine kinases. IRS-1 contains 14 potential tyrosine autophosphorylation sites. Shc is also a widely expressed tyrosine-phosphorylated, SH2-containing protein that may possess oncogenic potential. Interactions between Grb2 and the two phosphoproteins, IRS-1 and Shc, may play a crucial role in Ras protein activation and the control of specific downstream effector pathways in insulin-stimulated cells.

B. TRANSCRIPTIONAL EFFECTS OF INSULIN

Insulin regulates the transcriptional expression of a number of cellular genes.[483,484] Expression of distinct genes may be either stimulated or inhibited by the action of insulin, according to the type of cell and the predominant physiological conditions. Results from early studies on rat epididymal adipose tissue showed that insulin stimulates the incorporation of labeled precursors into total and specific fractions of RNA.[352,353] Insulin may have an important role in the regulation of gene expression in the liver. The hormone rapidly induces the expression of the hot spot 14 (S14) gene in the liver of diabetic rats.[485] S14 gene expression is also induced by thyroid hormone, but the normal function of the S14 protein is unknown. The level of a particular mRNA species (p33) in H4 rat hepatoma cells is induced by insulin.[486] PMA displays an insulin-like effect in H4 cells.[487] The regulation of p33 mRNA expression by insulin occurs at both the transcriptional and posttranscriptional levels.[488] Dexamethasone induces a marked increase of p33 mRNA in H4 cells. Insulin specifically decreases the level of albumin mRNA and inhibits albumin gene transcription in H4 cells.[489] The very marked reduction in cytoplasmic albumin mRNA observed in H4 cells treated with insulin, as compared with the decrease in transcription, indicates that the decrease in cytoplasmic albumin mRNA in the liver cells is due in part to posttranscriptional effects. In rat glioma cells, insulin regulates ferritin mRNA levels at the transcriptional level.[490]

Insulin may regulate the expression of genes encoding specific enzymes. In primary cultures of rat hepatocytes, regulation of glucokinase activity by insulin is controlled at the level of gene transcription.[491] The glucokinase gene is silent in the absence of insulin, and its expression is turned on by insulin and is acutely turned off by glucagon. When both hormones are present at maximal concentrations, the negative effect of glucagon prevails and the gene is repressed. The effect of glucagon in this system depends on the generation of cAMP and is mimicked by a cAMP analogue. Thus, a glucagon/cAMP-mediated repression mechanism is a key aspect in the regulation of glucokinase gene transcription in the hepatocyte. Glucokinase is an hexokinase isoenzyme that regulates the first step of glucose metabolism, the ATP-dependent phosphorylation of glucose to generate glucose 6-phosphate.

Expression of the gene for phosphoenolpyruvate decarboxykinase (PEPCK), an enzyme that governs the rate-limiting step of gluconeogenesis in the liver, is positively regulated by cAMP and glucocorticoids and negatively regulated by insulin. Inhibition of PEPCK gene expression by insulin is dominant since it predominates in the presence of cAMP and glucocorticoids. The PEPCK gene contains a cAMP-responsive element (CRE) and a glucocorticoid-responsive element (GRE) as well as a 15-bp insulin-responsive sequence (IRS).[492] The IRS is located within the PEPCK gene promoter and may be recognized by a specific activity which has been identified in liver extracts.

Insulin may contribute to the regulation of expression of certain hormone and growth factor genes. In endothelial cells, insulin stimulates endothelin (ET-1) gene expression.[493] Insulin regulates transcription of the growth hormone gene, and this effect is mediated by the insulin-induced synthesis of a DNA-binding protein which recognizes a sequence present on the upstream human growth hormone gene promoter.[494] The *trans*-acting insulin-induced protein has a molecular weight of 70 to 80 kDa and is responsive to insulin treatment.

1. Effects of Insulin of Proto-oncogene Expression

Insulin can induce changes in the expression of distinct proto-oncogenes in defined cellular systems. In cell lines established from BALB/c 3T3 mouse cells that constitutively express either the murine p53 or the human IGF-I gene, or both, but not in intact BALB/c 3T3 cells, the addition of insulin results in an increased transcriptional expression of the c-*myc* gene.[495] In rat hepatoma cells, insulin also regulates c-*myc* expression at the level of transcription.[496] Insulin induces c-K-*ras* gene expression in rat liver *in vivo* and in cultured normal rat hepatocytes.[497] The hormone stimulates a rapid but transient accumulation of c-*fos* gene mRNA in 3T3-L1 fibroblasts and adipocytes.[498] Expression of the c-H-*ras* gene is regulated by insulin, IGF, and EGF in murine fibroblasts.[499] Both insulin and IGF-I induce expression of c-*fos* mRNA in L6 rat skeletal muscle cells.[500]

The physiological effects of insulin in the liver may depend, at least in part, on alterations of proto-oncogene expression. In rat H4-IIE hepatoma cells, which proliferate in response to insulin, addition of the hormone results in increased rate of c-*fos* gene transcription and accumulation of c-*fos* mRNA.[501] This accumulation is dose dependent and reaches a maximum at 30 min. Induction of c-*fos* transcription in H4-IIE cells occurs as early as 5 min after the addition of insulin. This is the most rapid effect of insulin yet shown on the induction of gene expression. In H35 rat hepatoma cells that become quiescent under serum-deprived conditions or by treatment with dexamethasone, insulin induces DNA synthesis; and this induction is associated with an ordered increase in the levels of expression of c-*myc*, c-*fos*, c-K-*ras*, and c-H-*ras* gene transcripts.[502,503] Other growth factors, including EGF, do not affect H35 cells in terms of DNA synthesis or induction of c-*myc* mRNA expression. The phorbol ester PMA also induces c-*myc* and c-*fos* mRNA expression in H35 cells, but it does not induce DNA synthesis and would act on the expression of both proto-oncogenes through a mechanism which is different from that of insulin.

The activity of some oncoproteins may be altered by insulin at the posttranscriptional level. Insulin activates the kinase activity of Raf-1 by increasing its phosphorylation on serine.[504,505] Ras proteins are involved in the mechanism of insulin action, and their functions are exerted upstream of Raf-1. However, Ras activation by insulin does not involve phosphorylation of the Gap protein.[506] The functional significance of altered proto-oncogene expression in insulin-stimulated cells remains to be elucidated.

2. Mechanisms of Regulation of Gene Expression by Insulin

Insulin-induced regulation of gene expression may occur, at least in part, through the action of specific proto-oncogene products, in particular by activation of Ras proteins.[507] The c-Fos and c-Jun proteins, which are involved in the formation of AP-1 transcription activation complexes, could also have a role in the regulation of gene expression by insulin. Regulation of gene expression by insulin is associated with the presence in the regulated genes of *cis*-acting insulin response elements (IREs). Two independent IREs are present in the 5′-flanking side of the human gene coding for glyceraldehyde-3-phosphate dehydrogenase (GADPH).[508] These two IREs are sufficient to direct insulin-inducible GADPH gene expression. Exposure of differentiated 3T3-L1 cells to insulin increased the functional level of a nuclear protein selective for one of the two IREs. Similar changes were observed in the liver of rats subjected to certain nutritional manipulations *in vivo*.

C. EFFECTS OF INSULIN ON CELL PROLIFERATION AND DIFFERENTIATION

Insulin and the insulin-like growth factors, IGF-I and IGF-II, are importantly involved in the regulation of processes associated with DNA synthesis, cell proliferation, and cell differentiation. The three signaling agents may have a complementary role in adult tissues as well as in normal embryo development.[509]

1. Action of Insulin on DNA Synthesis

Insulin action is tightly associated with the initiation of DNA synthesis in different types of cells and tissues. In epithelial cells of mammary gland explants derived from 3-month-old virgin mice the effects of insulin are elicited, at least in part, by an insulin-dependent emergence of DNA polymerase activity.[510] Insulin and IGFs stimulate the incorporation of thymidine into DNA in cultured human skin fibroblasts.[511] In these cells, IGFs may be active at concentrations lower than those required for insulin. The combined action of insulin and IGFs fails to give additive results, suggesting the existence of a common mechanism for both types of polypeptides; whereas combined action of insulin and serum, or IGFs and serum, gives additive results, indicating that serum contains some factor(s) that are capable of complementing or reinforcing the action of insulin.

2. Mechanisms of the Mitogenic Action of Insulin

Insulin has intrinsic mitogenic activity.[512] The hormone plays a central role in the regulation of cell proliferation both *in vivo* and *in vitro*. Insulin is the only hormone required for the continued growth of CHO-K1 cells *in vitro*.[513] When these cells are incubated in the defined medium M-F12 without insulin for 48 to 72 h, the cells accumulate in G_1 phase of the cycle and respond to the addition of physiological concentrations of insulin with a marked (18-fold) increase in the rate of DNA synthesis after a lag of 8 to 18 h, entering in division after 24 h. This effect is initiated by the binding of insulin to its receptor and does not require RNA synthesis. In certain systems *in vitro,* insulin and IGF-I exhibit synergistic effects with growth factors such as EGF and PDGF in cell cycle control and stimulation of DNA synthesis.[514-516] Each factor (insulin, IGF-I, EGF, PDGF) may control discrete events within the cell cycle. Insulin and IGFs may be required for the action of other growth factors including, for example, the action of NGF for the induction of neurite formation in cultured human neuroblastoma cells.[517] Treatment of isolated fat cells with insulin stimulates transferrin uptake and causes redistribution of transferrin receptors from an internal microsomal compartment to the plasma membrane.[518]

The mitogenic effect of insulin can also be observed *in vivo*. The presence of insulin and certain growth factors is required for the optimal growth and development of many tissues *in vivo,* especially fetal tissues. The growth of fetal and adult tissues such as bone tissue may depend on insulin and/or IGF stimulation.[519] Insulin is important in the control of differentiation processes occurring at particular stages of ontogeny. Insulin, as well as glucagon and EGF, play important roles in the control of DNA synthesis and cell proliferation in various endodermally derived organs.[520]

Progesterone and insulin are capable of inducing the maturation of *Xenopus laevis* oocytes but they would act by different mechanisms. Insulin acts synergistically with progesterone to induce meiosis of *Rana pipiens* oocytes *in vitro*.[521] Ras proteins are probably involved in hormone-induced oocyte maturation.[522] Microinjection of a monoclonal antibody (Y13-259) against Ras proteins markedly reduces insulin-induced oocyte maturation but fails to inhibit progesterone induction.[523] Although both insulin and progesterone induce oocyte maturation through the production of a cytosolic meiosis- or maturation-promoting activity (MPA), some steps of the insulin action on this process (including the action of Ras proteins) are different from that induced by steroid hormones.

As mentioned earlier, Ras proteins may have an important role in the transductional mechanism of insulin action. The protein Grb2, which contains in its amino-terminal end one SH2 domain flanked by two SH3 domains, couples receptor tyrosine kinases to a protein, Sos, which activates the Ras signaling pathway.[481,482] However, in contrast to the EGF receptor that interacts directly with the Grb2-Sos complex, activation of the insulin receptor by insulin binding leads to phosphorylation of the two docking proteins IRS-1 and Shc.[524-526] The Grb2-Sos complex is probably involved in insulin-induced conversion of the inactive Ras-guanosine 5′-diphosphate (GDP) complex to the active Ras-GTP complex, which may result in activation of Raf-1 kinase, MAP kinase, and other protein kinases (Figure 1.1).

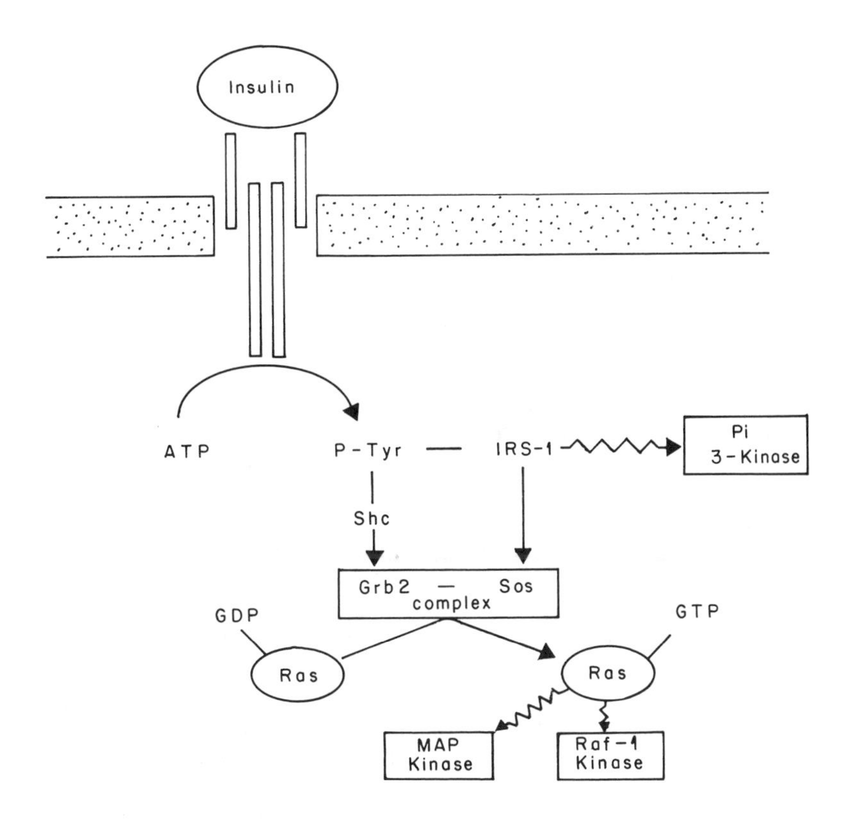

Figure 1.1 Hypothetical model of insulin-induced activation of Ras protein and serine/threonine kinases.

In addition to the Ras-Raf signaling pathway, other pathways may be involved in the mechanisms of the mitogenic action of insulin. The PI 3-kinase may be activated by the IRS-1 insulin receptor substrate, and it functions upstream of Raf and Ras in mediating insulin stimulation of cellular genes associated with a mitogenic response such as the c-*fos* proto-oncogene.[527] Certain protein phosphatases may have a role in the mitogenic action of insulin. SH2 domains contained in protein-tyrosine phosphatase may be phosphorylated by the insulin receptor tyrosine kinase and may bind to the carboxyl-terminus of insulin receptors.[528] Cell cycle progression in mouse fibroblasts stimulated by IGF-I and probably also by insulin depends on calcium entry into the cell, preceded by protein kinase C activation.[529] It is thus clear that many cellular pathways interacting in complex manners are involved in the mechanisms of the mitogenic action of insulin.

VI. ROLE OF INSULIN IN NEOPLASTIC PROCESSES

On the basis of studies performed *in vivo* and *in vitro* it has been recognized that most types of tumor cells depend on insulin for their proliferation. Insulin, as well as transferrin, is essential for the growth of both normal and transformed cells in serum-free media.[530] Malignant cells may require less insulin than normal cells for their growth in culture,[531] but the results from many studies indicate that most tumor cell lines require insulin as an essential component of the medium. Growth of a human breast cancer cell line (MCF-7) is sensitive to physiological concentrations of insulin, which produces a shortening of the G_1 transit time and a marked increase in the fraction of S phase cells by effects that are probably mediated via the insulin receptor.[532] Insulin and IGFs stimulate replication of the clonal cell line PC12, derived from a rat pheochromocytoma.[533] Anchorage-independent growth of murine melanoma cells in serum-free medium is dependent on insulin or melanocyte-stimulating hormone (MSH).[534] Insulin has also been recognized as a potent, specific growth factor in hepatoma cell lines.[535] Insulin stimulates serum-starved human hepatoma cells Hep3B to proliferate in a chemically defined medium, and somatostatin selectively inhibits the insulin-induced growth of Hep3B cells in a dose-dependent manner.[536] The mechanisms involved in such insulin actions are not understood but there is evidence that phospholipids may play a role in mediating the effects of insulin on growth of experimental tumors.[537]

Insulin may be required for the expression of a transformed phenotype. Addition of insulin to BALB/3T3 mouse cells markedly enhances the yield of transformation induced by x-irradiation.[538] This effect is dependent on the concentration of insulin added to the culture medium and is consistently observed in cell cultures irradiated at various dosage levels. Tumor cells producing insulin may exhibit resistance to insulin.[539]

A. DIABETES MELLITUS AND CANCER INCIDENCE

Mammary carcinomas induced in rats by administration of 7,12-dimethylbenz(a)anthracene (DMBA) or MNU are dependent on insulin; and most of these tumors regress, or grow more slowly, after the animals are made diabetic with administration of alloxan, streptozotocin, or diazoxide.[540-542] Prolonged insulin treatment may have remarkable effects on tumor growth in mice. Administration of insulin in pharmacological doses exerts a cocarcinogenic effect on a nonhormonally dependent tumor, squamous cell carcinoma induced in mice by the carcinogen, 3-methylcholanthrene (MCA).[543] The hormone changes tumor evolution, increasing DNA synthesis and inducing morphological alterations in the cancer cells, which affect the expression of differentiated characteristics. In contrast with these experimental observations, little is known about the relationship between diabetes and natural neoplastic diseases occurring in humans and other animal species. In a recent study, diabetes was shown to be a moderate risk factor for cancer of the uterus, vulva, and vagina, as well as the kidney and skin, but not the pancreas.[544] No association between type of treatment, including insulin, and cancer risk was noted. In contrast, the results obtained in another autopsy study indicated that death due to all cancers except cancer of the pancreas was much less common in the patients with diabetes, in both sexes and in each age group, as compared to a nondiabetic control group.[545] The reasons for such differences between diabetic and nondiabetic patients, as well as those of the results obtained in different studies, are unknown.

B. PRODUCTION OF INSULIN-LIKE SUBSTANCES BY TUMORS

A diversity of tumors may produce and secrete *in vivo* and/or *in vitro* substances immunologically cross-reactive with insulin. This phenomenon has been observed in human tumors such as renal adenocarcinoma,[546] carcinomas of the cervix and corpus uteri,[547] Hodgkin and non-Hodgkin lymphomas,[548,549] and mammary and bronchial carcinomas,[550] as well as in murine tumors such as myeloid leukemia and melanoma B16.[551-553] These observations suggest the possibility that insulin or insulin-like substances may act as an autocrine factor for the growth of a variety of tumors. Little is known, however, about the chemical structure of the insulin-like substances produced by non-β-cell tumors. These substances appear to be proteins related to insulin and are capable of inducing hypoglycemia, but they may be different from the pancreatic hormone.

An insulin-related factor (IRF) produced by the mouse teratoma-derived insulin-independent cell line 1246-3A binds to insulin receptors in the same tumor cells and acts in an autostimulatory manner.[554] The biological activities of IRF are similar to those of insulin. IRF was characterized as a heat-labile and acid-stable polypeptide with a molecular weight of 6000 and is composed of two chains connected by disulfide bonds which are necessary for the specific biological activity.[555] The amino acid composition of IRF is similar to that of mouse pancreatic insulins I and II. 1246-3A cells possess high-affinity receptors that bind IRF secreted by the cells into the medium, and the growth of 1246-3A cells is inhibited by an insulin-specific monoclonal antibody.[556,557] These data suggest that at least in some tumor cells insulin may act as an important autocrine factor favoring their selective growth.

C. POSTRECEPTOR DEFECTS OF INSULIN ACTION IN TUMOR CELLS

Little information is available at present about the possibility of postreceptor defects of insulin action occurring in neoplastic cells. In spontaneously transformed cloned rat hepatocytes, lack of a serum requirement for growth is accompanied by several defects in insulin action, including loss of insulin stimulation of protein synthesis and reduction in insulin ability to activate glycogen synthesis from glucose.[558] Postreceptor mechanisms appear to be the most important mechanisms involved in the mediation of insulin-induced desensitization phenomena occurring in a rat hepatoma cell line (H-35) exposed to the action of insulin.[559] Although insulin resistance observed in this hepatoma cell line is accompanied by a 50 to 60% decrease in insulin receptor number, this decrease cannot account for the marked impairment observed in the responses of these cells to insulin action, which should be attributed to postreceptor phenomena functioning abnormally in the tumor cells. In cultured rat hepatoma cells exhibiting different degrees of differentiation it was observed that the number of insulin receptors expressed at the cell surface is higher in the well-differentiated cells than in dedifferentiated cells, and that some postreceptor insulin effects such as the induction of tyrosine aminotransferase (TAT) activity are lost in the more differentiated tumor cells; whereas other effects occurring at the same level such as the conversion of the D form to the I form of glycogen synthase activity may be preserved.[560]

D. INSULIN ACTION AND THE INDUCTION OF DIFFERENTIATION IN NEOPLASTIC CELLS

Insulin and its receptor may have an important role in the processes of differentiation induced *in vitro* by several types of compounds. Induction of differentiation in human leukemia cell lines by calcitriol may be associated with expression of the insulin receptor.[561] Differentiation of the U-937 human monocyte-like cell line by calcitriol or by retinoic acid is accompanied by altered, although opposite, effects on insulin receptors.[562] Cytosolic Ca^{2+} is able to regulate the function of insulin receptors in U-937 cells induced to differentiation by calcitriol, while it remains without effect in the uninduced cells.[563] This phenomenon is synergistic with the action of phorbol esters on the binding of insulin. Retinoids are effective inhibitors of mammary gland carcinogenesis in the rat, and the addition of retinoic acid to media of organ cultures of rat mammary gland carcinomas blocks the stimulatory effect of insulin on DNA synthesis.[564]

E. ONCOGENIC POTENTIAL OF THE INSULIN RECEPTOR

The structure and function of the insulin receptor exhibit striking homologies with oncoproteins of the Src family, in particular with the Ros protein. Overexpression of normal insulin receptors can induce a ligand-dependent transformed phenotype in NIH/3T3 mouse fibroblasts transfected with a bovine papilloma virus/insulin receptor cDNA construct.[565] However, these cells are not capable of inducing tumor formation in nude mice, suggesting that in addition to insulin and its receptor, other factors may be necessary for their full neoplastic transformation capacity.

REFERENCES

1. **Philippe, J.,** Structure and pancreatic expression of the insulin and glucagon genes, *Endocr. Rev.,* 12, 252, 1991.
2. **Bell, G.I.,** Molecular defects in diabetes mellitus, *Diabetes,* 40, 413, 1991.
3. **Castaño, L. and Eisenbarth, G.S.,** Type-I diabetes: a chronic autoimmune disease of human, mouse, and rat, *Annu. Rev. Immunol.,* 8, 647, 1990.
4. **Campbell, I.L. and Harrison, L.C.,** Viruses and cytokines: evidence for multiple role in pancreatic β cell destruction in type 1 insulin-dependent diabetes mellitus, *J. Cell. Biochem.,* 40, 57, 1989.
5. **Dorman, J.S., LaPorte, R.E., Stone, R.A., and Trucco, M.,** Worldwide differences in the incidence of type I diabetes are associated with amino acid variation at position 57 of the HLA-DQ β chain, *Proc. Natl. Acad. Sci. U.S.A.,* 87, 7370, 1990.
6. **Clemmons, D.R.,** Role of peptide growth factors in development of macrovascular complications of diabetes, *Diabetes Care,* 14, 153, 1991.
7. **Straus, D.S.,** Effects of insulin on cellular growth and proliferation, *Life Sci.,* 29, 2131, 1981.
8. **Straus, D.S.,** Growth-stimulatory actions of insulin *in vitro* and *in vivo, Endocr. Rev.,* 5, 356, 1984.
9. **Hill, D.J. and Milner, R.D.G.,** Insulin as a growth factor, *Pediatr. Res.,* 19, 879, 1985.
10. **de Pablo, F., Scott, L.A., and Roth, J.,** Insulin and insulin-like growth factor I in early development: peptides, receptors and biological events, *Endocr. Rev.,* 11, 558, 1990.
11. **Czech, M.P.,** Insulin action, *Am. J. Med.,* 70, 142, 1981.
12. **Kaplan, S.A.,** The insulin receptor, *J. Pediatr.,* 104, 327, 1984.
13. **Rosen, O.M.,** After insulin binds, *Science,* 237, 1452, 1987.
14. **Kahn, C.R. and White, M.F.,** The insulin receptor and the molecular mechanism of insulin action, *J. Clin. Invest.,* 82, 1151, 1988.
15. **Nielsen, J.H., Linde, S., Welinder, B.S., Billestrup, N., and Madsen, O.D.,** Growth hormone is a growth factor for the differentiated pancreatic β-cell, *Mol. Endocrinol.,* 13, 165, 1989.
16. **Goodison, S., Kenna, S., and Ashcroft, S.J.H.,** Control of insulin gene expression by glucose, *Biochem. J.,* 285, 563, 1992.

17. **Hedeskov, C.J.,** Mechanism of glucose-induced insulin secretion, *Physiol. Rev.,* 60, 442, 1980.
18. **Zawalich, W.S.,** Modulation of insulin secretion from β-cells by phosphoinositide-derived second-messenger molecules, *Diabetes,* 37, 137, 1988.
19. **Zawalich, W.S., Diaz, V.A., and Zawalich, K.C.,** Influence of cAMP and calcium on [^{3}H]inositol efflux, inositol phosphate accumulation, and insulin release from isolated rat islets, *Diabetes,* 37, 1478, 1988.
20. **Zawalich, W.S. and Rasmussen, H.,** Control of insulin secretion: a model involving Ca^{2+}, cAMP and diacylglycerol, *Mol. Cell. Endocrinol.,* 70, 119, 1990.
21. **Jones, P.M. and Persaud, S.J.,** Arachidonic acid as a second messenger in glucose-induced insulin secretion from pancratic β-cells, *J. Endocrinol.,* 137, 7, 1993.
22. **Metz, S.A. and Dunlop, M.,** Stimulation of insulin release by phospholipase D. A potential role for endogenous phosphatidic acid in pancreatic islet function, *Biochem. J.,* 270, 427, 1990.
23. **Weir, G.C. and Bonner-Weir, S.,** Islets of Langerhans —the puzzle of intraislet interactions and their relevance to diabetes, *J. Clin. Invest.,* 85, 983, 1990.
24. **McDaniel, M.L., Hughes, J.H., Wolf, B.A., Easom, R.A., and Turk, J.W.,** Descriptive and mechanistic considerations of interleukin 1 and insulin secretion, *Diabetes,* 37, 1311, 1988.
25. **Campbell, I.L., Cutri, A., Wilson, A., and Harrison, L.C.,** Evidence for IL-6 production by and effects on the pancreatic β-cell, *J. Immunol.,* 143, 1188, 1989.
26. **Sandler, S., Bendtzen, K., Eizirik, D.L., and Welsh, M.,** Interleukin-6 affects insulin secretion and glucose metabolism of rat pancreatic islets *in vitro, Endocrinology,* 126, 1288, 1990.
27. **Southern, C., Schulster, D., and Green, I.C.,** Inhibition of insulin secretion from rat islets of Langerhans by interleukin-6 — An effect distinct from that of interleukin-1, *Biochem. J.,* 272, 243, 1990.
28. **Totsuka, Y., Tabuchi, M., Kojima, I., Etow, Y., Shibai, H., and Ogata, E.,** Stimulation of insulin secretion by transforming growth factor-β, *Biochem. Biophys. Res. Commun.,* 158, 1060, 1989.
29. **Koranyi, L., James, D.E., Kraegen, E.W., and Permutt, M.A.,** Feedback inhibition of insulin gene expression by insulin, *J. Clin. Invest.,* 89, 432, 1992.
30. **Rosenzweig, J.L., Havrankova, J., Lesniak, M.A., Brownstein, M., and Roth, J.,** Insulin is ubiquitous in extrapancreatic tissues of rats and humans, *Proc. Natl. Acad. Sci. U.S.A.,* 77, 572, 1980.
31. **Murakami, K., Taniguchi, H., and Baba, S.,** Presence of insulin-like immunoreactivity and its biosynthesis in rat and human parotid gland, *Diabetologia,* 22, 358, 1982.
32. **Raizada, M.K.,** Localization of insulin-like immunoreactivity in the neurons from primary cultures of rat brain, *Exp. Cell Res.,* 143, 351, 1983.
33. **Conlon, J.M., Reinecke, M., Thorndyke, M.C., and Falkmer, S.,** Insulin and other islet hormones (somatostatin, glucagon and PP) in the neuroendocrine system of some lower vertebrates and that of invertebrates — a minireview, *Horm. Metab. Res.,* 20, 406, 1988.
34. **Owerbach, D., Bell, G.I., Rutter, W.J., and Shows, T.B.,** The insulin gene is located on chromosome 11 in humans, *Nature (London),* 286, 82, 1980.
35. **Owerbach, D., Bell, G.I., Rutter, W.J., Brown, J.A., and Shows, T.B.,** The insulin gene is located on the short arm of chromosome 11 in humans, *Diabetes,* 30, 267, 1981.
36. **Chaganti, R.S.K., Jhanwar, S.C., Antonarakis, S.E., and Hayward, W.S.,** Germ-line chromosomal localization in chromosome 11p linkage: parathyroid hormone, β-globin, c-Ha-ras-1, and insulin, *Somatic Cell Mol. Genet.,* 11, 197, 1985.
37. **Zabel, B.U., Kronenberg, H.M., Bell, G.I., and Shows, T.B.,** Chromosome mapping of genes on the short arm of human chromosome 11: parathyroid hormone gene is at 11p15 together with the genes for insulin, c-Harvey-*ras* 1, and β-hemoglobin, *Cytogenet. Cell Genet.,* 39, 200, 1985.
38. **Glaser, T., Lewis, W.H., Bruns, G.A.P., Watkins, P.C., Rogler, C.E., Shows, T.B., Powers, V.E., Willard, H.F., Goguen, J.M., Simola, K.O.J., and Housman, D.E.,** The β-subunit of follicle-stimulating hormone is deleted in patients with aniridia and Wilms' tumour, allowing a further definition of the WAGR locus, *Nature (London),* 321, 882, 1986.
39. **Yunis, J.J. and Ramsay, N.K.C.,** Familial occurrence of the aniridia-Wilms' tumor syndrome with deletion 11p13-14.1, *J. Pediatr.,* 96, 1027, 1980.
40. **Kaneko, Y., Egues, M.C., and Rowley, J.D.,** Interstitial deletion of short arm of chromosome 11 limited to Wilms' tumor cells in a patient without aniridia, *Cancer Res.,* 41, 4577, 1981.
41. **Koufos, A., Hansen, M.F., Lampkin, B.C., Workman, M.L., Copeland, N.G., Jenkins, N.A., and Cavenee, W.K.,** Loss of alleles at loci on human chromosome 11 during genesis of Wilms' tumour, *Nature (London),* 309, 170, 1984.

42. **Orkin, S.H., Goldman, D.S., and Sallan, S.E.,** Development of homozygosity for chromosome 11p markers in Wilms' tumour, *Nature (London),* 309, 172, 1984.
43. **Fearon, E.R., Vogelstein, B., and Feinberg, A.P.,** Somatic deletion and duplication of genes on chromosome 11 in Wilms' tumours, *Nature (London),* 309, 176, 1984.
44. **Solomon, E.,** Recessive mutation in aetiology of Wilms' tumour, *Nature (London),* 309, 111, 1984.
45. **Reeve, A.E., Housiaux, P.J., Gardner, R.J.M., Chewings, W.E., Grindley, R.M., and Millow, L.J.,** Loss of a Harvey *ras* allele in sporadic Wilms' tumour, *Nature (London),* 309, 174, 1984.
46. **Rogler, C.E., Sherman, M., Su, C.Y., Shafritz, D.A., Summers, J., Shows, T.B., Henderson, A., and Kew, M.,** Deletion in chromosome 11p associated with a hepatitis B integration site in hepatocellular carcinoma, *Science,* 230, 319, 1985.
47. **Fearon, E.R., Feinberg, A.P., Hamilton, S.H., and Vogelstein, B.,** Loss of genes on the short arm of chromosome 11 in bladder cancer, *Nature (London),* 318, 377, 1985.
48. **Geiser, A.G. and Stanbridge, E.J.,** A review of the evidence for tumor suppressor genes, *Crit. Rev. Oncogenesis,* 1, 261, 1989.
49. **Bell, G.I., Swain, W.F., Pictet, R., Cordell, B., Goodman, H.M., and Rutter, W.J.,** Nucleotide sequence of a cDNA clone encoding human preproinsulin, *Nature (London),* 282, 525, 1979.
50. **Bell, G.I., Pictet, R.L., Rutter, W.J., Cordell, B., Tischer, E., and Goodman, H.M.,** Sequence of the human insulin gene, *Nature (London),* 284, 26, 1980.
51. **Sures, I., Goeddel, D.V., Gray, A., and Ullrich, A.,** Nucleotide sequence of human preproinsulin complementary DNA, *Science,* 208, 57, 1980.
52. **Bell, G.I., Pictet, R., and Rutter, W.J.,** Analysis of the regions flanking the human insulin gene and sequence of an Alu family member, *Nucleic Acids Res.,* 8, 4091, 1980.
53. **Ullrich, A., Dull, T.J., Gray, A., Brosius, J., and Sures, I.,** Genetic variation in the human insulin gene, *Science,* 209, 612, 1980.
54. **Steiner, D.F., Chan, S.J., Welsh, J.M., and Kwok, S.C.M.,** Structure and evolution of the insulin gene, *Annu. Rev. Genet.,* 19, 463, 1985.
55. **Bell, G.I., Selby, M.J., and Rutter, W.J.,** The highly polymorphic region near the human insulin gene is composed of simple tandemly repeating sequences, *Nature (London),* 295, 31, 1982.
56. **Ullrich, A., Dull, T.J., Gray, A., Philips, J.A., III, and Peter, S.,** Variation in the sequence and modification state of the human insulin gene flanking regions, *Nucleic Acids Res.,* 10, 2225, 1982.
57. **Rotwein, P., Chyn, R., Chirgwin, J., Cordell, B., Goodman, H.M., and Permutt, M.A.,** Polymorphism in the 5′-flanking region of the human insulin gene and its possible relation to type 2 diabetes, *Science,* 213, 117, 1981.
58. **Owerbach, D. and Nerup, J.,** Restriction fragment length polymorphism of the insulin gene in diabetes mellitus, *Diabetes,* 31, 275, 1982.
59. **Owerbach, D., Poulsen, S., Billesbolle, P., and Nerup, J.,** DNA insertion sequences near the insulin gene affect glucose regulation, *Lancet,* 1, 880, 1982.
60. **Rotwein, P.S., Chirgwin, J., Province, M., Knowler, W.C., Pettitt, D.J., Cordell, B., Goodman, H.M., and Permutt, M.A.,** Polymorphism in the 5′ flanking region of the human insulin gene: a genetic marker for non-insulin-dependent diabetes, *N. Engl. J. Med.,* 308, 65, 1983.
61. **Yokoyama, S.,** Polymorphism in the 5′-flanking region of the human insulin gene and the incidence of diabetes, *Am. J. Hum. Genet.,* 35, 193, 1983.
62. **Permutt, M.A., Rotwein, P., Andreone, T., Ward, W.K., and Porte, D., Jr.,** Islet β-cell function and polymorphism in the 5′-flanking region of the human insulin gene, *Diabetes,* 34, 311, 1985.
63. **Chakravarti, A., Elbein, S.C., and Perlmutt, M.A.,** Evidence for increased recombination near the human insulin gene: implication for disease association studies, *Proc. Natl. Acad. Sci. U.S.A.,* 83, 1045, 1986.
64. **Tager, H.S.,** Abnormal products of the human insulin gene, *Diabetes,* 33, 693, 1984.
65. **Vinik, A. and Bell, G.,** Mutant insulin syndromes, *Horm. Metab. Res.,* 20, 1, 1988.
66. **Nakagawa, S.H. and Tager, H.S.,** Role of the phenylalanine B25 side chain in directing insulin interaction with its receptor: steric and conformational effects, *J. Biol. Chem.,* 261, 7332, 1986.
67. **Rotwein, P., Naylor, S.L., and Chirgwin, J.M.,** Human insulin-related DNA sequences map to chromosomes 2 and 11, *Somatic Cell Mol. Genet.,* 12, 625, 1986.
68. **Lomedico, P., Rosenthal, N., Efstratiadis, A., Gilbert, W., Kolodner, R., and Tizard, R.,** The structure and evolution of the two nonallelic rat preproinsulin genes, *Cell,* 18, 545, 1979.
69. **Soares, M.B., Schon, E., Henderson, A., Karathanasis, S.K., Cate, R., Zeitlin, S., Chirgwin, J., and Efstratiadis, A.,** RNA-mediated gene duplication: the rat preproinsulin I gene is a functional retroposon, *Mol. Cell. Biol.,* 5, 2090, 1985.

70. **Todd, S., Yoshida, M.C., Fang, X.E., McDonald, L., Jacobs, J., Heinrich, G., Bell, G.I., Naylor, S.L., and Sakaguchi, A.Y.,** Genes for insulin I and II, parathyroid hormone, and calcitonin are on rat chromosome 1, *Biochem. Biophys. Res. Commun.,* 131, 1175, 1985.
71. **Kozak, M.,** Point mutations close to the AUG initiator codon affect the efficiency of translation of rat preproinsulin *in vivo, Nature (London),* 308, 241, 1984.
72. **Giddings, S.J. and Carnaghi, L.R.,** The two nonallelic rat insulin mRNAs and pre-mRNAs are regulated coordinately *in vivo, J. Biol. Chem.,* 263, 3845, 1988.
73. **Sample, C.E. and Steiner, D.F.,** Tissue-specific binding of a nuclear factor to the insulin gene promoter, *FEBS Lett.,* 222, 332, 1987.
74. **Hwung, Y.-P., Crowe, D.T., Wang, L.-H., Tsai, S.Y., and Tsai, M.-J.,** The COUP transcription factor binds to an upstream promoter element of the rat insulin II gene, *Mol. Cell. Biol.,* 8, 2070, 1988.
75. **Lakshmikumaran, M.S., D'Ambrosio, E., Laimins, L.A., Lin, D.T., and Furano, A.V.,** Long interspersed repeated DNA (LINE) causes polymorphism at the rat insulin 1 locus, *Mol. Cell. Biol.,* 5, 2197, 1985.
76. **Winter, W.E., Beppu, H., Maclaren, N.K., Cooper, D.L., Bell, G.I., and Wakeland, E.K.,** Restriction-fragment-length polymorphisms of 5′-flanking region of insulin I gene in BB and other rat strains. Absence of an association with IDDM, *Diabetes,* 36, 193, 1987.
77. **Laimins, L., Holmgren-König, M., and Khoury, G.,** Transcriptional "silencer" element in rat repetitive sequences associated with the rat insulin 1 gene locus, *Proc. Natl. Acad. Sci. U.S.A.,* 83, 3151, 1986.
78. **Meneses, P. and Ortiz, M.D.,** A protein extract from *Drosophila* with insulin-like activity, *Comp. Biochem. Physiol.,* 51A, 483, 1975.
79. **LeRoith, D., Lesniak, M.A., and Roth, J.,** Insulin in insects and annelids, *Diabetes,* 30, 70, 1981.
80. **Nagasawa, H., Kataoka, H., Isogai, A., Tamura, S., Suzuki, A., Ishizaki, H., Mizoguchi, A., Fujiwara, Y., and Susuki, A.,** Amino-terminal amino acid sequence of the silkworm prothoracicotrophic hormone: homology with insulin, *Science,* 226, 1344, 1984.
81. **Salem, M.A.M. and Phares, C.K.,** Insulin-like effects in the rat of the purified growth factor from *Spirometra mansonoides* plerocercoids, *Proc. Soc. Exp. Biol. Med.,* 285, 31, 1987.
82. **Chan, S.J., Cao, Q.-P., and Steiner, D.F.,** Evolution of the insulin superfamily: cloning of a hybrid insulin/insulin-like growth factor cDNA from amphioxus, *Proc. Natl. Acad. Sci. U.S.A.,* 87, 9319, 1990.
83. **Fawell, S.E. and Lenard, J.,** A specific insulin receptor and tyrosine kinase activity in the membranes of *Neurospora crassa, Biochem. Biophys. Res. Commun.,* 155, 59, 1988.
84. **LeRoith, D., Shiloach, J., Heffron, R., Rubinowitz, C., Tanenbaum, R., and Roth, J.,** Insulin-related material in microbes: similarities and differences from mammalian insulins, *Can. J. Biochem. Cell Biol.,* 63, 839, 1985.
85. **Leiter, E.H., Fewell, J.W., and Kuff, E.L.,** Glucose induces intracisternal type A retroviral gene transcription and translation in pancreatic β cells, *J. Exp. Med.,* 163, 87, 1986.
86. **Yamashita, S., Tobinaga, T., Ashizawa, K., Nagayama, Y., Yokota, A., Harakawa, S., Inoue, S., Hirayu, H., Izumi, M., and Nagataki, S.,** Glucose stimulation of proto-oncogene expression and deoxyribonucleic acid synthesis in rat islet cell line, *Endocrinology,* 123, 1825, 1988.
87. **Inagaki, N., Maekawa, T., Sudo, T., Ishii, S., Seino, Y., and Imura, H.,** c-Jun represses the human insulin promoter activity that depends on multiple cAMP response elements, *Proc. Natl. Acad. Sci. U.S.A.,* 89, 1045, 1992.
88. **Welsh, M., Welsh, N., Nilsson, T., Arkhammar, P., Pepinsky, R.B., Steiner, D.F., and Berggren, P.-O.,** Stimulation of pancreatic islet β-cell replication by oncogenes, *Proc. Natl. Acad. Sci. U.S.A.,* 85, 116, 1988.
89. **Hanahan, D.,** Heritable formation of pancreatic β-cell tumours in transgenic mice expressing recombinant insulin/simian virus 40 oncogenes, *Science,* 315, 115, 1985.
90. **Efrat, S., Baekkeskov, S., Lane, D., and Hanahan, D.,** Coordinate expression of the endogenous p53 gene in β cells of transgenic mice expressing hybrid insulin-SV40 T antigen genes, *EMBO J.,* 6, 2699, 1987.
91. **Bucchini, D., Ripoche, M.-A., Stinnakre, M.-G., Lorès, P., Monthioux, E., Absil, J., Lepesant, J.-A., Pictet, R., and Jami, J.,** Pancreatic expression of human insulin gene in transgenic mice, *Proc. Natl. Acad. Sci. U.S.A.,* 83, 2511, 1986.
92. **Jami, J.,** Expression du gène de l'insuline humaine chez les souris transgéniques, *C.R. Soc. Biol. (Paris),* 181, 475, 1987.

93. **Selden, R.F., Skoskewicz, M.J., Howie, K.B., Russell, P.S., and Goodman, H.M.,** Regulation of human insulin gene expression in transgenic mice, *Nature (London),* 321, 525, 1986.
94. **Adams, T.E., Alpert, S., and Hanahan, D.,** Non-tolerance and autoantibodies to a transgenic self antigen expressed in pancreatic β cells, *Nature (London),* 325, 223, 1987.
95. **Goldfine, I.D.,** The insulin receptor: molecular biology and transmembrane signaling, *Endocr. Rev.,* 8, 235, 1987.
96. **Gorden, P., Arakaki, R., Collier, E., and Carpentier, J.-L.,** Biosynthesis and regulation of the insulin receptor, *Yale J. Biol. Med.,* 62, 521, 1989.
97. **Rosen, O.M.,** Structure and function of insulin receptors, *Diabetes,* 38, 1508, 1989.
98. **Becker, A.B. and Roth, R.A.,** Insulin receptor structure and function in normal and pathological conditions, *Annu. Rev. Med.,* 41, 99, 1990.
99. **Olefsky, J.M.,** The insulin receptor: a multifunctional protein, *Diabetes,* 39, 1009, 1990.
100. **Häring, H.U.,** The insulin receptor: signalling mechanism and contribution to the pathogenesis of insulin resistance, *Diabetologia,* 34, 848, 1991.
101. **Walker, D.H., Kuppuswamy, D., Vivanathan, A., and Pike, L.J.,** Substrate specificity and kinetic mechanism of human placental insulin receptor/kinase, *Biochemistry,* 26, 1428, 1987.
102. **Wang, C., Goldfine, I.D., Fujita-Yamaguchi, Y., Gattner, H.-G., Brandenburg, D., and De Meyts, P.,** Negative and positive site-site interactions, and their modulation by pH, insulin analogs, and monoclonal antibodies, are preserved in the purified insulin receptor, *Proc. Natl. Acad. Sci. U.S.A.,* 85, 8400, 1988.
103. **O'Hare, T. and Pilch, P.F.,** Intrinsic kinase activity of the insulin receptor. The intact ($\alpha_2\beta_2$) insulin receptor from rat liver contains a kinase domain with greater intrinsic activity than the intact insulin receptor from human placenta, *Proc. Natl. Acad. Sci. U.S.A.,* 264, 602, 1989.
104. **Morrison, B.D., Feltz, S.M., and Pessin, J.E.,** Polylysine activates the insulin-dependent insulin receptor protein kinase, *J. Biol. Chem.,* 264, 9994, 1989.
105. **Yonezawa, K. and Roth, R.A.,** Various proteins modulate the kinase activity of the insulin receptor, *FASEB J.,* 4, 194, 1990.
106. **Yang-Feng, T.L., Francke, U., and Ullrich, A.,** Gene for human insulin receptor: localization to site on chromosome 19 involved in pre-B-cell leukemia, *Science,* 228, 728, 1985.
107. **Ebina, Y., Ellis, L., Jarnagin, K., Edery, M., Graf, L., Clauser, E., Ou, J., Masiarz, F., Kan, Y.W., Goldfine, I.D., Roth, R.A., and Rutter, W.J.,** The human insulin receptor cDNA: the structural basis for hormone-activated transmembrane signalling, *Cell,* 40, 747, 1985.
108. **Elbein, S.C., Corsetti, L., Ullrich, A., and Permutt, M.A.,** Multiple restriction fragment length polymorphisms at the insulin receptor locus: a highly informative marker for linkage analysis, *Proc. Natl. Acad. Sci. U.S.A.,* 83, 5223, 1986.
109. **Araki, E., Shimada, F., Uzawa, H., Mori, M., and Ebina, Y.,** Characterization of the promoter region of the human insulin receptor gene: evidence for promoter activity, *J. Biol. Chem.,* 262, 16186, 1987.
110. **Seino, S., Seino, M., Nishi, S., and Bell, G.I.,** Structure of the human insulin receptor gene and characterization of its promoter, *Proc. Natl. Acad. Sci. U.S.A.,* 86, 114, 1989.
111. **Seino, S., Seino, M., and Bell, G.I.,** Human insulin-receptor gene. Partial sequence and amplification of exons by polymerase chain reaction, *Diabetes,* 39, 123, 1990.
112. **Seino, S., Seino, M., and Bell, G.I.,** Human insulin-receptor gene, *Diabetes,* 39, 129, 1990.
113. **Ebina, Y., Edery, M., Ellis, L., Standring, D., Beaudoin, J., Roth, R.A., and Rutter, W.J.,** Expression of a functional human insulin receptor from a cloned cDNA in Chinese hamster ovary cells, *Proc. Natl. Acad. Sci. U.S.A.,* 82, 8014, 1985.
114. **Whittaker, J., Okamoto, A.K., Thys, R., Bell, G.I., Steiner, D.F., and Hofmann, C.A.,** High-level expression of human insulin receptor cDNA in mouse NIH 3T3 cells, *Proc. Natl. Acad. Sci. U.S.A.,* 84, 5237, 1987.
115. **Whittaker, J. and Okamoto, A.,** Secretion of soluble functional insulin receptors by transfected NIH3T3 cells, *J. Biol. Chem.,* 263, 3063, 1988.
116. **Ellis, L., Levitan, A., Cobb, M.H., and Ramos, P.,** Efficient expression in insect cells of a soluble, active human insulin receptor parotein-tyrosine kinase domain by use of a baculovirus vector, *J. Virol.,* 62, 1634, 1988.
117. **Rohilla, A.M.K., Anderson, C., Wood, W.M., and Berhanu, P.,** Insulin down-regulates the steady-state level of its receptor's messenger ribonucleic acid, *Biochem. Biophys. Res. Commun.,* 175, 527, 1991.

118. **Seino, S. and Bell, G.I.,** Alternative splicing of human insulin receptor messenger RNA, *Biochem. Biophys. Res. Commun.,* 159, 312, 1989.
119. **Mosthaf, L., Grako, K., Dull, T.J., Coussens, L., Ullrich, A., and McClain, D.A.,** Functionally distinct insulin receptors generated by tissue-specific alternative splicing, *EMBO J.,* 9, 2409, 1990.
120. **Petruzzelli, L., Herrera, R., Arenas-Garcia, R., Fernandez, R., Birnbaum, M.J., and Rosen, O.M.,** Isolation of a *Drosophila* genomic sequence homologous to the kinase domain of the human insulin receptor and detection of the phosphorylated *Drosophila* receptor with an anti-peptide antibody, *Proc. Natl. Acad. Sci. U.S.A.,* 83, 4710, 1986.
121. **Nishida, Y., Hata, M., Nishizuka, Y., Rutter, W.J., and Ebina, Y.,** Cloning of a *Drosophila* cDNA encoding a polypeptide similar to the human insulin receptor precursor, *Biochem. Biophys. Res. Commun.,* 141, 474, 1986.
122. **Garofalo, R.S. and Rosen, O.M.,** Tissue localization of *Drosophila melanogaster* insulin receptor transcripts during development, *Mol. Cell. Biol.,* 8, 1638, 1988.
123. **Thompson, K.L., Decker, S.J., and Rosner, M.R.,** Identification of a novel receptor in *Drosophila* for both epidermal growth factor and insulin, *Proc. Natl. Acad. Sci. U.S.A.,* 82, 8443, 1985.
124. **Janicot, M., Flores-Riveros, J.R., and Lane, M.D.,** The insulin-like growth factor 1 (IGF-1) receptor is responsible for mediating the effects of insulin, IGF-1, and IGF-2 in *Xenopus laevis* oocytes, *J. Biol. Chem.,* 266, 9382, 1991.
125. **Schaefer, E.M., Erickson, H.P., Federwisch, M., Wollmer, A., and Ellis, L.,** Structural organization of the human insulin receptor ectodomain, *J. Biol. Chem.,* 267, 23393, 1992.
126. **Yamada, K., Goncalves, E., Kahn, C.R., and Shoelson, S.E.,** Substitution of the insulin receptor transmembrane domain with the c-*neu*/*erbB2* transmembrane domain constitutively activates the insulin receptor kinase *in vitro, J. Biol. Chem.,* 267, 12452, 1992.
127. **Kathuria, S., Hartman, S., Grunfeld, C., Ramachandran, J., and Fujita-Yamaguchi, Y.,** Differential sensitivity of two functions of the insulin receptor to the associated proteolysis: kinase action and hormone binding, *Proc. Natl. Acad. Sci. U.S.A.,* 83, 8570, 1986.
128. **Chvatchko, Y., Gazzano, H., Van Obberghen, E., and Fehlmann, M.,** Subunit arrangement of insulin receptors in hepatoma cells, *Mol. Cell. Endocrinol.,* 36, 59, 1984.
129. **Maturo, J.M., III and Hollenberg, M.D.,** Distinct hydrodynamic forms of the insulin receptor: electrophoretic analysis of the R_I and R_{II} species, *Can. J. Physiol. Pharmacol.,* 63, 987, 1985.
130. **Böni-Schnetzler, M., Rubin, J.B., and Pilch, P.F.,** Structural requirement for the transmembrane activation of the insulin receptor kinase, *J. Biol. Chem.,* 261, 15281, 1986.
131. **Böni-Schnetzler, M., Scott, W., Waugh, S.M., DiBella, E., and Pilch, P.F.,** The insulin receptor. Structural basis for high affinity ligand binding, *J. Biol. Chem.,* 262, 8395, 1987.
132. **Böni-Schnetzler, M., Kaligian, A., DelVecchio, R., and Pilch, P.F.,** Ligand-dependent intersubunit association within the insulin receptor complex activates its intrinsic kinase activity, *J. Biol. Chem.,* 263, 6822, 1988.
133. **O'Hare, T. and Pilch, P.F.,** Separation and characterization of three insulin receptor species that differ in subunit composition, *Biochemistry,* 27, 5693, 1988.
134. **Wilden, P.A., Morrison, B.D., and Pessin, J.E.,** Wheat germ agglutinin stimulation of $\alpha\beta$ heterodimeric insulin receptor β-subunit autophosphorylation by noncovalent association into an $\alpha_2\beta_2$ heterodimeric state, *Endocrinology,* 124, 971, 1989.
135. **McElduff, A., Comi, R.J., and Grunberger, G.,** Structural difference of the insulin receptors from circulating monocytes and erythrocytes, *Biochem. Biophys. Res. Commun.,* 133, 1175, 1985.
136. **Newman, J.D. and Harrison, L.C.,** Homogeneous bivalent insulin receptor: purification using insulin coupled to 1,1′-carbonyldiimidazole-activated agarose, *Biochem. Biophys. Res. Commun.,* 132, 1059, 1985.
137. **Yip, C.C. and Jack, E.,** Insulin receptors are bivalent as demonstrated by photoaffinity labeling, *J. Biol. Chem.,* 267, 13131, 1992.
138. **Pang, D.T. and Shafer, J.A.,** Evidence that insulin receptor from human placenta has a high affinity for only one molecule of insulin, *J. Biol. Chem.,* 259, 8589, 1984.
139. **Sweet, L.J., Morrison, B.D., and Pessin, J.E.,** Isolation of functional $\alpha\beta$ heterodimers from the purified human placental $\alpha_2\beta_2$ heterotetrameric insulin receptor complex. A structural basis for insulin binding heterogeneity, *J. Biol. Chem.,* 262, 6939, 1987.
140. **Yip, C.C., Hsu, H., Patel, R.G., Hawley, D.M., Maddux, B.A., and Goldfine, I.D.,** Localization of the insulin-binding site to the cysteine-rich region of the insulin receptor α-subunit, *Biochem. Biophys. Res. Commun.,* 157, 321, 1989.

141. **Zhang, B. and Roth, R.A.,** Binding properties of chimeric insulin receptors containing the cysteine-rich domain of either the insulin-like growth factor I receptor or the insulin receptor related receptor, *Biochemistry,* 30, 5113, 1991.
142. **Hua, Q.X., Shoelson, S.E., Kochoyan, M., and Weiss, M.A.,** Receptor binding redefined by a structural switch in a mutant human insulin, *Nature (London),* 354, 238, 1991.
143. **Massagué, J. and Czech, M.P.,** Role of disulfides in the subunit structure of the insulin receptor: reduction of class I disulfides does not impair transmembrane signalling, *J. Biol. Chem.,* 257, 6729, 1982.
144. **Jarett, L. and Smith, R.M.,** Partial disruption of naturally occurring groups of insulin receptors on adipocyte plasma membranes by dithiothreitol and *N*-ethylmaleimide: the role of disulfide bonds, *Proc. Natl. Acad. Sci. U.S.A.,* 80, 1023, 1983.
145. **Maturo, J.M., III, Hollenberg, M.D., and Aglio, L.S.,** Insulin receptor: insulin-modulated interconversion between distinct molecular forms involving disulfide-sulfydryl exchange, *Biochemistry,* 22, 2579, 1983.
146. **Aglio, L.S., Maturo, J.M., III, and Hollenberg, M.D.,** Receptors for insulin and epidermal growth factor: interaction with organomercurial agarose, *J. Cell. Biochem.,* 28, 143, 1985.
147. **Pike, L.J., Eakes, A.T., and Krebs, E.G.,** Characterization of affinity-purified insulin receptor/kinase. Effects of dithiothreitol on receptor/kinase function, *J. Biol. Chem.,* 261, 3782, 1986.
148. **Hedo, J.A., Kahn, C.R., Hayashi, M., Yamada, K.M., and Kasuga, M.,** Biosynthesis and glycosylation of the insulin receptor: evidence for a single polypeptide precursor of the two major subunits, *J. Biol. Chem.,* 258, 10020, 1983.
149. **Lane, M.D., Ronnett, G., Slieker, L.J., Kohanski, R.A., and Olson, T.L.,** Post-translational processing and activation of insulin and EGF receptors, *Biochimie,* 67, 1069, 1985.
150. **Fehlmann, M., Peyron, J.-F., Samson, M., Van Obberghen, E., Brandenburg, D., and Brosette, N.,** Molecular association between major histocompatibility complex class I antigens and insulin receptors in mouse liver membranes, *Proc. Natl. Acad. Sci. U.S.A.,* 82, 8634, 1985.
151. **Phillips, M.L., Moule, M.L., Delovitch, T.L., and Yip, C.C.,** Class I histocompatibility antigens and insulin receptors: evidence for interactions, *Proc. Natl. Acad. Sci. U.S.A.,* 83, 3474, 1986.
152. **Sterkers, G., Zeliszewski, D., Chaussée, A.M., Deschamps, I., Font, M.P., Freidel, C., Hors, J., Betuel, H., Dausset, J., and Levy, J.P.,** HLA-DQ rather than HLA-DR region might be involved in dominant nonsusceptibility to diabetes, *Proc. Natl. Acad. Sci. U.S.A.,* 85, 6473, 1988.
153. **Hedo, J.A. and Gorden, P.,** Biosynthesis of the insulin receptor, *Horm. Metab. Res.,* 17, 487, 1985.
154. **Olson, T.S. and Lane, M.D.,** Post-translational acquisition of insulin binding activity by the insulin proreceptor. Correlation to recognition by autoimmune antibody, *J. Biol. Chem.,* 262, 6816, 1987.
155. **Lane, M.D., Slieker, L.J., Olson, T.S., and Martensen, T.M.,** Post-translational acquisition of ligand-binding and tyrosine kinase-domain function by the epidermal growth factor and insulin receptors, *J. Receptor Res.,* 7, 321, 1987.
156. **Olson, T.S., Bamberger, M.J., and Lane, M.D.,** Post-translational changes in tertiary and quaternary structure of the insulin proreceptor. Correlation with acquisition of function, *J. Biol. Chem.,* 263, 7342, 1988.
157. **Kobayashi, M., Sasaoka, T., Takata, Y., Ishibashi, O., Sugibayashi, M., Shigeta, Y., Hisatomi, A., Nakamura, E., Tamaki, M., and Teraoka, H.,** Insulin resistance by unprocessed insulin proreceptors point mutation at the cleavage site, *Biochem. Biophys. Res. Commun.,* 153, 657, 1988.
158. **Yoshimasa, Y., Seino, S., Whittaker, J., Kakehi, T., Kosaki, A., Kuzuya, H., Imura, H., Bell, G.I., and Steiner, D.F.,** Insulin-resistant diabetes due to a point mutation that prevents insulin proreceptor processing, *Science,* 240, 784, 1988.
159. **Forsayeth, J., Maddux, B., and Goldfine, I.D.,** Biosynthesis and processing of the human insulin receptor, *Diabetes,* 35, 837, 1986.
160. **Burant, C.F., Treutelaar, M.K., Block, N.E., and Buse, M.G.,** Structural differences between liver- and muscle-derived insulin receptors in rats, *J. Biol. Chem.,* 261, 14361, 1986.
161. **Bassas, L., de Pablo, F., Lesniak, M.A., and Roth, J.,** The insulin receptors of chick embryo show tissue-specific structural differences which parallel those of the insulin-like growth factor I receptors, *Endocrinology,* 121, 1468, 1987.
162. **Hedo, J.A., Collier, E., and Watkinson, A.,** Myristyl and palmityl acylation of the insulin receptor, *J. Biol. Chem.,* 262, 954, 1987.
163. **Schenker, E. and Kohanski, R.A.,** Conformational states of the insulin receptor, *Biochem. Biophys. Res. Commun.,* 157, 140, 1988.

164. **Forsayeth, J.R., Montemurro, A., Maddux, B.A., DePirro, R., and Goldfine, I.D.,** Effect of monoclonal antibodies on human insulin receptor autophosphorylation, negative cooperativity, and down-regulation, *J. Biol. Chem.,* 262, 4134, 1987.
165. **Thomopoulos, P.,** Les récepteurs insuliniques des cellules sanguines circulantes, *Diabete Métab.,* 7, 207, 1981.
166. **Donner, D.B. and Yonkers, K.,** Hormone-induced conformational changes in the hepatic insulin receptor, *J. Biol. Chem.,* 258, 9413, 1983.
167. **Lipson, K.E., Yamada, K., Kolhatkar, A.A., and Donner, D.B.,** Relationship between the affinity and proteolysis of the insulin receptor. Evidence that higher affinity receptors are preferentially degraded, *J. Biol. Chem.,* 261, 10833, 1986.
168. **Ridge, K.D., Hofmann, K., and Finn, F.M.,** ATP sensitizes the insulin receptor to insulin, *Proc. Natl. Acad. Sci. U.S.A.,* 85, 9489, 1988.
169. **Shoelson, S.E., White, M.F., and Kahn, C.R.,** Tryptic activation of the insulin receptor. Proteolytic truncation of the α-subunit releases the β-subunit from inhibitory control, *J. Biol. Chem.,* 263, 4852, 1988.
170. **Clark, S., Eckardt, G., Siddle, K., and Harrison, L.C.,** Changes in insulin-receptor structure associated with trypsin-induced activation of the receptor tyrosine kinase, *Biochem. J.,* 276, 27, 1991.
171. **Heffetz, D. and Zick, Y.,** Receptor aggregation is necessary for activation of the soluble insulin receptor kinase, *J. Biol. Chem.,* 261, 889, 1986.
172. **Kasuga, M., Karlsson, F.A., and Kahn, C.R.,** Insulin stimulates the phosphorylation of the 95,000-Dalton subunit of its own receptor, *Science,* 215, 185, 1982.
173. **Häring, H.-U., Kasuga, M., and Kahn, C.R.,** Insulin receptor phosphorylation in intact adipocytes, *Biochem. Biophys. Res. Commun.,* 108, 1538, 1982.
174. **Van Obberghen, E., Rossi, B., Kowalski, A., Gazzano, H., and Ponzio, G.,** Receptor-mediated phosphorylation of the hepatic insulin receptor: evidence that the M_r 95,000 receptor subunit is its own kinase, *Proc. Natl. Acad. Sci. U.S.A.,* 80, 945, 1983.
175. **Ballotti, R., Kowalski, A., White, M.F., Le Marchand-Brustel, Y., and Van Obberghen, E.,** Insulin stimulates tyrosine phosphorylation of its receptor β-subunit in intact rat hepatocytes, *Biochem. J.,* 241, 99, 1987.
176. **Tornqvist, H.E., Gunsalus, J.R., Nemenoff, R.A., Frackelton, A.R., Pierce, M.W., and Avruch, J.,** Identification of the insulin receptor tyrosine residues undergoing insulin-stimulated phosphorylation in intact rat hepatoma cells, *J. Biol. Chem.,* 263, 350, 1988.
177. **Kadowaki, T., Koyasu, S., Nishida, E., Tobe, K., Izumi, T., Takaku, F., Sakai, H., Yahara, I., and Kasuga, M.,** Tyrosine phosphorylation of common and specific sets of cellular proteins rapidly induced by insulin, insulin-like growth factor I, and epidermal growth factor in an intact cell, *J. Biol. Chem.,* 262, 7342, 1987.
178. **Herrera, R. and Rosen, O.M.,** Autophosphorylation of the insulin receptor *in vitro*. Designation of phosphorylation sites and correlation with receptor kinase activation, *J. Biol. Chem.,* 261, 11980, 1986.
179. **Kasuga, M., Zick, Y., Blithe, D.L., Crettaz, M., and Kahn, C.R.,** Insulin stimulates tyrosine phosphorylation of the insulin receptor in a cell-free system, *Nature (London),* 298, 667, 1982.
180. **Zick, Y., Whittaker, J., and Roth, J.,** Insulin stimulated phosphorylation of its own receptor: activation of a tyrosine-specific protein kinase that is tightly associated with the receptor, *J. Biol. Chem.,* 258, 3431, 1983.
181. **Kasuga, M., Fujita-Yamaguchi, Y., Blithe, D.L., and Kahn, C.R.,** Tyrosine-specific protein kinase activity is associated with the purified insulin receptor, *Proc. Natl. Acad. Sci. U.S.A.,* 80, 2137, 1983.
182. **Cobb, M.H. and Rosen, O.M.,** The insulin receptor and tyrosine protein kinase activity, *Biochim. Biophys. Acta,* 738, 1, 1983.
183. **Rees-Jones, R.W., Hendricks, S.A., Quarum, M., and Roth, J.,** The insulin receptor of rat brain is coupled to tyrosine kinase activity, *J. Biol. Chem.,* 259, 3470, 1984.
184. **Petruzzelli, L., Herrera, R., and Rosen, O.M.,** Insulin receptor is an insulin-dependent tyrosine protein kinase: copurification of insulin-binding activity and protein kinase activity to homogeneity from human placenta, *Proc. Natl. Acad. Sci. U.S.A.,* 81, 3327, 1984.
185. **Kohanski, R.A. and Lane, M.D.,** Kinetic evidence for activating and non-activating components of autophosphorylation of the insulin receptor protein kinase, *Biochem. Biophys. Res. Commun.,* 134, 1312, 1986.
186. **Yu, K.-T. and Czech, M.P.,** Tyrosine phosphorylation of insulin receptor β subunit activates the receptor tyrosine kinase in intact H-35 hepatoma cells, *J. Biol. Chem.,* 261, 4715, 1986.

187. **Stadtmauer, L. and Rosen, O.M.,** Phosphorylation of synthetic insulin receptor peptides by the insulin receptor kinase and evidence that the preferred sequence containing Tyr-1150 is phosphorylated *in vivo, J. Biol. Chem.,* 261, 10000, 1986.
188. **Ellis, L., Clauser, E., Morgan, D.O., Edery, M., Roth, R.A., and Rutter, W.J.,** Replacement of insulin receptor tyrosine residues 1162 and 1163 compromises insulin-stimulated kinase activity and uptake of 2-deoxyglucose, *Cell,* 45, 721, 1986.
189. **Blake, A.D., Hayes, N.S., Slater, E.E., and Strader, C.D.,** Insulin receptor desensitization correlates with attenuation of tyrosine kinase activity, but not of receptor endocytosis, *Biochem. J.,* 245, 357, 1987.
190. **Debant, A., Ponzio, G., Clauser, E., Contreres, J.O., and Rossi, B.,** Receptor cross-linking restores an insulin metabolic effect altered by mutation on tyrosine 1162 and tyrosine 1163, *Biochemistry,* 28, 14, 1989.
191. **Flores-Riveros, J.R., Sibley, E., Kastelic, T., and Lane, M.D.,** Substrate phosphorylation catalyzed by the insulin receptor tyrosine kinase. Kinetic correlation to autophosphorylation of specific sites in the β subunit, *J. Biol. Chem.,* 264, 21557, 1989.
192. **Reynet, C., Caron, M., Magré, J., Cherqui, G., Clauser, E., Picard, J., and Capeau, J.,** Mutation of tyrosine residues 1162 and 1163 of the insulin receptor affects hormone and receptor internalization, *Mol. Endocrinol.,* 4, 304, 1990.
193. **Zhang, B., Tavaré, J.M., Ellis, L., and Roth, R.A.,** The regulatory role of known tyrosine autophosphorylation sites of the insulin receptor kinase domain. An assessment by replacement with neutral and negatively charged amino acids, *J. Biol. Chem.,* 266, 990, 1991.
194. **Yu, K.-T. and Czech, M.P.,** Tyrosine phosphorylation of the insulin receptor β subunit activates the receptor-associated tyrosine kinase activity, *J. Biol. Chem.,* 259, 5277, 1984.
195. **Kasuga, M., Zick, Y., Blithe, D.L., Karlsson, F.A., Häring, H.U., and Kahn, C.R.,** Insulin stimulation of phosphorylation of the β subunit of the insulin receptor: formation of both phosphoserine and phosphotyrosine, *J. Biol. Chem.,* 257, 9891, 1982.
196. **Tavaré, J.M., Zhang, B., Ellis, L., and Roth, R.A.,** Insulin-stimulated serine and threonine phosphorylation of the human insulin receptor. An assessment of the role of serines 1305/1306 and threonine 1348 by their replacement with neutral or negatively charged amino acids, *J. Biol. Chem.,* 266, 21804, 1991.
197. **White, M.F., Takayama, S., and Kahn, C.R.,** Differences in the sites of phosphorylation of the insulin receptor *in vivo* and *in vitro, J. Biol. Chem.,* 260, 9470, 1985.
198. **Pang, D.T., Sharma, B.R., Shafer, J.A., White, M.F., and Kahn, C.R.,** Predominance of tyrosine phosphorylation of insulin receptors during the initial response of intact cells to insulin, *J. Biol. Chem.,* 260, 7131, 1985.
199. **Leef, J.W. and Larner, J.,** Insulin-mimetic effect of trypsin on the insulin receptor tyrosine kinase in intact adipocytes, *J. Biol. Chem.,* 262, 14837, 1987.
200. **Bollag, G.E., Roth, R.A., Beaudoin, J., Mochly-Rosen, D., and Koshland, D.E., Jr.,** Protein kinase C directly phosphorylates the insulin receptor *in vitro* and reduces its protein-tyrosine kinase activity, *Proc. Natl. Acad. Sci. U.S.A.,* 83, 5822, 1986.
201. **Takayama, S., White, M.F., and Kahn, C.R.,** Phorbol ester-induced serine phosphorylation of the insulin receptor decreases its tyrosine kinase activity, *J. Biol. Chem.,* 263, 3440, 1988.
202. **Jacobs, S. and Cuatrecasas, P.,** Phosphorylation of receptors for insulin and insulin-like growth factor I. Effects of hormones and phorbol esters, *J. Biol. Chem.,* 261, 934, 1986.
203. **Ballotti, R., Kowalski, A., Le Marchand-Brustel, Y., and Van Obberghen, E.,** Presence of an insulin-stimulated serine kinase in cell extracts from IM-9 cells, *Biochem. Biophys. Res. Commun.,* 139, 179, 1986.
204. **Roth, R.A. and Beaudoin, J.,** Phosphorylation of purified insulin receptor by cAMP kinase, *Diabetes,* 36, 123, 1987.
205. **Riedel, H., Dull, T.J., Schlessinger, J., and Ullrich, A.,** A chimaeric receptor allows insulin to stimulate tyrosine kinase activity of epidermal growth factor receptor, *Nature (London),* 324, 68, 1986.
206. **Morrison, B.D. and Pessin, J.E.,** Insulin stimulation of the insulin receptor kinase can occur in the complete absence of β subunit autophosphorylation, *J. Biol. Chem.,* 262, 2861, 1987.
207. **Chou, C.K., Dull, T.J., Russell, D.S., Gherzi, R., Lebwohl, D., Ullrich, A., and Rosen, O.M.,** Human insulin receptors mutated at the ATP-binding site lack protein tyrosine kinase activity and fail to mediate postreceptor effects of insulin, *J. Biol. Chem.,* 262, 1842, 1987.

208. **Desbois, C., Capeau, J., Hainault, I., Wicek, D., Reynet, C., Veissière, D., Caron, M., Picard, J., Guerre-Milo, M., and Cherqui, G.,** Differential role of insulin receptor autophosphorylation sites 1162 and 1163 in the long-term insulin stimulation of glucose transport, glycogenesis, and protein synthesis, *J. Biol. Chem.,* 267, 13488, 1992.
209. **Debant, A., Clauser, E., Ponzio, G., Filloux, C., Auzan, C., Contreres, J.-O., and Rossi, B.,** Replacement of insulin receptor tyrosine residues 1162 and 1163 does not alter the mitogenic effect of the hormone, *Proc. Natl. Acad. Sci. U.S.A.,* 85, 8032, 1988.
210. **Wilden, P.A., Backer, J.M., Kahn, C.R., Cahill, D.A., Schroeder, G.J., and White, M.F.,** The insulin receptor with phenylalanine replacing tyrosine-1146 provides evidence for separate signals regulating cellular metabolism and growth, *Proc. Natl. Acad. Sci. U.S.A.,* 87, 3358, 1990.
211. **Fujita-Yamaguchi, Y. and Kathuria, S.,** The monomeric $\alpha\beta$ form of the insulin receptor exhibits much higher insulin-dependent tyrosine-specific protein kinase activity than the intact $\alpha_2\beta_2$ form of the receptor, *Proc. Natl. Acad. Sci. U.S.A.,* 82, 6095, 1985.
212. **Deger, A., Krämer, H., Rapp, R., Koch, R. and Weber, U.,** The nonclassical insulin binding of insulin receptors from rat liver is due to the presence of two interacting α-subunits in the receptor complex, *Biochem. Biophys. Res. Commun.,* 135, 458, 1986.
213. **Fujita-Yamaguchi, Y., Sato, Y., and Kathuria, S.,** Removal of sialic acids from the purified insulin receptor results in enhanced insulin-binding and kinase activities, *Biochem. Biophys. Res. Commun.,* 129, 739, 1985.
214. **Herrera, R., Petruzzelli, L., Thomas, N., Bramson, H.N., Kaiser, E.T., and Rosen, O.M.,** An antipeptide antibody that specifically inhibits insulin receptor autophosphorylation and protein kinase activity, *Proc. Natl. Acad. Sci. U.S.A.,* 82, 7899, 1985.
215. **Taylor, R., Soos, M.A., Wells, A., Argyraki, M., and Siddle, K.,** Insulin-like and insulin-inhibitory effects of mononuclear antibodies for different epitopes on the human insulin receptors, *Biochem. J.,* 242, 123, 1987.
216. **Steele-Perkins, G. and Roth, R.A.,** Insulin-mimetic anti-insulin receptor monoclonal antibodies stimulate receptor kinase activity in intact cells, *J. Biol. Chem.,* 265, 9458, 1990.
217. **Morgan, D.O. and Roth, R.A.,** Acute insulin action requires insulin receptor kinase activity: introduction of an inhibitory monoclonal antibody into mammalian cells blocks the rapid effects of insulin, *Proc. Natl. Acad. Sci. U.S.A.,* 84, 41, 1987.
218. **Morgan, D.O., Ho, L., Korn, L.J., and Roth, R.A.,** Insulin action is blocked by a monoclonal antibody that inhibits the insulin receptor kinase, *Proc. Natl. Acad. Sci. U.S.A.,* 83, 328, 1986.
219. **Yu, K.T., Pessin, J.E., and Czech, M.P.,** Regulation of insulin receptor kinase by multisite phosphorylation, *Biochimie,* 67, 1081, 1985.
220. **Grunberger, G., Zick, Y., Roth, J., and Gorden, P.,** Protein kinase activity of the insulin receptor in human circulating and cultured mononuclear cells, *Biochem. Biophys. Res. Commun.,* 115, 560, 1983.
221. **Simpson, I.A. and Hedo, J.A.,** Insulin receptor phosphorylation may not be a prerequisite for acute insulin action, *Science,* 223, 1301, 1984.
222. **Tamura, S., Fujita-Yamaguchi, Y., and Larner, J.,** Insulin-like effect of trypsin on the phosphorylation of rat adipocyte insulin receptor, *J. Biol. Chem.,* 258, 14749, 1983.
223. **Gottschalk, W.K.,** The pathway mediating insulin's effects on pyruvate dehydrogenase bypasses the insulin receptor tyrosine kinase, *J. Biol. Chem.,* 166, 8814, 1991.
224. **Forsayeth, J.R., Caro, J.F., Sinha, M.K., Maddux, B.A., and Goldfine, I.D.,** Monoclonal antibodies to the human insulin receptor that activate glucose transport but not insulin receptor kinase activity, *Proc. Natl. Acad. Sci. U.S.A.,* 84, 3448, 1987.
225. **Ponzio, G., Dolais-Kitabgi, J., Louvard, D., Gautier, N., and Rossi, B.,** Insulin and rabbit anti-insulin receptor antibodies stimulate additively the intrinsic receptor kinase activity, *EMBO J.,* 6, 333, 1987.
226. **Ganguly, S.,** Description of a human placental anti-insulin receptor polyclonal antibody that activates insulin binding to the receptor *in vitro, Oncogene,* 3, 341, 1988.
227. **Gherzi, R., Russell, D.S., Taylor, S.I., and Rosen, O.M.,** Reevaluation of the evidence that an antibody to the insulin receptor is insulinmimetic without activating the protein tyrosine kinase activity of the receptor, *J. Biol. Chem.,* 262, 16900, 1987.
228. **Ueno, A., Arakaki, N., Takeda, Y., and Fujio, H.,** Inhibition of tyrosine autophosphorylation of the solubilized insulin receptor by an insulin-stimulating peptide derived from bovine serum albumin, *Biochem. Biophys. Res. Commun.,* 144, 11, 1987.

229. **Malchoff, C.D., Messina, J.L., Gordon, V., Tamura, S., and Larner, J.,** Inhibition of insulin receptor phosphorylation by indomethacin, *Mol. Cell. Biochem.,* 69, 83, 1985.
230. **Horn, R.S., Lystad, E., Adler, A., and Walaas, O.,** Evidence that insulin and guanosine triphosphate regulate dephosphorylation of the β-subunit of the insulin receptor in sarcolemma membranes isolated from skeletal muscle, *Biochem. J.,* 234, 527, 1986.
231. **Strout, H.V., Vicario, P.P., Saperstein, R., and Slater, E.E.,** A protein phosphotyrosine phosphatase distinct from alkaline phosphatase with activity against the insulin receptor, *Biochem. Biophys. Res. Commun.,* 151, 633, 1988.
232. **Hashimoto, N., Feener, E.P., Zhang, W.-R., and Goldstein, B.J.,** Insulin receptor protein-tyrosine phosphatases. Leukocyte common antigen-related phosphatase rapidly deactivates the insulin receptor kinase by preferential dephosphorylation of the receptor regulatory domain, *J. Biol. Chem.,* 267, 13811, 1992.
233. **Grunberger, G., Zick, Y., and Gorden, P.,** Defect in phosphorylation of insulin receptors in cells from an insulin-resistant patient with normal insulin binding, *Science,* 223, 932, 1984.
234. **Grunberger, G., Comi, R.J., Taylor, S.I., and Gorden, P.,** Tyrosine kinase activity of the insulin receptor of patients with type A extreme insulin resistance: studies with circulating mononuclear cells and cultured lymphocytes, *J. Clin. Endocrinol. Metab.,* 59, 1152, 1984.
235. **Le Marchand-Brustel, Y., Gremeaux, T., Ballotti, R., and Van Obberghen, E.,** Insulin receptor tyrosine kinase is defective in skeletal muscle of insulin-resistant obese mice, *Nature (London),* 315, 676, 1985.
236. **Taylor, S.I., Cama, A., Kadowaki, H., Kadowaki, T., and Accili, D.,** Mutations of the human insulin receptor gene, *Trends Endocrinol. Metab.,* 1, 134, 1990.
237. **Taylor, S.I., Kadowaki, T., Kadowaki, H., Accili, D., Cama, A., and McKeon, C.,** Mutations in insulin-receptor gene in insulin-resistant patients, *Diabetes Care,* 13, 257, 1990.
238. **Makino, H., Taira, M., Shimada, F., Hashimoto, N., Suzuki, Y., Nozaki, O., Hatanaka, Y., and Yoshida, S.,** Insulin receptor gene mutation: a molecular genetical and functional analysis, *Cell. Signal.,* 4, 351, 1992.
239. **Taira, M., Taira, M., Hashimoto, N., Shimada, F., Suzuki, Y., Kanatsuka, A., Nakamura, F., Ebina, Y., Tatibana, M., Makino, H., and Yoshida, S.,** Human diabetes associated with a deletion of the tyrosine kinase domain of the insulin receptor, *Science,* 245, 63, 1989.
240. **Shimada, F., Taira, M., Suzuki, Y., Hashimoto, N., Nozaki, O., Taira, M., Tatibana, M., Ebina, Y., Tawata, M., Onaya, T., Makino, H., and Yoshida, S.,** Insulin-resistant diabetes associated with partial deletion of insulin-receptor gene, *Lancet,* 335, 1179, 1990.
241. **Odawara, M., Kadowaki, T., Yamamoto, R., Shibasaki, Y., Tobe, K., Accili, D., Bevins, C., Mikami, Y., Matsuura, N., Akanuma, Y., Takaku, F., Taylor, S.I., and Kasuga, M.,** Human diabetes associated with a mutation in the tyrosine kinase domain of the insulin receptor, *Science,* 245, 66, 1989.
242. **Klinkhamer, M.P., Groen, N.A., van der Zon, G.C.M., Lindhout, D., Sandkuyl, L.A., Krans, H.M.J., Möller, W., and Maassen, J.A.,** A leucine-to-proline mutation in the insulin receptor in a family with insulin resistance, *EMBO J.,* 8, 2503, 1989.
243. **Longo, N., Shuster, R.C., Griffin, L.D., and Elsas, L.J.,** Insulin-receptor autophosphorylation and kinase activity are constitutively increased in fibroblasts cultured from a patient with heritable insulin resistance, *Biochem. Biophys. Res. Commun.,* 167, 1229, 1990.
244. **Longo, N., Langley, S.D., Griffin, L.D., and Elsas, L.J.,** Activation of glucose transport by a natural mutation in the human insulin receptor, *Proc. Natl. Acad. Sci. U.S.A.,* 90, 60, 1993.
245. **Seemanová, E., Rüdiger, H.W., and Dreyer, M.,** Autosomal dominant insulin resistance syndrome due to a postbinding defect, *Am. J. Med. Genet.,* 44, 705, 1992.
246. **Rüdiger, H.W., Ahrens, P., Dreyer, M., Frorath, B., Löffel, C., and Schmidt-Preuss, U.,** Impaired insulin-induced RNA synthesis secondary to a genetically defective insulin receptor, *Hum. Genet.,* 69, 76, 1985.
247. **Hoepffner, H.-J., Dreyer, M., Reimers, U., Schmidt-Preuss, U., Koepp, H.-P., and Rüdiger, H.W.,** A new familial syndrome with impaired function of three related peptide growth factors, *Hum. Genet.,* 83, 209, 1989.
248. **Perrotti, N., Taylor, S.I., Richert, N.D., Rapp, U.R., Pastan, I.H., and Roth, J.,** Immunoprecipitation of insulin receptors from cultured human lymphocytes (IM-9 cells) by antibodies to $pp60^{src}$, *Science,* 227, 761, 1985.

249. **Perrotti, N., Grunberger, G., Richert, N.D., and Taylor, S.I.,** Immunological similarity between the insulin receptor and the protein encoded by the *src* oncogene, *Endocrinology,* 118, 2349, 1986.
250. **White, M.F., Werth, D.K., Pastan, I., and Kahn, C.R.,** Phosphorylation of the solubilized insulin receptor by the gene product of the Rous sarcoma virus, $pp60^{src}$, *J. Cell. Biochem.,* 26, 169, 1984.
251. **Yu, K.-T., Werth, D.K., Pastan, I.H., and Czech, M.P.,** *src* kinase catalyzes the phosphorylation and activation of the insulin receptor kinase, *J. Biol. Chem.,* 260, 5838, 1985.
252. **Wong, T.W. and Goldberg, A.R.,** Kinetics and mechanism of angiotensin phosphorylation by the transforming gene product of Rous sarcoma virus, *J. Biol. Chem.,* 259, 3127, 1984.
253. **Fukami, Y. and Lipmann, F.,** Purification of the Rous sarcoma virus *src* kinase by casein-agarose affinity chromatography, *Proc. Natl. Acad. Sci. U.S.A.,* 82, 321, 1985.
254. **Ullrich, A., Bell, J.R., Chen, E.Y., Herrera, R., Petruzzelli, L.M., Dull, T.J., Gray, A., Coussens, L., Liao, Y.-C., Tsubokawa, M., Mason, A., Seeburg, P.H., Grunfeld, C., Rosen, O.M., and Ramachandran, J.,** Human insulin receptor and its relationship to the tyrosine kinase family of oncogenes, *Nature (London),* 313, 756, 1985.
255. **Neckameyer, W.S., Shibuya, M., Hsu, M.-T., and Wang, L.-H.,** Proto-oncogene c-*ros* codes for a molecule with structural features common to those of growth factor receptors and displays tissue-specific and developmentally regulated expression, *Mol. Cell. Biol.,* 6, 1478, 1986.
256. **Boulton, T.G., Gregory, J.S., Jong, S.-M., Wang, L.-H., Ellis, L., and Cobb, M.H.,** Evidence for insulin-dependent activation of S66 and microtubule-associated protein-2 kinase via a human insulin receptor/v-Ros hybrid, *J. Biol. Chem.,* 265, 2713, 1990.
257. **Chen, J., Heller, D., Poon, B., Kang, L., and Wang, L.-H.,** The proto-oncogene c-*ros* codes for a transmembrane tyrosine protein kinase sharing sequence and structural homology with *sevenless* protein of *Drosophila melanogaster, Oncogene,* 6, 257, 1991.
258. **Tessarollo, L., Nagarajan, L., and Parada, L.F.,** c-*ros:* the vertebrate homolog of the *sevenless* tyrosine kinase receptor is tightly regulated during organogenesis in mouse embryonic development, *Development,* 115, 11, 1992.
259. **Ellis, L., Morgan, D.O., Jong, S.-M., Wang, L.-H., Roth, R.A., and Rutter, W.J.,** Heterologous transmembrane signaling by a human insulin receptor-v-ros hybrid in Chinese hamster ovary cells, *Proc. Natl. Acad. Sci. U.S.A.,* 84, 5101, 1987.
260. **Wang, L.-H., Lin, B., Jong, S.-M.J., Dixon, D., Ellis, L., Roth, R.A., and Rutter, W.J.,** Activation of transforming potential of the human insulin receptor gene, *Proc. Natl. Acad. Sci. U.S.A.,* 84, 5725, 1987.
261. **Poon, B., Dixon, D., Ellis, L., Roth, R.A., Rutter, W.J., and Wang, L.-H.,** Molecular basis of the activation of the tumorigenic potential of Gag-insulin receptor chimeras, *Proc. Natl. Acad. Sci. U.S.A.,* 88, 877, 1991.
262. **Marshall, S. and Olefsky, J.M.,** Characterization of insulin-induced receptor loss and evidence for internalization of the insulin receptor, *Diabetes,* 30, 746, 1981.
263. **Carpentier, J.-L., Van Obberghen, E., Gorden, P., and Orci, L.,** Binding, membrane redistribution, internalization and lysosomal association of (^{125}I)anti-insulin receptor antibody in IM-9-cultured human lymphocyte, *Exp. Cell Res.,* 134, 81, 1981.
264. **Lane, M.D.,** The regulation of insulin receptor level and activity, *Nutr. Rev.,* 39, 417, 1981.
265. **Berhanu, P., Olefsky, J.M., Tsai, P., Thamm, P., Saunders, D., and Brandenburg, D.,** Internalization and molecular processing of insulin receptors in isolated rat adipocytes, *Proc. Natl. Acad. Sci. U.S.A.,* 79, 4069, 1982.
266. **Hwang, D.L., Papolan, T., Barseghian, G., Josefsberg, Z., and Lev-Ran, A.,** Absence of down-regulation of insulin receptors in human breast cancer cells (MCF-7) cultured in serum-free medium: comparison with epidermal growth factor receptor, *J. Receptor Res.,* 5, 27, 1985.
267. **Treadway, J.L., Whittaker, J., and Pessin, J.E.,** Regulation of the insulin receptor kinase by hyperinsulinism, *J. Biol. Chem.,* 264, 15136, 1989.
268. **Rouiller, D.G. and Gorden, P.,** Homologous down-regulation of the insulin receptor is associated with increased receptor biosynthesis in cultured human lymphocytes (IM-9 line), *Proc. Natl. Acad. Sci. U.S.A.,* 84, 126, 1987.
269. **McDonald, A.R., Maddux, B.A., Okabayashi, Y., Wong, K.Y., Hawley, D.M., Logsdon, C.D., and Goldfine, I.D.,** Regulation of insulin-receptor mRNA levels by glucocorticoids, *Diabetes,* 36, 779, 1987.

270. **Maasen, J.A., Krans, H.M.J., and Möller, W.,** The effect of insulin, serum and dexamethasone on mRNA levels for the insulin receptor in the human lymphoblastoic cell line IM-9, *Biochim. Biophys. Acta,* 930, 72, 1987.
271. **Iwama, N., Nomura, M., Kajimoto, Y., Imano, E., Kubota, M., Watarai, T., Kawamori, R., Shichiri, M., and Kamada, T.,** Effect of dexamethasone on the synthesis and degradation of insulin receptor mRNA in cultured IM-9 cells, *Diabetologia,* 30, 899, 1987.
272. **Correze, C., Pierre, M., Thibout, H., and Toru-delbauffe, D.,** Autophosphorylation of the insulin receptor in rat adipocytes is modulated by thyroid hormone status, *Biochem. Biophys. Res. Commun.,* 126, 1061, 1985.
273. **Tanti, J.-F., Grémeaux, T., Rochet, N., Van Obberghen, E., and Le Marchand-Brustel, Y.,** Effect of cyclic AMP-dependent protein kinase on insulin receptor tyrosine kinase activity, *Biochem. J.,* 245, 19, 1987.
274. **Takayama, S., White, M.F., Lauris, V., and Kahn, C.R.,** Phorbol esters modulate insulin receptor phosphorylation and insulin action in cultured hepatoma cells, *Proc. Natl. Acad. Sci. U.S.A.,* 81, 7797, 1984.
275. **Häring, H., Kirsch, D., Obermaier, B., Ermel, B., and Machicao, F.,** Tumor promoting phorbol esters increase the K_m of the ATP-binding site of the insulin receptor kinase from rat adipocytes, *J. Biol. Chem.,* 261, 3869, 1986.
276. **Hofmann, C.A., Lotan, R.M., Ku, W.W., and Oeltmann, T.N.,** Insulin-ricin *B* hybrid molecules mediate an insulin-associated effect on cells which do not bind insulin, *J. Biol. Chem.,* 258, 11774, 1983.
277. **Hofmann, C.A., Crettaz, M., Bruns, P., Hessel, P., and Hadawi, G.,** Cellular responses elicited by insulin mimickers in cells lacking detectable plasma membrane insulin receptors, *J. Cell. Biochem.,* 27, 401, 1985.
278. **Miller, D.S.,** Stimulation of RNA and protein synthesis by intracellular insulin, *Science,* 240, 506, 1988.
279. **Soler, A.P., Thompson, K.A., Smith, R.M., and Jarett, L.,** Immunological demonstration of the accumulation of insulin, but not insulin receptors, in nuclei of insulin-treated cells, *Proc. Natl. Acad. Sci. U.S.A.,* 86, 6640, 1989.
280. **Rakowicz-Szulczynska, E.M., Otwiaska, D., and Koprowski, H.,** Plasma membrane-mediated nuclear uptake and chromatin binding of insulin in tumor cell lines, *Mol. Carcinogenesis,* 3, 150, 1990.
281. **McElduff, A., Poronnik, P., Baxter, R.C., and Williams, P.,** A comparison of the insulin and insulin-like growth factor I receptors from rat brain and liver, *Endocrinology,* 122, 1933, 1988.
282. **Brennan, W.A., Jr.,** Developmental aspects of the rat brain insulin receptor: loss of sialic acid and fluctuation in number characterize fetal development, *Endocrinology,* 122, 2364, 1988.
283. **Häring, H.U., White, M.F., Kahn, C.R., Kasuga, M., Lauris, V., Fleischmann, R.F., Murray, M., and Pawelek, J.,** Abnormality of insulin binding and receptor phosphorylation in an insulin-resistant melanoma cell culture, *J. Cell Biol.,* 99, 900, 1984.
284. **Kahn, R., Murray, M., and Pawelek, J.,** Inhibition of proliferation of Cloudman S91 melanoma cells by insulin and characterization of some insulin-resistant variants, *J. Cell. Physiol.,* 103, 109, 1980.
285. **McElduff, A., Grunberger, G., and Gorden, P.,** An alteration in apparent molecular weight of the insulin receptor from the human monocyte cell line U-937, *Diabetes,* 34, 686, 1985.
286. **Lev-Ran, A., Carr, B.I., Hwang, D.L., and Roitman, A.,** Binding of epidermal growth factor and insulin and the autophosphorylation of their receptors in experimental primary hepatocellular carcinomas, *Cancer Res.,* 46, 4656, 1986.
287. **Hwang, D.L., Lev-Ran, A., and Tay, Y.-C.,** Hepatocarcinogens induce decrease in mRNA transcripts of receptors for insulin and epidermal growth factor in the rat liver, *Biochem. Biophys. Res. Commun.,* 146, 87, 1987.
288. **Lotan, R.M. and Lender, M.,** Insulin receptor regulation in minimal deviation hepatoma cell line, *Cancer Biochem. Biophys.,* 8, 173, 1986.
289. **Kaplan, G.C., Pillion, D.J., Rutter, W.J., Kim, H., and Barker, P.E.,** Insulin receptor overexpression in a human pre-B acute lymphocytic leukemia cell line with a t(91;19) chromosome translocation near the INSR locus, *Biochem. Biophys. Res. Commun.,* 159, 1275, 1989.
290. **Chaplinski, T.J., Bennett, T.E., and Caro, J.F.,** Alteration in insulin receptor expression accompanying differentiation of HL-60 leukemia cells, *Cancer Res.,* 46, 1203, 1986.

291. **Auberger, P., Falquerho, L., Contreres, J.O., Pages, G., Le Cam, G., Rossi, B., and Le Cam, A.,** Characterization of a natural inhibitor of the insulin receptor tyrosine kinase: cDNA cloning, purification, and anti-mitogenic activity, *Cell,* 58, 631, 1989.
292. **Colombo, B.M., Falquerho, L., Manenti, G., Dragani, T.A., and Le Cam, A.,** Expression of the pp63 gene encoding the insulin receptor tyrosine kinase inhibitor in proliferating liver and in liver tumors, *Biochem. Biophys. Res. Commun.,* 180, 967, 1991.
293. **Jarett, L.,** Insulin uptake and receptor reutilization (Symposium Summary), *Fed. Proc.,* 42, 2553, 1983.
294. **Hollenberg, M.D.,** Receptor dynamics and insulin action, in *Insulin: Its Receptor and Diabetes,* Hollenberg, M.D., Ed., Marcel Dekker, New York, 1985, 85.
295. **Russell, D.S., Gherzi, R., Johnson, E.L., Chou, C.-K., and Rosen, O.M.,** The protein-tyrosine kinase activity of the insulin receptors is necessary for insulin-mediated receptor down-regulation, *J. Biol. Chem.,* 262, 11833, 1987.
296. **Backer, J.M., Kahn, C.R., and White, M.F.,** Tyrosine phosphorylation of the insulin receptor during insulin-stimulated internalization in rat hepatoma cells, *J. Biol. Chem.,* 264, 1694, 1989.
297. **Hari, J. and Roth, R.A.,** Defective internalization of insulin and its receptor in cells expressing mutated insulin receptors lacking kinase activity, *J. Biol. Chem.,* 262, 15341, 1987.
298. **Backer, J.M., Kahn, C.R., and White, M.F.,** Tyrosine phosphorylation of the insulin receptor is not required for receptor internalization: studies in 2,4-dinitrophenol-treated cells, *Proc. Natl. Acad. Sci. U.S.A.,* 86, 3209, 1989.
299. **Iacopetta, B., Carpentier, J.-L., Pozzan, T., Lew, D.P., Gorden, P., and Orci, L.,** Role of intracellular calcium and protein kinase C in the endocytosis of transferrin and insulin by HL-60 cells, *J. Cell Biol.,* 103, 851, 1986.
300. **Fehlmann, M., Carpentier, J.-L., Van Obberghen, E., Freychet, P., Thamm, P., Saunders, D., Brandenburg, D., and Orci, L.,** Internalized insulin receptors are recycled to the cell surface in rat hepatocytes, *Proc. Natl. Acad. Sci. U.S.A.,* 79, 5921, 1982.
301. **Hueckstaedt, T., Olefsky, J.M., Brandenburg, D., and Heidenreich, K.A.,** Recycling of photoaffinity-labeled insulin receptors in rat adipocytes. Dissociation of insulin-receptor complexes is not required for receptor recycling, *J. Biol. Chem.,* 261, 8655, 1986.
302. **Smith, R.M. and Jarett, L.,** Receptor-mediated endocytosis and intracellular processing of insulin: ultrastructural and biochemical evidence for cell-specific heterogeneity and distinction from nonhormonal ligands, *Lab. Invest.,* 58, 613, 1988.
303. **Levy, J.R. and Olefsky, J.M.,** The trafficking and processing of insulin and insulin receptors in cultured rat hepatocytes, *Endocrinology,* 121, 2075, 1987.
304. **Duckworth, W.C.,** Insulin degradation: mechanisms, products, and significance, *Endocr. Rev.,* 9, 319, 1988.
305. **Duckworth, W.C., Hamel, F.G., and Peavy, D.E.,** Hepatic metabolism of insulin, *Am. J. Med.,* 85 (Suppl. 5A), 71, 1988.
306. **Smith, R.M., Seely, B.L., Shah, N., Olefsky, J.M., and Jarett, L.,** Tyrosine kinase-defective insulin receptors undergo insulin-induced microaggregation but do not concentrate in coated pits, *J. Biol. Chem.,* 266, 17522, 1991.
307. **Sonne, O.,** Receptor-mediated degradation of insulin in isolated rat adipocytes. Formation of a degradation product slightly smaller than insulin, *Biochim. Biophys. Acta,* 927, 106, 1987.
308. **Duckworth, W.C., Hamel, F.G., Peavy, D.E., Liepnieks, J.J., Ryan, M.P., Hermodson, M.A., and Frank, B.H.,** Degradation products of insulin generated by hepatocytes and by insulin protease, *J. Biol. Chem.,* 263, 1826, 1988.
309. **Affholter, J.A., Fried, V.A., and Roth, R.A.,** Human insulin-degrading enzyme shares structural and functional homologies with *E. coli* protease III, *Science,* 242, 1415, 1988.
310. **Affholter, J.A., Hsieh, C.-L., Francke, U., and Roth, R.A.,** Insulin-degrading enzyme: stable expression of the human complementary DNA, characterization of its protein product, and chromosomal mapping of the human and mouse genes, *Mol. Endocrinol.,* 4, 1125, 1990.
311. **Smith, R.M., Laudenslager, N.H., Shah, N., and Jarett, L.,** Insulin binding and processing by H4IIEC3 hepatoma cells: ultrastructural and biochemical evidence for a unique route of internalization and processing, *J. Cell. Physiol.,* 130, 428, 1987.
312. **King, G.L. and Johnson, S.M.,** Receptor-mediated transport of insulin across endothelial cells, *Science,* 227, 1583, 1985.

313. **Goldfine, I.D., Clawson, G.A., Smuckler, E.A., Purrello, F., and Vigneri, R.,** Action of insulin at the nuclear envelope, *Mol. Cell. Biochem.,* 48, 3, 1982.
314. **Purrello, F., Vigneri, R., Clawson, G.A., and Goldfine, I.D.,** Insulin stimulation of nucleoside triphosphate activity in isolated nuclear envelopes, *Science,* 216, 1005, 1982.
315. **Purrello, F., Burnham, D.B., and Goldfine, I.D.,** Insulin regulation of protein phosphorylation in isolated rat liver nuclear envelopes: potential relationship to mRNA metabolism, *Proc. Natl. Acad. Sci. U.S.A.,* 80, 1189, 1983.
316. **Rakowicz-Szulczynska, E.M., Rodeck, U., Herlyn, M., and Koprowski, H.,** Chromatin binding of epidermal growth factor, nerve growth factor, and platelet-derived growth factor in cells bearing the appropriate surface receptors, *Proc. Natl. Acad. Sci. U.S.A.,* 83, 3728, 1986.
317. **Ueda, M., Robinson, F.W., Smith, M.M., and Kono, T.,** Effects of monensin on insulin processing in adipocytes: evidence that the internalized insulin-receptor complex has some physiological activities, *J. Biol. Chem.,* 260, 3941, 1985.
318. **Podlecki, D.A., Smith, R.M., Koo, M., Tsai, P., Hueckenstaedt, T., Brandenburg, D., Lasher, R.S., Jarett, L., and Olefsky, J.M.,** Nuclear translocation of the insulin receptor. A possible mediator of insulin's long term effects, *J. Biol. Chem.,* 262, 3362, 1987.
319. **Cayanis, E., Sarangarajan, R., Lombes, M., Nahon, E., Edelman, I.S., and Erlanger, B.F.,** Identification of an epitope shared by the DNA-binding domain of glucocorticoid receptor and the B chain of insulin, *Proc. Natl. Acad. Sci. U.S.A.,* 86, 2138, 1989.
320. **Morgan, D.O., Ellis, L., Rutter, W.J., and Roth, R.A.,** Antibody-induced down regulation of a mutated insulin receptor lacking an intact cytoplasmic domain, *Biochemistry,* 26, 2959, 1987.
321. **Berhanu, P., Rohilla, A.M.K., and Rutter, W.J.,** Replacement of the human insulin receptor transmembrane and cytoplasmic domains by corresponding domains of the oncogene product v-*ros* leads to accelerated internalization, degradation, and down-regulation, *J. Biol. Chem.,* 265, 9505, 1990.
322. **Kohanski, R.A., Frost, S.C., and Lane, M.D.,** Insulin-dependent phosphorylation of the insulin receptor-protein kinase and activation of glucose transport in 3T3-L1 adipocytes, *J. Biol. Chem.,* 261, 12272, 1986.
323. **Planchart, A. and Barros-Pita, J.C.,** Role of divalent cations in cold and ouabain sensitive glucose uptake of adipose tissue stimulated by insulin, *Acta Cient. Venez.,* 28, 385, 1977.
324. **McDonald, J.M., Bruns, D.E., and Jarett, L.,** Ability of insulin to increase calcium uptake by adipocyte endoplasmic reticulum, *J. Biol. Chem.,* 253, 3504, 1978.
325. **Hobson, C.H., Upton, J.D., Loten, E.G., and Rennie, P.I.C.,** Is extracellular calcium required for insulin action?, *J. Cyclic Nucleotide Res.,* 6, 179, 1980.
326. **Malchoff, D.M. and Bruns, D.E.,** Dissociation of insulin's effects on cell metabolism and on subcellular calcium transport systems of 3T3-L1 adipocytes, *Biochem. Biophys. Res. Comm.,* 100, 501, 1981.
327. **Walaas, O. and Horn, R.S.,** The controversial problem of insulin action, *Trends Pharmacol. Sci.,* 2, 196, 1981.
328. **Chan, K.-M. and McDonald, J.M.,** Identification of an insulin-sensitive calcium-stimulated phosphoprotein in rat adipocyte plasma membranes, *J. Biol. Chem.,* 257, 7443, 1982.
329. **Pershadsingh, H.A. and McDonald, J.M.,** Hormone-receptor coupling and the molecular mechanism of insulin action in the adipocyte: a paradigm for Ca^{2+} homeostasis in the initiation of the insulin-induced metabolic cascade, *Cell Calcium,* 5, 111, 1984.
330. **Cheng, K., Thompson, M., Craig, J., Schwartz, C., Locher, E., and Larner, J.,** Cell membrane signals in the mechanism of insulin action, *Ann. Clin. Lab. Sci.,* 14, 78, 1984.
331. **Sung, C.K.,** Insulin receptor signaling through non-tyrosine kinase pathways: evidence from anti-receptor antibodies and insulin receptor mutants, *J. Cell. Biochem.,* 48, 26, 1992.
332. **Lönnroth, P., Davies, J.I., Lönnroth, I., and Smith, U.,** The interaction between the adenylate cyclase system and insulin-stimulated glucose transport, *Biochem. J.,* 243, 789, 1987.
333. **Smith, C.J., Vasta, V., Degerman, E., Belfrage, P., and Manganiello, V.C.,** Hormone-sensitive cyclic GMP-inhibited cyclic AMP phosphodiesterase in rat adipocytes. Regulation of insulin- and cAMP-dependent activation by phosphorylation, *J. Biol. Chem.,* 266, 13385, 1991.
334. **Gabbay, R.A. and Lardy, H.A.,** Insulin inhibition of hepatic cAMP-dependent protein kinase: decreased affinity of protein kinase for cAMP and possible differential regulation of intrachain sites 1 and 2, *Proc. Natl. Acad. Sci. U.S.A.,* 84, 2218, 1987.

335. **Weber, H.W., Chung, F.-Z., Day, K., and Appleman, M.M.,** Insulin stimulation of cyclic AMP phosphodiesterase is independent from the G-protein pathways involved in adenylate cyclase regulation, *J. Cyclic Nucleot. Protein Phosphoryl.,* 11, 345, 1987.
336. **Graves, C.B., Goewert, R.R., and McDonald, J.M.,** The insulin receptor contains a calmodulin-binding domain, *Science,* 230, 827, 1985.
337. **Graves, C.B., Gale, R.D., Laurino, J.P., and McDonald, J.M.,** The insulin receptor and calmodulin. Calmodulin enhances insulin-mediated receptor kinase activity and insulin stimulates phosphorylation of calmodulin, *J. Biol. Chem.,* 261, 10429, 1986.
338. **Colca, J.R., DeWald, D.B., Pearson, J.D., Palazuk, B.J., Laurino, J.P., and McDonald, J.M.,** Insulin stimulates the phosphorylation of calmodulin in intact adipocytes, *J. Biol. Chem.,* 262, 11399, 1987.
339. **Sacks, D.B. and McDonald, J.M.,** Insulin-stimulated phosphorylation of calmodulin by rat liver insulin receptor preparations, *J. Biol. Chem.,* 263, 2377, 1988.
340. **Laurino, J.P., Colca, J.R., Pearson, J.D., DeWald, D.B., and McDonald, J.M.,** The *in vitro* phosphorylation of calmodulin by the insulin receptor tyrosine kinase, *Arch. Biochem. Biophys.,* 265, 8, 1988.
341. **Wong, E.C.C., Sacks, D.B., Laurino, J.P., and McDonald, J.M.,** Characteristics of calmodulin phosphorylation by the insulin receptor kinase, *Endocrinology,* 123, 1830, 1988.
342. **Sacks, D.B., Fujita-Yamaguchi, Y., Gale, R.D., and McDonald, J.M.,** Tyrosine-specific phosphorylation of calmodulin by the insulin receptor kinase purified from human placenta, *Biochem. J.,* 263, 803, 1989.
343. **Klip, A., Ramlal, T., and Koivisto, U.-M.,** Stimulation of Na^+/H^+ exchange by insulin and phorbol ester during differentiation of 3T3-L1 cells. Relation to hexose uptake, *Endocrinology,* 123, 296, 1988.
344. **Goodman, S.A., Esau, B., and Koontz, J.W.,** Insulin and phorbol myristic acetate induce ornithine decarboxylase in Reuber H35 rat hepatoma cells by different mechanisms, *Arch. Biochem. Biophys.,* 266, 343, 1988.
345. **Oksenberg, D., Dieckmann, B.S., and Greenberg, P.L.,** Functional interactions between colony-stimulating factors and the insulin family hormones for human myeloid leukemic cells, *Cancer Res.,* 50, 6471, 1990.
346. **Luttrell, L., Kilgour, E., Larner, J., and Romero, G.,** A pertussis toxin-sensitive G-protein mediates some aspects of insulin action on BC_3H-1 murine myocytes, *J. Biol. Chem.,* 265, 16873, 1990.
347. **Larner, J., Galasko, G., Cheng, K., DePaoli-Roach, A.A., Huang, L., Daggy, P., and Kellogg, J.,** Generation by insulin of a chemical mediator that controls protein phosphorylation and dephosphorylation, *Science,* 206, 1408, 1979.
348. **Stevens, E.V.J. and Husbands, D.R.,** Insulin-dependent production of low-molecular-weight compounds that modify key enzymes in metabolism, *Comp. Biochem. Physiol.,* 81B, 1, 1985.
349. **Malchoff, C.D., Huang, L., Gillespie, N., Villar Palasi, C., Schwartz, C.F.W., Cheng, K., Hewlett, E.L., and Larner, J.,** A putative mediator of insulin action which inhibits adenylate cyclase and adenosine 3′,5′-monophosphate-dependent protein kinase: partial purification from rat liver: site and kinetic mechanism of action, *Endocrinology,* 120, 1327, 1987.
350. **Farese, R.V.,** Lipid-derived mediators in insulin action, *Proc. Soc. Exp. Biol. Med.,* 195, 312, 1990.
351. **Farese, R.V., Standaert, M.L., Arnold, T., Yu, B., Ishizuka, T., Hoffman, J., Vila, M., and Cooper, D.R.,** The role of protein kinase C in insulin action, *Cell. Signal.,* 4, 133, 1992.
352. **Pimentel, E., Gonzalez, C.A., and Gonzalez, F.,** Biochemical effects of insulin on subcellular fractions from rat adipose tissue, *Acta Endocrinol.,* Suppl. 173, 121, 1973.
353. **Pimentel, E., Gonzalez, C.A., and Gonzalez-Mujica, F.,** Effects of insulin and glucose on subcellular fractions from rat adipose tissue, *Acta Diabetol. Lat.,* 11, 206, 1974.
354. **Honeyman, T.W., Strohsnitter, W., Scheid, C.R., and Schimmel, R.J.,** Phosphatidic acid and phosphatidylinositol labelling in adipose tissue, *Biochem. J.,* 212, 489, 1983.
355. **Narayanan, U., Keuker, C., and Hilf, R.,** Membrane-associated phosphatidylinositol kinase of R3230AC mammary tumors and normal mammary glands and effects of insulin on tumor enzyme activity, *Cancer Res.,* 48, 6727, 1988.
356. **Taylor, D., Ushing, R.J., Blackmore, P.F., Prpic, V., and Exton, J.H.,** Insulin and epidermal growth factor do not affect phosphoinositide metabolism in rat liver plasma membranes and hepatocytes, *J. Biol. Chem.,* 260, 2011, 1985.
357. **Sakai, M. and Wells, W.W.,** Action of insulin on the subcellular metabolism of polyphosphoinositides in isolated rat hepatocytes, *J. Biol. Chem.,* 261, 10058, 1986.

358. **Thakker, J.K., DiMarchi, R., MacDonald, K., and Caro, J.F.,** Effect of insulin and insulin-like growth factors I and II on phosphatidylinositol and phosphatidylinositol 4,5-bisphosphate breakdown in liver from humans with and without type II diabetes, *J. Biol. Chem.,* 264, 7169, 1989.
359. **L'Allemain, G. and Pouysségur, J.,** EGF and insulin action in fibroblasts. Evidence that phosphoinositide hydrolysis is not an essential mitogenic signalling pathway, *FEBS Lett.,* 197, 344, 1986.
360. **Besterman, J.M., Watson, S.P., and Cuatrecasas, P.,** Lack of association of epidermal growth factor-, insulin-, and serum-induced mitogenesis with stimulation of phosphoinositide degradation in BALB/c 3T3 fibroblasts, *J. Biol. Chem.,* 261, 723, 1986.
361. **Pennington, S.R. and Martin, B.R.,** Insulin-stimulated phosphoinositide metabolism in isolated fat cells, *J. Biol. Chem.,* 260, 11039, 1985.
362. **Carrascosa, J.M., Schleicher, E., Maier, R., Hackenberg, C., and Wieland, O.H.,** Separation of the protein-tyrosine kinase and phosphatidylinositol kinase activities of the human placental insulin receptor, *Biochim. Biophys. Acta,* 971, 170, 1988.
363. **Romero, G., Luttrell, L., Rogol, A., Zeller, K., Hewlett, E., and Larner, J.,** Phosphatidylinositol-glycan anchors of membrane proteins: potential precursors of insulin mediators, *Science,* 240, 509, 1988.
364. **Saltiel, A.R. and Cuatrecasas, P.,** Insulin stimulates the generation from hepatic plasma membranes of modulators derived from an inositol glycolipid, *Proc. Natl. Acad. Sci. U.S.A.,* 83, 5793, 1986.
365. **Saltiel, A.R., Fox, J.A., Sherline, P., and Cuatrecasas, P.,** Insulin-stimulated hydrolysis of a novel glycolipid generates modulators of cAMP phosphodiesterase, *Science,* 233, 967, 1986.
366. **Saltiel, A.R., Sherline, P., and Fox, J.A.,** Insulin-stimulated diacylglycerol production results from the hydrolysis of a novel phosphatidylinositol glycan, *J. Biol. Chem.,* 262, 1116, 1987.
367. **Fox, J.A., Soliz, N.M., and Saltiel, A.R.,** Purification of a phosphatidylinositol-glycan-specific phospholipase C from liver plasma membranes: a possible target of insulin action, *Proc. Natl. Acad. Sci. U.S.A.,* 84, 2663, 1987.
368. **Saltiel, A.R.,** Second messengers of insulin action, *Trends Endocrinol. Metab.,* 1, 158, 1990.
369. **Kelly, K.L., Kiechle, F.L., and Jarett, L.,** Insulin stimulation of phospholipid methylation in isolated rat adipocyte plasma membranes, *Proc. Natl. Acad. Sci. U.S.A.,* 81, 1089, 1984.
370. **Standaert, M.L. and Pollet, R.J.,** Insulin-glycerolipid mediators and gene expression, *FASEB J.,* 2, 2453, 1988.
371. **Standaert, M.L., Farese, R.V., Cooper, D.R., and Pollet, R.J.,** Insulin-induced glycerolipid mediators and the stimulation of glucose transport in BC3H-1 myocytes, *J. Biol. Chem.,* 263, 8696, 1988.
372. **Suzuki, S., Sugawara, K., Satoh, Y., and Toyota, T.,** Insulin stimulates the generation of two putative insulin mediators, inositol-glycan and diacylglycerol, in BC3H-1 myocytes, *J. Biol. Chem.,* 266, 8115, 1991.
373. **Cherqui, G., Reynet, C., Caron, M., Melin, B., Wicek, D., Clauser, E., Capeau, J., and Picard, J.,** Insulin receptor tyrosine residues 1162 and 1163 control insulin stimulation of myristoyl-diacylglycerol generation and subsequent activation of glucose transport, *J. Biol. Chem.,* 265, 21254, 1990.
374. **Hesketh, J.E. and Campbell, G.P.,** Effects of insulin, pertussis toxin and cholera toxin on protein synthesis and diacylglycerol production in 3T3 fibroblasts: evidence for a G-protein mediated activation of phospholipase C in the insulin signal mechanism, *Biosci. Rep.,* 7, 533, 1987.
375. **Etindi, R. and Fain, J.N.,** Insulin does not activate a phosphoinositide-specific phospholipase C in adipocytes, *Mol. Cell. Endocrinol.,* 67, 149, 1989.
376. **Farese, R.V., Larson, R.E., and Sabir, M.A.,** Insulin acutely increases phospholipids in the phosphatidate-inositide cycle in rat adipose tissue, *J. Biol. Chem.,* 257, 4042, 1982.
377. **Farese, R.V., Konda, T.S., Davis, J.S., Standaert, M.L., Pollet, R.J., and Cooper, D.R.,** Insulin rapidly increases diacylglycerol by activating *de novo* phosphatidic acid synthesis, *Science,* 236, 586, 1987.
378. **Grunberger, G.,** Interplay between insulin signalling and protein kinase C, *Cell. Signal.,* 3, 171, 1991.
379. **Farese, R.V., Standaert, M.L., Arnold, T., Yu, B.Z., Ishikura, T., Hoffman, J., Vila, M., and Cooper, D.R.,** The role of protein kinase C in insulin action, *Cell. Signal.,* 4, 133, 1992.
380. **Cooper, D.R., Konda, T.S., Standaert, M.L., Davis, J.S., Pollet, R.J., and Farese, R.V.,** Insulin increases membrane and cytosolic protein kinase C activity in BC3H-1 myocytes, *J. Biol. Chem.,* 262, 3633, 1987.
381. **Gomez, M.L., Medrano, E.E., Cafferatt, E.G.A., and Tellez-Iñon, M.T.,** Protein kinase C is differentially regulated by thrombin, insulin, and epidermal growth factor in human mammary tumor cells, *Exp. Cell Res.,* 175, 74, 1988.

382. **Walaas, S.I., Horn, R.S., Adler, A., Albert, K.A., and Walaas, O.,** Insulin increases membrane protein kinase C activity in rat diaphragm, *FEBS Lett.*, 220, 311, 1987.
383. **Egan, J.J., Saltis, J., Wek, S.A., Simpson, I.A., and Londos, C.,** Insulin, oxytocin, and vasopressin stimulate protein kinase C activity in adipocyte plasma membranes, *Proc. Natl. Acad. Sci. U.S.A.*, 87, 1052, 1990.
384. **Cooper, D.R., Watson, J.E., Hernandez, H., Yu, B., Standaert, M.L., Ways, D.K., Arnold, T.T., Ishizuka, T., and Farese, R.V.,** Direct evidence for protein kinase C involvement in insulin-stimulated hexose uptake, *Biochem. Biophys. Res. Commun.*, 188, 142, 1992.
385. **Ishizuka, T., Yamamoto, M., Kajita, K., Nagashima, T., Yasuda, K., Miura, K., Cooper, D.R., and Farese, R.V.,** Insulin stimulates novel protein kinase C in rat adipocytes, *Biochem. Biophys. Res. Commun.*, 183, 814, 1992.
386. **Farese, R.V., Standaert, M.L., Francois, A.J., Ways, K., Arnold, T.P., Hernandez, H., and Cooper, D.R.,** Effects of insulin and phorbol esters on subcellular distribution of protein kinase C isoforms in rat adipocytes, *Biochem. J.*, 288, 319, 1992.
387. **Vaartjes, W.J. and de Haas, C.G.M.,** Acute effects of tumor-promoting phorbol esters on hepatic intermediary metabolism, *Biochem. Biophys. Res. Commun.*, 129, 721, 1985.
388. **Messina, J.L. and McCann, J.A.,** Interaction of insulin and phorbol esters on the regulation of DNA synthesis in rat hepatoma cells, *Biochem. Biophys. Res. Commun.*, 172, 759, 1990.
389. **Grunberger, G. and Gorden, P.,** Affinity alteration of insulin receptor induced by a phorbol ester, *Am. J. Physiol.*, 243, E319, 1982.
390. **Grunberger, G., Iacopetta, B., Carpentier, J.-L., and Gorden, P.,** Diacylglycerol modulation of insulin receptor from cultured human mononuclear cells. Effects on binding and internalization, *Diabetes,* 35, 1364, 1986.
391. **Lavan, B.E., Kuhné, M.R., Garner, G.W., Anderson, D., Reedijk, M., Pawson, T., and Lienhard, G.E.,** The association of insulin-elicited phosphotyrosine proteins with *src* homology 2 domains, *J. Biol. Chem.*, 267, 11631, 1992.
392. **Okamoto, M., Hayashi, T., Kono, S., Inoue, G., Kubota, M., Okamoto, M., Kuzuya, H., and Imura, H.,** Specific activity of phosphatidylinositol 3-kinase is increased by insulin stimulation, *Biochem. J.*, 290, 327, 1993.
393. **Giorgetti, S., Ballotti, R., Kowalski-Chavel, A., Tartare, S., and Van Obberghen, E.,** The insulin and insulin-like growth factor I receptor substrate IRS-1 associates with and activates phosphatidylinositol 3-kinase *in vitro, J. Biol. Chem.*, 268, 7358, 1993.
394. **Sung, C.K. and Goldfine, I.D.,** Phosphatidylinositol-3-kinase is a nontyrosine phosphorylated member of the insulin receptor signalling complex, *Biochem. Biophys. Res. Commun.*, 189, 1024, 1992.
395. **Yonezawa, K., Ueda, H., Hara, K., Nishida, K., Ando, A., Chavanieu, A., Matsuba, H., Shii, K., Yokono, K., Fukui, Y., Calas, B., Grigorescu, F., Dhand, R., Gout, I., Otsu, M., Waterfield, M.D., and Kasuga, M.,** Insulin-dependent formation of a complex containing an 85-kDa subunit of phosphatidylinositol 3-kinase and tyrosine-phosphorylated insulin receptor substrate 1, *J. Biol. Chem.*, 267, 25958, 1992.
396. **Hayashi, H., Kamohara, S., Nishioka, Y., Kanai, F., Miyake, N., Fukui, Y., Shibasaki, F., Takenawa, T., and Ebina, Y.,** Insulin treatment stimulates the tyrosine phosphorylation of the α-type 85-kDa subunit of phosphatidylinositol 3-kinase *in vivo, J. Biol. Chem.*, 267, 22575, 1992.
397. **Hayashi, H., Nishioka, Y., Kamohara, S., Kanai, F., Ishii, K., Fukui, Y., Shibasaki, F., Tanekawa, T., Kido, H., Katsunuma, N., and Ebina, Y.,** The α-type 85-kDa subunit of phosphatidylinositol 3-kinase is phosphorylated at tyrosines 368, 580, and 607 by the insulin receptor, *J. Biol. Chem.*, 268, 7107, 1993.
398. **Lavan, G.E. and Lienhard, G.E.,** The insulin-elicited 60-kDa phosphotyrosine protein in rat adipocytes is associated with phosphatidylinositol 3-kinase, *J. Biol. Chem.*, 268, 5921, 1993.
399. **Matuoka, K., Fukami, K., Nakanishi, O., Kawai, S., and Takenawa, T.,** Mitogenesis in response to PDGF and bombesin abolished by microinjection of antibody to PIP_2, *Science,* 239, 640, 1988.
400. **Czech, M.P.,** Insulin action and the regulation of hexose transport, *Diabetes,* 29, 399, 1980.
401. **Moore, R.D.,** Effects of insulin upon ion transport, *Biochim. Biophys. Acta,* 737, 1, 1983.
402. **Bessman, S.P.,** The metabolic actions of insulin do not depend upon transport of glucose into the cell, *Biochem. Med. Metab. Biol.*, 48, 194, 1992.
403. **Lane, M.D., Flores-Riveros, J.R., Hresko, R.C., Kaestner, K.H., Liao, K., Janicot, M., Hoffman, R.D., McLenithan, J.C., Kastelic, T., and Christy, R.J.,** Insulin-receptor tyrosine kinase and glucose transport, *Diabetes Care,* 13, 565, 1990.

404. **Ramlal, T., Sarabia, V., Bilan, P.J., and Klip, A.,** Insulin-mediated translocation of glucose transporters from intracellular membranes to plasma membranes: sole mechanism of stimulation of glucose transport in L6 muscle cells, *Biochem. Biophys. Res. Commun.,* 157, 1329, 1988.
405. **Calderhead, D.M., Kitagawa, K., Tanner, L.I., Holman, G.D., and Lienhard, G.E.,** Insulin regulation of the two glucose transporters in 3T3-L1 adipocytes, *J. Biol. Chem.,* 265, 13800, 1990.
406. **Calderhead, D.M., Kitagawa, K., Tanner, L.I., Holman, G.D., and Lienhard, G.E.,** Insulin regulation of the two glucose transporters in 3T3-L1 adipocytes, *J. Biol. Chem.,* 265, 13800, 1990.
407. **Sivitz, W.I., DeSautel, S.L., Kayano, T., Bell, G.I., and Pessin, J.E.,** Regulation of glucose transporter messenger RNA in insulin-deficient states, *Nature (London),* 340, 72, 1989.
408. **Sargeant, R.J. and Paquet, M.R.,** Effect of insulin on the rates of synthesis and degradation of GLUT1 and GLUT4 glucose transporters in 3T3-L1 adipocytes, *Biochem. J.,* 290, 913, 1993.
409. **Flores-Riveros, J.R., McLenithan, J.C., Ezaki, O., and Lane, D.M.,** Insulin down-regulates expression of the insulin-responsive glucose transporter (GLUT4) gene: effects on transcription and mRNA turnover, *Proc. Natl. Acad. Sci. U.S.A.,* 90, 512, 1993.
410. **Bird, T.A., Davies, A., Baldwin, S.A., and Saklatvala, J.,** Interleukin 1 stimulates hexose transport in fibroblasts by increasing the expression of glucose transporters, *J. Biol. Chem.,* 265, 13578, 1990.
411. **Omatsukanbe, M. and Kitasato, H.,** Insulin and noradrenaline independently stimulate the translocation of glucose transporters from intracellular stores to the plasma membrane in mouse brown adipocytes, *FEBS Lett.,* 314, 246, 1992.
412. **James, D.E., Strube, M., and Mueckler, M.,** Molecular cloning and characterization of an insulin-regulatable glucose transporter, *Nature (London),* 338, 83, 1989.
413. **Orci, L., Ravazzola, M., Baetens, D., Inman, L., Amherdt, M., Peterson, R.G., Newgard, C.B., Johnson, J.H., and Unger, R.H.,** Evidence that down-regulation of β-cell glucose transporters in non-insulin-dependent diabetes may be the cause of diabetic hyperglycemia, *Proc. Natl. Acad. Sci. U.S.A.,* 87, 9953, 1990.
414. **Slot, J.W., Moxley, R., Geuze, H.J., and James, D.E.,** No evidence for expression of the insulin-regulatable glucose transporter in endothelial cells, *Nature (London),* 346, 369, 1990.
415. **White, M.K., Rall, T.B., and Weber, M.J.,** Differential regulation of glucose transporter isoforms by the *src* oncogene in chicken embryo fibroblasts, *Mol. Cell. Biol.,* 11, 4448, 1991.
416. **Häring, H.U., White, M.F., Kahn, C.R., Ahmad, Z., DePaoli-Roach, A.A., and Roach, P.J.,** Interaction of the insulin receptor kinase with serine/threonine kinases *in vitro, J. Cell. Biochem.,* 28, 171, 1985.
417. **Goldstein, B.J.,** Protein-tyrosine phosphatases and the regulation of insulin action, *J. Cell. Biochem.,* 48, 33, 1992.
418. **Cicirelli, M.F., Tonks, N.K., Diltz, C.D., Weiel, J.E., Fischer, E.H., and Krebs, E.G.,** Microinjection of a protein-tyrosine-phosphatase inhibits insulin action in *Xenopus* oocytes, *Proc. Natl. Acad. Sci. U.S.A.,* 87, 5514, 1990.
419. **Dent, P., Lavoinne, A., Nakielny, S., Caudwell, F.B., Watt, P., and Cohen, P.,** The molecular mechanism by which insulin stimulates glycogen synthesis in mammalian skeletal muscle, *Nature (London),* 348, 302, 1990.
420. **Kadowaki, T., Koyasu, S., Nishida, E., Sakai, H., Takaku, F., Yahara, I., and Kasuga, M.,** Insulin-like growth factors, insulin, and epidermal growth factor cause rapid cytoskeletal reorganization in KB cells. Clarification of the roles of type I insulin-like growth factor receptors and insulin receptors, *J. Biol. Chem.,* 261, 16141, 1988.
421. **McRoberts, J.A., Aranda, R., Riley, N., and Kang, H.,** Insulin regulates the paracellular permeability of cultured intestinal epithelial cell monolayers, *J. Clin. Invest.,* 85, 1127, 1990.
422. **Siegfried, B.A., Reo, N.V., Ewy, C.S., Shalwitz, R.A., Ackerman, J.J.H., and McDonald, J.M.,** Effects of hormone and glucose administration on hepatic glucose and glycogen metabolism *in vivo:* a ^{13}C NMR study, *J. Biol. Chem.,* 260, 16137, 1985.
423. **Bessman, S.P., Mohan, C., and Zaidise, I.,** Intracellular site of insulin action: mitochondrial Krebs cycle, *Proc. Natl. Acad. Sci. U.S.A.,* 83, 5067, 1986.
424. **Vaheri, A., Ruoslahti, E., Hovi, T., and Norling, S.,** Stimulation of density-inhibited cell cultures by insulin, *J. Cell. Physiol.,* 81, 355, 1973.
425. **Ross, M. and Ballard, F.J.,** Regulation of protein metabolism and DNA synthesis by fibroblast growth factor in BHK-21 cells, *Biochem. J.,* 249, 363, 1988.
426. **Moley, J.F., Morrison, S.D., and Norton, J.A.,** Insulin reversal of cancer cachexia in rats, *Cancer Res.,* 45, 4925, 1985.

427. **Roth, R.A., Zhang, B., Ching, J.E., and Kovacina, K.,** Substrates and signalling complexes: the tortured path to insulin action, *J. Cell. Biochem.,* 48, 12, 1992.
428. **Lawrence, J.C.,** Signal transduction and protein phosphorylation in the regulation of cellular metabolism by insulin, *Annu. Rev. Physiol.,* 54, 177, 1992.
429. **Sommercorn, J., Mulligan, J.A., Lozeman, F.J., and Krebs, E.G.,** Activation of casein kinase II in response to insulin and to epidermal growth factor, *Proc. Natl. Acad. Sci. U.S.A.,* 84, 8834, 1987.
430. **Chan, C.P., McNall, S.J., Krebs, E.G., and Fisher, E.H.,** Stimulation of protein phosphatase activity by insulin and growth factors in 3T3 cells, *Proc. Natl. Acad. Sci. U.S.A.,* 85, 6257, 1988.
431. **Liao, K., Hoffman, R.D., and Lane, D.M.,** Phosphotyrosyl turnover in insulin signaling. Characterization of two membrane-bound pp15 protein tyrosine phosphatases from 3T3-L1 adipocytes, *J. Biol. Chem.,* 266, 6544, 1991.
432. **O'Brien, R.M., Houslay, M.D., Milligan, G., and Siddle, K.,** The insulin receptor tyrosyl kinase phosphorylates holomeric forms of the guanine nucleotide regulatory proteins G_1 and G_0, *FEBS Lett.,* 212, 281, 1987.
433. **Häring, H.U., White, M.F., Machicao, F., Ermel, B., Schleicher, E., and Obermeier, B.,** Insulin rapidly stimulates phosphorylation of a 46-kDa membrane protein on tyrosine residues as well as phosphorylation of several soluble proteins in intact cells, *Proc. Natl. Acad. Sci. U.S.A.,* 84, 113, 1987.
434. **Caro, J.F., Shafer, J.A., Taylor, S.I., Raju, S.M., Perrotti, N., and Sinha, M.K.,** Insulin stimulated protein phosphorylation in human plasma liver membranes: detection of endogenous or plasma membrane associated substrates for insulin receptor kinase, *Biochem. Biophys. Res. Commun.,* 149, 1008, 1987.
435. **Fanciulli, M., Paggi, M.G., Mancini, A., Del Carlo, C., Floridi, A., Taylor, S.I., and Perrotti, N.,** pp120: a common endogenous substrate for insulin and IGF-I receptor-associated tyrosine kinase activity in the highly malignant AS-30D rat hepatoma cells, *Biochem. Biophys. Res. Commun.,* 160, 168, 1989.
436. **Klee, U. and Singh, T.J.,** Insulin-stimulated tyrosine phosphorylation of a 43 kDa protein in rat liver membranes, *Mol. Cell. Biochem.,* 108, 19, 1991.
437. **Margolis, R.N., Taylor, S.I., Seminara, D., and Hubbard, A.L.,** Identification of pp120, an endogenous substrate for the hepatocyte insulin receptor kinase, as an integral membrane glycoprotein of the bile canalicular domain, *Proc. Natl. Acad. Sci. U.S.A.,* 85, 7256, 1988.
438. **Rosen, O.M., Rubin, C.S., Cobb, M.H., and Smith, C.J.,** Insulin stimulates the phosphorylation of ribosomal protein S6 in a cell-free system derived from 3T3-L1 adipocytes, *J. Biol. Chem.,* 256, 3630, 1981.
439. **Thomas, G., Martin-Pérez, J., Siegmann, M., and Otto, A.M.,** The effect of serum, EGF, $PGF_{2\alpha}$, and insulin on S6 phosphorylation and the initiation of protein and DNA synthesis, *Cell,* 30, 235, 1982.
440. **Martin-Pérez, J., Siegmann, M., and Thomas, G.,** EGF, $PGF_{2\alpha}$, and insulin induce the phosphorylation of identical S6 peptides in Swiss mouse 3T3 cells: effect of cAMP on early sites of phosphorylation, *Cell,* 36, 287, 1984.
441. **Blenis, J., Spivack, J.G., and Erikson, R.L.,** Phorbol ester, serum, and Rous sarcoma virus transforming gene product induce similar phosphorylations of ribosomal protein S6, *Proc. Natl. Acad. Sci. U.S.A.,* 81, 6408, 1984.
442. **Trevillyan, J.M., Perisic, O., Traugh, J.A., and Byus, C.V.,** Insulin- and phorbol ester-stimulated phosphorylation of ribosomal protein S6, *J. Biol. Chem.,* 260, 3041, 1985.
443. **Ballou, L.M., Siegmann, M., and Thomas, G.,** S6 kinase in quiescent Swiss mouse 3T3 cells is activated by phosphorylation in response to serum treatment, *Proc. Natl. Acad. Sci. U.S.A.,* 85, 7154, 1988.
444. **Decker, S.,** Phosphorylation of ribosomal protein S6 in avian sarcoma virus-transformed chicken embryo fibroblasts, *Proc. Natl. Acad. Sci. U.S.A.,* 78, 4112, 1981.
445. **Hecht, L.B. and Straus, D.S.,** Insulin-sensitive, serum-sensitive protein kinase activity that phosphorylates ribosomal protein S6 in cultured fibroblast-melanoma hybrid cells, *Endocrinology,* 119, 470, 1986.
446. **Tabarini, D., Heinrich, J., and Rosen, O.M.,** Activation of S6 kinase activity in 3T3-L1 cells by insulin and phorbol ester, *Proc. Natl. Acad. Sci. U.S.A.,* 82, 4369, 1985.
447. **Cobb, M.H., Burr, J.G., Linder, M.E., Gray, T.B., and Gregory, J.S.,** Similar ribosomal protein S6 kinase activity is found in insulin-treated 3T3-L1 cells and chick embryo fibroblasts transformed by Rous sarcoma virus, *Biochem. Biophys. Res. Commun.,* 137, 702, 1986.

448. **Cobb, M.H.,** An insulin-stimulated ribosomal protein S6 kinase in 3T3-L1 cells, *J. Biol. Chem.,* 261, 12994, 1986.
449. **Sturgill, T.W., Ray, L.B., Erikson, E., and Maller, J.L.,** Insulin-stimulated MAP-2 kinase phosphorylates and activates ribosomal protein S6 kinase II, *Nature (London),* 334, 715, 1988.
450. **Olivier, A.R., Ballou, L.M., and Thomas, G.,** Differential regulation of S6 phosphorylation by insulin and epidermal growth factor in Swiss mouse 3T3 cells: insulin activation of type 1 phosphatase, *Proc. Natl. Acad. Sci. U.S.A.,* 85, 4720, 1988.
451. **Kulkarni, R.K. and Straus, D.S.,** Insulin-mediated phosphorylation of ribosomal protein S6 in mouse melanoma cells and melanoma x fibroblast hybrid cells in relation to cell proliferation, *Biochim. Biophys. Acta,* 762, 542, 1983.
452. **Nilsen-Hamilton, M., Hamilton, R.T., Allen, W.R., and Potter-Perigo, S.,** Synergistic stimulation of S6 ribosomal protein phosphorylation and DNA synthesis by epidermal growth factor and insulin in quiescent 3T3 cells, *Cell,* 31, 237, 1982.
453. **Bernier, M., Laird, D.M., and Lane, M.D.,** Insulin-activated tyrosine phosphorylation of a 15-kilodalton protein in intact 3t3-L1 adipocytes, *Proc. Natl. Acad. Sci. U.S.A.,* 84, 1844, 1987.
454. **Yamada, K., Lipson, K.E., Marino, M.W., and Donner, D.B.,** Effect of growth hormone on protein phosphorylation in isolated rat hepatocytes, *Biochemistry,* 26, 715, 1987.
455. **Lawrence, J.C., Jr., Hiken, J.F., Inkster, M., Scott, C.W., and Mumby, M.C.,** Insulin stimulates the generation of an adipocyte phosphoprotein that is isolated with a monoclonal antibody against the regulatory subunit of bovine heart cAMP-dependent protein kinase, *Proc. Natl. Acad. Sci. U.S.A.,* 83, 3649, 1986.
456. **Goren, J.P., Neufeld, E., and Boland, D.,** A 180,000 molecular weight glycoprotein substrate of the insulin receptor tyrosine kinase is present in human placenta and in rat liver, muscle, heart and brain plasma membrane preparations, *Cell. Signal.,* 2, 537, 1990.
457. **Kletzien, R.F. and Day, P.,** Modulation of the G_0 to S phase transit time by insulin: potential involvement of protein phosphorylation, *J. Cell. Physiol.,* 105, 533, 1980.
458. **Rees-Jones, R.W. and Taylor, S.I.,** An endogenous substrate for the insulin receptor-associated tyrosine kinase, *J. Biol. Chem.,* 260, 4461, 1985.
459. **Sadoul, J.-L., Peyron, J.-F., Ballotti, R., Debant, A., Fehlmann, M., and Van Obberghen, E.,** Identification of a cellular 110,000-Da protein substrate for the insulin-receptor kinase, *Biochem. J.,* 227, 887, 1985.
460. **Goren, H.J. and Boland, D.,** The 180,000 molecular weight plasma membrane insulin receptor substrate is a protein tyrosine phosphatase and is elevated in diabetic plasma membranes, *Biochem. Biophys. Res. Commun.,* 180, 463, 1991.
461. **Karasik, A., Pepinsky, R.B., and Kahn, C.R.,** Insulin and epidermal growth factor stimulates phosphorylation of a 170-kDa protein in intact hepatocytes immunologically related to lipocortin 1, *J. Biol. Chem.,* 263, 18558, 1988.
462. **White, M.F., Maron, R., and Kahn, C.R.,** Insulin rapidly stimulates tyrosine phosphorylation of a M_r-185,000 protein in intact cells, *Nature (London),* 318, 183, 1985.
463. **Keller, S.R., Kitagawa, K., Aebersold, R., Lienhard, G.E., and Garner, C.W.,** Isolation and characterization of the 160,000-Da phosphotyrosyl protein, a putative participant in insulin signaling, *J. Biol. Chem.,* 266, 12817, 1991.
464. **Rothenberg, P.L., Lane, W.S., Karasik, A., Backer, J., White, M., and Kahn, C.R.,** Purification and partial sequence analysis of pp185, the major cellular substrate of the insulin receptor tyrosine kinase, *J. Biol. Chem.,* 266, 8302, 1991.
465. **Izumi, T., White, M.F., Kadowaki, T., Takaku, F., Akanuma, Y., and Kasuga, M.,** Insulin-like growth factor I rapidly stimulates tyrosine phosphorylation of a M_r 185,000 protein in intact cells, *J. Biol. Chem.,* 262, 1282, 1987.
466. **Sun, X.J., Rothenberg, P., Kahn, C.R., Backer, J.M., Araki, E., Wilden, P.A., Cahill, D.A., Goldstein, B.J., and White, M.F.,** Structure of the insulin receptor substrate IRS-1 defines a unique signal transduction protein, *Nature (London),* 352, 73, 1991.
467. **Kelly, K.L. and Ruderman, N.B.,** Insulin-stimulated phosphatidylinositol 3-kinase. Association with a 185-kDa tyrosine-phosphorylated protein (IRS-1) and localization in a low density membrane vesicle, *J. Biol. Chem.,* 268, 4391, 1993.
468. **Fleischmann, R.D. and Pawelek, J.M.,** Evidence that a 90-kDa phosphoprotein, an associated kinase, and a specific phosphatase are involved in the regulation of Cloudman melanoma cell proliferation by insulin, *Proc. Natl. Acad. Sci. U.S.A.,* 82, 1007, 1985.

469. **Sacks, D.B., Davis, H.W., Crimmins, D.L., and McDonald, J.M.,** Insulin-stimulated phosphorylation of calmodulin, *Biochem. J.,* 286, 211, 1992.
470. **Graves, C.B. and McDonald, J.M.,** Insulin and phorbol ester stimulate phosphorylation of a 40-kDa protein in adipocyte plasma membranes, *J. Biol. Chem.,* 260, 11286, 1985.
471. **Morley, J.E., Levine, A.S., Brown, D.M., and Handwerger, B.S.,** Calmodulin levels in diabetic mice, *Biochem. Biophys. Res. Commun.,* 108, 1418, 1982.
472. **Akiyama, T., Kadowaki, T., Nishida, E., Kadooka, T., Ogawara, H., Fukami, Y., Sakai, H., Takaku, F., and Kasuga, M.,** Substrate specificities of tyrosine-specific protein kinases toward cytoskeletal proteins *in vitro, J. Biol. Chem.,* 261, 14797, 1986.
473. **Kadowaki, T., Fujita-Yamaguchi, Y., Nishida, E., Takaku, F., Akiyama, T., Kathuria, S., Akanuma, Y., and Kasuga, M.,** Phosphorylation of tubulin and microtubule-associated proteins by the insulin receptor kinase, *J. Biol. Chem.,* 260, 4016, 1985.
474. **Ray, L.B. and Sturgill, T.W.,** Rapid stimulation by insulin of a serine/threonine kinase in 3T3-L1 adipocytes that phosphorylates microtubule-associated protein 2 *in vitro, Proc. Natl. Acad. Sci. U.S.A.,* 84, 1502, 1987.
475. **Wandossell, F., Serrano, L., and Avila, J.,** Phosphorylation of alpha-tubulin carboxyl-terminal tyrosine prevents its incorporation into microtubules, *J. Biol. Chem.,* 262, 8268, 1987.
476. **Ray, L.B. and Sturgill, T.W.,** Insulin-stimulated microtubule-associated protein kinase is phosphorylated on tyrosine and threonine *in vivo, Proc. Natl. Acad. Sci. U.S.A.,* 85, 3753, 1988.
477. **Fujita-Yamaguchi, Y., Kathuria, S., Xu, Q.-Y., McDonald, J.M., Nakano, H., and Kamata, T.,** *In vitro* tyrosine phosphorylation studies on RAS proteins and calmodulin suggest that polylysine-like basic peptides or domains may be involved in interactions between insulin receptor kinase and its substrate, *Proc. Natl. Acad. Sci. U.S.A.,* 86, 7306, 1989.
478. **Sacks, D.B., Glenn, K.C., and McDonald, J.M.,** The carboxyl terminal segment of the c-Ki-*ras* 2 gene product mediates insulin-stimulated phosphorylation of calmodulin and stimulates insulin-independent autophosphorylation of the insulin receptor, *Biochem. Biophys. Res. Commun.,* 161, 399, 1989.
479. **Luttrell, L.M., Luttrell, D.K., Parsons, S.J., and Rogol, A.D.,** Insulin and phorbol ester induce distinct phosphorylations of $pp60^{c\text{-}src}$ in the BC3H-1 murine myocyte cell line, *Oncogene,* 4, 317, 1989.
480. **Egan, S.E., Giddings, B.W., Brooks, M.W., Buday, L., Sizeland, A.M., and Weinberg, R.A.,** Association of Sos Ras exchange protein with Grb2 is implicated in tyrosine kinase signal transduction and transformation, *Nature (London),* 363, 45, 1993.
481. **Li, N., Batzer, A., Daly, R., Yajnik, V., Skolnik, E., Chardin, P., Bar-Sagi, D., Margolis, B., and Schlessinger, J.,** Guanine-nucleotide-releasing factor hSos1 binds to Grb2 and links receptor tyrosine kinases to Ras signalling, *Nature (London),* 363, 85, 1993.
482. **Skolnik, E.Y., Lee, C.-H., Batzer, A., Vicentini, L.M., Zhou, M., Daly, R., Myers, M.J., Jr., Backer, J.M., Ullrich, A., and Schlessinger, J.,** The SH2/SH3 domain-containing protein GRB2 interacts with tyrosine-phosphorylated IRS1 and Shc: implications for insulin control of *ras* signalling, *EMBO J.,* 12, 1929, 1993.
483. **Meisler, M.H. and Howard, G.,** Effects of insulin on gene transcription, *Annu. Rev. Physiol.,* 51, 701, 1989.
484. **O'Brien, R.M. and Granier, D.K.,** Regulation of gene expression by insulin, *Biochem. J.,* 278, 609, 1991.
485. **Jump, D.B., Bell, A., Lepar, G., and Hu, D.,** Insulin rapidly induces rat liver S14 gene transcription, *Mol. Endocrinol.,* 4, 1655, 1990.
486. **Messina, J.L., Hamlin, J., and Larner, J.,** Effects of insulin alone on the accumulation of a specific mRNA in rat hepatoma cells, *J. Biol. Chem.,* 260, 16418, 1985.
487. **Messina, J.L., Hamlin, J., and Larner, J.,** Positive interaction between insulin and phorbol esters on the regulation of a specific messenger ribonucleic acid in rat hepatoma cells, *Endocrinology,* 121, 1227, 1987.
488. **Messina, J.L.,** Insulin and dexamethasone regulation of a rat hepatoma messenger ribonucleic acid: insulin has a transcriptional and posttranscriptional effect, *Endocrinology,* 124, 754, 1989.
489. **Straus, D.S. and Takemoto, C.D.,** Insulin negatively regulates albumin mRNA at the transcriptional and post-transcriptional level in rat hepatoma cells, *J. Biol. Chem.,* 262, 1955, 1987.
490. **Yokomori, N., Iwasa, Y., Aida, K., Inoue, M., Tawata, M., and Onaya, T.,** Transcriptional regulation of ferritin messenger ribonucleic acid levels by insulin in cultured rat glioma cells, *Endocrinology,* 128, 1474, 1991.

491. **Iynedjian, P.B., Jotterand, D., Nouspikel, T., Asfari, M., and Pilot, P.-R.,** Transcriptional induction of glucokinase gene by insulin in cultured liver cells and its repression by the glucagon-cAMP system, *J. Biol. Chem.,* 264, 21824, 1989.
492. **O'Brien, R.M., Lucas, P.C., Forest, C.D., Magnusson, M.A., and Granner, D.K.,** Identification of a sequence in the PEPCK gene that mediates a negative effect of insulin on transcription, *Science,* 249, 533, 1990.
493. **Oliver, F.J., Delarubia, G., Feener, E.P., Lee, M.E., Loeken, M.R., Shiba, T., Quertermous, T., and King, G.L.,** Stimulation of endothelin-1 gene expression by insulin in endothelial cells, *J. Biol. Chem.,* 266, 23251, 1991.
494. **Prager, D., Gebremedhin, S., and Melmed, S.,** An insulin-induced DNA-binding protein for the human growth hormone gene, *J. Clin. Invest.,* 85, 1680, 1990.
495. **Gai, X., Rizzo, M.-G., Valpreda, S., and Baserga, R.,** Regulation of c-*myc* mRNA levels by insulin or platelet-poor plasma, *Oncogene Res.,* 5, 111, 1989.
496. **Messina, J.L.,** Inhibition and stimulation of c-*myc* gene transcription by insulin in rat hepatoma cells. Insulin alters the intragenic pausing of c-*myc* transcription, *J. Biol. Chem.,* 266, 17955, 1991.
497. **Chan, S.O., Wong, S.S.C., and Yeung, D.C.Y.,** Insulin induction of c-Ki-*ras* in rat liver and in cultured normal rat hepatocytes, *Comp. Biochem. Physiol.,* B104, 341, 1993.
498. **Stumpo, D.J. and Blackshear, P.J.,** Insulin and growth factor effects on c-*fos* expression in normal and protein kinase C-deficient 3T3-L1 fibroblasts and adipocytes, *Proc. Natl. Acad. Sci. U.S.A.,* 83, 9453, 1986.
499. **Lu, K., Levine, R.A., and Campisi, J.,** c-*ras*-Ha gene expression is regulated by insulin or insulinlike growth factor and epidermal growth factor in murine fibroblasts, *Mol. Cell. Biol.,* 9, 3411, 1989.
500. **Ong, J., Yamashita, S., and Melmed, S.,** Insulin-like growth factor I induces c-*fos* messenger ribonucleic acid in L6 rat skeletal muscle cells, *Endocrinology,* 120, 353, 1987.
501. **Messina, J.L.,** Insulin's regulation of c-*fos* gene transcription in hepatoma cells, *J. Biol. Chem.,* 265, 11700, 1990.
502. **Taub, R., Roy, A., Dieter, R., and Koontz, J.,** Insulin as a growth factor in rat hepatoma cells. Stimulation of proto-oncogene expression, *J. Biol. Chem.,* 262, 10893, 1987.
503. **Cook, P.W., Weintraub, W.H., Swanson, K.T., Machen, T.E., and Firestone, G.L.,** Glucocorticoids confer normal serum/growth factor-dependent growth regulation to Fu5 rat hepatoma cells *in vitro*. Sequential expression of cell cycle-regulated genes without changes in intracellular calcium or pH, *J. Biol. Chem.,* 263, 19296, 1988.
504. **Kovacina, K.S., Yonezawa, K., Brautigan, D.L., Tonks, N.K., Rapp, U.R., and Roth, R.A.,** Insulin activates the kinase activity of the Raf-1 proto-oncogene by increasing its serine phosphorylation, *J. Biol. Chem.,* 265, 12115, 1990.
505. **Blackshear, P.J., McNeill Haupt, D., App, H., and Rapp, U.R.,** Insulin activates the Raf-1 protein kinase, *J. Biol. Chem.,* 265, 12131, 1990.
506. **Porras, A., Nebreda, A.R., Benito, M., and Santos, E.,** Activation of Ras by insulin in 3T3 L1 cells does not involve GTPase-activating protein phosphorylation, *J. Biol. Chem.,* 267, 21124, 1992.
507. **Burgering, B.M.T., Medema, R.H., Maassen, J.A., Van de Wetering, M.L., Van der Eb, A.J., McCormick, F., and Bos, J.L.,** Insulin stimulation of gene expression mediated by $p21^{ras}$ activation, *EMBO J.,* 10, 1103, 1991.
508. **Nasrin, N., Ercolani, L., Denaro, M., Kong, X.F., Kang, I., and Alexander, M.,** An insulin response element in the glyceraldehyde-3-phosphate dehydrogenase gene binds a nuclear protein induced by insulin in cultured cells and by nutritional manipulations *in vivo, Proc. Natl. Acad. Sci. U.S.A.,* 87, 5273, 1990.
509. **Girbau, M., Gomez, J.A., Lesniak, M.A., and de Pablo, F.,** Insulin and insulin-like growth factor I both stimulate metabolism, growth, and differentiation in the postneurula chick embryo, *Endocrinology,* 121, 1477, 1987.
510. **Lockwood, D.H., Voytovich, A.E., Stockdale, F.E., and Topper, Y.J.,** Insulin-dependent DNA polymerase and DNA synthesis in mammary epithelial cells *in vitro, Proc. Natl. Acad. Sci. U.S.A.,* 58, 658, 1967.
511. **Rechler, M.M., Podskalny, J.M., Goldfine, I.D., and Wells, C.A.,** DNA synthesis in human fibroblasts: stimulation by insulin and by nonsuppressible insulin-like activity (NSILA-S), *J. Clin. Endocrinol. Metab.,* 39, 512, 1974.
512. **Petrides, P.E. and Bohlen, P.,** The mitogenic activity of insulin: an intrinsic property of the molecule, *Biochem. Biophys. Res. Commun.,* 95, 1138, 1980.

513. **Mamounas, M., Gervin, D., and Englesberg, E.,** The insulin receptor as a transmitter of a mitogenic signal in Chinese hamster ovary CHO-K1 cells, *Proc. Natl. Acad. Sci. U.S.A.,* 86, 9294, 1989.
514. **O'Keefe, E.J. and Pledger, W.J.,** A model of cell cycle control: sequential events regulated by growth factors, *Mol. Cell. Endocrinol.,* 31, 167, 1983.
515. **Shipley, G.D., Childs, C.B., Volkenant, M.E., and Moses, H.L.,** Differential effects of epidermal growth factor, transforming growth factor, and insulin on DNA and protein synthesis and morphology in serum-free cultures of AKR-2B cells, *Cancer Res.,* 44, 710, 1984.
516. **Sand, T.-E. and Christoffersen, T.,** Temporal requirement for epidermal growth factor and insulin in the stimulation of hepatocyte DNA synthesis, *J. Cell. Physiol.,* 131, 141, 1987.
517. **Recio-Pinto, E., Lang, F.F., and Ishii, D.N.,** Insulin and insulin-like growth factor II permit nerve growth factor binding and the neurite formation response in cultured neuroblastoma cells, *Proc. Natl. Acad. Sci. U.S.A.,* 81, 2562, 1984.
518. **Davis, R.J., Corvera, S., and Czech, M.P.,** Insulin stimulates cellular iron uptake and causes the redistribution of intracellular transferrin receptors to the plasma membrane, *J. Biol. Chem.,* 261, 8708, 1986.
519. **Hickman, J. and McElduff, A.,** Insulin promotes growth of the cultured rat osteosarcoma cell line UMR-106-01: an osteoblast-like model, *Endocrinology,* 124, 701, 1989.
520. **Scheving, L.A., Scheving, L.E., Tsai, T.H., and Pauly, J.E.,** Circadian stage-dependent effects of insulin and glucagon on incorporation of (^{3}H)thymidine into deoxyribonucleic acid in the esophagus, stomach, duodenum, jejunum, ileum, caecum, colon, rectum, and spleen of the adult female mouse, *Endocrinology,* 111, 308, 1982.
521. **Lessman, C.A. and Schuetz, A.W.,** Insulin induction of meiosis of *Rana pipiens* oocytes relation to endogenous progesterone, *Gamete Res.,* 6, 95, 1982.
522. **Birchmeier, C., Broek, D., and Wigler, M.,** RAS proteins can induce meiosis in *Xenopus* oocytes, *Cell,* 43, 615, 1985.
523. **Deshpande, A.K. and Kung, H.-F.,** Insulin induction of *Xenopus laevis* oocyte maturation is inhibited by monoclonal antibody against p21 *ras* proteins, *Mol. Cell. Biol.,* 7, 1285, 1987.
524. **Kovacina, K.S. and Roth, R.A.,** Identification of SHC as a substrate of the insulin receptor kinase distinct from the GAP-associated 62 kDa tyrosine phosphoprotein, *Biochem. Biophys. Res. Commun.,* 192, 1303, 1993.
525. **Baltensperger, K., Kozma, L.M., Cherniack, A.D., Klarlund, J.K., Chawla, A., Banerjee, U., and Czech, M.P.,** Binding of the Ras activator Son of Sevenless to insulin receptor substrate-1 signaling complexes, *Science,* 260, 1950, 1993.
526. **Skolnik, E.Y., Batzer, A., Li, N., Lee, C.-H., Lowenstein, E., Mohammadi, M., Margolis, B., and Schlessinger, J.,** The function of GRB2 in linking the insulin receptor to Ras signaling pathways, *Science,* 260, 1953, 1993.
527. **Yamauchi, K., Holt, K., and Pessin, J.E.,** Phosphatidylinositol 3-kinase functions upstream of Ras and Raf in mediating insulin stimulation of c-*fos* transcription, *J. Biol. Chem.,* 268, 14597, 1993.
528. **Maegawa, H., Ugi, S., Ibayashi, O., Tachikawaide, N., Tanaka, Y., Takagi, Y., Kikkawa, R., Shigeta, Y., and Kashiwagi, A.,** Src homology 2 domains of protein tyrosine phosphatase are phosphorylated by insulin receptor kinase and bind to the COOH-terminus of insulin receptors *in vitro, Biochem. Biophys. Res. Commun.,* 194, 208, 1993.
529. **Kojima, I., Mogami, H., Shibata, H., and Ogata, E.,** Role of calcium entry and protein kinase c in the progression of insulin-like growth factor-I in Balb/c 3T3 cells, *J. Biol. Chem.,* 268, 10003, 1993.
530. **Zirvi, K., Chee, D.O., and Hill, G.J.,** Continuous growth of human tumor cell lines in serum-free media, *In Vitro Cell. Devel. Biol.,* 22, 369, 1986.
531. **Cohen, N.D. and Hilf, R.,** Influence of insulin on growth and metabolism of 7,12-dimethylbenz(a)anthracene-induced mammary tumors, *Cancer Res.,* 34, 3245, 1974.
532. **Gross, G.E., Boldt, D.H., and Osborne, C.K.,** Perturbation by insulin of human breast cancer cell cycle kinetics, *Cancer Res.,* 44, 3570, 1984.
533. **Dahmer, M.K. and Perlman, R.L.,** Insulin and insulin-like growth factors stimulate deoxyribonucleic acid synthesis in PC12 pheochromocytoma cells, *Endocrinology,* 122, 2109, 1988.
534. **Bregman, M.D., Abdel Malek, Z.A., and Meyskens, F.L., Jr.,** Anchorage-independent growth of murine melanoma in serum-less media is dependent on insulin or melanocyte-stimulating hormone, *Exp. Cell Res.,* 157, 419, 1985.
535. **Koontz, J.W. and Iwahashi, M.,** Insulin as a potent, specific growth factor in a rat hepatoma cell line, *Science,* 211, 947, 1981.

536. **Chou, C.-K., Ho, L.-T., Ting, L.-P., Hu, C., Su, T.-S., Chang, W.-C., Suen, C.-S., Huan, M.-Y., and Chang, C.,** Selective suppression of insulin-induced proliferation of cultured human hepatoma cells by somatostatin, *J. Clin. Invest.,* 79, 175, 1987.
537. **Narayanan, U., Ribes, J.A., and Hilf, R.,** Effects of streptozotocin-induced diabetes and insulin on phospholipid content of R3230AC mammary tumor cells, *Cancer Res.,* 45, 4833, 1985.
538. **Umeda, M., Tanaka, K., and Ono, T.,** Effect of insulin on the transformation of BALB/3T3 cells by X-ray irradiation, *Jpn. J. Cancer Res.,* 74, 864, 1983.
539. **Sener, A. and Malaisse, W.J.,** Resistance to insulin of tumoral insulin-producing cells, *FEBS Lett.,* 193, 150, 1985.
540. **Heuson, J.C. and Legros, N.,** Influence of insulin deprivation on growth of the 7,12-dimethylbenz(a)anthracene-induced mammary carcinoma in rats subjected to alloxan diabetes and food restriction, *Cancer Res.,* 31, 226, 1972.
541. **Shafie, S.M. and Hilf, R.,** Insulin receptor levels and magnitude of insulin-induced responses in 7,12-dimethylbenz(a)anthracene-induced mammary tumors in rats, *Cancer Res.,* 41, 826, 1981.
542. **Berger, M.R., Fink, M., Feichter, G.E., and Janetschek, P.,** Effects of diazoxide-induced reversible diabetes on chemically induced autochthonous mammary carcinomas in Sprague-Dawley rats, *Int. J. Cancer,* 35, 395, 1985.
543. **Lupulescu, A.P.,** Effect of prolonged insulin treatment on carcinoma formation in mice, *Cancer Res.,* 45, 3288, 1985.
544. **O'Mara, B.A., Byers, T., and Schenfeld, E.,** Diabetes mellitus and cancer risk: a multisite case-control study, *J. Chronic Dis.,* 38, 435, 1985.
545. **Joron, G.E., Laryea, E., Jaeger, D., and Macdonald, L.,** Cause of death in 1144 patients with diabetes mellitus: an autopsy study, *Can. Med. Assoc. J.,* 134, 759, 1986.
546. **Pavelic, K. and Popovic, M.,** Insulin and glucagon secretion by renal adenocarcinoma, *Cancer,* 48, 98, 1981.
547. **Pavelic, K., Bolanca, M., Vecek, N., Pavelic, J., Marotti, T., and Vuk-Pavlaovic, S.,** Carcinomas of the cervix and corpus uteri in humans: stage-dependent blood levels of substance(s) immunologically cross-reactive with insulin, *J. Natl. Cancer Inst.,* 68, 891, 1982.
548. **Pavelic, K., Odavic, M., Peric, B., Hrsak, I., and Vuk-Pavlovic, S.,** Correlation of substance(s) immunologically cross-reactive with insulin, glucose and growth hormone in Hodgkin's lymphoma patients, *Cancer Lett.,* 17, 81, 1982.
549. **Pavelic, K. and Vuk-Pavlovic, S.,** C-peptide does not parallel increases of serum levels of substances immunologically cross-reactive with insulin in non-Hodgkin's lymphoma patients, *Blood,* 61, 925, 1983.
550. **Pavelic, L., Pavelic, K., and Vuk-Pavlovic, S.,** Human mammary and bronchial carcinomas: *in vivo* and *in vitro* secretion of substances immunologically cross-reactive with insulin, *Cancer,* 53, 2467, 1984.
551. **Bajzer, Z., Pavelic, K., and Vuk-Pavlovic, S.,** Growth self-incitement in murine melanoma B16: a phenomenological model, *Science,* 225, 930, 1984.
552. **Vuk-Pavlovic, Z., Pavelic, K., and Vuk-Pavlovic, S.,** Modulation of *in vitro* growth of murine myeloid leukemia by an autologous substance immunochemically cross-reactive with insulin and antiinsulin serum, *Blood,* 67, 1031, 1986.
553. **Vuk-Pavlovic, S., Opara, E.C., Levanat, S., Vrbanec, D., and Pavelic, K.,** Autocrine tumor growth regulation and tumor-associated hypoglycemia in murine melanoma B16 *in vivo, Cancer Res.,* 46, 2208, 1986.
554. **Yamada, Y. and Serrero, G.,** Characterization of an insulin-related factor secreted by a teratoma cell line, *Biochem. Biophys. Res. Commun.,* 135, 533, 1986.
555. **Yamada, Y. and Serrero, G.,** Purification of an insulin-related factor secreted by a teratoma-derived mesodermal cell line, *J. Biol. Chem.,* 262, 209, 1987.
556. **Yamada, Y. and Serrero, G.,** Autocrine growth induced by the insulin-related factor in the insulin-independent teratoma cell line 1246-3A, *Proc. Natl. Acad. Sci. U.S.A.,* 85, 5936, 1988.
557. **Gazzano, H. and Serrero, G.,** Identification of insulin receptors on the insulin-independent variant 1246-3A cell line, *J. Cell. Physiol.,* 136, 348, 1988.
558. **Petersen, B. and Blecher, M.,** Insulin receptors and functions in normal and spontaneously transformed cloned rat hepatocytes, *Exp. Cell Res.,* 120, 119, 1979.
559. **Krett, N.L., Heaton, J.H., and Gelehrter, T.D.,** Insulin resistance in H-35 rat hepatoma cells is mediated by post-receptor mechanisms, *Mol. Cell. Endocrinol.,* 32, 91, 1983.

560. **Crettaz, M. and Kahn, C.R.,** Analysis of insulin action using differentiated and dedifferentiated hepatoma cells, *Endocrinology,* 113, 1201, 1983.

561. **Chaplinski, T.J. and Bennett, T.E.,** Induction of insulin receptor expression of human leukemic cells by 1α,25 dihydroxyvitamin D_3, *Leukemia Res.,* 11, 37, 1987.

562. **Rouis, M., Thomopoulos, P., Louache, F., Testa, U., Hervy, C., and Titeux, M.,** Differentiation of U-937 human monocyte-like cell line by 1α,25-dihydroxyvitamin D_3 or by retinoic acid, *Exp. Cell Res.,* 157, 539, 1985.

563. **Rouis, M., Thomopoulos, P., Cherier, C., and Testa, U.,** Inhibition of insulin receptor binding by A23187: synergy with phorbol esters, *Biochem. Biophys. Res. Commun.,* 130, 9, 1985.

564. **Welsch, C.W., DeHoog, J.V., Scieszka, K.M., and Aylsworth, C.F.,** Retinoid feeding, hormone inhibition, and/or immune stimulation and the progression of *N*-methyl-*N*-nitrosourea-induced rat mammary carcinoma: suppression by retinoids of peptide hormone-induced tumor cell proliferation *in vivo* and *in vitro, Cancer Res.,* 44, 166, 1984.

565. **Giorgino, F., Belfiore, A., Milazzo, G., Costantino, A., Maddux, B., Whittaker, J., Goldfine, I.D., and Vigneri, R.,** Overexpression of insulin receptors in fibroblast and ovary cells induces a ligand-mediated transformed phenotype, *Mol. Endocrinol.,* 5, 542, 1991.

Chapter 2

Insulin-Like Growth Factors

I. INTRODUCTION

The insulin-like growth factors (IGFs) or somatomedins are polypeptide hormones that exhibit close homologies to insulin and possess potent anabolic and mitogenic effects both *in vivo* and *in vitro*.[1-10] Both insulin and the IGFs have an important role in developmental processes.[11] The IGFs may also be involved in the pathogenesis of diabetes mellitus and other chronic diseases.[12]

The IGFs are constituents of a complex of compounds present in serum and designated as nonsuppressible insulin-like activity (NSILA).[13] Insulin-like growth factor I (IGF-I) and insulin-like growth factor II (IGF-II) are members of a family of hormones that mediate many of the growth-promoting actions of growth hormones.[14] A substance, operationally termed "sulfation factor", involved in mediating the effects of growth hormone on somatic growth is represented by the somatomedins.[15] The IGFs are required for normal fetal and postnatal growth and development; and somatomedin activity, as determined by a radioreceptor assay, is present in human embryos.[16] IGF-II/somatomedin A may have a more important role in the fetal and neonatal organism, while IGF-I/somatomedin C is the predominant form of IGF in the adult life. However, both IGF-I and IGF-II play important roles in fetal development.[17]

The IGFs have potent mitogenic effects on a wide diversity of cell types. In cultured mouse fibroblasts, IGFs behave as progression factors, inducing DNA synthesis and cell proliferation when the cells are rendered competent by the previous treatment with platelet-derived growth factor (PDGF). IGFs are involved in the control of progression through the G_1 phase of the cell cycle and can also stimulate DNA synthesis in competent cells primed with factors such as epidermal growth factor (EGF) (primed competent cells). The effect of IGFs on the entry of cells into the S phase is apparently exerted at a posttranscriptional level.[18] The mitogenic effects of IGFs have been demonstrated also *in vivo*. Injection of IGF-I to hypophysectomized frogs induces DNA synthesis in mitosis in tissues such as the lens epithelium.[19] In addition to their mitogenic effects, IGFs have most important roles in processes associated with cellular differentiation during prenatal and postnatal life. These effects have been characterized in defined cellular systems *in vitro*. Terminal differentiation of myoblasts in culture may depend on the interaction between stimulatory growth factors represented by IGFs and TGFs and the inhibitory factors represented by TGFs.[20] The differentiation-inducing properties of these factors may be dissociated from their mitogenic effects. Autocrine secretion of IGF-II appears to play a central role in the "spontaneous" differentiation of skeletal muscle cells incubated in low serum medium.[21]

The mitogenic effects of IGF-I may be different from those of IGF-II, according to the cell type and the predominant physiological conditions. IGF-I, but not IGF-II, is a strong stimulator of DNA synthesis in cultured mouse keratinocytes.[22] Insulin has a similar mitogenic effect on keratinocytes, but this effect is mediated through the IGF-I receptor, to which insulin can bind with low affinity. IGF-I may be involved in the regulation of transcription and replicating enzyme induction necessary for DNA synthesis.[23] In certain types of cells, for example, in thyroid cells, IGFs may act as growth modulators by an autocrine or paracrine mechanism.[24] IGFs are present in the serum of all mammalian species as well as in chicken, turtle, and frog serum.[25] The amount of IGF extracted from serum of different species does not correlate with the size of the animals.

II. STRUCTURE AND BIOSYNTHESIS OF IGFs

The two types of insulin-like growth factors, IGF-I and IGF-II, have different structures and are recognized by distinct receptors on the cell surface. They are synthesized by a number of cell types and may act in an autocrine, paracrine, or endocrine fashion.

A. THE IGF GENES

Analysis of the human IGF genes showed the existence of a complex exon-intron structure and the presence of multiple promoters.[26] The IGF-II gene *(IGF2)* is located on human chromosome 11p15, in close proximity to the loci for insulin *(INS)* and the c-H-*ras*-1 proto-oncogene;[27] and the IGF-I gene *(IGF1)* is located on human chromosome 12q23.[28] The 5′ end of the *IGF2* gene is within 12.6 kbp of the

3′ end of the *INS* gene, and both genes are separated by an *Alu* sequence.[29] The possible biological significance of these syntenies is unknown.

1. Structure and Expression of the *IGF1* Gene

Full-size cDNA sequences coding for IGF-I have been isolated from human DNA libraries.[30-32] The complete nucleotide sequence of the high molecular weight human IGF-I mRNA was determined.[33] The human *IGF1* gene spans a region of more than 95 kb of chromosomal DNA and contains five exons interrupted by four introns. The DNA sequence of exons 1 through 4 encodes a 195-amino acid precursor polypeptide, while exons 1, 2, 3, and 5 code for a 153-amino acid polypeptide. These two polypeptides are generated by alternative RNA processing of the primary *IGF1* transcripts and both of them contain the 70-amino acid IGF-I protein.[32] As the distinct carboxyl-terminal extensions of each IGF-I precursor molecule reside on separate exons, these results suggest that tissue-specific factors may play a role in IGF-I biosynthesis by influencing RNA splicing or protein processing and that the extension peptides themselves may have discrete biological functions. The human *IGF1* gene contains two promoters that are regulated in a cell type-specific manner.[34]

cDNAs encoding mouse liver IGF-I have been isolated and sequenced.[35] Alternative RNA splicing events result in the synthesis of two types of mouse IGF-I precursor, termed prepro-IGF-IA and prepro-IGF-IB, which differ in the size and sequence of the carboxyl-terminal peptide. The sequences of mouse and human prepro-IGF-IA are highly conserved and exhibit 94% identity.

cDNA encoding IGF-I was isolated from a rat kidney library.[36] The predicted polypeptide sequence shows 96 and 99% homology with the human and mouse prepro-IGF-I protein, respectively. Under stringent conditions, rat IGF-I cDNA hybridizes with three mRNA species which are detectable in all tissues from intact adult rats and for which expression is dependent on the presence of growth hormone. The proximal promoter region of the rat IGF-I gene, which is involved in the regulation of IGF-I gene expression, was also cloned and characterized.[37]

2. Structure and Expression of the *IGF2* Gene

The *IGF2* gene is situated close to human chromosome region 11p15, only 1.4 kb from the 3′ end of the insulin gene, and consists of at least seven exons spanning a total region of 14 kb.[38,39] cDNA sequences of the human *IGF2* gene with capability of coding for the IGF-II polypeptide precursor molecule have been constructed.[40] The cloned human *IGF2* gene has been expressed in *Escherichia coli* in the form of a biologically active protein.[41] The 5′-untranslated region of a cDNA corresponding to a human placental library was found to be different from that of adult human liver.[42] Three promoters have been identified in the human *IGF2* gene.[43] Differential initiation of transcription can occur at the three distinct promoter sites, resulting in the synthesis of mRNA species of different length which are expressed in a tissue- and development-specific manner. IGF-II promoter 1 is not active in fetal tissues, but only in adult human liver, and gives rise to a 5.3-kb mRNA species. The sequence of the 5′-noncoding region of human IGF-II mRNA exhibits extensive variability not only from one tissue to another but also within the mRNA population of a particular organ such as the liver or the placenta.[44] Initiation of transcription and processing of the primary transcript of the human *IGF2* gene are apparently not regulated by single mechanisms, as is usually observed for eukaryotic genes. The biological significance of these peculiarities is unknown. Expression of the *IGF2* gene is repressed by the WT1 product of the Wilms' tumor suppressor gene.[45] Sequences from the *IGF2* and other genes linked on human chromosome 11p are lost in the tumor cells of patients with various histological types of advanced lung carcinomas.[46]

cDNAs encoding the mouse IGF-II precursor polypeptide have been isolated from a placental library and sequenced.[47] Mouse prepro-IGF-II is predicted to be composed of 180 amino acid residues and shows 84 and 97% identity with human and rat IGF-II precursors, respectively. In contrast to the adult human liver, prepro-IGF-II mRNA transcripts are not detected in the adult mouse liver.

The rat *IGF2* gene extends over 12 kbp and contains two 5′-noncoding exons and three protein-coding exons.[48] The two 5′ exons represent alternative 5′ regions of different mRNA molecules and are expressed from two distinct promoters. The two promoters are used with different efficiencies but exhibit similar tissue-specific expression and regulation with developmental age. The high heterogeneity of IGF-I gene transcripts observed in the rat is due not only to the use of two different promoters, each transcribing alternative 5′-noncoding exons, but also to the use of alternate sequential polyadenylation sites.[49] The biological significance of developmentally regulated IGF-II RNA heterogeneity and differential splicing of the rat IGF-II gene transcripts is not understood. The relative constancy of mRNA transcribed from the single rat IGF-II gene in normal cells is altered in tumor cells such as rat ascites hepatomas, thereby indicating that each IGF-II mRNA level can be regulated independently.[50]

B. STRUCTURE OF IGFS

The primary structures of IGF-I and IGF-II indicate that these growth factors and insulin have extensive homology and may have diverged from a common ancestor molecule.[51,52] Like insulin, IGFs contain B and A domains; the amino-terminal part of the B domain is connected via smaller peptide, the C domain, to the A domain. In addition, and unlike proinsulin, the IGFs contain an extension peptide called the D domain.

1. Structure of IGF-I

Human IGF-I consists of a single chain of 70 amino acids with three disulfide bridges which is synthesized as a 130-amino acid precursor. The primary structure of rat IGF-I and its biological and immunologic activities in comparison with human IGF-I were determined after isolation and purification of the protein from rat serum.[53]

DNA and protein synthesis techniques have been applied to the study of structural requirements of IGF-I for exerting its biological effects. A synthetic gene coding for an IGF-I analogue replacing the only methionine of the protein at position 59 was constructed.[54] The artificial protein was found to be 60% as active as native human IGF-I in a radioimmunoassay and 50% as potent as native IGF-I in a radioreceptor assay, and displayed mitogenic activity for BALB/c 3T3 cells. A constructed IGF-I molecule prepared by solid-phase synthesis, in which the amino-terminal portion is truncated (destripeptide-IGF-I), is about 7 times more potent than IGF-I in stimulating protein synthesis in L6 myoblasts.[55] The mechanism of the increased biological potency of the destripeptide-IGF-I is unknown. The higher potency of the truncated form of IGF-I is not associated with an increased ability to compete for IGF-I binding to L6 myoblasts. Intact and amino-terminally shortened forms of IGF-I are capable of inducing mammary gland differentiation and development.[56] A modified IGF-I lacking the amino-terminal tripeptide Gly-Pro-Glu is normally present in bovine colostrum and possesses enhanced biological activity compared with the unmodified IGF-I molecule.[57] Studies with a constructed hybrid insulin-like molecule in which the A chain moiety was the sequence corresponding to the A domain of IGF-I and the B chain moiety was that of insulin indicated that the growth-promoting activity of IGF-I is associated with the A domain of the molecule and that the amino-terminal B domain does not contribute directly to growth factor activity but is responsible for recognition by carrier proteins.[58] Analysis of IGF-I structural analogues in which one or more of the three tyrosine residues (Tyr-24, Tyr-31, and Tyr-60) were replaced with nonaromatic residues indicated that all of them are involved in the high-affinity binding of IGF-I to the IGF-I receptor (type I IGF receptor), while Tyr-60 is important for maintaining binding to the IGF-II receptor (type II IGF receptor).[59]

2. Structure of IGF-II

Human IGF-II is a single-chain polypeptide of 7.5 kDa and 67 amino acids with three disulfide bridges which is synthesized as a 180-amino acid precursor. Human proIGF-II has an 89-amino acid carboxyl-terminal extension of unknown function.[29,60] A similar structure is contained in the rat IGF-II precursor polypeptide. IGF-II is synthesized in the rat liver as a 20-kDa precursor polypeptide which is processed to intermediate and mature IGF-II species at the time of secretion or shortly thereafter.[60-62] The mature 7.8-kDa IGF-II molecule corresponds to the amino-terminus of the 20-kDa IGF-II precursor.[63] The artificial synthesis of biologically active human IGF-II by the solid-phase method has been reported.[64]

Evidence has been obtained in favor of the existence of a second human gene coding for IGF-II which may be responsible for the presence of a variant form of IGF-II polypeptide in human serum.[65] Two mRNAs coding for IGF-II, one of 4 kb and the other of 1.2 kb, have been detected in the cytoplasm of an IGF-II-producing cell line (BRL-3A) derived from normal rat liver.[66] Although the physiological role of the larger (4 kb) IGF-II mRNA has not been elucidated, both molecular species of IGF-II mRNAs are regulated coordinately with developmental age, being higher in liver from neonatal rats but not detectable in liver from older animals. Recently, an IGF-II variant has been purified from Cohn fraction IV_1 of human plasma.[67] The amino-terminal sequence of the first 35 amino acids of this variant showed a replacement of Ser-29 of IGF-II with a tetrapeptide present in the variant.

3. Evolutionary Conservation of IGFs

The evolutionary conservation of amino acid sequences in IGFs indicates the biological importance of these hormones. Not only the mature IGF proteins but also their precursors are well conserved. The predicted amino acid sequences of a cDNA encoding rat prepro-IGF-I exhibits 96 and 99% homology with the human and mouse prepro-IGF-I, respectively.[36]

IGFs and insulin are close relatives in evolution. A region of the IGF molecules, the A region, exhibits 43% amino acid homology with the A chain of the insulin molecule; and another region of the IGF molecules, termed B, exhibits 41% homology with the B chain of the insulin molecule. IGFs present in the serum of all mammals give cross-reactions in a radioimmunoassay established for human IGF-I.[24] Since IGF antibodies recognize mostly a region of IGF molecule, the C region, this region must have been conserved during evolution. In contrast, only guinea pig and monkey sera contain material that cross-reacts in the radioimmunoassay for human IGF-II, indicating that the C region of the IGF-II molecule has undergone considerable evolutionary alterations. The C region consists of 12 amino acids in IGF-I and 8 amino acids in IGF-II.

C. PRODUCTION OF IGFs IN NORMAL TISSUES

The IGFs are synthesized in many tissues, especially in fetal tissues including heart, lung, kidney, liver, pancreas, spleen, small intestine, colon, brain, and pituitary gland.[29,68] The human produces IGF-I and IGF-II at rates of 10 and 13 mg/d, respectively.[69] However, the abundance of IGF-I and IGF-II synthesis in each tissue varies. The abundance of IGF-I mRNA in polyadenylated RNAs from adult rat liver is 10- to 50-fold higher than in other tissues, which indicates that in the adult rat the liver is a major site of IGF-I synthesis and source of IGF-I circulating in the blood.[36,68,70] IGF-I is synthesized also by gonadal cells. The production of IGF-I by immature rat Sertoli cells is stimulated by thyroid hormone.[71]

Expression of IGF-II transcripts in rat tissues is subjected to developmental regulation.[72] Hepatic IGF-II mRNA levels decrease gradually during the rat postnatal development, reaching adult levels at 3 weeks of age.[73] Higher abundance of IGF-II mRNAs in rat fetal tissues compared with adult tissues supports the hypothesis that IGF-II is a fetal somatomedin.[74] In rat plasma, immunoreactive IGF-II is elevated at birth and decreases 20- to 100-fold by 21 d of age. Tissue concentrations of IGF-II mRNA show variations, being higher in muscle, skin, liver, lung, thymus, and intestine. Low but significant levels of IGF-II mRNA are present in neural tissues from fetal and neonatal rats whereas no IGF-II mRNA could be detected in spleen and pancreas. In each tissue, the highest level of expression of the *IGF2* gene occurs at the youngest ages of the animals. In the adult rat, the highest level of IGF-II expression is present in the brain.[36] IGF-II mRNA is translated and prepro-IGF-II processed in different fetal tissues.[75] The highest levels of IGF-II are observed in liver, limb, lung, intestine, and brain; lower levels are observed in heart and kidney. The specific functions of IGF-II at the level of particular organs and tissues are not understood. IGF-II is secreted into milk, but the possible biological significance of this secretion is unknown.[76]

Both IGF-I and IGF-II are synthesized and retained in bone, where they can stimulate DNA, collagen, and noncollagenous protein synthesis.[77] IGF-II produced by human bone cells may be involved in bone formation in a paracrine and/or autocrine fashion. IGF-II is probably identical to a factor called skeletal growth factor which is implicated in bone volume regulatory mechanisms.[78]

IGFs are synthesized in the human placenta, which suggests that they may have an autocrine/paracrine role in the regulation of placental growth.[79] Insulin-related mRNA sequences, possibly including both insulin and IGFs, are transcribed in the normal human placenta, where they may represent up to 0.1% of the total polyadenylated RNA.[80] Placentas from diabetic women express much more of these sequences which may be involved in stimulating the growth of the human fetus. IGFs may be involved not only in growth stimulation but also in enhancing differentiation processes occurring in specific cell types.[81]

IGFs may have important functions in the nervous system. IGF-I is synthesized in fetal rat astrocytes and acts as a growth promoter for the same cells by activation of the type I IGF receptor tyrosine kinase.[82] IGF-I may act through autocrine or paracrine mechanisms to stimulate astroglial cell growth during normal brain development. IGF-II is present in the human cerebrospinal fluid and is apparently synthesized in the brain.[83] The brain is the only tissue in the adult rat where 4.7- and 3.9-kb IGF-II mRNAs are consistently observed. Five distinct size classes of IGF-II have been separated from the human brain on the basis of their immunoreactivity, the smallest component having a molecular weight of 7.5 kDa, identical to that of purified IGF-II from human serum. The highest concentrations of IGF-II occur in the anterior pituitary gland. The structure of the larger forms of IGF-II-like immunoreactive material and the function of IGF-II in the brain are still unknown.

D. REGULATION OF IGF SYNTHESIS

Different types of endogenous and exogenous factors contribute to regulating the synthesis and physiological actions of IGFs. EGF, insulin, and growth hormone contribute to the regulation of IGF-I production by human and mouse embryonic tissues as well as by human adult tumor-derived and transformed cells.[84,85] Expression of the *IGF1* gene partially depends on growth hormone, which is capable of inducing a rapid activation of *IGF1* gene transcription *in vivo*.[86] Administration of a single injection of growth hormone to hypophysectomized rats may result in a dose-dependent increase in *IGF1* transcripts in various tissues. However, in the liver and to a lesser extent in the kidney, IGF-I mRNAs are detectable in hypophysectomized rats, suggesting that in these tissues there is some degree of growth hormone-independent *IGF1* expression. IGF synthesis is regulated by growth hormone at the transcriptional level in the mouse liver, but in other organs and tissues, including the pancreas, this expression may be independent of growth hormone.[87] Another member of the growth hormone family of proteins, prolactin, may also be involved in regulating the transcriptional activity of the IGF-I gene in the liver. Injection of ovine prolactin into hypophysectomized rats results in a marked increase in the level of IGF-I mRNA in the liver and serum levels of IGF-I.[88]

In addition to growth hormone and prolactin, other hormones and growth factors may be involved in the regulation of IGF synthesis and secretion. Administration of adrenocorticotropin (ACTH) to rats results in decreased levels of adrenal IGF-I and IGF-II mRNA.[89] The levels of adrenal IGF-I and IGF-II mRNA do not change during the period of rapid organ growth that follows unilateral adrenalectomy. In human fetal adrenal cell cultures, both tumor necrosis factor-α (TNF-α) and interferon-γ (IFN-γ) inhibit IGF-II gene expression.[90] Transcriptional expression of the *IGF2* gene may be inhibited by glucocorticoids.[73] In bone tissue (mouse calvaria), the synthesis of IGF-I and IGF-II is differentially regulated by parathyroid hormone (PTH), transforming growth factor-β (TGF-β), and calcitriol.[91] The mechanisms associated with the regulation of IGF production are little understood. The synthesis of IGF-I in osteoblast-enriched cultures can be induced by cyclic adenosine-3′,5′-monophosphate (cAMP).[92]

E. IGF-RELATED FACTORS

Factors related to IGFs are produced in several organs and tissues, but their structural and functional relationships with IGFs remain incompletely defined.

1. Multiplication-Stimulating Activity

A factor called multiplication-stimulating activity (MSA) has been isolated from fetal rat liver and has been detected at high concentrations in fetal rat serum.[93] The cellular receptors for growth and metabolic activities of MSA are separate from the insulin receptors.[94] Rat MSA is probably identical to human IGF-II.[95]

2. Lentropin

Lentropin, a factor which is present in the vitreous humor of the eye and which is capable of stimulating lens fiber differentiation from chicken embryo lens epithelial cells *in vitro,* shares a functional and immunologic relationship with IGF-I.[96] Lentropin may be closely related to IGF-I, and an activity present in fetal bovine serum that promotes lens fiber cell differentiation is probably similar or identical to bovine IGF-I. Both type I IGF receptors and insulin receptors are present on lens epithelial cells. The source of lentropin found in vitreous humor has not been identified.

3. 15K IGF-II

A form of IGF-II with a molecular weight of 15,000, termed 15K IGF-II, was purified to homogeneity from the Cohn fraction IV_{1-4} of human plasma.[97] The amino acid composition of 15K IGF-II indicates that its carboxyl-terminal region may be different from that predicted from the analysis of human IGF-II cDNA clones. 15K IGF-II binds to IGF receptors and exhibits mitogenic properties. The precise structural and functional relationship between 15K IGF-II, the 10-kDa form of IGF-II previously purified from human serum,[65] and the classic 7.5-kDa form of the protein is not clear at present.

F. PRODUCTION OF IGFs IN TUMOR TISSUES

IGFs and IGF-like polypeptides are synthesized by various tumor tissues. IGFs or IGF-related proteins are secreted by some but not all human mammary carcinoma cell lines.[98,99] These cell lines also produce other growth factors such as PDGF and TGF-α. The rat epithelial-like cell line 18,54-SF, which can grow in serum- and hormone-free medium, secretes a polypeptide which is identical with IGF-II.[100] Certain human primary tumors exhibit high levels of IGF-I. An increased IGF-I content has been found in different types of primary human lung tumors.[101] Normal human lung tissue also contains IGF-I, and it is not known whether the increased IGF-I content of lung tumors is due to primary *in situ* production of the growth factor or is secondary to overproduction of other growth factors (such as PDGF) capable of stimulating IGF-I synthesis. Enhanced levels of IGF-I and IGF-II mRNAs have been detected in human colon carcinomas as well as in the majority of liposarcomas.[102] No significant levels of EGF or TGF-α were detected in the same tumors. A high level of expression of *IGF2* mRNA has been detected in the cells of Wilms' tumors.[74,103] Human adrenal pheochromocytomas contain high amounts of immunoreactive IGF-II, as compared to normal human adrenal medulla.[104] The high amounts of IGF-II protein present in pheochromocytomas are not reflected by a corresponding increase in IGF-II mRNA; whereas the opposite situation is seen in Wilms' tumors, where immunoreactive IGF-II is in the same range as in nontumor tissues despite increased expression of IGF-II mRNA. Both IGF-I and IGF-II mRNAs are frequently detected in rat medullary thyroid carcinomas in different stages of differentiation, but these mRNAs are usually not present in anaplastic medullary thyroid carcinomas.[105] The possible role of IGF production in the origin and/or development of tumors is not understood. It is clear, however, that IGFs do not function as universal autocrine factors in tumors, since they are not produced in all types of tumors or even in all tumors of the same type.

G. HUMAN DISEASES ASSOCIATED WITH ALTERED LEVELS OF IGFs IN SERUM OR URINE

The only clinical disorder in which an elevation of immunoreactive IGF-I level in blood has been unambiguously found is acromegaly,[3] a disease characterized by hypersecretion of growth hormone associated with a pituitary adenoma. Determination of serum IGF-I levels in acromegaly may serve as a useful diagnostic index when other clinical tests do not permit a clear-cut interpretation. Similarly, it is of value in assessing the success of therapy for acromegaly. Several diseases have been found to be accompanied by decreased serum levels of IGF-I and/or IGF-II, including pituitary dwarfism associated with growth hormone deficiency and the Laron syndrome. This syndrome is characterized by dwarfism associated with decreased IGF-I and IGF-II levels despite elevated growth hormone levels, and is attributed to an incapacity to produce and secrete IGFs under the influence of growth hormone. Treatment with growth hormone is usually ineffective in patients with the Laron syndrome.

IGF-I is present in human urine.[106] About one third of the IGF-I immunoreactivity in urine is free and the remainder is in a high molecular weight (43 kDa) form. The IGF-I values (corrected for urinary creatinine concentrations) in the urine of acromegalic patients are elevated, while those of patients with hypopituitarism are low, indicating that urinary IGF-I values are growth hormone dependent.

H. DEGRADATION OF IGFs

Both IGF-I and IGF-II are degraded by enzymes which are present in different types of cells. Two specific IGF-degrading systems are associated with kidney membranes.[107] These degrading systems are independent of the IGF receptor systems; are separated from but related to the insulin degrading systems; and are represented by enzymes which are apparently specific for IGF-I and IGF-II, respectively.

III. IGF-BINDING PROTEINS

Unlike insulin, the IGFs do not circulate as free polypeptide hormones but are noncovalently bound to specific carrier proteins in serum and other extracellular fluids as well as in certain normal and tumor tissues.[108-110] There are at least four specific forms of IGF-binding proteins (IGF-BPs) in blood, and these forms are secreted by a number of cell types. The rates of IGF-BP secretion and clearance appear to be regulated by hormones and growth factors. The major role of IGF-BPs is to alter the interaction of IGFs with their cell surface receptors, but other functions such as transport of IGFs out of the vasculature and modulation of target cell actions may also be important.

A. STRUCTURE AND FUNCTION OF IGF-BPs

The amino acid sequence of plasma or tissue IGF-BPs does not show similarity to that of cell surface IGF type I or type II receptors. Regions of the IGF-I molecule which are involved in binding to human IGF-BPs are distinct from type I receptor binding sites.[111] Structural studies suggest that IGF-BPs are multifunctional proteins that regulate not only transport through blood and entry into the extravascular space, but also presentation of IGFs to cell surface receptors. IGF-BPs may act as a reservoir, releasing continuously low amounts of IGF-I, and the IGF-I-BP complex is a better mitogen than temporary large concentrations of free IGF-I.[112] IGF-BPs released into assay buffer during binding assay by human fibroblasts and glioblastoma cells exert a modulating effect on IGF-I binding to the cell surface.[113] Other mechanisms may be additionally involved in the regulation of IGF-I biological activity. A covalent cross-linking of IGF-I to a specific inhibitor present in human serum has been demonstrated.[114] Human osteosarcoma cells release into the medium an IGF-BP which inhibits IGF action and proliferation of bone cells in culture.[115,116] The possible physiological role of IGF-BP inhibitors is unknown.

Initial analysis of IGF-BP forms demonstrated that IGF-BPs are represented by two major molecular forms, IGF-BP-1 and IGF-BP-2. The IGF-BP-1 corresponds to a small (25.3 kDa), growth hormone-independent protein which was detected in the human plasma, the amniotic fluid, the placenta, and the HepG2 hepatoma cell line. IGF-BP-1 derived from amniotic fluid may function as a cell growth inhibitor.[117] However, IGF-BP-1 can stimulate DNA synthesis in human granulosa cells and fetal skin fibroblasts.[118] Depending on the target cells, IGF-BP-1 can either stimulate or inhibit DNA synthesis. IGF-BP-1 is phosphorylated by human endometrial stromal cells and multiple protein kinases *in vitro*.[119] Phosphorylation of IGF-BP-1 is an important modification, and binding studies show that the phosphorylated form of IGF-BP-1 secreted by HepG2 human hepatoma cells has a higher affinity for IGF-I than the dephosphorylated form.[120] The gene for IGF-BP-1 is located on human chromosome region 7p14-p12.[121]

The IGF-BP-2 is a large protein present in human plasma. Insulin is involved in the regulation of IGF-BP-2 mRNA expression in rat hepatocytes.[122] A third form of IGF-BP, IGF-BP-3, associates with an acid-labile protein of 88 kDa to form a 150-kDa complex that binds IGF-I and IGF-II with high affinity and is the major carrier of the IGFs in the circulation postnatally. The human gene for IGF-BP-3 has been characterized.[123] cDNAs encoding two new types of human IGF-BPs (IGF-BP-4 and IGF-BP-5) have been isolated and the amino acid sequences were deduced.[124] The IGF-BP-4 and IGF-BP-5 are proteins of approximately 30 kDa. Other forms of IGF-BPs have been identified recently, including a form present in the culture medium conditioned by the simian virus 40 (SV40)-transformed human fibroblast cell line AG 2804.[125] A growth inhibitor purified from serum-free conditioned medium of the HT29 human colon carcinoma cell line was identified as the recently cloned human IGF-BP-4.[126]

IGF-carrier protein complexes of 150 kDa may stabilize and maintain a circulating reservoir of IGFs. These complexes differ from free IGF in several aspects, including loss of the acute insulin-like actions characteristic of free IGFs and inhibition of IGF access to cell membrane receptors. Circulating IGF-carrier complexes are growth hormone dependent.[127] Some forms of these complexes are elevated in acromegaly, where hypersecretion of growth hormone causes increased IGF-I levels; and are diminished in patients with genetic or idiopathic growth hormone deficiency and defects of the growth hormone receptor (Laron's syndrome), where both IGF-I and IGF-II are decreased, as well as in Pygmy adults and children who have an isolated deficiency of IGF-I.[128]

B. PRODUCTION AND SECRETION OF IGF-BPs

IGF-BPs are produced by a diversity of cell types and may display different functions in different organs and tissues. A preparation of IGF-BP purified from human amniotic fluid potentiates DNA synthesis and proliferation of several cell types in response to IGF-I.[129] IGF-BP-1 may enhance IGF-I response and has been identified as one of the most highly expressed immediate-early genes in liver regeneration.[130] The results suggest that IGF-I and IGF-BP-1 interact with hepatocytes or nonparenchymal liver cells and may act in a paracrine and/or autocrine fashion for maintaining normal liver architecture during regeneration. Porcine granulosa cells secrete IGF-BP-2 and IGF-BP-3, suggesting that these cells may be a source of follicular fluid IGF-BPs and that IGF-BPs may act as modulators of the IGF autocrine/paracrine system in the ovary.[131] Cultured human, bovine, and rodent endothelial cells also produce IGF-BPs and release IGF-BPs into the medium.[132,133] IGF-BPs could be used for IGF functions within the endothelial cells and/or for IGF storage and transport across the endothelium. The IGF-BPs produced by endothelial cells may have intrinsic biological activity which is retained when the IGF-binding domain is occupied by IGF-I.[134] IGF-BPs may be secreted into the bloodstream and circulate as a serum-binding protein or as a building block for larger serum IGF-BPs. In the absence of serum, human skin fibroblasts in monolayer culture secrete at least three types of IGF-BPs.[135] The primary structure of IGF-BPs produced by human HepG2 hepatoma cells or present in human amniotic fluid has been defined from cloned cDNAs.[136,137]

Hormones and growth factors are involved in the regulation of IGF-BP expression. EGF stimulates production of the IGF-BP-1 protein in human granulosa-luteal cells as well as in human hepatoma cells.[138,139] In contrast, insulin inhibits transcription of the IGF-BP-1 gene.[140] IGF-BP-3 production by human skin fibroblasts is stimulated by TGF-β.[141] Treatment of hypopituitary children with growth hormones can result in alterations in the plasma levels of unoccupied IGF-BPs.[142] Expression of the genes for IGF-BP-2 and IGF-BP-3 in human proliferative endometrium is under the influence of estradiol, and in the secretory endometrium is under the influence of both estradiol and progesterone.[143]

Tumor cells may secrete various molecular forms of IGF-BPs. Human breast cancer cell lines secrete heterogeneous IGF-BPs that are structurally and immunologically distinct from the small IGF-BP found in plasma, amniotic fluid, and HepG2-conditioned media.[144] IGF-BPs are also secreted by small-cell lung cancer cell lines.[145] Human central nervous system tumor cells produce a variety of IGF-BP types *in vitro* as well as *in vivo*.[146] Since IGF receptors are expressed in a diversity of neoplastic cells and these cells may respond to IGFs, IGF-BPs produced by tumor cells may be important modulators of IGF activity in tumor cell growth.

IV. FUNCTIONS OF IGFs

IGFs are importantly involved in the mechanisms that control the proliferation and/or differentiation of many types of cells. The physiological properties of IGF-I are different from those of IGF-II. IGF-I is more growth hormone dependent and more mitogenic than IGF-II, which is more insulin-like in its actions and is present in the blood at levels about three times greater than those of IGF-I. A monoclonal antibody to IGF-I is capable of blocking the stimulation of DNA synthesis by human plasma or calf serum.[147] IGF-I/somatomedin C is the mediator of growth hormone in many, but probably not all, peripheral tissues. Available evidence strongly suggests that the effects of growth hormone on somatic cell growth and development are mediated by IGF-I.

A. INSULIN AND IGFs IN EMBRYONIC AND FETAL DEVELOPMENT

Insulin and IGFs are importantly involved in embryonic and fetal development as well as in the postnatal growth of vertebrates.[11] Genes encoding receptors for insulin and IGF-I are expressed in *Xenopus laevis* oocytes and embryos,[148] and insulin and/or IGF-I may be important in amphibian oogenesis and early embryogenesis. The growth and early development of the chick embryo are regulated in a complementary and partially overlapping form by insulin and IGF-I.[149] IGF-I gene expression is tightly regulated during early organogenesis in the chick embryo.[150] Insulin and the IGFs are expressed during mammalian development. Although IGF-II has been considered as the predominant somatomedin during the fetal period, both IGF-I and IGF-II are expressed in mammalian embryos and are involved in the regulation of early fetal development.[17] IGF-I and IGF-II are present in the tissues or body fluids of fetuses from different mammalian species, and in some cases their expression is markedly elevated in comparison to postnatal levels.[93] IGF-I is involved in the modulation of erythropoietin-stimulated proliferation and differentiation of erythroid progenitors in the fetal liver.[151,152] IGF-I can stimulate normal somatic growth in growth hormone-deficient transgenic mice.[153] The developmental patterns of expression of the IGFs and IGF-BPs in the rhesus monkey and human are similar.[154] Synthesis of IGF-BPs in the liver is regulated during primate development at the level of transcription.

The differentiation-inducing effect of IGFs can be studied in selected systems *in vitro*. The ability of murine preadipocytes 3T3-L1 to differentiate into adipocytes is suppressed in a medium depleted of hormones and growth factors and can be restored by the addition of physiological amounts of IGF-I or pharmacological doses of insulin.[155] Insulin or IGF-I is essential for the induction of a variety of RNAs in 3T3-L1 adipocytes, suggesting that they may control many aspects of the differentiation program of these cells. IGF-I may also be involved in the terminal differentiation of muscle cells, which is associated with the production of myogenin; and this process can be specifically inhibited by the use of an antisense oligodeoxyribonucleotide to myogenin mRNA.[156] Transcriptional expression of the IGF-II gene is activated during myoblast differentiation.[157] IGF-II produced by endoderm cells, in particular visceral endoderm cells, may serve as an early embryonic growth factor.[158] Embryonal cells such as the human embryonic lung fibroblast cell line WI-38 produces IGF-I and expresses the type I IGF receptor, as well as a number of IGF-binding proteins, which suggests that the proliferation of these cells depends on an autocrine mechanism involving IGF-I and its receptor.[159] It seems likely that IGF-I-associated autocrine mechanisms are also involved in the growth and development of embryonic and fetal tissues *in vivo*.

B. ROLE OF IGFs IN REPRODUCTIVE FUNCTIONS

IGFs are expressed in the gonads and may be involved in the control of gametogenesis and reproductive functions in both sexes. IGF-II is present in the human seminal plasma and ovarian follicular fluid, suggesting a role for IGF-II in the regulation of gonadal cell growth.[160,161] The IGFs may have an important role in spermatogenesis by selectively stimulating spermatogonial, but not meiotic, DNA synthesis during this process.[162]

IGFs may have a role in ovarian follicular development.[163] They are secreted by granulosa cells *in vitro,* and ovarian follicular levels of IGFs increase during follicular growth *in vivo.*[164] Granulosa cells possess high-affinity, low-capacity binding sites which prefer IGF-I over IGF-II or insulin.[165] The three factors stimulate granulosa cell differentiation, as measured by the accumulation of estrogen and progesterone and the induction of gonadotropin receptors on the cell surface. There are fundamental differences in the cellular localization and hormonal regulation of ovarian IGF gene expression in that IGF-II gene expression (unlike IGF-I) is theca interstitial (rather than granulosa) cell specific, and is subject to down- (as opposed to up-) regulation in response to estrogenic stimulation.[166] In contrast, type I and type II IGF receptors exist on both somatic cell types of the ovary. These observations suggest that IGF-II of theca interstitial cell origin not only may play an autocrine role but also may serve as one of several signals through which this ovarian androgen-producing cell may communicate in a paracrine fashion with the adjacent granulosa cell compartment. Two forms of IGF-BPs are produced and secreted by granulosa cells, suggesting a role for these proteins in the IGF-associated autocrine/paracrine system in the ovary.[130] The IGF-BPs may have an inhibitory role on granulosa cell steroidogenesis, acting by sequestration of endogenously produced IGF-I and by direct interaction with the cell to elicit the inhibition.[167]

Biochemical, morphological, and structural changes occurring in the uterus and conceptuses may be correlated with IGF-I production and secretion. Estrogen and progesterone may be involved in the regulation of IGF-I gene expression and in the synthesis and secretion of IGF-I in the pig uterus.[168] However, the relative contributions of serum and conceptus-derived estrogen in these processes during the pig preimplantation period remain to be elucidated.

C. INTERACTION BETWEEN IGFs AND OTHER GROWTH FACTORS

IGFs do not act in an independent manner but they contribute to establish, through complex interactions with other hormones and growth factors, the regulatory infrastructure required for the proliferation and differentiation of normal cells. Particularly important are the interactions between insulin and IGFs. The construction of synthetic hybrid insulin-IGF molecules may contribute to the definition of the specific domains of IGFs involved in growth-promoting activity.[169]

Complex interactions between IGFs and other hormones and growth factors are important for the regulation of cell proliferation and/or differentiation processes. For example, interactions between IGFs and EGF are important for regulating the proliferation of normal rat epithelial cells maintained in a serum-free culture system *in vitro.*[170] IGF-I is produced by cultured porcine aortic smooth muscle cells, and the synergistical action of IGF-I and PDGF induces in these cells an increase in DNA synthesis that exceeds the sum of the individual effects obtained when either growth factor is added alone.[171] IGF-I has a growth-promoting effect on sparse cultures of human fetal fibroblasts; however, other growth factors such as PDGF and EGF have similar effects on the same cells and the mitogenic effects of IGF-I, PDGF, and EGF on the fibroblast cells appear to be additive.[172] These results suggest that the three factors have unique mechanisms of action and/or that they each act at different points of the cell cycle.

D. ROLE OF IGF-I IN HORMONE ACTION

The important role of IGF-I/somatomedin C as a mediator of growth hormone action is well recognized. However, it is not yet clear whether all the physiological actions of growth hormone are mediated by IGF-I. The regulation of cell proliferation by growth hormone may require a synergistic interaction with some locally produced growth factors.[173]

The actions of hormones and growth factors on some of their target cells may depend on the local production or presence of IGF-I. Expression of IGF-I is induced by estrogen in the uterus of ovariectomized prepubertal rats.[88] Estrogen may have no direct effects on the proliferation of at least some of target cells, and IGF-I may function as a mediator of estrogen action in these cells. IGF-I may be necessary for the trophic effects of insulin on cartilage growth *in vitro.*[174] The growth-promoting effect of EGF on BALB/c 3T3 mouse fibroblasts requires the activation of an autocrine loop involving IGF-I and its receptor.[175]

IGF-I could have a general role in the growth response to different exogenous stimuli, including hormones and growth factors.

E. OTHER FUNCTIONS OF IGFs

The IGFs exert multiple metabolic and physiological effects on different organs and tissues. IGF-I is one of the factors involved in bone remodelation, stimulating osteoclastic bone resorption through a direct or indirect action of supporting the generation and activation of osteoclasts.[176] IGF-II may play an important role in early bone development in the rat.[177] Whereas IGF-I mRNA is expressed in trabecular bone, cortical bone, and periosteum, but not in cartilage, IGF-II mRNA is highly expressed in cartilage and periosteum and much less in bone. These findings are not consistent with the generally accepted assumption of local IGF-I production by chondrocytes as mediators of growth hormone stimulation of longitudinal bone growth. Rather, growth hormone would have a direct effect on chondrocytes. IGF-I present in chondrocytes of the growth plate would not be locally produced, and IGF-II could be involved in bone longitudinal growth at later stages of development.

IGF-I has an effect on the chemotaxis of both aortic and retinal capillary endothelial cells *in vitro* and may have an important role in neovascularization processes *in vivo*.[178] Certain steps of hematopoietic processes may depend on the presence of IGFs. IGF-I or insulin is required for the formation of erythroid colony development (formation of CFU-E) *in vitro*.[179] A substance apparently identical to IGF-II is able to stimulate thymidine incorporation in cultures of fetal calf liver erythroid cells.[180] It appears that IGF-II is captured in the extracellular matrix and may act in a way similar to that of fibroblast growth factor (FGF), interleukin 3 (IL-3), or colony-stimulating factor 2 (CSF-2). IGF-II may have an important role in the nervous system and can stimulate motor nerve regeneration.[181]

V. IGF RECEPTORS

The cellular actions of IGFs are initiated by their binding to specific receptors.[182,183] Membrane receptors for IGF-I and IGF-II are also called type I and type II IGF receptors, respectively. These receptors are present in a wide diversity of mammalian organs and tissues, including placenta and kidney.[107,184] IGF-I receptors (type I IGF receptors) have been detected in the human lymphocytic cell line IM-9, and autologous regulation of IGF-I receptor expression is observed when these cells are incubated with the growth factor.[185] No IGF-I-specific binding sites are present in the adult human liver, which indicates that if IGF-I has any effect on this organ it is by interacting with other receptors.[186] Type I and type II IGF receptors are structurally different from the insulin receptor. However, insulin and IGFs may partially interact with the receptors for each other. The insulin receptor is phosphorylated in response to treatment of HepG2 cells with IGF-I.[187] In general, IGF-I and insulin act mostly through their homologous receptors; and IGF-II acts by cross-reacting with both the insulin and the type I (heterologous) receptors.[188] The physiological role of the type II IGF receptor in IGF signal transduction is not clear at present. Autocrine mechanisms of IGF-I action require the interaction of secreted IGF-I with its receptor on the cell surface.[189]

A. TYPE I IGF RECEPTOR

The IGF-I receptor (type I IGF receptor) has been purified from several sources, including human placental membranes.[190-192] The mature, glycosylated type I receptor is a transmembrane protein with a structure which is similar to that of the insulin receptor. It is composed of two 125-kDa α subunits and two 90-kDa β subunits, which are associated in the form of a heterotetrameric disulfide-linked complex.[193,194] The type I IGF receptor binds IGF-I with higher affinity than IGF-II and can also recognize insulin, but with a lower affinity. The binding of IGF-I to its receptor is regulated by multiple factors, including monovalent ions. There is an inverse relationship between NaCl concentration and the specific binding of [^{125}I]IGF-I to monolayers of human skin fibroblasts.[195] Most probably, Na^+ and other ions induce conformational changes in the IGF-I receptor as well as in IGF-binding proteins associated with the cell surface.

Insulin may interact on the cell surface either with its own receptor or with the IGF-I receptor (type I IGF receptor). Subunit structure, sizes of the protein portions of α and β subunits, amino acid composition of the subunits, and functional properties of the type I receptor are similar to those of the insulin receptor.[191] The assembly *in vitro* of hybrid human insulin/IGF-I receptors, composed of an insulin $\alpha\beta$ heterodimer half receptor and an IGF-I heterodimer $\alpha\beta$ half receptor, has been reported.[196] A constructed molecule containing the insulin A chain and a hybrid insulin and IGF-I B chain exhibited affinity for both the insulin receptor and the IGF-I receptor.[197] The study of chimeric insulin receptors containing the cysteine-rich domain of the IGF-I receptor indicates that this region can confer binding to both IGF-I and IGF-II.[198] The native type I IGF receptor can be distinguished from the insulin receptor by its ability to bind IGF-I with 100- to 1000-fold higher affinity than it binds insulin, as well as by the differential inhibitory effects of mono- and polyclonal antibodies.[199] The structural determinants of human IGF-I required to maintain binding to the type I IGF receptor and to the type II IGF receptor and serum-binding proteins are different.[200] The study of chimeric insulin-IGF-I receptors indicates that the major determinant for IGF-I binding specificity is located within the cysteine-rich region (amino acids 131 to 315) of the type I receptor, whereas the corresponding region of the insulin receptor is of minor importance for insulin specificity.[201] Despite their close evolutionary relationship, insulin and IGF-I have developed distinct modes of physical interaction with their cognate receptors within binding pockets that appear to exhibit the same overall conformation.

The type I IGF receptor gene maps to human chromosome region 15q25-q26. The complete primary structure of the human type I IGF receptor has been deduced from cloned cDNA.[202,203] The cDNA sequence predicts a 1367-amino acid precursor including a 30-residue signal peptide which is removed from the nascent chain. The type I receptor exhibits extensive similarity with the insulin receptor not only at the level of the precursor and subunit size, but also at the level of structural topology and primary sequence. The ligand-binding pockets of insulin and type I IGF receptors are formed by the extracellular α subunits and some extracellular portions of the β subunits. The 350-kDa heterotetrameric type I receptor complex spans the cell membrane via two β subunit domains, leaving 195 amino acid portions of the β subunits protruding from the cell surface to which the entire extracellular α subunits are attached by disulfide bonds. There is evidence suggesting the existence of an alternate form of the β subunit of the type I IGF receptor.[204] This alternate form of the receptor would be generated by a particular type of mRNA transcript, but the possible physiological significance of this form of IGF-I receptor is not known.

IGF-I binding to its receptor stimulates tyrosine kinase activity of the receptor molecule, which is necessary for activation of the IGF-I-stimulated signal transduction cascade.[205] The receptor β subunit is phosphorylated on tyrosine immediately after IGF-I binding.[206] An antipeptide antibody to the IGF-I receptor amino acid sequence 1232 to 1246 inhibits the receptor kinase activity.[207] The IGF-I receptor is a substrate for the insulin-activated insulin receptor kinase in intact L6 skeletal muscle cells.[208] This result suggests that some of the biological actions of insulin could be mediated by the IGF-I receptor as a result of its phosphorylation by the insulin receptor kinase.

Functionally active IGF-I receptors have been expressed at high levels in stable transfectants of Chinese hamster ovary (CHO) cells.[209] The expressed receptor was capable of mediating rapid (glucose uptake), intermediate (glycogen synthesis), and long-term (DNA synthesis) effects of IGF-I. The type I IGF receptor expressed in transfected cells has the same potency at stimulating these three responses as the human insulin receptor expressed in the same parental CHO cells, which suggests that there are no inherent differences in the abilities of these two receptors to mediate various responses and that the different physiological roles of insulin and IGF-I are determined by the tissue distribution of their respective cellular receptors and/or the pharmacodynamics of the two hormones. However, studies on the expression of chimeric receptor molecules consisting of insulin receptor extracellular ligand-binding sequences linked to the intracellular domain of the IGF-I receptor by the transmembrane domain of either the insulin or IGF-I receptor suggest that the receptors for IGF-I and insulin may mediate short-term responses similarly, but display distinct characteristics in their long-term mitogenic signaling potentials.[210]

The IGF-I receptor expressed in CHO cells transfected with an IGF-I cDNA clone recognizes IGF-II with high affinity, suggesting that the cellular effects of IGF-II are, at least in part, mediated by the IGF-I receptor. The available evidence indicates that the type I IGF receptor is able to transduce — although with different efficiency — the mitogenic action of insulin, IGF-I, and IGF-II and that there are intrinsic differences among human cells in the receptor through which insulin stimulates DNA synthesis; whereas in some cell types insulin exerts its mitogenic effect through activation of its own receptor, in other cells it can have this effect through activation of the type I IGF receptor.[211] The supraphysiological doses of insulin required for optimum growth of cells such as normal rat mammary epithelial cells *in vitro* are probably due to its acting weakly through an IGF-I-mediated growth promoting mechanism.[212] The mitogenic effect of IGF-II is mediated solely through activation of the type I IGF receptor.

There is more than one molecular species of type I IGF receptor. The human placenta contains two distinct species of type I IGF receptors.[213] Two species of type I receptor, termed type IA and type IB receptors, have been detected in human placenta, lymphoid cells, and hepatoma tissue by means of the monoclonal antibody 5D9.[199] The 5D9 antibody recognizes the type IA receptor which binds insulin with approximately 10 times higher affinity than the type IB receptor which is not recognized by the 5D9 antibody. Both species of the type I receptor may coexist in the same cell but it is not known whether they arise from a posttranslational modification of the receptor molecule, or whether there are two different genes for the type I receptor or a single gene that directs the synthesis of two distinct mRNAs by differential processing. Studies with antibodies directed against distinct epitopes in the β subunits of the IGF-I receptor suggest the existence in the rat brain of two distinct types of receptor β subunits which may differ in their amino acid sequence.[214] The peak of IGF-I receptor expression coincides with a period of rat brain development during which active neurogenesis is occurring, i.e., 16 to 18 d of gestation. An IGF-I receptor purified from fetal rat skeletal muscle contains a β subunit of 105 kDa, in contrast to the β subunit of the same tissue present in adult rats.[215] The fetal receptor appears to account for the high tyrosine kinase activity of fetal muscle and may be an important mediator of responses to both insulin and IGF-I during early development of the rat. Expression of this receptor decreases markedly during the first 2 weeks of postnatal life. Truncated molecular species of IGF-I receptors, some of them with higher biological potency, have been identified in the porcine uterus, but it is not known whether they are natural products or are formed by limited proteolysis of the intact species during the isolation process.[216]

A 35-kDa protein (35K protein) secreted by human fibroblasts maintained in culture *in vitro* acts as a modulator of the binding of IGF-I to its receptor.[217] The nature of the interaction between the 35K protein and the type I receptor, as well as the biological significance of the 35K protein, is unknown. A 34-kDa protein (34K protein) synthesized and secreted by the human endometrium during the secretory phase may have a significant role in inhibiting IGF-I binding to its receptor and may thus regulate the biological action of IGF-I in the endometrium.[218,219] The 34K IGF-I-binding protein could act as a local modulator of the IGF-I action at the interface between the decidua and the placenta. However, the physiological role of the 35K and 34K proteins involved in the regulation of the functional properties of the IGF-I receptor, as well as their precise relationship, remains to be established. A binding activity with a molecular weight of 28 to 33 kDa, termed BP-25, was detected in the human breast cancer cell lines MDA-MB 231 and Hs578T.[220] The precise physiological significance of BP-25 is also unknown but it could be involved in the regulation of breast cancer cell growth.

A monoclonal antibody against IGF-I is capable of inhibiting the mitogenic effect of IGF-I but not insulin, and also blocks the stimulation of DNA synthesis by human plasma or calf serum.[147] The growth-promoting actions of insulin in most tissues seem to be solely mediated by the binding of the hormone to its high-affinity receptor;[221,222] although in cells such as cultured human skin fibroblasts and breast cells, type I IGF receptors may mediate the growth-stimulating action of insulin.[94,223,224] In the IGF-I-responsive MG-63 human osteosarcoma cell line, IGF-I is internalized after binding to its receptor on the cell surface and the endocytosed hormone is degraded in lysosomes.[225] This internalization is mediated through the type I IGF receptor and not through the type II IGF receptor or the IGF-binding proteins which are also present in MG-63 cells. It is unknown whether the type I IGF receptor is internalized and recycled.

B. TYPE II IGF RECEPTOR

The type II IGF receptor is structurally different from the dimeric type I receptor and is represented by a monomeric structure of 250 to 270 kDa. The receptor has been purified to homogeneity from rat placenta liver and from human chondrosarcoma cells.[226-228] Glycosylation of the type II receptor protein is required for the acquisition of IGF-II binding activity.[229] The type II IGF receptor, as well as the insulin and type I IGF receptors, exhibits a molecular weight which is lower in the brain than in the liver; this is due to differences in the N-linked glycosylation patterns.[230] The possible biological significance of such differences is unknown. The type II IGF receptor was purified from the IGF-II-producing cell line 18,54-SF, and an antibody was generated which is capable of both immunoprecipitating the type II receptor and blocking IGF-II binding to its receptor.[231] The antibody specifically inhibits IGF-II binding not only to 18,54-SF cells, but also to rat placental membranes. The presence of abundant type II receptors on an IGF-II-secreting cell line is consistent with an autocrine role for IGF-II in selected cells.

Insulin may increase the expression of IGF-II receptors on the cell surface;[232] and both IGF-I purified from human plasma and recombinant IGF-I should bind to the type II receptor, although with different affinities.[233] However, the type II IGF receptor does not display a significant affinity for either insulin or IGF-I.[234,235] The major actions of IGF-II are exerted through its binding to the IGF-I receptor, and the major function of the IGF-II receptor would be related to limiting growth by degrading IGF-II.

Upon IGF-II binding, type II receptors are internalized and are subsequently recycled rapidly back to the cell surface.[236] However, type II receptors can be internalized and can recycle in the absence of the ligand.[237] Multiple factors are involved in the expression of type II receptors on the cell surface. Both cell surface and intracellular type II IGF receptors are regulated by cell density in primary cultures of adult rat hepatocytes.[238] Insulin modulates type II receptor expression by inducing a constitutive recycling of the receptor through a pathway which is independent from that of the glucose transport recycling.[236,239] Incubation of insulin-treated adipocytes with the acidotropic agent chloroquine results in interference with type II receptor recycling to the cell surface. Significant amounts of circulating type II IGF receptor are normally present in rat serum and are subjected to developmental regulation.[240] The origin and role of circulating type II IGF receptors are not known.

The role of the type II IGF receptor in signal transduction is not clear. The mitogenic effects of IGF-II are mediated solely through activation of the type I IGF receptor.[211] In human hepatoma cells HepG2, the type II IGF receptor can mediate an insulin-like biological response such as stimulation of glycogen synthesis.[241] This fact is interesting because, in contrast to the insulin receptor, the type II receptor is devoid of tyrosine kinase activity. Unexpectedly, the type II IGF receptor has been identified recently as a receptor for mannose-6-phosphate.[242,243] The identity was confirmed by biochemical and immunologic methods. However, the type II IGF receptor responds dissimilarly to the two distinct high-affinity ligands, IGF-II and mannose-6-phosphate.[244] In contrast to the hormone, the sugar does not induce an interaction of the type II receptor with G proteins in phospholipid vesicles or in native cellular membranes.

The type II IGF receptor is implicated in the delivery of more than 50 lysosomal enzymes to the lysosome. In human breast cancer cells, the lysosomal enzyme cathepsin D and IGF-II recognize the same receptor molecule on distinct, but interacting sites.[245] A consequence of such interaction is that cathepsin D can modulate the mitogenic activity of IGF-II. These results suggest that IGF-II is a full agonist, whereas cathepsin D is a partial agonist/antagonist of the type II IGF receptor system.

The type II IGF receptor is abundantly present in many types of cells, including human T lymphocytes.[246] Type II IGF receptor mRNA and protein are expressed at high levels during rat embryogenic development.[247] The physiological significance of this expression is suggested by the fact that it is principally localized to tissues expressing high levels of IGF-II mRNA, such as the heart and major vessels, and commences at the same time as IGF-II expression. No expression of type II IGF receptors occurs in the developing central nervous system or in peripheral nervous tissues. A strong downregulation of type II IGF receptor expression occurs postnatally. High levels of type II IGF receptors present in embryonic and fetal tissues could serve to stabilize local concentrations of IGF-II at required values by endocytosing excessive amounts of locally synthesized growth factor. Different patterns of production of IGF-I and IGF-II occur during rat embryonic development; and the early and widespread expression of type I IGF receptor gene, in contrast to the relatively limited and localized pattern of IGF-I gene expression, is consistent with the view that the type I receptor may mediate the effects of both IGF-II and IGF-I during embryogenesis.[248] A striking example of genomic imprinting is observed in the mouse, where the paternal gene for IGF-II receptor is poorly transcribed.[249] The hepatocytes of regenerating rat liver express increased levels of type II IGF receptors.[250]

Type II receptors are expressed in continuous cell lines such as BRL-3A cells derived from an adult rat liver, which produce IGF-II and are capable of proliferating in serum-free conditions, suggesting an autocrine function for IGF-II in these cell lines.[234] However, the type II IGF receptor present in rat hepatoma H35 cells (which are cells that express abundant type II receptors but have no detectable type I receptors), as well as the type II IGF receptor present in fetal and postnatal human fibroblasts, do not directly mediate the mitogenic action of IGF-II.[251-253] This action is mediated via the type I and not the type II IGF receptor. In H35 cells, both IGF-II and MSA stimulate metabolic responses (amino acid transport and tyrosine aminotransferase and glycogen synthase activities) to the same magnitude as insulin but only at high concentrations, which suggests that these peptides are acting through the insulin receptor.[254] Insulin and IGF-II enhance erythroid colony formation by human bone marrow cells, and this stimulation may be explained by activation of a common receptor or postreceptor system.[255]

The type II IGF receptor is phosphorylated on tyrosine by a tyrosine kinase in adipocyte plasma membranes.[256] Furthermore, IGF-II stimulates phosphorylation of the type II IGF receptor in monolayer cultures of rat embryo fibroblasts and rat liver cells; but since IGF-II interacts with both type I and type II IGF receptors, it is not clear which receptor mediates the phosphorylation.[257] The bulk of this phosphorylation occurs on serine residues. In the human placenta there is a unique, high-affinity binding site for IGF-II; and the available evidence suggests that this site is closely associated with or, more probably, is physically located on the α subunit of the type I IGF receptor.[258] The site of the type I receptor with higher affinity for IGF-II was designated IB and the site of the same receptor with higher affinity for IGF-I was designated the IA site.

In IM-9 human lymphoblastoid cells, the receptor for IGF-II has the characteristics of a type I receptor and may be either an atypical insulin receptor or a unique form of type I receptor.[259] The type II receptor plays little or no role in mediating the actions of the two somatomedins, IGF-I and IGF-II, in L6 myoblasts.[260] In conclusion, the type II IGF receptor would not act as a direct mediator of at least some of the biological actions of IGF-II.

C. IGF RECEPTORS IN DEVELOPMENT AND DIFFERENTIATION

Regulation of IGF receptor expression has an important role in development and differentiation. The IGF receptors antedate and initially predominate over insulin receptors in chick embryonic development.[261] The IGF-II receptor is a major protein in some fetal rat tissues and its expression is subjected to developmental regulation, suggesting that it plays an important role in fetal growth and development.[262] The high concentrations of both IGF-II and the IGF-II/mannose-6-phosphate receptor during fetal life raise interesting questions about the functional interaction of a growth factor with a receptor that binds numerous lysosomal enzymes. Studies on developing mouse embryonic limb bud in organ culture suggest that in differentiated tissues IGF-II may no longer act as a classical growth factor.[263] Preadipocytes express high levels of receptors for both IGF-I and IGF-II and are more responsive to IGF-I than to insulin in both acute and delayed types of response.[264] After their differentiation into adipocytes, the levels of IGF receptors decrease and the insulin receptors increase. These results support the idea that the IGFs primarily regulate undifferentiated cells, whereas insulin primarily regulates differentiated cells. The IGF-I receptor is required for the proliferation of hematopoietic cells.[265] Expression of IGF-II receptors may have an important role in liver regeneration. Partial hepatectomy in the rat leads to a rapid increase in the expression of IGF-II receptors at the cell surface, possibly as a result of increased receptor translocation, followed by a later increase in receptor mRNA and protein synthesis.[266]

D. REGULATION OF IGF RECEPTOR EXPRESSION

The physiological factors involved in the regulation of IGF receptor expression are little known, but hormones and growth factors may have an important role in this regulation. Both growth hormone and insulin are able to increase the number of IGF-II receptors in adipose cells from hypophysectomized rats.[267] Growth hormone also plays a role in maintaining the number of IGF-II receptors and modulates the stimulatory effect of insulin on IGF-II binding. Expression of IGF-II receptor mRNA is developmentally regulated in rat tissues at different stages of growth.[268] Calcitriol increases the number of IGF-I receptor sites in the mouse osteoblastic cell line MC3T3-E1, which is able to secrete IGF-I.[269] These results suggest a paracrine or autocrine system of IGF-I in bone cells, especially osteoblasts.

VI. POSTRECEPTOR MECHANISMS OF ACTION OF IGFs

IGF-I appears to act through postreceptor mechanisms similar to that of insulin, whereas the cellular effects of IGF-II would be exerted through other mechanisms. Both somatomedins may influence DNA, RNA, and protein synthesis in the target cells. High doses of somatomedins, in particular IGF-I, may result in insulin-like effects. Injection of supraphysiological doses of IGF-I produces in humans a hypoglycemia which is clinically indistinguishable from that induced by insulin.[270] The levels of epinephrine, norepinephrine, growth hormone, glucagon, and cortisol respond similarly to both agents, IGF-I and insulin.

A. MECHANISMS OF ACTION OF IGF-I

The cellular and biochemical mechanisms of action of IGF-I are very complex. Some of these mechanisms are dependent on the cell cycle. For example, in BALB/c 3T3 fibroblasts IGF-I-induced stimulation of glucose uptake is observed in both quiescent and primed competent cells, whereas IGF-I-induced DNA synthesis is detected only in primed competent cells.[271] In KB human epidermoid carcinoma cells, IGF-I and IGF-II as well as insulin and EGF cause a rapid membrane ruffling.[206] The morphological change is observed within 1 min after the addition of the growth factors and is accompanied by microfilament reorganization.

1. Adenylyl Cyclase Activity

The possible role of the adenylyl cyclase system in the mechanism of action of somatomedins is poorly understood. In crude membrane preparations of rat adipocytes, hepatocytes, and spleen lymphocytes, as well as in chicken cartilage, IGF-I inhibits the basal activity of adenylyl cyclase.[272] A similar effect is observed when the same preparations are exposed to epinephrine, parathormone, or prostaglandin (PGE_1). The possible physiological significance of these changes remains to be elucidated.

2. Tyrosine Phosphorylation

After binding to the type I receptor on the cell surface, IGF-I stimulates phosphorylation of tyrosine residues on the β subunits of both its own receptor and the insulin receptor.[273-276] The autophosphorylated receptor acquires tyrosine kinase activity with specificity for cellular substrates which include 185- and 240-kDa proteins (p185 and p240) that may also be phosphorylated by the activated insulin receptor.[206,277] The p185 and p240 substrates are shared with the activated insulin receptor kinase but their functions are unknown. However, studies using selected tyrosine-containing polymers indicate that the substrate specificity and enzymatic action of the insulin and IGF-I receptor are partially distinct.[278] In intact cells, the IGF-I receptor is phosphorylated not only on tyrosine but also on serine and threonine residues.[279] The enzyme catalyzing this phosphorylation, which is stimulated by phorbol esters, has not been characterized but it seems to be different from protein kinase C.

IGF-I stimulates the proliferation of highly metastatic NL-17 mouse colon carcinoma cells, and this effect is tightly associated with tyrosine phosphorylation of two proteins of 150 and 160 kDa.[280] The cellular mechanism of IGF-I action may also include tyrosyl phosphorylation of nuclear proteins.[281] IGF-I stimulation of murine mesangial cells may result in the rapid phosphorylation of nuclear proteins on tyrosine residues, including the c-Jun protein. Since the c-Jun protein is an essential component of the transcription factor AP-1, its phosphorylation may result in regulatory changes in gene expression.

3. Calcium Influx and G Protein Activation

Calcium influx may play a critical role in the mitogenic action of IGF-I. Relatively low concentrations of IGF-I stimulate calcium influx in primed competent BALB/c 3T3 cells by activating a calcium-permeable cation channel via the IGF-I receptor.[282] Pertussis toxin blocks IGF-I-mediated Ca^{2+} influx and DNA synthesis in BALB/c 3T3 cells, suggesting the involvement of a pertussis toxin-sensitive GTP-binding protein (probably G_1) in the cellular mechanisms of IGF-I action.[283] These results suggest that Ca^{2+} influx and a G protein coupled with the IGF-I receptor are involved in the mitogenic action of IGF-I.

4. Phospholipid Metabolism

The effect of IGF-I on phospholipid metabolism remains little characterized. In primed competent BALB/c 3T3 mouse cells, IGF-I stimulates the production of 1,2-diacylglycerol via multiple pathways and induces phosphatidylcholine breakdown by a mechanism involving protein kinase C.[284]

5. Regulation of Gene Expression

IGF-I is involved in the regulation of gene expression at both the transcriptional and posttranscriptional levels.[285] Some labile proteins may be involved in the regulation of the transcriptional activity of these genes. Regulation of other genes by IGF-I may be exerted at the level of transcript stability. The growth-promoting effects of IGF-I are associated with an increase in cellular RNA content; and it has been shown that exposure of cells to IGF-I can activate transcription from the ribosomal DNA promoter, which is involved in regulating ribosomal RNA synthesis.[286]

6. Proto-oncogene Products

The IGFs may participate in the regulation of proto-oncogene expression in particular cellular systems. Both IGF-I and IGF-II induce a rapid and transient induction of c-*fos* mRNA in murine osteoblast-like cells.[287] Expression of c-*fos* mRNA is induced by insulin and IGF-I in L6 rat skeletal muscle cells.[288] IGF-I induces transient expression of c-*fos* transcripts in cultured rat thyroid cells FRTL5.[289] An even stronger induction of c-*fos* is produced in the same cells by the addition of calf serum. IGF-I and additional factors contained in serum increase c-*fos* mRNA levels, at least in part, by activation of the c-*fos* promoter. In BALB/c 3T3 mouse fibroblasts, IGF-I stimulates transcription of the c-*jun* proto-oncogene.[290] With the c-Fos protein, the c-Jun protein forms the AP-1 complex which is involved in regulation of gene transcription. In the human neuroblastoma cell line SH-SY5Y, IGF-I regulates c-*myc* and GTPase-activating protein 43 (GAP-43) mRNA expression.[291] The role of proto-oncogene products in the mechanisms of action of IGF-I in different types of cells is not understood. However, as is true for other growth factors with receptors possessing tyrosine kinase activity, Ras proteins may have a role in the IGF-I intracellular signaling pathways. The protein Grb2, which contains in its amino-terminal end one SH2 domain flanked by two SH3 domains, couples receptor tyrosine kinases to a protein, Sos, which activates the Ras signaling pathway.[292,293]

B. MECHANISMS OF ACTION OF IGF-II

The postreceptor mechanisms of action of IGF-II may be, at least in part, different from those of IGF-I. The mechanisms of IGF-II may include changes in cell membrane G proteins and calcium influx as well as alterations in phospholipid metabolism.

1. G Proteins

A 40-kDa pertussis toxin-sensitive protein involved in the mechanism of action of IGF-II in BALB/c 3T3 fibroblasts may be represented by the α subunit of a G_1 protein.[294] This signaling system is activated in quiescent cultured cells primed with either competence factors such as PDGF and EGF or the viral K-Ras protein, which may function downstream of the signaling pathway of the competence growth factors. PDGF and EGF or the v-K-Ras oncoprotein may allow the interaction of IGF-II with the G protein signaling pathway.

2. Calcium Influx

After binding to its cell surface receptor, IGF-II increases the $[Ca^{2+}]_i$ in primed competent BALB/c 3T3 cells (previously treated with EGF) by stimulating a sustained calcium influx.[295,296] This mechanism is totally dependent on extracellular calcium and involves pertussis toxin-sensitive GTP-binding protein. The IGF-II-stimulated calcium influx may be causally related to the mitogenic effect of IGF-II. The action of IGF-II on the $[Ca^{2+}]_i$ is cell cycle specific. While EGF and IGF-II, either alone or in combination, do not affect the $[Ca^{2+}]_i$ in quiescent cells, the combination of EGF and IGF-II generates a calcium signal in cells rendered competent by treatment with PDGF. A sustained calcium influx may be a message of IGF-II action causally related to the induction of DNA synthesis and cell proliferation.

3. Phospholipid Metabolism

In preparations of basolateral membranes isolated from canine kidney incubated with rat IGF-II, the growth factor activates phospholipase C and induces an increase in the concentration of inositol trisphosphate and diacylglycerol.[297] This effect may result in activation of protein kinase C, phosphorylation of basolateral membrane proteins on serine and threonine residues, and increased Na^+/H^+ exchange across the brush-border membrane. At the same time, the influx of calcium would be stimulated.

4. Regulation of Gene Expression

The phenotypic changes induced by IGF-II may be associated, at least in part, with changes in the expression of specific genes. Unfortunately, the genes regulated by IGF-II remain little characterized.

VII. ROLE OF INSULIN AND IGFs IN NEOPLASIA

Insulin and IGFs may be important in the growth regulation of some, but not all, human and nonhuman malignancies.[298,299] The role of insulin and IGFs in tumor growth *in vivo* is little understood, but receptors for these hormones are present in many different types of tumor cells.[300,301] Similar or identical receptors are present in cell lines derived from a diversity of human or nonhuman tumors. It has been suggested that increased amounts of insulin and IGF receptors in the tumor cells may be a factor contributing to the hypoglycemia observed in certain cancer patients.[302] Overexpression of the human IGF-I receptor can promote neoplastic transformation in a ligand-dependent manner.[303]

A. PRIMARY TUMORS

Primary human and nonhuman tumors may produce insulin and/or IGFs and may contain type I and/or type II IGF receptors. Of course, most insulinomas are characterized by the secretion of high amounts of insulin, which may result in hypoglycemia and may be helpful to the diagnosis of such tumors by the finding of high amounts of circulating insulin either in basal conditions or after appropriate stimulation.

In addition to insulin, high levels of proinsulin and C-peptide are frequently present in the blood of patients with insulinoma.[304]

Relatively high amounts of IGFs are produced by some primary human tumors. A 40- to 100-fold increase in the levels of IGF-II mRNA was detected in about one fifth of primary human liver cancers, whereas in benign liver tumors and liver cirrhosis IGF-II gene expression was comparable to that of normal adult liver.[305] Interestingly, the primary liver cancers expressed 6-kb IGF-II gene transcripts which are found in fetal liver, whereas 5.3-kb transcripts previously detected in normal adult liver were mainly present in benign liver tumors and cirrhosis. Reexpression of fetal-type IGF-II transcripts in liver cancer may reflect an undifferentiated state of the tumor cells. A high level of IGF-II expression in the liver tumor cells could contribute to hepatocarcinogenic processes by an autocrine mechanism. IGF-II produced by fibroblasts may function as a paracrine growth factor in some human breast tumors.[306,307] IGF-I plasma concentrations were found to be increased in breast cancer patients when compared with a control population.[308] High contents of IGF-I and IGF-II have been detected in freshly obtained surgical specimens from human tumors of smooth muscle origin, both of the benign type (leiomyomas) and of the malignant type (leiomyosarcomas).[309,310] A role for IGF-II, but not IGF-I, in the development of human smooth muscle tumors has been suggested.

Variable amounts of receptors for insulin and IGFs are present in a diversity of human primary tumors as well as in the metastases from these tumors. Whereas in some tumors the receptors for insulin may predominate, other tumors contain normal or elevated levels of somatomedin receptors. Insulin receptors are present at relatively high levels in mammary cancers, raising the possibility that insulin may have a role in the growth of these tumors.[311] IGF-I receptors have been detected in plasma membrane preparations from primary breast and colon carcinoma specimens obtained at surgery.[312] IGF-I activity as well as IGF-I and IGF-II receptors are present in most human primary breast cancer specimens and breast cancer cell lines.[313,314] IGF receptors are present in over 90% of primary breast cancers and correlate positively to estrogen and progesterone receptors in breast cancer tissue, but no correlation exists in these malignant tissues between IGF receptor and EGF receptor.[315-318] Amplification of the IGF-I receptor gene, which occurs rarely in breast cancer, may be associated with overexpression of the receptor.[319] These results suggest that IGF-I may act via endocrine, paracrine, or autocrine mechanisms for the stimulation of breast tumor growth. However, no association has been found between IGF-I receptors in breast cancers and lymph node status, tumor size, or differentiation grade of the tumor. IGF-I receptors are also present in about half of the tissue specimens from benign human breast tumors.[320]

The human endometrium, either normal or neoplastic, exhibits high-affinity IGF-I binding sites.[321] The fact that IGF-I binding activity increases as the histological grade of the malignant endometrial tissue increases suggests that IGF-I may play a role in supporting the growth of this neoplastic tissue.

The possible role of insulin and IGFs in the growth of primary human lung tumors is unknown. IGF-I was detected in primary and metastatic tissue and in cell lines of small-cell lung carcinomas (SCLC), but not in non-small-cell lung carcinomas (NSCLC).[322,323] IGF-I could function as an autocrine growth factor in the growth of SCLC. However, the results from another study of 75 primary lung cancers using immunohistochemical methods and quantitative receptor autoradiographic techniques showed that IGF-I is positive in all cases of NSCLC but not in SCLC.[324] The reason for such contradictory results is unknown. Specific binding sites for IGF-I were detected in all types of human lung tumors examined. Several SCLC cell lines secrete both IGF-I and IGF-I-binding proteins, while NSCLC cell lines secrete IGF-I-binding proteins only.[325] The levels of IGF-I in the blood of patients with SCLC or NSCLC are usually normal, but most patients with lung cancer have elevated blood levels of IGF-I-binding proteins. IGF-I binding sites are present in SCLC cell lines as well as in fresh cell samples from these tumors.[313] Stimulation of thymidine incorporation into DNA is observed when SCLC cell lines are incubated in the presence of IGF-I, and inhibition of DNA synthesis occurs when the cells are exposed to IGF-I-specific monoclonal antibodies. Since IGF-I is present in primary and metastatic SCLC tumor tissue and in most SCLC cell lines, it appears that IGF-I may function, at least in certain cases, as an autocrine growth factor for SCLCs.

Insulin and/or IGF may have an important role in the growth of human sarcomas. IGF-I receptors are present in primary osteogenic sarcomas as well as in the human osteosarcoma cell line NG-63, where IGF-I acts as a potent stimulator of cell growth.[326] IGF-II is implicated in the growth and motility of human rhabdomyosarcoma cells, which are highly invasive *in vivo*.[327] These cells express high levels of transcripts for both IGF-II and type I IGF receptors. The polyanionic compound suramin can inhibit the growth of human rhabdomyosarcoma cells by interrupting an IGF-II-associated autocrine growth loop.[328] Rhabdomyosarcomas, the most common soft tissue sarcoma of childhood, appear to arise from developing striated muscle-forming cells. It is unclear whether the production of elevated amounts of IGF-II by rhabdomyosarcomas simply reflects the embryonal muscle origin of the tumor cells or whether a disordered regulation of IGF-II production is a primary event that may lead, through the action of some additional lesions, to the expression of a transformed phenotype.

Primary human endocrine tumors may express insulin and IGFs as well as insulin and IGF receptors. Expression of the IGF-II receptor is increased in primary human thyroid neoplasms.[329] Medullary thyroid carcinomas, which are characterized by the secretion of calcitonin and serotonin, and carcinoid tumors, which originate from enterochromaffin cells in the small bowel and produce serotonin and tachykinin peptides, produce IGF-I and may also express in some cases IGF-I receptors; this suggests an autocrine mechanism for the growth of these tumors.[330,331] Insulin and IGF-I may exert regulatory effects on the growth of primary adrenal tumors, including adrenocortical carcinomas and adenomas associated with Cushing's or Conn's syndrome or adrenogenital syndrome, and pheochromocytomas.[332] Insulin and IGF-I may act as autocrine factors in adrenocortical carcinomas. In addition, IGF-I may play a role in the function of adrenocortical carcinoma tissues. Human ovarian cancer cell lines as well as primary ovarian tumors may express IGF-I, IGF-I-binding proteins, and IGF-I receptors, suggesting a role for IGF-I in the growth of these tumors.[333] However, a similar expression occurs in normal ovarian tissue as well.

Human neural tumors may express relatively high levels of insulin and/or IGF receptors. Primary human glioblastomas were found to express functional IGF-I receptors.[334] In an independent series, increased concentrations of IGF-II receptors, but not of receptors for insulin or IGF-I, were detected in cell membranes prepared from primary human glioblastomas, as compared to normal brain tissue.[335] Moreover, IGFs may be produced by the neural tumors since they were identified in the cytosol of glioblastoma tissues. Neuroblastomas are highly malignant tumors that arise in cells of neural crest origin destined to become chromaffin tissue or neurons of the peripheral nervous system. These tumors account for up to half of all cancers in infants. Examination of neuroblastoma tumor tissues using *in situ* hybridization histochemical analysis showed that the IGF-II gene was expressed in 5/21 neuroblastomas, but was detectable in cells of nonmalignant tissues including adrenal cortical cells, stromal fibroblasts, and eosinophils in all 21 tumors.[336] Locally produced IGF-II may stimulate the growth of primary neuroblastomas through an autocrine mechanism in some cases and through a paracrine mechanism in other cases.

Type I IGF receptors were detected in Wilms' tumors (a malignant renal neoplasm of childhood), and these receptors are activated by IGF binding *in vitro*.[337] An antibody with specificity to the IGF-I receptor inhibited growth of Wilms' tumor in cell cultures and heterotransplants in athymic mice.[338]

Insulin binding in primary human tumors (breast carcinoma, colonic adenocarcinoma, gastric carcinoma, adrenocortical carcinoma, lymphoma, and pheochromocytoma) may be very similar to that of normal tissues in both binding site concentration and affinity.[298] Insulin receptors from human brain tumors of glial origin are indistinguishable from the normal peripheral receptors of target tissues for insulin.[339] Similar mechanisms may influence insulin binding in neoplastic and nonneoplastic tissue, and insulin receptors may be preserved during malignant transformation.

Resting normal human lymphocytes do not express insulin receptors, but the receptors emerge on the cell surface upon stimulation with mitogens and other substances.[340,341] In human leukemia, insulin receptors are present only on immature cells, although some cases of poorly differentiated malignant lymphoma show no insulin receptors in the tumor cells.[342] Lymphoblasts from patients with ALL exhibit insulin-binding sites, the number of receptors per cell being higher in the null-cell type than in the T-cell type; whereas in 6 of 14 patients with CLL, the tumor cells lacked insulin receptors and in the remaining 8 patients with CLL the number of insulin binding sites per cell were low.[343] In a study on the binding of insulin to lymphoblasts from 46 children with leukemia, including 35 cases of ALL, variable levels of insulin receptors were detected but no correlation with several different clinical parameters was found.[344]

IGF-II overexpression may be associated with demethylation of the IGF-II gene in some tumors, but amplification or deletion of the IGF-I and IGF-II genes were not detected in 40 primary tumors from adult patients.[345] Loss of heterozygosity or an imbalance of the two alleles of the IGF-II gene on human chromosome 11p15 was found in childhood tumors, including neuroblastoma, nephroblastoma, and adrenocortical carcinoma. The possible role of this alteration in the origin or development of childhood tumors is unknown.

Hepatitis and hepatocellular carcinomas may be etiologically associated with infection with the woodchuck hepatitis virus (WHV) in woodchucks. High levels of IGF-II mRNA were detected in 45% of hepatocellular carcinomas arising in woodchucks with persistent WHV infection.[346] Analysis of WHV RNA in the same carcinomas revealed that the tumors with high levels of IGF-II mRNA contained low or undetectable levels of WHV RNA, and vice versa the tumors with low levels of IGF-II mRNA contained high levels of WHV RNA. Integrated WHV DNA was present in hepatocellular carcinomas from both groups, but viral DNA-replicating forms were present predominantly in the carcinomas with low levels of IGF-II mRNA. Whereas the latter carcinomas were highly differentiated, tumors with high levels of IGF-II mRNA were more anaplastic. The exact role of IGF-II in woodchuck hepatocellular carcinomas remains to be determined.

B. EXPERIMENTAL TUMORS

The structural and functional characteristics of insulin and IGF receptors present in experimental tumors may be similar, at least in some cases, to those of the respective normal tissues. The insulin-responsive rat mammary adenocarcinoma R3230AC exhibits responses comparable to those of normal tissues in receptor binding affinity, specificity for ligand, and downregulation by insulin.[347] However, certain hormonal perturbations of the R3230AC tumor-bearing hosts may be associated with alterations in insulin binding as well as with basal and insulin-induced tyrosine kinase activities in the transplanted tumor. The tumor from diabetic rats displays increased responsiveness of insulin receptor kinase activity to insulin *in vitro,* compared to tumors from intact or insulin-treated diabetic rats.

Insulin and glucagon receptors are present in the plasma membranes of rat Morris hepatomas of varying growth rates.[348] The binding of insulin, however, is diminished in Morris hepatomas when compared to plasma membranes from rat normal liver, which is apparently due to either a decrease in binding affinity or a change in site-site interactions. A significant correlation exists between the binding of insulin and the growth rate of different Morris hepatomas. Administration of the carcinogenic agent 2-acetylaminofluorene (2-AAF) to rats for 2 d or more results in decreased insulin receptor number and insulin receptor-associated kinase activity in the liver, whereas EGF receptors may be unchanged in number and kinase activity for the ligand.[349]

Oncogene-induced neoplastic transformation may be associated with increased expression and/or altered function of insulin and/or IGF receptors. Rous sarcoma virus (RSV)-induced transformation of chicken embryo fibroblasts (CEF) cells is associated with constitutive phosphorylation of the IGF-I receptor β subunit on tyrosine, suggesting that the v-Src oncoprotein may alter the regulation of cell proliferation and differentiation through a constitutive mitogenic signal elicited by a continued expression of an activated IGF-receptor.[350] Transfection of a plasmid vector expressing a *ts* mutant of the oncogenic virus SV40 T antigen into mouse 3T3 cells results in induction of IGF-I mRNA and protein expression.[351] Moreover, the SV40 T antigen requires a functional IGF-I receptor to stimulate cellular growth in low concentrations of serum or PDGF. Mouse keratinocytes transfected with a plasmid constitutively expressing the c-*fos* proto-oncogene exhibit a marked potentiation of the mitogenic response to IGF-I.[352]

C. TUMOR CELL LINES

Insulin and IGFs receptors are present at variable levels in most of the cultured neoplastic cell lines examined, but only some of these cell lines synthesize insulin or IGF mRNA and protein. No IGF-I mRNA was detected in human breast or colon cancer cell lines analyzed with a sensitive and specific RNAse protection assay using an antisense RNA probe as well as with *in situ* hybridization histochemistry. In contrast to the cell lines, IGF-I mRNA was easily detectable in breast tissues but it is produced by the stromal cells and not by normal or malignant epithelial cells. These findings suggest that IGF-I may function in the normal and malignant breast as either a paracrine stimulator of epithelial cells or an autocrine stimulator of stromal cells.[353] IGF-I transcripts were also detected in three neuroepithelioma cell lines as well as in one of two Ewing's sarcoma cell lines.

Insulin and IGF-binding capacity may be decreased, normal, or increased in tumor cell lines when compared with the respective nontransformed counterparts.[354-356] The number of insulin-binding sites in seven different Burkitt's lymphoma cell lines containing c-*myc* gene translocations was found to be decreased by >90%, compared to lymphoblastoid cells of normal karyotype, with no change in receptor affinity.[357] Structural analysis revealed that the tumor cells had an increase in a precursor form (mol wt 210,000) of the insulin receptor, suggesting that its decreased expression is associated with defective processing of the receptor precursor molecule. Some of the Burkitt tumor cell lines exhibited reduced expression of class I major histocompatibility complex (MHC) antigens, and an inverse correlation was found between the expression of the c-*myc* proto-oncogene and both insulin receptors and class I MHC antigens. As unactivated (resting) B lymphocytes do not express insulin receptors, the reduced expression of insulin receptors in Burkitt's tumor cells may reflect their less activated phenotype compared to that of lymphoblastoid cells.

The effects of insulin or IGFs on tumor cell lines are variable and depend on the particular type of cell. In the teratoma (teratocarcinoma) cell line Tera-2, which was isolated from a lung metastasis of a primary human testicular teratoma, addition of IGFs to the cells grown in serum-free medium results in stimulation of the population multiplication.[358] Tera-2 cells possess functional IGF-I receptors, and the effects of insulin and IGF-I on their proliferation are possibly mediated by these receptors.[359] However, IGFs are not able to induce substantial change in the thymidine-labeling index of Tera-2 cells; and their effect on population multiplication appears to be exerted by increasing cell survival.

The expression and function of insulin receptors in human breast cancer cell lines have been characterized.[360] IGF-I and to a lesser extent insulin and IGF-II are potent mitogenic agents for macrophage chemotactic factor 7 (MCF-7) and T47D human breast cancer cell lines maintained in a serum-free culture medium.[361] IGF-II may be an autocrine/paracrine growth factor for certain breast cancer cell lines, and this effect would take place through an activation of the IGF-I receptor.[362] IGF-I receptors are present in different amounts in different breast cancer cell lines.[363] The mitogenic action of insulin on these cells is mediated, at least in part, by IGF receptors.[223,364-366] Insulin is mitogenic to T47D cells only at supraphysiological concentrations, probably due to its limited binding to the IGF receptors. Absence of downregulation of insulin receptors is observed in the human breast cancer cell line MCF-7 cultured in serum-free medium.[367] Estrogen alone is hardly mitogenic for MCF-7 cells, and these cells require insulin or IGFs for growth.[368] At suboptimal insulin concentrations, estrogen acts synergistically with insulin, possibly by inducing an autocrine production of growth factors by the MCF-7 cells. Expression of IGF-II in MCF-7 cells may decrease estrogen sensitivity of the cells under both anchorage-dependent and -independent growth conditions.[369] IGF-II may function in an autocrine manner to stimulate breast cancer cell growth and may play a role in loss of steroid sensitivity in cancer cells.

In addition to breast cancer cell lines, cell lines derived from other types of human tumors may contain insulin and/or IGF receptors and may respond to the respective physiological ligand. SCLC cell lines may synthesize an IGF-I precursor molecule, express IGF-I immunoreactivity, possess IGF-I receptors, and exhibit IGF-I-mediated growth stimulation; moreover, the basal growth of these cells is inhibited by a monoclonal antibody specific to the IGF-I receptor.[370,371] These findings suggest that IGF-I may be an important autocrine growth factor for SCLCs. IGF-II receptors may be expressed by SCLC cell lines, but IGF-II exhibits a higher affinity for the IGF-I receptor than for the IGF-II receptor itself.[372] The function of the IGF-II receptor in normal and tumor cells remains to be elucidated.

The human pancreatic carcinoma cell line MIA-PaCa 2 produces IGF-I and TGF-α, and possesses receptors for both growth factors.[373] The molecular size of IGF-I produced by MIA-PaCa 2 cells is similar to that of authentic IGF-I; and an antibody to the IGF-I receptor inhibits the growth of MIA-PaCa 2 cells, suggesting that IGF-I is associated with an autocrine loop which stimulates the growth. Addition of IGF-I to the culture medium stimulates the growth of the pancreatic carcinoma cells.

Human hepatoma cell lines express transcripts of the genes encoding IGF-I, PDGF-1, and PDGF-2/c-*sis* proteins and their respective receptors, suggesting that autocrine regulation may be an important mechanism for the maintenance of these cells.[374] IGF receptors present in hepatoma cells may have altered functional properties. Although the IGF-II receptors present in rat hepatoma cell lines are similar to those of primary rat hepatocytes in their size and ligand-binding characteristics, they are defective in regulation by cell density.[375]

High-affinity type I IGF receptors are expressed in the human choriocarcinoma cell lines JEG-3 and BeWo.[376] A 34-kDa (34K) IGF-binding protein synthesized and secreted by human decidua, but not placenta, inhibits the binding and action of IGFs in JEG-3 cells.[377]

IGF-I can stimulate the growth of both normal and neoplastic neural cells through binding to its surface receptor. B104 rat neuroblastoma cells possess abundant IGF-I receptors; and the factor requirement of these cells for proliferation *in vitro* is satisfied by IGF-I through binding to these receptors, which are similar or identical to those present in nonneuronal tissues.[378] The human neuroblastoma cell line SH-SY5Y expresses both IGF-II and its receptor, suggesting that IGF-II mediates autocrine growth of this cell line.[379] The human retinoblastoma cell line Y79 contains both insulin-specific mRNA and insulin-binding sites, suggesting an insulin-dependent autocrine mechanism for the growth of these cells.[380] Most human tumor cell lines of neuroectodermal origin with a chromosome translocation t(11;22), including primitive neuroectodermal tumors, Ewing's sarcoma, and esthesioneuroblastoma, express IGF-I transcripts, whereas cell lines without the t(11;22) translocation, including primitive neuroectodermal tumors and neuroblastoma, do not express IGF-I transcripts.[381] In addition to IGF-I, neuroblastoma cell lines may always express IGF-II transcripts. The primitive neuroectodermal tumor cell line CHP-100, which carries the t(11;22), was found to secrete both IGF-I and the IGF-binding protein, IGFBP-2. In addition, CHP-100 cells were found to express type I IGF receptor mRNA; and blockade of this receptor by a specific monoclonal antibody inhibited serum-free growth. These results demonstrate that interruption of an autocrine loop may result, at least in certain cases, in inhibition of tumor cell growth.

Both IGF-I and its receptor are present in rat medullary thyroid carcinoma cell lines, suggesting a role for IGF-I as an autocrine growth factor for this tumor.[382] Medullary thyroid carcinoma cells are derived from thyroid C cells and are known to produce a variety of neuropeptides including calcitonin, calcitonin-related peptide, neurotensin, and somatostatin. Calcitonin and the related peptide fragment CCAP have been used as markers for the diagnosis and follow-up of human medullary thyroid carcinomas.[304]

Insulin and IGF-I receptors are present in human leukemia and lymphoma cell lines. The receptors exhibit cell cycle-specific regulation, which favors the concept that leukemic cells remain hormone- and growth factor-responsive to a certain degree and may be dependent on the stimulatory action of insulin and/or IGF-I during certain stages of the oncogenic process.[383] However, the role of insulin, IGFs, and other growth factors in leukemogenesis is not clear. Although insulin and IGF-I can stimulate growth of some cultured human leukemia cells, the presence of insulin of IGF-I receptors alone does not predict hormone-induced mitogenic responses.[384]

Insulin receptors are abundantly present in the cell line IM-9, which has cells that are B-type lymphoblasts derived from a patient with multiple myeloma.[385] In IM-9 cells the majority of IGF-II binding activity is due to an atypical high-affinity insulin receptor or to a unique type I IGF receptor.[259] In contrast, very low levels of insulin receptors were detected in a human Hodgkin's disease cell line as well as in some murine lymphoid tumor cell lines.[386] The number of insulin receptors in Burkitt's lymphoma-derived Daudi and Raji cell lines is reduced by over 95%, compared to that of the IM-9 cell line.[387] Insulin receptor-negative cell lines may possess receptors for IGF-I and/or IGF-II, and these receptors can mediate some, but not all, the insulin-like effects of IGFs. In acute nonlymphocytic leukemia (ANLL) cell lines that possess both insulin and IGF-I and IGF-II receptors, IGF-I and IGF-II can stimulate glycogen synthesis; however, in Hodgkin's disease cells lacking insulin receptors no effect of IGFs on glycogen synthesis is observed, which indicates that at least some of the insulin-like effects of IGFs in ANLL cells are mediated through the insulin receptor.[388] On the other hand, IGF-I and IGF-II stimulate DNA synthesis in ANLL cell lines possessing insulin receptors as well as in the Hodgkin's disease cell line lacking these receptors.

In most neoplastic cell lines, insulin and IGF receptors are of apparently normal structure. However, abnormalities in receptor affinity,[389] suggesting possible changes in the structure of insulin receptors, have been detected in some cell lines. An alteration of insulin receptor molecular weight has been detected in the human monocytic cell line U-937.[390] Significant differences have been detected between IGF receptors expressed on cell lines established from human brain tumors of glial origin (astrocytoma grades III and IV) and those characterized in normal adult brain of human and rat.[391] The properties of IGF receptors expressed by human glioma cells are similar to those found in fetal astrocytes. Abnormalities of insulin binding and receptor phosphorylation have been described in the insulin-resistant mouse melanoma cell line Cloudman S91.[392] This cell line is defective in both its affinity for insulin and its autophosphorylation properties; one site of autophosphorylation is absent in the altered receptor, which would make it unable to stimulate cell growth. Growth stimulation of melanoma cells by insulin may be mediated, at least in part, via the type I IGF receptor.[393]

D. ROLE OF INSULIN AND IGFs IN TUMOR CELL GROWTH AND METASTASIZATION

Insulin and IGFs may have a central role in the complex processes associated with the growth and metastasization of tumor cells. It has been shown that episome-based antisense cDNA transcription of IGF-I induces loss of tumorigenicity of rat fibroblastoma cells.[394] IGF-I and to a lesser extent insulin and IGF-II stimulate cell motility in the highly metastatic human melanoma cell line A2058.[395,396] These cells have high-affinity binding sites for IGF-I, and the effects of IGF-I on A2058 cells are predominantly chemotactic. The IGF-I receptor appears to mediate a strong motility response to a factor, the autocrine motility factor (AMF), that may be secreted from secondary target organs and may act as a "homing" signal for extravasating tumor cells. The stimulatory effect of AMF on cell motility is additive to that of IGF-I. AMF stimulates phosphoinositide metabolism in A2058 cells via a pertussis toxin-sensitive G protein signal transduction pathway.[397]

E. INSULIN AND IGF RECEPTORS IN THE INDUCTION OF DIFFERENTIATION OF NEOPLASTIC CELLS

Induction of differentiation of neoplastic hematopoietic cell lines may be associated with marked changes in the expression of insulin and/or IGF receptors. The human erythroleukemia cell line K-562 has abundant type I IGF receptors, and these receptors decrease as a function of hemin-induced cell differentiation.[398] Differentiation of K-562 cells leads mainly to the formation of early erythroblasts that can synthesize embryonic hemoglobin.

In the human promyelocytic leukemia cell line HL-60, induction to differentiation by different agents (calcitriol, retinoic acid, dimethyl sulfoxide [DMSO], phorbol ester) is accompanied by an increase in the number of plasma membrane insulin receptors.[399] In contrast, type I IGF receptors decrease when HL-60 cells are induced to differentiate into macrophage-like cells by treatment with 12-*O*-tetradecanoylphorbol-13-acetate (TPA) or calcitriol.[400] This cell line produces a peptide, or peptides, with insulin-like activity which is distinct from insulin or IGFs and which may possibly play a role in the growth of HL-60 cells.[401]

In the U-937 human monocyte-like cell line, induction of differentiation by calcitriol is accompanied by increased insulin binding, whereas induction of differentiation by retinoic acid decreases the hormone binding.[402] Abnormal insulin binding and altered plasma membrane physical properties have been described in a Friend erythroleukemia cell clone resistant to differentiation induced by dimethyl sulfoxide (DMSO).[403] Phorbol esters enhance the phosphorylation of both insulin and IGF-I receptors in a human lymphocyte cell line (IM-9) and modulate insulin receptor phosphorylation and insulin action in cultured hepatoma cells.[404,405]

REFERENCES

1. **Baxter, R.C.,** The somatomedins: insulin-like growth factors, *Adv. Clin. Chem.,* 25, 49, 1986.
2. **Zapf, J. and Froesch, E.R.,** Insulin-like growth factors/somatomedins: structure, secretion, biological actions and physiological role, *Horm. Res.,* 24, 121, 1986.
3. **Zapf, J. and Froesch, E.R.,** Pathophysiological and clinical aspects of the insulin-like growth factors, *Horm. Res.,* 24, 160, 1986.
4. **D'Ercole, A.J.,** Somatomedins/insulin-like growth factors and fetal growth, *J. Dev. Physiol.,* 9, 481, 1987.
5. **Daughaday, W.H. and Rotwein, P.,** Insulin-like growth factor-I and factor-II. Peptide, messenger ribonucleic acid and gene structures, serum, and tissue concentrations, *Endocr. Rev.,* 10, 68, 1989.
6. **Humbel, R.F.,** Insulin-like growth factors I and II, *Eur. J. Biochem.,* 190, 445, 1990.
7. **Rutanen, E.M. and Pekonen, F.,** Insulin-like growth factors and their binding proteins, *Acta Endocrinol.,* 123, 7, 1990.
8. **Sara, V.R. and Hall, K.,** Insulin-like growth factors and their binding proteins, *Physiol. Rev.,* 70, 591, 1990.
9. **Quin, J.D.,** The insulin-like growth factors, *Q. J. Med.,* 82, 81, 1992.
10. **Cohick, W.S. and Clemmons, D.R.,** The insulin-like growth factors, *Annu. Rev. Physiol.,* 55, 131, 1993.
11. **de Pablo, F., Scott, L.A., and Roth, J.,** Insulin and insulin-like growth factor I in early development: peptides, receptors, and biological events, *Endocr. Rev.,* 11, 558, 1990.
12. **Bach, L.A. and Rechler, M.M.,** Insulin-like growth factors and diabetes, *Diabetes Metab. Rev.,* 8, 229, 1992.
13. **Zapf, J., Rinderknecht, E., Humbel, R.E., and Froesch, E.R.,** Nonsuppressible insulin-like activity (NSILA) from human serum: recent accomplishments and their physiological implications, *Metabolism,* 27, 1803, 1978.
14. **Phillips, L.S. and Vassilopoulou-Sellin, R.,** Somatomedins, *N. Engl. J. Med.,* 302, 371, 1980.
15. **Daughaday, W.H., Hall, K., Raben, M.S., Salmon, W.D., Van den Brande, J.L., and Van Wyk, J.J.,** Somatomedin: proposed designation for sulphation factor, *Nature (London),* 235, 107, 1972.
16. **Sara, V.R., Hall, K., Rodeck, C.H., and Wetterberg, L.,** Human embryonic somatomedin, *Proc. Natl. Acad. Sci. U.S.A.,* 78, 3175, 1981.
17. **Rotwein, P., Pollock, K.M., Watson, M., and Milbrandt, J.D.,** Insulin-like growth factor gene expression during rat embryonic development, *Endocrinology,* 121, 2141, 1987.
18. **Campisi, J. and Pardee, A.B.,** Post-transcriptional control of the onset of DNA synthesis by an insulin-like growth factor, *Mol. Cell. Biol.,* 4, 1807, 1984.

19. **Rothstein, H., Van Wyk, J.J., Hayden, J.H., Gordon, S.R., and Weinsieder, A.,** Somatomedin C: restoration *in vivo* of cycle traverse in G_0/G_1 blocked cells of hypophysectomized animals, *Science,* 208, 410, 1980.
20. **Florini, J.R. and Magri, K.A.,** Effects of growth factors on myogenic differentiation, *Am. J. Physiol.,* 256, C701, 1989.
21. **Florini, J.R., Magri, K.A., Ewton, D.Z., James, P.L., Grindstaff, K., and Rotwein, P.S.,** "Spontaneous" differentiation of skeletal myoblasts is dependent upon autocrine secretion of insulin-like growth factor-II, *J. Biol. Chem.,* 266, 15917, 1991.
22. **Ristow, H.-J. and Messmer, T.O.,** Basic fibroblast growth factor and insulin-like growth factor I are strong mitogens for cultured mouse keratinocytes, *J. Cell. Physiol.,* 137, 277, 1988.
23. **Yang, H.C. and Pardee, A.B.,** Insulin-like growth factor I regulation of transcription and replicating enzyme induction necessary for DNA synthesis, *J. Cell. Physiol.,* 127, 410, 1986.
24. **Bachrach, L.K., Eggo, M.C., Hintz, R.L., and Burrow, G.N.,** Insulin-like growth factors in sheep thyroid cells: action, receptors and production, *Biochem. Biophys. Res. Commun.,* 154, 861, 1988.
25. **Zangger, I., Zapf, J., and Froesch, E.R.,** Insulin-like growth factor I and II in 14 animal species and man as determined by three radioligand assays and two bioassays, *Acta Endocrinol.,* 114, 107, 1987.
26. **Sussenbach, J.S., Steenbergh, P.H., and Holthuizen, P.,** Structure and expression of the human insulin-like growth factor genes, *Growth Regul.,* 2, 1, 1992.
27. **Morton, C.C., Byers, M.G., Nakai, H., Bell, G.I., and Millow, L.J.,** Human genes for insulin-like growth factors I and II and epidermal growth factor are located on 12q22-q24, 11p15, and 4q25-q27, respectively, *Cytogenet. Cell Genet.,* 41, 245, 1986.
28. **Mathew, S., Murty, V.V.V.S., Hunziker, W., and Chaganti, R.S.K.,** Subregional mapping of 13 single-copy genes on the long arm of chromosome 12 by fluorescence *in situ* hybridization, *Genomics,* 14, 775, 1992.
29. **Bell, G.I., Gerhard, D.S., Fong, N.M., Sanchez-Pescador, R., and Rall, L.B.,** Isolation of the human insulin-like growth factor genes: insulin-like growth factor II and insulin genes are contiguous, *Proc. Natl. Acad. Sci. U.S.A.,* 82, 6450, 1985.
30. **Rotwein, P.,** Two insulin-like growth factor I messenger RNAs are expressed in human liver, *Proc. Natl. Acad. Sci. U.S.A.,* 83, 77, 1986.
31. **Le Bouc, Y., Dreyer, D., Jaeger, F., Binoux, M., and Sondermeyer, P.,** Complete characterization of the human IGF-I nucleotide sequence isolated from a newly constructed adult liver cDNA library, *FEBS Lett.,* 196, 108, 1986.
32. **Rotwein, P., Pollock, K.M., Didier, D.K., and Krivi, G.G.,** Organization and sequence of the human insulin-like growth factor I gene. Alternative RNA processing produces two insulin-like growth factor I precursor peptides, *J. Biol. Chem.,* 261, 4828, 1986.
33. **Steenbergh, P.H., Koonenremst, A.M.B.C., Cleutjens, C.B.J.M., and Sussenbach, J.S.,** Complete nucleotide sequence of the high molecular weight human IGF-I messenger RNA, *Biochem. Biophys. Res. Commun.,* 175, 507, 1991.
34. **Jansen, E., Steenbergh, P.H., Vanschaik, F.M.A., and Sussenbach, J.S.,** The human IGF-I gene contains two cell type-specifically regulated promoters, *Biochem. Biophys. Res. Commun.,* 187, 1219, 1992.
35. **Bell, G.I., Stempien, M.M., Fong, N.M., and Rall, L.B.,** Sequences of liver cDNAs encoding two different mouse insulin-like growth factor I precursors, *Nucleic Acids Res.,* 14, 7873, 1986.
36. **Murphy, L.J., Bell, G.I., Duckworth, M.L., and Friesen, H.G.,** Identification, characterization, and regulation of a rat complementary deoxyribonucleic acid which encodes insulin-like growth factor-I, *Endocrinology,* 121, 684, 1987.
37. **Werner, H., Stannard, B., Bach, M.A., LeRoith, D., and Roberts, C.T., Jr.,** Cloning and characterization of the proximal promoter region of the rat insulin-like growth factor I (IGF-I) gene, *Biochem. Biophys. Res. Commun.,* 169, 1021, 1990.
38. **de Pagter-Holthuizen, P., Höppener, J.W.M., Jansen, M., Geurts van Kessel, A.H.M., van Ommen, G.J.B., and Sussenbach, J.S.,** Chromosomal localization and preliminary characterization of the human gene encoding insulin-like growth factor II, *Hum. Genet.,* 69, 170, 1985.
39. **de Pagter-Holthuizen, P., Jansen, M., van Schaik, F.M.A., van der Kammen, R., Oosterwijk, C., Van den Brande, J.L., and Sussenbach, J.S.,** The human insulin-like growth factor II gene contains two development-specific promoters, *FEBS Lett.,* 214, 259, 1987.

40. **Bell, G.I., Merryweather, J.P., Sanchez-Pescador, R., Stempien, M.M., Priestley, L., Scott, J., and Rall, L.B.,** Sequence of a cDNA clone encoding human preproinsulin-like growth factor II, *Nature (London),* 310, 775, 1984.
41. **Furman, T.C., Epp, J., Hsiung, H.M., Hoskins, J., Long, G.L., Mendelsohn, L.G., Schoner, B., Smith, D.P., and Smith, M.C.,** Recombinant human insulin-like growth factor II expressed in *Escherichia coli, Biotechnology,* 5, 1047, 1987.
42. **Shen, S.-J., Daimon, M., Wang, C.-Y., Jansen, M., and Ilan, J.,** Isolation of an insulin-like growth factor II cDNA with a unique 5′ untranslated region from human placenta, *Proc. Natl. Acad. Sci. U.S.A.,* 85, 1947, 1988.
43. **de Pagter-Holthuizen, P., Jansen, M., van der Kammen, R.A., van Schaik, F.M.A., and Sussenbach, J.S.,** Differential expression of the human insulin-like growth factor II gene. Characterization of the IGF-II mRNAs and an mRNA encoding a putative IGF-II-associated protein, *Biochim. Biophys. Acta,* 950, 282, 1988.
44. **Le Bouc, Y., Noguiez, P., Sondermeijer, P., Dreyer, D., Girard, F., and Binoux, M.,** A new 5′-non-coding region for human placental insulin-like growth factor II mRNA expression, *FEBS Lett.,* 222, 181, 1987.
45. **Drummond, I.A., Madden, S.L., Rohvernutter, P., Bell, G.I., Sukhatve, V.P., and Rauscher, F.J.,** Repression of the insulin-like growth factor II gene by the Wilms tumor suppressor WT1, *Science,* 257, 674, 1992.
46. **Shiraishi, M., Morinaga, S., Noguchi, M., Shimosato, Y., and Sekiya, T.,** Loss of genes on the short arm of chromosome 11 in human lung carcinomas, *Jpn. J. Cancer Res.,* 78, 1302, 1987.
47. **Stempien, M.M., Fong, N.M., Rall, L.B., and Bell, G.I.,** Sequence of a placental cDNA encoding the mouse insulin-like growth factor II precursor, *DNA,* 5, 357, 1986.
48. **Frunzio, R., Chiarotti, L., Brown, A.L., Graham, D.E., Rechler, M.M., and Bruni, C.B.,** Structure and expression of the rat insulin-like growth factor II (rIGF-II) gene, *J. Biol. Chem.,* 261, 17138, 1986.
49. **Chiariotti, L., Brown, A.L., Frunzio, R., Clemmons, D.R., Rechler, M.M., and Bruni, C.B.,** Structure of the rat insulin-like growth factor II transcriptional unit: heterogeneous transcripts are generated from two promoters by use of multiple polyadenylation sites and differential ribonucleic acid splicing, *Mol. Endocrinol.,* 2, 1115, 1988.
50. **Ueno, T., Takahashi, K., Matsuguchi, T., Endo, H., and Miyamoto, M.,** Transcriptional deviation of the rat insulin-like growth factor II gene initiated at three alternative leader exons between neonatal tissues and ascites hepatomas, *Biochim. Biophys. Acta,* 950, 411, 1988.
51. **Rinderknecht, E. and Humbel, R.E.,** The amino acid sequence of human insulin-like growth factor I and its structural homology with proinsulin, *J. Biol. Chem.,* 253, 2769, 1978.
52. **Rinderknecht, E. and Humbel, R.E.,** Primary structure of human insulin-like growth factor II, *FEBS Lett.,* 89, 283, 1978.
53. **Tamura, K., Kobayashi, M., Ishii, Y., Tamura, T., Hashimoto, K., Nakamura, S., Niwa, M., and Zapf, J.,** Primary structure of rat insulin-like growth factor-I and its biological activities, *J. Biol. Chem.,* 264, 5616, 1989.
54. **Peters, M.A., Lau, E.P., Snitman, D.L., Van Wyk, J.J., Underwood, L.E., Russell, W.E., and Svoboda, M.E.,** Expression of a biologically active analogue of somatomedin-C/insulin-like growth factor I, *Gene,* 35, 83, 1985.
55. **Ballard, F.J., Francis, G.L., Ross, M., Bagley, C.J., May, B., and Wallace, J.C.,** Natural and synthetic forms of insulin-like growth factor-1 (IGF-1) and the potent derivative, destripeptide IGF-1: biological activities and receptor binding, *Biochem. Biophys. Res. Commun.,* 149, 398, 1987.
56. **Ruan, W.F., Newman, C.B., and Kleinberg, D.L.,** Intact and amino-terminally shortened forms of insulin-like growth factor-I induce mammary gland differentiation and development, *Proc. Natl. Acad. Sci. U.S.A.,* 89, 10872, 1992.
57. **Francis, G.L., Upton, F.M., Ballard, F.J., McNeil, K.A., and Wallace, J.C.,** Insulin-like growth factors 1 and 2 in bovine serum, *Biochem. J.,* 251, 95, 1988.
58. **Chen, Z.Z., Schwartz, G.P., Zong, L., Burke, G.T., Chanley, J.D., and Katsoyannis, P.G.,** Determinants of growth-promoting activity reside in the A-domain of insulin-like growth factor I, *Biochemistry,* 27, 6105, 1988.
59. **Bayne, M.L., Applebaum, J., Chicchi, G.C., Miller, R.E., and Cascieri, M.A.,** The roles of tyrosines 24, 31, and 60 in the high affinity binding of insulin-like growth factor-I to the type 1 insulin-like growth factor receptor, *J. Biol. Chem.,* 265, 15648, 1990.

60. **Dull, T.J., Gray, A., Hayflick, J.S., and Ullrich, A.,** Insulin-like growth factor II precursor gene organization in relation to insulin gene family, *Nature (London),* 310, 777, 1984.
61. **Whitfield, H.J., Bruni, C.B., Frunzio, R., Terrell, J.E., Nissley, S.P., and Rechler, M.M.,** Isolation of a cDNA clone encoding rat insulin-like growth factor-II precursor, *Nature (London),* 312, 277, 1984.
62. **Yang, Y.W.-H., Romanus, J.A., Liu, T.-Y., Nissley, S.P., and Rechler, M.M.,** Biosynthesis of rat insulin-like growth factor II. I. Immunochemical demonstration of a 20-kilodalton biosynthetic precursor of rat insulin-like growth factor II in metabolically labeled BRL-3A rat liver cells, *J. Biol. Chem.,* 260, 2570, 1985.
63. **Yang, Y.W.-H., Rechler, M.M., Nissley, S.P., and Coligan, J.E.,** Biosynthesis of rat insulin-like growth factor II. II. Localization of mature rat insulin-like growth factor II (7484 daltons) to the amino terminus of the 20-kilodalton biosynthetic precursor by radiosequence analysis, *J. Biol. Chem.,* 260, 2578, 1985.
64. **Li, C.H., Yamashiro, D., Hammonds, R.G., Jr., and Westphal, M.,** Synthetic insulin-like growth factor II, *Biochem. Biophys. Res. Commun.,* 127, 420, 1985.
65. **Zumstein, P.P., Lüthi, C., and Humbel, R.E.,** Amino acid sequence of a variant pro-form of insulin-like growth factor II, *Proc. Natl. Acad. Sci. U.S.A.,* 82, 3169, 1985.
66. **Graham, D.E., Rechler, M.M., Brown, A.L., Frunzio, R., Romanus, J.A., Bruni, C.B., Whitfield, H.J., Nissley, S.P., Seelig, S., and Berry, S.,** Coordinate developmental regulation of high and low molecular weight mRNAs for rat insulin-like growth factor II, *Proc. Natl. Acad. Sci. U.S.A.,* 83, 4519, 1986.
67. **Hampton, B., Burgess, W.H., Marshak, D.R., Cullen, K.J., and Perdue, J.F.,** Purification and characterization of an insulin-like growth factor II variant from human plasma, *J. Biol. Chem.,* 264, 19155, 1989.
68. **Lund, P.K., Moats-Staats, B.M., Hynes, M.A., Simmons, J.G., Jansen, M., D'Ercole, A.J., and Van Wyk, J.J.,** Somatomedin-C/insulin-like growth factor-I and insulin like growth factor-II mRNAs in rat fetal and adult tissues, *J. Biol. Chem.,* 261, 14539, 1986.
69. **Guler, H.-P., Zapf, J., Schmid, C., and Froesch, E.R.,** Insulin-like growth factors I and II in healthy man. Estimations of half-lives and production rates, *Acta Endocrinol.,* 121, 753, 1989.
70. **Murphy, L.J., Bell, G.I., and Friesen, H.G.,** Tissue distribution of insulin-like growth factor I and II messenger ribonucleic acid in the adult rat, *Endocrinology,* 120, 1279, 1987.
71. **Palmero, S., Prati, M., Barreca, A., Minuto, F., Giordano, G., and Fugassa, E.,** Thyroid hormone stimulates the production of insulin-like growth factor I (IGF-I) by immature rat Sertoli cells, *Mol. Cell. Endocrinol.,* 68, 61, 1990.
72. **Brown, A.L., Graham, D.E., Nissley, S.P., Hill, D.J., Strain, A.J., and Rechler, M.M.,** Developmental regulation of insulin-like growth factor II mRNA in different rat tissues, *J. Biol. Chem.,* 261, 13144, 1986.
73. **Levinovitz, A. and Norstedt, G.,** Developmental and steroid hormonal regulation of insulin-like growth factor II expression, *Mol. Endocrinol.,* 3, 7979, 1989.
74. **Scott, J., Cowell, J., Robertson, M.E., Priestley, L.M., Wadey, R., Hopkins, B., Pritchard, J., Bell, G.I., Rall, L.B., Graham, C.F., and Knott, T.J.,** Insulin-like growth factor-II gene expression in Wilms' tumour and embryonic tissue, *Nature (London),* 317, 260, 1985.
75. **Romanus, J.A., Yang, Y.W.-H., Adams, S.O., Sofair, A.N., Tseng, L.Y.-H., Nissley, S.P., and Rechler, M.M.,** Synthesis of insulin-like growth factor II (IGF-II) in fetal rat tissues: translation of IGF-II ribonucleic acid and processing of pre-pro-IGF-II, *Endocrinology,* 122, 709, 1988.
76. **Prosser, C.G. and Fleet, I.R.,** Secretion of insulin-like growth factor II into milk, *Biochem. Biophys. Res. Commun.,* 183, 1230, 1992.
77. **McCarthy, T.L., Centrella, M., and Canalis, E.,** Insulin-like growth factor (IGF) and bone, *Connect. Tissue Res.,* 20, 277, 1989.
78. **Mohan, S., Jennings, J.C., Linkhart, T.A., and Baylink, D.J.,** Primary structure of human skeletal growth factor: homology with human insulin-like growth factor-II, *Biochim. Biophys. Acta,* 966, 44, 1988.
79. **Fant, M., Munro, H., and Moses, A.C.,** An autocrine/paracrine role for insulin-like growth factors in the regulation of human placental growth, *J. Clin. Endocrinol. Metab.,* 63, 499, 1986.
80. **Liu, K.-S., Wang, C.-Y., Mills, N., Gyves, M., and Ilan, J.,** Insulin-related genes expressed in human placenta from normal and diabetic pregnancies, *Proc. Natl. Acad. Sci. U.S.A.,* 82, 3668, 1985.

81. **Schmid, C., Steiner, T., and Froesch, E.R.,** Insulin-like growth factor I supports differentiation of cultured osteoblast-like cells, *FEBS Lett.,* 173, 48, 1984.
82. **Ballotti, R., Nielsen, F.C., Pringle, N., Kowalski, A., Richardson, W.D., Van Obberghen, E., and Gammeltoft, S.,** Insulin-like growth factor I in cultured rat astrocytes: expression of the gene, and receptor tyrosine kinase, *EMBO J.,* 6, 3633, 1987.
83. **Haselbacher, G.K., Schwab, M.E., Pasi, A., and Humbel, R.E.,** Insulin-like growth factor II (IGF II) in human brain: regional distribution of IGF II and of higher molecular mass forms, *Proc. Natl. Acad. Sci. U.S.A.,* 82, 2153, 1985.
84. **Atkison, P.R., Bala, R.M., and Hollenberg, M.D.,** Somatomedin-like activity from cultured embryo-derived cells: partial characterization and stimulation of production by epidermal growth factor (urogastrone), *Can. J. Biochem. Cell Biol.,* 62, 1335, 1984.
85. **Atkison, P.R., Hayden, L.J., Bala, R.M., and Hollenberg, M.D.,** Production of somatomedin-like activity by human adult tumor-derived, transformed, and normal cell cultures and by cultured rat hepatocytes: effects of culture conditions and of epidermal growth factor (urogastrone), *Can. J. Biochem. Cell Biol.,* 62, 1343, 1984.
86. **Bichell, D.P., Kikuchi, K., and Rotwein, P.,** Growth hormone rapidly activates insulin-like growth factor I gene transcription *in vivo, Mol. Endocrinol.,* 6, 1899, 1992.
87. **Mathews, L.S., Norstedt, G., and Palmiter, R.D.,** Regulation of insulin-like growth factor I gene expression by growth hormone, *Proc. Natl. Acad. Sci. U.S.A.,* 83, 9343, 1986.
88. **Murphy, L.J., Murphy, L.C., and Friesen, H.G.,** Estrogen induces insulin-like growth factor-I expression in the rat uterus, *Mol. Endocrinol.,* 1, 445, 1987.
89. **Townsend, S.F., Dallman, M.F., and Miller, W.L.,** Rat insulin-like growth factor-I and -II mRNAs are unchanged during compensatory adrenal growth but decrease during ACTH-induced adrenal growth, *J. Biol. Chem.,* 265, 22117, 1990.
90. **Ilvesmaki, V., Jaatela, M., Saksela, E., and Voutilainen, R.,** Tumor necrosis factor-α and interferon-γ inhibit insulin-like growth factor-II gene expression in human fetal adrenal cell cultures, *Mol. Cell. Endocrinol.,* 91, 59, 1993.
91. **Linkhart, T.A. and Keffer, M.J.,** Differential regulation of insulin-like growth factor-I (IGF-I) and IGF-II release from cultured neonatal mouse calvaria by parathyroid hormone, transforming growth factor-β, and 1,25-dihydroxyvitamin D_3, *Endocrinology,* 128, 1511, 1991.
92. **McCarthy, T.L., Centrella, M., and Canilis, E.,** Cyclic AMP induces insulin-like growth factor I synthesis in osteoblast-enriched cultures, *J. Biol. Chem.,* 265, 15353, 1990.
93. **Moses, A.C., Nissley, S.P., Short, P.A., Rechler, M.M., White, R.M., Knight, A.B., and Higa, O.Z.,** Increased levels of multiplication-stimulatory activity, an insulin-like growth factor, in fetal rat serum, *Proc. Natl. Acad. Sci. U.S.A.,* 77, 3649, 1980.
94. **King, G.L., Kahn, C.R., Rechler, M.M., and Nissley, S.P.,** Direct demonstration of separate receptors for growth and metabolic activities of insulin and multiplication-stimulating activity (an insulin-like growth factor) using antibodies to the insulin receptor, *J. Clin. Invest.,* 66, 130, 1980.
95. **Marquardt, H., Todaro, G.J., Henderson, L.E., and Oroszlan, S.,** Purification and primary structure of a polypeptide with multiplication stimulating activity from rat liver cell cultures: homology with human insulin-like growth factor II, *J. Biol. Chem.,* 256, 6859, 1981.
96. **Beebe, D.C., Silver, M.H., Belcher, K.S., Van Wyk, J.J., Svoboda, M.E., and Zelenka, P.S.,** Lentropin, a protein that controls lens fiber formation is related functionally and immunologically to the insulin-like growth factors, *Proc. Natl. Acad. Sci. U.S.A.,* 84, 2327, 1987.
97. **Gowan, L.K., Hampton, B., Hill, D.J., Schlueter, R.J., and Perdue, J.F.,** Purification and characterization of a unique high molecular weight form of insulin-like growth factor II, *Endocrinology,* 121, 449, 1987.
98. **Huff, K.K., Kaufman, D., Gabbay, K.H., Spencer, E.M., Lippman, M.E., and Dickson, R.B.,** Secretion of an insulin-like growth factor-I-related protein by human breast cancer cells, *Cancer Res.,* 46, 4613, 1986.
99. **Peres, R., Betsholtz, C., Westermark, B., and Heldin, C.-H.,** Frequent expression of growth factors for mesenchymal cells in human mammary carcinoma cell lines, *Cancer Res.,* 47, 3425, 1987.
100. **Tanaka, H., Asami, O., Hayano, T., Sasaki, I., Yoshitake, Y., and Nishikawa, K.,** Identification of a family of insulin-like growth factor II secreted by cultured rat epithelial-like cell line 18,54-SF: application of a monoclonal antibody, *Endocrinology,* 124, 870, 1989.

101. **Minuto, F., Del Monte, P., Barreca, A., Fortini, P., Cariola, G., Catrambone, G., and Giordano, G.,** Evidence for an increased somatomedin-C/insulin-like growth factor I content in primary human lung tumors, *Cancer Res.,* 46, 985, 1986.
102. **Tricoli, J.V., Rall, L.B., Karakousis, C.P., Herrera, L., Petrelli, N.J., Bell, G.I., and Shows, T.B.,** Enhanced levels of insulin-like growth factor messenger RNA in human colon carcinomas and liposarcomas, *Cancer Res.,* 46, 6169, 1986.
103. **Reeve, A.E., Eccles, M.R., Wilkins, R.J., Bell, G.I., and Millow, L.J.,** Expression of insulin-like growth factor-II transcripts in Wilms' tumour, *Nature (London),* 317, 258, 1985.
104. **Haselbacher, G.K., Irminger, J.-C., Zapf, J., Ziegler, W.H., and Humbel, R.E.,** Insulin-like growth factor II in human adrenal pheochromocytomas and Wilms' tumors: expression at the mRNA and protein level, *Proc. Natl. Acad. Sci. U.S.A.,* 84, 1104, 1987.
105. **Höppener, J.W.M., Steenbergh, P.H., Slebos, R.J.C., de Pagter-Holtzhuizen, P., Roos, B.A., Jansen, M., Van den Brande, J.L., Sussenbach, J.S., Jansz, H.S., and Lips, C.J.M.,** Expression of insulin-like growth factor-I and -II genes in rat medullary thyroid carcinoma, *FEBS Lett.,* 215, 122, 1987.
106. **Hizuka, N., Takano, K., Tanaka, I., Asakawa, K., Miyakawa, M., Horikawa, R., and Shizume, K.,** Demonstration of insulin-like growth factor I in human urine, *J. Clin. Endocrinol. Metab.,* 64, 1309, 1987.
107. **Bhaumick, B. and Bala, R.M.,** Binding and degradation of insulin-like growth factors I and II by rat kidney membrane, *Endocrinology,* 120, 1439, 1987.
108. **Baxter, R.C.,** Insulin-like growth factor (IGF) binding proteins — The role of serum IGFBPs in regulating IGF availability, *Acta Paediatr. Scand.,* Suppl. 372, 107, 1991.
109. **Rechler, M.M. and Brown, A.L.,** Insulin-like growth factor binding proteins — Gene structure and expression, *Growth Regul.,* 2, 55, 1992.
110. **Clemmons, D.R.,** IGF binding proteins — Regulation of cellular actions, *Growth Regul.,* 2, 80, 1992.
111. **Clemmons, D.R., Cascieri, M.A., Camacho-Hubner, C., McCusker, R.H., and Bayne, M.L.,** Discrete alterations of the insulin-like growth factor-I molecule which alter its affinity for insulin-like growth factor-binding proteins result in changes in bioactivity, *J. Biol. Chem.,* 265, 12210, 1990.
112. **Blum, W.F., Jenne, E.W., Reppin, F., Kietzmann, K., Ranke, M.B., and Bierich, J.R.,** Insulin-like growth factor I (IGF-I)-binding protein complex is a better mitogen than free IGF-I, *Endocrinology,* 125, 766, 1989.
113. **McCusker, R.H., Camacho-Hubner, C., Bayne, M.L., Cascieri, M.A., and Clemmons, D.R.,** Insulin-like growth factor (IGF) binding to human fibroblasts and glioblastoma cells: the modulating effect of cell released IGF binding proteins (IGFBPs), *J. Cell. Physiol.,* 144, 244, 1990.
114. **Ooi, G.T. and Herington, A.C.,** Covalent cross-linking of insulin-like growth factor-1 to a specific inhibitor from human serum, *Biochem. Biophys. Res. Commun.,* 137, 411, 1986.
115. **Mohan, S., Bautista, C., Wergedal, J., and Baylink, D.J.,** Isolation of an inhibitory insulin-like growth factor (IGF) binding protein from bone cell-conditioned medium: a potential local regulator of IGF action, *Proc. Natl. Acad. Sci. U.S.A.,* 86, 8338, 1989.
116. **Campbell, P.G. and Novak, J.F.,** Insulin-like growth factor binding protein (IGFBP) inhibits IGF action on human osteosarcoma cells, *J. Cell. Physiol.,* 149, 293, 1991.
117. **Liu, L., Brinkman, A., Blat, C., and Harel, L.,** IGFBP-1, an insulin like growth factor binding protein, is a cell growth inhibitor, *Biochem. Biophys. Res. Commun.,* 174, 673, 1991.
118. **Koistinen, R., Itkonen, O., Selenius, P., and Seppälä, M.,** Insulin-like growth factor-binding protein-1 inhibits binding of IGF-I on fetal skin fibroblasts but stimulates their DNA synthesis, *Biochem. Biophys. Res. Commun.,* 173, 408, 1990.
119. **Frost, R.A. and Tseng, L.,** Insulin-like growth factor-binding protein-1 is phosphorylated by cultured human endometrial stromal cells and multiple protein kinases *in vitro, J. Biol. Chem.,* 266, 18082, 1991.
120. **Jones, J.I., D'Ercole, A.J., Camacho-Hubner, C., and Clemmons, D.R.,** Phosphorylation of insulin-like growth factor (IGF)-binding protein 1 in cell culture and *in vivo:* effects of affinity for IGF-I, *Proc. Natl. Acad. Sci. U.S.A.,* 88, 7481, 1991.
121. **Ekstrand, J., Ehrenberg, E., Stern, I., Stellan, B., Zech, L., and Luthman, H.,** The gene for insulin-like growth factor-binding protein-1 is localized to human chromosomal region 7p14-p12, *Genomics,* 6, 413, 1990.

122. **Böni-Schnetzler, M., Schmid, C., Mary, J.-L., Zimmerli, B., Meier, P.J., Zapf, J., Schwander, J., and Froesch, E.R.,** Insulin regulates the expression of the insulin-like growth factor binding protein 2 mRNA in rat hepatocytes, *Mol. Endocrinol.,* 4, 1320, 1990.
123. **Cubbage, M.L., Suwanichkul, A., and Powell, D.R.,** Insulin-like growth factor binding protein-3. Organization of the human chromosomal gene and demonstration of promoter activity, *J. Biol. Chem.,* 265, 12642, 1990.
124. **Kiefer, M.C., Masiarz, F.R., Bauer, D.M., and Zapf, J.,** Identification and molecular cloning of two new 30-kDa insulin-like growth factor binding proteins isolated from adult human serum, *J. Biol. Chem.,* 266, 9043, 1991.
125. **Martin, J.L., Willets, K.E., and Baxter, R.C.,** Purification and properties of a novel insulin-like growth factor-II binding protein from transformed human fibroblasts, *J. Biol. Chem.,* 265, 4124, 1990.
126. **Culouscou, J.-M. and Shoyab, M.,** Purification of a colon cancer cell growth inhibitor and its identification as an insulin-like growth factor binding protein, *Cancer Res.,* 51, 2813, 1991.
127. **White, R.M., Nissley, S.P., Moses, A.C., Rechler, M.M., and Johnsonbaugh, R.E.,** The growth-hormone dependence of a somatomedin-binding protein in human serum, *J. Clin. Endocrinol. Metab.,* 53, 49, 1981.
128. **Hardouin, S., Gourmelen, M., Noguiez, P., Seurin, D., Roghani, M., Le Bouc, Y., Povoa, G., Merimee, T.J., Hossenlopp, P., and Binoux, M.,** Molecular forms of serum insulin-like growth factor (IGF)-binding proteins in man: relationships with growth hormone and IGFs and physiological significance, *J. Clin. Endocrinol. Metab.,* 69, 1291, 1989.
129. **Elgin, R.G., Busby, W.H., Jr., and Clemmons, D.R.,** An insulin-like growth factor (IGF) binding protein enhances the biologic response to IGF-I, *Proc. Natl. Acad. Sci. U.S.A.,* 84, 3254, 1987.
130. **Mohn, K.L., Melby, A.E., Tewari, D.S., Laz, T.M., and Taub, R.,** The gene encoding rat insulinlike growth factor-binding protein 1 is rapidly and highly induced in regenerating liver, *Mol. Cell. Biol.,* 11, 1393, 1991.
131. **Mondschein, J.S., Smith, S.A., and Hammond, J.M.,** Production of insulin-like growth factor binding proteins (IGFBPs) by porcine granulosa cells: identification of IGFBP-2 and -3 and regulation by hormones and growth factors, *Endocrinology,* 127, 2298, 1990.
132. **Bar, R.S., Harrison, L.C., Baxter, R.C., Boes, M., Dake, B.L., Booth, B., and Cox, A.,** Production of IGF-binding proteins by vascular endothelial cells, *Biochem. Biophys. Res. Commun.,* 148, 734, 1987.
133. **Bar, R.S., Booth, B.A., Boes, M., and Dake, B.L.,** Insulin-like growth factor-binding proteins from vascular endothelial cells: purification, characterization, and intrinsic biological activities, *Endocrinology,* 125, 1910, 1989.
134. **Booth, B.A., Bar, R.S., Boes, M., Dake, B.L., Bayne, M., and Cascieri, M.,** Intrinsic bioactivity of insulin-like growth factor-binding proteins from vascular endothelial cells, *Endocrinology,* 127, 2630, 1990.
135. **Martin, J.L. and Baxter, R.C.,** Insulin-like growth factor-binding proteins (IGF-BPs) produced by human skin fibroblasts: immunological relationship to other human IGF-BPs, *Endocrinology,* 123, 1907, 1988.
136. **Lee, Y.-L., Hintz, R.L., James, P.M., Lee, P.D.K., Shively, J.E., and Powell, D.R.,** Insulin-like growth factor (IGF) binding protein complementary deoxyribonucleic acid from human HEP G2 hepatoma cells: predicted protein sequence suggests an IGF binding domain different from those of the IGF-I and IGF-II receptors, *Mol. Endocrinol.,* 2, 404, 1988.
137. **Brewer, M.T., Stetler, G.L., Squires, C.H., Thompson, R.C., Busby, W.H., and Clemmons, D.R.,** Cloning, characterization, and expression of a human insulin-like growth factor binding protein, *Biochem. Biophys. Res. Commun.,* 152, 1289, 1988.
138. **Angervo, M., Koistinen, R., and Seppala, M.,** Epidermal growth factor stimulates production of insulin-like growth factor-binding protein-1 in human granulosa-luteal cells, *J. Endocrinol.,* 134, 127, 1992.
139. **Angervo, M.,** Epidermal growth factor enhances insulin-like growth factor binding protein-1 synthesis in human hepatoma cells, *Biochem. Biophys. Res. Commun.,* 189, 1177, 1992.
140. **Powell, D.R., Suwanichkul, A., Cubbage, M.L., DePaolis, L.A., Snuggs, M.B., and Lee, P.D.K.,** Insulin inhibits transcription of the human gene for insulin-like growth factor-binding protein-1, *J. Biol. Chem.,* 266, 18868, 1991.
141. **Martin, J.L. and Baxter, R.C.,** Transforming growth factor-β stimulates production of insulin-like growth factor-binding protein-3 by human skin fibroblasts, *Endocrinology,* 128, 1425, 1991.

142. **Hintz, R.L., Liu, F., Rosenfeld, R.G., and Kemp, S.F.,** Plasma somatomedin-binding proteins in hypopituitarism: changes during growth hormone therapy, *J. Clin. Endocrinol. Metab.,* 53, 100, 1981.
143. **Giudice, L.C., Milkowski, D.A., Lamson, G., Rosenfeld, R.G., and Irwin, J.C.,** Insulin-like growth factor binding proteins in human endometrium: steroid-dependent messenger ribonucleic acid expression and protein synthesis, *J. Clin. Endocrinol. Metab.,* 72, 779, 1991.
144. **De Leon, D.D., Wilson, D.M., Bakker, B., Lamson, G., Hintz, R.L., and Rosenfeld, R.G.,** Characterization of insulin-like growth factor binding proteins from human breast cancer cells, *Mol. Endocrinol.,* 3, 567, 1989.
145. **Jaques, G., Kiefer, P., Rotsch, M., Hennig, C., Göke, R., Richter, G., and Havemann, K.,** Production of insulin-like growth factor binding proteins by small-cell lung cancer cell lines, *Exp. Cell Res.,* 184, 396, 1989.
146. **Unterman, T.G., Glick, R.P., Waites, G.T., and Bell, S.C.,** Production of insulin-like growth factor-binding proteins by human central nervous system tumors, *Cancer Res.,* 51, 3030, 1991.
147. **Russell, W.E., Van Wyk, J.J., and Pledger, W.J.,** Inhibition of the mitogenic effects of plasma by a monoclonal antibody to somatomedin C, *Proc. Natl. Acad. Sci. U.S.A.,* 81, 2389, 1984.
148. **Scavo, L., Shuldiner, A.R., Serrano, J., Dashner, R., Roth, J., and DePablo, F.,** Genes encoding receptors for insulin and insulin-like growth factor-I are expressed in *Xenopus* oocytes and embryos, *Proc. Natl. Acad. Sci. U.S.A.,* 88, 6214, 1991.
149. **Girbau, M., Gomez, J.A., Lesniak, M.A., and de Pablo, F.,** Insulin and insulin-like growth factor I both stimulate metabolism, growth, and differentiation in the postneurula chick embryo, *Endocrinology,* 121, 1477, 1987.
150. **Serrano, J., Shuldiner, A.R., Roberts, C.T., Jr., LeRoith, D., and de Pablo, F.,** The insulin-like growth factor I (IGF-I) gene is expressed in chick embryos during early organogenesis, *Endocrinology,* 127, 1547, 1990.
151. **Congote, L.F.,** Effects of insulin-like growth factor I, platelet-derived growth factor, fibroblast growth factor, and transforming growth factor-beta on thymidine incorporation into fetal liver cells, *Exp. Hematol.,* 15, 936, 1987.
152. **Akahane, K., Tojo, A., Tobe, K., Kasuga, M., Urabe, A., and Takaku, F.,** Binding properties and proliferative potency of insulin-like growth factor I in fetal mouse liver cells, *Exp. Hematol.,* 15, 1068, 1987.
153. **Behringer, R.R., Lewin, T.M., Quaife, C.J., Palmiter, R.D., Brinster, R.L., and D'Ercole, A.J.,** Expression of insulin-like growth factor I stimulates normal somatic growth in growth hormone-deficient transgenic mice, *Endocrinology,* 127, 1033, 1990.
154. **Liu, F., Powell, D.R., Styne, D.M., and Hintz, R.L.,** Insulin-like growth factors (IGFs) and IGF-binding proteins in the developing rhesus monkey, *J. Clin. Endocrinol. Metab.,* 72, 905, 1991.
155. **Smith, P.J., Wise, L.S., Berkowitz, R., Wan, C., and Rubin, C.S.,** Insulin-like growth factor-I is an essential regulator of the differentiation of 3T3-L1 adipocytes, *J. Biol. Chem.,* 263, 9402, 1988.
156. **Florini, J.R. and Ewton, D.Z.,** Highly specific inhibition of IGF-I-stimulated differentiation by an antisense olygodeoxyribonucleotide to myogenin mRNA. No effects on other actions of IGF-I, *J. Biol. Chem.,* 265, 13435, 1990.
157. **Kou, K. and Rotwein, P.,** Transcriptional activation of the insulin-like growth factor II gene during myoblast differentiation, *Mol. Endocrinol.,* 7, 291, 1993.
158. **Nagarajan, L., Anderson, W.B., Nissley, S.P., Rechler, M.M., and Jetten, A.M.,** Production of insulin-like growth factor II (MSA) by endoderm-like cells derived from embryonal carcinoma cells: possible mediator of embryonic cell growth, *J. Cell. Physiol.,* 124, 199, 1985.
159. **Moats-Staats, B.M., Retsch-Bogart, G.Z., Price, W.A., Jarvis, H.W., D'Ercole, A.J., and Stiles, A.D.,** Insulin-like growth factor-I (IGF-I) antisense oligodeoxynucleotide mediated inhibition of DNA synthesis by WI-38 cells: evidence for autocrine actions of IGF-I, *Mol. Endocrinol.,* 7, 171, 1993.
160. **Baxter, R.C., Martin, J.L., and Handelsman, D.J.,** Identification of human semen insulin-like growth factor-I/somatomedin C immunoreactivity and binding protein, *Acta Endocrinol.,* 106, 420, 1984.
161. **Ramasharma, K., Cabrera, C.M., and Li, C.H.,** Identification of insulin-like growth factor-II in human seminal and follicular fluids, *Biochem. Biophys. Res. Commun.,* 140, 536, 1986.
162. **Soder, O., Bang, P., Wahab, A., and Parvinen, M.,** Insulin-like growth factors selectively stimulate spermatogonial, but not meiotic, deoxyribonucleic acid synthesis during rat spermatogenesis, *Endocrinology,* 131, 2344, 1992.

163. **Giudice, L.C.,** Insulin-like growth factors and ovarian follicular development, *Endocr. Rev.,* 13, 641, 1992.
164. **Hammond, J.M., Baranao, J.L.S., Skaleris, D., Knight, A.B., Romanus, J.A., and Rechler, M.M.,** Production of insulin-like growth factors by ovarian granulosa cells, *Endocrinology,* 117, 2553, 1985.
165. **Davoren, J.B., Kasson, B.G., Li, C.H., and Hsueh, A.J.W.,** Specific insulin-like growth factor (IGF) I- and II-binding sites on rat granulosa cells: relation to IGF action, *Endocrinology,* 119, 2155, 1986.
166. **Hernandez, E.R., Roberts, C.T., Jr., Hurwitz, A., LeRoith, D., and Adashi, E.Y.,** Rat ovarian insulin-like growth factor II gene expression is theca-interstitial cell-exclusive: hormonal regulation and receptor distribution, *Endocrinology,* 127, 3249, 1990.
167. **Bicsak, T.A., Shimonaka, M., Malkowski, M., and Ling, N.,** Insulin-like growth factor-binding protein (IGF-BP) inhibition of granulosa cell function: effect on cyclic adenosine 3′,5′-monophosphate, deoxyribonucleic acid synthesis, and comparison with the effect of an IGF-I antibody, *Endocrinology,* 126, 2184, 1990.
168. **Simmen, R.C.M., Simmen, F.A., Hofig, A., Farmer, S.J., and Bazer, F.W.,** Hormonal regulation of insulin-like growth factor gene expression in pig uterus, *Endocrinology,* 127, 2166, 1990.
169. **Joshl, S., Ogawa, H., Burke, G.T., Tseng, L.Y.-H., Rechler, M.M., and Katsoyannis, P.G.,** Structural features involved in the biological activity of insulin and the insulin-like growth factors: A^{27}insulin/B IGF-I, *Biochem. Biophys. Res. Commun.,* 133, 423, 1985.
170. **Ethier, S.P., Kudla, A., and Cundiff, K.C.,** Influence of hormone and growth factor interactions on the proliferative potential of normal rat mammary epithelial cells *in vitro, J. Cell. Physiol.,* 132, 161, 1987.
171. **Clemmons, D.R.,** Exposure to platelet-derived growth factor modulates the porcine aortic smooth muscle cell response to somatomedin C, *Endocrinology,* 117, 77, 1985.
172. **Conover, C.A., Rosenfeld, R.G., and Hintz, R.L.,** Hormonal control of the replication of human fetal fibroblasts: role of somatomedin C/insulin-like growth factor I, *J. Cell. Physiol.,* 128, 47, 1986.
173. **Ashcom, G., Gurland, G., and Schwartz, J.,** Growth hormone synergizes with serum growth factors in inducing c-*fos* transcription in 3T3-F442A cells, *Endocrinology,* 131, 1915, 1992.
174. **Alarid, E.T., Schlechter, N.L., Russell, S.M., and Nicoll, C.S.,** Evidence suggesting that insulin-like growth factor I is necessary for the trophic effects of insulin on cartilage growth *in vitro, Endocrinology,* 130, 2305, 1992.
175. **Pietrzkowski, Z., Sell, C., Lammers, R., Ullrich, A., and Baserga, R.,** Roles of insulinlike growth factor 1 (IGF-1) and the IGF-1 receptor in epidermal growth factor-stimulated growth of 3T3 cells, *Mol. Cell. Biol.,* 12, 3883, 1992.
176. **Mochizuki, H., Hakeda, Y., Wakatsuki, N., Usui, N., Akashi, S., Sato, T., Tanaka, K., and Kumegawa, M.,** Insulin-like growth factor-I supports formation and activation of osteoclasts, *Endocrinology,* 131, 1075, 1992.
177. **Shinar, D.M., Endo, N., Halperin, D., Rodan, G.A., and Weinreb, M.,** Differential expression of insulin-like growth factor-I (IGF-I) and IGF-II messenger ribonucleic acid in growing rat bone, *Endocrinology,* 132, 1158, 1993.
178. **Grant, M., Jerdan, J., and Merimee, T.J.,** Insulin-like growth factor-I modulates endothelial cell chemotaxis, *J. Clin. Endocrinol. Metab.,* 65, 370, 1987.
179. **Sawada, K., Krantz, S.B., Dessypris, E.N., Koury, S.T., and Sawyer, S.T.,** Human colony-forming units-erythroid do not require accessory cells, but do require direct interaction with insulin-like growth factor I and/or insulin for erythroid development, *J. Clin. Invest.,* 83, 1701, 1989.
180. **Li, Q., Blacher, R., Esch, F., and Congote, L.F.,** A heparin-binding erythroid cell stimulating factor from fetal bovine serum has the N-terminal sequence of insulin-like growth factor II, *Biochem. Biophys. Res. Commun.,* 166, 557, 1990.
181. **Near, S.L., Whalen, L.R., Miller, J.A., and Ishii, D.N.,** Insulin-like growth factor-II stimulates motor nerve regeneration, *Proc. Natl. Acad. Sci. U.S.A.,* 89, 11716, 1992.
182. **Nissley, S.P., Haskell, J.F., Sasaki, N., De Vroede, M.A., and Rechler, M.M.,** Insulin-like growth factor receptors, *J. Cell Sci.,* Suppl. 3, 39, 1985.
183. **Neely, E.K., Beukers, M.W., Oh, Y., Cohen, P., and Rosenfeld, R.G.,** Insulin-like growth factor receptors, *Acta Paediatr. Scand.,* Suppl. 372, 116, 1991.
184. **Jonas, H.A., Newman, J.D., and Harrison, L.C.,** An atypical insulin receptor with high affinity for insulin-like growth factors copurified with placental insulin receptors, *Proc. Natl. Acad. Sci. U.S.A.,* 83, 4124, 1986.

185. **Rosenfeld, R.G. and Hintz, R.L.,** Characterization of a specific receptor for somatomedin C (SM-C) on cultured human lymphocytes: evidence that SM-C modulates homologous receptor concentration, *Endocrinology,* 107, 1841, 1980.
186. **Caro, J.F., Poulos, J., Ittoop, O., Pories, W.J., Flickinger, E.G., and Sinha, M.K.,** Insulin-like growth factor I binding in hepatocytes from human liver, human hepatoma, and normal, regenerating, and fetal rat liver, *J. Clin. Invest.,* 81, 976, 1988.
187. **Duronio, V.,** Insulin receptor is phosphorylated in response to treatment of HepG2 cells with insulin-like growth factor I, *Biochem. J.,* 270, 27, 1990.
188. **Kadowaki, T., Koyasu, S., Nishida, E., Sakai, H., Takaku, F., Yahara, I., and Kasuga, M.,** Insulin-like growth factors, insulin, and epidermal growth factor cause rapid cytoskeletal reorganization in KB cells. Clarification of the role of type I insulin-like growth factor receptors and insulin receptors, *J. Biol. Chem.,* 261, 16141, 1986.
189. **Dai, Z., Stiles, A.D., Moats-Staats, B., Van Wyk, J.J., and D'Ercole, A.J.,** Interaction of secreted insulin-like growth factor-I (IGF-I) with cell surface receptors is the dominant mechanism of IGF-I's autocrine actions, *J. Biol. Chem.,* 267, 19565, 1992.
190. **Bhaumick, B., Bala, R.M., and Hollenberg, M.D.,** Somatomedin receptor of human placenta: solubilization, photolabeling, partial purification, and comparison with insulin receptor, *Proc. Natl. Acad. Sci. U.S.A.,* 78, 4279, 1981.
191. **Fujita-Yamaguchi, Y., LeBon, T.R., Tsubokawa, M., Henzel, W., Kathuria, S., Koyal, D., and Ramachandran, J.,** Comparison of insulin-like growth factor I receptor and insulin receptor purified from human placental membranes, *J. Biol. Chem.,* 261, 16727, 1986.
192. **Feltz, S.M., Swanson, M.L., Wemmie, J.A., and Pessin, J.E.,** Functional properties of an isolated $\alpha\beta$ heterodimeric human placenta insulin-like growth factor 1 receptor complex, *Biochemistry,* 27, 3234, 1988.
193. **Massagué, J. and Czech, M.P.,** The subunit structures of two distinct receptors for insulin-like growth factors I and II and their relationship to the insulin receptor, *J. Biol. Chem.,* 257, 5038, 1982.
194. **Pilch, P.F., O'Hare, T., Rubin, J., and Böni-Schnetzler, M.,** The ligand binding subunit of the insulin-like growth factor 1 receptor has properties of a peripheral membrane protein, *Biochem. Biophys. Res. Commun.,* 136, 45, 1986.
195. **Moses, A.C., Usher, P., Ikari, N., King, P.P., Tramontano, D., and Flier, J.S.,** Multiple factors influence insulin-like growth factor-I binding to human skin fibroblasts, *Endocrinology,* 125, 867, 1989.
196. **Treadway, J.L., Morrison, B.D., Goldfine, I.D., and Pessin, J.E.,** Assembly of insulin/insulin-like growth factor-1 hybrid receptors *in vitro, J. Biol. Chem.,* 264, 21450, 1989.
197. **Cara, J.F., Mirmira, R.G., Nakagawa, S.H., and Tager, H.S.,** An insulin-like growth factor I/insulin hybrid exhibiting high potency for intereaction with the type I insulin-like growth factor and insulin receptors of placental plasma membranes, *J. Biol. Chem.,* 265, 17820, 1990.
198. **Zhang, B. and Roth, R.A.,** Binding properties of chimeric insulin receptors containing the cysteine-rich domain of either the insulin-like growth factor I receptor or the insulin receptor related receptor, *Biochemistry,* 30, 5113, 1991.
199. **Morgan, D.O. and Roth, R.A.,** Identification of a monoclonal antibody which can distinguish between two distinct species of the type I receptor for insulin-like growth factor, *Biochem. Biophys. Res. Commun.,* 138, 1341, 1986.
200. **Cascieri, M.A., Chicchi, G.G., Applebaum, J., Hayes, N.S., Green, B.G., and Bayne, M.L.,** Mutants of human insulin-like growth factor I with reduced affinity for the type 1 insulin-like growth factor receptor, *Biochemistry,* 27, 3229, 1988.
201. **Schumacher, R., Mosthaf, L., Schlessinger, J., Brandenburg, D., and Ullrich, A.,** Insulin and insulin-like growth factor-1 binding specificity is determined by distinct regions of their cognate receptors, *J. Biol. Chem.,* 266, 19288, 1991.
202. **Ullrich, A., Gray, A., Tam, A.W., Yang-Feng, T., Tsubokawa, M., Collins, C., Henzel, W., Le Bon, T., Kathuria, S., Chen, E., Jacobs, S., Francke, U., Ramachandran, J., and Fujita-Yamaguchi, Y.,** Insulin-like growth factor I receptor primary structure: comparison with insulin receptor suggests structural determinants that define specificity, *EMBO J.,* 5, 2503, 1986.
203. **Abbott, A.M., Bueno, R., Pedrini, M.T., Murray, J.M., and Smith, R.J.,** Insulin-like growth factor I receptor gene structure, *J. Biol. Chem.,* 267, 10759, 1992.
204. **Yee, D., Lebovic, G.S., Marcus, R.R., and Rosen, N.,** Identification of an alternate type I insulin-like growth factor receptor β subunit mRNA transcript, *J. Biol. Chem.,* 264, 21439, 1989.

205. **Kato, H., Faria, T.N., Stannard, B., Roberts, C.T., Jr., and LeRoith, D.,** Role of tyrosine kinase activity in signal transduction by the insulin-like growth factor-I (IGF-I) receptor. Characterization of kinase-deficient IGF-I receptors and the action of an IGF-I-mimetic antibody (αIR-3), *J. Biol. Chem.,* 268, 2655, 1993.
206. **Kadowaki, T., Koyasu, S., Nishida, E., Tobe, K., Izumi, T., Takaku, F., Sakai, H., Yahara, I., and Kasuga, M.,** Tyrosine phosphorylation of common and specific sets of cellular proteins rapidly induced by insulin, insulin-like growth factor I, and epidermal growth factor in an intact cell, *J. Biol. Chem.,* 262, 7342, 1987.
207. **Kaliman, P., Baron, V., Gautier, N., and Van Obberghen, E.,** Antipeptide antibody to the insulin-like growth factor I receptor sequence 1232–1246 inhibits the receptor kinase activity, *J. Biol. Chem.,* 267, 10645, 1992.
208. **Beguinot, F., Smith, R.J., Kahn, C.R., Maron, R., Moses, A.C., and White, M.F.,** Phosphorylation of insulin-like growth factor I receptor by insulin receptor tyrosine kinase in intact cultured skeletal muscle cells, *Biochemistry,* 27, 3222, 1988.
209. **Steele-Perkins, G., Turner, J., Edman, J.C., Hari, J., Pierce, S.B., Stover, C., Rutter, W.J., and Roth, R.A.,** Expression and characterization of a functional human insulin-like growth factor I receptor, *J. Biol. Chem.,* 263, 11486, 1988.
210. **Lammers, R., Gray, A., Schlessinger, J., and Ullrich, A.,** Differential signalling potential of insulin- and IGF-1-receptor cytoplasmic domains, *EMBO J.,* 8, 1369, 1989.
211. **Furlanetto, R.W., DiCarlo, J.N., and Wisehart, C.,** The type II insulin-like growth factor receptor does not mediate deoxyribonucleic acid synthesis in human fibroblasts, *J. Clin. Endocrinol. Metab.,* 64, 1142, 1987.
212. **Deeks, S., Richards, J., and Nandi, S.,** Maintenance of normal rat mammary epithelial cells by insulin and insulin-like growth factor 1, *Exp. Cell Res.,* 174, 448, 1988.
213. **Jonas, H.A. and Harrison, L.C.,** The human placenta contains two distinct binding and immunoreactive species of insulin-like growth factor-I receptors, *J. Biol. Chem.,* 260, 2288, 1985.
214. **Garofalo, R.S. and Rosen, O.M.,** Insulin and insulinlike growth factor 1 (IGF-1) receptors during central nervous system development: expression of two immunologically distinct IGF-1 receptor β subunits, *Mol. Cell. Biol.,* 9, 2806, 1989.
215. **Alexandrides, T.K. and Smith, R.J.,** A novel fetal insulin-like growth factor (IGF) I receptor. Mechanism for increased IGF I- and insulin-stimulated tyrosine kinase activity in fetal muscle, *J. Biol. Chem.,* 264, 12922, 1989.
216. **Ogasawara, M., Karey, K.P., Marquardt, H., and Sirbasku, D.A.,** Identification and purification of truncated insulin-like growth factor I from porcine uterus. Evidence for high biological potency, *Biochemistry,* 28, 2710, 1989.
217. **Clemmons, D.R., Han, V.K.M., Elgin, R.G., and D'Ercole, A.J.,** Alterations in the synthesis of a fibroblast surface associated 35K protein modulates the binding of somatomedin-C/insulin-like growth factor I, *Mol. Endocrinol.,* 1, 339, 1987.
218. **Rutanen, E.-M., Pekonen, F., and Mäkinen, T.,** Soluble 34K binding protein inhibits the binding of insulin-like growth factor I to its cell receptors in human secretory phase endometrium: evidence for autocrine/paracrine regulation of growth factor action, *J. Clin. Endocrinol. Metab.,* 66, 173, 1988.
219. **Pekonen, F., Suikkari, A.-M., Mäkinen, T., and Rutanen, E.-M.,** Different insulin-like growth factor binding species in human placenta and decidua, *J. Clin. Endocrinol. Metab.,* 67, 1250, 1988.
220. **Yee, D., Favoni, R.E., Lupu, R., Cullen, K.J., Lebovic, G.S., Huff, K.K., Lee, P.D.K., Lee, Y.L., Powell, D.R., Dickson, R.B., Rosen, N., and Lippman, M.E.,** The insulin-like growth factor binding protein BP-25 is expressed in human breast cancer cells, *Biochem. Biophys. Res. Commun.,* 158, 38, 1989.
221. **Massagué, J., Blinderman, L.A., and Czech, M.P.,** The high affinity insulin receptor mediates growth stimulation in rat hepatoma cells, *J. Biol. Chem.,* 257, 13958, 1982.
222. **Czech, M.P., Oppenheimer, C.L., and Massagué, J.,** Interrelationships among receptor structures for insulin and peptide growth factors, *Fed. Proc.,* 42, 2598, 1983.
223. **Furlanetto, R.W. and DiCarlo, J.N.,** Somatomedin-C receptors and growth effects in human breast cells maintained in long-term tissue culture, *Cancer Res.,* 44, 2122, 1984.
224. **Flier, J.S., Usher, P., and Moses, A.C.,** Monoclonal antibody to the type I insulin-like growth factor (IGF-I) receptor blocks IGF-I receptor-mediated DNA synthesis: clarification of the mitogenic mechanisms of IGF-I and insulin in human skin fibroblasts, *Proc. Natl. Acad. Sci. U.S.A.,* 83, 664, 1986.

225. **Furlanetto, R.W.,** Receptor-mediated endocytosis and lysosomal processing of insulin-like growth factor I by mitogenically responsive cells, *Endocrinology,* 122, 2044, 1988.
226. **Oppenheimer, C.L. and Czech, M.P.,** Purification of type II insulin-like growth factor receptor from rat placenta, *J. Biol. Chem.,* 258, 8539, 1983.
227. **Cooper, J.L. and Smith, G.L.,** Insulin-like growth factor II binding to cultured human chondrosarcoma cells, *Proc. Soc. Exp. Biol. Med.,* 179, 68, 1985.
228. **Scott, C.D. and Baxter, R.C.,** Purification and immunological characterization of the rat liver insulin-like growth factor-II receptor, *Endocrinology,* 120, 1, 1987.
229. **MacDonald, R.G. and Czech, M.P.,** Biosynthesis and processing of the type II insulin-like growth factor receptor in H-35 hepatoma cells, *J. Biol. Chem.,* 260, 11357, 1985.
230. **McElduff, A., Poronnik, P., and Baxter, R.C.,** The insulin-like growth factor-II (IGF II) receptor from rat brain is of lower apparent molecular weight than the IGF II receptor from rat liver, *Endocrinology,* 121, 1306, 1987.
231. **Rosenfeld, R.G., Hodges, D., Pham, H., Lee, P.D.K., and Powell, D.R.,** Purification of the insulin-like growth factor II (IGF-II) receptor from an IGF-II-producing cell line, and generation of an antibody which both immunoprecipitates and blocks the type 2 IGF receptor, *Biochem. Biophys. Res. Commun.,* 138, 304, 1986.
232. **Oka, Y., Mottola, C., Oppenheimer, C.L., and Czech, M.P.,** Insulin activates the appearance of insulin-like growth factor II receptors on the adipocyte cell surface, *Proc. Natl. Acad. Sci. U.S.A.,* 81, 4028, 1984.
233. **Rosenfeld, R.G., Conover, C.A., Hodges, D., Lee, P.D.K., Misra, P., Hintz, R.L., and Li, C.H.,** Heterogeneity of insulin-like growth factor-I affinity for the insulin-like growth factor-II receptor: comparison of natural, synthetic and recombinant DNA-derived insulin-like growth factor-I, *Biochem. Biophys. Res. Commun.,* 143, 199, 1987.
234. **Lee, P.D.K., Hodges, D., Hintz, R.L., Wyche, J.H., and Rosenfeld, R.G.,** Identification of receptors for insulin-like growth factor II in two insulin-like growth factor II producing cell lines, *Biochem. Biophys. Res. Commun.,* 134, 595, 1986.
235. **Barenton, B., Guyda, H.J., Goodyer, C.G., Polychronakos, C., and Posner, B.I.,** Specificity of insulin-like growth factor binding to type-II IGF receptors in rabbit mammary gland and hypophysectomized rat liver, *Biochem. Biophys. Res. Commun.,* 149, 555, 1987.
236. **Oka, Y., Rozek, L.M., and Czech, M.P.,** Direct demonstration of rapid insulin-like growth factor II receptor internalization and recycling in rat adipocytes. Insulin stimulates ^{125}I-insulin-like growth factor II degradation by modulating the IGF-II receptor recycling process, *J. Biol. Chem.,* 260, 9435, 1985.
237. **Oka, Y. and Czech, M.P.,** The type II insulin-like growth factor receptor is internalized and recycled in the absence of ligand, *J. Biol. Chem.,* 261, 9090, 1986.
238. **Scott, C.D. and Baxter, R.C.,** Insulin-like growth factor-II receptors in cultured rat hepatocytes: regulation by cell density, *J. Cell. Physiol.,* 133, 532, 1987.
239. **Oka, Y., Kasuga, M., Kanazawa, Y., and Takaku, F.,** Insulin induces chloroquine-sensitive recycling of insulin-like growth factor II but not of glucose transporters in rat adipocytes, *J. Biol. Chem.,* 262, 17480, 1987.
240. **Kiess, W., Greenstein, L.A., White, R.M., Lee, L., Rechler, M.M., and Nissley, S.P.,** Type II insulin-like growth factor receptor is present in rat serum, *Proc. Natl. Acad. Sci. U.S.A.,* 84, 7720, 1987.
241. **Hari, J., Pierce, S.B., Morgan, D.O., Sara, V., Smith, M.C., and Roth, R.A.,** The receptor for insulin-like growth factor II mediates an insulin-like response, *EMBO J.,* 6, 3367, 1987.
242. **Roth, R.A.,** Structure of the receptor for insulin-like growth factor II: the puzzle amplified, *Science,* 239, 1269, 1988.
243. **Polychronakos, C., Guyda, H.J., and Posner, B.I.,** Mannose 6-phosphate increases the affinity of its cation-dependent receptor for insulin-like growth factor II by displacing inhibitory endogenous ligands, *Biochem. Biophys. Res. Commun.,* 157, 632, 1988.
244. **Okamoto, T., Nishimoto, I., Murayama, Y., Ohkuni, Y., and Ogata, E.,** Insulin-like growth factor-II/mannose 6-phosphate receptor is incapable of activating GTP-binding proteins in response to mannose 6-phosphate, but capable in response to insulin-like growth factor-II, *Biochem. Biophys. Res. Commun.,* 168, 1201, 1990.
245. **Mathieu, M., Rochefort, H., Barenton, B., Prebois, C., and Vignon, F.,** Interactions of cathepsin-D and insulin-like growth factor-II (IGF-II) on the IGF-II/mannose-6-phosphate receptor in human breast cancer cells and possible consequences on mitogenic activity of IGF-II, *Mol. Endocrinol.,* 4, 1327, 1990.

246. **Brown, T.J., Ercolani, L., and Ginsberg, B.H.,** Demonstration of receptors for insulin-like growth factor-II on human T lymphocytes, *J. Receptor Res.,* 5, 297, 1985.
247. **Senior, P.V., Byrne, S., Brammar, W.J., and Beck, F.,** Expression of the IGF-II/mannose-6-phosphate receptor mRNA and protein in the developing rat, *Development,* 109, 67, 1990.
248. **Bondy, C.A., Werner, H., Roberts, C.T., Jr., and LeRoith, D.,** Cellular pattern of insulin-like growth factor-I (IGF-I) and type I IGF receptor gene expression in early organogenesis: comparison with IGF-II gene expression, *Mol. Endocrinol.,* 4, 1386, 1990.
249. **Haig, D. and Graham, C.,** Genomic imprinting and the strange case of the insulin-like growth factor II receptor, *Cell,* 64, 1045, 1991.
250. **Scott, C.D. and Baxter, R.C.,** Insulin-like growth factor II/mannose-6-phosphate receptors are increased in hepatocytes from regenerating rat liver, *Endocrinology,* 126, 2543, 1990.
251. **Mottola, C. and Czech, M.P.,** The type II insulin-like growth factor receptor does not mediate increased DNA synthesis in H-35 hepatoma cells, *J. Biol. Chem.,* 259, 12705, 1984.
252. **Conover, C.A., Misra, P., Hintz, R.L., and Rosenfeld, R.G.,** Effect of an anti-insulin-like growth factor I receptor antibody on insulin-like growth factor II stimulation of DNA synthesis in human fibroblasts, *Biochem. Biophys. Res. Commun.,* 139, 501, 1986.
253. **Conover, C.A., Rosenfeld, R.G., and Hintz, R.L.,** Insulin-like growth factor II binding and action in human fetal fibroblasts, *J. Cell. Physiol.,* 133, 560, 1987.
254. **Krett, N.L., Heaton, J.H., and Gelehrter, T.D.,** Mediation of insulin-like growth factor actions by the insulin receptor in H-35 rat hepatoma cells, *Endocrinology,* 120, 401, 1987.
255. **Dainiak, N. and Kreczko, S.,** Interactions of insulin, insulinlike growth factor II, and platelet-derived growth factor in erythropoietic culture, *J. Clin. Invest.,* 76, 1237, 1985.
256. **Corvera, S., Whitehead, R.E., Mottola, C., and Czech, M.P.,** The insulin-like growth factor II receptor is phosphorylated by a tyrosine kinase in adipocyte plasma membranes, *J. Biol. Chem.,* 261, 7675, 1986.
257. **Haskell, J.F., Nissley, S.P., Rechler, M.M., Sasaki, N., Greenstein, L., and Lee, L.,** Evidence for the phosphorylation of the type II insulin-like growth factor receptor in cultured cells, *Biochem. Biophys. Res. Commun.,* 130, 793, 1985.
258. **Casella, S.J., Han, V.K., D'Ercole, A.J., Svoboda, M.E., and Van Wyk, J.J.,** Insulin-like growth factor II binding to the type I somatomedin receptor: evidence for two high affinity binding sites, *J. Biol. Chem.,* 261, 9268, 1986.
259. **Misra, P., Hintz, R.L., and Rosenfeld, R.G.,** Structural and immunological characterization of insulin-like growth factor II binding to IM-9 cells, *J. Clin. Endocrinol. Metab.,* 63, 1400, 1986.
260. **Ewton, D.Z., Falen, S.L., and Florini, J.R.,** The type II insulin-like growth factor (IGF) receptor has low affinity for IGF-I analogs: pleiotropic actions of IGFs on myoblasts are apparently mediated by the type I receptor, *Endocrinology,* 120, 115, 1987.
261. **Bassas, L., de Pablo, F., Lesniak, M.A., and Roth, J.,** Ontogeny of receptors for insulin-like peptides in chick embryo tissues: early dominance of insulin-like growth factor over insulin receptors in brain, *Endocrinology,* 117, 2321, 1985.
262. **Sklar, M.M., Kiess, W., Thomas, C.L., and Nissley, S.P.,** Developmental expression of the tissue insulin-like growth factor II/mannose 6-phosphate receptor in the rat. Measurement by quantitative immunoblotting, *J. Biol. Chem.,* 264, 16733, 1989.
263. **Bhaumick, B. and Bala, R.M.,** Receptors for insulin-like growth factors I and II in developing embryonic mouse limb bud, *Biochim. Biophys. Acta,* 927, 117, 1987.
264. **Shimizu, M., Torti, F., and Roth, R.A.,** Characterization of the insulin and insulin-like growth factor receptors and responsitivity of a fibroblast/adipocyte cell line before and after differentiation, *Biochem. Biophys. Res. Commun.,* 137, 552, 1986.
265. **Reiss, K., Porcu, P., Sell, C., Pietrzkowski, Z., and Baserga, R.,** The insulin-like growth factor-1 receptor is required for the proliferation of hemopoietic cells, *Oncogene,* 7, 2243, 1992.
266. **Scott, C.D., Ballesteros, M., and Baxter, R.C.,** Increased expression of insulin-like growth factor-II/mannose-6-phosphate receptor in regenerating rat liver, *Endocrinology,* 127, 2210, 1990.
267. **Lönnroth, P., Assmundsson, K., Edén, S., Enberg, G., Gause, I., Hall, K., and Smith, U.,** Regulation of insulin-like growth factor II receptors by growth hormone and insulin in rat adipocytes, *Proc. Natl. Acad. Sci. U.S.A.,* 84, 3619, 1987.
268. **Ballesteros, M., Scott, C.D., and Baxter, R.C.,** Developmental regulation of insulin-like growth factor II/mannose 6-phosphate receptor mRNA in the rat, *Biochem. Biophys. Res. Commun.,* 172, 775, 1990.

269. **Kurose, H., Yamaoika, K., Okada, S., Nakajima, S., and Seino, Y.,** 1,25-dihydroxyvitamin D_3 (1,25-$(OH)_2D_3$) increases insulin-like growth factor I (IGF-I) receptors in clonal osteoblastic cells. Study on interaction of IGF-I and 1,25-$(OH)_2D_3$, *Endocrinology,* 126, 2088, 1990.
270. **Guler, H.-P., Zapf, J., and Froesch, E.R.,** Short-term metabolic effects of recombinant human insulin-like growth factor I in healthy adults, *N. Engl. J. Med.,* 317, 137, 1987.
271. **Kojima, I., Kitaoka, M., and Ogata, E.,** Studies on the cell cycle-dependency of the actions of insulin-like growth factor-I in Balb/c 3T3 cells, *J. Cell. Physiol.,* 143, 529, 1990.
272. **Tell, G.P.E., Cuatrecasas, P., Van Wyk, J.J., and Hintz, R.L.,** Somatomedin: inhibition of adenylate cyclase activity in subcellular membranes of various tissues, *Science,* 180, 312, 1973.
273. **Rubin, J.B., Shia, M.A., and Pilch, P.F.,** Stimulation of tyrosine-specific phosphorylation *in vitro* by insulin-like growth factor I, *Nature (London),* 305, 438, 1983.
274. **Zick, Y., Sasaki, N., Rees-Jones, R.W., Grunberger, G., Nissley, S.P., and Rechler, M.M.,** Insulin-like growth factor-I (IGF-I) stimulates tyrosine kinase activity in purified receptors from a rat liver cell line, *Biochem. Biophys. Res. Commun.,* 119, 6, 1984.
275. **Sasaki, N., Rees-Jones, R.W., Zick, Y., Nissley, S.P., and Rechler, M.M.,** Characterization of insulin-like growth factor I-stimulated tyrosine kinase activity associated with the beta-subunit of type I insulin-like growth factor receptors of rat liver cells, *J. Biol. Chem.,* 260, 9793, 1985.
276. **Yu, K.-T., Peters, M.A., and Czech, M.P.,** Similar control mechanisms regulate the insulin and type I insulin-like growth factor receptor kinases. Affinity-purified insulin-like growth factor I receptor kinase is activated by tyrosine phosphorylation of its beta subunit, *J. Biol. Chem.,* 261, 11341, 1986.
277. **Izumi, T., White, M.F., Kadowaki, T., Takaku, F., Akanuma, Y., and Kasuga, M.,** Insulin-like growth factor I rapidly stimulates tyrosine phosphorylation of a M_r 185,100 protein in intact cells, *J. Biol. Chem.,* 262, 1282, 1987.
278. **Sahal, D., Ramachandran, J., and Fujita-Yamaguchi, Y.,** Specificity of tyrosine protein kinases of the structurally related receptors for insulin and insulin-like growth factor I: Tyr-containing synthetic polymers as specific inhibitors or substrates, *Arch. Biochem. Biophys.,* 260, 416, 1988.
279. **Jacobs, S. and Cuatrecasas, P.,** Phosphorylation of receptors for insulin and insulin-like growth factor I: effects of hormones and phorbol esters, *J. Biol. Chem.,* 261, 934, 1986.
280. **Yamori, T., Iizuka, Y., Takayama, Y., Nishiya, S., Iwashita, S., Yamazaki, A., Takatori, T., and Tsuruo, T.,** Insulin-like growth factor I rapidly induces tyrosine phosphorylation of a M_r 150,000 and a M_r 160,000 protein in highly metastatic mouse colon carcinoma 26 NL-17 cells, *Cancer Res.,* 51, 5859, 1991.
281. **Oemar, B.S., Law, N.M., and Rosenzweig, S.A.,** Insulin-like growth factor-1 induces tyrosyl phosphorylation of nuclear proteins, *J. Biol. Chem.,* 266, 24241, 1991.
282. **Kojima, I., Matsunaga, H., Kurokawa, K., Ogata, E., and Nishimoto, I.,** Calcium influx: an intracellular message of the mitogenic action of insulin-like growth factor-I, *J. Biol. Chem.,* 263, 16561, 1988.
283. **Nishimoto, I., Ogata, E., and Kojima, I.,** Pertussis toxin inhibits the action of insulin-like growth factor-I, *Biochem. Biophys. Res. Commun.,* 148, 403, 1987.
284. **Kojima, I., Kitaoka, M., and Ogata, E.,** Insulin-like growth factor-I stimulates diacylglycerol production via multiple pathways in Balb/c 3T3 cells, *J. Biol. Chem.,* 265, 16846, 1990.
285. **Zumstein, P. and Stiles, C.D.,** Molecular cloning of gene sequences that are regulated by insulin-like growth factor I, *J. Biol. Chem.,* 262, 11252, 1987.
286. **Surmacz, E., Kaczmarek, L., Ronning, O., and Baserga, R.,** Activation of the ribosomal DNA promoter in cells exposed to insulinlike growth factor I, *Mol. Cell. Biol.,* 7, 657, 1987.
287. **Merriman, H.L., La Tour, D., Linkhart, T.A., Mohan, S., Baylink, D.J., and Strong, D.D.,** Insulin-like growth factor-I and insulin-like growth factor-II induce c-fos in mouse osteoblastic cells, *Calcif. Tissue Int.,* 46, 258, 1990.
288. **Ong, J., Yamashita, S., and Melmed, S.,** Insulin-like growth factor I induces c-*fos* messenger ribonucleic acid in L6 rat skeletal muscle cells, *Endocrinology,* 120, 353, 1987.
289. **Damante, G., Cox, F., and Rapoport, B.,** IGF-I increases c-fos expression in FRTL5 rat thyroid cells by activating the c-fos promoter, *Biochem. Biophys. Res. Commun.,* 151, 1194, 1988.
290. **Chiou, S.T. and Chiang, W.C.,** Insulin-like growth factor I stimulates transcription of the c-*jun* proto-oncogene in Balb/c 3T3 cells, *Biochem. Biophys. Res. Commun.,* 183, 524, 1992.
291. **Sumantran, V.N. and Feldman, E.L.,** Insulin-like growth factor I regulates c-*myc* and GAP-43 messenger ribonucleic acid expression in SH-SY5Y human neuroblastoma cells, *Endocrinology,* 132, 2017, 1993.

292. **Egan, S.E., Giddings, B.W., Brooks, M.W., Buday, L., Sizeland, A.M., and Weinberg, R.A.,** Association of Sos Ras exchange protein with Grb2 is implicated in tyrosine kinase signal transduction and transformation, *Nature (London),* 363, 45, 1993.
293. **Li, N., Batzer, A., Daly, R., Yajnik, V., Skolnik, E., Chardin, P., Bar-Sagi, D., Margolis, B., and Schlessinger, J.,** Guanine-nucleotide-releasing factor hSos1 binds to Grb2 and links receptor tyrosine kinases to Ras signalling, *Nature (London),* 363, 85, 1993.
294. **Okamoto, T., Asano, T., Harada, S., Ogata, E., and Nishimoto, I.,** Regulation of transmembrane signal transduction of insulin-like growth factor II by competence type growth factors or viral *ras* p21, *J. Biol. Chem.,* 266, 1085, 1991.
295. **Nishimoto, I., Ohkuni, Y., Ogata, E., and Kojima, I.,** Insulin-like growth factor II increases cytoplasmic free calcium in competent Balb/c 3T3 cells treated with epidermal growth factor, *Biochem. Biophys. Res. Commun.,* 142, 275, 1987.
296. **Nishimoto, I., Hata, Y., Ogata, E., and Kojima, I.,** Insulin-like growth factor II stimulates calcium influx in competent BALB/c 3T3 cells primed with epidermal growth factor. Characteristics of calcium influx and involvement of GTP-binding protein, *J. Biol. Chem.,* 262, 12120, 1987.
297. **Rogers, S.A. and Hammerman, M.R.,** Insulin-like growth factor II stimulates production of inositol trisphosphate in proximal tubular basolateral membranes from canine kidney, *Proc. Natl. Acad. Sci. U.S.A.,* 85, 4037, 1988.
298. **Cullen, K.J., Yee, D., and Rosen, N.,** Insulinlike growth factors in human malignancy, *Cancer Invest.,* 9, 443, 1991.
299. **MacAulay, V.M.,** Insulin-like growth factors and cancer, *Br. J. Cancer,* 65, 311, 1992.
300. **Benson, A.E. and Holdaway, I.M.,** Insulin receptors in human cancer, *Br. J. Cancer,* 44, 917, 1981.
301. **Wong, M. and Holdaway, I.M.,** Insulin binding by normal and neoplastic colon tissue, *Int. J. Cancer,* 35, 335, 1985.
302. **Stuart, C.A., Prince, M.J., Peters, E.J., Smith, F.E., Townsend, C.M., III, and Poffenbarger, P.L.,** Insulin receptor proliferation: a mechanism for tumor-associated hypoglycemia, *J. Clin. Endocrinol. Metab.,* 63, 879, 1986.
303. **Kaleko, M., Rutter, W.J., and Miller, A.D.,** Overexpression of the human insulinlike growth factor I receptor promotes ligand-dependent neoplastic transformation, *Mol. Cell. Biol.,* 10, 464, 1990.
304. **Pimentel, E.,** Peptide hormone precursors, subunits, and fragments as human tumor markers, in *Human Tumor Markers,* Ting, W.S., Chen, J.-S., and Schwartz, M.K., Eds., Excerpta Medica, Amsterdam, 1989, 105.
305. **Cariani, E., Lasserre, C., Seurin, D., Hamelin, B., Kemeny, F., Franco, D., Czech, M. P., Ullrich, A., and Brechot, C.,** Differential expression of insulin-like growth factor II mRNA in human primary liver cancers, benign liver tumors, and liver cirrhosis, *Cancer Res.,* 48, 6844, 1988.
306. **Yee, D., Cullen, K.J., Paik, S., Perdue, J.F., Hampton, B., Schwartz, A., Lippman, M.E., and Rosen, N.,** Insulin-like growth factor II mRNA expression in human breast cancer, *Cancer Res.,* 48, 6691, 1988.
307. **Cullen, K.J., Smith, H.S., Hill, S., Rosen, N., and Lippman, M.E.,** Growth factor messenger RNA expression by human breast fibroblasts from benign and malignant lesions, *Cancer Res.,* 51, 4978, 1991.
308. **Peyrat, J.P., Bonneterre, J., Hecquet, B., Vennin, P., Louchez, M.M., Fournier, C., Lefebvre, J., and Demaille, A.,** Plasma insulin-like growth factor-1 (IGF-1) concentrations in human breast cancer, *Eur. J. Cancer,* 29A, 492, 1993.
309. **Höppener, J.W.M., Mosselman, S., Roholl, P.J.M., Lambrechts, C., Slebos, R.J.C., de Pagter-Holthuizen, P., Lips, C.J.M., Jansz, H.S., and Sussenbach, J.S.,** Expression of insulin-like growth factor-I and -II genes in human smooth muscle tumours, *EMBO J.,* 7, 1379, 1988.
310. **Gloudemans, T., Prinsen, I., Van Unnik, J.A.M., Lips, C.J.M., Den Oetter, W., and Sussenbach, J.S.,** Insulin-like growth factor gene expression in human smooth muscle tumors, *Cancer Res.,* 50, 6689, 1990.
311. **Papa, V., Pezzino, V., Costantino, A., Belfiore, A., Giuffrida, D., Frittitta, L., Vannelli, G.B., Brand, R., Goldfine, I.D., and Vigneri, R.,** Elevated insulin receptor content in human breast cancer, *J. Clin. Invest.,* 86, 1503, 1990.
312. **Pollak, M.N., Perdue, J.F., Margolese, R.G., Baer, K., and Richard, M.,** Presence of somatomedin receptors on primary human breast and colon carcinomas, *Cancer Lett.,* 38, 223, 1987.
313. **Foekens, J.A., Portengen, H., Janssen, M., and Klijn, J.G.M.,** Insulin-like growth factor-1 receptors and insulin-like growth factor-1-like activity in human primary breast cancer, *Cancer,* 63, 2139, 1989.

314. **Cullen, K.J., Yee, D., Sly, W.S., Perdue, J., Hampton, B., Lippman, M.E., and Rosen, N.,** Insulin-like growth factor receptor expression and function in human breast cancer, *Cancer Res.,* 50, 48, 1990.
315. **Pekonen, F., Partanen, S., Mäkinen, T., and Rutanen, E.-M.,** Receptors for epidermal growth factor and insulin-like growth factor I and their relation to steroid receptors in human breast cancer, *Cancer Res.,* 48, 1343, 1988.
316. **Peyrat, J.-P., Bonneterre, J., Beuscart, R., Djiane, J., and Demaille,** Insulin-like growth factor 1 receptors in human breast cancer and their relation to estradiol and progesterone receptors, *Cancer Res.,* 48, 6429, 1988.
317. **Foekens, J.A., Portengen, H., van Putten, W.L.J., Trapman, A.M.A.C., Reubi, J.-C., Alexieva-Figusch, J., and Klijn, J.G.M.,** Prognostic value of receptors for insulin-like growth factor 1, somatostatin, and epidermal growth factor in human breast cancer, *Cancer Res.,* 49, 7002, 1989.
318. **Bonneterre, J., Peyrat, J.P., Beuscart, R., and Demaille, A.,** Prognostic significance of insulin-like growth factor 1 receptors in human breast cancer, *Cancer Res.,* 50, 6931, 1990.
319. **Berns, E.M.J.J., Klijn, J.G.M., Van Staveren, I.I., Portengen, H., and Foekens, J.A.,** Sporadic amplification of the insulin-like growth factor 1 receptor gene in human breast tumors, *Cancer Res.,* 52, 1036, 1992.
320. **Peyrat, J.P., Bonneterre, J., Laurent, J.C., Louchez, M.M., Amrani, S., Leroy-Martin, B., Vilain, M.O., Delobelle, A., and Demaille, A.,** Presence and characterization of insulin-like growth factor I receptors in human benign breast disease, *Eur. J. Cancer Clin. Oncol.,* 24, 1425, 1988.
321. **Talavera, F., Reynolds, R.K., Roberts, J.A., and Menon, K.M.J.,** Insulin-like growth factor I receptors in normal and neoplastic human endometrium, *Cancer Res.,* 50, 3019, 1990.
322. **Macaulay, V.M., Teale, J.D., Everard, M.J., Joshi, G.P., Smith, I.E., and Millar, J.L.,** Somatomedin-C/insulin-like growth factor-I is a mitogen for human small cell lung cancer, *Br. J. Cancer,* 57, 91, 1988.
323. **Macaulay, V.M., Everard, M.J., Teale, J.D., Trott, P.A., Van Wyk, J.J., Smith, I.E., and Millar, J.L.,** Autocrine function for insulin-like growth factor I in human small cell lung cancer cell lines and fresh tumor cells, *Cancer Res.,* 50, 2511, 1990.
324. **Shigematsu, K., Kataoka, Y., Kamio, T., Kurihara, M., Niwa, M., and Tsuchiyama, H.,** Partial characterization of insulin-like growth factor I in primary lung cancers using immunohistochemical and receptor autoradiographic techniques, *Cancer Res.,* 50, 2481, 1990.
325. **Reeve, J.G., Payne, J.A., and Bleehen, N.M.,** Production of immunoreactive insulin-like growth factor-I (IGF-I) and IGF-I binding proteins by human lung tumours, *Br. J. Cancer,* 61, 727, 1990.
326. **Pollak, M.N., Polychronakos, C., and Richards, M.,** Insulinlike growth factor I: a potent mitogen for human osteogenic sarcoma, *J. Natl. Cancer Inst.,* 82, 301, 1990.
327. **El-Badry, O.M., Minniti, C., Kohn, E.C., Houghton, P.J., Daughaday, W.H., and Helman, L.J.,** Insulin-like growth factor II acts as an autocrine growth and motility factor in human rhabdomyosarcoma tumors, *Cell Growth Differ.,* 1, 325, 1990.
328. **Minniti, C.P., Maggi, M., and Helman, L.J.,** Suramin inhibits the growth of human rhabdomyosarcoma by interrupting the insulin-like growth factor II autocrine growth loop, *Cancer Res.,* 52, 1830, 1992.
329. **Yashiro, T., Tsushima, T., Murakami, H., Obara, T., Fujimoto, Y., Shizume, K., and Ito, K.,** Insulin-like growth factor-II (IGF-II)/mannose-6-phosphate receptors are increased in primary human thyroid neoplasms, *Eur. J. Cancer,* 27, 699, 1991.
330. **Nilsson, O., Wängberg, B., Wigander, A., and Ahlman, H.,** Immunocytochemical evidence for the presence of IGF-I and IGF-I receptors in human endocrine tumours, *Acta Physiol. Scand.,* 144, 211, 1992.
331. **Nilsson, O., Wängberg, B., Theodorsson, E., Skottner, A., and Ahlman, H.,** Presence of IGF-I in human midgut carcinoid tumours — An autocrine regulator of carcinoid tumour growth?, *Int. J. Cancer,* 51, 195, 1992.
332. **Kamio, T., Shigematsu, K., Kawai, K., and Tsuchiyama, H.,** Immunoreactivity and receptor expression of insulinlike growth factor I and insulin in human adrenal tumors, *Am. J. Pathol.,* 138, 83, 1991.
333. **Yee, D., Morales, F.R., Hamilton, T.C., and Von Hoff, D.D.,** Expression of insulin-like growth factor I, its binding proteins, and its receptor in ovarian cancer, *Cancer Res.,* 51, 5107, 1991.
334. **Merrill, M.J. and Edwards, N.A.,** Insulin-like growth factor-I receptors in human glial tumors, *J. Clin. Endocrinol. Metab.,* 71, 199, 1990.
335. **Sara, V.R., Prisell, P., Sjögren, B., Persson, L., Boethius, J., and Enberg, G.,** Enhancement of insulin-like growth factor 2 receptors in glioblastoma, *Cancer Lett.,* 32, 229, 1986.

336. **El-Badry, O.M., Helman, L.J., Chatten, J., Steinberg, S.M., Evans, A.E., and Israel, M.A.,** Insulin-like growth factor II-mediated proliferation of human neuroblastoma, *J. Clin. Invest.,* 87, 648, 1991.
337. **Gansler, T., Allen, K.D., Burant, C.F., Inabnett, T., Scott, A., Buse, M.G., Sens, D.A., and Garvin, A.J.,** Detection of type 1 insulinlike growth factor (IGF) receptors in Wilms' tumors, *Am. J. Pathol.,* 130, 431, 1988.
338. **Gansler, T., Furlanetto, R., Gramling, T.S., Robinson, K.A., Blocker, N., Buse, M.G., Sens, D.A., and Garvin, A.J.,** Antibody to type I insulinlike growth factor receptor inhibits growth of Wilms' tumor in culture and in athymic mice, *Am. J. Pathol.,* 135, 961, 1989.
339. **Grunberger, G., Lowe, W.L., Jr., McElduff, A., and Glick, R.P.,** Insulin receptor of human cerebral gliomas: structure and function, *J. Clin. Invest.,* 77, 997, 1986.
340. **Helderman, J.H., Reynolds, T.C., and Strom, T.B.,** The insulin receptors as a universal marker of activated lymphocytes, *Eur. J. Immunol.,* 8, 589, 1978.
341. **Helderman, J.H.,** Role of insulin in the intermediary metabolism of the activated thymic derived lymphocytes, *J. Clin. Invest.,* 67, 1636, 1981.
342. **Chen, P., Kwan, S., Hwang, T., Chiang, B.N., and Chou, C.-K.,** Insulin receptors on leukemia and lymphoma cells, *Blood,* 62, 251, 1983.
343. **Thomopoulos, P. and Marie, J.P.,** Insulin receptors in acute and chronic lymphoid leukaemias, *Eur. J. Clin. Invest.,* 10, 387, 1980.
344. **Pui, C.-H. and Costlow, M.E.,** Clinical and biologic correlates of insulin binding by leukemia lymphoblastas, *Leukemia Res.,* 9, 843, 1985.
345. **Schneid, H., Seurin, D., Noguiez, P., and Le Bouc, Y.,** Abnormalities of insulin-like growth factor (IGF-I and IGF-II) genes in human tumor tissue, *Growth Regul.,* 2, 45, 1992.
346. **Fu, X.-X., Su, C.Y., Lee, Y., Hintz, R., Biempica, L., Snyder, R., and Rogler, C.E.,** Insulinlike growth factor II expression and oval cell proliferation associated with hepatocarcinogenesis in woodchuck hepatitis virus carriers, *J. Virol.,* 62, 3422, 1988.
347. **Hilf, R., Livingston, J.N., and Crofton, D.H.,** Effects of diabetes and sex steroid hormones on insulin receptor tyrosine kinase activity in R3230AC mammary adenocarcinomas, *Cancer Res.,* 48, 3741, 1988.
348. **Pezzino, V., Vigneri, R., Siperstein, M.D., and Goldfine, I.D.,** Insulin and glucagon receptors in Morris hepatomas of varying growth rates, *Cancer Res.,* 39, 1443, 1979.
349. **Hwang, D.L., Roitman, A., Carr, B.I., Barseghian, G., and Lev-Ran, A.,** Insulin and epidermal growth factor receptors in rat liver after administration of the hepatocarcinogen 2-acetylaminofluorene: ligand binding and autophosphorylation, *Cancer Res.,* 46, 1955, 1986.
350. **Kozma, L.M. and Weber, M.J.,** Constitutive phosphorylation of the receptor for insulinlike growth factor I in cells transformed by the *src* oncogene, *Mol. Cell. Biol.,* 10, 3626, 1990.
351. **Porcu, P., Ferber, A., Pietrzkowski, Z., Roberts, C.T., Adamo, M., LeRoith, D., and Baserga, R.,** The growth-stimulatory effect of simian virus 40 T antigen requires the interaction of insulinlike growth factor 1 with its receptor, *Mol. Cell. Biol.,* 12, 5069, 1992.
352. **Reiss, M., Radin, A.I., and Weisberg, T.F.,** Constitutive expression of the c-*fos* proto-oncogene in murine keratinocytes: potentiation of the mitogenic response to insulin-like growth factor 1, *Cancer Res.,* 50, 6641, 1990.
353. **Yee, D., Paik, S., Lebovic, G.S., Marcus, R.R., Favoni, R.E., Cullen, K.J, Lippman, M.E., and Rosen, N.,** Analysis of insulin-like growth factor I gene expression in malignancy: evidence for a paracrine role in human breast cancer, *Mol. Endocrinol.,* 3, 509, 1989.
354. **Hoffmann, S.S. and Kolodny, G.M.,** Insulin receptors in 3T3 fibroblasts: relationship to growth phase, transformation and differentiation into new cell types, *Exp. Cell Res.,* 107, 293, 1977.
355. **Petersen, B. and Blecher, M.,** Insulin receptors and functions in normal and spontaneously transformed cloned rat hepatocytes, *Exp. Cell Res.,* 120, 119, 1979.
356. **Hofmann, C., Marsh, J.W., Miller, B., and Steiner, D.F.,** Cultured hepatoma cells as a model system for studying insulin processing and biological responsiveness, *Diabetes,* 29, 865, 1980.
357. **Newman, J.D. and Harrison, L.C.,** Insulin receptor expression in Burkitt lymphoma cell lines, *Int. J. Cancer,* 44, 467, 1989.
358. **Biddle, C., Li, C.H., Schofield, P.N., Tate, V.E., Hopkins, B., Engstrom, W., Huskisson, N.S., and Graham, C.F.,** Insulin-like growth factors and the multiplication of Tera-2, a human teratoma-derived cell line, *J. Cell Sci.,* 90, 475, 1988.

359. **Weima, S.M., Stet, L.T., van Rooijen, M.A., van Buul-Offers, S.C., van Zoelen, E.J.J., de Laat, S.W., and Mummery, C.L.,** Human teratocarcinoma cells express functional insulin-like growth factor I receptors, *Exp. Cell Res.,* 184, 427, 1989.
360. **Milazzo, G., Giorgino, F., Damante, G., Sung, C., Stampfer, M.R., Vigneri, R., Goldfine, I.D., and Belfiore, A.,** Insulin receptor expression and function in human breast cancer cell lines, *Cancer Res.,* 52, 3924, 1992.
361. **Karey, K.P. and Sibasku, D.A.,** Differential responsiveness of human breast cancer cell lines MCF-7 and T47D to growth factors and 17 β-estradiol, *Cancer Res.,* 48, 4083, 1988.
362. **Osborne, C.K., Coronado, E.B., Kitten, L.J., Arteaga, C.I., Fuqua, S.A.W., Ramasharma, K., Marshall, M., and Li, C.H.,** Insulin-like growth factor-II (IGF-II): a potential autocrine/paracrine growth factor for human breast cancer acting via the IGF-I receptor, *Mol. Endocrinol.,* 3, 1701, 1989.
363. **Peyrat, J.P., Bonneterre, J., Dusanter-Fourt, I., Leroy-Koy-Martin, B., Djiane, J., and Demaille, A.,** Characterization of insulin-like growth factor 1 receptors (IGF1-R) in human breast cancer cell lines, *Bull. Cancer,* 76, 311, 1989.
364. **Myal, Y., Shiu, R.P.C., Bhaumick, B., and Bala, M.,** Receptor binding and growth-promoting activity of insulin-like growth factors in human breast cancer cells (T-47-D) in culture, *Cancer Res.,* 44, 5486, 1984.
365. **De Leon, D.D., Bakker, B., Wilson, D.M., Hintz, R.L., and Rosenfeld, R.G.,** Demonstration of insulin-like growth factor (IGF-I and -II) receptors and binding protein in human breast cancer cell lines, *Biochem. Biophys. Res. Commun.,* 152, 398, 1988.
366. **Pollak, M.N., Polychronakos, C., Yousefi, S., and Richard, M.,** Characterization of insulin-like growth factor I (IGF-I) receptors of human breast cancer cells, *Biochem. Biophys. Res. Commun.,* 154, 326, 1988.
367. **Hwang, D.L., Papoian, T., Barseghian, G., Josefsberg, Z., and Lev-Ran, A.,** Absence of down-regulation of insulin receptors in human breast cancer cells (MCF-7) cultured in serum-free medium: comparison with epidermal growth factor, *J. Receptor Res.,* 5, 27, 1985.
368. **van der Burg, B., Rutteman, G.R., Blankenstein, M.A., de Laat, S.W., and van Zoelen, E.J.J.,** Mitogenic stimulation of human breast cancer cells in a growth factor-defined medium: synergistic action of insulin and estrogen, *J. Cell. Physiol.,* 134, 101, 1988.
369. **Daly, R.J., Harris, W.H., Wang, D.Y., and Darbre, P.D.,** Autocrine production of insulin-like growth factor II using an inducible expression system results in reduced estrogen sensitivity of MCF-7 human breast cancer cells, *Cell Growth Differ.,* 2, 457, 1991.
370. **Jaques, G., Rotsch, M., Wegmann, C., Worsch, U., Maasberg, M., and Havemann, K.,** Production of immunoreactive insulin-like growth factor I and response to exogenous IGF-I in small cell lung cancer cell lines, *Exp. Cell Res.,* 176, 336, 1988.
371. **Nakanishi, Y., Muishine, J.L., Kasprzyk, P.G., Natale, R.B., Maneckjee, R., Avis, I., Treston, A.M., Gazdar, A.F., Minna, J.D., and Cuttitta, F.,** Insulin-like growth factor-I can mediate autocrine proliferation of human small cell lung cancer cell lines *in vitro, J. Clin. Invest.,* 82, 354, 1988.
372. **Schardt, C., Rotsch, M., Erbil, C., Göke, R., Richter, G., and Havemann, K.,** Characterization of insulin-like growth factor II receptors in human small cell lung cancer cell lines, *Exp. Cell Res.,* 204, 22, 1993.
373. **Ohmura, E., Okada, M., Onoda, N., Kamiya, Y., Murakami, H., Tsushima, T., and Shizume, K.,** Insulin-like growth factor I and transforming growth factor alpha as autocrine growth factors in human pancreatic cancer cell growth, *Cancer Res.,* 50, 103, 1990.
374. **Tsai, T.-F., Yauk, Y.-K., Chou, C.-K., Ting, L.-P., Chang, C., Hu, C., Han, S.-H., and Su, T.-S.,** Evidence of autocrine regulation in human hepatoma cell lines, *Biochem. Biophys. Res. Commun.,* 153, 39, 1988.
375. **Scott, C.D., Taylor, J.E., and Baxter, R.C.,** Differential regulation of insulin-like growth factor-II receptors in rat hepatocytes and hepatoma cells, *Biochem. Biophys. Res. Commun.,* 151, 815, 1988.
376. **Ritvos, O., Rutanen, E.-M., Pekonen, F., Jalkanen, J., Suikkari, A.-M., and Ranta, T.,** Characterization of functional type I insulin-like growth factor receptors from human choriocarcinoma cells, *Endocrinology,* 122, 395, 1988.
377. **Ritvos, O., Ranta, T., Jalkanen, J., Suikkari, A.-M., Voutilainen, R., Bohn, H., and Rutanen, E.-M.,** Insulin-like growth factor (IGF) binding protein from human decidua inhibits the binding and biological action of IGF-I in cultured choriocarcinoma cells, *Endocrinology,* 122, 2150, 1988.

378. **Orlowski, C.C., Chernausek, S.D., and Akeson, R.,** Actions of insulin-like growth factor-I on the B104 neuronal cell line: effects on cell replication, receptor characteristics, and influence of secreted binding protein on ligand binding, *J. Cell. Physiol.*, 139, 469, 1989.
379. **Martin, D.M. and Feldman, E.L.,** Regulation of insulin-like growth factor-II expression and its role in autocrine growth of human neuroblastoma cells, *J. Cell. Physiol.*, 155, 290, 1993.
380. **Das, A., Pansky, B., Budd, G.C., and Reid, T.W.,** Human retinoblastoma-Y79 cells contain both insulin-specific messenger RNA and insulin-binding sites, *Neurosci. Lett.*, 121, 231, 1991.
381. **Yee, D., Favoni, R.E., Lebovic, G.S., Lombana, F., Powell, D.R., Reynolds, C.P., and Rosen, N.,** Insulin-like growth factor I expression by tumors of neuroectodermal origin with the t(11;22) chromosomal translocation. A potential autocrine growth factor, *J. Clin. Invest.*, 86, 1806, 1990.
382. **Okimura, Y., Kitajima, N., Uchiyama, T., Yagi, H., Abe, H., Shakutsui, S., and Chihara, K.,** Insulin-like growth factor I (IGF-I) production and the presence of IGF-I receptors in rat medullary thyroid carcinoma cell line 6-23 (clone 6), *Biochem. Biophys. Res. Commun.*, 161, 589, 1989.
383. **Hartmann, W., Hitzler, H., Schlickenrieder, J.H.M., Zapf, J., Heit, W., Gaedicke, G., and Vetter, U.,** Heterogeneity of insulin and insulin-like growth factor I binding in a human Burkitt type ALL cell line during the cell cycle and in three Burkitt type ALL sublines, *Leukemia*, 2, 241, 1988.
384. **Sinclair, J., McClain, D., and Taetle, R.,** Effects of insulin and insulin-like growth factor I on growth of human leukemia cells in serum-free and protein-free medium, *Blood*, 72, 66, 1988.
385. **Kaplan, S.A.,** The insulin receptor, *J. Pediatr.*, 104, 327, 1984.
386. **Straus, D.S. and Pang, K.J.,** Insulin receptors on cultured murine lymphoid tumor cell lines, *Mol. Cell. Biochem.*, 47, 161, 1982.
387. **Newman, J.D., Campbell, I.L., Maher, F., and Harrison, L.C.,** Insulin receptor expression in the Burkitt lymphoma cells Daudi and Raji, *Mol. Endocrinol.*, 3, 597, 1989.
388. **Vetter, U., Schlickenrieder, J.H.M., Zapf, J., Hartmann, W., Heit, W., Hitzler, H., Byrne, P., Gaedicke, G., Heinze, E., and Teller, W.M.,** Human leukemic cells: receptor binding and biological effects of insulin and insulin-like growth factors, *Leukemia Res.*, 10, 1201, 1986.
389. **Maturo, J.M., III and Hollenberg, M.D.,** Insulin receptors in transformed fibroblasts and in adipocytes: a comparative study, *Can. J. Biochem.*, 57, 497, 1979.
390. **McElduff, A.M., Grunberger, G., and Gorden, P.,** An alteration in apparent molecular weight of the insulin receptors from the human monocyte cell line U-937, *Diabetes*, 34, 686, 1985.
391. **Gammeltoft, S., Ballotti, R., Kowalski, A., Westermark, B., and Van Obberghen, E.,** Expression of two types of receptor for insulin-like growth factors in human malignant glioma, *Cancer Res.*, 48, 1233, 1988.
392. **Häring, H.U., White, M.F., Kahn, C.R., Kasuga, M., Lauris, V., Fleischmann, R., Murray, M., and Pawelek, J.,** Abnormality of insulin binding and receptor phosphorylation in an insulin-resistant melanoma cell line, *J. Cell Biol.*, 99, 900, 1984.
393. **Rodeck, U., Herlyn, M., and Koprowski, H.,** Interactions between growth factor receptors and corresponding monoclonal antibodies in human tumors, *J. Cell. Biochem.*, 35, 315, 1987.
394. **Trojan, J., Blossey, B.K., Johnson, T.R., Rudin, S.D., Tykocinsky, M., Ilan, J., and Ilan, J.,** Loss of tumorigenicity of rat glioblastoma directed by episome-based antisense cDNA transcription of insulin-like growth factor I, *Proc. Natl. Acad. Sci. U.S.A.*, 89, 4874, 1992.
395. **Stracke, M.L., Kohn, E.C., Aznavoorian, S.A., Wilson, L.L., Salomon, D., Krutzsch, H.C., Liotta, L.A., and Schiffmann, E.,** Insulin-like growth factors stimulate chemotaxis in human melanoma cells, *Biochem. Biophys. Res. Commun.*, 153, 1076, 1988.
396. **Stracke, M.L., Engel, J.D., Wilson, L.W., Rechler, M.M., Liotta, L.A., and Schiffmann, E.,** The type I insulin-like growth factor receptor is a motility receptor in human melanoma cells, *J. Biol. Chem.*, 264, 21544, 1989.
397. **Kohn, E.C., Liotta, L.A., and Schiffmann, E.,** Autocrine motility factor stimulates a three-fold increase in inositol trisphosphate in human melanoma cells, *Biochem. Biophys. Res. Commun.*, 166, 757, 1990.
398. **Hizuka, N., Sukegawa, I., Takano, K., Asakawa, K., Horikawa, R., Tsushima, T., and Shizume, K.,** Characterization of insulin-like growth factor I receptors on human erythroleukemia cell line (K-562 cells), *Endocrinol. Jpn.*, 34, 81, 1987.
399. **Yamanouchi, T., Tsushima, T., Murakami, H., Sato, Y., Shizume, K., Oshimi, K., and Mizoguchi, H.,** Differentiation of human promyelocytic leukemia cells is accompanied by an increase in insulin receptors, *Biochem. Biophys. Res. Commun.*, 108, 414, 1982.

400. **Sukegawa, I., Hizuka, N., Takano, K., Asakawa, K., and Shizume, K.,** Decrease in IGF-I binding sites on human promyelocytic leukemia cell line (HL-60) with differentiation, *Endocrinol. Jpn.,* 34, 365, 1987.
401. **Yamanouchi, T., Tsushima, T., Kasuga, M., and Takaku, F.,** Variables that regulate the production of insulin-like peptide(s) in human leukemia cell line (HL-60), *Biochem. Biophys. Res. Commun.,* 129, 293, 1985.
402. **Rouis, M., Thomopoulos, P., Louache, F., Testa, U., Hervy, C., and Titeux, M.,** Differentiation of U-937 human monocyte-like cell line by 1 alpha,25-dihydroxyvitamin D_3 or by retinoic acid: opposite effects on insulin receptors, *Exp. Cell Res.,* 157, 539, 1985.
403. **Simon, I., Brown, T.J., and Ginsberg, B.H.,** Abnormal insulin binding and membrane physical properties of a Friend erythroleukemia clone resistant to dimethylsulfoxide-induced differentiation, *Biochim. Biophys. Acta,* 803, 39, 1984.
404. **Jacobs, S., Sahyoun, N.E., Saltiel, A.R., and Cuatrecasas, P.,** Phorbol esters stimulate the phosphorylation of receptors for insulin and somatomedin C, *Proc. Natl. Acad. Sci. U.S.A.,* 80, 6211, 1983.
405. **Takayama, S., White, M.F., Lauris, V., and Kahn, C.R.,** Phorbol esters modulate insulin receptor phosphorylation and insulin action in cultured hepatoma cells, *Proc. Natl. Acad. Sci. U.S.A.,* 81, 7797, 1984.

Chapter 3

Epidermal Growth Factor

I. INTRODUCTION

Epidermal growth factor (EGF) is a 6-kDa polypeptide that exhibits potent mitogenic activity in a diversity of cell types *in vivo* and *in vitro.*[1-9] EGF was first isolated from submaxillary glands of male mice, where it is present in very high amounts. It is also synthesized and secreted by many other different types of cells *in vivo* and *in vitro,* including human fibroblasts and salivary adenocarcinoma cells in culture.[10,11] EGF may be considered, at least in some aspects, as a hormone because it circulates with the blood and can act at far distant sites.[12] The normal serum levels of EGF in the mouse are of approximately 1 ng/ml (1.7×10^{-10} *M*). These levels are regulated by hormones such as thyroxine and testosterone and may show variation according to sex and stage of development.[13,14]

In addition to blood, EGF is present in various body fluids, including urine, milk, colostrum, saliva, seminal fluid, pancreatic juice, cerebrospinal fluid, and amniotic fluid.[5] Urinary EGF is not derived from blood but may be originated in the kidney. A clear peak of urinary EGF excretion is observed in the second year of life, but the physiological meaning of this peak is unknown.[15] High molecular weight forms of EGF are present in only trace quantities in normal urine but may be markedly elevated in the urine of some cancer patients.[16] Intact EGF, EGF fragments, and EGF-related peptides have been isolated from human urine as well as from the urine of normal and tumor-bearing mice.[17,18]

The precise physiological role of EGF has not been defined, but EGF can either stimulate or inhibit the proliferation and/or differentiation of a wide variety of cells, depending on the cell type and its state of differentiation. In addition, EGF may be involved in the regulation of important metabolic pathways. The role of EGF in oncogenic processes is very complex and is poorly understood at present.[19] Studies with malignant cells cultured *in vitro* suggest that EGF may have, in a manner similar to TGF-β, a dual role in the growth of tumor cells, being capable of exerting both growth stimulating and growth inhibiting effects.[20] The effects of EGF on tumor growth *in vivo* are little known.

II. EGF STRUCTURE AND EGF-RELATED PROTEINS

Mouse EGF is a 6045-Da single-chain polypeptide of 53 amino acids with three disulfide bonds.[21] Human EGF, also called urogastrone because it is a potent inhibitor of gastric acid secretion, has been isolated from human urine and its amino acid sequence is similar to that of mouse EGF, differing in 16 positions.[22,23] Mouse and human EGF are biologically equipotent and both bind to human fibroblast receptors with similar affinities.[24,25] The three-dimensional structural arrangements of mouse and rat EGF are different. Whereas two of the eight aromatic residues of mouse EGF are internalized in the protein, rat EGF is folded in such a way that all of its hydrophobic aromatic residues are solvent-exposed.[26] In spite of these differences, mouse and rat EGF have similar structural receptor-binding regions, and rat EGF competes for binding to the same cellular receptor and elicits the same cellular responses as does mouse EGF. Mouse EGF was synthesized by the solid-phase method.[27] The synthetic polypeptide is identical to the natural material in amino acid composition, chromatographic behavior, receptor binding, and stimulation of DNA synthesis. Synthetic fragments of different parts of the EGF molecule showed diminished biological activity and some fragments were completely inactive, but the study of such fragments may be important for establishing structure-function relationships of the EGF molecule. Aromaticity at position 37 of human EGF is not obligatory for activity.[28] Two forms of mouse EGF have been detected by liquid chromatography, α-EGF and β-EGF, and it has been shown that β-EGF represents the des-asparaginyl-1 form of the polypeptide and that both forms are essentially equipotent as mitogens.[29] Mouse and human EGF seem to derive from a primitive protein related to certain blood coagulation factors.[30]

A. AMPHIREGULIN

A protein exhibiting striking structural and functional homology with EGF is amphiregulin. Human amphiregulin is a glycosylated polypeptide of 84 amino acids which was initially purified from the conditioned medium of the human breast carcinoma cell line MCF-7.[30-34] The 40-amino acid carboxyl-terminal segment shows significant homology with both EGF and transforming growth factor-α (TGF-α).

Amphiregulin partially competes with EGF for binding to the EGF receptor. However, in human epithelial cells the action of amphiregulin may be solely mediated by its interaction with the EGF receptor.[35] Amphiregulin has been detected in the nucleus of epithelial cells, but its possible function in this location is unknown.[36]

Amphiregulin is found in the human placenta and ovaries. Normal human keratinocytes produce a growth factor, the keratinocyte-derived growth factor (KAF or KDGF), which is identical with amphiregulin.[37] Human keratinocytes and mammary epithelial cell cultures express amphiregulin mRNA, but this RNA is expressed at low levels in the normal human epidermis. Markedly elevated levels of amphiregulin gene transcripts are present in benign hyperproliferative skin epithelial lesions (psoriasis) as well as in malignant epithelial lesions such as colon, gastric, and mammary carcinoma.[38] There is evidence that amphiregulin may act as an autocrine growth stimulator for colonic carcinoma cells and may perform a growth regulatory function in normal and malignant colon *in vivo*.[39] In general, amphiregulin may be considered as a bifunctional growth modulator which can promote the growth of normal epithelial cells and inhibits the growth of some carcinoma cells. However, the precise physiological role of amphiregulin remains to be elucidated.

B. OTHER MAMMALIAN EGF-RELATED GROWTH FACTORS

Factors immunologically related to EGF are produced by different types of cells but the biological significance of these factors is unknown. Fetal mouse salivary mesenchymal cells secrete a protein of 15 kDa that is immunologically related to EGF.[40] The EGF-like protein contributes to the anchorage-independent growth of normal rat kidney (NRK) cells in soft agarose, suggesting that its structure and function are similar to those of TGFs. An EGF-like immunoreactive material present in platelets is released in high amounts during blood coagulation.[41] A 67-kDa molecule is the main storage form of EGF in human platelets, and at least two independent mechanisms exist for its release.[42] EGF-like mitogens are produced and secreted by cultured cells from bovine pituitary glands as well as by L929 mouse fibroblasts cultured in serum-free medium.[43,44] A factor called keratinocyte growth factor, isolated from term human placenta, is distinct from EGF and is more active than EGF in stimulating [^{3}H]thymidine incorporation and protein synthesis in cultured human keratinocytes.[45] The keratinocyte growth factors are structurally related to fibroblast growth factors (FGFs). A mitogen called heparin-binding EGF-like growth factor (HB-EGF) was purified from the conditioned medium of the human monocyte/macrophage-like cell line U-937.[46] The HB-EGF secreted by U-937 cells contains at least 86 amino acids and its carboxyl-terminal half shares approximately 40% sequence identity with human EGF.[47] The amino-terminal part of HB-EGF contains a highly hydrophilic stretch of amino acid residues that may represent a heparin-binding region. HB-EGF may serve as a receptor mediating the binding and uptake of diphtheria toxin into monkey cells.[48] Characterization of rat and mouse cDNA clones encoding HB-EGF showed that the factor is expressed in a number of tissues, including lung, skeletal muscle, brain, and heart.[49] HB-EGF is derived from a transmembrane precursor and is a potent mitogen for smooth muscle cells, fibroblasts, and keratinocytes.

C. EGF-RELATED GENES IN INVERTEBRATES

A cDNA clone, termed uEGF-1, isolated from the sea urchin *Strongylocentrotus purpuratus,* recognizes transcripts that are preferentially expressed in embryonic ectoderm.[50] Sequencing of the cDNA revealed that it can encode a protein with strong homology to EGF and other EGF-related proteins. The homology is restricted to a 38-amino acid cysteine-rich repeat to these proteins, which include vaccinia virus growth factor (VVGF), TGF-α, tissue-type plasminogen activator, mammalian clotting factors IX and X, low-density lipoprotein (LDL) receptor, and the *Notch* product of *Drosophila melanogaster.* The temporal pattern of expression of transcripts recognized by the uEGF-1 cDNA clone is consistent with a function of the gene product during early sea urchin development.

Significant structural homologies exist between EGF, TGF-α, and the homeotic neurogenic locus *Notch* locus of *Drosophila.*[51] *Notch* encodes a protein that exhibits homology to EGF and mammalian proteins containing EGF-like repeated sequences.[52] In addition to *Notch,* other neurogenic genes of the insect exhibit homology to EGF but the physiological role of these genes is unknown.[53] The *Drosophila* gene *Serrate (Ser)* encodes a membrane protein with extracellular domain containing two cysteine-rich regions, one of which is organized in a tandem array of 14 EGF-like repeats.[54] The *Ser* gene has a complex pattern of expression during the development of *Drosophila.* Another *Drosophila* gene, *sos,* functions in signaling pathways that are initiated by the Sev and EGF receptor tyrosine kinases; and its product is similar to the CDC25 protein of the yeast *Saccharomyces cerevisiae,* which is a Ras-activating factor involved in guanine nucleotide exchange.[55] Ras proteins have an important role in EGF signal transduction in vertebrates.

The product of *lin-12,* a homeotic gene controlling certain binary decisions during the development of the nematode *Caenorhabditis elegans* is homologous to a set of mammalian proteins that includes EGF.[56] The *let-23* gene required for vulva induction in the nematode encodes a receptor-like protein with tyrosine kinase activity which is a member of the EGF receptor family.[57] The normal ligand of this receptor is unknown but an interaction between the ligand and the receptor would be required for vulval development in the nematode. The *let-23* gene product acts upstream of the *let-60 ras* gene, suggesting the existence of a link between a tyrosine kinase receptor and Ras proteins in a pathway related to cell-type determination.

D. EGF-RELATED SEQUENCES IN POXVIRUSES

A 140-amino acid polypeptide encoded by one of the early genes of a cytolytic poxvirus, the vaccinia virus (VV), is closely related to both EGF and TGF-α and may be able to bind the EGF cellular receptor and act as a growth factor.[58-60] The factor, called vaccinia virus growth factor (VVGF), is found in the medium of vaccinia virus-infected cells, is recognized by antibodies to mouse EGF, and is capable of stimulating autophosphorylation of the EGF receptor.[61,62] In this way, the VV protein could mediate binding of the virus to the surface of cells expressing EGF receptors and could also stimulate growth of the infected cells and/or the neighboring cells. Interestingly, EGF receptor occupancy inhibits vaccinia virus infection.[63] Antibodies against the EGF receptor reduce the ability of VV to infect cells productively, indicating that the virus binds to cells via the EGF receptor.[64] The VVGF gene is present twice, once at each end of the VV genome within the inverted terminal repeat. A constructed VV variant lacking more than half of VVGF genes has an attenuated phenotype following intracranial and intradermal inoculations into mice and rabbits, respectively.[65] Thus, expression of the VVGF gene is important to the virulence of VV.

Another member of the poxvirus family is the Shope fibroma virus (SFV), which induces benign fibromas in adult rabbits and invasive atypical fibrosarcomas in newborn rabbits and immunosuppressed adult rabbits.[66] Interestingly, SFV also encodes a gene product, the Shope fibroma virus growth factor (SFGF), which exhibits significant homology with EGF and TGF-α.[67] A 55-amino acid peptide, comprising the carboxyl portion (residues 26 to 80) of SFGF, shares the biological activities of EGF in mouse cells both *in vitro* and *in vivo*.[68,69] The myxoma virus (MV), which is a poxvirus of the leporipoxvirus subgroup involved in the etiology of myxomatosis, encodes an 85-amino acid EGF-like growth factor, the myxoma virus growth factor (MVGF).[70] A recombinant virus between SFV and MV, the malignant rabbit fibroma virus (MRFV), also elaborates a factor that is capable of competing effectively with EGF; this induces an inhibition of EGF-stimulated EGF binding to its receptor, which results in inhibition of EGF-stimulated proliferation on susceptible target cells.[71] The homologies existing between poxviruses and EGF suggest that the capture of cellular genes by viruses (especially genes related to peptide growth factors), is not limited to retroviruses, where they are transduced as viral oncogenes, but may also occur with DNA viruses and may be more common than it has been suspected until now.

III. THE EGF GENE AND EGF BIOSYNTHESIS

The murine gene encoding EGF was assigned to chromosome 3 by interspecific somatic cell hybridization analysis.[72] The 5′-flanking region of the EGF gene was isolated from a mouse cDNA library, and its nucleotide sequence was determined.[73] Sequence comparisons indicated regions of the gene which may be involved in hormonal regulation of its expression.

The human gene for EGF is located on chromosome 4, at region 4q25-q27.[74] The human gene has been cloned and expressed in *Escherichia coli,* which resulted in the synthesis and secretion of human EGF by the manipulated bacteria.[75] The human EGF gene was chemically synthesized by phosphite-coupling procedures; and its expression in yeast yielded a single-chain polypeptide of 53 amino acids that was shown to induce biological effects characteristic of EGF action, including promotion of epithelial cell proliferation and inhibition of gastric acid secretion.[76] Studies with synthetic peptides corresponding to fragments of the amino acid EGF sequence showed that residues 20 to 31 define a primary receptor-binding site of EGF, whereas the amino- and carboxyl-terminal regions of the EGF molecule would provide the necessary conformational stability for binding of the middle region.[77] Human EGF produced in genetically manipulated *E. coli* is identical to native human EGF in primary sequence, receptor binding, and stimulation of EGF receptor kinase activity.[78] Selective qualitative alterations introduced in the EGF protein synthesized in bacteria may be useful for structure-function analysis.

A. THE EGF PRECURSOR

EGF is biosynthesized in mammals from a large precursor.[79-81] The EGF mRNA contains about 4,750 nucleotides and predicts the synthesis of an EGF precursor (prepro-EGF) of 1,217 amino acids (130,000 Da). The amino-terminal segment of the EGF precursor contains seven peptides with sequences similar but not identical to EGF. The polypeptides p788 and p789, which are considered as reliable markers for transformation of human fibroblasts, show an amino acid composition which is remarkably similar to that of residues 630 to 680 of the EGF precursor.[82] Mouse National Institutes of Health (NIH)/3T3 fibroblasts transfected with an expression vector carrying the human EGF precursor gene synthesize a 150- to 180-kDa EGF precursor polypeptide which is membrane associated and glycosylated, and is partially secreted into the conditioned medium.[83]

The levels of the EGF precursor protein in the mouse kidney are unexpectedly high, being only twofold less than in the mouse submaxillary gland in spite of the fact that the kidney contains less mature EGF than the submaxillary gland.[81] The relatively low levels of EGF in the kidney may reflect differences between the processing of the precursor in this tissue and the submaxillary gland. In the kidney, the highest amounts of EGF precursor are found in the distal tubules, which are involved in the fine regulation of urine composition. The function of the EGF precursor at this site is unknown. Pro-EGF and high molecular weight EGF-like peptides are present in adult mouse urine.[84]

B. STRUCTURAL HOMOLOGIES OF THE EGF PRECURSOR

Homology has been detected between a portion of prepro-EGF (amino acid residues 63 to 880) and a 26-kDa polypeptide, termed p788, which is considered as a reliable marker for neoplastic transformation in human fibroblasts.[82] After carcinogen-induced transformation, the rate of p788 synthesis is markedly elevated. However, elevated synthesis of p788 and a related polypeptide, p789, appears to be unrelated to the degree of tumorigenicity of transformed cell lines.

Protein C, a vitamin K-dependent plasma protein that exerts a regulatory function in blood coagulation, has two domains that exhibit homology to the EGF precursor.[85] A Ca^{2+}-binding site is present in the EGF homology region of protein C.

Another interesting sequence homology exists between the mouse EGF precursor and a region of the bovine LDL receptor, suggesting a common ancestor gene for both proteins.[86] Deletion analysis has shown that one function of the EGF precursor-like region in the LDL receptor is to accelerate acid-dependent ligand dissociation and thereby to facilitate receptor recycling.[87]

Structural homology has been observed between the EGF precursor and a peptide sequence encoded by the second exon of the atrial natriuretic factor (ANF) precursor.[88] ANF, a polypeptide hormone produced in the heart (cardiac atrium), plays an important role in regulating blood pressure and extracellular fluid volume.

The mouse EGF precursor protein contains three regions of sequence homology with the v-Mos oncoprotein and the normal c-Mos protein.[89] Similarity is greatest between the carboxyl-terminal region of the v-Mos protein and a part of the cytoplasmic domain of the EGF precursor. Similarities are also observed between two regions of the murine c-Mos protein sequence and parts of the extracellular domain of the EGF precursor.[89] These structural relationships suggest that the gene of the EGF precursor and the c-*mos* proto-oncogene may have evolved from a common ancestor. However, the evolutionary and biological significance of structural homologies existing between the EGF precursor and other proteins remain to be determined.

IV. PRODUCTION AND BIOLOGICAL EFFECTS OF EGF

Both the sites of production and the biological actions of EGF or EGF-related peptides show an extremely wide diversity, and many questions yet remain to be answered in relation to these important aspects. Different types of cells may show specific or unspecific responses to EGF.

A. MITOGENIC AND DEVELOPMENTAL ACTIONS OF EGF

Cellular proliferation and differentiation during developmental processes implicate an intimate interaction between the cells and the extracellular environment; and many growth factors, including EGF, may be involved in modulating these complex interactions.[90] EGF-like activity is present, for example, in the developing chick embryo.[91] Very little EGF activity is detectable in the chick embryo prior to day 8, a peak of activity appears over days 10 and 12, the activity falls to undetectable levels during days 14 to 17, and a later rise is observed over days 18 to 20 of embryonic life (hatching is at day 21). There is no information on the site of synthesis or distribution of the chick embryonic EGF, and it is also not known whether the observed changes in EGF activity represent an endocrine or a paracrine type of EGF secretion. The physiological role of EGF-like activity in the developing chick embryo is not understood, but there is evidence that EGF promotes chick embryonic angiogenesis.[92]

For an unknown reason, extremely high concentrations of EGF are found in the submaxillary (submandibular) glands of adult male mice.[93,94] Expression of the EGF gene may be controlled in a different manner in the mouse kidney and the submaxillary gland. Whereas submaxillary EGF mRNA levels are about 16-fold higher in the mouse male than in the female, the kidney levels of this RNA are 2- to 4-fold higher in the mouse female than in the male.[95] Moreover, renal EGF mRNA concentrations are less responsive to hormones than those in the submaxillary gland. The production of EGF is decreased in old mice but it can be induced by treatment with T3 or androgen to the same levels found in young mice.[96] The production of EGF is decreased in the submaxillary gland of diabetic mice but it can be restored to normal by insulin treatment.[97] Sialoadenectomy in mice results in a rapid and marked decrease in plasma EGF levels; and this is accompanied by a decrease in sperm production, which may be completely reversed by EGF replacement.[98] EGF may thus play a role in male reproductive function by stimulating the meiotic phase of spermatogenesis.

EGF may act as a potent mitogen and anabolic agent for a variety of tissues of ectodermal and endodermal origin, being involved in wound healing.[99] Topical application of biosynthetic EGF accelerates the rate of partial-thickness wounds in humans.[100] Stimulation of wound healing and collagen production by EGF is due to increased fibroblast proliferation, and not to increased expression of type I and type III procollagen genes.[101]

EGF may have an important role in embryogenic processes. Human and murine EGF, as well as TGF-α, are active in promoting eyelid opening in newborn mice.[102] In the mouse, EGF is involved in the control of secondary palate formation by a developmentally regulated process which includes a quantitative modulation of glycosaminglycan synthesis.[103] In the rabbit, EGF is involved in the development of the gastrointestinal tract and is capable of exerting a trophic effect on stomach, small intestine, and pancreas.[104] In the neonatal period of the rabbit, EGF can promote hepatic growth and maturation; exogenous EGF is capable of inducing precocious maturation of the liver, and, depending on the route of administration, of stimulating liver growth.[105]

In certain organs and tissues the mitogenic action of EGF may be supplemented, or perhaps replaced, by other mitogenic substances. Gastrin-releasing peptide (GRP), which is analogous to bombesin (a tetradecapeptide discovered in amphibian skin), is particularly abundant in the human fetal lung where it is produced in neuroendocrine cells and may have important mitogenic effects which are independent on the presence of EGF.[106] In other tissues the growth-promoting effects of EGF may be counterbalanced by the action of growth inhibitors. For example, the potent mitogenic effect of EGF in cultured rat aortic smooth muscle cells is inhibited by TGF-β.[107] Alterations in the delicate balance between the opposite effects of EGF and TGF-β on the proliferation of vascular smooth muscle cells may be important in atherogenesis.

Old animals exhibit a poor mitogenic response to EGF. The responsiveness of cultured hepatocytes from 22- to 24-month-old rats to EGF is diminished in relation to DNA synthesis, as compared to 2- or 3-month-old rat hepatocytes.[108] Interestingly, c-*myc* proto-oncogene expression is induced by EGF to a similar extent in the liver cells from both old and young animals. These results suggest that the old cells maintain their response to EGF at the transcriptional level, but are blocked from entering the S phase of the cycle. Furthermore, the results show that expression of c-*myc* is not sufficient by itself to stimulate DNA synthesis.

B. METABOLIC AND PHYSIOLOGIC ACTIONS OF EGF

In addition to its mitogenic and developmental effects, EGF displays a wide diversity of important metabolic and physiological activities.[109] These activities include stimulation of ion transport, enhancement of endogenous protein phosphorylation, alterations in cell morphology, and stimulation of DNA synthesis.

EGF displays, in addition to its stimulating actions, a variety of inhibitory effects, including a marked inhibition of gastric acid secretion; and for this reason EGF purified from human urine was termed urogastrone. EGF exerts an influence on the contractile function of the stomach, which depends in part on the production of prostaglandins and the presence of extracellular calcium.[110] A similar influence is exerted by EGF on the smooth muscle system of other organs such as the trachea.[111]

The mouse liver contains a very high number of EGF receptors; and EGF may exert important effects in this organ, especially in the control of hepatocyte homeostasis, glucose metabolism, and gluconeogenesis.[112] EGF may have a role in the regulation of glycogen synthase, a key enzyme involved in glycogen metabolism, through phosphorylation and dephosphorylation mechanisms.[113] It has been suggested that EGF mimics the action of insulin in the liver, but this hypothesis is misleading. In hepatocytes from starved rats, EGF stimulates the incorporation of piruvate into glucose, but does not augment either the incorporation of glucose into glycogen or the oxidation of glucose to CO_2.[114] EGF can also regulate protein metabolism, including protein breakdown in cell systems such as the human epidermoid carcinoma cell line A431.[115]

EGF has striking effects on vascular functions, especially on arterial contractility.[116] Binding sites for EGF are present in the tunica media of coronary arteries, supporting the hypothesis that EGF plays a role in vascular function by acting directly on smooth muscle cells of arterial tissue.[117] The synthesis of elastin is inhibited by EGF in chick aortic smooth muscle cells.[118] Studies in anesthetized dogs demonstrated a potent vasodilatory effect of EGF which may result from its direct action on arterioles, suggesting a potential role for EGF (and possibly also for TGF-α) in the regulation of blood flow *in vivo*.[119,120] These effects of EGF are probably mediated by prostaglandins and are specifically abolished by indomethacin.

EGF has important effects on the kidney. Both EGF and TGF-α (which recognizes the EGF receptor) are required for kidney tubulogenesis in matrigel cultures maintained in serum-free medium.[121] EGF stimulates prostaglandin synthesis in cultured kidney cells (Madin-Darby canine kidney [MDCK] cells).[122] Brief contact of EGF with the porcine renal epithelial cell line LLC-PK_1 results in a mitogenic effect.[123] LLC-PK_1 cells are polar cells, and exposure of the basolateral membrane of these cells to EGF is necessary for mitogenesis; exposure of the apical surface to EGF is ineffective.

EGF stimulates bone resorption in neonatal mouse calvaria in organ culture via a prostaglandin-mediated mechanism.[124] Both EGF and TGF-α stimulate bone resorption by increasing the proliferation of osteoclast precursors, which leads to an increased numbers of osteoclasts.[125]

EGF has important effects on the production and/or action of different hormones. Synthesis of hormones induced by EGF may occur by different mechanisms, involving or not involving new protein synthesis. Prolactin mRNA expression induced by EGF in GH4C1 rat pituitary tumor cells occurs via a protein synthesis-dependent pathway, as indicated by the abolishing effects of inhibitors such as cycloheximide and puromycin, suggesting a requirement for the synthesis of an intermediary regulatory protein.[126] In contrast, prolactin synthesis induced in GH4C1 cells by thyrotropin-releasing hormone (TRH), 12-*O*-tetradecanoylphorbol-13-acetate (TPA), forskolin, or dibutyryl cyclic adenosine-3′,5′-monophosphate (cAMP) is independent of new protein synthesis. The action of thyroid hormone in different types of cells is modulated by EGF. Low concentrations of EGF decrease nuclear thyroid hormone receptors and thyroid hormone responses in GH4C1 cells, including the stimulation of growth hormone synthesis.[127] EGF inhibits radioiodine uptake but stimulates DNA synthesis in newborn rat thyroids grown in nude mice.[128] In primary cultures of dog and bovine thyroid cells, both TSH and EGF stimulate thyroid cell proliferation; however, while TSH maintains thyroglobulin mRNA levels close to their starting value, EGF inhibits the expression of the thyroglobulin gene and renders thyroglobulin mRNA levels undetectable.[129] EGF and TRH act similarly on a clonal pituitary cell strain, modulating hormone production and inhibiting cell proliferation.[130] Production of EGF in the mouse thyroid gland is increased by thyroxine administration *in vivo*.[131]

EGF may have important effects on gonadal functions. According to the predominant physiological conditions, EGF may have either stimulatory or inhibitory effects on steroidogenic processes occurring in freshly cultured Leydig cells from the testis of prepubertal and adult rats and mice.[132] The effects of EGF on Leydig cells are partially dependent on the presence of luteinizing hormone (LH) and are exerted through mechanisms that are apparently different from those of LHRH and the Sertoli cell factor (SCF), a paracrine factor produced by Sertoli cells.

EGF may be involved in ovarian folliculogenesis, influencing the growth and differentiation of rat granulosa cells.[133] EGF and the EGF receptor are expressed in the follicular and stromal cells of the human ovary, and there marked changes occur in their levels during the course of follicular growth and regression.[134] Although EGF is a potent mitogen for granulosa cells, EGF attenuates follicle-stimulating hormone (FSH)-mediated differentiation of these cells, suggesting that it promotes the proliferation of granulosa cells at the expense of their differentiation.[135] The proliferation of granulosa cells induced by

EGF is associated with the expression of FSH-binding sites. EGF can inhibit human chorionic gonadotropin (hCG)-induced estradiol and progesterone secretion and ovulation in the perfused rabbit ovary, which may be due to attenuation of cell-to-cell communication in the ovarian response to hCG.[136] TGF-β enhances the induction of EGF receptors by FSH and augments the inhibitory actions of EGF during granulosa cell differentiation.[137] Granulosa cells are less able to produce cAMP and cAMP-induced LH receptors in the presence of EGF and TGF-β *in vitro*. The precise role of TGF-β, EGF, and other growth factors during granulosa cell differentiation *in vivo* is unknown.

Receptors for EGF are present in smooth muscle-containing tissue from vascular and nonvascular origin, and EGF can modulate the contractility of isolated arterial strips and nonvascular smooth muscle-containing preparations. There is evidence that EGF is involved in the regulation of ocular ciliary muscle tension. This effect, as studied *in vitro,* is concentration dependent and does not require extracellular calcium.[138]

EGF immunoreactive material has been detected in the central nervous system of the rat, especially in the forebrain and midbrain structures of pallidal areas of the brain, where it could act as either a neurotransmitter or neuromodulator.[139] EGF may possibly act as a mitogenic agent during development of the central nervous system. In addition, EGF may act as a neurite elongation and maintenance factor for particular neurons of the central nervous system.[140] Highly purified mouse EGF enhances the survival and process outgrowth in neurons derived from the subneocortical telencephalon of neonatal rat brain.

EGF has important physiological effects in the placenta and the mammary gland. Secretion of mouse placental lactogen I *in vitro* is stimulated by EGF, whereas EGF inhibits the secretion of placental lactogen II.[141] EGF causes morphological differentiation, but not proliferation, of human trophoblast cells in culture.[142] The EGF-induced differentiation of trophoblasts results in increased secretion of hCG and placental lactogen by the syncytial cells.

EGF may have a role in the development and function of the mammary gland during puberty and pregnancy.[143] The effects of EGF on the development of the mammary gland have been shown *in vivo* by the use of slow-releasing cholesterol-based EGF pellets inserted directly into the gland.[144] TGF-α has effects similar to those of EGF in this system, which requires the systemic presence of estrogen and progesterone in the animal. Milk production partially depends on the action of EGF. Lactogenic hormones increase the content of EGF mRNA in mouse mammary glands.[145] In HC11 cells, which are normal mammary epithelial cells of murine origin that maintain differentiation-specific functions, activation of the EGF and c-Erb-B2 receptor kinases (but not the PDGF receptor kinase) allows the cells to respond optimally to lactogenic hormones such as glucocorticoids and prolactin.[146] EGF and/or EGF-like proteins are present in normal human and bovine milk.[147,148] A polypeptide present in human milk is identical to EGF and to urogastrone from human urine.[149] Milk EGF is selectively transported across the ileal epithelium in suckling rats and may display local effects through binding to specific receptors present in the intestine.[150] The possibility should be considered that EGF ingested with breast milk can be absorbed, at least in part, from the digestive tract into the circulation to affect other tissues of the newborn human or rodent.

The importance of EGF in the growth and function of the mammary gland is indicated by the fact that sialoadenectomy of mice decreases milk production and increases offspring mortality during the lactation period.[151] EGF administration to sialoadenectomized pregnant mice increases survival rate of offspring to almost a normal level. Since the mouse submaxillary gland is a rich source of EGF, these results are more easily explained by assuming that the gland has an endocrine function associated with EGF secretion. Retinoic acid enhances the mitogenic effect of EGF in mouse mammary gland in culture, which may occur through an increased binding of EGF to mammary cells.[152]

It has been suggested that EGF may inhibit hormone-induced milk protein synthesis by the mammary gland, including the synthesis of casein and the α-lactalbumin activity. The inhibitory effect of EGF on α-lactalbumin activity is enhanced by retinoic acid in a dose-dependent manner.[152] However, these inhibitory effects of EGF may be artifactual responses due to the presence of insulin in the culture medium at unphysiologically high concentrations.[153]

C. REGULATION OF EGF PRODUCTION

The factors normally involved in the regulation of EGF production and secretion by different organs and tissues remain little characterized. Heterologous regulation of the production of EGF by other hormones and growth factors may be of great importance and may operate in different manners according to the different organs and tissues. In the mouse submandibular gland of hypothyroid mice, thyroid hormone increases EGF gene expression without affecting cellular proliferation.[154] In the mouse uterus, expression of the EGF gene is restricted to the luminal and glandular epithelia and is upregulated by estrogens.[155] EGF receptors in the uterus are expressed before the synthesis of EGF. The functions of EGF in the uterus are unknown but EGF may have an important role in the early stages of pregnancy.

V. THE EGF RECEPTOR

The physiological actions of EGF are initiated by its binding to the EGF receptor located on the cell surface.[156-165] This binding stimulates the autophosphorylation of the receptor on tyrosine residues and the intrinsic tyrosine-specific kinase activity of the receptor. The activated EGF receptor protein kinase has affinity for a diversity of cellular protein substrates. The physiological effects of EGF, including mitogenesis, are tightly associated with the specific kinase activity of the EGF receptor. Inhibition of the specific kinase activity of the EGF receptor by compounds derived from erbstatin (a substance isolated from the medium of an actinomycete) results in inhibition of EGF-dependent proliferation of EGF-sensitive cells.[166,167]

The membrane receptor for EGF has been purified from different sources including human placenta[168-170] and mouse liver, as well as from the human epidermoid carcinoma cell line A431, which expresses a very high number of EGF receptors. Paradoxically, the proliferation of A431 cells is inhibited by EGF.[171] Monoclonal antibodies to EGF receptors have been extensively used for the isolation and purification of EGF receptors from different types of cells.[172-178] The majority of these antibodies recognize oligosaccharide determinants in the EGF receptor molecule. High-affinity EGF binding is specifically reduced by a monoclonal antibody, and appears necessary for early responses.[179] One antibody of the immunoglobulin (Ig)M type inhibits EGF binding to its receptor and mimics the biological effects of EGF.[172] Only one EGF-binding site is contained in each EGF receptor protein molecule.[180] Local aggregation of EGF receptor complexes is required for receptor activation by EGF.[177,181] High-affinity EGF receptors on the surface of A431 cells are preferentially localized in the vicinity of membrane-associated cytoskeletal filaments.[182,183] EGF induces an association of the receptor to the cytoskeleton in A431 cells.[184] Interaction of ligand-activated EGF receptors with the cytoskeleton is related to receptor clustering.[185]

A. THE EGF RECEPTOR GENE

The EGF receptor in human cells is encoded by a gene located on chromosome 7, at region 7p12-p14.[186,187] This gene is expressed in different types of human cell lines.[188] The complete nucleotide sequence of a cDNA from the normal human EGF receptor gene has been determined, which allowed deduction of the primary structure of the human EGF receptor protein.[187] The primary structure of the chicken EGF receptor protein was also deduced from the nucleotide sequence of a cDNA corresponding to the chicken EGF receptor.[189] The mouse EGF receptor gene resides on chromosome 11, which also contains the loci for α-globin, colony-stimulating factor 2 (CSF-2), and interleukin 3 (IL-3).[190] The possible biological significance of these associations is unknown. The EGF receptor gene and the c-*erb*-B1 proto-oncogene are identical genes.[191,192]

1. Amplification and Rearrangement of the EGF Receptor Gene

Amplification, rearrangement, and enhanced expression of the EGF receptor gene occur in A431 cells.[193] These cells may contain 15 to 25 copies of the EGF receptor gene, and cDNA copies of the EGF receptor gene have been obtained from them.[194-196] This cDNA is homologous to a variety of RNAs which are overproduced in A431 cells,[197] but it is not known whether these RNAs are translated, and whether overproduction of some particular species of EGF receptor-related RNAs and/or proteins may be associated with the appearance of a malignant phenotype in A431 cells and perhaps also in other cells. The mechanisms responsible for generation of EGF receptor-related transcripts in A431 cells are unknown but could be associated with structural genomic changes or with alterations in RNA transcription or RNA-processing events. Aberrant EGF receptor-coding RNAs may be created in A431 cells by gene rearrangement within chromosome 7, resulting in a fusion of the 5′ portion of the EGF receptor gene to an unidentified region of genomic DNA.[198]

2. The Promoter Region of the EGF Receptor Gene

The 5′-flanking promoter region of the human EGF receptor gene has some unusual characteristics. In contrast to most other eukaryotic genes, this promoter contains neither a TATA box nor a CAAT box, but has an extremely high G+C content and contains five CCGCCC repeats and four (TCC)TCCTCCTCC repeats. This region is situated close to, or within, a DNase I-hypersensitive site in human A431 cells.[199] The promoter region of the EGF receptor shows a striking similarity with the promoter of the human c-H-*ras* proto-oncogene, and this similarity may be relevant to the molecular mechanisms by which the expression of such growth control genes is regulated.[200] An S1 nuclease-sensitive site is located 80 to 100 bp upstream from the major transcription initiation site of the EGF receptor gene.[201] Generally, differences in sensitivity to nucleases are associated with alterations in chromatin structure and consequent changes in gene expression. The existence of the S1 nuclease-sensitive site within the promoter region of the EGF receptor gene suggests that nuclear *trans*-acting factors interact with this site to stimulate transcription of the gene. The transcription factor Sp1, which has a molecular weight of approximately 500 kDa, and other DNA-binding proteins are required for the optimal expression of the EGF receptor gene in A431 cells.[202] In addition, transcription of the EGF receptor is stimulated by a 270-kDa protein which exhibits a weak or indirect interaction with the DNA template. A 36-bp element in the TATA-less, G+C-rich 5′ region of the EGF receptor gene mediates inductive responses to EGF, phenylmercuric acetate (PMA), and cAMP.[203] This element displays many characteristics of an enhancer, acting in an orientation-independent manner. A 50-kDa nuclear protein present in HeLa cells binds with high affinity and specificity to this element.

Restriction fragment length polymorphisms (RFLPs) have been found in the EGF receptor gene. Examination of the EGF receptor gene from human normal and tumor tissue samples for structural alterations by digestion with *Hin*dIII restriction endonuclease and Southern blot analysis showed the existence of multiple RFLPs.[204] Rearrangement and amplification of the EGF receptor gene occurred infrequently in the tissues examined, either normal or neoplastic.

B. STRUCTURE AND FUNCTION OF THE EGF RECEPTOR

The EGF receptor is a transmembrane glycoprotein of 175 kDa consisting of a single polypeptide chain of 1186 amino acids and N-linked carbohydrates. As deduced from a cDNA encoding the EGF receptor precursor, the receptor molecule can be divided into two functional domains: an extracellular domain of 621 amino acids containing the hormone-binding site and a cytoplasmic domain of 542 amino acids containing the sequence related to tyrosine kinase activity.[196] The two domains are linked by a transmembrane region constituted by a stretch of 23 predominantly hydrophobic amino acids. A striking feature of the EGF receptor is that the cytoplasmic domain contains 9 cysteine residues, a value well within the range of most cytoplasmic proteins, but the extracellular domain contains 51 cysteine residues. Most of these extracellular cysteines are concentrated within two regions of approximately 170 amino acids, and there is evidence indicating the presence of sulfhydryl groups in the EGF receptor.[205] The purified EGF receptor can be reconstituted into artificial phospholipid bilayers by using a detergent-dialysis method.[206] In this system the EGF receptor is uniformly oriented within the bilayer, with the EGF-binding domain facing outside the liposomes, and the incorporated receptor is functional in binding EGF and a monoclonal antireceptor antibody.

1. Functional Domains of the EGF Receptor

A model for the ATP-binding site of the EGF receptor and other structurally related proteins (the v-*erb*-B and c-*src* oncogene products, mammalian cyclic AMP-dependent protein kinase, and the cell division protein CDC28) has been proposed on the basis of the conservation of certain amino acid residues.[207] The amino acid residues corresponding to the nucleotide-binding site of the EGF receptor have been identified.[208] Monoclonal antibodies to the EGF receptor present in A431 cells act as inhibitors of EGF binding to its receptor on the cell surface and are antagonists of EGF-stimulated tyrosine kinase activity.[209] A monoclonal antibody to the EGF receptor from A431 cells may function as a noncompetitive agonist of EGF action.[177]

The extracellular portion of the EGF receptor protein comprises four domains (I to IV); and domain III, which is flanked by the two cysteine-rich domains II and IV, would contribute most of the interactions that define binding specificity for EGF.[210] Binding of EGF to the extracellular domain of the receptor not only relieves an inhibitory constraint but also induces a conformation optimal for tyrosine kinase activity.[211] Full activity appears to require a Mn^{2+}-binding site within the tyrosine kinase core domain. Interaction of EGF with its receptor would also require the formation of a ternary complex involving a different cell surface molecule.[212]

Functional studies have been made with the antibody Ab 2913, which binds the intracellular kinase portion of the EGF receptor.[213] The antibody Ab 2913 is able to activate tyrosine kinase in A431 cell membrane preparations, stimulating receptor autophosphorylation as well as the phosphorylation of exogenous substrates. In a manner similar to EGF, Ab 2913 causes receptor autophosphorylation on tyrosine residues only, and tryptic peptide maps suggest that the antibody and EGF stimulate phosphorylation of the same amino acid residues. The EGF receptor kinase is degraded in conjunction with the external portion of the receptor molecule in lysosomes, and the kinase fragment is not translocated to the nucleus.[213] Ab 2913 recognizes the EGF receptor in a diversity of cell lines and is able to immunoprecipitate the v-Erb-B oncoprotein from avian erythroblastosis virus (AEV)-infected chicken fibroblasts.

Interaction of the EGF receptor with its cellular substrates is effected through specific amino acid sequences contained in the carboxyl-terminal region of the receptor molecule. Two known substrates of the EGF receptor, phospholipase C-γ and the GTPase-activating protein (GAP) of the Ras effector loop, bind to phosphotyrosine-containing regions of the activated EGF receptor by direct interaction involving their SH2/SH3 domains.[214] Tyrosine phosphorylation of the EGF receptor carboxyl-terminus is probably important for binding of the physiological ligand.

2. Molecular Forms of the EGF Receptor

Three different products encoded by the single chicken c-*erb*-B gene are represented by proteins of 300, 170, and 95 kDa.[215] The 95-kDa protein is synthesized from a low molecular weight c-*erb*-B gene transcript that exclusively encodes the ligand-binding domain of the receptor. The majority of this form of the EGF receptor produced by chicken embryo fibroblast (CEF) cells is released into the culture medium, where it may participate in ligand binding. A secreted, truncated EGF receptor peptide of 70 kDa encoded by a 2.6-kb transcript of the avian EGF receptor gene, generated by alternative splicing, binds TGF-α and can block ligand-dependent transformation (soft agar colony formation in CEF cells).[216] Truncated, soluble forms of growth factor receptors are widely distributed and may play important regulatory functions.

Different molecular and functional forms of the EGF receptor found in different cell types are probably due to posttranscriptional modifications since the receptor is encoded by a single gene. The EGF receptor expressed in activated mammalian lymphocytes may be qualitatively different from the receptor expressed in other types of cells. The EGF-related growth factors TGF-α and VVGF, which stimulate cell growth via binding to the EGF receptor, do not compete for the EGF receptor expressed in murine lymphocytes induced to produce interferon-γ (IFN-γ).[217] The EGF receptor-associated tyrosine kinase activity plays an important role in EGF-induced contraction of gastric smooth muscle cells;[218] and guinea pig gastric smooth muscle cells maintained in culture express an EGF receptor subtype with higher affinity for TGF-α, in comparison to the receptor expressed in nonmuscle tissues.[219] TGF-α is more potent than EGF in enhancing anchorage-independent growth of T3M4 human pancreatic carcinoma cells; and this effect may be due to the expression in these cells of a variant EGF receptor which is encoded by an allele that contains a nucleotide change at position 1749, determining the presence of a lysine instead of an arginine in the EGF receptor protein.[220] The same substitution was found in three other human pancreatic cancer cell lines, as well as in RL95-2 human endometrial carcinoma cells and A549 human lung cancer cells.

The precise functional significance of qualitatively distinct EGF receptors expressed by different types of normal or tumor cells is unknown. Some variants of the EGF receptor may be due to posttranslational modifications. The presence of three functionally different types of EGF receptors in HeLa cells may result from phosphorylation of the receptor and the formation of a ternary complex.[221]

3. Transmission of the EGF-Elicited Transmembrane Signal

Transmembrane signaling of the EGF receptor probably depends on the intrinsic tyrosine kinase activity of the receptor molecule. However, the molecular mechanisms involved in the transmission of the EGF-elicited signal across the membrane are poorly understood. Two models can be considered for the signal transduction: an intramolecular model and an intermolecular model.[222] In an intramolecular model it is assumed that EGF binding to the receptor induces a conformational change in the extracellular domain of the receptor molecule, which is transmitted in some manner through the transmembrane region to the kinase domain which is consequently activated. In an intermolecular model it is assumed that receptor-receptor interactions which are mediated by EGF binding lead to activation of the receptor kinase. This type of mechanism bypasses the requirement for a conformational change to be transmitted through the hydrophobic stretch connecting the two functional domains (extracellular and cytoplasmic). Testing the predictions of each model in an *in vitro* system composed of a detergent-solubilized EGF receptor led to results which were compatible with an intermolecular mechanism rather than with an intramolecular mechanism for EGF receptor activation.[223] Studies with EGF receptor mutants with altered transmembrane regions, prepared by *in vitro* site-directed mutagenesis, also lent support to an intermolecular mechanism or receptor activation.[224]

Clustering of EGF receptor molecules on the cell surface is required for signal transduction and subsequent generation of biological responses in EGF-sensitive cells. Both EGF receptor monomers and oligomers (probably dimers) exist in solution but binding of EGF to its receptor would favor a rapid and reversible oligomerization of receptor molecules. The key regulatory step in the activation of the EGF receptor is the EGF-dependent noncovalent conversion of receptor monomers into dimers, which would induce receptor autophosphorylation.[225] Dimerization of EGF receptors may have an important role in the mechanisms underlying receptor activation. Heterodimerization with defective EGF receptors functions as a dominant negative mutation suppressing the activation and response of normal receptors by formation of unproductive heterodimers.[226] However, the results of studies with detergent-solubilized EGF receptor in monomeric form indicated that an individual receptor molecule has the potential to function as a transmembrane signal transducer, lending support to the idea of an intramolecular mechanism for EGF receptor activation.[227] The ligand-binding domain of the EGF receptor is not required for dimerization of the receptor.[228] In contrast to the c-Src and c-Fms tyrosine kinases, the carboxyl-terminus of the EGF receptor does not have a negative regulatory function and may rather facilitate transmission of receptor-dependent signals for cell proliferation.[229]

Transmembrane signaling by the activated EGF receptor is acutely regulated by protein kinase C. Activation of protein kinase C by treatment of cells with phorbol esters causes an inhibition of the high-affinity binding of EGF to its cell surface receptor and inhibits the tyrosine kinase activity of the receptor molecule, a process called transmodulation. The molecular mechanism of EGF receptor transmodulation is not understood but this process may cause desensitization of the EGF receptor. The transmodulated receptor EGF is capable of signal transduction, although through an altered pathway that involves the action of a protein kinase (termed T669) which phosphorylates the receptor at the Thr-669 residue.[230]

4. Evolutionary Conservation of the EGF Receptor

The EGF receptor structure is highly conserved in different vertebrate species (human, baboon, dog, rat, mouse, frog, and chicken), which indicates an essential function in different cell types. The chicken EGF receptor, as deduced from the nucleotide sequence of a cDNA clone containing its complete coding sequence, is a 170-kDa glycoprotein.[210] NIH/3T3 cells devoid of endogenous EGF receptors transfected with appropriate cDNA constructs expressed the chicken EGF receptor protein. Murine EGF binds to the chicken EGF receptor with 100-fold lower affinity than to the human receptor molecule, but surprisingly human TGF-α binds equally well or even better to the chicken EGF receptor than to the human EGF receptor. Moreover, TGF-α stimulates DNA synthesis 100-fold better than EGF in NIH/3T3 cells that expressed the chicken EGF receptor. There is no information, however, concerning the structure or even the existence and tissue distribution of EGF and TGF-α in birds.

Two distinct *erb*-B-related genes encoding EGF receptor-like proteins are present in the genome of *Xiphophorus* fishes.[231,232] The protein product of at least one of these two genes possesses tyrosine kinase activity, is developmentally regulated, and is involved in the development of melanomas in the fish.

EGF receptor-related DNA sequences are present in invertebrates, including *Drosophila melanogaster*.[233] Such sequences show varying degrees of homology to members of the *src* gene family, including the oncogenes v-*src*, v-*abl*, v-*fes*, v-*fps*, v-*yes*, and v-*fms*. Use of alternative 5′ exons and tissue-specific expression is observed in the *Drosophila* EGF receptor homolog transcripts.[234] The complete nucleotide sequence of the *Drosophila* EGF receptor homolog gene has been determined.[235] The protein encoded by this gene has, as the human EGF receptor, three distinct domains: an extracellular putative EGF-binding domain, a hydrophobic transmembrane region, and a cytoplasmic kinase domain. The overall amino acid homology between the human and the *Drosophila* receptor is 41% in the extracellular domain and 55% in the kinase domain. Two cysteine regions, a hallmark of the human ligand-binding domain, have been conserved; and it is clear from the sequence that the extracellular and the cytoplasmic domains of the receptor have been part of the same molecule for over 800 million years.[235] The *Drosophila* receptor has affinity for both EGF and insulin.[236-238] The EGF-binding protein of *Drosophila* may be an enzyme with insulin-degrading properties. However, the exact physiological role of the *Drosophila* EGF receptor has not been characterized, and a specific EGF clone has not been isolated from the *Drosophila* genome. *Ellipse (Elp)* alleles are dominant mutations of the *Drosophila* EGF receptor homolog for which expression is associated with alterations in the spacing pattern of the ommatidia, the units that compose the insect eye.[239]

C. BIOSYNTHESIS OF THE EGF RECEPTOR

The biosynthesis of EGF receptor, as studied in human A431 cells, is initiated in the form of a 70-kDa protein which is subjected to different posttranslational modifications,[159,240-244] including the addition of seven or more N-linked high-mannose oligosaccharide chains onto a 138-kDa molecule to form a 160-kDa intermediate. The oligosaccharide chains are subsequently modified by addition of terminal sugars, including fucose and sialic acid, to give the mature 175-kDa form of the receptor.[245] The external EGF-binding domain of the receptor contains about 30% carbohydrate,[180] and blood group-related antigens are expressed on the carbohydrate chains of the receptor.[246] In addition to high-mannose oligosaccharides, the receptor contains N-linked backbone structures of oligosaccharide chains and peripheral monosaccharides conferring blood group and other polymorphic antigen properties. Glycosylation is necessary for the acquisition of ligand-binding capacity of the receptor.[244,247] It is not known whether glycosylation *per se* is required for ligand binding or must precede a later processing step where binding activity is acquired, but the acquisition of binding activity occurs relatively late in the processing pathway of the receptor precursor. Posttranslational glycosylation of the receptor is not required for its autophosphorylation associated with the acquisition of tyrosine kinase activity.[248]

EGF receptor biosynthesis in normal human fibroblasts is similar to that found in neoplastic human A431 cells.[249,250] Biosynthesis of the receptor can be reduced by the expression of constructed molecular vectors containing EGF receptor-specific antisense RNA sequences.[251] Expression of the v-Src oncoprotein can result in inhibition of EGF receptor biosynthesis.[252]

D. EGF RECEPTOR EXPRESSION

The EGF receptor is expressed by various types of cells. This expression correlates more to cell lineage and specific stages of differentiation than to cell proliferation.[253] The EGF receptor is not expressed in hematopoietic cells.

EGF receptors are expressed in both adult and fetal tissues, including human fetal membranes,[254] and their number per cell increases during embryo development.[255] All types of cells from human amnion, chorion, decidua, and placenta contain EGF receptors.[256] The levels of EGF receptor mRNA and protein increase in human placentas throughout the gestational period.[257] In the placenta, receptors for EGF are almost exclusively expressed in the syncytiotrophoblast, which represents a nonmitotic, end-stage derivative of the cytotrophoblast.[258] Expression of the EGF receptor in the human syncytiotrophoblast parallels the expression of hCG in the same tissue. In contrast, high levels of expression of the c-*myc* gene are expressed in the cytotrophoblast but not in the syncytiotrophoblast. The EGF receptor is expressed in the sexually immature avian oviduct, but receptor upregulation does not occur during estrogen-promoted oviduct growth and differentiation.[259] Functional EGF receptors are present in the rat uterus,[260] but the possible role of EGF in uterine physiology is unknown. A similar ignorance exists in relation to many other organs and tissues where EGF receptors have been identified and in which the precise role of EGF in regulatory phenomena is not understood. EGF receptors are present in luteal, thecal, and granulosa cells of the rat ovary, suggesting that EGF may play an important role in the regulation of different ovarian functions.[261]

In the rat liver, a marked change in EGF binding is observed following birth.[262] The roles of EGF and its receptor in hepatocyte proliferation *in vivo* are not understood. Freshly isolated hepatocytes from rats have a weak proliferative response to EGF, even though they express large numbers of both high- and low-affinity EGF receptors.[263] Expression of EGF receptors on cultured rabbit chondrocytes increases by treatment of the cells with retinoic acid and IL-1β, which inhibits proteoglycan synthesis; and decreases when the cells are treated with parathyroid hormone (PTH) or dibutyryl cAMP, which stimulates proteoglycan synthesis.[264] Thus, EGF receptor expression may be a negative marker for chondrocyte differentiation. EGF receptors are not expressed by some mouse 3T3 fibroblast variants which do not respond to EGF and do not contain EGF-related antigens or mRNA.[265]

1. Regulation of EGF Receptor Expression

Homologous and heterologous regulation of EGF receptor expression is observed in a wide diversity of cell types *in vivo* and *in vitro*. Biosynthesis of EGF receptor mRNA and protein in a rat hepatic cell line (WB cells) and a human breast cancer cell line is stimulated by EGF.[266,267] Exposure of KB human carcinoma cells to either EGF or phorbol ester stimulates the synthesis of EGF receptor mRNA and protein.[268] Addition of cycloheximide (an inhibitor of protein synthesis) together with EGF further enhances EGF receptor mRNA accumulation, suggesting that EGF modulation of EGF receptor expression takes place, at least in part, at a posttranscriptional level.

Hormones and growth factors other than EGF may be involved in the regulation of EGF receptor expression in various cell types. Heterologous regulation of EGF receptor expression is observed, for example, in the aortic muscle cell line A10, in which arginine-vasopressin and 5-hydroxytryptamine cause a reduction in the amount of EGF binding to the cells.[269] Retinoic acid suppresses EGF receptor synthesis in human epidermoid carcinoma ME180 cells.[270] This effect is exerted at the transcriptional level and is due to the interaction of the retinoic acid receptor-γ (RAR-γ) with a distinct element contained in the EGF receptor gene promoter.

In addition to hormones and growth factors, other agents of either endogenous or exogenous origin, including infectious agents and drugs, may be involved in regulating EGF receptor expression. Treatment of mouse 3T3 cells and human A431 cells with protamine results in increased binding activity of the EGF receptor and EGF-stimulated mitogenic response.[271] The effect of protamine is associated with increased expression of EGF receptors on the cell surface and could be attributed to an activation of cryptic or inactive receptors present in the plasma membrane. It is well known that basic proteins such as protamine and histones may act as powerful stimulators of cell growth.

The biochemical mechanisms involved in homologous and heterologous regulation of EGF receptor expression and function are poorly understood. Gangliosides may contribute to modulation of EGF receptor signal transduction,[272] and calmodulin inhibits the EGF receptor tyrosine kinase.[273] Rat Sertoli cells secrete a growth factor that can block EGF binding to its receptor.[274] Infection of a variety of cell types by group C human adenoviruses results in downregulation of the EGF receptor, which may be due to a protein product of the virus that induces internalization and degradation of the EGF receptor.[275]

2. Quantitative Variations in EGF Receptor Expression

Variable amounts of EGF receptors are expressed on the surface of different types of cells. Most cells contain about 10^5 EGF receptors per cell, but the human epidermoid carcinoma cell line A431 contains as many as 2×10^6 EGF receptors per cell and has been used for this reason as an excellent source for the purification and characterization of these receptors.[180] The EGF receptor gene is amplified 15- to 20-fold in A431 cells, yet amplification cannot explain the high level of expression of EGF receptors in these cells because it does not lead to concomitant increase of 5.8- and 10.5-kb mRNA transcripts which are responsible for the synthesis of the receptor.[196] In contrast to normal cells, A431 cells synthesize a 2.8-kb transcript at levels 100-fold greater than that of the larger mRNA species. This mRNA encodes a 70-kDa secreted polypeptide representing almost the entire extracellular domain of the receptor. Since this truncated receptor is secreted, it cannot account for the overexpression of functionally intact EGF receptors in A431 cells. EGF inhibits the growth of A431 cells maintained in monolayer culture but stimulates the growth of these cells in soft agar.[25] The latter effect of EGF is antagonized by TGF-β.

The human breast cancer cell line MDA-468 is characterized by a very high number of EGF receptors associated with amplification and overexpression of the EGF receptor gene.[276] From 22 grade III or IV human astrocytoma cell lines examined, one cell line (SK-MG-3) exhibited an unusually high number of specific EGF-binding sites associated with amplification and rearrangement of the EGF receptor gene.[277] Double minute chromosomes were detected in SK-MG-3 cells but no homogeneously staining regions were observed. No abnormal EGF receptor-related mRNA species were detected in the same line. Of 10 primary human glioblastomas graded as glioblastoma multiform (astrocytoma grade III or IV), 4 exhibited an elevated EGF receptor level associated with amplification of EGF receptor gene sequences.[194] The rarity of EGF receptor gene amplification in astrocytoma cell lines selected and propagated in monolayer suggests that the amplification does not represent an advantage and may even lead to counterselection, at least under certain conditions of *in vitro* culture.[277]

The growth of both A431 and MDA-468 cells may be stimulated *in vitro* by low concentrations of EGF (between 0.1 and 10 p*M*), but is inhibited at higher concentrations (0.1 to 10 n*M*) which stimulate most other cells. An MDA-468 clone selected for resistance to EGF-induced growth inhibition shows a number of receptors within the normal range, which suggests that a correlation between EGF receptor number and EGF-induced proliferative response may be a general phenomenon.[276] Sustained high levels of EGF achieved *in vivo* by the administration of testosterone to female athymic mice do not affect the growth of solid A431 tumors in the animals, whereas low levels of EGF stimulate growth of the tumor.[278] These data suggest that the mechanism involved in the inhibition of A431 cell growth *in vitro* does not operate *in vivo* and that the effect of EGF *in vivo* is associated with growth promotion in both normal and tumor cells.

3. Functional Heterogeneity of EGF Receptors

EGF receptors are functionally heterogeneous and occur in both high- and low-affinity states in A431 cells and other types of cells. Two classes of EGF receptors have been identified, for example, in PC12 rat pheochromocytoma cells: one high-affinity class with 7,600 sites per cell and another low-affinity class with 62,000 sites per cell.[279] The two subpopulations of EGF receptors may respond in different fashion to agents such as TGF-β.[280] Two distinct subtypes of EGF receptors have been identified in porcine and bovine coronary artery preparations, in which EGF elicits a contractile response.[281] These EGF receptor subtypes can be distinguished by their sensitivity to indomethacin and dexamethasone, their desensitization properties, and the order of potencies for EGF and TGF-α. Nonetheless, the two different receptor systems both require the influx of extracellular Ca^{2+}.

High-affinity receptors mediate the growth-related effects of EGF and can be distinguished from the major population of low-affinity receptors by their reduced rate of lateral diffusion.[282] The molecular basis of this heterogeneity is not understood. In particular, it is not known whether a small proportion of a heterogeneous population of receptors associate with an effector molecule to give multiple EGF receptor classes in equilibrium or a heterogeneous population of EGF receptors preexists but responds independently and in different ways to EGF.

4. Mutant Cells with Decreased EGF Receptor Expression

Mutants of the human KB carcinoma cell line (a subclone of HeLa cells) may be resistant to EGF due to decreased EGF receptor expression.[283] Decreased levels of EGF receptor mRNA are expressed in the mutant KB cells. However, the 2- to 3-fold decrease in the numbers of EGF receptors observed in the mutant KB cells are not sufficient to explain the 30- to 1000-fold resistance to EGF exhibited by these cells, suggesting that additional functional alterations are present in the mutant cells.

5. EGF Receptor-Related Polypeptides

Polypeptides related to the EGF receptors may be produced by some types of cells but their precise structure and function are unknown. In addition to the mature EGF receptor located on the cell membrane, A431 cells synthesize and secrete a 105-kDa EGF receptor-related peptide which is not derived from the mature receptor but is separately produced by the cell.[284] The synthesis of this polypeptide may occur from a distinct mRNA generated via alternate splicing. In primary cultures of adult rat hepatocytes the 175-kDa form of the EGF receptor is progressively replaced by low molecular weight (130 and 105 kDa) forms of the receptor.[285] Although the number of the EGF-binding sites per cell do not change during 3 weeks of culture, there is a decrease in EGF-binding affinity which could be attributed to changes in the structure and/or phosphorylation patterns of the receptor.

6. Expression of the EGF Receptor in Heterologous Cells

Transfection of a human EGF receptor cDNA into EGF receptor-negative NIH/3T3 mouse fibroblasts results in expression of functional EGF receptor protein which confers responsiveness to the mitogenic action of EGF as well as an EGF-dependent expression of a transformed phenotype.[286,287] The transforming potential of retroviral vectors carrying a human EGF receptor gene is directly related to the number of EGF receptors expressed on the surface by the transfected murine cells.[288] The vigorous proliferation elicited by EGF in NIH/3T3 cells transfected with a human EGF receptor gene is sustained by a spectrum of early intracellular signals that include phosphoinositide hydrolysis and increases in $[Ca^{2+}]_i$ and pH_i.[289] By contrast, expression of the human EGF receptor alone is not sufficient to induce or maintain proliferation in IL-3-dependent, freshly isolated bone marrow-derived mouse cells exposed to EGF, although it can do so in established hematopoietic cell lines.[290] Heterologous expression of the EGF receptor in the Burkitt lymphoma cell line Namalwa after infection with a recombinant retroviral vector also results in the processing of EGF receptor molecules that are expressed in a functionally mature form on the cell surface.[291]

7. Absence of Functional EGF Receptors

Functional EGF receptors are absent from certain cell line clones. Variant Swiss-Webster 3T3 mouse cell lines unable to develop a mitogenic response to EGF stimulation are characterized by the absence of functionally active EGF receptors.[292] Attempts to complement by cocultivation the EGF-nonresponsive phenotype of independently isolated variants of these variant cell lines were unsuccessful. The molecular defect responsible for the absence of EGF receptors in these cells remains to be characterized.

E. PHOSPHORYLATION AND PROCESSING OF THE EGF RECEPTOR

Three functional sites can be distinguished in the EGF receptor: the EGF-binding site, the protein kinase catalytic site, and the major autophosphorylation site.[293] The functional properties of the receptor depend on its state of phosphorylation at different amino acid residues.[294] In cells not exposed to EGF, the EGF receptor is already phosphorylated at several sites on tyrosine, threonine, and serine residues located in the carboxyl-terminal portion of the molecule; and phosphorylation at these sites increases following EGF binding, which additionally induces phosphorylation at threonine in the amino-terminal, EGF-binding domain of the molecule.[295] All the autophosphorylation sites of the EGF receptor, as well as those of the related Erb-B2 protein, are located in the carboxyl-terminal portion of the protein.[296] At least some phosphorylations of the receptor occur during its process of maturation, before the addition of high-mannose oligosaccharide which determines the formation of a 160-kDa form of receptor precursor.[249] The tyrosine kinase activity of the EGF receptor, which probably depends on its state of phosphorylation, is similar in the liver of adult and senescent mouse.[297]

As other hormone and growth factor receptors located on the cell surface, the EGF receptor undergoes autophosphorylation upon ligand binding. The results of studies utilizing mixed populations of EGF receptor molecules altered by *in vitro* site-directed mutagenesis suggest that autophosphorylation of solubilized EGF receptors can occur by an intermolecular process involving cross phosphorylation of the receptors, which is probably facilitated by receptor oligomerization.[298]

1. Phosphorylation of the EGF Receptor on Tyrosine

After ligand binding, the EGF receptor is autophosphorylated on tyrosine,[299,300] which may result in the acquisition of tyrosine kinase activity with affinity for different substrates.[301,302] EGF receptor autophosphorylation occurs immediately (<10 s) after EGF binding.[303] Three major *in vitro* autophosphorylation sites near the carboxyl-terminus of the EGF receptor have been assigned: Tyr-1068 (site P3), Tyr-1148 (site P2), and Tyr-1173 (site P1).[304] Each of these residues is immediately preceded by an amino acid with an acidic side chain. P1 is selectively phosphorylated *in vivo,* while sites P2 and P3 are autophosphorylation sites used to a lesser extent *in vivo*. Studies with site-directed mutagenesis (phenylalanine for tyrosine at position 1173) showed that the self-phosphorylation site P1 can act as a competitive/alternate substrate for the EGF receptor and that this region at the carboxyl-terminus of the molecule is important for the modulation of its maximal biological activity.[305]

In contrast to the insulin receptor and the viral oncoproteins possessing tyrosine protein kinase activity, the EGF receptor region that is active as a protein kinase lacks the major site(s) of tyrosine phosphorylation.[306] This fact indicates that tyrosine phosphorylation within the catalytic domain of the receptor molecule is not important for the protein-tyrosine kinase activity associated with the EGF receptor. A highly conserved tyrosine residue at codon 845 within the kinase domain of the human EGF receptor may not be required for transforming activity.[307] Not an increase, but a decrease in the tyrosine kinase activity of the EGF receptor has been observed in A431 cells after a brief treatment with EGF.[308] It has been suggested that this decrease may represent a feedback mechanism by which responsiveness to the growth factor is regulated. Tumor necrosis factor-α (TNF-α) can modulate EGF receptor tyrosine phosphorylation and protein kinase activity in human tumor cells.[309,310] Since the TNF-α receptor does not possess intrinsic tyrosine kinase activity, the phosphorylation of EGF receptors induced by TNF-α must be due to the secondary activation of tyrosine kinases.

The physiological significance of EGF receptor phosphorylation on tyrosine residues is not completely understood. EGF-induced tyrosine phosphorylation of the receptor occurs to a similar extent and on identical tryptic peptides in preparations from cells of various *in vitro* ages, including senescent cells that become unable to synthesize DNA in response to mitogenic stimuli.[311] The fully phosphorylated EGF receptor and EGF receptor protected from autophosphorylation by the binding of antibodies have equal tyrosine-specific kinase activity toward exogenous synthetic peptide substrates.[312] Monoclonal antibodies acting as partial EGF receptor agonists are capable of stimulating the EGF receptor tyrosine kinase activity both in membrane preparations and in intact cells, but these antibodies fail to induce other EGF-like effects including stimulation of DNA synthesis.[178] Over a wide range of stoichiometry of phosphorylation, there is no apparent change in either EGF binding to the receptor or receptor tyrosine kinase activity.[301] A truncated form of the EGF receptor, lacking the carboxyl-terminal 20 kDa of the molecule which contains all of the three major autophosphorylation sites, is still capable of transducing a mitogenic signal in Chinese hamster ovary (CHO) cells.[313] Constructed EGF receptor mutants in which individual autophosphorylation sites have been replaced by phenylalanine residues are as active as wild-type receptors in mediating various EGF responses, including stimulation of DNA synthesis, when expressed in NIH/3T3 cells lacking endogenous EGF receptors.[314,315] The mutant receptors do not show altered kinase activity and the EGF-binding affinity of the receptor is not affected by the mutations. Mitogenic signaling by the EGF receptor may occur by a mechanism not involving binding of SH2 domain-containing proteins to the activated receptor. Thus, EGF receptor autophosphorylation on tyrosine may play some as yet unidentified role in the function of the EGF receptor.

Membrane lipids, especially glycosphingolipids, can modulate the EGF-dependent tyrosine phosphorylation of the EGF receptor.[316] Sphingosine increases the affinity and number of EGF receptors on the surface of A431 cells and causes an increase in the affinity of the receptor.[317,318] A ganglioside, neuraminyllactosylceramide, detected in the membrane of A431 carcinoma cells and B16 melanoma cells strongly enhances tyrosine kinase activity associated with the EGF receptor.[319] In general, there is evidence that the lipid microenvironment is important in regulating the expression and function of growth factor receptors at the level of the plasma membrane, including receptor-receptor interactions.

2. Phosphorylation of the EGF Receptor on Serine and Threonine

In addition to tyrosine, the EGF receptor may be phosphorylated *in vitro* on serine and threonine by several enzymes, including cAMP-dependent protein kinase.[320] The significance of this reaction to the conditions occurring in the intact animal is suggested by the analysis of the *in vivo* phosphorylated receptor, which shows the presence of phosphoserine and phosphothreonine in addition to phosphotyrosine.[300] Adenosine 5′-triphosphate (ATP) can modulate the EGF receptor through phosphorylation of the receptor molecule on serine and threonine residues.[321] The affinity of the receptor for EGF is modulated, via altered receptor phosphorylation, by Ca^{2+} and protein kinase C activity.[322,323] The EGF receptor of A431 cells is phosphorylated by the type III subspecies of protein kinase C.[324] Functional consequences of protein kinase C-catalyzed phosphorylation at threonine in the EGF receptor kinase are a reduction of the receptor affinity for its ligand and an inhibition of EGF-stimulated tyrosine kinase activity.[294,301,325,326] However, studies using site-directed mutagenesis techniques suggest that the mechanism of regulation of apparent affinity of the EGF receptor may be independent of the major sites of serine and threonine phosphorylation of the receptor.[327]

Phorbol ester-activated protein kinase C phosphorylates the EGF receptor at Thr-654, a residue located within a basic sequence of nine amino acids close to the cytoplasmic face of the plasma membrane.[328] Thr-654 is at a position where it can be involved in the modulation of signaling between the internal kinase domain and the external EGF-binding domain. Phosphorylation of the receptor at Thr-654 induced by protein kinase C results in reduction of the tyrosine kinase activity of the receptor molecule.[323,329] Phosphorylation at Thr-654 inhibits ligand-induced internalization and downregulation of the EGF receptor.[330] Alternative sites of phosphorylation may be important for the regulation of the functional properties of the receptor. Phosphorylation of Thr-654 may not be required for the negative regulation of the EGF receptor by the calcium ionophore A23187 and the nonphorbol tumor promoter thapsigargin, a natural plant product that was purified from the roots of *Thapsia garganica* (Apiaceae).[331] Protein kinase C-dependent phosphorylation of the unoccupied EGF receptor at Thr-654 and EGF binding regulate functional receptor loss, but this change occurs as a result of independent mechanisms.[332]

A major site of phosphorylation of the EGF receptor after ligand binding is Thr-669.[333] Phosphorylation of the EGF receptor on Thr-669 depends on the activity of the microtubule-associated protein 2 (MAP-2) kinase and a related enzyme, the ERT kinase.[334,335] The ERT kinase recognizes the consensus primary sequence Pro-Leu-Ser/Thr-Pro and is also involved in the phosphorylation of the human c-Myc protein on Ser-62 and the rat c-Jun protein on Ser-246. The activity of the MAP-2 and ERT kinases is stimulated by EGF and is transmodulated by PDGF and phorbol ester. Phosphorylation of the EGF receptor at Thr-669 and Ser-671 may mediate its interaction with a specific tyrosine kinase substrate and is required for efficient ligand-induced internalization of the receptor.[336] Another important site of regulation of the EGF receptor by phosphorylation is located on the carboxyl-terminal subdomain of the receptor and is represented by Ser 1046/7.[337] This site is a substrate for the calcium- and calmodulin-dependent protein kinase II (CAM kinase II), and its phosphorylation causes a negative regulation of tyrosine kinase activity associated with the receptor protein.

Phosphorylation of the EGF receptor on serine and threonine partially depends on pathways involving EGF receptor-induced tyrosine phosphorylation. The isoflavone compound genistein, which is an specific inhibitor of tyrosine kinases *in vitro,* inhibits not only the tyrosine kinase activity of the EGF receptor in intact A431 cells but also the EGF-stimulated serine and threonine phosphorylation of the receptor.[338] The state of phosphorylation of the EGF receptor depends not only on phosphorylating enzymes but also on active dephosphorylating processes. Certain hormones or growth factors may contribute to the regulation of these processes. Calcineurin purified from bovine brain catalyzes complete dephosphorylation of phosphotyrosine and phosphoserine residues in the human placental EGF receptor.[339]

F. INTERNALIZATION AND DEGRADATION OF THE EGF RECEPTOR COMPLEX

The EGF receptor complex is internalized into the cell and is, at least partially, fused with lysosomes, which results in the production of proteolytic fragments and simultaneous downregulation of the receptors located at the cell surface.[340-345] The internalization and downregulation of the human EGF receptor are regulated by receptor carboxyl-terminal tyrosine residues.[346]

The internalization/degradation events include a sequential translocation of surface bound EGF: first to a plasma membrane-associated fraction, which is a transient form that is still attached to the plasma membrane but not accessible to the extracellular fluid, and then to an intravesicular fraction which represents the endosomic form of EGF.[347] Several intermediates in the degradation of EGF have been detected in KB human epidermoid carcinoma cells. Some proteolysis of EGF may already occur in the endocytic vesicules while the final proteolytic products are formed after arrival of EGF in the lysosomal compartment.[348,349] Recycling or inactivation of the EGF receptor may not occur during downregulation, and EGF receptors may be rapidly degraded in the presence of EGF.[350] EGF-induced proteolytic processing of the EGF receptor appears to occur primarily at the amino-terminal EGF-binding domain of the receptor molecule which would be oriented within the interior of endocytic or lysosomal vesicles and would be exposed to intravesicular proteases.[351] Truncation of receptor amino-terminal sequences may contribute to constitutive activation of its signaling mechanism. Thus proteolysis of the EGF receptor may have a role in receptor function.

The physiological factors involved in EGF receptor degradation are little known. Cathepsins are the major proteases that cleave the EGF receptor, and Ras proteins may possess inhibitory activity against cathepsins.[352,353] Cleavage of EGF receptors by cathepsin L is inhibited by Ras protein in a dose-dependent fashion. These results raise the possibility that Ras proteins can suppress the degradation of growth-related proteins such as EGF receptors and thereby affect cell proliferation and/or differentiation. However, the presence of a functional EGF receptor may not be required for *ras*-mediated transformation.[354] Other growth factors may affect the internalization and/or degradation of the EGF receptor in different types of cells. TGF-β inhibits EGF receptor endocytosis and downregulation in cultured fetal rat hepatocytes.[355]

In addition to the association with the degradative pathway, internalization of EGF receptors serves to shuttle the intact EGF receptor complex across particular types of polarized cells such as MDCK cells.[356] This process, called transcytosis, occurs also in other systems, for example, in the receptor-mediated transport of insulin across endothelial cells.[357] Transcytosis may serve to unidirectionally transport intact protein molecules across epithelial cell barriers. However, the role of receptors in this process is not clear, and there is evidence that the EGF receptor mediates EGF uptake but not transcytosis.[358]

EGF binding induces a decrease in the population of mature EGF receptors but has no discernible effect on receptor synthesis or stability, suggesting that the reduction in EGF-binding capacity observed during downregulation is produced solely by a change in the rate of degradation of receptors.[250] Comparison of different cell lines which are either responsive or nonresponsive to EGF indicated that the presence of specific EGF receptors may not correlate with cell growth, but differences in the processing of EGF after binding to the receptor were apparent in these cell lines.[359]

Internalized EGF is proteolytically processed to a number of high molecular weight products in PANC-1 human pancreatic carcinoma cells and other types of cells.[360] In PANC-1 cells, EGF undergoes only limited processing, being able to bypass the cellular degradative pathways and being rebound to the cell after its slow release into the culture medium.[361] In A431 cells, only 15 to 20% of the EGF receptors are associated with the plasma membrane; while 80 to 85% of them are found in an intracellular location, associated with lysosome-like structures, a tubulo-vesicular system, the rough endoplasmic reticulum, and the nuclear envelope.[362] Immunofluorescence visualization using antibodies directed to kinase and extracellular domains of the EGF receptor showed their perinuclear localization and recycling back to the cell surface.[363] In different cell lines, labeled EGF is found to be tightly bound to chromatin after 1 h of incubation.[364,365] Moreover, binding of the EGF receptor to isolated chromatin is inhibited by monoclonal antibodies specific for the EGF receptor. These monoclonal antibodies are taken up by the cell, translocated to the nucleus, and bound to the chromatin receptor instead of the original ligand (EGF). Since the antibodies do not stimulate DNA synthesis and inhibition of epidermoid carcinoma cell growth may occur without an induction of tyrosine kinase activity, it seems likely that this activity does not represent the only critical point in cell induction to proliferation. Binding sites for EGF are present in nuclear fractions purified from rat liver, and these sites fluctuate in number during liver regeneration after partial hepatectomy.[366] The nuclear EGF-binding protein is similar in size and ligand affinity to the plasma membrane receptor and is recognized by EGF receptor-specific monoclonal antibodies.[367] The possible physiological significance of nuclear EGF-binding sites is not understood, however. Enzymatic processing of EGF or the EGF receptor complex within intracellular vesicles is not necessary for an EGF-induced response, including the specific induction of RNA synthesis.[368]

The processes related to the internalization and degradation of EGF receptor complexes have been additionally studied by different experimental approaches, including the effect of phorbol esters, studies with monoclonal antibodies, studies with monensin and methylamine, site-directed mutagenesis procedures, and use of specific inhibitors of the EGF receptor kinase. Results obtained with these approaches are discussed next.

In contrast to EGF-induced EGF receptor internalization, phorbol ester-induced internalization of the EGF receptor does not cause delivery of the EGF receptors to lysosomes; however, the receptors reappear on the cell surface, even in the continuous presence of phorbol ester, as a consequence of receptor recycling.[369] Phorbol ester may be involved in inducing EGF receptor internalization through the phosphorylation of a specific threonine residue.

Treatment of mouse fibroblasts with concanavalin A results in stabilization of the EGF receptor complexes at the cell surface. These stabilized complexes allow an acute response to EGF (RNA synthesis), but are not able to stimulate later responses such as mitogenesis.[370] The internalization and processing of the hormone-receptor complex appear to be necessary to produce the mitogenic signal. Studies with monoclonal antibodies directed against the EGF receptor also suggest that the internalization of ligand-receptor complexes is necessary for EGF to exert its mitogenic effects.[172] However, this conclusion is contradicted by the results of other studies. The use of some monoclonal antibodies directed against the EGF receptor indicates that internalization of the receptor in A431 cells is not necessarily coupled to its phosphorylation. Monoclonal antibodies that specifically bind to human EGF receptors are internalized without stimulating receptor phosphorylation.[371] Monoclonal antibodies which behave as partial EGF agonists in that they are capable of stimulating the receptor tyrosine kinase activity both in membrane preparations and in intact cells can also induce receptor clustering, but fail to induce effects including stimulation of DNA synthesis.[178]

Monensin and the lysosomotropic amines chloroquine and methylamine prevent degradation of EGF and cause intracellular accumulation of EGF receptors, blocking the EGF-induced mitogenic response.[372] These results suggest that sequestration of the EGF-receptor complexes in a cytosolic compartment may be a primary signal required for correct intracellular transport of EGF and for events that might lead to generation of a mitogenic response. The EGF receptor of A431 cells is associated with the cytoskeleton both at the cell surface and at intracellular sites, and this structural association of the receptor may contribute to the modulation of the activity and substrate specificity of the EGF receptor kinase.[373] Both the primary amine, methylamine, and the ionophore, monensin, disrupt intracellular pH gradients but they appear to act by different mechanisms. In contrast to the case with the primary amines, EGF material is not lost from the cells (mouse BALB/c 3T3 fibroblasts) when chased in the continued presence of the drug in the medium.[374]

Mutational analysis of cloned EGF receptor sequences has been used to investigate the function of different receptor domains.[375] Deletion of cytoplasmic sequences abolished high, but not low, affinity-binding sites and impairs the ability of the protein to internalize into cells. A four-amino acid insertion mutation at residue 708 abolished the tyrosine kinase activity of the receptor; however, the mutant receptor exhibited both the high-and low-affinity states, internalized efficiently, and was able to cause cells to undergo DNA synthesis in response to EGF.[375] Thus, a critical role for the tyrosine kinase activity of the EGF receptor may, at least in this case, not be required for an EGF-induced mitogenic response. An artificially generated mutant EGF receptor in which Lys-721, a key residue in the ATP-binding site, was replaced by alanine, was properly processed and expressed at the surface of NIH/3T3 cells but did not possess tyrosine kinase activity, and was not downregulated after EGF binding.[376] Comparison of the internalization and intracellular sorting of normal EGF receptor and a mutant tyrosine kinase-negative EGF receptor separately expressed in NIH/3T3 cells lacking endogenous EGF receptor indicated that tyrosine phosphorylation of the receptor within a distinct intracellular vacuole, the multivesicular body, provides a sorting signal that selects the normal EGF receptor molecule for degradation.[377] These results suggest that degradation of EGF receptors after endocytosis is due to the specific endogenous kinase activity of the receptor. EGF receptor mutants with either a defective kinase or large cytoplasmic domain deletions may exhibit antiproliferative and antioncogenic potencies in intact cells in culture.[378] Thus signaling-defective EGF receptor mutants may have antioncogenic activity. Further studies using different mutated versions of the EGF receptor may contribute to a better characterization of the specific functional roles of discrete regions of the receptor molecule.

The functional relevance of the specific kinase activity of the EGF receptor may be defined more clearly with the use of effective inhibitors of this activity such as synthetic 4-hydroxycinnamamide derivatives.[379] It is unknown, however, whether these derivatives are specific inhibitors of the EGF receptor tyrosine kinase or can also inhibit other similar kinases, including those associated with certain oncoproteins.

G. THE *ERB*-B ONCOPROTEIN AND THE EGF RECEPTOR

Avian erythroblastosis virus (AEV) induces erythroblastosis when injected into chickens, and can also occasionally induce sarcomas or carcinomas.[380] AEV can also induce transformation of immature erythroid cells and fibroblasts in culture. AEV-transformed embryonic erythroid cells can undergo differentiation.[381] The transforming sequences of AEV correspond to the v-*erb* oncogene, of which the cellular counterpart, c-*erb,* is present in the genome of all vertebrate species including humans and fishes.[382] AEV carries two different oncogenes, v-*erb*-A and v-*erb*-B, derived from separate, unlinked DNA sequences in the cellular genome.[383] The v-*erb*-A and v-*erb*-B oncogenes can cooperate in the induction of a transformed phenotype in chicken erythroblasts and fibroblasts.[384] An AEV variant, AEV-H, transduces only the v-*erb*-B oncogene and is capable of inducing both erythroblastosis and sarcomas in chickens.[385] The *erb*-B gene is frequently transduced by another virus, the Rous-associated virus type 1 (RAV-1), which can also induce a rapid-onset erythroblastosis when inoculated into 1-week-old chickens.[386]

A strong structural homology has been observed between the v-Erb-B oncoprotein and the EGF receptor purified from the A431 human epidermoid carcinoma cell line and normal human placenta cells.[387] Moreover, the c-*erb*-B proto-oncogene and the EGF receptor gene are located on the same region of the human and mouse chromosomes 7 and 11, respectively,[388] suggesting their possible identity. The 3′-coding domain of the human EGF receptor gene cloned from A431 cells has striking homology to the v-*erb*-B oncogene. The products of both the v-*erb*-B oncogene and the c-*erb*-B proto-oncogene may be involved in oncogenic processes. In a manner similar to that of the v-*erb*-B oncogene, nonmutated c-*erb*-B proto-oncogenes expressed from constructed vectors can cause erythroblastosis and renal adenocarcinomas in chickens.[389] The viral oncoprotein, v-Erb-B, and the product of the normal c-*erb*-B1 proto-oncogene, the c-Erb-B1 or c-Erb-B protein, are intimately related. The normal c-Erb-B protein and the EGF receptor are, in fact, identical molecules.

1. Structure and Function of the v-*erb*-B Oncoprotein

The product of the v-*erb*-B oncogene is a 68-kDa glycoprotein, $gp68^{v\text{-}erb\text{-}B}$, which is modified to a 74-kDa protein, $gp74^{v\text{-}erb\text{-}B}$, located on the cell surface.[390,391] The orientation of the v-Erb-B oncoprotein in the membrane has been characterized.[392] The use of glycoprotein processing inhibitors has shown that incorrectly glycosylated v-Erb-B oncoprotein is inserted normally into the plasma membrane and is still capable of exerting its oncogenic activity on susceptible cells.[393] The v-Erb-B oncoprotein is a truncated form of the EGF receptor, with deletion of the amino-terminal end. Site-specific antibodies to the v-Erb-B protein precipitate the EGF receptor.[394] v-Erb-B contains only the transmembrane and tyrosine kinase domains of the EGF receptor and lacks most of the extracellular domain responsible for EGF binding.[196,387] Thus, v-Erb-B corresponds to a growth factor receptor which has lost the regulatory binding domain but has retained the membrane-locating and biochemical-activating domains. The v-Erb-B protein represents a kind of unregulated receptor which is expressed constitutively in the activated state.

Construction of chimeric genes and artificial mutants has been useful to elucidate some of the structural requirements for v-*erb*-B-induced transformation. A chimeric gene construct encoding the extracellular and the transmembrane domain of the human EGF receptor protein joined to sequences coding for the cytoplasmic domain of the v-Erb-B protein was transported to the surface of Rat-1 fibroblasts and bound EGF, and its autophosphorylation activity was stimulated by the ligand.[395] Expression of the chimeric EGF receptor/v-*erb*-B gene in Rat-1 cells resulted in anchorage-independent growth and EGF-induced focus formation in monolayer cultures. Mitogenic and transforming activities of chimeric EGF receptors carrying v-*erb*-B cytoplasmic sequences may be associated with constitutive signaling activity without autophosphorylation.[396]

A constructed *gag-erb*-B fused gene encoding the v-Erb-B protein modified at its amino-terminus by the large Gag viral polypeptide sequence was able to induce transformation of chicken embryo fibroblasts (CEF cells).[397] Expression of a truncated EGF receptor lacking the extracellular ligand-binding domain was associated with EGF-independent activation of the intrinsic tyrosine kinase activity and transformation of immortalized rodent fibroblasts.[398] The transformed phenotype became enhanced by further truncation of the carboxyl-terminal domain of the EGF receptor containing two tyrosine autophosphorylation sites. Overexpression of EGF receptors with an intact extracellular region in transfected Rat-1 cells was associated with EGF-dependent neoplastic transformation.

Some AEV isolates contain v-*erb*-B oncogenes that lack codons for the immediate carboxyl-terminus of the EGF receptor; and the biological properties of such variant viruses, including tumor-inducing ability and specificity, are different from those of the prototype retrovirus.[399] Thus, differences in the transforming potential of v-*erb*-B-carrying viruses correlate with differences in *erb*-B sequences encoding the carboxyl-terminal domain of the EGF receptor protein. Removal of the carboxyl-terminal tyrosine residue of the receptor, implicated in modulation of the kinase activity, does not lead to a fully transformed phenotype in transfected fibroblasts, whereas a point mutation in the kinase domain was found to be sufficient for sarcomagenic activation of the receptor molecule.[400] The product of a chimeric oncogene containing the transmembrane glycosylated domain of the v-Erb-B protein linked to the kinase catalytic domain of the v-*src* protein was capable of transforming cultured fibroblasts to an oncogenic phenotype closely resembling that induced by the parent v-*erb*-B oncogene.[401] The chimeric oncogene failed to transform erythroid progenitor cells, which is consistent with evidence indicating that the carboxyl-terminal domain of the v-*erb*-B oncoprotein is involved in erythroid target cell specificity.

In contrast to the truncated intracytoplasmic version of the receptor corresponding to the expression of the v-*erb*-B oncogene, A431 human carcinoma cells synthesize high amounts of 2.8-kb mRNA transcripts which encode a secreted polypeptide representing almost the entire extracellular EGF-binding domain of the receptor.[196] Since this abnormal version of the EGF receptor is secreted, it cannot account for the high number of functionally intact EGF receptors that are expressed in A431 cells.

A mutated form of the v-Erb-B protein lacking two thirds of the extracellular region retained ability to transform cells to an oncogenic phenotype.[402] The transmembrane domain of v-Erb-B could also be deleted without completely abolishing the ability of this protein to transform avian fibroblasts.[403] The results of studies with constructed mutants of the c-Erb-B protein indicate that the transforming potential of this protein in fibroblastic cells can be activated by multiple mechanisms, including carboxyl-terminal truncation, internal deletion, and point mutations.[399]

2. Phosphorylation of the v-Erb-B Oncoprotein

The v-*erb*-B oncoprotein is phosphorylated primarily on serine and threonine, but it contains also minor amounts of phosphotyrosine, and protein phosphorylation at tyrosine residues is induced by the v-Erb-B protein *in vivo* and *in vitro*.[404,405] Both the EGF receptor and the v-Erb-B oncoprotein possess tyrosine kinase activity. Site-specific antibodies to the Src-homologous domain of the v-Erb-B protein neutralize the tyrosine kinase activity of the EGF receptor.[406,407] AEV-transformed CEF cells exhibit increased phosphorylation of a number of cellular proteins on tyrosine residues, including 36- and 42-kDa proteins.[404] The major *in vivo* tyrosine autophosphorylation site is Tyr-1173 (site P1), which is located 14 residues from the carboxyl-terminus of the receptor molecule and is not found in the v-Erb-B protein.[304] v-Erb-B is fused at the carboxyl-terminus to four residues of the viral Env protein. Therefore, v-Erb-B represents a protein which is terminated prematurely with respect to the normal cellular EGF receptor sequences from which it is derived.

Antibodies to a synthetic oligopeptide have been used as a probe for the kinase activity of the avian EGF receptor and v-Erb-B protein.[408] Inhibition of the tyrosine kinase activity of the protein can be achieved by using antibodies generated against a synthetic peptide corresponding to amino acid residues 285 to 296 of the predicted v-Erb-B protein sequence, which corresponds to the tyrosine kinase domain of both v-Erb-B and the EGF receptor.[409] The v-Erb-B protein does not possess intrinsic kinase activity with specificity for inositol phospholipids.[410]

Structural domains of the v-Erb-B protein required for fibroblast transformation have been defined by site-directed mutagenesis.[411] AEV genomes bearing lesions induced within the v-Erb-B kinase domain exhibit a drastically decreased ability to transform avian fibroblasts. In contrast, mutations in the extracellular domain or at the extreme carboxyl-terminus of v-Erb-B have no effect on AEV-mediated fibroblast transformation. Thus, an intact protein kinase domain would be required for the preservation of the acutely transforming potential of the AEV oncogene product. However, the precise relationship between the kinase activity of the v-Erb-B protein and the induction of transformation is not totally clear. Analysis of *ts* mutants and nonconditional host range mutants of AEV demonstrated that there is no simple correlation between the autophosphorylation activity of the v-Erb-B protein and the transformation ability of various AEV mutants.[412] Although the tyrosine kinase activity of v-Erb-B may be central to AEV-induced transformation, it would be in itself insufficient.

A single amino acid substitution in the v-Erb-B protein confers a thermolabile phenotype in *ts* 167 AEV-transformed HD6 erythroid precursor cells.[413] The change is located in the center of the tyrosine kinase domain, corresponding to amino acid position 826 of the human EGF receptor sequence. There is a correlation between alterations in the kinetics of Erb-B tyrosine kinase activity and the oncogenic activation of the protein.[414] Mutation at the Erb-B phosphorylation site Ser-477/478 may cause increased oncogenic potential of the protein without altering its kinase activity.[415] Ser-477/478 may be a negative regulatory phosphorylation site at the carboxyl-terminal domain of the protein and may act in some way to suppress the transforming effects of the wild-type Erb-B protein.

H. HOMOLOGIES BETWEEN THE EGF RECEPTOR AND OTHER PROTEINS

The c-Erb-B1/EGF receptor protein is closely related to other normal cellular proteins, in particular to the transmembrane proteins c-Neu/Erb-B2 and amphiregulin.

1. The c-Neu/Erb-B2 Protein

The c-*erb*-B2/*HER*-2 gene is the human counterpart of the rat c-*neu* proto-oncogene. The promoter of this gene contains sequences involved in regulatory responses to extracellular signals including EGF, TPA, dibutyryl cAMP, and retinoic acid.[416,417] The product of the c-*neu*/*erb*-B2 gene is a glycoprotein of 185 kDa, $p185^{c\text{-}neu}$ or $p185^{c\text{-}erb\text{-}B2}$, which is closely related in its structure to the EGF receptor.[418] c-Neu/Erb-B2 is a transmembrane protein which consists of an extracellular binding domain, a transmembrane anchoring segment, and an intracytoplasmic tyrosine kinase domain. The c-Neu/Erb-B2 protein is expressed at low levels in differentiated keratinocytes, but not in the basal cell layer of the human skin.[419] Relatively high levels of expression of c-Neu/Erb-B2 are observed in human fetal epithelial cells.[420,421] There is an inverse relationship between the expression of the c-Neu/Erb-B2 and EGF protein in human female genital tract tissues and placenta.[422] c-Neu/Erb-B2 protein is expressed by other cell types, including Schwann cells during peripheral nerve development as well as in Wallerian degeneration.[423] Immunoreactivity for the c-Neu/Erb-B2 protein is observed in the nucleus of Schwann cells, but the possible physiological significance of this expression is unknown.

The c-Neu/Erb-B2 protein possesses tyrosine kinase activity and probably binds a growth factor.[424] Several putative Erb-B2 ligands have been proposed. A glycoprotein of 30 kDa (gp30), secreted by the human breast cancer cell line MDA-MB 231, binds independently to the Erb-B2 and the EGF receptor.[425] The gp30 protein is capable of inducing differentiation of human breast cancer cell lines expressing the Erb-B2 receptor.[426] Proteins of 25 kDa secreted by activated mouse macrophages or purified from bovine kidney have been proposed to be a ligand for the Erb-B2 receptor.[427,428] A protein, termed Neu-activating factor (NAF), purified from medium conditioned by the human T-cell line ATL-2 interacts with the extracellular domain of the Erb-B2 receptor and stimulates its tyrosine kinase activity.[429] NAF is capable of stimulating the growth of cells bearing the Erb-B2 receptor. A protein of 75 kDa purified from SKBr-3 human breast cancer cells binds and activates the Erb-B2 receptor.[430] A protein of 45 kDa, heregulin, purified from the conditioned medium of the MDA-MB-231 human breast cancer cell line activates phosphorylation of the Erb-B2 receptor on tyrosine.[431] Heregulin gene transcripts have been found in several normal tissues and cell lines. A protein of 44 kDa, termed Neu differentiation factor (NDF), secreted by *ras*-transformed rat cells can stimulate the phosphorylation of the c-Neu/Erb-B2 receptor.[432,433] NDF contains an EGF domain and an Ig homology unit. A number of mitogens for Schwann cells with molecular weights between 34 and 45 kDa, termed glial growth factors (GGFs), may be ligands of the Erb-B2 receptor.[434] The precise role of these proteins in the activation of the Erb-B2 receptor in intact animals is not understood. Further studies are required for a better characterization of the physiological Erb-B2 receptor ligand(s). Experiments with chimeric EGF receptor/c-Neu protein indicate that the signal pathways of the EGF receptor and c-Neu/Erb-B2 protein are similar, including certain early gene responses.[435-437] The chimeric protein is able to stimulate the glucose transport and ornithine decarboxylase (ODC) activity and induces the Fos/Jun transcription factor complex.

Autophosphorylation of the c-Neu/Erb-B2 protein is required for its mitogenic action and high-affinity substrate coupling.[438] Two autophosphorylation sites contained in c-Neu/Erb-B2 correspond to tyrosine residues which are also contained in the EGF receptor.[439] Mutagenic analysis of c-Neu shows that the carboxyl-terminal domain, which contains the major autophosphorylation site, Tyr-1248, regulates transformation negatively; and autophosphorylation eliminates this negative regulation.[440] c-Neu protein expressed by different types of tumor cells is a substrate for EGF, insulin, and phorbol ester-induced phosphorylation.[441-443] Studies on cell lines expressing both the EGF receptor and c-Neu protein indicated an intermolecular association of both proteins in the form of a heterodimeric structure on the cell surface and suggested that c-Neu is directly involved in regulating the function of the EGF receptor.[444] The heterodimer-containing cells express EGF receptors with unusually high affinity for the ligand and high levels of tyrosine kinase activity. Coexpression of EGF receptors and high levels of c-Neu may lead to transformation of rodent fibroblasts, and anti-EGF receptor and anti-Neu monoclonal antibodies can inhibit the tumorigenic growth of the transformed cells implanted into nude mice.[445]

Point mutations of the c-*neu* gene can activate its oncogenic potential. Expression of the activated *neu* oncogene, but not the normal proto-oncogene, in NIH/3T3 cells is associated with enhanced levels of c-*jun* and c-*fos* expression and loss of growth factor-induced immediate-early gene responses.[446] Protein kinase C is involved in the oncogenic signaling of the active (mutant) Erb-B2 protein that leads to Jun/Fos-mediated transcriptional activation in the nucleus of NIH/3T3 cells.[447] Other types of murine cells are resistant to *neu*-induced transformation. FDC-P2 myeloid cells expressing an activated c-*neu* gene can grow in the absence of the specific growth factor, IL-3, but do not proliferate indefinitely and do not form colonies in soft agar medium.[448] Mutationally activated c-Neu/Erb-B2 proteins may transform cells by mimicking ligand-induced activation.[449]

The EGF receptor and c-Neu/Erb-B2 proteins may be expressed independently in benign and malignant breast tissues.[450] Analysis of Erb-B2 expression by immunohistochemistry or other methods may be useful as a prognostic marker in breast cancer.[451,452] Erb-B2 overexpression, associated or not with amplification of the gene sequences, may occur in advanced stages of the disease and may be indicative of a poor prognosis. The *erb*-B2 gene is usually not overexpressed in hyperplastic or displastic breast lesions. Breast cancers negative for the estrogen receptor and positive for the Erb-B2 protein have a relatively poor prognosis.[453] Expression of *erb*-B2 mRNA in human breast cancer cell lines may be regulated by estrogen, but is uncoupled from estrogen-stimulated growth of these cells.[454] The oncogenic potential of the unactivated Erb-B2 protein is suggested by the fact its overexpression in the mammary epithelium of female transgenic mice results in the appearance of focal mammary tumors that metastasize with high frequency after a long latency.[455]

2. The Erb-B3 Protein

A third member of the *erb*-B/EGF receptor gene family is termed *erb*-B3 or *HER3*.[456] Cloning and sequence analysis of the *erb*-B3 gene isolated from the A431 human adenocarcinoma cell line indicate that it encodes a 180-kDa transmembrane receptor with close similarity to the EGF receptor and Erb-B2 proteins. The Erb-3 protein possesses tyrosine kinase activity and is normally expressed in cells of epithelial and neuroectodermal origin. The Erb-B3 receptor-associated tyrosine kinase is activated upon binding of an unidentified ligand, which is probably represented by a growth factor. Constitutive activation of the Erb-B3 kinase has been observed in some human breast tumor cell lines.[457]

I. FACTORS INVOLVED IN ALTERED EGF RECEPTOR EXPRESSION

Many endogenous and exogenous factors are capable of producing alterations in the number, affinity, phosphorylation, or activity of EGF receptors.[458] Expression of the EGF receptor depends on the influence of both homologous and heterologous factors. Regulatory peptides, hormones, growth factors, and neurotransmitters, may participate in the transmodulation of EGF receptor number and/or affinity. EGF receptor levels in the liver, the uterus, and the mammary gland may show significant changes according to various physiological states including fasting and hormonal changes.[459-462] In the mouse mammary gland, EGF receptor levels reflect the dynamic state of the organ: being high at an early age when the proliferative capacity of the gland is high, declining as the proliferative capacity approximates a static state in mature virgin mice, and increasing again to reach the highest levels when the mammary gland undergoes rapid growth during midpregnancy.[460]

Expression of EGF receptor mRNA in the regenerating rat liver is subjected to regulatory mechanisms acting at the transcriptional and/or posttranscriptional levels.[463] After partial hepatectomy, transcripts of the EGF receptor gene are dramatically decreased in 6- and 12-h regenerating liver, show a marked increase at 24 h, and decrease after 72 h. Similar induction of the hepatic EGF receptor is observed *in vivo* by the action of EGF or as a consequence of partial hepatectomy.[464]

1. Cell Membrane Components

Cell surface glycolipids (gangliosides) may be involved in the regulation of EGF-dependent phosphorylation of the receptor.[465] While some ganglioside derivatives act as inhibitors or the EGF receptor kinase activity, other derivatives may strongly promote phosphorylation of the receptor on tyrosine residues. These modifications include de-*N*-acetylation of gangliosides in either the sialic acid or ceramide moiety of the molecule. They may be important for the modulation of the EGF receptor-associated triggering of a positive or negative transmembrane signal.

2. Steroid Hormones

Steroid hormones may contribute to regulating the expression of EGF receptors in their target cells. Estrogen and antiestrogens modulate EGF receptor level in the human breast cancer cell line MCF-7.[466] Expression of EGF and growth hormone receptors in the human breast cancer cell line T-47D is regulated by progestins, but not by testosterone, estradiol, or cortisol.[467] Regulation of EGF receptor expression by progestin (medroxyprogesterone acetate) in T-47D cell may occur, at least in part, at the transcriptional level.[468] However, the results obtained with ovariectomized mice indicate that ovarian hormones are probably not required for the expression of EGF receptors in the mammary gland *in vivo*.[469] An EGF receptor-mediated pathway remains intact in the mammary gland epithelium in the absence of ovarian steroids, and local availability of either EGF or TGF-α is sufficient to stimulate normal ductal growth.

Expression of EGF receptors in the rat prostate is modulated by androgens.[470] Glucocorticoids increase both the high-affinity EGF-binding sites and the EGF-induced receptor phosphorylation in HeLa cells, which is associated with an enhanced rate of cellular growth.[471] Treatment of the human breast cancer cell line BT-20 with calcitriol results in a progressive increase in the number of EGF receptors per cell.[472] Estrogens regulate acutely the levels of EGF receptors in the uterus, which raises the possibility that events coupled to this receptor may play a role in estrogen-stimulated growth.[461] Estrogen-induced regulation of EGF receptor expression in the uterus occurs at the transcriptional level.[473]

3. Peptide Hormones

Pituitary hormones and insulin may be involved in regulating the expression of EGF receptors in specific tissues. Growth hormone administered *in vivo* induces the expression of EGF receptors in mouse liver.[474] EGF receptor expression, as well as responsiveness to the mitogenic action of EGF, is modulated in the thyroid gland *in vitro* by thyroid-stimulating hormone (TSH) through a cyclic AMP-dependent process.[475] In rats with diabetes induced by either alloxan or streptozotocin, the expression of EGF receptors in the liver is decreased; however, it is not known whether this decrease is caused by the hypoinsulinemia, the hyperglycemia, or a combined action of both factors.[476] Binding of EGF to its receptor is decreased in the liver of genetically diabetic (C57BL/KsJ *db/db*) mice as well as in mice with streptozotocin-induced diabetes.[477] This decrease involves alterations in the level of EGF receptor mRNA and can be restored to almost normal levels by daily administration of insulin.

4. Thyroid Hormone

Thyroid hormone is involved in the regulation of EGF receptor levels *in vivo,* the levels being markedly reduced in hypothyroid animals.[459] EGF receptor levels in both normal mammary glands and spontaneous breast tumors in mice are diminished in hypothyroid animals, and this reduction is reversed by treatment with L-thyroxine.[478] It is possible that at least some effects of thyroid hormone on cell growth and differentiation are mediated through alterations in EGF receptor levels.

5. Growth Factors

Growth factors other than EGF may be involved in an heterologous modulation of EGF receptor expression, which may occur through alteration of EGF receptor phosphorylation or other mechanisms. Transmodulation of EGF receptors by platelet-derived growth factor (PDGF) occurs by a mechanism distinct from that of the phorbol ester PMA.[479,480] In density-arrested BALB/c 3T3 cells, PDGF causes a rapid, protein synthesis-independent reduction in EGF-binding capacity. The effects of PDGF on EGF receptor phosphorylation may exhibit specificity for different types of cells, including cells such as human skin fibroblasts and human lung fibroblasts.[481]

EGF receptors are expressed in the PC12 rat pheochromocytoma cell line.[279] PC12 cells also possess nerve growth factor (NGF) receptors and exhibit neuronal differentiation in response to NGF, which is accompanied by a marked decrease in the expression of EGF receptors on the cell surface due to decreased receptor biosynthesis.[482] The NGF-induced downregulation of EGF receptors in PC12 cells may be part of the mechanism by which the differentiating PC12 cells become insensitive to mitogens and exhibit cessation of DNA synthesis and cell division.

Different interferons may have different effects on the expression of the EGF receptor gene in particular types of cells. Treatment of A431 cells with IFN-α or IFN-β represses the expression of the EGF receptor gene without affecting the cellular morphology, but 48 h of treatment of the same cells with IFN-γ results in a significant increase in the level of expression of EGF receptor mRNA and a dramatic change in morphology and viability of the cells.[483] Treatment of MDBK cells with human IFN-α_2 results in a dose-dependent inhibition of cell growth which is associated with a reduction in EGF binding to its receptor on the cell surface.[484]

TGF-α, which is structurally related to EGF and exerts cellular actions through its binding to the EGF receptor, modulates EGF receptor expression; and this effect involves complex synergistic interactions with heterologous hormones and growth factors including TGF-β, thyroid hormone, and retinoic acid.[485] TGF-β contributes to the control of EGF receptor levels in cultured NRK fibroblasts and MDA-468 human breast carcinoma cells.[486,487] Accumulation of EGF receptor mRNA is induced by TGF-β in MDA-468 cells, which may contribute to amplification of the EGF-induced growth inhibitory response of these cells. TNF-α, which stimulates the growth of cultured human fibroblasts, increases the binding of EGF to these cells.[488] The increased EGF receptor expression is due to an increase in receptor synthesis, which increases the number of EGF-binding sites without altering the receptor-binding affinity.

Peptides of the bombesin family may cause a rapid and temperature-dependent reduction in the affinity of the EGF receptor for its ligand in Swiss 3T3 cells.[489] The action of bombesin-like peptides on EGF receptors is apparently mediated by protein kinase C-induced phosphorylation of the receptor. Endothelial cell-derived growth factor (ECDGF), which was isolated from PC13 embryonal carcinoma cells, modulates the activity of the EGF receptor in mouse fibroblasts apparently through protein kinase C-induced phosphorylation.[490]

6. Changes of EGF Receptors in Cultured Cells

Mouse embryo cells arrested in G1 due to nutrient deficiency show a reduction in EGF binding of 10 to 20% of that observed under other conditions.[491] This effect appears to be due to decreased receptor number and could represent the mechanism by which cells are able to enter a quiescent state when faced with the possibility of becoming deficient in essential low molecular weight nutrients. No changes in the number or kinase activity of the EGF receptor have been observed in the liver of senescent mice when compared with the liver of adult mice.[297] Apparently, EGF receptor retains normal structure and function in aging cells.[492] The structure and function of the EGF receptor appears to be unchanged in senescent human fibroblasts in culture, and the reason for the lack of proliferative response of senescent cells to EGF remains unknown. Dimethyl sulfoxide (DMSO), which may produce important effects on the growth and differentiation of cells *in vitro,* stimulates protein-tyrosine kinase activity of the EGF receptor.[493]

Infection of cultured cells with various agents may result in altered EGF receptor expression. Persistent infection of mouse BALB/c 3T3 cells with type 3 reovirus is associated with a 70 to 90% decrease in the number of EGF receptors and altered response to exogenous EGF.[494] It is thus clear that a multitude of agents is involved in the regulation of EGF receptor expression in different types of cells, but the precise biological significance of these regulatory effects is unknown.

VI. EGF RECEPTORS IN TUMOR CELLS

Various types of tumor cells exhibit amplification of EGF receptor gene sequences, and changes in the levels of expression of EGF receptors are frequently found in neoplastic cells. It has been suggested the EGF receptor may be a target for cancer therapy.[495] However, the precise significance of EGF receptor alterations in relation to tumor growth and development remains to be elucidated.

A. AMPLIFICATION OF THE EGF RECEPTOR GENE IN TUMOR CELLS

High amounts of EGF receptors (30 to 100 times more than normal), associated with amplification and overexpression of the EGF receptor gene, are found in the human epidermoid carcinoma cell line A431 as well as in the MDA-468 human breast cancer cell line.[276] MDA-468 cells express high numbers of EGF receptors on the cell surface, and their growth is inhibited by exogenous EGF. The EGF receptor gene was found to be amplified 10- to 14-fold in the BT-20 breast cancer cell line.[496] The growth of cell lines with a high level of EGF receptor gene amplification and overexpression is inhibited by EGF at concentrations that stimulate most other cell lines. EGF-resistant variants isolated from A431 cells that express fewer EGF receptors have a greater potential for differentiation *in vitro* than parental cells.[497] However, the mechanism of EGF-induced inhibition of A431 cell growth does not seem to be directly related to the high number of EGF receptors expressed on the surface of these cells. A variant of A431 cells, A431R-1, exhibits a number of EGF receptors which are similar to those of the parental cells, but the variant cells proliferate in both monolayer and soft agar cultures in response to EGF.[498] The reasons for the differences between A431 cells and A431R-1 cells are not clear but the EGF receptors of the variant A431R-1 cells can be modulated by agents known to modulate the binding of EGF to the parental A431 cells.

High incidence of amplification of the EGF receptor gene has been detected in different human squamous carcinoma cell lines,[496] but the amplification is much less frequent in primary squamous and nonsquamous cell human carcinomas.[500] Malignant squamous cells with amplified EGF receptor genes may more readily adapt to growth in tissue culture, leading to establishment as cell lines, than lines without this amplification. Overexpression of EGF receptors was found in cultured human squamous carcinomas of the skin, oral cavity, and esophagus.[501] In these cells EGF action is associated not with stimulatory but with growth inhibitory effects. EGF receptor overexpression may be a hallmark of squamous cell carcinomas.[502] It is unknown, however, whether the elevated levels of EGF receptors existing in these cells are causally associated with the development of the malignant phenotype and at which stage EGF receptor overexpression occurs. The cloned human hepatoma cell line Li-7A also possesses a very high number of EGF receptors per cell, and its growth is inhibited by EGF.[504]

Primary human tumor cells may contain amplified EGF receptor gene sequences. Amplification of the EGF receptor gene was found in 3 of 21 uncultured tissue samples from primary breast carcinomas.[503] Overexpression of the receptor protein and elevated levels of the kinase activity associated with this protein were found in the cases with EGF receptor gene amplification. Amplification (10- to 30-fold) and overexpression of the EGF receptor gene, associated with unbalanced (1;7) chromosome translocation, were detected in both peripheral blood and bone marrow cells of three patients with myelodysplastic syndromes.[505] The origin of this amplification was not clear, however, since these patients had been treated with cytotoxic and anti-inflammatory drugs; and it is known that drugs blocking DNA synthesis can cause gene amplification. EGF receptor gene amplification or rearrangement was not detected in 17 fresh tumor samples obtained from patients with various forms of head and neck cancer.[506]

B. EXPRESSION OF EGF RECEPTORS IN HUMAN TUMOR CELLS

EGF receptors are frequently present in primary human tumors as well as in cell lines derived from different types of human tumors such as pulmonary, gastric, bone, oral, laryngeal, and cervical cancers.[507,508] EGF receptors are also present in normal human fibroblasts. The number of EGF receptors per cell is highly variable among different primary tumors as well as in tumor cell lines; in some lines the receptors are readily detectable, whereas in other lines the number of these receptors are below the limits of detection.[188] Some cell lines express elevated numbers of EGF receptors.[253] Human tumor cell lines and primary human tumors may contain increased amounts of the EGF receptors without possessing amplified EGF receptor gene sequences.[509,510] EGF receptors are overexpressed in more than half of head and neck tumors, but only a minority of these tumors contain amplified EGF receptor genes.[511]

1. EGF Receptors in Mammary Tumors

The possible role of EGF and its binding to the specific receptors in breast cancer development has been the focus of a number of recent clinical studies.[512] EGF receptors are present in some human breast cancer cell lines, but in other similar lines they are apparently absent. There is no clear-cut correlation between EGF binding and EGF-induced mitogenic activity in breast cancer cell lines with EGF receptors,[513] and even cell lines expressing very high amounts of EGF receptors per cell may not respond *in vitro* to EGF.[514] It has been proposed that tumor-derived growth factors induced by estradiol may act via the EGF receptor in hormone-dependent mammary carcinoma cells.[515] In some human mammary cancer cell lines, a positive correlation may exist between protein kinase C, estrogen receptors, and EGF receptors.[516] However, in MCF-7 human breast cancer cells treated with phorbol ester there is an inverse regulation of the expression of estrogen receptor mRNA and EGF receptor mRNA.[517]

Variants of the MDA-468 human breast cancer cell line, which is characterized by amplification of the EGF receptor gene, may lose this amplification and may exhibit normal numbers of the EGF receptor on the cell surface. These variants are resistant to EGF-induced growth inhibition but retain their tumorigenic potential in nude mice, although the tumors generated by them show a growth rate that is lower than that of tumors derived from the parental cell line.[518] Human breast cancer cell lines such as T47D cells possess high-affinity receptors for calcitriol, and exposure of these cells to calcitriol results in a dose-dependent decrease in the number of EGF receptors expressed on the cell surface.[519]

Qualitative changes of the EGF receptor protein may be found in some breast cancer cell lines. The MDA-MB231 human breast cancer cell line contains EGF receptors with a molecular weight which is slightly larger than that of the receptors from normal human fibroblasts.[520] The difference may be due to posttranslational modification such as glycosylation or phosphorylation. Turnover of EGF receptors in MDA-MB231 cells does not increase upon EGF binding.

Studies of primary human breast tumor samples may be more relevant to the clinical situation than the study of breast cancer cell lines. Significant levels of EGF receptors have been detected in approximately half of the biopsy specimens from unselected primary and metastatic human breast cancer tumors.[521-523] A higher incidence of EGF receptors (up to 91%) was found in other studies.[524,525] It was suggested that the presence of EGF receptors in the tumor cells may be an indicator of poor prognosis in patients with breast cancer.[526-528] In a group of 221 primary breast cancers, the relapse-free survival and overall survival were found to be shorter for EGF receptor-positive vs. EGF receptor-negative tumors.[529] Coexpression of the c-*neu*/*erb*-B2 proto-oncogene and the EGF receptor in the tumor tissue had an additive adverse effect in this series, and it was concluded that the EGF receptor status may be a useful marker for lack of response to endocrine therapy of breast cancer. In another series of 135 primary breast cancer patients, the relapse-free survival rate of EGF receptor-positive patients was found to be worse than that of EGF receptor-negative patients, particularly in node-positive cases.[530] However, no significant association has been found in other clinical studies between EGF receptor expression and lymph node status, tumor size, or differentiation grade of the tumor.[525]

EGF receptor expression in primary breast cancer has been compared to estrogen and progesterone receptor expression.[531,532] In breast cancer biopsies, EGF binding may correlate inversely with the estrogen receptor and progesterone receptor contents and with the age of the patients.[533-535] No association was found between the magnitude of EGF binding in individual tumors and either estrogen or progesterone receptor levels. Studies with transfection of a plasmid expressing high levels of the EGF receptor into the estrogen receptor-positive ZR 75-1 human breast cancer cell line indicate that an increase in EGF receptor expression alone may not be sufficient to induce an estrogen-independent phenotype in human breast cancer cells.[536] Elevated levels of EGF receptors found in primary breast cancer can occur in the presence or absence of gene amplification and can be associated with overexpression of the c-*neu*/*erb*-B2 proto-oncogene.[537] In a minority of mammary tumors both genes are amplified in the primary tumor and metastatic lesions from the same patient.

The possible role of the interaction between EGF and its receptor in human breast cancer development and evolution remains to be characterized. The EGF receptor status may be independent of tumor growth and morphological prognostic parameters.[527] EGF receptor binding in breast cancer tissue and adjacent normal breast tissue may not show statistically significant differences.[534] The ensemble of these results suggests that EGF receptors present in breast cancer tissue may not have a central role in the regulation of tumor growth. However, further studies are required for a proper evaluation of the role of EGF and its receptor in human mammary tumorigenic processes.

2. EGF Receptors in Tumors of the Digestive Tract

Human esophageal carcinoma cell lines have fewer, but higher affinity EGF receptors, and are tolerant to increased levels of EGF.[538] Other human esophageal carcinoma cell lines produce both EGF and TGF-α and express high levels of EGF receptors, suggesting that these two factors may act in an autocrine manner for the growth of esophageal carcinomas.[539] Furthermore, anti-EGF and anti-TGF-α monoclonal antibodies inhibit DNA synthesis in the cultured esophageal cancer cells. Overproduction of EGF receptors was detected in esophageal squamous cell carcinomas.[540] The tumor cells of some patients with esophageal cancer have amplified EGF receptor gene sequences, but considerable variation in the abundance of EGF receptor transcripts is observed in both the tumor and adjacent nontumor specimens.[541]

EGF receptors and their physiological ligands, EGF and TGF-α, are present in human gastric carcinoma cell lines as well as in many primary gastric carcinomas.[542,543] However, the same receptors are present in parietal cells of normal human gastric mucosa. In an immunohistochemical study of 122 gastric carcinomas the incidence of EGF receptor positivity was about 4% in early tumors and 34% in advanced tumors, regardless of the histological type.[544] Stomach cancer patients with synchronous expression of EGF and its receptor in the tumor would have a poor prognosis.[545] EGF and its receptor may play a role in gastric tumor progression through an autocrine and/or paracrine mechanism.

EGF receptor immunoreactivity was observed in 47 of 61 (77%) human colon carcinomas, but no correlation was found with tumor staging or histological type.[542] In a study on 19 samples of primary human colorectal carcinoma and adjacent mucosa no significant difference in the expression of EGF receptors was found between the tumor and the normal mucosa.[546] Similar results were obtained in another study of 55 samples from colorectal carcinoma and normal colon mucosa tissue.[547] At least in some cases, the levels of EGF receptor expression are lower in carcinomatous than in normal colorectal tissue.[548] Differentiation-inducing agents modulate the expression of EGF receptors in colon carcinoma cell lines.[549]

EGF receptors are present in the normal human liver and may be expressed at decreased levels in human primary hepatomas.[550,551] There is an almost ubiquitous expression of EGF receptors in primary pancreatic cancer and chronic pancreatitis.[552] Receptors for EGF as well as the EGF and TGF-α proteins are expressed in the acinar and duct cells of normal human pancreas, and are overexpressed in tumors of the organ.[553] High levels of expression of EGF receptors were found in four human pancreatic carcinoma cell lines; and in three of these lines structural alterations were detected in chromosome 7, although none of the cell lines exhibited an amplification of the EGF receptor gene.[554] Pancreatic cancer cells overexpressing EGF receptors may utilize endogenously produced TGF-α as a superagonist in inducing anchorage-independent growth *in vitro*.[555]

3. EGF Receptors in Lung Tumors

High levels of EGF receptors have been detected in cell lines derived from nonsmall cell lung cancers (non-SCLC) but not in small-cell lung cancer (SCLC) cell lines.[556,557] All histological types of primary non-SCLC express similarly high concentrations of EGF receptors when compared to normal lung tissue.[558] EGF receptor gene amplification was detected in two human lung adenocarcinoma cell lines; whereas three other adenocarcinoma cell lines, as well as two lines of squamous cell carcinoma and one line of small-cell lung carcinoma, did not exhibit amplification of the gene.[559] Some lung cancer cell lines are negative for EGF receptor expression.

Monoclonal antibodies have been applied to the analysis of EGF receptor expression in lung tumors. Immunohistochemical analysis of human lung tumors with monoclonal antibodies against the EGF receptor showed that the receptor is not present in SCLCs, whereas the majority of non-SCLCs is positive.[560] In a study of human lung carcinomas using the monoclonal antibody MAb 528, which is specific for the external domain of the EGF receptor, distinctive patterns of anti-EGF receptor reactivity correlated with tumor histopathology.[561] EGF receptor expression was detected in epidermoid and large-cell lung carcinomas as well as in some lung adenocarcinomas, but was not found in SCLCs. It was suggested that immunohistochemical detection of EGF receptors may be applied to distinguish epidermoid and large-cell lung carcinomas, as well as some adenocarcinomas of the lung, from SCLCs.

Primary human lung tumors frequently exhibit an increase in EGF receptor binding and autophosphorylation.[562] Studies with an EGF receptor-specific monoclonal antibody and the immunoperoxidase technique showed that the receptor is expressed in non-SCLCs and that squamous lung carcinomas exhibit a more highly intense staining than nonsquamous tumors.[563] Moreover, intensity of staining for the EGF receptor may be related to the stage of spread of the tumors. However, a study of 152 archival specimens of non-SCLCs did not indicate any significant relationship between the EGF receptor status and the clinical outcome of the disease.[564] The patients with 50% or more EGF receptor-positive tumor cells tended to have an improved survival, but the results failed to reach statistical significance. In a series of 51 primary non-SCLCs, EGF receptor concentrations were higher in the tumors than in normal lung tissues from the same patient, but no clear relationships were found between EGF receptor concentrations or positivity rates and several prognostic parameters.[565] The binding characteristics of EGF receptors were similar in normal and tumor lung tissue; and no differences were found in EGF receptor concentrations between squamous cell carcinomas, adenocarcinomas, and large cell carcinomas. Since EGF receptor concentrations are very high in fetal lung, the increased expression of EGF receptors in lung cancer may be a consequence of tissue dedifferentiation and could contribute to regulation of the lung tumor growth by an autocrine mechanism. Differential expression of EGF receptors and their ligands (EGF, TGF-α, and amphiregulin) in 114 non-NSCL tumor samples and adjacent benign lung tissue did not indicate any correlation with either disease-free or overall survival.[566] Further studies are required for an evaluation of the biological significance of EGF receptor expression in human lung tumors and the possible application of EGF receptor analysis in the clinical evaluation of lung cancer patients.

A plasma membrane glycoprotein of 160 kDa (gp160), similar in several respects to the EGF receptor, was detected in a large proportion of human lung tumors.[567] The gp160 protein is recognized by the monoclonal antibody 5E8 and is distinct from the EGF receptor molecule. The human lung cancer cell line Calu-3, which is negative for the EGF receptor, is positive for gp160. Protein gp160 serves as a substrate for protein kinase, but its normal function was not characterized.

4. EGF Receptors in Brain Tumors

High levels of EGF receptors may be present in human brain tumors, especially in tumors of nonneuronal origin such as glioblastomas and meningiomas, whereas the levels present in neuroblastomas are similar to those observed in the brain of patients who died from diseases not related to the central nervous system.[194,568,569] The EGF receptor gene may be amplified and overexpressed in more than half of malignant gliomas, but not in meningiomas.[570] Moreover, in primary meningiomas there is no clear correlation between EGF receptor levels and the histopathology of tumors.[571] In contrast, astrocytomas exemplify a progression of malignancy from benign (low grade astrocytomas) toward an increased malignant phenotype (glioblastomas), and in these tumors there is a correlation between the amounts of EGF receptors and the degree of malignancy. Multidrug-resistant human neuroblastoma cell lines, obtained by selection with vincristine or actinomycin D, may have increased levels of EGF receptor expression compared to the drug-sensitive parental cells.[572] The multidrug-resistant neuroblastoma cells do not contain amplified EGF receptor genes, and reversion to drug sensitivity is accompanied by a return to the parental level of EGF receptor.

Human malignant glioma (glioblastoma) cell lines frequently express EGF receptor mRNA and protein.[573] In addition, most of these lines express PDGF mRNA and PDGF receptors, suggesting the operation of two autocrine loops in the malignant glioma cells: one related to EGF and the other to PDGF. Structural alterations of the EGF receptor gene have been detected in xenografts of human glioma cell lines containing amplified EGF receptor genes.[574] EGF receptor proteins present in different types of human tumor cell lines may exhibit qualitative alterations. EGF receptors in the human glioblastoma cell line SF268 may differ from those in epidermoid carcinoma cell line A431.[575] SF268 cells contain two- to fivefold amplification of the EGF receptor gene, which gives rise to abundant quantities of surface EGF receptors that bound EGF with high affinity; however, binding of the ligand fails to elicit cellular DNA synthesis and the receptor is enzymatically inactive, not exhibiting tyrosine kinase activity.[576] The amplified EGF receptor gene of SF268 cells may contain a mutation that affects one or more functions of the receptor. The malignant human glioma cell line D-298 MG contains an amplified and rearranged EGF receptor that is characterized by an in-frame deletion of 83 amino acids in the extracellular domain of the receptor.[577] Both EGF and TGF-α bound to the mutant receptor with high affinity and enhanced the kinase activity of the receptor. Overexpression of the mutant EGF receptor by amplification of the gene sequences may provide a growth advantage to D-298 MG cells. The mutant receptor was capable of transducing EGF-stimulated glioma cell proliferation and invasiveness in an *in vitro* model. Primary human malignant gliomas may contain amplified EGF receptor gene sequences.[578,579]

Three molecules with EGF-binding capacity (mol wt 170, 150, and 125 kDa) were identified in biopsies from human meningiomas.[580] Expression of such putative EGF receptors is modulated by PDGF, but the biological significance of the mixed population of EGF receptors in tumors such as human meningiomas is unknown. The cells of two cases of glioblastoma multiforme containing amplified c-*erb*-B genes exhibited EGF receptors about 30 kDa shorter than the normal 170-kDa receptor due to short deletions within the c-*erb*-B/EGF receptor sequence.[581] The 140-kDa abnormal EGF receptor contained in these cells showed a constitutive elevation of protein-tyrosine kinase activity in the absence of its ligand, which may have contributed to the tumorigenic process. The KE cell line derived from a human glioblastoma tumor specimen express a 190-kDa protein (p190) which represents an endogenous structurally and functionally altered EGF receptor.[582] The p190 protein synthesized by KE cells exhibits an alteration in the amino-terminal region and contains an activated tyrosine kinase that cannot be modulated by the addition of isolated growth factors. The possible role of EGF receptor expression in the malignant behavior of transformed cells is unknown but the effects of an antibody specific to the EGF receptor on glioma cell lines suggest that in brain tumors expressing an increased number of EGF receptors, EGF of an EGF-like ligand such as TGF-α may selectively facilitate expansive tumor growth and tumor cell invasion.[583]

An apparent loss or lack of expression of EGF receptors on the cell surface is observed in certain tumor cells. EGF receptors are present in normal human pituitary cells but may be absent in human pituitary adenomas of both the hormone-secreting and nonsecreting types.[584]

The possible clinical value of the analysis of EGF receptor expression in intracranial tumors remains controversial. In a study of 55 primary cerebral tumors and 131 breast tumors it was concluded that the EGF receptor protein tends to be expressed in malignant tumors of poorer prognosis, but the extent of this expression is variable and not sufficiently consistent to be of prognostic significance.[585]

5. EGF Receptors in Melanoma

The EGF receptor is expressed by human melanoma cells containing an extra copy of chromosome 7 (where the receptor gene is normally located).[586] Cells of benign pigmented lesions (nevi) or radial growth phase (nonmetastatic) primary melanoma do not express EGF receptors and do not have extra copies of chromosome 7. The possible role of EGF receptor expression in melanoma cell growth is unknown, but the anti-EGF receptor antibody can partially suppress spontaneous metastasis of human melanoma M24met tumors in athymic nude mice.[587]

6. EGF Receptors in Sarcomas

Increased levels of EGF receptors may be expressed in some human sarcomas.[588] EGF receptors are present in cell lines established from human osteosarcoma and giant tumors of the bone, but their characteristics are comparable or identical to those of normal cells; and the same receptors are present in normal bone, which suggests that EGF may be involved in normal bone metabolism.[589] Kaposi's sarcoma-derived cells exhibit a low mitogenic response to both EGF and TGF-α, but this alteration is not due to a reduced expression of EGF receptor mRNA and protein in these cells.[590]

7. EGF Receptors in Urogenital and Gonadal Tumors

EGF may have a role in the growth of urogenital tumors. Primary cultured adult rat urothelial cells secrete into the medium EGF or an EGF-like substance which is able to stimulate the growth of rat bladder stroma cells in a coculture system, and this effect is probably mediated by EGF receptors.[591] EGF receptors are present at highly variable levels in a diversity of human urogenital tumors. Elevated levels of EGF receptor mRNA and protein have been found in more than one half of surgical specimens of renal cell carcinoma cancers, but amplification of the EGF receptor gene is only occasionally observed in these tumors.[592-594] Human renal tumors may also display high levels of IGF-I receptors and may overexpress TGF-α mRNA and protein in comparison to adjacent normal kidney tissue. Variations in EGF receptor levels have been observed in bladder cancer when invasive and superficial tumors have been compared.[595] Receptors for EGF are present in the human urinary bladder carcinoma cell line 5637, in which growth is regulated by EGF.[596] A study on primary bladder tumor membranes demonstrated the presence of EGF receptors in the tumors.[597] Invasive bladder tumors exhibited higher EGF receptor levels than superficial tumors or normal bladder mucosa.

EGF receptors are expressed in human prostatic tumors. Comparison of the concentrations of EGF receptors in primary prostate cancer and benign prostatic hyperplasia showed that expression of EGF receptors is higher in the hyperplastic than in the carcinoma tissue.[598] In prostate carcinoma, the expression of EGF receptors varies according to the histological grade of the cancer; well-differentiated tumors have more abundant EGF receptors than poorly differentiated tumors.

EGF receptors are present in normal human endometrium, and a survey of 37 primary endometrial adenocarcinomas indicated a progressive decrease of EGF receptor expression in cancers of increasing grade.[599] The human choriocarcinoma cell line JEG-3 expresses EGF receptors, and the JEG-3 cells respond to addition of EGF with a marked increase in the secretion of hCG into the medium.[589] Occupancy of the EGF receptors present in the human endometrial carcinoma cell line RL95-2 by the ligand results not in stimulation but in inhibition of cell proliferation.[600]

Increased expression of EGF and insulin-like growth factor I (IGF-I) receptors has been found in ovarian adenocarcinoma tissues.[601] A number of these tumors (between 20 to 30%) contain amplified c-*myc* and c-*erb*-B-2 gene sequences. Analysis of biopsy specimens of gynecological malignancies indicated that EGF receptors were present in a number of ovarian and uterine tumors.[602] All vulvar epidermoid carcinomas and uterine sarcomas analyzed were EGF receptor positive. Higher EGF receptor levels were found in metastatic than in primary ovarian tumors from this series, and the EGF receptor was expressed mainly in well or moderately differentiated endometrial tumors. EGF receptors are more likely to be present in malignant ovaries than in benign ovarian tumors or normal ovaries.[603]

8. EGF Receptors in Thyroid Tumor Cells

Examination of EGF receptors in primary-cultured human thyroid cells from 9 adenomas and 18 differentiated thyroid carcinomas indicated that total binding of [^{125}I]-EGF to these cells does not differ from that of the adjacent nonneoplastic counterparts.[604] Both nonneoplastic and neoplastic thyroid cells possess EGF receptors with two components, and the association constant of the carcinoma cells high-affinity EGF receptor component was found to be lower than that of adjacent nonneoplastic thyroid cells as well as that of thyroid adenoma cells. Addition of EGF to the cultured cells showed that the increase in [^{3}H]thymidine incorporation was similar in neoplastic and nonneoplastic thyroid cells. Addition of TSH caused an increase in thymidine incorporation by thyroid adenoma cells, but not by carcinoma or nonneoplastic cells. Combined treatment with EGF and TSH additively promoted thyroid adenoma cell growth only.

9. EGF Receptors in Skin Tumors

Basal cells of normal human skin synthesize DNA, undergo mitosis, and then differentiate into upper squamous layers. Expression of EGF receptors in the normal skin as well as in keratoacanthoma, solar keratosis, Bowen's disease, squamous cell carcinoma basal cell carcinoma, and Paget's disease is associated with cell proliferation.[605] In contrast, the expression of Ras proteins in the same tissues is associated with cell differentiation.

C. EXPRESSION OF EGF RECEPTORS IN EXPERIMENTAL TUMORS

The tumorigenic potential of the EGF receptor/c-*erb*-B proto-oncogene is indicated by the results of experiments in which the EGF receptor gene has been inserted into a retrovirus vector. NIH/3T3 fibroblasts transfected with this vector or infected with the corresponding rescued retrovirus develop a fully transformed phenotype *in vitro* that requires both functional EGF receptor expression and the presence of EGF in the growth medium.[286] Oncogenic viruses, carcinogenic agents, and tumor promoters may produce important changes in EGF receptors.

1. Acute Transforming Retroviruses

Acute transforming retroviruses may induce a rapid and profound decrease in EGF receptor levels.[606] Rous sarcoma virus (RSV)-induced transformation of rodent fibroblasts is associated with a marked decrease in the expression of EGF receptors on the cell surface.[607] The v-*src* and v-*fps* oncoproteins, which possess tyrosine kinase activity, may lead to a rapid decrease in EGF receptors through activation of the EGF-dependent growth regulatory system.[608] A similar decrease in the expression of EGF receptors is observed *in vivo* when renal tumors are induced with Kirsten murine sarcoma virus (K-MuSV).[609] The presence of EGF receptors may be required for the expression of malignant properties by cells transformed by acute retroviruses. A-MuLV-transformed fibroblasts lacking the EGF receptor are not tumorigenic in nude mice.[610]

2. Oncogenic DNA Viruses

Infection of cells with Simian virus 40 (SV40) may produce important quantitative and qualitative changes in EGF receptors, which may be associated with changes in the oligosaccharide portion of the receptor molecule.[611] SV40-transformed mouse fibroblasts would secrete autocrine EGF-like growth factors, which may contribute to mask the expression of EGF receptors on the cell surface and may alter the receptor regulation by retinoic acid.[612] Transformation of REF cells by adenovirus Ad5 results in the loss of thyroid hormone inducibility of Na^+/K^+-ATPase and a decreased binding of EGF to cell membrane receptors.[613] Reversion to anchorage-dependent growth, flat morphology, and nontumorigenicity of rat brain cells infected with adenovirus Ad2 is accompanied by increased expression of EGF receptors on the cell surface and decreased production of TGF-α.[614] It is not known whether these complex changes are causally associated with the process of neoplastic transformation or are instead the consequence of this process.

3. Chemical Carcinogens

Hepatocellular carcinomas induced in rats by acetylaminofluorene (AAF) or diethylnitrosamine (DENA) are characterized by a marked decrease in EGF receptor number and/or affinity, as well as in autophosphorylation of the receptors, as compared to peritumorous or normal tissues.[615-617] Similar changes may occur in the insulin receptors. These changes are associated with reduced levels of mRNAs for EGF and insulin receptors.[618] Although carcinogen administration causes an initial alteration in the

EGF and insulin receptors of the majority of liver cells, most of them recover their receptor function; and only a small minority become truly initiated and retain the changed characteristics up to the tumor stage where they are fixed in an irreversible form.[618] Since the changes occur in both the EGF and insulin receptors, the possibility of a common underlying mechanism for alteration of the receptors during chemical carcinogenesis should be considered.

Decreased levels of EGF receptors are observed in chemically transformed cells at saturation density *in vitro*.[619] Gradual decrease in EGF receptor expression is observed in the course of spontaneous neoplastic progression of whole Chinese hamster embryo (CHE) lineages during serial passage in culture.[620]

D. BIOLOGICAL SIGNIFICANCE AND MECHANISMS OF EGF RECEPTOR ALTERATIONS IN MALIGNANT CELLS

The role of EGF receptors in the origin and/or maintenance of a transformed phenotype remains poorly understood. It has been suggested that a high concentration of EGF receptors on the cell surface may facilitate the growth of tumor cells both *in vitro* and *in vivo*.[621] However, it is clear that overexpression or aberrant expression of EGF receptors is not universally involved in carcinogenic processes. A decrease in EGF receptors is not a general phenomenon of neoplastic transformation since it is not observed in cells transformed by DNA viruses or by most types of chemical carcinogens.[622] TGF-β induces not a decrease but a rapid increase in the number of EGF receptors on the plasma membrane.[486]

Different molecular mechanisms may be involved in the modulation of EGF receptor expression on the surface of tumor cells.[623] The decreased levels of EGF receptors observed in transformed cells could result either from an alteration in the processes of synthesis, degradation, and/or location of functional receptors; or from the production in transformed cells of endogenous EGF or a related substance that binds to the EGF receptor.[609] Changes in the pattern of DNA methylation may have a role in the modulation of EGF receptor expression. Hypomethylation of the EGF receptor gene and the c-*myc* gene has been detected in human hepatocellular carcinomas as well as in fetal liver.[624] Structurally altered EGF receptors may occur in some types of transformed cells. Activation of c-*erb*-B in avian leukosis virus (ALV)-induced erythroblastosis leads to the expression of a truncated EGF receptor with tyrosine kinase activity.[625]

Some malignant cells may be capable of synthesizing EGF receptors but the receptor molecules may not be expressed on the cell surface. Undifferentiated OC15 mouse embryonal carcinoma stem cells synthesize EGF receptors with specific kinase activity, but these receptors are not expressed on the cell surface until after differentiation.[626] A possible explanation for this alteration is the presence of conformational changes in membrane components surrounding the receptors in transformed cells. Cells transformed by acute retroviruses or viral oncogenes may be characterized by the apparent absence of EGF-binding sites on the cell surface. Rat kidney cells transformed by K-MuSV lose the ability to bind EGF.[609] However, the presence of EGF receptors in these cells can be demonstrated by immunoprecipitation with a monoclonal antibody to the receptor; and, furthermore, two-dimensional peptide mapping indicates that the EGF receptor present in the transformed cells has apparently the same structure as its normal counterpart.[627] The loss of EGF receptor activity on the surface of v-K-*ras*-transformed NRK cells is not due to improper insertion of EGF receptors into the membrane, but is associated with an increased phosphorylation of the EGF receptors on threonine and serine.[628] This modification may be mediated by protein kinase C since v-K-*ras*-transformed cells are characterized by a constitutively high level of expression of this enzyme in the absence of stimulators such as phorbol esters. Calmodulin antagonists induce decreased binding of EGF to SV40-transformed human fibroblasts (WI38 cells) in a dose-dependent manner, but this effect is not observed in normal human cells.[629] The effect observed in transformed cells would be due to a decrease in the affinity of the plasma membrane EGF receptor for the EGF molecule when calmodulin activity is inhibited.

A direct relationship between the altered EGF receptor expression and cellular tumorigenic properties appears to be nonexistent.[630] Although the growth of A431 cells implanted in athymic mice may be inhibited by monoclonal antibodies against the EGF receptor,[631,632] the mechanism of this inhibition is unclear. Moreover, the effect of the EGF receptor-specific monoclonal antibody MAb 425 on the growth of A431 cells depends on its concentration; whereas concentrations equal or higher than 1 n*M* induce growth inhibition, low concentrations of MAb 425 may stimulate cell growth.[633] Monoclonal antibodies specific to the human EGF receptor may be unable to block autocrine growth stimulation in TGF-secreting melanoma cells.[634]

The disparate variations in the number of cellular EGF receptors under different conditions make rather unlikely the possibility that the process of transformation should always be mediated by an activation of the EGF-dependent cellular proliferation regulatory system. However, cellular processes beyond EGF binding may be important for the EGF-induced regulation of cell growth in normal or malignant cells. While both malignant and normal human urothelial cells bind EGF equally well *in vitro,* only urothelial transitional carcinoma cell lines respond to EGF in two parameters relevant to neoplasia: growth and induction of ODC activity.[635]

1. Tumor Promoters and EGF Receptor Expression

A profound decrease in EGF receptor levels occurs in cells exposed to potent tumor promoters that activate protein kinase C such as phorbol esters.[636] These compounds share several biological activities with EGF, stimulating the proliferation of mammary gland epithelium and preventing DNA fragmentation induced by serum deprivation in cultured fibroblasts.[637,638] Phorbol esters inhibit acid secretion by the stomach *in vivo* in a manner which is similar to EGF.[639] However, it is not known whether this action of phorbol esters is mediated by the EGF receptor.

The functional properties of EGF receptors may be modified by tumor promoters. The decrease in EGF receptor expression observed in phorbol ester-treated A431 cells is associated with altered EGF receptor phosphorylation on tyrosine residues.[325,640,641] Phorbol diesters, as well as other tumor promoters (indole alkaloids and polyacetates), inhibit EGF-stimulated tyrosine phosphorylation of the receptor, which correlates with loss of EGF binding to the high-affinity EGF receptor.[640] The phorbol esters TPA and PMA may induce increased serine and threonine phosphorylation within the binding domain of the EGF receptor.[249,295,628,641,642] The major site of phorbol ester-induced EGF receptor phosphorylation is a threonine residue located on the amino-terminal, EGF-binding domain of the receptor molecule.[124] Phorbol esters cause the phosphorylation of EGF receptors in normal human fibroblasts at a unique site, Thr-654, which is located nine amino acids away from the predicted transmembrane domain of the receptor on the cytoplasmic side of the plasma membrane.[643] Thr-654 may be involved in regulating the tyrosine kinase activity of the EGF receptor as well as the binding of EGF. Inhibition of the EGF receptor affinity for its ligand caused by phorbol esters correlates with phosphorylation of Thr-654 but not other sites on the receptor.[644]

The effects of tumor promoters on the phosphorylation and affinity of EGF receptors may be exerted by substitution for diacylglycerol at the level of protein kinase C activation. A strong correlation has been observed between the potency of diacylglycerol analogues to modulate protein kinase C activity *in vitro* and their capacity to mimic the action of phorbol ester *in vivo.*[645] Phorbol esters may regulate EGF receptor expression through a mechanism involving changes in protein kinase C activity, but this regulation is not necessarily associated with changes in the oligomeric state of the receptor.[646] Protein kinase C activation does not represent a universal mechanism for regulation of EGF receptor expression by tumor promoters. The liver tumor promoter phenobarbital decreases EGF binding in rat hepatocytes, but this response is mediated by a pathway which is independent of protein kinase C activation.[647] A distinct class of tumor promoters represented by palytoxin (a water-soluble compound of mol wt 2681 isolated from a coelenterate from the genus *Palythoa*) exerts a negative regulation on the EGF receptor expression through a pathway that is not dependent on protein kinase C.[648]

The pattern of cellular response to phorbol ester-induced decrease in EGF receptor expression may depend on physiological conditions related to the cycling state of the cell. Addition of TPA with low serum concentration to quiescent cultures of BALB/c 3T3 cells causes a rapid decrease in the subsequent binding and accumulation of EGF without stimulating cell cycle traverse, but a marked increase in DNA synthesis is observed when TPA is added directly to spent medium or to fresh medium containing insulin.[649]

2. Effects of Oncoproteins on EGF Receptor Expression

The transforming action of oncoproteins may be exerted, at least in part, through activation of growth factor receptors. The EGF receptor becomes constitutively activated and phosphorylated on tyrosine in cells transformed by the v-*src* oncogene.[650] Moreover, in cells coexpressing the v-Src oncoprotein and the EGF receptor, phospholipase C-γ is constitutively phosphorylated. Overexpression of normal or mutated c-H-*ras* genes in noninvasive bladder cancer cells (human papillary transitional cell carcinoma cell lines) confer on these cells an invasive phenotype *in vivo,* and this effect is associated with increased EGF receptor expression.[651] Certain oncoproteins may affect intracellular signaling pathways through the altered expression of EGF receptors.

VII. POSTRECEPTOR MECHANISMS OF ACTION OF EGF

The postreceptor mechanisms involved in EGF action are exceedingly complex, and there is an almost complete ignorance about their sequential operation in either normal or transformed cells. Many changes are observed in sensitive cells exposed to EGF, including alterations at the level of the plasma membrane, the cytoplasm, and the nucleus. EGF not only can modulate the functions of cells directly exposed to the factor but also can have important physiological effects on neighbor cells. It is capable of altering gap junctional intercellular communication in cells such as normal human epidermal keratinocytes grown in a defined medium.[652] Intercellular communication is an important mechanism by which cells exchange regulatory ions and molecules to control tissue homeostasis, cell growth, and differentiation as well as synchronization of tissue reactions and tissue regeneration.

The intracellular mediators or second messengers responsible for the pleiotropic effects induced by EGF are little known.[653] The postreceptor mechanisms that allow a specific response to EGF may be present even in cells in which the EGF receptor is absent. Introduction by means of an expression vector of an EGF receptor gene into a nontumorigenic hematopoietic cell line (the 32D myeloic cell line), which is devoid of EGF receptors and is absolutely dependent on IL-3 for its proliferation and survival, confers to the manipulated cells the ability to express EGF receptors and to utilize EGF for transduction of a mitogenic signal.[654] It is thus evident that 32D hematopoietic cells, in spite of the absence of EGF receptors, contain and express all of the intracellular components of the EGF signaling pathway necessary to evoke a mitogenic response and sustain continuous proliferation.

A. THE ADENYLYL CYCLASE SYSTEM

The role of the adenylyl cyclase system in the mechanism of action of EGF is little known. In several epithelial cell lines, EGF may cause a rapid modulation of cAMP metabolism.[655] The concentrations of cAMP in response to a variety of cAMP-elevating agents are increased in the EGF-stimulated epithelial cells. In bovine luteal cell membranes, ligand-activated EGF receptors interact with the adenylyl cyclase system and act as modulators of this system.[656] The EGF receptor could interact with a component of the adenylyl cyclase system, which may result in stimulation of cAMP synthesis and accumulation of cAMP. In cultured rat cardiac myocytes EGF can also stimulate cAMP accumulation.[657] It seems likely that at least some of the physiological effects of EGF may be mediated by cAMP-dependent protein kinases.

B. POLYAMINE SYNTHESIS

ODC plays a key role in the synthesis of polyamines. Treatment of A431 human carcinoma cells with EGF results in markedly stimulated ODC activity despite the inhibition of mitogenesis produced by EGF in these cells.[658] Therefore, an alteration of polyamine metabolism does not seem to provide an explanation for the aberrant mitogenic response of some cells to EGF. The use of ODC inhibitors suggests that the mechanism of EGF-induced inhibition of mitogenesis in A431 cells is operative at a step distal to the initial activation of ODC by EGF.

C. ION FLUXES AND DISTRIBUTION

At least in some cellular systems, changes in monovalent and/or divalent ion fluxes and intracellular Ca^{2+} distribution may be involved in the mechanisms of EGF action. EGF induces an elevation of the pH_i in chicken granulosa cells.[659] In Swiss mouse 3T3 cells, the action of EGF is not associated with changes in the $[Ca^{2+}]_i$, but EGF induces intracellular alkalinization through stimulation of the Na^+/H^+ exchange.[660] In T51B epithelial cells derived from rat liver, EGF-induced DNA synthesis requires extracellular Ca^{2+} influx.[661] Treatment of rat hepatocytes with EGF *in vitro* results in a rise in the $[Ca^{2+}]_i$, and the initial phase of this increase is not affected by reducing the external Ca^{2+} concentration.[662] In serum-deprived Rat-1 cells, EGF stimulates Ca^{2+} influx via channels which are not voltage dependent.[663] Addition of EGF to human A431 cells induces a rapid but transient rise in the $[Ca^{2+}]_i$ by stimulating the entry of Ca^{2+} into the cell through voltage-independent Ca^{2+} channels located in the plasma membrane.[664,665] This, in turn, causes the activation of Ca^{2+}-dependent K^+ channels, which results in a delayed membrane hyperpolarization and leads to the activation of a second class of Ca^{2+} channels that are sensitive to membrane hyperpolarization. There is an absolute requirement for EGF receptor-associated tyrosine kinase in EGF-induced activation of Ca^{2+} channels.[666] The EGF-induced activation of Ca^{2+} channels may involve stimulation of phospholipase A_2 activity, probably as a consequence of EGF receptor tyrosine kinase activity.[667] Subsequent 5-lipoxygenase-mediated production of leukotriene C_4 may serve as the second messenger in the activation of Ca^{2+} channels by EGF. Intracellular redistribution of Ca^{2+} may be associated with alterations in phosphoinositide metabolism. The calmodulin antagonist — trifluoperazine — is a potent inhibitor of EGF-induced tyrosine phosphorylation of the EGF receptor, although this effect may not be due to calmodulin antagonism since another calmodulin antagonist — W7 — has only a slight effect on EGF receptor phosphorylation.[668]

The functional activity of the Na^+/H^+ antiporter, which has been characterized as a 110-kDa glycoprotein,[669] may be regulated by EGF in its target cells. In intestinal cells, EGF upregulates Na^+/H^+ exchange activity.[670] Mitogenic stimulation of human and hamster cell lines with EGF resulted in increased phosphorylation of the Na^+/H^+ antiporter on serine residues. However, the precise role of Na^+/H^+ antiport activation in EGF action is not clear. Studies with the human breast cancer cell line MDA-468, which contains an amplification of the EGF receptor gene and expresses very high numbers of EGF receptors, indicate that although activation of antiport and cytoplasmic alkalinization takes place in response to EGF, it is not a prerequisite for augmented gene expression, growth inhibition, or growth stimulation.[671] Activation of Na^+/H^+ exchange and cytoplasmic alkalinization may not be necessary for EGF-induced changes in cell proliferation. Further studies are required for a proper evaluation of the role of Na^+/H^+ antiporter activation in the mechanisms of action of EGF in different types of cells.

D. ACTIVATION OF G PROTEINS

The membrane-associated guanosine triphosphate (GTP)-regulatory proteins (G proteins) are essential signal-coupling components involved in the cellular response to many hormones, growth factors, and neurotransmitters. Studies with pertussis toxin, which acts as a potent and selective inhibitor of G protein activity, show that G proteins are involved in at least one pathway of the growth-modulating response to EGF in the MDA-468 human breast cancer cell line.[652] This pathway is, however, not responsible for the activation of c-*myc* and c-*fos* proto-oncogene expression observed in EGF-stimulated MDA-468 cells. In isolated perfused rat hearts, EGF stimulates adenylyl cyclase activity via a G protein.[673] The G protein which mediates the actions of EGF may be G_s or a G_s-like protein.

E. ACTIVATION OF RAS PROTEINS

Cellular Ras proteins have an important role in the transduction of the EGF signal from the activated cell surface receptor to the biochemical machinery of the cell. Activation of Ras proteins is associated GTP binding; this depends on the opposite actions of the GTPase-activating protein (GAP), which negatively regulate Ras activity by increasing GTP hydrolysis to GDP, and the guanine nucleotide exchange factor, which stimulates the dissociation of Ras-bound GDP.

The GAP protein is phosphorylated by the activated EGF receptor on Tyr-460, a residue adjacent to SH2 domains.[674] However, in Rat-1 fibroblasts EGF leads to a rapid activation of Ras proteins by a mechanism involving modulation of Ras guanine nucleotide exchange factor activity, without significant changes in GAP activity.[675] In proliferating rat parotid gland acinar cells, EGF stimulates the formation of active GTP-Ras complexes by increasing the activity of the Ras guanine nucleotide exchange factor.[676] EGF receptors generate Ras-GTP complexes more efficiently than insulin receptors, which may explain the stronger mitogenic activity of EGF in certain types of cells.[677]

The Grb2 protein, which contains one SH2 domain flanked by two SH3 domains, plays a key role in signal transmission from the ligand-activated EGF receptor to the intracellular machinery.[678,679] Grb2 associates with the EGF receptor via its SH2 domain and acts as a bridge to the Sos protein through its SH3 domains.[680-684] Sos, a guanine nucleotide-releasing factor for Ras, is essential for the control of Ras activity. In Rat-1 fibroblasts, Sos protein forms a complex with the activated EGF receptor and Grb2 in the plasma membrane.[685] The SH2 domain of Grb2 may bind to the Tyr-1068 autophosphorylation site of the EGF receptor, and the SH3 domains of Grb2 bind to the carboxyl-terminal domain of Sos. Activation of Sos by Grb2, which may be affected by phosphorylation, increases the amount of GTP-bound form of Ras, which in some manner may result in mitogenesis or other effects (Figure 3.1).

F. PHOSPHOINOSITIDE METABOLISM

There are conflicting results on the role of phosphoinositides in the cellular mechanism of action of EGF. Phosphatidylinositol (PI) kinase activity is not intrinsic to the EGF receptor molecule.[686] However, the receptor is associated with PI kinases in various types of cells. A PI kinase activity is associated with the EGF receptor in a mouse fibroblast cell line expressing human EGF receptors, and this activity increases with treatment of the cells with physiologically relevant concentrations of EGF.[687] Antiphosphotyrosine antibody immunoprecipitates PI 3-kinase activity in cells exposed to EGF,[688] suggesting a role for EGF in the regulation of an enzyme involved in phosphatidylinositol phosphorylation at the D3 position of the

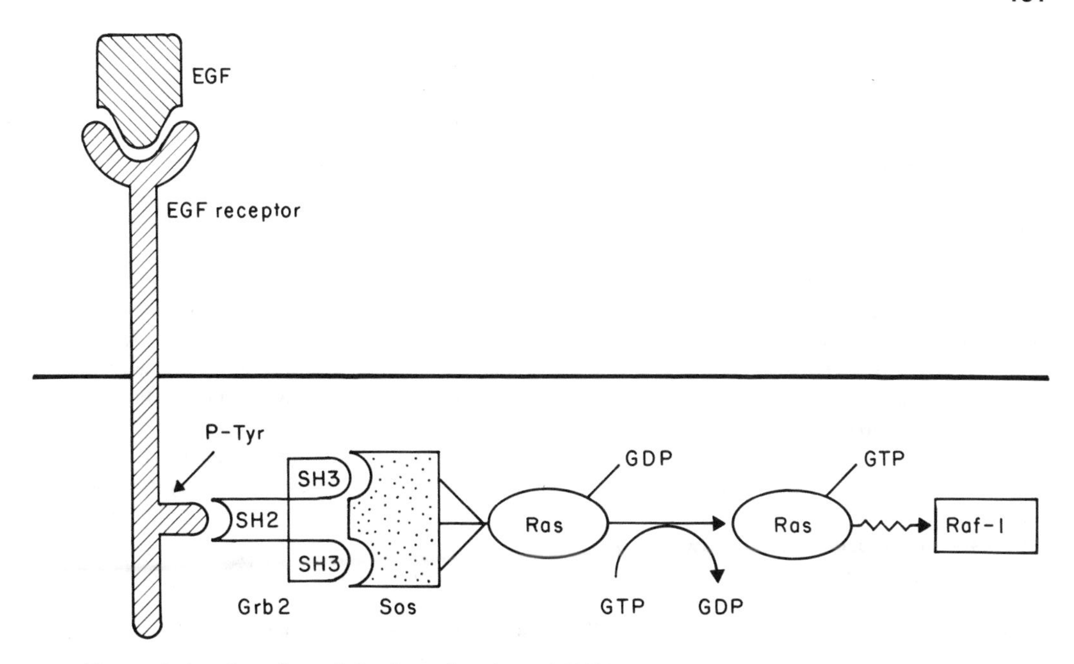

Figure 3.1. Coupling of the ligand-activated EGF receptor to Ras-signaling pathways.

inositol ring. Other enzymes, PI 4-kinase and PI 5-kinase, which phosphorylate the inositol ring at position D4 and D5, respectively, are associated with the region of the EGF receptor located between the inner membrane face and the ATP-binding site.[689] PI kinases may be associated with and tyrosine phosphorylated by the EGF receptor as part of the mechanism coordinating responses between signal transduction pathways. However, EGF-induced phosphorylation of PI kinases on tyrosine may not be sufficient to activate the enzymes.

Different types of cells under different conditions may show a diversity of changes in phosphoinositide metabolism in response to EGF. In cells that overexpress the EGF receptor, EGF rapidly stimulates the production of inositol 1,4,5-trisphosphate and increases the incorporation of phosphate into phosphatidylinositol 4-monophosphate, phosphatidylinositol 4,5-bisphosphate, and phosphatidic acid as early as 15 s after the addition of EGF.[690-692] In A431 cells, these changes are associated with activation of the respective PI kinases, which involves processes dependent on phosphorylation of tyrosine residues.[693] PI kinase activity is increased in the particulate fraction prepared from A431 cells that have been pretreated with EGF.[694] These metabolic changes are accompanied by a concomitant rise in cytosolic Ca^{2+} which is apparently due to mobilization of Ca^{2+} from intracellular stores.[326,695] The effects of EGF on phosphoinositide metabolism and Ca^{2+} intracellular redistribution in a human hepatocellular carcinoma-derived cell line phospholipase C (PLC/PRF/5) are similar to those observed in A435 cells.[696] The effects of EGF on phosphoinositide turnover and $[Ca^{2+}]_i$ are greatly augmented in Rat-1 cells by TGF-β.[697] In cultured rat hepatocytes, both EGF and angiotensin II stimulate a rapid increase in the production of inositol 1,4,5-trisphosphate and an increase in $[Ca^{2+}]_i$.[662] Treatment of hepatocytes with phorbol ester prior to the addition of EGF blocked these effects. An almost complete inhibition of EGF-induced phosphoinositide hydrolysis is observed in A431 cells treated with phorbol ester.[698]

Phosphorylation and activation of phospholipase C may have a role in the mechanism of EGF action. Treatment of A431 cells with EGF results in stimulation of the phosphorylation of phospholipase C on both tyrosine and serine residues, and this phosphorylation is independent of EGF receptor internalization and extracellular calcium.[699] Activation of phospholipase C induced by EGF is inhibited by erbstatin, a tyrosine kinase inhibitor.[700] The EGF receptor, but not the insulin receptor, can directly phosphorylate the PLC-II/PLC-γ form of the enzyme in an efficient and selective manner *in vitro* and *in vivo,* suggesting that activation of PLC-II may have a role in the EGF receptor-signaling mechanism.[701-704]

Both EGF and PDGF promote a rapid translocation of PLC-II/PLC-γ from the cytosol to the plasma membrane.[705,706] Activation of PLC induced by EGF may result in stimulation of phosphoinositide turnover and formation of inositol phosphates. As a consequence of these changes, both protein kinase C and diacylglycerol kinase may be activated, which results in 1,2-diacylglycerol breakdown to phosphatidic acid.[707-710] In turn, 1,2-diacylglycerol can modulate binding and phosphorylation of the EGF receptor, probably through activation of protein kinase C.[322,711,712] Treatment of 3T3 cells with diacylglycerol results in a rapid decrease in the affinity of the EGF receptor for the ligand.[713] The mechanism of EGF-induced phospholipase C activation is unknown but it may be associated with phosphorylation of the enzyme on tyrosine by a primary or secondary action of the activated EGF receptor tyrosine kinase.[714]

In certain cellular systems the action of EGF is not associated with changes in phosphoinositide metabolism, or the changes induced by EGF in this metabolism may not be responsible for the cellular actions of EGF. Neither insulin nor EGF stimulate the synthesis of phosphatidylinositol, phosphatidylinositol-4-phosphate, or phosphatidylinositol-4,5-bisphosphate in rat liver plasma membranes or intact hepatocytes, suggesting that the insulin- and EGF-stimulated receptor kinases do not act on phosphoinositides in the liver.[715,716] EGF alone or in combination with insulin stimulates growth-arrested CHL fibroblasts to undergo DNA replication and to divide without significant activation of phosphatidylinositol turnover.[717] The ability of insulin and IGF-I to mimic the effect of EGF on labeling of phosphatidylinositol-3,4-bisphosphate without eliciting any of the diverse actions of EGF on the differentiated functions of MA-10 Leydig tumor cells argues against a role for phosphatidylinositol-3,4-bisphosphate as a mediator of the actions of EGF on these cells.[718] EGF alone may not stimulate phosphoinositide degradation in BALB/c 3T3 mouse fibroblasts at high concentrations.[719] In contrast to bombesin, the mitogenic effect of EGF in Swiss 3T3 cells does not involve activation of phospholipase C, increased production of 1,2-diacylglycerol, and protein kinase C activation.[660] Studies using a monoclonal antibody specific for phosphatidylinositol-4,5-bisphosphate clearly show that increased phosphatidylinositol breakdown is not crucially required for the mitogenic action of EGF on NIH/3T3 cells.[720] In rat renal cortical slices, EGF stimulates incorporation of phosphate into phosphatidic acid and phosphoinositides, but does not affect phosphoinositide breakdown by a mechanism involving phospholipase C.[721] On the other hand, studies on cells expressing mutant EGF receptors suggest that formation of a complex between the autophosphorylated EGF receptor and phospholipase C is necessary for activation of the enzyme *in vivo*.[722]

Phosphoinositide metabolic pathways involving enzymes other than phospholipase C may undergo alterations in cells stimulated by EGF. In cultured human skin fibroblasts and mouse Swiss 3T3 cells, EGF stimulates a rapid and sustained accumulation of 1,2-diacylglycerol via phospholipase D-catalyzed hydrolysis of phosphatidylcholine.[723,724] Changes in phosphoinositide metabolism may result in alterations in monovalent ion fluxes and pH_i. In chicken granulosa cells, EGF elevates the pH_i by a mechanism that would involve activation of protein kinase C.[725] Further studies are required for a better characterization of the precise role of phosphoinositide metabolism in the mechanisms of action of EGF, but the available results suggest that different types of cells may exhibit different responses of this metabolism to EGF stimulation.

G. ARACHIDONATE METABOLISM AND PROSTAGLANDIN BIOSYNTHESIS

EGF signal transduction may involve the metabolism of arachidonic acid to oxygenated metabolites. Metabolism of arachidonic acid to prostaglandins occurs in EGF-stimulated murine BALB/c 3T3 cells and is necessary for c-*myc* gene expression in these cells.[726] In contrast, PDGF-dependent mitogenesis in BALB/c 3T3 cells does not involve arachidonic acid metabolism and increased production of prostaglandins. EGF increases the production of prostaglandin endoperoxide (PGE_2) by human amnion cells.[727] The production of prostacyclin by aortic smooth muscle cells is stimulated by EGF, but this effect depends on a concurrent stimulus capable of increasing the $[Ca^{2+}]_i$.[728]

In vasopressin-treated rat glomerular mesangial cells, a major effect of EGF is to enhance free arachidonate release and PGE_2 production by activation of phospholipase A_2.[729-731] The mechanism involved in modulation of phospholipase A_2 activity by EGF in the kidney is unknown, but it appears to be independent of alterations in phospholipase C, Ca^{2+}, and protein kinase C. The tyrosine kinase activity of the EGF receptor is necessary for phospholipase A_2 activation.[732] Studies with human amnion cells indicate that the EGF-induced increased secretion of PGE_2 by these cells is due to a mechanism that involves induction of prostaglandin endoperoxide synthase (PGH_2 synthase) and is dependent on the presence of nonesterified arachidonic acid.[733] However, in gastric smooth muscle cells EGF may not act through activation of phospholipase A_2, but rather via the metabolism of diacylglycerol by diacylglycerol lipase, thereby liberating arachidonic acid for the synthesis of prostanoids involved in muscle contraction.[734]

Linoleic and arachidonic acid metabolism may play a central role in transducing the EGF mitogenic signal in Syrian hamster embryo (SHE) fibroblasts.[735] The major product of linoleate in these cells is

13-hydroxyoctadecadienoic acid (HODE). Biosynthesis of HODE is enhanced upon EGF stimulation of quiescent SHE cells, and inhibition of EGF receptor-associated tyrosine kinase activity blocks EGF-stimulated biosynthesis of HODE. The primary products of arachidonate in SHE cells are PGE_2 and PGF_2, which are formed via the cyclooxygenase pathway, but inhibition of cyclooxygenase activity does not alter the EGF mitogenic response in SHE cells. Interestingly, the linoleate products do not enhance the EGF mitogenic effect in variant SHE cells that have lost tumor suppressor gene function.

H. PHOSPHORYLATION OF CELLULAR PROTEINS

Activation of the tyrosine kinase activity intrinsic to the EGF receptor is crucial for the transmission of the EGF signal.[156-163] Phosphorylation of proteins with different intracellular locations is stimulated as a consequence of the formation of a complex between EGF and its cellular receptor,[736] but it is difficult to determine which of these proteins are direct substrates of the EGF-stimulated kinase activity of the receptor and which are phosphorylated by other, secondarily activated kinases. In addition to the stimulation of protein phosphorylation on tyrosine residues, EGF stimulates the phosphorylation of cellular proteins on serine and threonine residues, suggesting the activation of a cascade of different protein kinases that may be indirectly augmented upon transmission of the transducing signal represented by an activated EGF receptor.[737,738] The EGF receptor itself can be phosphorylated on serine and threonine residues in a reaction which requires both EGF and cAMP-dependent protein kinase.[739]

1. Membrane Proteins

EGF rapidly induces a series of structural and functional changes on the surface of responsive cells.[740] Some of the cellular proteins that are phosphorylated as a consequence of the formation of the EGF receptor complex are membrane components. The band 3 protein, which is one of the major integral membrane proteins of the human erythrocyte (and probably also of other cell types) is phosphorylated on tyrosine residues *in vitro* by the purified EGF receptor and v-Src kinases.[741] The physiological relevance of band 3 glycoprotein phosphorylation by the EGF receptor kinase is not known since there are few EGF receptors on human erythrocytes. A 59-kDa protein that is phosphorylated on tyrosine and serine residues by EGF in rat liver membranes was represented by the c-Src protein or another similar kinase.[742] Src-related kinases may have a role in EGF signal transduction.

2. Calpactins

The EGF receptor kinase participates in the phosphorylation of a 36-kDa protein (p36), which is located on the inner aspect of the plasma membrane.[743] The p36 protein is present in fibroblastic and endothelial cells in all of the mouse and rat tissues examined; but it is present at very low or undetectable levels in skeletal and smooth muscle cells, erythrocytes, and lymphocytes.[744] The 36p protein may be identical with the β subunit of G proteins which play a key role in the stimulation (G_s) or inhibition (G_i) of adenylyl cyclase.[745] EGF kinase substrates identified and isolated from human placenta are recognized by polyclonal antisera against the retinal rod outer segment G protein, transducin.[746] Protein p36 is also a cellular substrate for the v-Src oncoprotein;[747] and both protein kinases, v-Src and the EGF receptor, phosphorylate the protein exclusively on tyrosine residues.[748] Protein p36 is identical with calpactin I (lipocortin II), a member of a family of calcium-phospholipid-binding proteins. The cellular protein p35 is also a substrate for the activated EGF receptor kinase.[749]

A substrate of the EGF receptor kinase in A431 cells as well as in normal cells, including membranes from human placental cells, is a 35-kDa protein which is also a member of the family of calcium-phospholipid-binding proteins, including membranes from human placental cells.[750-752] The 35-kDa substrate of EGF receptor is identical with calpactin II/lipocortin I.[753] Lipocortins are misnomers since glucocorticoids may not have an effect on the synthesis or secretion of calpactins (proteins p35 and p36).[754]

3. Ribosomal Protein S6

An important substrate for EGF-stimulated phosphorylation is the 40S ribosomal protein S6.[755] Phosphorylation of S6, which may result in the activation of this kinase,[756] is stimulated by other hormones and growth factors in different tissues, including insulin in liver cells and human chorionic gonadotropin (hCG) in Leydig cells.[757,758] S6 phosphorylation is also stimulated by oncoproteins possessing tyrosine kinase activity such as the v-Src and v-Fps proteins.[759] Phosphorylation of S6 involves the activation of a specific kinase.[760] A small amount of S6 phosphorylation induced by action of EGF and other mitogens may be related to the activity of a cAMP-dependent protein kinase.[761] A potential mediator of EGF-induced

S6 phosphorylation is the cAMP-independent, Ca^{2+}-independent protein kinase, protease-activated kinase II (PAK-II).[762] The mechanisms involved in regulation of PAK-II activity *in vivo* are unknown. The mitogen-activated S6 kinase, which is distinct from other known protein kinases,[763] is directly responsible for EGF-induced phosphorylation of the ribosomal protein S6. In addition to protein kinases, EGF modulates S6 phosphorylation through activation of specific phosphatases.[764] Phosphorylation of S6 could contribute to the stimulation of cell proliferation under particular physiological conditions,[765] although a direct regulation of DNA synthesis through S6 phosphorylation seems unlikely.

4. Cytoskeletal Proteins

EGF, as well as insulin and IGFs, causes rapid membrane ruffling and alterations in the cytoskeletal structure of susceptible normal and transformed cells associated with microfilament reorganization but not with microtubule reorganization.[766] The molecular mechanisms responsible for these changes are not understood but microfilament- and microtubule-associated proteins are important substrates for the activated EGF receptor kinase as well as for the insulin receptor and v-Src tyrosine kinases.[767] The particular cytoskeletal substrates of these kinases may be different, however. The EGF receptor kinase phosphorylates the β but not the α subunit of fodrin on tyrosine residues, and this phosphorylation is markedly inhibited by F-actin. Microtubule associated protein MAP-2 is one of the best cytoskeletal substrates for the EGF receptor kinase. This kinase is also able to phosphorylate the β subunit of tubulin.

5. Enzymes

The complex biochemical actions of EGF may be exerted, in part, through the phosphorylation of cellular enzymes. Addition of EGF to the media of cells proliferating in culture may either enhance or depress the specific activities of different enzymes.[768] These changes may or may not involve new protein synthesis. The EGF receptor phosphorylates several glycolytic enzymes, including key regulatory enzymes of the glycolytic pathway.[769] EGF or EGF-like factors could contribute to the high rate of glycolysis observed in different tumors.[770] EGF stimulates the phosphorylation of pyruvate kinase in freshly isolated rat hepatocytes, suggesting a role for EGF in the regulation of glycolysis.[771]

Activation of MAP kinases may be important for the mitogenic activity of EGF. In dog thyroid epithelial cells (thyrocytes), EGF induces p42 and p44 MAP kinase phosphorylation on tyrosine, serine, and threonine residues.[772] In contrast, MAP kinases are not activated in thyroid cells stimulated through the TSH/cAMP mitogenic pathway. Stimulation of Chang liver or monkey kidney cells with EGF results in translocation of nonhistone protein to the nucleus, which may be due to inhibition of the lysosomal degradation of these proteins.[773] It is thus clear that different mitogens may stimulate mitogenesis in the same type of cells by using different intracellular pathways.

6. Hormones, Growth Factors, and Their Receptors

Peptide hormones such as gastrin-17 and growth hormone can be phosphorylated by EGF-stimulated protein kinase activity.[774,775] The possible physiological role of these modifications is not understood. Steroid hormone receptors may be substrates for EGF-induced phosphorylation. In cultured human breast epithelial cells, EGF induces phosphorylation of the glucocorticoid receptor on both tyrosine and serine residues.[776] The progesterone receptor is phosphorylated exclusively on tyrosine residues by the activated EGF and insulin receptor kinases, but not by the PDGF receptor kinase.[777] Different sites of the progesterone receptor are phosphorylated by these kinases. The possible physiological role of the EGF-stimulated phosphorylation of steroid hormone receptors is unknown.

An action of EGF in its target cells may consist of the induction of a redistribution of specific cell membrane components, in particular, transferrin receptors. EGF induces a rapid and transient increase in the expression of transferrin receptors at the cell surface with no change in total cellular receptor content, which suggests that the newly appearing receptors are temporarily unable to be internalized.[778] In malignant human epithelial cell lines, EGF induces rapid and variable changes in the expression and phosphorylation of transferrin receptors.[779] However, there is no consistent relationship between EGF-induced alterations in transferrin receptors and surface transferrin receptor expression in these cell lines.

7. Other Cellular Protein Substrates

In human A431 cells, EGF enhances the phosphorylation of acidic cytosol proteins with molecular weight of 17, 27, 34, and 80 kDa.[780] The same proteins are phosphorylated by phorbol ester (TPA) and dibutyryl cAMP, suggesting that they may be substrates for protein kinase C and cAMP-dependent protein kinases. In other cell types the major increase in tyrosine phosphorylation by the action of EGF (or other growth factors and mitogens) occurs on a 42-kDa protein, and to a lesser extent on a 40-kDa protein.[781] Interestingly, EGF-stimulated phosphorylation of the 42-kDa cytosolic protein is dependent on protein kinase C activity but occurs on tyrosine residues, indicating that protein kinase C is capable of stimulating the activity of cellular tyrosine kinases, probably by phosphorylation of these enzymes on serine/threonine residues.[782] The tyrosine kinases activated by phosphorylations related to EGF-stimulated protein kinase C activity could include products of the Src family.

Stimulation of cultured dog thyroid epithelial cells with EGF increases the phosphorylation of five polypeptides, including two 42-kDa proteins phosphorylated on tyrosine, threonine, and/or serine residues.[783] The phorbol ester TPA, which mimics all the effects of EGF in dog thyroid cells, induces the phosphorylation of a set of proteins in these cells that overlap, at least in part, those phosphorylated by EGF. On the other hand, the mitogenic effects of TSH on dog thyroid cells are exerted mainly through activation of the cAMP-dependent pathway and do not include phosphorylation of 42-kDa proteins. Thus, different mitogenic agents can use different independent metabolic pathways for cell stimulation.

8. Oncoproteins

The ligand-activated EGF receptor kinase may have a role in the phosphorylation of oncoproteins. The c-Raf-1 protein kinase is a substrate *in vivo* for phosphorylation on serine induced by the EGF receptor.[784] EGF receptor tyrosine kinase activity, but neither EGF receptor autophosphorylation nor internalization, is required for c-Raf-1 phosphorylation by the ligand-activated EGF receptor. The c-Neu/Erb-B2 protein is phosphorylated on tyrosine as well as on serine and threonine residues by the EGF receptor kinase in several types of cells, including MKN-7 human adenocarcinoma cells and SK-BR-3 mammary tumor cells.[441,442]

EGF may be involved in regulating the phosphorylation of Ras proteins. A synthetic peptide containing the autophosphorylation site of the v-H-Ras oncoprotein is phosphorylated on tyrosine by the ligand-stimulated receptor kinase of A431 cell membranes.[785] Membranes isolated from NRK cells transformed by Harvey murine sarcoma virus (H-MuSV) exhibit increased phosphorylation of the v-H-Ras protein when EGF is added to the incubation medium.[786] Insulin has a similar effect on the phosphorylation of the v-Ras protein. The important role of Ras proteins in the mechanism of EGF signal transduction was discussed previously.

I. TRANSCRIPTIONAL AND TRANSLATIONAL EFFECTS OF EGF

EGF may be involved in the regulation of RNA and protein synthesis in its target cells. However, the precise significance of the induction of different RNAs and proteins by EGF in different types of cells is frequently enigmatic.

A number of RNAs and proteins are induced by EGF in the target cells. Stimulation of quiescent rat fibroblasts by either EGF or serum results in increase of different molecular species of mRNAs, the most abundant being those coding for proteins with extensive homology to the glycolytic enzymes lactate dehydrogenase, enolase, and triose phosphate isomerase.[787] It is not clear, however, whether EGF induction of these mRNAs (and presumably of the corresponding proteins) is directly related to the triggering of the particular mitogenic signal or is a consequence of the general growth response. A 55-kDa phosphoprotein is secreted by cultured rat kidney fibroblasts stimulated by EGF.[788] Not only positive regulatory elements, but also negative regulatory elements contained in the 5′-flanking region of EGF-responsive genes such as the calcyclin gene may be regulated by EGF.[789]

In AKR-2B mouse embryo cells, EGF induces the synthesis of mRNAs transcribed from virus-like 30 (VL30) sequence elements, which are closely related to sequences present in integrated retroviruses and certain classes of transposable genetic elements.[790,791] Expression of the proliferation-associated gene VL30 is rapidly induced by EGF in the mouse cell line RVL-3.[792] Expression of VL30 is also induced by activators of protein kinase C such as TPA and *sn*-1,2-dioctanoylglycerol. High concentrations of a mRNA species, pTR1 RNA, have been detected in rat fibroblasts transformed by the polyoma virus, the v-*src* oncogene, or the mutant EJ c-H-*ras* oncogene; and the same RNA is rapidly induced by the addition of EGF to the culture medium of normal rat fibroblasts.[787] Both the growth factor and the oncoproteins control pTR1 gene expression at the transcriptional level. pTR1 RNA is not present in serum-stimulated cells and is apparently not a participant of the general cellular growth response. The function of the pTR1-derived protein has not been established, but the results indicate that growth factors and oncoproteins may act by controlling the transcriptional activity of similar or identical genes.

Differential hybridization screening of cDNA libraries may allow the detection of mRNAs specifically induced by growth factors. Expression of a gene, termed cMG1, is transiently activated by EGF in the

RIE-1 rat intestinal epithelial cell line.[793] Analysis of cDNA derived from cMG1 mRNA showed that the cMG1 gene encodes a polypeptide of 338 amino acids which exhibits no homology to any other protein, with the exception of the product of the TIS11 gene, a TPA-induced sequence in Swiss 3T3 cells. The two genes, cMG1 and TIS11, may code for a type of early response genes.

Expression of the secreted protease, transin, is stimulated by EGF at the transcriptional level.[794] As demonstrated with the use of antisense c-*fos* mRNA, EGF-induced expression of the transin gene in NIH/3T3 fibroblasts occurs despite an equivalent inhibition of endogenous c-*fos* mRNA levels.[795] In contrast, PDGF stimulation of transin mRNA is blocked by a selective reduction in the levels of the c-Fos protein. The stimulatory effect of both EGF and PDGF on transin gene transcription involves factors recognizing a nucleotide sequence, TGAGTCA, which is found in the transin promoter and is known to be a binding site for Fos-Jun/AP-1 protein complexes. Transin synthesis in EGF-stimulated Rat-1 fibroblasts is blocked at the transcriptional level by TGF-β.[796] Transin is also induced by various oncoproteins and TGF-α, as well as by phorbol ester, cAMP, and cytochalasins. Transin shows high structural homology to, and may be the rat homolog of, the human protease stromelysin, for which the principal substrates are proteoglycans and fibronectin. A second gene existing in the rat genome, the transin-2 gene, codes for a protein which shows homology with human fibroblast collagenase.[797] Expression of the transin-2 gene is under different control in REF cells and is not stimulated by EGF, TGF-α, cAMP, or cytochalasin D. However, both genes (transin and transin-2) are expressed in different types of transformed cells, and their protease products could play a role in tumor invasion. Expression of the transin gene is stimulated in the mouse skin by a classical initiation-promotion protocol which induces the appearance of squamous cell carcinomas.[798] A single application of phorbol ester to the normal mouse epidermis elicits transient increase in transin mRNA.

EGF induces in mouse embryo cells a rapid increase in actin mRNA levels, which could involve modulation of a specific, labile repressor of actin gene transcription.[799,800] Actin or its higher ordered derivatives are essential components of microfilaments; these structures are involved in important cellular processes, including morphogenesis, motility, and mitosis. Microfilaments may play a necessary role in the initiation of DNA synthesis in response to EGF.[801] An altered microfilament organization is found in cells transformed by diverse biological and chemical agents.

Secretion of specific proteins may be stimulated by EGF in certain cellular systems. Treatment of NRK-49F cells with EGF alone, or with different combinations of EGF plus TGF-β, modulates the secretion of a major nonglycosylated phosphoprotein of 69 kDa (pp69).[802] This protein is phosphorylated on serine residues and is secreted only by nontransformed NRK-49f cells. An antibody against pp69 recognizes, in addition to pp69, another phosphoprotein of 62 kDa (pp62) that is secreted by spontaneously transformed NRK-49F cells. A precursor-product relationship may exist between pp69 secreted by nontransformed NRK-49F cells and pp62 secreted by these cells when they become neoplastically transformed.

EGF induces tyrosine hydroxylase mRNA and protein in cells such as the rat PCG2 pheochromocytoma cell line.[803] Sequences in the 5′-flanking region of the tyrosine hydroxylase gene contain the information required for the modulation of its expression by EGF, suggesting a role for EGF in the synthesis of neurotransmitters. Tyrosine hydroxylase catalyzes the conversion of tyrosine to 3,4-dihydroxyphenylalanine, which is the initial step in the biosynthesis of catecholamines. The major sites of synthesis of tyrosine hydroxylase and catecholamines are the adrenal medulla, the sympathetic gangli, and certain defined regions of the brain.

The action of EGF on enzyme induction may in some cases be indirect, by amplifying the effect of other hormones and growth factors. EGF enhances in a dose-dependent manner the induction of tyrosine aminotransferase (TAT) and tryptophan oxygenase in primary cultures of adult rat hepatocytes without itself having any effect on these enzymes in the absence of glucocorticoids.[804] EGF may have inhibitory effects on the synthesis of certain RNAs and proteins. In a primary culture system of the dog thyroid gland, EGF decreases thyroglobulin mRNA and protein synthesis to undetectable levels, affecting profoundly the morphology of the gland and its capacity for iodide trapping.[805] EGF may have a bifunctional effect on the expression of casein and α-lactalbumin genes in explants from rodent mammary glands *in vitro;* and the animal species (mouse or rat) as well as the presence of other hormones and growth factors (insulin, aldosterone, corticosterone, prolactin) may greatly influence the type of action, either stimulatory or inhibitory, of EGF on casein gene expression.[806] EGF inhibits casein production and the accumulation of casein mRNA induced by insulin, cortisol, and prolactin in primary cultures of mammary epithelial cells from pregnant mice.[807] Various cAMP derivatives counteract the inhibitory effect of EGF, which suggests a possible modulatory function of the cyclic nucleotide in regulation of casein production at either the transcriptional or the translational level. In contrast, cAMP does not reverse the stimulatory effect of EGF on mammary cell proliferation, which indicates that cAMP selectively counteracts the effect of EGF on the differentiation of the mammary gland. The action of cAMP on casein production is observed in the mammary gland organ culture system, but not in mammary cell cultures; this suggests that EGF may not act directly on mammary epithelial cells, but that its effects are mediated through the paracrine action on nonepithelial cells of the gland such as fat or mesenchymal cells which are present in tissue explants but not in cell culture systems.

The mechanisms involved in the transcriptional effects of EGF are little understood but probably include the activation of certain transcription factors. Short *cis*-active nucleotide sequences of the rat prolactin or M-MuLV genes can transfer transcriptional regulation by both EGF and phorbol esters to fusion genes.[808] These sequences act in a position- and orientation-independent manner, thus behaving in a manner similar to that of enhancers, being capable of binding some *trans*-acting factors. A possible mechanism for EGF-induced modification of genomic functions is the regulation of DNA topoisomerases and DNA-nicking activity. EGF binding to mouse and human cells enhances topoisomerase activity, and this activity in the nucleus corresponds with DNA synthesis in the cells.[809] Purified EGF receptors of human and murine origin can nick supercoiled double-stranded DNA in an ATP-dependent fashion. However, DNA-nicking activity is not intrinsic to the EGF receptor but is mediated by a 100-kDa cellular protein.[810] Other growth factors may modulate the transcriptional action of EGF. The transcriptional response of specific genes from AKR-2B mouse fibroblasts to EGF can be stably altered as a consequence of exposure to TGF-β.[811] EGF and TGF-β exert a strong synergistic effect on specific RNA induction. The positive or negative cooperative effects of different hormones and growth factors at the transcriptional level may explain, in part, the importance of the microenvironment on the action of these signaling agents.

J. EGF-INDUCED PROTO-ONCOGENE EXPRESSION

EGF regulates the expression of distinct proto-oncogenes in its target cells, and some of the physiological effects of EGF may be exerted through the products of these genes. There are cell- and species-specific differences in the effects of EGF on proto-oncogene expression. Transcriptional expression of the c-*fos* and c-*myc* genes is increased by EGF in TSH-stimulated primary cultures of porcine thyrocytes.[812] In contrast, TSH alone does not induce c-*fos* and c-*myc* expression and does not stimulate the growth of cultured thyrocytes. Both EGF and TGF-α (which utilizes the EGF receptor) induce c-*myc* and c-*fos* gene expression in CH3/10T1/2 mouse fibroblasts.[813] In NIH/3T3 cells, EGF induces a rapid increase of c-*myc* mRNA levels.[814,815] EGF, as well as insulin and IGF, regulate c-H-*ras* gene expression in murine fibroblasts.[816] Both EGF and TGF-α are mitogenic for immature astrocytes and trophic for developing brain neurons; and they induce c-*fos* gene expression in mature retinal Müller cells *in vivo,* as shown by injection of the growth factors into the vitreous cavity of adult rabbits.[817] In resting EL2 rat fibroblasts, stimulation with EGF results in induction of c-*fos* mRNA after 15 min, whereas c-*myc* mRNA is detectable 5 h after EGF stimulation.[818] EGF superinduces the c-*fos* gene in EL2 cells, as compared with the induction caused by other growth factors.[819]

In addition to c-*fos* and c-*myc,* EGF stimulates c-*jun* transcription in rat fibroblasts.[820] EGF-stimulated Rat-1 fibroblasts express both c-*fos* and c-*jun* genes, and this expression is required for the synthesis of stromelysin (transin) by the rat cells.[821] Stromelysin is an enzyme that belongs to the family of metal-dependent proteinases involved in degradation of the extracellular matrix. The expression of these enzymes correlates to tumor progression. EGF-treated EL2 cells acquire some properties reminiscent of transformed cells, including growth in soft agar and loss of contact inhibition. The *jun*-B gene is the only member of the *jun* gene family that is induced by EGF in transfected P19 embryonal carcinoma cells expressing EGF receptors, and the upregulation of *jun*-B gene expression by EGF is mediated by the inverted repeat present in its promoter and is dependent on the Jun-D gene product.[822]

Other growth factors may interact with EGF in the induction of proto-oncogene expression. Treatment of MDA-468 human mammary carcinoma cells with EGF results in accumulation of c-*myc* mRNA, and this effect is inhibited by the presence of TGF-β.[823] Both EGF and TGF-β exert growth inhibitory effects on MDA-468 cells but TGF-β alone induces little change on c-*myc* expression in these cells. In another study it was found that EGF regulates moderately and to different extents the transcription of the c-*erb*-B1/EGF receptor gene and c-*neu*/*erb*-B2 gene in MDA-468 cells, but c-*myc,* c-H-*ras,* and c-*fps* gene transcription is not altered.[824] TGF-β_1 modulates the EGF-dependent transcription of the c-*erb*-B1/EGF receptor and c-*neu*/*erb*-B2 genes in MDA-468 cells. Thyroid hormone may exert synergistic control on the action of EGF, alone or in combination with TGF-β_1, on the transcription of the two genes. Both EGF and NGF induce rapid transient changes in proto-oncogene expression in PC12 rat pheochromocytoma cells.[825]

Induction of c-*fos* and/or c-*myc* gene expression by EGF may not correlate with the proliferation of certain types of cells. EGF induces transient expression of c-*fos* and c-*myc* mRNA and protein in the human epidermoid carcinoma cell line A431, in spite of the fact that these cells respond to EGF with a decreased growth rate.[826] EGF induces c-*fos* and c-*myc* (but not p53) gene expression in human breast cancer MDA-468 cells, which are growth-inhibited by EGF, as well as in the MDA-468-S4 variant of these cells, which is dependent on EGF for cell proliferation.[672,827] These results indicate that induction of c-*fos* and c-*myc* by EGF is not strictly correlated with proliferative activity but should be attributed to the primary interaction of EGF with its receptor. Moreover, c-*fos* and c-*myc* gene expression is also induced by cyanide bromide-cleaved EGF (CNBr-EGF), a molecule with no mitogenic activity but still capable of triggering some of the early cellular actions of EGF.[826] Strong induction of c-*fos* and, to a lesser extent, of c-*myc* is produced by the phorbol ester TPA and by the calcium ionophore A23187, suggesting that protein kinase C may be involved in proto-oncogene activation by growth factors. Combined treatment with EGF and other growth factors may result in complex changes in proto-oncogene expression. A dual effect on c-*erb*-B and c-H-*ras* expression is observed in A431 cells treated with a combination of EGF and IFN-γ.[828]

The mechanisms involved in regulation of proto-oncogene expression by EGF are little understood. In canine MDCK cells, EGF-induced expression of the c-*myc* and c-*fos* genes is not correlated to the extent of EGF receptor autophosphorylation or EGF-stimulated DNA synthesis, but may involve alternative mechanisms such as protein kinase C or phospholipase C activation.[829] In contrast to c-*fos* and c-*myc* expression induced by PDGF or bombesin, which depends on increased phosphoinositide turnover and activation of protein kinase C, induction of c-*fos* and c-*myc* by EGF in mouse fibroblasts is independent of this metabolic pathway.[830] EGF induces c-*myc* RNA in Swiss 3T3 cells without eliciting phosphoinositide turnover, protein kinase C activation, or Ca^{2+} mobilization.[831] Induction of c-*fos* and c-*jun* gene expression by EGF in mouse NIH/3T3 cells does not depend on an increase in the $[Ca^{2+}]_i$ or activation of protein kinase C.[832] On the other hand, induction of c-*fos* and c-*myc* expression by EGF or calcium ionophore may be cAMP dependent.[833] However, in rat fibroblasts stimulation of stromelysin (transin) mRNA expression by EGF is dependent on induction of c-*fos* and c-*jun* genes and activation of protein kinase C.[821] Thus, the effects of EGF on gene expression are cell specific and may involve the operation of different metabolic pathways. In primary cultures of rat hepatocytes, stimulation of c-*myc* gene expression by EGF may be mediated by the prostaglandins E_2 and F_{2a}.[834]

EGF-induced c-*fos* gene transcription would be accompanied by increased binding of a regulatory factor to an enhancer sequence contained in the c-*fos* gene promoter.[835] The increased binding would not result from increased synthesis of the regulatory factor but rather from an activation mechanism involving a preformed inactive factor. However, the regulation of c-*fos* gene expression is complex, and the exact role of this putative regulatory factor has not been elucidated. The enhancer-binding activity does not decrease with time as transcription decreases in EGF-treated A431 cells, and stimulation of c-*fos* expression induced by TPA or calcium ionophore A23187 is not accompanied by an increased level of binding activity. Moreover, HeLa cells do not show detectable c-*fos* gene transcription in spite of the fact that they contain significant amounts of the binding activity.[835] Both EGF and TPA rapidly induce the phosphorylation of a complexed and chromatin-associated 33-kDa nuclear protein in C3H10T1/2 mouse cells.[836] This effect is selectively abolished by 2-aminopurine, a compound which has been reported to block serum-stimulated c-*fos* and c-*myc* gene induction. The 33-kDa phosphoprotein (pp33) may have a crucial role in c-*fos* gene induction in response to EGF and TPA.

At least in certain types of cells, a site in the c-*fos* gene which binds the transcriptional activator AP-1 can mediate induction by EGF and phorbol ester (TPA).[837] In KB human epidermoid carcinoma cells, both EGF and an anti-EGF receptor monoclonal antibody (MoAB 225) induce c-*fos* gene expression predominantly through posttranscriptional mechanisms.[838] In general, the mechanisms and biological significance of induction of proto-oncogene expression by EGF and other growth factors are little known.

The results of studies on EGF-stimulated expression of proto-oncogenes using cells cultured *in vitro* should be interpreted with caution. Addition of EGF to short-term primary cultures of rat hepatocytes resulted in the entry of cells into DNA synthesis 20 h after inclusion of the mitogen, and this effect was associated with altered c-H-*ras* gene expression.[839] The levels of c-*myc* mRNA also showed a dramatic increase in the cultured cells, but this increase occurred at similar levels with or without the addition of EGF to the cultured cells. These results suggest that the isolation of cells for culture and/or the culture conditions *in vitro* may produce adaptative changes in c-*myc* expression.

K. EFFECTS OF EGF ON DNA SYNTHESIS, CELL PROLIFERATION, AND CELL DIFFERENTIATION

EGF is involved in the regulation of DNA synthesis and cell proliferation as well as in the expression of differentiated functions in defined cellular systems both *in vitro* and *in vivo*. EGF is a potent mitogenic factor involved in regulation of the proliferation of different cell types, especially of epithelial cells both *in vitro* and *in vivo*. There may be differences among animal species in relation to the physiological requirements for an EGF-induced mitogenic response. While cultured mouse fibroblasts require to be primed by a "competence factor" such as PDGF or FGF in order to become responsive to growth induced by a "progression factor" such as EGF or IGF-I, rat embryo fibroblasts may be mitogenically stimulated by the action of EGF alone.[814] EGF-induced growth of EL2 rat embryo fibroblasts is associated with the production of at least four inducible-secreted proteins (mol wt 29 to 68 kDa) which are apparently required for DNA synthesis.[840]

Regeneration of epidermal tissue in split-thickness wounds and partial-thickness burns is enhanced by topical treatment with biosynthetic EGF.[841] EGF also promotes the proliferation of smooth muscle cells and is capable of reversing the antiproliferative effects of heparin on these cells.[842] Depending on the type of cell and the local or general physiological conditions, EGF may act not as a stimulator but as an inhibitor of cell proliferation; this is demonstrated by the growth-inhibitory action of EGF on A431 human epidermoid carcinoma cells, which express an extremely high number of EGF receptors on the surface. However, EGF causes a marked synthesis of DNA and an accumulation of nonhistone proteins in A431 cells, without progression into mitosis.[843] The reason for such discrepancy between DNA synthesis and cell division in EGF-stimulated A431 cells is unknown. It may be hypothesized that the cell cycle in EGF-stimulated cells is controlled by two surface signals: one related to stimulation of the biosynthetic events associated with the G_1 and S phases of the cycle; and the other leading to progression into mitosis, which would not be generated or may be rather inhibited by EGF in A431 cells.

The proliferative response of cells to EGF depends on their state of growth and differentiation. In rat intestinal epithelial (RIE)-1 cells, the effects of EGF on cell proliferation depend on the cell population density, being stimulatory in dense cultures and inhibitory at low population densities.[844] Senescence of cultured cells is associated with decreased responsiveness to EGF.[845]

The mechanisms responsible for the different effects of EGF on cellular proliferation are little understood. An effect of EGF on replicating cells such as HeLa, A431, and D HER 14 may consist of transient inhibition of the transition from G_2 to M phase of the cycle.[846] Primary cultures of adult rat hepatocytes in serum-free medium optimally respond to EGF-induced cell proliferation at low extracellular Ca^{2+} concentrations.[847] Hormones and growth factors modulate the stimulation of DNA synthesis and cell proliferation induced by EGF. Insulin at relatively high concentrations and IGFs at physiological concentrations synergistically enhance EGF-stimulated DNA in the murine keratinocyte cell line MK.[848] In contrast, TGF-β and dexamethasone inhibit in a reversible manner the stimulation of DNA synthesis elicited by EGF in MK cells. Administration of EGF to intact mice results in stimulation of DNA synthesis and phosphatidylinositol lipid turnover in the uterus, mimicking the effects of estrogen.[849] There may exist interactions between the EGF and estrogen-signaling pathways.

EGF has important effects on cell differentiation. Depending on the type of cell and the physiological conditions, EGF may either stimulate or inhibit cell differentiation. The growth of normal and neoplastic human urothelial cells in a defined medium is not stimulated by EGF, but addition of EGF to these cells results in the induction of morphological changes characteristic of terminal differentiation *in vitro*.[850] Rat adipocyte precursor EGF-binding sites and their differentiation are inhibited by EGF and TGF-α in a dose-dependent fashion through mechanisms which are independent of cell growth.[851]

Binding of EGF to its receptor on the cell surface is essential for the cellular response, but the specificity of this response may crucially depend on the type of cell and the physiological conditions. Stabilized complexes of EGF and its receptor on the cell surface can stimulate RNA synthesis but not mitogenesis, suggesting that internalization and processing of the EGF-receptor complex is necessary to produce the mitogenic signal.[370] In the RL95-2 human endometrial carcinoma cell line, occupancy of the EGF receptor by the normal ligand results not in stimulation, but in inhibition of cell proliferation.[600] The reason for this response is unknown but it indicates that postreceptor mechanisms are crucially involved in the cellular response to EGF action. EGF-induced inhibition of A431 human carcinoma cell proliferation is due to cell cycle arrest in both the G_1 and S/G_2 phases, which is preceded by inhibition of protein synthesis.[852] However, although the overall rate of cellular protein synthesis is decreased, the relative synthetic rates of some proteins in EGF-treated cells are increased, while others are decreased. Stimulation or inhibition of cellular proliferation in response to EGF may depend on the differential expression of particular types of proteins. In EGF-sensitive cells, EGF-induced phosphorylation of the EGF receptor may be necessary but insufficient in itself to trigger mitogenesis. The TNR9 variant line of Swiss 3T3 cells does not respond mitogenically to TPA but shows this response when exposed to EGF. TPA binding, protein kinase C activation, and EGF receptor phosphorylation, however, proceed normally in the nonmitogenic variant.[853] EGF and other growth factors do not depend solely on the protein kinase C pathway to elicit their mitogenic effect.

Calcium ions may play a central role in EGF-stimulated cell proliferation, probably by acting as an intracellular mediator. EGF reduces by over 50-fold the extracellular Ca^{2+} requirement for multiplication of cultured skin fibroblasts.[854,855] Polyamines may also be involved in the cellular mechanisms of action of EGF in relation to the control of cell proliferation. EGF causes a marked and transient increase of ODC activity in cultured epidermis of chick embryos.[856] EGF-inducible ODC activity is comparable in young and senescent human fibroblast cells in culture, and thus the lack of proliferative response of senescent cells to EGF cannot be explained on this basis.[492] Cyclin may be involved in the control of cell proliferation by EGF. EGF inhibits the proliferation of A431 cells; and the inhibition is paralleled by a decreased synthesis of cyclin, both phenomena being absent in A431 cell variants resistant to the growth inhibitory effect of EGF.[857]

In addition to stimulatory signals, the net result of EGF on cell proliferation may depend on the operation of negative signals. In general, the control of cellular proliferation may depend on the opposed actions of growth factors and growth inhibitors. There is evidence of the existence in monkey cells of a 48-kDa growth inhibitor protein capable of antagonizing EGF-mediated effects on stimulation of DNA synthesis through mechanisms depending on RNA synthesis.[858] Synergistic or antagonistic effects on cellular proliferation are observed between EGF and other hormones and growth factors, which may depend on the physiological condition of the cells. In cultures of WI38 cells at low density, EGF and dexamethasone alone generate only a small growth response, whereas together there is a marked synergistic effect.[859] In contrast, no synergism between EGF and dexamethasone is observed in senescent WI38 cells. The mechanism by which dexamethasone synergistically enhances EGF-stimulated growth of WI38 cells remains unclear, but does not appear to involve changes in the number or affinity of EGF receptors. Insulin is required for EGF-induced DNA synthesis in defined cellular systems, for example, in primary cultures of rat hepatocytes.[860] DNA synthesis is stimulated in both adult and young rat hepatocytes in primary culture by EGF in the presence of insulin. The initiation time and magnitude of DNA synthesis are altered by the age of the donor and the substratum on which the cells are explanted. Moreover, insulin and EGF exert different effects on hepatocytes in the prereplicative period.[861] IGF-I and its receptor may have an important role in the mechanism of action of EGF in certain types of cells. The growth-promoting action of EGF on mouse fibroblasts requires an autocrine loop involving the interaction of endogenously produced IGF-I with its receptor on the cell surface.[862] Pretreatment of human fibroblasts with interferon abolishes the mitogenic effect of EGF without affecting either the EGF receptor binding or the downregulation of the EGF receptor and without blocking early events (increased amino acid transport) in the course of EGF-stimulated thymidine incorporation.[863] In primary cultures of adult rat hepatocytes, EGF stimulates DNA synthesis, and this effect is greatly enhanced by norepinephrine which reduces binding of EGF to its receptor at the cell surface.[864] It is thus clear that interaction of EGF with other hormones and growth factors may have great influence on the control of cell proliferation in different tissues and that this interaction may involve receptor and/or postreceptor mechanisms.

Oncoproteins may either diminish or abolish the requirement of EGF for cellular proliferation. Induction of N-Ras protein expression in NIH/3T3 mouse fibroblasts by means of a molecular construct containing a normal N-*ras* proto-oncogene attached to a steroid hormone-inducible transcription unit resulted in induction of DNA synthesis, and this effect could be augmented in the presence of EGF.[865] C3H/10T1/2 mouse fibroblasts overexpressing the chicken c-Src protein exhibited a severalfold enhanced DNA synthesis and mitogenic response to EGF, but not other growth factors, relative to that of the parent line.[866,867] c-Src would interact with the mitogenic signal transduction pathway of EGF in some event distal to EGF binding.

Clonal BALB/c mouse epidermal keratinocyte (BALB/MK) cells have an absolute requirement for nanomolar concentrations of EGF for their proliferation. Infection of these cells with BALB-MuSV or K-MuSV retroviruses induces a complete abrogation of their requirement for EGF,[868] indicating that v-Ras proteins confer to epidermal cells the rapid acquisition of EGF-independent growth. Members of both the *ras* and *src* oncogene families may supplant the EGF requirement of BALB/MK-2 keratinocytes and may induce alterations in the terminal differentiation processes of these cells.[869] The presence of a functional EGF receptor may not be required for oncogene-induced neoplastic transformation. Murine

NR6 cells, which lack functional EGF receptors, acquire a fully malignant phenotype upon transfection with a plasmid carrying an activated human EJ c-H-*ras* gene.[867] Although RSV-induced transformation of mouse and rat fibroblasts is associated with a marked decrease in the expression of the EGF receptor on the cell surface, NR6 cells can be transformed by the v-*src* oncogene.[607] Any changes induced by oncoproteins on the levels of EGF receptor expression should be considered as symptomatic rather than as necessary for the loss of growth control.

VIII. ROLE OF EGF IN ONCOGENESIS

The influence of EGF in oncogenic processes occurring in humans and other animal species is not clear. EGF or EGF-like proteins are synthesized by some, but not all human tumors, even of the same type. The possible influence of EGF on tumor growth is unknown but it seems likely that EGF may regulate, in conjunction with other factors or with particular proto-oncogene products, the growth of some tumors. Under serum-free conditions *in vitro,* most fresh samples of human malignant tumors, primary or metastatic, exhibit an *in vitro* growth response to EGF.[870] Serum-free growth of human tumors independent of EGF is usually very limited. In certain conditions EGF may display oncoprotein-like properties. Expression of a constructed vector encoding the EGF precursor polypeptide induces transformation of NIH/3T3 mouse fibroblasts.[871]

A. PRODUCTION OF EGF OR EGF-LIKE PROTEINS BY TUMOR CELLS

EGF or EGF-like proteins are produced by several types of primary tumors and tumor cell lines. Human gastric cancer cells, strain MKN-45, synthesize and secrete a large amount of an EGF-like material, EGF-LI, into the culture medium.[872] Among human gastric carcinoma cell lines, EGF-LI is produced mainly by those derived from poorly differentiated tumors, and treatment of these cells with retinoic acid may reduce the synthesis and secretion of the EGF-related material.[873] Immunohistochemical and radioimmunoassay studies have shown that while EGF immunoactivity is present in early human gastric carcinomas, EGF-positive tumor cells are detected in only one fifth of advanced gastric carcinomas and one third of scirrhous carcinomas.[874] Patients with EGF-positive gastric carcinomas may have a poorer prognosis than those with EGF-negative tumors, but the possible role of EGF in the evolution of these tumors is difficult to evaluate.

Hormone-responsive tumor cells may produce EGF mRNA and protein or EGF-related peptides. Analysis of different human breast cancer cell lines with a cDNA probe specific for the human EGF precursor indicated that whereas EGF expression is not detectable in some of these lines (MCF-7, BT 20, and HBL 100), high levels of EGF mRNA are present in the T47D cell line, and these levels are increased when the cells are pretreated with progestins.[875] Treatment of the human breast cancer cell line MCF-7 with 17 β-estradiol induces the synthesis of large amounts of EGF-related polypeptides which are secreted into the culture medium.[876-878] However, the role of such polypeptides in the metabolism of MCF-7 cells is unknown. The study of EGF expression in samples from primary human breast cancers may be more informative than that obtained from cell lines. EGF expression is detectable in a high percentage of human breast cancer biopsies, and is more likely to be expressed at detectable levels in breast tumors also expressing either estrogen receptors or progesterone receptors.[879] The EGF gene is also expressed by normal human breast tissue.

Autocrine mechanisms involving EGF or EGF-like growth factor and its receptor may be implicated in the growth of certain tumor cells. Cell lines derived from rat mammary tumor cells (RMT/RMC cells) exhibit independence from either single or multiple growth factors, and examination of the conditioned medium of some of these lines indicates that independence of EGF is mediated by an autocrine mechanism involving an EGF-like activity.[880,881] This activity may be represented by amphiregulin or a similar substance capable of interacting with the EGF receptor. The RMT cells that express growth factor independence *in vitro* may frequently exhibit neoplastic potential *in vivo.*

B. ONCOGENE-INDUCED ABROGATION OF EGF REQUIREMENT

Infection of cells by acute retroviruses, or transfection with cloned viral oncogenes or activated proto-oncogenes, may result in partial or total abrogation of the requirement for the exogenous supply of EGF. Expression of the v-Ras oncoprotein in normal human keratinocytes results in EGF-independent growth.[882] Mouse cells transfected with either an EJ c-H-*ras* gene alone or a combination of c-H-*ras* plus c-*myc* lose their dependence on EGF for DNA synthesis, and the cultures become committed to S phase in serum-free medium supplemented with insulin alone.[883] Introduction of an activated c-H-*ras* gene in normal human cells (mesothelial cells or fibroblasts) abrogates their requirement for the exogenous supply of EGF.[884] However, the human cells transfected with the mutant c-H-*ras* gene are not neoplastically transformed and depend on the exogenous supply of serum mitogens and cortisol for their growth.

Overexpression of an oncogene or proto-oncogene may potentiate the mitogenic effects of EGF. BALB/MK murine keratinocytes expressing a transcriptionally activated human c-*myc* proto-oncogene exhibit a markedly increased rate of DNA synthesis in response to EGF.[885] Transfection of liver cells from adult rats with a c-*myc* gene by electroporation results in potentiation of DNA synthesis and cell proliferation in response to EGF, and this potentiation is dependent on the amount of c-*myc* DNA transfected.[886] The mechanism by which the c-*myc* gene regulates EGF activity is unknown.

C. INFLUENCE OF EGF ON CARCINOGENIC PROCESSES

EGF can act in certain systems as a tumor-promoting agent, perhaps through its capacity to induce an increased rate of SCE.[887] It is difficult, however, to evaluate the role of EGF in carcinogenic processes, especially in those occurring under natural conditions. The basal cell layer of normal urothelium is richly endowed with cell surface receptors, and superficial cells of premalignant and malignant urothelium express abundant EGF receptors. Intravesical administration of EGF induces ODC activity and DNA synthesis in the rat bladder.[888] Since relatively high amounts of EGF are excreted with the urine and thus incubate with bladder epithelial cells continuously, these studies suggest that urinary EGF may play a role in bladder tumor development and/or growth.

In cultured cells, EGF may display opposite effects according to different experimental protocols. EGF promotes radiation-induced cell transformation in the C3H10T1/2 system,[889] but EGF is capable of suppressing 3-methylcholanthrene (MCA)-induced transformation in the same system.[890] The growth of KB human epidermoid carcinoma cells is reversibly inhibited by EGF; and this inhibition is augmented by TGF-β, although TGF-β itself does not inhibit the growth of KB cells.[891] KB cells cultured in the presence of EGF show relatively high motility and grow dispersely as single cells, whereas the cells cultured in the absence of EGF grow in clusters. In nontumorigenic mouse MSK-C3H-NU epidermal keratinocytes, treatment with either phorbol ester or EGF results in the reversible induction of an anchorage-independent phenotype.[892] When applied to mouse skin following a carcinogenic dose of MCA, EGF can reduce the latent period prior to tumor appearance and can increase the frequency of tumors.[893] Monoclonal antibodies raised against the extracellular domain of the EGF receptor inhibit the growth *in vitro* of human oral epidermoid carcinoma (KB) cells, which exhibit elevated levels of EGF receptors.[894] Moreover, the monoclonal antibodies, alone or in combination with antineoplastic drugs, inhibit the growth of KB cells transplanted in different modes into nude mice.

The submaxillary gland is a rich source of EGF in mice, and its surgical removal (sialoadenectomy) markedly reduces the incidence of mammary tumors in the virgin female animals.[895] Long-term treatment of the sialoadenectomized mice with EGF increases the incidence of mammary tumors. Sialoadenectomy of mammary tumor-bearing mice causes a rapid and sustained cessation of tumor growth, but EGF administration quickly restores the rate of tumor growth to the usual level. Implantation of spontaneous mouse mammary tumors in female nude mice is decreased by sialoadenectomy and increased by treatment with EGF.[896] Moreover, treatment of sialoadenectomized mice with EGF increases the incidence of successful implantation of mammary tumors. The results of these studies suggest that EGF produced by the submaxillary glands may play a crucial role in mouse mammary tumorigenesis.

Most primary human tumors of nonhematopoietic origin (about 80%) respond to exogenous EGF with increased growth rate *in vitro*.[897] The majority of breast carcinomas have little *in vitro* growth without exogenous EGF. It is thus clear that EGF is involved in the growth of different types of tumor cells *in vitro* and probably also *in vivo*. However, the precise role of EGF in tumorigenesis remains to be established.

D. EGF-INDUCED DIFFERENTIATION OF TUMOR CELLS

Activation of the EGF receptor-associated tyrosine kinase activity may be capable of inducing the differentiation of certain types of tumor cells. Transfection of a functional EGF receptor gene into undifferentiated murine P19 embryonal carcinoma cells and binding of EGF to the expressed receptor lead to neuronal differentiation of the P19 cells.[898] The differentiation is associated with c-*jun* proto-oncogene induction.

E. REGULATION OF ONCOGENIC VIRUS EXPRESSION BY EGF

Nucleotide sequences capable of being recognized by EGF may be contained in viruses with oncogenic potential. Certain types of human papillomaviruses (HPVs) have been implicated in the origin and/or development of human malignancies, including cervical cancer.[899] HPV-16 and HPV-18 immortalize human squamous epithelial cells *in vitro*. Functional regulatory elements contained in the long control region (LCR) of the virus can bind proteins of both viral and cellular origin; and one of these elements, the EGFRE, has a silencer activity and is regulated by EGF in an HPV-16-immortalized human keratinocyte cell line.[900] EGF regulation of HPV-16 gene expression is dissociated from EGF-induced cell proliferation and enhanced c-*myc* proto-oncogene expression.

REFERENCES

1. **Carpenter, G. and Cohen, S.,** Epidermal growth factor, *Annu. Rev. Biochem.,* 48, 193, 1979.
2. **Schlessinger, J., Schreiber, A.B., Levi, A., Lax, I., Libermann, T., and Yarden, Y.,** Regulation of cell proliferation by epidermal growth factor, *CRC Crit. Rev. Biochem.,* 14, 93, 1982.
3. **Hollenberg, M.D. and Armstrong, G.D.,** Epidermal growth factor-urogastrone and its receptor, in *Polypeptide Hormone Receptors,* Posner, B.I., Ed., Marcel Dekker, New York, 1985, 201.
4. **Carpenter, G.,** Epidermal growth factor: biology and receptor metabolism, *J. Cell Sci.,* Suppl. 3, 1, 1985.
5. **Carpenter, G. and Zendegui, J.G.,** Epidermal growth factor, its receptor, and related proteins, *Exp. Cell Res.,* 164, 1, 1986.
6. **Stoscheck, C.M. and King, L.E., Jr.,** Functional and structural characteristics of EGF and its receptor and their relationship to transforming proteins, *J. Cell. Biochem.,* 31, 135, 1986.
7. **Carpenter, G. and Cohen, S.,** Epidermal growth factor, *J. Biol. Chem.,* 265, 7709, 1990.
8. **Fisher, D.A. and Lakshmanan, J.,** Metabolism and effects of epidermal growth factor and related growth factors in mammals, *Endocr. Rev.,* 11, 418, 1990.
9. **Laurence, D.J.R. and Gusterson, B.A.,** The epidermal growth factor. A review of structural and functional relationships in the normal organism and in cancer cells, *Tumor Biol.,* 11, 229, 1990.
10. **Kurobe, M., Furukawa, S., and Hayashi, K.,** Synthesis and secretion of an epidermal growth factor (EGF) by human fibroblast cells in culture, *Biochem. Biophys. Res. Commun.,* 131, 1080, 1985.
11. **Sato, M., Yoshida, H., Hayashi, Y., Miyakami, K., Bando, T., Yanagawa, T., Yura, Y., Azuma, M., and Ueno, A.,** Expression of epidermal growth factor and transforming growth factor-β in a human salivary gland adenocarcinoma cell line, *Cancer Res.,* 45, 6160, 1985.
12. **Hollenberg, M.D.,** Epidermal growth factor: a polypeptide acquiring hormonal status, *PAABS Rev.,* 5, 265, 1976.
13. **Perheentupa, J., Lakshmanan, J., Hoath, S.B., and Fisher, D.A.,** Hormonal modulation of mouse plasma concentration of epidermal growth factor, *Acta Endocrinol.,* 107, 571, 1984.
14. **Perheentupa, J., Lakshmanan, J., Hoath, S.B., Beri, U., Kim, H., Macaso, T., and Fisher, D.A.,** Epidermal growth factor measurements in mouse plasma: method, ontogeny, and sex difference, *Am. J. Physiol.,* 248, E391, 1985.
15. **Mattila, A.-L., Perheentupa, J., Pesonen, K., and Viinikka, L.,** Epidermal growth factor in human urine from birth to puberty, *J. Clin. Endocrinol. Metab.,* 61, 997, 1985.
16. **Stromberg, K., Hudgins, W.R., Dorman, L.S., Henderson, L.E., Sowder, R.C., Sherrell, B.J., Mount, C.D., and Orth, D.N.,** Human brain tumor-associated urinary high molecular weight transforming growth factor: a high weight form of epidermal growth factor, *Cancer Res.,* 47, 1190, 1987.
17. **Twardzik, D.R., Kimball, E.S., Sherwin, S.A., Ranchalis, J.E., and Todaro, G.J.,** Comparison of growth factors functionally related to epidermal growth factor in the urine of normal and human tumor-bearing athymic mice, *Cancer Res.,* 45, 1934, 1985.
18. **Mounts, C.D., Lukas, T.J., and Orth, D.N.,** Purification and characterization of epidermal growth factor (β-urogastrone) and epidermal growth factor fragments from large volumes of human urine, *Arch. Biochem. Biophys.,* 240, 33, 1985.
19. **Stoscheck, C.M. and King, L.E., Jr.,** Role of epidermal growth factor in carcinogenesis, *Cancer Res.,* 46, 1030, 1986.
20. **Lee, K., Tanaka, M., Hatanaka, M., and Kuze, F.,** Reciprocal effects of epidermal growth factor and transforming growth factor β on the anchorage-dependent and -independent growth of A431 epidermoid carcinoma cells, *Exp. Cell Res.,* 173, 156, 1987.

21. **Savage, C.R., Jr., Inagami, T., and Cohen, S.,** The primary structure of epidermal growth factor, *J. Biol. Chem.,* 247, 7612, 1972.
22. **Gregory, H.,** Isolation and structure of urogastrone and its relationship to epidermal growth factor, *Nature (London),* 257, 325, 1975.
23. **Gregory, H. and Preston, B.M.,** The primary structure of human urogastrone, *Int. J. Pept. Protein Res.,* 9, 107, 1977.
24. **Hollenberg, M.D. and Gregory, H.,** Human urogastrone and mouse epidermal growth factor share a common receptor site in cultured human fibroblasts, *Life Sci.,* 20, 267, 1976.
25. **Hollenberg, M.D. and Gregory, H.,** Epidermal growth factor-urogastrone: biological activity and receptor binding derivatives, *Mol. Pharmacol.,* 17, 314, 1980.
26. **Mayo, K.H., Schaudies, P., Savage, C.R., De Marco, A., and Kaptein, R.,** Structural characterization and exposure of aromatic residues in epidermal growth factor from the rat, *Biochem. J.,* 239, 13, 1986.
27. **Heath, W.F. and Merrifield, R.B.,** A synthetic approach to structure-function relationships in the murine epidermal growth factor molecule, *Proc. Natl. Acad. Sci. U.S.A.,* 83, 6367, 1986.
28. **Engler, D.A., Hauser, M.R., Cook, J.S., and Nigoyi, S.K.,** Aromaticity at position-37 in human epidermal growth factor is not obligatory for activity, *Mol. Cell. Biol.,* 11, 2425, 1991.
29. **DiAugustine, R.P., Walker, M.P., Klapper, D.G., Grove, R.I., Willis, W.D., Harvan, D.J., and Hernandez, O.,** β-Epidermal growth factor is the des-asparaginyl[1] form of the polypeptide, *J. Biol. Chem.,* 260, 2807, 1985.
30. **Doolittle, R.F., Feng, D.F., and Johnson, M.S.,** Computer-based characterization of epidermal growth factor precursor, *Nature (London),* 307, 558, 1984.
31. **Shoyab, M., McDonald, V.M., Bradley, J.G., and Todaro, G.J.,** Amphiregulin: a bifunctional growth-modulating glycoprotein produced by the phorbol 12-myristate 13-acetate-treated human breast carcinoma cell line MCF-7, *Proc. Natl. Acad. Sci. U.S.A.,* 85, 6528, 1988.
32. **Shoyab, M., Plowman, G.D., McDonald, V.M., Bradley, J.G., and Todaro, G.J.,** Structure and function of human amphiregulin: a member of the epidermal growth factor family, *Science,* 243, 1074, 1989.
33. **Plowman, G.D., Green, J.M., McDonald, V.L., Neubauer, M.G., Disteche, C.M., Todaro, G.J., and Shoyab, M.,** The amphiregulin gene encodes a novel epidermal growth factor-related protein with tumor-inhibitory activity, *Mol. Cell. Biol.,* 10, 1969, 1990.
34. **Plowman, G.D., Whitney, G.S., Neubauer, M.G., Green, J.M., McDonald, V.L., Todaro, G.J., and Shoyab, M.,** Molecular cloning and expression of an additional epidermal growth factor receptor-related gene, *Proc. Natl. Acad. Sci. U.S.A.,* 87, 4905, 1990.
35. **Johnson, G.R., Kannan, B., Shoyab, M., and Stromberg, K.,** Amphiregulin induces tyrosine phosphorylation of the epidermal growth factor receptor and $p185^{erbB2}$. Evidence that amphiregulin acts exclusively through the epidermal growth factor receptor at the surface of human epithelial cells, *J. Biol. Chem.,* 268, 2924, 1993.
36. **Johnson, G.R., Saeki, T., Auersperg, N., Gordon, A.W., Shoyab, M., Salomon, D.S., and Stromberg, K.,** Response to and expression of amphiregulin by ovarian carcinoma and normal ovarian surface epithelial cells: nuclear localization of endogenous amphiregulin, *Biochem. Biophys. Res. Commun.,* 180, 481, 1991.
37. **Cook, P.W., Mattox, P.A., Keeble, W.W., Pittelkow, M.R., Plowman, G.D., Shoyab, M., Adelman, J.P., and Shipley, G.D.,** A heparin sulfate-regulated human keratinocyte autocrine factor is similar or identical to amphiregulin, *Mol. Cell. Biol.,* 11, 2547, 1991.
38. **Cook, P.W., Pittelkow, M.R., Keeble, W.W., Graves-Deal, R., Coffey, R.J., Jr., and Shipley, G.D.,** Amphiregulin messenger RNA is elevated in psoriatic epidermis and gastrointestinal carcinomas, *Cancer Res.,* 52, 3224, 1992.
39. **Johnson, G.R., Saeki, T., Gordon, A.W., Shoyab, M., Salomon, D.S., and Stromberg, K.,** Autocrine action of amphiregulin in a colon carcinoma cell line and immunocytochemical localization of amphiregulin in human colon, *J. Cell Biol.,* 118, 741, 1992.
40. **Ram, T.G., Venkateswaran, V., Oliver, S.A., and Hosick, H.L.,** A transforming growth factor related to epidermal growth factor is expressed by fetal mouse salivary mesenchymal cells in culture, *Biochem. Biophys. Res. Commun.,* 175, 37, 1991.
41. **Kurobe, M., Tokida, N., Furukawa, S., and Hayashi, K.,** Some properties of human epidermal growth factor (hEGF)-like immunoreactive material originating from platelets during blood coagulation, *Biochem. Biophys. Res. Commun.,* 13, 729, 1986.

42. **Pesonen, K., Viinikka, L., Myllylä, G., Kiuru, J., and Perheentupa, J.,** Characterization of material with epidermal growth factor immunoreactivity in human serum and platelets, *J. Clin. Endocrinol. Metab.,* 68, 486, 1989.
43. **Kudlow, J.E. and Kobrin, M.S.,** Secretion of epidermal growth factor-like mitogens by cultured cells from bovine anterior pituitary glands, *Endocrinology,* 115, 911, 1984.
44. **Lage, A., Pérez, R., Valdés, D., and Gavilondo, J.,** Estudios sobre el factor de crecimiento epidérmico. III. Obtención de moléculas EGF equivalentes, a partir del medio condicionado de células L929, *Interferón Biotecnol.,* 2, 199, 1985.
45. **Chiu, M.L. and O'Keefe, E.J.,** Placental keratinocyte growth factor: partial purification and comparison with epidermal growth factor, *Arch. Biochem. Biophys.,* 269, 75, 1989.
46. **Higashiyama, S., Abraham, J.A., Miller, J., Fiddes, J.C., and Klagsbrun, M.,** A heparin-binding growth factor secreted by macrophage-like cells that is related to EGF, *Science,* 251, 936, 1991.
47. **Higashiyama, S., Lau, K., Besner, G.E., Abraham, J.A., and Klagsbrun, M.,** Structure of heparin-binding EGF-like growth factor. Multiple forms, primary structure, and glycosylation of the mature protein, *J. Biol. Chem.,* 267, 6205, 1992.
48. **Naglich, J.G., Metherall, J.E., Russell, D.W., and Eidels, L.,** Expression cloning of a diphtherial toxin receptor —Identity with a heparin-binding EGF-like growth factor precursor, *Cell,* 69, 1051, 1992.
49. **Abraham, J.A., Damm, D., Bajardi, A., Miller, J., Klagsbrun, M., and Ezekowitz, R.A.B.,** Heparin-binding EGF-like growth factor: characterization of rat and mouse cDNA clones, domain conservation across species, and transcript expression in tissues, *Biochem. Biophys. Res. Commun.,* 190, 125, 1993.
50. **Hursh, D.A., Andrews, M.E., and Raff, R.A.,** A sea urchin gene encodes a polypeptide homologous to epidermal growth factor, *Science,* 237, 1487, 1987.
51. **Burgess, A.W.,** Growth factors and their receptors: specific roles in development, *Bioessays,* 6, 79, 1987.
52. **Wharton, K.A., Johansen, K.M., Xu, T., and Artavanis-Tsakonas, S.,** Nucleotide sequence from the neurogenic locus Notch implies a gene product that shares homology with proteins containing EGF-like repeats, *Cell,* 43, 110, 1985.
53. **Knust, E., Dietrich, U., Tepass, U., Bremer, K.A., Weigel, D., Vässin, H., and Campos-Ortega, J.A.,** EGF homologous sequences encoded in the genome of *Drosophila melanogaster,* and their relation to neurogenic genes, *EMBO J.,* 6, 761, 1987.
54. **Thomas, U., Speicher, S.A., and Knust, E.,** The *Drosophila* gene *Serrate* encodes an EGF-like transmembrane protein with a complex expression patter in embryos and wing discs, *Development,* 111, 749, 1991.
55. **Bonfini, L., Karlovich, C.A., Dasgupta, C., and Banerjee, U.,** The *Son of sevenless* gene product: a putative activator of Ras, *Science,* 255, 603, 1992.
56. **Greenwald, I.,** *lin*-12, a nematode homeotic gene, is homologous to a set of mammalian proteins that includes epidermal growth factor, *Cell,* 43, 583, 1985.
57. **Aroian, R.V., Koga, M., Mendel, J.E., Ohshima, Y., and Sternberg, P.W.,** The *let-23* gene necessary for *Caenorhabditis elegans* vulval induction encodes a tyrosine kinase of the EGF receptor subfamily, *Nature (London),* 348, 693, 1990.
58. **Blomquist, M.C., Hunt, L.T., and Barker, W.C.,** Vaccinia virus 19-kilodalton protein: relationship to several mammalian proteins, including two growth factors, *Proc. Natl. Acad. Sci. U.S.A.,* 81, 7363, 1984.
59. **Brown, J.P., Twardzik, D.R., Marquardt, H., and Todaro, G.J.,** Vaccinia virus encodes a polypeptide homologous to epidermal growth factor and transforming growth factor, *Nature (London),* 313, 491, 1985.
60. **Twardzik, D.R., Brown, J.P., Ranchalis, J.E., Todaro, G.J., and Moss, B.,** Vaccinia virus-infected cells release a novel polypeptide functionally related to transforming and epidermal growth factors, *Proc. Natl. Acad. Sci. U.S.A.,* 82, 5300, 1985.
61. **Stroobant, P., Rice, A.P., Gullick, W.J., Cheng, D.J., Kerr, I.M., and Waterfield, M.D.,** Purification and characterization of vaccinia virus growth factor, *Cell,* 42, 383, 1985.
62. **King, C.S., Cooper, J.A., Moss, B., and Twardzik, D.R.,** Vaccinia virus growth factor stimulates tyrosine protein kinase activity of A431 cell epidermal growth factor receptors, *Mol. Cell. Biol.,* 6, 332, 1986.

63. **Eppstein, D.A., Marsh, Y.V., Schreiber, A.B., Newman, S.R., Todaro, G.J., and Nestor, J.J., Jr.,** Epidermal growth factor receptor occupancy inhibits vaccinia virus infection, *Nature (London),* 318, 663, 1985.
64. **Marsh, Y.V. and Eppstein, D.A.,** Vaccinia virus and the EGF receptor: a portal for infectivity?, *J. Cell. Biochem.,* 34, 239, 1987.
65. **Buller, R.M.L, Chakrabarti, S., Cooper, J.A., Twardzik, D.R., and Moss, B.,** Deletion of the vaccinia virus growth factor gene reduces virus virulence, *J. Virol.,* 62, 866, 1988.
66. **Smith, J.W., Tevethia, S.S., Levy, B.M., and Rawls, W.E.,** Comparative studies on host responses to Shope fibroma virus in adult and newborn rabbits, *J. Natl. Cancer Inst.,* 50, 1529, 1973.
67. **Chang, W., Upton, C., Hu, S.-L., Purchio, A.F., and McFadden, G.,** The genome of Shope fibroma virus, a tumorigenic poxvirus, contains a growth factor gene with sequence similarity to those encoding epidermal growth factor and transforming growth factor α, *Mol. Cell. Biol.,* 7, 535, 1987.
68. **Lin, Y.-Z., Caporaso, G., Chang, P.-Y., He, X.-H., and Tam, J.P.,** Synthesis of a biological active tumor growth factor from the predicted DNA sequence of Shope fibroma virus, *Biochemistry,* 27, 5640, 1988.
69. **Ye, Y., Lin, Y.-Z., and Tam, T.P.,** Shope fibroma virus growth factor exhibits epidermal growth factor activities in newborn mice, *Biochem. Biophys. Res. Commun.,* 154, 497, 1988.
70. **Upton, C., Macen, J.L., and McFadden, G.,** Mapping and sequencing of a gene from myxoma virus that is related to those encoding epidermal growth factor and transforming growth factor α, *J. Virol.,* 61, 1271, 1987.
71. **Strayer, D.S. and Leibowitz, J.L.,** Inhibition of epidermal growth factor-induced cellular proliferation, *Am. J. Pathol.,* 128, 203, 1987.
72. **Zabel, B.U., Eddy, R.L., Lalley, P.A., Scott, J., Bell, G.I., and Shows, T.B.,** Chromosomal locations of the human and mouse genes for precursors of epidermal growth factor and the β subunit of nerve growth factor, *Proc. Natl. Acad. Sci. U.S.A.,* 82, 469, 1985.
73. **Pascall, J.C. and Brown, K.D.,** Structural analysis of the 5′-flanking sequence of the mouse epidermal growth factor gene, *J. Mol. Endocrinol.,* 1, 5, 1988.
74. **Morton, C.C., Byers, M.G., Nakai, H., Bell, G.I., and Shows, T.B.,** Human genes for insulin-like growth factors I and II and epidermal growth factor are located on 12q22-24.1, 11p15, 4q25-q27, respectively, *Cytogenet. Cell Genet.,* 41, 245, 1986.
75. **Oka, T., Sakamoto, S., Miyoshi, K.-I., Fuwa, T., Yoda, K., Yamasaki, M., Tamura, G., and Miyake, T.,** Synthesis and secretion of human epidermal growth factor by *Escherichia coli, Proc. Natl. Acad. Sci. U.S.A.,* 82, 7212, 1985.
76. **Urdea, M.S., Merryweather, J.P., Mullenbach, G.T., Coit, D., Heberlein, U., Valenzuela, P., and Barrk, P.J.,** Chemical synthesis of a gene for human epidermal growth factor urogastrone and its expression in yeast, *Proc. Natl. Acad. Sci. U.S.A.,* A80, 7461, 1983.
77. **Komoriya, A., Hortsch, M., Meyers, C., Smith, M., Kanety, H., and Schlessinger, J.,** Biologically active synthetic fragments of epidermal growth factor: localization of a major receptor-binding region, *Proc. Natl. Acad. Sci. U.S.A.,* 81, 1351, 1984.
78. **Engler, D.A., Matsunami, R.K., Campion, S.R., Stringer, C.D., Stevens, A., and Nigoyi, S.K.,** Cloning of authentic human epidermal growth factor as a bacterial secretory protein and its initial structure-function analysis by site-directed mutagenesis, *J. Biol. Chem.,* 263, 12384, 1988.
79. **Scott, J., Urdea, M., Quiroga, M., Sanchez-Pescador, R., Fong, N., Selby, M., Rutter, W.J., and Bell, G.I.,** Structure of a mouse submaxillary messenger RNA encoding epidermal growth factor and seven related proteins, *Science,* 221, 236, 1983.
80. **Gray, A., Dull, T.J., and Ullrich, A.,** Nucleotide sequence of epidermal growth factor cDNA predicts a 128,000-molecular weight protein precursor, *Nature (London),* 303, 722, 1983.
81. **Rall, L.B., Scott, J., Bell, G.I., Crawford, R.J., Penschow, J.D., Niall, H.D., and Coghlan, J.P.,** Mouse prepro-epidermal growth factor synthesis by the kidney and other tissues, *Nature (London),* 313, 228, 1985.
82. **Burbeck, S., Latter, G., Metz, E., and Leavitt, J.,** Neoplastic human fibroblast proteins are related to epidermal growth factor precursor, *Proc. Natl. Acad. Sci. U.S.A.,* 81, 5360, 1984.
83. **Mroczkowski, B., Reich, M., Whittaker, J., Bell, G.I., and Cohen, S.,** Expression of human epidermal growth factor precursor cDNA in transfected mouse NIH 3T3 cells, *Proc. Natl. Acad. Sci. U.S.A.,* 85, 126, 1988.

84. **Lakshmanan, J., Salido, E.C., Lam, R., Barajas, L., and Fisher, D.A.,** Identification of pro-epidermal growth factor and high molecular weight epidermal growth factors in adult mouse urine, *Biochem. Biophys. Res. Commun.,* 173, 902, 1990.
85. **Öhlin, A.-K. and Stenflo, J.,** Calcium-dependent interaction between the epidermal growth factor precursor-like region of human protein C and a monoclonal antibody, *J. Biol. Chem.,* 262, 13798, 1987.
86. **Russell, D.W., Schneider, W.J., Yamamoto, T., Luskey, K.L., Brown, M.S., and Goldstein, J.L.,** Domain map of the LDL receptor: sequence homology with the epidermal growth factor precursor, *Cell,* 37, 577, 1984.
87. **Davis, C.G., Goldstein J.L., Südhof, T.C., Anderson, R.G.W., Russell, D.W., and Brown, M.S.,** Acid-dependent ligand dissociation and recycling of LDL receptor mediated by growth factor homology region, *Nature (London),* 326, 760, 1987.
88. **Hayashida, H. and Miyata, T.,** Sequence similarity between epidermal growth factor precursor and atrial natriuretic factor precursor, *FEBS Lett.,* 185, 125, 1985.
89. **Baldwin, G.S.,** Epidermal growth factor precursor is related to the translation product of the Moloney sarcoma virus oncogene *mos, Proc. Natl. Acad. Sci. U.S.A.,* 82, 1921, 1985.
90. **Baley, P., Lützelschwabl, I., Scott-Burden, T., Küng, W., and Eppenberger, U.,** Modulation of extracellular-matrix synthesized by cultured stromal cells from normal human breast tissue by epidermal growth factor, *J. Cell. Biochem.,* 43, 111, 1990.
91. **Mesiano, S., Browne, C.A., and Thorburn, G.D.,** Detection of endogenous epidermal growth factor-like activity in the developing chick embryo, *Dev. Biol.,* 110, 23, 1985.
92. **Stewart, R., Nelson, J., and Wilson, D.J.,** Epidermal growth factor promotes chick embryonic angiogenesis, *Cell Biol. Int. Rep.,* 13, 957, 1989.
93. **Cohen, S.,** Isolation of a mouse submaxillary gland protein accelerating incisor eruption and eyelid opening in the new-born animal, *J. Biol. Chem.,* 237, 1555, 1962.
94. **Turkington, R.W., Males, J.L., and Cohen, S.,** Synthesis and storage of epithelial-epidermal growth factor in submaxillary gland, *Cancer Res.,* 31, 252, 1971.
95. **Gubits, R.M., Shaw, P.A., Gresik, E.W., Onetti-Muda, A., and Barka, T.,** Epidermal growth factor gene expression is regulated differentially in mouse kidney submandibular gland, *Endocrinology,* 119, 1382, 1986.
96. **Gresik, E.W., Wenk-Salamone, K., Onetti-Muda, A., Gubits, R.M., and Shaw, P.A.,** Effect of advanced age on the induction by androgen or thyroid hormone of epidermal growth factor and epidermal growth factor mRNA in the submandibular glands of C57BL/6 male mice, *Mech. Ageing Dev.,* 34, 175, 1986.
97. **Kasayama, S., Ohba, Y., and Oka, T.,** Epidermal growth factor deficiency associated with diabetes mellitus, *Proc. Natl. Acad. Sci. U.S.A.,* 86, 7644, 1989.
98. **Tsutsumi, O., Kurachi, H., and Oka, T.,** A physiological role of epidermal growth factor in male reproductive function, *Science,* 233, 975, 1986.
99. **Buckley, A., Davidson, J.M., Kamerath, C.D., Wolt, T.B., and Woodward, S.C.,** Sustained release of epidermal growth factor accelerates wound repair, *Proc. Natl. Acad. Sci. U.S.A.,* 82, 7340, 1985.
100. **Brown, G.L., Nanney, L.B., Griffen, J., Cramer, A.B., Yancey, J.M., Curtsinger, L.J., III, Holtzin, L., Schultz, G.S., Jurkiewicz, M.J., and Lynch, J.B.,** Enhancement of wound healing by topical treatment with epidermal growth factor, *N. Engl. J. Med.,* 321, 76, 1989.
101. **Laato, M., Hähäri, V.-M., Niinikoski, J., and Vuorio, E.,** Epidermal growth factor increases collagen production in granulation tissue by stimulation of fibroblast proliferation and not by activation of procollagen genes, *Biochem. J.,* 247, 385, 1987.
102. **Smith, J.M., Sporn, M.B., Roberts, A.B., Derynk, R., Winkler, M.E., and Gregory, H.,** Human transforming growth factor-α causes precocious eyelid opening in newborn mice, *Nature (London),* 315, 515, 1985.
103. **Turley, E.A., Hollenberg, M.D., and Pratt, R.M.,** Effect of epidermal growth factor/urogastrone on glycosaminoglycan synthesis and accumulation *in vitro* in the developing mouse palate, *Differentiation,* 28, 279, 1985.
104. **O'Loughlin, E.V., Chung, M., Hollenberg, M., Hayden, J., Zahavi, I., and Gall, D.G.,** Effect of epidermal growth factor on ontogeny of the gastrointestinal tract, *Am. J. Physiol.,* 249, G674, 1985.
105. **Opleta, K., O'Loughlin, E.V., Shaffer, E.A., Hayden, J., Hollenberg, M., and Gall, D.G.,** Effect of epidermal growth factor on growth and postnatal development of the rabbit liver, *Am. J. Physiol.,* 253, G622, 1987.

106. **Willey, J.C., Lechner, J.F., and Harris, C.C.,** Bombesin and the C-terminal tetradecapeptide of gastrin-releasing peptide are growth factors for normal human bronchial epithelial cells, *Exp. Cell Res.,* 153, 245, 1984.
107. **Ouchi, Y., Hirosumi, J., Watanabe, M., Hattori, A., Nakamura, T., and Orimo, H.,** Inhibitory effect of transforming growth factor-β on epidermal growth factor-induced proliferation of cultured rat aortic smooth muscle cells, *Biochem. Biophys. Res. Commun.,* 157, 301, 1988.
108. **Sawada, N.,** Hepatocytes from old rats retain responsiveness of c-*myc* expression to EGF in primary culture but do not enter S phase, *Exp. Cell Res.,* 181, 584, 1989.
109. **Hollenberg, M.D.,** Epidermal growth factor urogastrone: new targets for the ligand and for its receptor, *Proc. West. Pharmacol. Soc.,* 29, 479, 1986.
110. **Muramatsu, I., Itoh, H., Lederis, K., and Hollenberg, M.D.,** Distinctive actions of epidermal growth factor-urogastrone in isolated smooth muscle preparations from guinea pig stomach: differential inhibition by indomethacin, *J. Pharmacol. Exp. Ther.,* 245, 625, 1988.
111. **Patel, P., Itoh, H., Lederis, K., and Hollenberg, M.D.,** Contraction of guinea pig trachea by epidermal growth factor-urogastrone, *Can. J. Physiol. Pharmacol.,* 66, 1308, 1988.
112. **Soley, M. and Hollenberg, M.D.,** Epidermal growth factor (urogastrone)-stimulated gluconeogenesis in isolated mouse hepatocytes, *Arch. Biochem. Biophys.,* 255, 136, 1987.
113. **Chan, C.P. and Krebs, E.G.,** Epidermal growth factor stimulates glycogen synthase activity in cultured cells, *Proc. Natl. Acad. Sci. U.S.A.,* 82, 4563, 1985.
114. **Moreno, F., Pastor-Anglada, M., Hollenberg, M.D., and Soley, M.,** Effects of epidermal growth factor (urogastrone) on gluconeogenesis, glucose oxidation, and glycogen synthesis in isolated rat hepatocytes, *Biochem. Cell Biol.,* 67, 724, 1989.
115. **Ballard, F.J.,** Regulation of protein breakdown by epidermal growth factor in A431 cells, *Exp. Cell Res.,* 157, 172, 1985.
116. **Muramatsu, I., Hollenberg, M.D., and Lederis, K.,** Vascular actions of epidermal growth factor-urogastrone: possible relationship to prostaglandin production, *Can. J. Physiol. Pharmacol.,* 63, 994, 1985.
117. **Gan, B. and Hollenberg, M.D.,** Autoradiographic localization of binding sites for epidermal growth factor-urogastrone (EGF-URO) in coronary arteries, *Eur. J. Pharmacol.,* 167, 407, 1989.
118. **Ichiro, T., Tajija, S., and Nishikawa, T.,** Preferential inhibition of elastin synthesis by epidermal growth factor in chick aortic smooth muscle cells, *Biochem. Biophys. Res. Commun.,* 168, 850, 1990.
119. **Gan, B.S., MacCannell, K.L., and Hollenberg, M.D.,** Epidermal growth factor-urogastrone causes vasodilatation in the anesthetized dog, *J. Clin. Invest.,* 80, 199, 1987.
120. **Gan, B.S., Hollenberg, M.D., MacCannell, K.L., Lederis, K., Winkler, M.E., and Derynck, R.,** Distinct vascular actions of epidermal growth factor-urogastrone and transforming growth factor-α, *J. Pharmacol. Exp. Ther.,* 242, 331, 1987.
121. **Taub, M., Wang, Y., Szczesny, T.M., and Kleinman, H.K.,** Epidermal growth factor or transforming growth factor α is required for kidney tubulogenesis in matrigel cultures in serum-free medium, *Proc. Natl. Acad. Sci. U.S.A.,* 87, 4002, 1990.
122. **Levine, L. and Hassid, A.,** Epidermal growth factor stimulates prostaglandin biosynthesis by canine kidney (MDCK) cells, *Biochem. Biophys. Res. Commun.,* 76, 1181, 1977.
123. **Mullin, J.M. and McGinn, M.T.,** Epidermal growth factor-induced mitogenesis in kidney epithelial cells (LLC-PK_1), *Cancer Res.,* 48, 4886, 1988.
124. **Tashjian, A.H., Jr. and Levine, L.,** Epidermal growth factor stimulates prostaglandin production and bone resorption in cultured mouse calvaria, *Biochem. Biophys. Res. Commun.,* 85, 966, 1978.
125. **Takahashi, N., MacDonald, B.R., Hon, J., Winkler, M.E., Derynck, R., Mundy, G.R., and Roodman, G.D.,** Recombinant human transforming growth factor-α stimulates the formation of osteoclast-like cells in long-term human marrow cultures, *J. Clin. Invest.,* 78, 894, 1986.
126. **Gilchrist, C.A. and Shull, J.D.,** Epidermal growth factor induces prolactin messenger RNA in GH_4C_1 cells via a protein synthesis-dependent pathway, *Mol. Cell. Endocrinol.,* 92, 201, 1993.
127. **Kaji, H. and Hinkle, P.J.,** Epidermal growth factor decreases thyroid hormone receptors and attenuates thyroid hormone responses in GH_4C_1 cells, *Endocrinology,* 120, 537, 1987.
128. **Ozawa, S. and Spaulding, S.W.,** Epidermal growth factor inhibits radioiodine uptake but stimulates deoxyribonucleic acid synthesis in newborn rat thyroid grown in nude mice, *Endocrinology,* 127, 604, 1990.

129. **Vassart, G., Bacolla, A., Brocas, H., Christophe, D., de Martynoff, G., Leriche, A., Mercken, A., Parma, J., Pohl, V., Targovnik, H., and Van Heuverswyn, B.**, Structure, expression and regulation of the thyroglobulin gene, *Mol. Cell. Endocrinol.*, 40, 89, 1985.
130. **Schonbrunn, A., Kransoff, M., Westendorf, J.M., and Tashjian, A.H.**, Epidermal growth factor and thyrotropin-releasing hormone act similarly on a clonal pituitary cell strain: modulation of hormone production and inhibition of cell proliferation, *J. Cell. Physiol.*, 85, 786, 1980.
131. **Ozawa, S., Sheflin, L.G., and Spaulding, S.W.**, Thyroxine increases epidermal growth factor levels in the mouse thyroid *in vivo, Endocrinology,* 128, 1396, 1991.
132. **Verhoeven, G. and Cailleau, J.**, Stimulatory effects of epidermal growth factor on steroidogenesis in Leydig cells, *Mol. Cell. Endocrinol.*, 47, 99, 1986.
133. **Bendell, J.J. and Dorrington, J.H.**, Epidermal growth factor influences growth and differentiation of rat granulosa cells, *Endocrinology,* 127, 533, 1990.
134. **Maruo, T., Ladines-Llave, C.A., Samoto, T., Matsuo, H., Manalo, A.S., Ito, H., and Mochizuki, M.**, Expression of epidermal growth factor and its receptor in the human ovary during follicular growth and regression, *Endocrinology,* 132, 924, 1993.
135. **May, J.V., Buck, P.A., and Schomberg, D.W.**, Epidermal growth factor enhances [^{125}]iodo-follicle-stimulating hormone binding by cultured porcine granulosa cells, *Endocrinology,* 120, 2413, 1987.
136. **Endo, K., Atlas, S.J., Rone, J.D., Zanagnolo, R.V.L., Kuo, T.-C., Dharmarajan, A.M., and Wallach, E.E.**, Epidermal growth factor inhibits follicular response to human chorionic gonadotropin: possible role of cell to cell communication in the response to gonadotropin, *Endocrinology,* 130, 186, 1992.
137. **Feng, P., Catt, K.J., and Knecht, M.**, Transforming growth factor β regulates the inhibitory actions of epidermal growth factor during granulosa cell differentiation, *J. Biol. Chem.*, 261, 14167, 1986.
138. **Yamaguchi, K., Lederis, K., and Hollenberg, M.D.**, Contraction of porcine ocular ciliary muscle by epidermal growth factor-urogastrone, *Eur. J. Pharmacol.*, 191, 245, 1990.
139. **Fallon, J.H., Seroogy, K.B., Loughlin, S.E., Morrison, R.S., Bradshaw, R.A., Knauer, D.J., and Cunningham, D.D.**, Epidermal growth factor immunoreactive material in the central nervous system: location and development, *Science,* 224, 1107, 1984.
140. **Morrison, R.S., Kornblum, H.I., Leslie, F.M., and Bradshaw, R.A.**, Trophic stimulation of cultured neurons from neonatal rat brain by epidermal growth factor, *Science,* 238, 72, 1987.
141. **Yamaguchi, M., Ogren, L., Endo, H., Thordarson, G., Kensing, R., and Talamantes, F.**, Epidermal growth factor stimulates mouse placental lactogen I but inhibits mouse placental lactogen II secretion *in vitro, Proc. Natl. Acad. Sci. U.S.A.*, 89, 11396, 1992.
142. **Morrish, D.W., Bhardwaj, D., Dabbagh, L.K., Marusyk, H., and Siy, O.**, Epidermal growth factor induces differentiation and secretion of human chorionic gonadotropin and placental lactogen in normal human placenta, *J. Clin. Endocrinol. Metab.*, 65, 1282, 1987.
143. **Oka, T., Kurachi, H., Yoshimura, M., Tsutsumi, O., Cossu, M.F., and Taga, M.**, Study of the growth factors for the mammary gland: epidermal growth factor and mesenchyme-derived growth factor, *Nucl. Med. Biol.*, 14, 353, 1987.
144. **Vonderhaar, B.K.**, Local effects of EGF, α-TGF, and EGF-like growth factors on lobuloalveolar development of the mouse mammary gland *in vivo, J. Cell. Physiol.*, 132, 581, 1987.
145. **Fenton, S.E. and Sheffield, L.G.**, Lactogenic hormones increase epidermal growth factor messenger RNA content of mouse mammary glands, *Biochem. Biophys. Res. Commun.*, 181, 1063, 1991.
146. **Taverna, D., Groner, B., and Hynes, N.**, Epidermal growth factor receptor, platelet growth factor receptor, and c-erbB-2 receptor activation all promote growth but have distinctive effects upon mouse mammary epithelial cell differentiation, *Cell Growth Differ.*, 2, 145, 1991.
147. **Hirata, Y., Nishimura, T., Uchihashi, M., and Fujita, T.**, Partial purification and characterization of epidermal growth factor in human breast milk, *Endocrinol. Jpn.*, 33, 433, 1986.
148. **Iacopetta, B.J., Grieu, F., Horisberger, M., and Sanahara, G.I.**, Epidermal growth factor in human and bovine milk, *Acta Paediatr.*, 81, 287, 1992.
149. **Petrides, P.E., Hosang, M., Shooter, E., Esch, F.S., and Böhlen, P.**, Isolation and characterization of epidermal growth factor from human milk, *FEBS Lett.*, 187, 89, 1985.
150. **Gonnella, P.A., Siminoski, K., Murphy, R.A., and Neutra, M.R.**, Transepithelial transport of epidermal growth factor by absorptive cells of suckling rat ileum, *J. Clin. Invest.*, 80, 22, 1987.
151. **Okamoto, S. and Oka, T.**, Evidence for physiological function of epidermal growth factor: pregestational sialoadenectomy of mice decreases milk production and increases offspring mortality during lactation period, *Proc. Natl. Acad. Sci. U.S.A.*, 81, 6059, 1984.

152. **Komura, H., Wakimoto, H., Chu-Fung, C., Terakawa, N., Aono, T., Tanizawa, O., and Matsumoto, K.,** Retinoic acid enhances cell responses to epidermal growth factor in mouse mammary gland in culture, *Endocrinology,* 118, 1530, 1986.
153. **Sankaran, L. and Topper, Y.J.,** Is EGF a physiological inhibitor or mouse mammary casein synthesis? Unphysiological responses to pharmacological levels of hormones, *Biochem. Biophys. Res. Commun.,* 146, 121, 1987.
154. **Fujieda, M., Murata, Y., Hayashi, H., Kambe, F., Matsui, N., and Seo, H.,** Effect of thyroid hormone on epidermal growth factor gene expression in mouse submandibular gland, *Endocrinology,* 132, 121, 1993.
155. **Huet-Hudson, Y.M., Chakraborty, C., De, S.K., Suzuki, Y., Andrews, G.K., and Dey, S.K.,** Estrogen regulates the synthesis of epidermal growth factor in mouse uterine epithelial cells, *Mol. Endocrinol.,* 4, 510, 1990.
156. **Carpenter, G.,** The biochemistry and physiology of the receptor-kinase for epidermal growth factor, *Mol. Cell. Endocrinol.,* 31, 1, 1983.
157. **Carpenter, G.,** Properties of the receptor for epidermal growth factor, *Cell,* 37, 357, 1984.
158. **Thompson, D.M. and Gill, G.N.,** The EGF receptor: structure, regulation and potential role in malignancy, *Cancer Surv.,* 4, 767, 1985.
159. **Soderquist, A.M. and Carpenter, G.,** Biosynthesis and metabolic degradation of receptors for epidermal growth factor, *J. Membr. Biol.,* 90, 97, 1986.
160. **Carpenter, G.,** Receptors for epidermal growth factor and other polypeptide mitogens, *Annu. Rev. Biochem.,* 56, 881, 1987.
161. **Schlessinger, J.,** The epidermal growth factor receptor as a multifunctional allosteric protein, *Biochemistry,* 27, 3119, 1988.
162. **Maihle, N.J. and Kung, H.-J.,** C-*erb*B and the epidermal growth-factor receptor: a molecule with dual identity, *Biochim. Biophys. Acta,* 948, 287, 1989.
163. **Velu, T.J.,** Structure, function and transforming potential of the epidermal growth factor receptor, *Mol. Cell. Endocrinol.,* 70, 205, 1990.
164. **Hayman, M.J. and Enrietto, P.J.,** Cell transformation by the epidermal growth factor receptor and v-*erb*B, *Cancer Cells,* 3, 302, 1991.
165. **Iwashita, S. and Kobayashi, M.,** Signal transduction system for growth factor receptors associated with tyrosine kinase activity: epidermal growth factor receptor signalling and its regulation, *Cell. Signal.,* 4, 123, 1992.
166. **Yaish, P., Gazit, A., Gilon, C., and Levitzki, A.,** Blocking of EGF-dependent cell proliferation by EGF receptor kinase inhibitors, *Science,* 242, 933, 1988.
167. **Umezawa, K., Hori, T., Tajima, H., Imoto, M., Isshiki, K., and Takeuchi, T.,** Inhibition of epidermal growth factor-induced DNA synthesis by tyrosine kinase inhibitors, *FEBS Lett.,* 260, 198, 1990.
168. **Hock, R.A., Nexo, E., and Hollenberg, M.D.,** Isolation of the human placenta receptor for epidermal growth factor-urogastrone, *Nature (London),* 277, 403, 1979.
169. **Hock, R.A. and Hollenberg, M.D.,** Characterization of the receptor for epidermal growth factor-urogastrone in human placenta membranes, *J. Biol. Chem.,* 255, 10731, 1980.
170. **Hock, R.A., Nexo, E., and Hollenberg, M.D.,** Solubilization and isolation of the human placenta receptor for epidermal growth factor-urogastrone, *J. Biol. Chem.,* 255, 10737, 1980.
171. **Gill, G.N. and Lazar, C.S.,** Increased phosphotyrosine content and inhibition of proliferation in epidermal growth factor-treated A431 cells, *Nature (London),* 293, 305, 1981.
172. **Schreiber, A.B., Libermann, T.A., Lax, I., Yarden, Y., and Schlessinger, J.,** Biological role of epidermal growth factor receptor clustering, *J. Biol. Chem.,* 258, 846, 1983.
173. **Richert, N.D., Willingham, M.C., and Pastan, I.,** Epidermal growth factor receptor: characterization of a monoclonal antibody specific for the receptor of A431 cells, *J. Biol. Chem.,* 258, 8902, 1983.
174. **Gooi, H.C., Schlessinger, J., Lax, L., Yarden, Y., Libermann, T.A., and Feizi, T.,** Monoclonal antibody reactive with the human epidermal-growth-factor receptor recognizes the blood-group-A antigen, *Biosci. Rep.,* 3, 1045, 1983.
175. **Parker, P.J., Young, S., Gullick, W.J., Mayes, E.L.V., Bennett, P., and Waterfield, M.D.,** Monoclonal antibodies against the human epidermal growth factor receptor from A431 cells: isolation, characterization, and use in the purification of active epidermal growth factor receptor, *J. Biol. Chem.,* 259, 9906, 1984.

176. **Yarden, Y., Harari, I., and Schlessinger, J.,** Purification of an active EGF receptor kinase with monoclonal antireceptor antibodies, *J. Biol. Chem.,* 260, 315, 1985.
177. **Fernandez-Pol, J.A.,** Epidermal growth factor receptor of A431 cells: characterization of a monoclonal anti-receptor antibody noncompetitive agonist of epidermal growth factor action, *J. Biol. Chem.,* 260, 5003, 1985.
178. **Defize, L.H.K., Moolenaar, W.H., van der Saag, P.T., and de Laat, S.W.,** Dissociation of cellular responses to epidermal growth factor using anti-receptor monoclonal antibodies, *EMBO J.,* 5, 1187, 1986.
179. **Bellot, F., Moolenaar, W., Kris, R., Mirakhur, B., Verlaan, I., Ullrich, A., Schlessinger, J., and Felder, S.,** High-affinity epidermal growth factor binding is specifically reduced by a monoclonal antibody, and appears necessary for early responses, *J. Cell Biol.,* 110, 491, 1990.
180. **Weber, W., Bertics, P.J., and Gill, G.N.,** Immunoaffinity purification of the epidermal growth factor receptor: stoichiometry of binding and kinetics of self-phosphorylation, *J. Biol. Chem.,* 259, 14631, 1984.
181. **Schechter, Y., Hernaez, L., Schlessinger, J., and Cuatrecasas, P.,** Local aggregation of hormone-receptor complexes is required for activation by epidermal growth factor, *Nature (London),* 278, 835, 1979.
182. **Wiegant, F.A.C., Blok, F.J., Defize, L.H.K., Linnemans, W.A.M., Verkley, A.J., and Boonstra, J.,** Epidermal growth factor receptors associated to cytoskeletal elements of epidermoid carcinoma (A431) cells, *J. Cell Biol.,* 103, 87, 1986.
183. **Roy, L.M., Gittinger, C.K., and Landreth, G.E.,** Characterization of the epidermal growth factor receptor associated with cytoskeletons of A431 cells, *J. Cell. Physiol.,* 140, 295, 1989.
184. **van Bergen en Henegouwen, P.M.P., Defize, L.H.K., de Kroon, J., van Damme, H., Verkleij, A.J., and Boonstra, J.,** Ligand-induced association of epidermal growth factor receptor to the cytoskeleton of A431 cells, *J. Cell. Biochem.,* 39, 455, 1989.
185. **van Belzen, N., Spaargaren, M., Verkleij, A.J., and Boonstra, J.,** Interaction of epidermal growth factor receptors with the cytoskeleton is related to receptor clustering, *J. Cell. Physiol.,* 145, 365, 1990.
186. **Kondo, I. and Shimizu, N.,** Mapping of the human gene for epidermal growth factor receptor (EGFR) on the p13-q22 region of chromosome 7, *Cytogenet. Cell Genet.,* 35, 9, 1983.
187. **Merlino, G.T., Ishii, S., Whang-Peng, J., Knutsen, T., Xu, Y.-H., Clark, A.J.L., Stratton, R.H., Wilson, R.K., Ma, D.P., Roe, B.A., Hunts, J.H., Shimizu, N., and Pastan, I.,** Structure and localization of genes encoding aberrant and normal epidermal growth factor receptor RNAs from A431 human carcinoma cells, *Mol. Cell. Biol.,* 5, 1722, 1985.
188. **Xu, Y.-H., Richert, N., Ito, S., Merlino, G.T., and Pastan, I.,** Characterization of epidermal growth factor receptor gene expression in malignant and normal human cell lines, *Proc. Natl. Acad. Sci. U.S.A.,* 81, 7308, 1984.
189. **Lax, I., Johnson, A., Howk, R., Sap, J., Bellot, F., Winkler, M., Ullrich, A., Vennstrom, B., Schlessinger, J., and Givol, D.,** Chicken epidermal growth factor (EGF) receptor: cDNA cloning, expression in mouse cells, and differential binding of EGF and transforming growth factor α, *Mol. Cell. Biol.,* 8, 1970, 1988.
190. **Silver, J., Whitney, J.B., III, Kozak, C., Hollis, G., and Kirsch, I.,** *ErbB* is linked to the α-globin locus on mouse chromosome 11, *Mol. Cell. Biol.,* 5, 1784, 1985.
191. **Hayman, M.J.,** *erb*-B: growth factor receptor turned oncogene, *Trends Genet.,* 2, 260, 1986.
192. **Maihle, N.J. and Kung, H.-J.,** c-*erbB* and the epidermal growth-factor receptor: a molecule with dual identity, *Biochim. Biophys. Acta,* 948, 287, 1989.
193. **Merlino, G.T., Xu, Y.-H., Ishii, S., Clark, A.J.L., Semba, K., Toyoshima, K., Yamamoto, T., and Pastan, I.,** Amplification and enhanced expression of the epidermal growth factor receptor gene in A431 human carcinoma cells, *Science,* 224, 417, 1984.
194. **Libermann, T.A., Nusbaum, H.R., Razon, N., Kris, R., Lax, I., Soreq, H., Whittle, N., Waterfield, M.D., Ullrich, A., and Schlessinger, J.,** Amplification, enhanced expression and possible rearrangement of EGF receptor gene in primary human brain tumours of glial origin, *Nature (London),* 313, 144, 1985.
195. **Lin, C.R., Chen, W.S., Kruiger, W., Stolarsky, L.S., Weber, W., Evans, R.M., Verma, I.M., Gill, G.N., and Rosenfeld, M.G.,** Expression cloning of human EGF receptor complementary DNA: gene amplification and three related messenger RNA products in A431 cells, *Science,* 224, 843, 1984.

196. **Ullrich, A., Coussens, L., Hayflick, J.S., Dull, T.J., Gray, A., Tam, A.W., Lee, J., Yarden, Y., Libermann, T.A., Schlessinger, J., Downward, J., Mayes, E.L.V., Whittle, N., Waterfield, M.D., and Seeburg, P.H.,** Human epidermal growth factor cDNA sequence and aberrant expression of the amplified gene in A431 epidermoid carcinoma cells, *Nature (London),* 309, 418, 1984.
197. **Xu, Y.-H., Ishii, S., Clark, A.J.L., Sullivan, M., Wilson, R.K., Ma, D.P., Roe, B.A., Merlino, G.T., and Pastan, I.,** Human epidermal growth factor receptor cDNA is homologous to a variety of RNAs overproduced in A431 carcinoma cells, *Nature (London),* 309, 806, 1984.
198. **Merlino, G.T., Ishii, S., Whang-Peng, J., Knutsen, T., Xu, Y.-H., Clark, A.J.L., Stratton, R.H., Wilson, R.K., Ma, D.P., Roe, B.A., Hunts, J.H., Shimizu, N., and Pastan, I.,** Structure and localization of genes encoding aberrant and normal epidermal growth factor receptor RNAs from A431 human carcinoma cells, *Mol. Cell. Biol.,* 5, 1722, 1985.
199. **Ishii, S., Xu, Y.-H., Stratton, R.H., Roe, B.A., Merlino, G.T., and Pastan, I.,** Characterization and sequence of the promoter region of the human epidermal growth factor receptor gene, *Proc. Natl. Acad. Sci. U.S.A.,* 82, 4920, 1985.
200. **Ishii, S., Merlino, G.T., and Pastan, I.,** Promoter region of the human Harvey *ras* proto-oncogene: similarity of the EGF receptor proto-oncogene promoter, *Science,* 230, 1378, 1985.
201. **Johnson, A.C., Jinno, Y., and Merlino, G.T.,** Modulation of epidermal growth factor receptor proto-oncogene transcription by a promoter site sensitive to S1 nuclease, *Mol. Cell. Biol.,* 8, 4174, 1988.
202. **Kageyama, R., Merlino, G.T., and Pastan, I.,** Epidermal growth factor (EGF) receptor gene transcription. Requirement for Sp1 and an EGF receptor-specific factor, *J. Biol. Chem.,* 263, 6329, 1988.
203. **Hudson, L.G., Thompson, K.L., Xu, J., and Gill, G.N.,** Identification and characterization of a regulated promoter element in the epidermal growth factor receptor gene, *Proc. Natl. Acad. Sci. U.S.A.,* 87, 7536, 1990.
204. **Lee, J.S., Ro, J.S., Eisbruch, A., Shtalrid, M., Ferrell, R.E., Gutterman, J.U., and Blick, M.,** Multiple restriction fragment length polymorphisms of the human epidermal growth factor receptor gene, *Cancer Res.,* 48, 4045, 1988.
205. **Aglio, L.S., Maturo, J.M., III, and Hollenberg, M.D.,** Receptors for insulin and epidermal growth factor: interaction with organomercurial agarose, *J. Cell. Biochem.,* 28, 143, 1985.
206. **Panayotou, G.N., Magee, A.I., and Geisow, M.J.,** Reconstitution of the epidermal growth factor receptor in artificial lipid bilayers, *FEBS Lett.,* 183, 321, 1985.
207. **Sternberg, M.J.E. and Taylor, W.R.,** Modelling the ATP-binding site of oncogene products, the epidermal growth factor receptor and related proteins, *FEBS Lett.,* 175, 387, 1984.
208. **Russo, M.W., Lukas, T.J., Cohen, S., and Staros, J.V.,** Identification of residues in the nucleotide binding site of the epidermal growth factor receptor/kinase, *J. Biol. Chem.,* 260, 5205, 1985.
209. **Gill, G.N., Kawamoto, T., Cochet, C., Le, A., Sato, J.D., Masui, H., McLeod, C., and Mendelsohn, J.,** Monoclonal anti-epidermal growth factor receptor antibodies which are inhibitors of epidermal growth factor binding and antagonists of epidermal growth factor-stimulated tyrosine protein kinase activity, *J. Biol. Chem.,* 259, 7755, 1984.
210. **Lax, I., Burgess, W.H., Bellot, F., Ullrich, A., Schlessinger, J., and Givol, D.,** Localization of a major receptor-binding domain for epidermal growth factor by affinity labeling, *Mol. Cell. Biol.,* 8, 1831, 1988.
211. **Wedegaertner, P.B. and Gill, G.N.,** Activation of the purified protein tyrosine kinase domain of the epidermal growth factor receptor, *J. Biol. Chem.,* 264, 11346, 1989.
212. **Mayo, K.H., Nunez, M., Burke, C., Starbuck, C., Lauffenburger, D., and Savage, C.R., Jr.,** Epidermal growth factor receptor binding is not a simple one-step process, *J. Biol. Chem.,* 264, 17838, 1989.
213. **Beguinot, L., Werth, D., Ito, S., Richert, N., Willingham, M.C., and Pastan, I.,** Functional studies on the EGF receptor with an antibody that recognizes the intracellular portion of the receptor, *J. Biol. Chem.,* 261, 1801, 1986.
214. **Margolis, B., Li, N., Koch, A., Mohammadi, M., Hurwitz, D.R., Zilberstein, A., Ullrich, A., Pawson, T., and Schlessinger, J.,** The tyrosine phosphorylated carboxyterminus of the EGF receptor is a binding site for GAP and PLC-γ, *EMBO J.,* 9, 4375, 1990.
215. **Maihle, N.J., Flickinger, T.W., Raines, M.A., Sanders, M.L., and Kung, H.J.,** Native avian c-*erbB* gene expresses a secreted protein product corresponding to the ligand-binding domain of the receptor, *Proc. Natl. Acad. Sci. U.S.A.,* 88, 1825, 1991.

216. **Flickinger, T.W., Maihle, N.J., and Kung, H.-J.,** An alternatively processed mRNA from the avian *c-erbB* gene encodes a soluble, truncated form of the receptor that can block ligand-dependent transformation, *Mol. Cell. Biol.,* 12, 883, 1992.
217. **Abdullah, N.A., Torres, B.A., Basu, M., and Johnson, H.M.,** Differential effects of epidermal growth factor, transforming growth factor-α, and vaccinia virus growth factor in the positive regulation of IFN-γ production, *J. Immunol.,* 143, 113, 1989.
218. **Yang, S.-G., Saifeddine, M., and Hollenberg, M.D.,** Tyrosine kinase inhibitors and the contractile action of epidermal growth factor-urogastrone and other agonists in gastric smooth muscle, *Can. J. Physiol. Pharmacol.,* 70, 85, 1992.
219. **Yang, S.-G. and Hollenberg, M.D.,** Distinct receptors for epidermal growth factor-urogastrone in cultured gastric smooth muscle cells, *Am. J. Physiol.,* 260, G827, 1991.
220. **Moriai, T., Kobrin, M.S., and Korc, M.,** Cloning of a variant epidermal growth factor receptor, *Biochem. Biophys. Res. Commun.,* 191, 1034, 1993.
221. **Berkers, J.A.M., van Bergen en Henegouwen, P.M.P., and Boonstra, J.,** Three classes of epidermal growth factor receptors on HeLa cells, *J. Biol. Chem.,* 266, 922, 1991.
222. **Yarden, Y. and Schlessinger, J.,** Self-phosphorylation of epidermal growth factor receptor: evidence for a model of intermolecular allosteric activation, *Biochemistry,* 26, 1434, 1987.
223. **Yarden, Y. and Schlessinger, J.,** Epidermal growth factor induces rapid, reversible aggregation of the purified epidermal growth factor receptor, *Biochemistry,* 26, 1443, 1987.
224. **Kashles, O., Szapary, D., Bellot, F., Ullrich, A., Schlessinger, J., and Schmidt, A.,** Ligand-induced stimulation of epidermal growth factor receptor mutants with altered transmembrane regions, *Proc. Natl. Acad. Sci. U.S.A.,* 85, 9567, 1988.
225. **Böni-Schnetzler, M. and Pilch, P.F.,** Mechanism of epidermal growth factor receptor autophosphorylation and high-affinity binding, *Proc. Natl. Acad. Sci. U.S.A.,* 84, 7832, 1987.
226. **Kashles, O., Yarden, Y., Fischer, R., Ullrich, A., and Schlessinger, J.,** A dominant negative mutation suppresses the function of normal epidermal growth factor receptors in heterodimerization, *Mol. Cell. Biol.,* 11, 1454, 1991.
227. **Koland, J.G. and Cerione, R.A.,** Growth factor control of epidermal growth factor receptor kinase activity via an intramolecular mechanism, *J. Biol. Chem.,* 263, 2230, 1988.
228. **Kwatra, M.M., Bigner, D.D., and Cohn, J.A.,** The ligand binding domain of the epidermal growth factor receptor is not required for receptor dimerization, *Biochim. Biophys. Acta,* 1134, 178, 1992.
229. **Velu, T.J., Vass, W.C., Lowy, D.R., and Beguinot, L.,** Functional heterogeneity of proto-oncogene tyrosine kinases: the C terminus of the human epidermal growth factor receptor facilitates cell proliferation, *Mol. Cell. Biol.,* 9, 1772, 1989.
230. **Northwood, I.C. and Davis, R.J.,** Signal transduction by the epidermal growth factor receptor after functional desensitization of the receptor tyrosine protein kinase activity, *Proc. Natl. Acad. Sci. U.S.A.,* 87, 6107, 1990.
231. **Zechel, C., Schleenbecker, U., Anders, A., and Anders, F.,** v-*erb*B related sequences in *Xiphophorus* that map to melanoma determining Mendelian loci and overexpress in a melanoma cell line, *Oncogene,* 3, 605, 1988.
232. **Wittbrodt, J., Adam, D., Malitschek, B., Mäueler, W., Raulf, F., Telling, A., Robertson, S.M., and Schartl, M.,** Novel putative receptor tyrosine kinase encoded by the melanoma-inducing *Tu* locus in *Xiphophorus, Nature (London),* 341, 415, 1989.
233. **Wadsworth, S.C., Vincent, W.S., III, and Bilodeau-Wentworth, D.,** A *Drosophila* genomic sequence with homology to human epidermal growth factor receptor, *Nature (London),* 314, 178, 1985.
234. **Schejter, E.D., Segal, D., Glazer, L., and Shilo, B.-Z.,** Alternative 5′ exons and tissue-specific expression of the *Drosophila* EGF receptor homolog transcripts, *Cell,* 46, 1091, 1986.
235. **Livneh, E., Glazer, L., Segal, D., Schlessinger, J., and Shilo, B.-Z.,** The *Drosophila* EGF receptor gene homolog: conservation of both hormone binding and kinase domains, *Cell,* 40, 599, 1985.
236. **Thompson, K.L., Decker, S.J., and Rosner, M.R.,** Identification of a novel receptor in *Drosophila* for both epidermal growth factor and insulin, *Proc. Natl. Acad. Sci. U.S.A.,* 82, 8443, 1985.
237. **Garcia, J.V., Stoppelli, M.P., Decker, S.J., and Rosner, M.R.,** An insulin epidermal growth factor-binding protein from *Drosophila* has insulin-degrading activity, *J. Cell Biol.,* 108, 177, 1989.
238. **Zak, N.B. and Shilo, B.Z.,** Biochemical properties of the *Drosophila* EGF receptor homolog (DER) protein, *Oncogene,* 5, 1589, 1990.

239. **Baker, N.E. and Rubin, G.M.,** Effect on eye development of dominant mutations in *Drosophila* homologue of the EGF receptor, *Nature (London),* 340, 150, 1989.
240. **Carlin, C.R. and Knowles, B.B.,** Biosynthesis of the epidermal growth factor receptor in human epidermoid carcinoma-derived A431 cells, *J. Biol. Chem.,* 259, 7902, 1984.
241. **Lane, M.D., Ronnett, G., Slieker, L.J., Kohanski, R.A., and Olson, T.L.,** Post-translational processing and activation of insulin and EGF receptors, *Biochimie,* 67, 1069, 1985.
242. **Carlin, C.R. and Knowles, B.B.,** Biosynthesis and glycosylation of the epidermal growth factor receptor in human tumor-derived cell lines A431 and Hep 3B, *Mol. Cell. Biol.,* 6, 257, 1986.
243. **Kay, D.G., Lai, W.H., Uchihashi, M., Khan, M.N., Posner, B.I., and Bergeron, J.J.M.,** Epidermal growth factor receptor kinase translocation and activation *in vivo, J. Biol. Chem.,* 261, 8473, 1986.
244. **Lane, M.D., Slieker, L.J., Olson, T.S., and Martensen, T.M.,** Post-translational acquisition of ligand binding- and tyrosine kinase-domain function by the epidermal growth factor and insulin receptors, *J. Receptor Res.,* 7, 321, 1987.
245. **Mayes, E.L.V. and Waterfield, M.D.,** Biosynthesis of the epidermal growth factor receptor in A431 cells, *EMBO J.,* 3, 531, 1984.
246. **Childs, R.A., Gregoriou, M., Scudder, P., Thorpe, S.J., Rees, A.R., and Feizi, T.,** Blood group-active chains on the receptor for epidermal growth factor of A431 cells, *EMBO J.,* 3, 2227, 1984.
247. **Slieker, L.J. and Lane, M.D.,** Post-translational processing of the epidermal growth factor receptor: glycosylation-dependent acquisition of ligand-binding capacity, *J. Biol. Chem.,* 260, 687, 1985.
248. **Slieker, L.J., Martensen, T.M., and Lane, M.D.,** Biosynthesis of the epidermal growth factor receptor: post-translational glycosylation-independent acquisition of tyrosine kinase autophosphorylation activity, *Biochem. Biophys. Res. Commun.,* 153, 96, 1988.
249. **Decker, S.J.,** Effects of epidermal growth factor and 12-*O*-tetradecanoylphorbol-13-acetate on metabolism of the epidermal growth factor receptor in normal human fibroblasts, *Mol. Cell. Biol.,* 4, 1718, 1984.
250. **Stoscheck, C.M., Soderquist, A.M., and Carpenter, G.,** Biosynthesis of the epidermal growth factor receptor in cultured human cells, *Endocrinology,* 116, 528, 1985.
251. **Hunts, J., Merlino, G., Pastan, I., and Shimizu, N.,** Reduction of EGF receptor synthesis by antisense RNA vectors, *FEBS Lett.,* 206, 319, 1986.
252. **Wasilenko, W.J., Nori, M., Testerman, N., and Weber, M.J.,** Inhibition of epidermal growth factor receptor biosynthesis caused by the *src* oncogene product, $pp60^{v\text{-}src}$, *Mol. Cell. Biol.,* 10, 1254, 1990.
253. **Real, F.X., Rettig, W.J., Chesa, P.G., Melamed, M.R., Old, L.J., and Mendelsohn, J.,** Expression of epidermal growth factor receptor in human cultured cells and tissues: relationship to cell lineage and stage of differentiation, *Cancer Res.,* 46, 4726, 1986.
254. **Rao, C.V., Carman, F.R., Chegini, N., and Schultz, G.S.,** Binding sites for epidermal growth factor in human fetal membranes, *J. Clin. Endocrinol. Metab.,* 58, 1034, 1984.
255. **Nexo, E., Hollenberg, M.D., Figueroa, A., and Pratt, R.M.,** Detection of epidermal growth factor-urogastrone and its receptor during fetal mouse development, *Proc. Natl. Acad. Sci. U.S.A.,* 77, 2782, 1980.
256. **Chegini, N. and Rao, C.V.,** Epidermal growth factor binding to human amnion, chorion, decidua, and placenta from mid-and term pregnancy: quantitative light microscopic autoradiographic studies, *J. Clin. Endocrinol. Metab.,* 61, 529, 1985.
257. **Chen, C.-F., Kurachi, H., Fujita, Y., Terakawa, N., Miyake, A., and Tanizawa, O.,** Changes in epidermal growth factor receptor and its messenger ribonucleic acid levels in human placenta and isolated trophoblast cells during pregnancy, *J. Clin. Endocrinol. Metab.,* 67, 1171, 1988.
258. **Maruo, T. and Mochizuki, M.,** Immunohistochemical localization of epidermal growth factor receptor and myc oncogene product in human placenta: implication for trophoblast proliferation and differentiation, *Am. J. Obstet. Gynecol.,* 165, 721, 1987.
259. **Simmen, F.A.,** Expression of the avian c-*erb* B (EGF receptor) proto-oncogene during estrogen-promoted oviduct growth, *Biochim. Biophys. Acta,* 910, 182, 1987.
260. **Mukku, V.R. and Stancel, G.M.,** Receptors for epidermal growth factor in the rat uterus, *Endocrinology,* 117, 149, 1985.
261. **Chabot, J.-G., St-Arnaud, R., Walker, P., and Pelletier, G.,** Distribution of epidermal growth factor receptors in the rat ovary, *Mol. Cell. Endocrinol.,* 44, 99, 1986.
262. **Hoath, S.B., Pickens, W.L., Bucuvalas, J.C., and Suchy, F.J.,** Characterization of hepatic epidermal growth factor receptors in the developing rat, *Biochim. Biophys. Acta,* 930, 107, 1987.

263. **Wollenberg, G.K., Harris, L., Farber, E., and Hayes, M.A.,** Inverse relationship between epidermal growth factor induced proliferation and expression of high affinity surface epidermal growth factor receptors in rat hepatocytes, *Lab. Invest.,* 60, 254, 1989.
264. **Kinoshita, A., Takinawa, M., and Suzuki, F.,** Demonstration of receptors for epidermal growth factor on cultured rabbit chondrocytes and regulation of their expression by various growth and differentiation factors, *Biochem. Biophys. Res. Commun.,* 183, 14, 1992.
265. **Schneider, C.A., Lim, R.W., Terwilliger, E., and Herschman, H.R.,** Epidermal growth factor-nonresponsive 3T3 variants do not contain epidermal growth factor receptor-related antigen or mRNA, *Proc. Natl. Acad. Sci. U.S.A.,* 83, 333, 1986.
266. **Kudlow, J.E., Cheung, J.-Y.C., and Bjorge, J.D.,** Epidermal growth factor stimulates the synthesis of its own receptor in a human breast cancer cell line, *J. Biol. Chem.,* 261, 4134, 1986.
267. **Earp, H.S., Austin, K.S., Blaisdell, J., Rubin, R.A., Nelson, K.G., Lee, L.W., and Grisham, J.W.,** Epidermal growth factor (EGF) stimulate EGF receptor synthesis, *J. Biol. Chem.,* 261, 4777, 1986.
268. **Clark, A.J.L., Ishii, S., Richert, N., Merlino, G.T., and Pastan, I.,** Epidermal growth factor regulates the expression of its own receptor, *Proc. Natl. Acad. Sci. U.S.A.,* 82, 8374, 1985.
269. **Blay, J. and Hollenberg, M.D.,** Heterologous regulation of EGF receptor function in cultured aortic smooth muscle cells, *Eur. J. Pharmacol.,* 172, 1, 1989.
270. **Zheng, Z.-S., Polakowska, R., Johnson, A., and Goldsmith, L.A.,** Transcriptional control of epidermal growth factor receptor by retinoic acid, *Cell Growth Differ.,* 3, 225, 1992.
271. **Lokeshwar, V.B., Huang, S.S., and Huang, J.S.,** Protamine enhances epidermal growth factor (EGF)-stimulated mitogenesis by increasing cell surface EGF receptor number. Implications for existence of cryptic EGF receptors, *J. Biol. Chem.,* 264, 19318, 1989.
272. **Weis, F.M.B. and Davis, R.J.,** Regulation of epidermal growth factor receptor signal transduction. Role of gangliosides, *J. Biol. Chem.,* 265, 12059, 1990.
273. **Jose, E.S., Benguria, A., Geller, P., and Villalobo, A.,** Calmodulin inhibits the epidermal growth factor receptor tyrosine kinase, *J. Biol. Chem.,* 267, 15237, 1992.
274. **Holmes, S.D., Spotts, G., and Smith, R.G.,** Rat Sertoli cells secrete a growth factor that blocks epidermal growth factor (EGF) binding to its receptor, *J. Biol. Chem.,* 261, 4076, 1986.
275. **Carlin, C.R., Tollefson, A.E., Brady, H.A., Hoffman, B.L., and Wold, W.S.M.,** Epidermal growth factor receptor is down-regulated by a 10,400 MW protein encoded by the E3 region of adenovirus, *Cell,* 57, 135, 1989.
276. **Filmus, J., Pollak, M.N., Cailleau, R., and Buick, R.N.,** MDA-468, a human breast cancer cell line with a high number of epidermal growth factor (EGF) receptors, has an amplified EGF receptor gene and is growth inhibited by EGF, *Biochem. Biophys. Res. Commun.,* 128, 898, 1985.
277. **Filmus, J., Pollak, M.N., Cairncross, J.G., and Buick, R.N.,** Amplified, overexpressed and rearranged epidermal growth factor receptor gene in a human astrocytoma cell line, *Biochem. Biophys. Res. Commun.,* 131, 207, 1985.
278. **Ginsburg, E. and Vonderhaar, B.K.,** Epidermal growth factor stimulates the growth of A431 tumors in athymic mice, *Cancer Lett.,* 28, 143, 1985.
279. **Boonstra, J., Mummery, C.L., van der Saag, P.T., and de Laat, S.W.,** Two receptor classes for epidermal growth factor on pheochromocytoma cells, distinguishable by temperature, lectins, and tumor promoters, *J. Cell. Physiol.,* 123, 347, 1985.
280. **Assoian, R.K.,** Biphasic effects of type β transforming growth factor on epidermal growth factor receptors in NRK fibroblasts: functional consequences for epidermal growth factor stimulated mitosis, *J. Biol. Chem.,* 260, 9613, 1985.
281. **Gan, B.S. and Hollenberg, M.D.,** Distinct coronary artery receptor systems for epidermal growth factor-urogastrone, *J. Pharmacol. Exp. Ther.,* 252, 1277, 1990.
282. **Rees, A.R., Gregoriou, M., Johnson, P., and Garland, P.B.,** High affinity epidermal growth factor receptors on the surface of A431 cells have restricted lateral diffusion, *EMBO J.,* 3, 1843, 1984.
283. **Hwang, J., Richert, N., Pastan, I., and Gottesman, M.M.,** Mutant KB cells with decreased EGF receptor expression: biochemical characterization, *J. Cell. Physiol.,* 133, 127, 1987.
284. **Weber, W., Gill, G.N., and Spiess, J.,** Production of an epidermal growth factor receptor-related protein, *Science,* 224, 294, 1984.
285. **O'Connor-McCourt, M., Soley, M., Hayden, L.J., and Hollenberg, M.D.,** Receptors for epidermal growth factor (urogastrone) and insulin in primary cultures of rat hepatocytes maintained in serum-free medium, *Biochem. Cell Biol.,* 64, 803, 1986.

286. **Velu, T.J., Beguinot, L., Vass, W.C., Willingham, M.C., Merlino, G.T., Pastan, I., and Lowy, D.R.,** Epidermal growth factor-dependent transformation by a human EGF receptor proto-oncogene, *Science,* 238, 1408, 1987.
287. **Di Fiore, P.P., Pierce, J.H., Fleming, T.P., Hazan, R., Ullrich, A., King, C.R., Schlessinger, J., and Aaronson, S.A.,** Overexpression of the human EGF receptor confers an EGF-dependent transformed phenotype to NIH 3T3 cells, *Cell,* 51, 1063, 1987.
288. **Velu, T.J., Beguinot, L., Vass, W.C., Zhang, K., Pastan, I., and Lowy, D.R.,** Retroviruses expressing different levels of the normal epidermal growth factor receptor: biological properties and new bioassay, *J. Cell. Biochem.,* 39, 153, 1989.
289. **Pandiella, A., Beguinot, L., Velu, T.J., and Meldolesi, J.,** Transmembrane signalling at epidermal growth factor receptors overexpressed in NIH 3T3 cells. Phosphoinositide hydrolysis, cytosolic Ca^{2+} increase and alkalinization correlate with epidermal-growth-factor-induced cell proliferation, *Biochem. J.,* 254, 223, 1988.
290. **von Rüden, T. and Wagner, E.F.,** Expression of functional human EGF receptor on murine bone marrow cells, *EMBO J.,* 7, 2749, 1988.
291. **Oval, J., Hershberg, R., Gansbacher, B., Gilboa, E., Schlessinger, J., and Taetle, R.,** Expression of functional epidermal growth factor receptors in a human hematopoietic cell line, *Cancer Res.,* 51, 150, 1991.
292. **Terwilliger, E. and Herschman, H.R.,** 3T3 variants unable to bind epidermal growth factor cannot complement in co-culture, *Biochem. Biophys. Res. Commun.,* 118, 60, 1984.
293. **Basu, M., Biswas, R., and Das, M.,** 42,000-molecular weight EGF receptor has protein kinase activity, *Nature (London),* 311, 477, 1984.
294. **Bertics, P.J., Weber, W., Cochet, C., and Gill, G.N.,** Regulation of the epidermal growth factor receptor by phosphorylation, *J. Cell. Biochem.,* 29, 195, 1985.
295. **Chinkers, M. and Garbers, D.L.,** Phorbol ester-induced threonine phosphorylation of the human epidermal growth factor receptor occurs within the EGF binding domain, *Biochem. Biophys. Res. Commun.,* 123, 618, 1984.
296. **Margolis, B.L., Lax, I., Kris, R., Dombalagian, M., Honegger, A.M., Howk, R., Givol, D., Ullrich, A., and Schlessinger, J.,** All autophosphorylation sites of epidermal growth factor (EGF) receptor and HER2/*neu* are located in their carboxyl-terminal tails. Identification of a novel site in EGF receptor, *J. Biol. Chem.,* 264, 10667, 1989.
297. **Finocchiaro, L., Komano, O., and Loeb, J.,** Epidermal growth factor stimulated protein kinase shows similar activity in liver of senescent and adult mice, *FEBS Lett.,* 187, 96, 1985.
298. **Honegger, A.M., Kris, R.M., Ullrich, A., and Schlessinger, J.,** Evidence that autophosphorylation of solubilized receptors for epidermal growth factor is mediated by intermolecular cross-phosphorylation, *Proc. Natl. Acad. Sci. U.S.A.,* 86, 925, 1989.
299. **Ushiro, H. and Cohen, S.,** Identification of phosphotyrosine as a product of epidermal growth factor-activated protein kinase in A-431 cell membranes, *J. Biol. Chem.,* 255, 8363, 1980.
300. **Hunter, T. and Cooper, J.A.,** Epidermal growth factor induces rapid tyrosine phosphorylation of proteins in A431 human tumor cells, *Cell,* 24, 741, 1981.
301. **Downward, J., Waterfield, M.D., and Parker, P.J.,** Autophosphorylation and protein kinase C phosphorylation of the epidermal growth factor receptor. Effect on tyrosine kinase activity and ligand binding affinity, *J. Biol. Chem.,* 260, 14538, 1985.
302. **Bertics, P.J. and Gill, G.N.,** Self-phosphorylation enhances the protein-tyrosine kinase activity of the epidermal growth factor receptor, *J. Biol. Chem.,* 260, 14642, 1985.
303. **Kadowaki, T., Koyasu, S., Nishida, E., Tobe, K., Izumi, T., Takaku, F., Sakai, H., Yahara, I., and Kasuga, M.,** Tyrosine phosphorylation of common and specific sets of cellular proteins rapidly induced by insulin, insulin-like growth factor I, and epidermal growth factor in an intact cell, *J. Biol. Chem.,* 262, 7342, 1987.
304. **Downward, J., Parker, P., and Waterfield, M.D.,** Autophosphorylation sites on the epidermal growth factor receptor, *Nature (London),* 311, 483, 1984.
305. **Bertics, P.J., Chen, W.S., Hubler, L., Lazar, C.S., Rosenfeld, M.G., and Gill, G.N.,** Alteration of epidermal growth factor receptor activity by mutation of its primary carboxyl-terminal site of tyrosine self-phosphorylation, *J. Biol. Chem.,* 263, 3610, 1988.
306. **Chinkers, M. and Brugge, J.S.,** Characterization of structural domains of the human epidermal growth factor receptor obtained by partial proteolysis, *J. Biol. Chem.,* 259, 11534, 1984.

307. **Gotoh, N., Tojo, A., Hino, M., Yazaki, Y., and Shibuya, M.,** A highly conserved tyrosine residue at codon 845 within the kinase domain is not required for the transforming activity of human epidermal growth factor receptor, *Biochem. Biophys. Res. Commun.,* 186, 768, 1992.
308. **Chinkers, M. and Garbers, D.L.,** Suppression of protein tyrosine kinase activity of the epidermal growth factor receptor by epidermal growth factor, *J. Biol. Chem.,* 261, 8295, 1986.
309. **Donato, N.J., Gallick, G.E., Steck, P.A., and Rosenblum, M.G.,** Tumor necrosis factor modulates epidermal growth factor receptor phosphorylation and kinase activity in human tumor cells, *J. Biol. Chem.,* 264, 20474, 1989.
310. **Donato, N.J., Rosenblum, M.G., and Steck, P.A.,** Tumor necrosis factor regulates tyrosine phosphorylation on epidermal growth factor receptors in A431 carcinoma cells: evidence for a distinct mechanism, *Cell Growth Differ.,* 3, 259, 1992.
311. **Brooks, K.M., Phillips, P.D., Carlin, C.R., Knowles, B.B., and Cristofalo, V.J.,** EGF-dependent phosphorylation of the EGF receptor in plasma membranes isolated from young and senescent WI-38 cells, *J. Cell. Physiol.,* 133, 523, 1987.
312. **Gullick, W.J., Downward, J., and Waterfield, M.D.,** Antibodies to the autophosphorylation sites of the epidermal growth factor receptor protein-tyrosine kinase as probes of structure and function, *EMBO J.,* 4, 2869, 1985.
313. **Clark, S., Cheng, D.J., Hsuan, J.J., Haley, J.D., and Waterfield, M.D.,** Loss of three major auto phosphorylation sites in the EGF receptor does not block the mitogenic action of EGF, *J. Cell. Physiol.,* 134, 421, 1988.
314. **Honegger, A., Dull, T.J., Bellot, F., Van Obberghen, E., Szapary, D., Schmidt, A., Ullrich, A., and Schlessinger, J.,** Biological activities of EGF-receptor mutants with individually altered autophosphorylation sites, *EMBO J.,* 7, 3045, 1988.
315. **Decker, S.J.,** Transmembrane signaling by epidermal growth factor receptors lacking autophosphorylation sites, *J. Biol. Chem.,* 268, 9176, 1993.
316. **Bremer, E.G., Schlessinger, J., and Hakomori, S.,** Ganglioside-mediated modulation of cell growth. Specific effects of G_{M3} on tyrosine phosphorylation of the epidermal growth factor receptor, *J. Biol. Chem.,* 261, 2434, 1986.
317. **Faucher, M., Gironès, N., Hannun, Y.A., Bell, R.M., and Davis, R.J.,** Regulation of the epidermal growth factor receptor phosphorylation state by sphingosine in A431 human epidermoid carcinoma cells, *J. Biol. Chem.,* 263, 5319, 1988.
318. **Davis, R.J., Gironès, N., and Faucher, M.,** Two alternative mechanisms control the interconversion of functional states of the epidermal growth factor receptor, *J. Biol. Chem.,* 263, 5373, 1988.
319. **Hanai, N., Dohi, T., Nores, G.A., and Hakomori, S.,** A novel ganglioside, de-*N*-acetyl-GM_3 ($II^3NeuNH_2LacCer$), acting as a strong promoter for epidermal growth factor receptor kinase and as a stimulator for cell growth, *J. Biol. Chem.,* 263, 6296, 1988.
320. **Rackoff, W.R., Rubin, R.A., and Earp, H.S.,** Phosphorylation of the hepatic EGF receptor with cAMP-dependent protein kinase, *Mol. Cell. Endocrinol.,* 34, 113, 1984.
321. **Hosoi, K. and Edidin, M.,** Exogenous ATP and other nucleoside phosphates modulate epidermal growth factor receptors of A-431 epidermoid carcinoma cells, *Proc. Natl. Acad. Sci. U.S.A.,* 86, 4510, 1989.
322. **Fearn, J.C. and King, A.C.,** EGF receptor affinity is regulated by intracellular calcium and protein kinase C, *Cell,* 40, 991, 1985.
323. **King, C.S. and Cooper, J.A.,** Effects of protein kinase C activation after epidermal growth factor binding on epidermal growth factor receptor phosphorylation, *J. Biol. Chem.,* 261, 10073, 1986.
324. **Ido, M., Sekiguchi, K., Kikkawa, U., and Nishizuka, Y.,** Phosphorylation of the EGF receptor from A431 epidermoid carcinoma cells by three distinct types of protein kinase C, *FEBS Lett.,* 219, 215, 1987.
325. **Cochet, C., Gill, G.N., Meisenhelder, J., Cooper, J.A., and Hunter, T.,** C-kinase phosphorylates the epidermal growth factor receptor and reduces its epidermal growth factor-stimulated tyrosine protein kinase activity, *J. Biol. Chem.,* 259, 2553, 1984.
326. **Pandiella, A., Vicentini, L.M., and Meldolesi, J.,** Protein kinase C-mediated feed back inhibition of the Ca^{2+} response at the EGF receptor, *Biochem. Biophys. Res. Commun.,* 149, 145, 1987.
327. **Countaway, J.L., McQuilkin, P., Gironès, N., and Davis, R.J.,** Multisite phosphorylation of the epidermal growth factor receptor. Use of site-directed mutagenesis to examine the role of serine/threonine phosphorylation, *J. Biol. Chem.,* 265, 3407, 1990.

328. **Hunter, T., Ling, N., and Cooper, J.A.,** Protein kinase C phosphorylation of the EGF receptor at a threonine residue close to the cytoplasmic face of the plasma membrane, *Nature (London),* 311, 480, 1984.
329. **Whiteley, B. and Glaser, L.,** Epidermal growth factor (EGF) promotes phosphorylation at threonine-654 of the EGF receptor: possible role of protein kinase C in homologous regulation of the EGF receptor, *J. Cell Biol.,* 103, 1355, 1986.
330. **Lund, K.A., Lazar, C.S., Chen, W.S., Walsh, B.J., Welsh, J.B., Herbst, J.J., Walton, G.M., Rosenfeld, M.G., Gill, G.N., and Willey, H.S.,** Phosphorylation of the epidermal growth factor receptor at threonine 654 inhibits ligand-induced internalization and down regulation, *J. Biol. Chem.,* 265, 20517, 1990.
331. **Friedman, B., van Amsterdam, J., Fujiki, H., and Rosner, M.R.,** Phosphorylation at threonine-654 is not required for negative regulation of the epidermal growth factor receptor by non-phorbol tumor promoters, *Proc. Natl. Acad. Sci. U.S.A.,* 86, 812, 1989.
332. **Lin, C.R., Chen, W.S., Lazar, C.S., Carpenter, C.D., Gill, G.N., Evans, R.M., and Rosenfeld, M.E.,** Protein kinase C phosphorylation at Thr 654 of the unoccupied EGF receptor and EGF binding regulate functional receptor loss by independent mechanisms, *Cell,* 44, 839, 1986.
333. **Countaway, J.L., Northwood, I.C., and Davis, R.J.,** Mechanism of phosphorylation of the epidermal growth factor receptor at threonine 669, *J. Biol. Chem.,* 264, 10828, 1989.
334. **Northwood, I.C., Gonzalez, F.A., Wartmann, M., Raden, D.L., and Davis, R.J.,** Isolation and characterization of two growth factor-stimulated protein kinases that phosphorylate the epidermal growth factor receptor at threonine 669, *J. Biol. Chem.,* 266, 15266, 1991.
335. **Alvarez, E., Northwood, I.C., Gonzalez, F.A., Latour, D.A., Seth, A., Abate, C., Curran, T., and Davis, R.J.,** Pro-Leu-Ser/Thr-Pro is a consensus primary sequence for substrate protein phosphorylation. Characterization of the phosphorylation of c-*myc* and c-*jun* proteins by an epidermal growth factor receptor threonine 669 protein kinase, *J. Biol. Chem.,* 266, 15277, 1991.
336. **Heisermann, G.J., Wiley, H.S., Walsh, B.J., Ingraham, H.A., Fiol, C.J., and Gill, G.N.,** Mutational removal of the Thr^{669} and Ser^{671} phosphorylation sites alters substrate specificity and ligand-induced internalization of the epidermal growth factor receptor, *J. Biol. Chem.,* 265, 12820, 1990.
337. **Countaway, J.L., Nairn, A.C., and Davis, R.J.,** Mechanism of desensitization of the epidermal growth factor receptor protein-tyrosine kinase, *J. Biol. Chem.,* 267, 1129, 1992.
338. **Akiyama, T., Ishida, J., Nakagawa, S., Ogawara, H., Watanabe, S., Itoh, N., Shibuya, M., and Fukami, Y.,** Genistein, a specific inhibitor of tyrosine-specific protein kinases, *J. Biol. Chem.,* 262, 5592, 1987.
339. **Pallen, C.J., Valentine, K.A., Wang, J.H., and Hollenberg, M.D.,** Calcineurin-mediated dephosphorylation of the human placental membrane receptor for epidermal growth factor urogastrone, *Biochemistry,* 24, 4727, 1985.
340. **Das, M. and Fox, C.F.,** Molecular mechanism of mitogen action: processing of receptor induced by epidermal growth factor, *Proc. Natl. Acad. Sci. U.S.A.,* 75, 2644, 1978.
341. **O'Connor-McCourt, M. and Hollenberg, M.D.,** Receptors, acceptors, and the action of polypeptide hormones: illustrative studies with epidermal growth factor (urogastrone), *Can. J. Biochem. Cell Biol.,* 61, 670, 1983.
342. **Beguinot, L., Lyal, R.M., Willingham, M.C., and Pastan, I.,** Down-regulation of the epidermal growth factor receptor in KB cells is due to receptor internalization and subsequent degradation in lysosomes, *Proc. Natl. Acad. Sci. U.S.A.,* 81, 2384, 1984.
343. **Matrisian, L.M., Planck, S.R., and Magun, B.E.,** Intracellular processing of epidermal growth factor. I. Acidification of ^{125}I-epidermal growth factor in intracellular organelles, *J. Biol. Chem.,* 259, 3047, 1984.
344. **Planck, S.R., Finch, J.S., and Magun, B.E.,** Intracellular processing of epidermal growth factor. II. Intracellular cleavage of the COOH-terminal region of ^{125}I-epidermal growth factor, *J. Biol. Chem.,* 259, 3053, 1984.
345. **Wiley, H.S., Van Nostrand, W., McKinney, D.N., and Cunningham, D.D.,** Intracellular processing of epidermal growth factor and its effect on ligand-receptor interactions, *J. Biol. Chem.,* 260, 5290, 1985.
346. **Helin, K. and Beguinot, L.,** Internalization and down-regulation of the human epidermal growth factor receptor are regulated by carboxyl-terminal tyrosines, *J. Biol. Chem.,* 266, 8363, 1991.
347. **Hertel, C. and Perkins, J.P.,** Sequential appearance of epidermal growth factor in plasma membrane-associated and intracellular vesicles during endocytosis, *J. Biol. Chem.,* 262, 11407, 1987.

348. **Haigler, H.T., Wiley, H.S., Moehring, J.M., and Moehring, T.J.,** Altered degradation of epidermal growth factor in a diphtheria toxin-resistant clone of KB cells, *J. Cell. Physiol.,* 124, 322, 1985.
349. **Miller, K., Beardmore, J., Kanety, H., Schlessinger, J., and Hopkins, C.R.,** Localization of the epidermal growth factor (EGF) receptor within the endosome of EGF-stimulated epidermoid carcinoma (A431) cells, *J. Cell Biol.,* 102, 500, 1986.
350. **Stoscheck C.M. and Carpenter, G.,** Characterization of the metabolic turnover of epidermal growth factor receptor protein in A-431 cells, *J. Cell. Physiol.,* 120, 296, 1984.
351. **Decker, S.J.,** Epidermal growth factor-induced truncation of the epidermal growth factor receptor, *J. Biol. Chem.,* 264, 17641, 1989.
352. **Hiwasa, T., Sakiyama, S., Yokoyama, S., Ha, J.-M., Fujita, J., Noguchi, S., Bando, Y., Kominami, E., and Katunuma, N.,** Inhibition of cathepsin L-induced degradation of epidermal growth factor receptors by c-Ha-*ras* gene products, *Biochem. Biophys. Res. Commun.,* 151, 78, 1988.
353. **Hiwasa, T., Sakiyama, S., Yokoyama, S., Ha, J.-M., Noguchi, S., Bando, Y., Kominami, E., and Katanuma, N.,** Degradation of epidermal growth factor receptors by cathepsin L-like protease: inhibition of the degradation by c-Ha-*ras* gene products, *FEBS Lett.,* 233, 367, 1988.
354. **McKay, I.A., Malone, P., Marshall, C.J., and Hall, A.,** Malignant transformation of murine fibroblasts by a human c-Ha-*ras*-1 oncogene does not require a functional epidermal growth factor receptor, *Mol. Cell. Biol.,* 6, 3382, 1986.
355. **Baskin, G., Schenker, S., Frosto, T., and Henderson, G.,** Transforming growth factor β-1 inhibits epidermal growth factor receptor endocytosis and down-regulation in cultured fetal rat hepatocytes, *J. Biol. Chem.,* 266, 13238, 1991.
356. **Maratos-Flier, E., Kao, C.-Y.Y., Verdin, E.M., and King, G.L.,** Receptor-mediated vectorial transcytosis of epidermal growth factor by Madin-Darby canine kidney cells, *J. Cell Biol.,* 105, 1595, 1987.
357. **King, G.L. and Johnson, S.M.,** Receptor-mediated transport of insulin across endothelial cells, *Science,* 227, 1583, 1985.
358. **Brandli, A.W., Adamson, E.D., and Simons, K.,** Transcytosis of epidermal growth factor. The epidermal growth factor receptor mediates uptake but not transcytosis, *J. Biol. Chem.,* 266, 8560, 1991.
359. **Schaudies, R.P., Harper, R.A., and Savage, C.R., Jr.,** ^{125}I-EGF binding to responsive and nonresponsive cells in culture: loss of cell-associated radioactivity relates to growth induction, *J. Cell. Physiol.,* 124, 493, 1985.
360. **Schaudies, R.P., Gorman, R.M., Savage, C.R., Jr., and Poretz, R.D.,** Proteolytic processing of epidermal growth factor within endosomes, *Biochem. Biophys. Res. Commun.,* 143, 710, 1987.
361. **Korc, M. and Magun, B.E.,** Recycling of epidermal growth factor in a human pancreatic carcinoma cell line, *Proc. Natl. Acad. Sci. U.S.A.,* 82, 6172, 1985.
362. **Carpentier, J.-L., Rees, A.R., Gregoriou, M., Kris, R., Schlessinger, J., and Orci, L.,** Subcellular distribution of the external and internal domains of the EGF receptor in A-431 cells, *Exp. Cell Res.,* 166, 312, 1986.
363. **Murthy, U., Basu, M., Sen-Majumdar, A., and Das, M.,** Perinuclear location and recycling of epidermal growth factor receptor kinase: immunofluorescence visualization using antibodies directed to kinase and extracellular domains, *J. Cell Biol.,* 103, 333, 1986.
364. **Rakowicz-Szulczynska, E.M., Rodeck, U., Herlyn, M., and Koprowski, H.,** Chromatin binding of epidermal growth factor, nerve growth factor, and platelet-derived growth factor in cells bearing the appropriate surface receptors, *Proc. Natl. Acad. Sci. U.S.A.,* 83, 3728, 1986.
365. **Rakowicz-Szulczynska, E.M., Otwiaska, D., Rodeck, U., and Koprowski, H.,** Epidermal growth factor (EGF) and monoclonal antibody to cell surface EGF receptor bind to the same chromatin receptor, *Arch. Biochem. Biophys.,* 268, 456, 1989.
366. **Ichii, S., Yoshida, A., and Yoshikawa, Y.,** Binding sites for epidermal growth factor in nuclear fraction from rat liver, *Endocrinol. Jpn.,* 35, 567, 1988.
367. **Marti, U., Burwen, S.J., Wells, A., Barker, M.E., Huling, S., Feren, A.M., and Jones, A.L.,** Localization of epidermal growth factor receptor in hepatocyte nuclei, *Hepatology,* 13, 15, 1991.
368. **Matrisian, L.M., Rodland, K.D., and Magun, B.E.,** Disruption of intracellular processing of epidermal growth factor by methylamine inhibits epidermal growth factor-induced DNA synthesis but not early morphological or transcriptional events, *J. Biol. Chem.,* 262, 6908, 1987.
369. **Beguinot, L., Hanover, J.A., Ito, S., Richert, N.D., Willingham, M.C., and Pastan, I.,** Phorbol esters induce transient internalization without degradation of unoccupied epidermal growth factor receptors, *Proc. Natl. Acad. Sci. U.S.A.,* 82, 2774, 1985.

370. **Wakshull, E.M. and Wharton, W.,** Stabilized complexes of epidermal growth factor and its receptor on the cell surface stimulate RNA synthesis but not mitogenesis, *Proc. Natl. Acad. Sci. U.S.A.,* 82, 8513, 1985.
371. **Sunada, H., Magun, B.E., Mendelsohn, J., and MacLeod, C.L.,** Monoclonal antibody against epidermal growth factor receptor is internalized without stimulating receptor phosphorylation, *Proc. Natl. Acad. Sci. U.S.A.,* 83, 3825, 1986.
372. **King, A.C.,** Monensin, like methylamine, prevents degradation of ^{125}I-epidermal growth factor, causes intracellular accumulation of receptors and blocks the mitogenic response, *Biochem. Biophys. Res. Commun.,* 124, 585, 1984.
373. **Landreth, G.E., Williams, L.K., and Rieser, G.D.,** Association of the epidermal growth factor receptor kinase with the detergent-insoluble cytoskeleton of A431 cells, *J. Cell Biol.,* 101, 1341, 1985.
374. **Cooper, J.L., Selinfreund, R., Wakshull, E., and Wharton, W.,** Interaction between monensin and lysosomotropic amines in the regulation of the processing of epidermal growth factor by BALB/c 3T3 cells, *Mol. Cell. Biochem.,* 73, 1, 1987.
375. **Prywes, R., Livneh, E., Ullrich, A., and Schlessinger, J.,** Mutations in the cytoplasmic domain of EGF receptor affect EGF binding and receptor internalization, *EMBO J.,* 5, 2179, 1986.
376. **Honegger, A.M., Dull, T.J., Felder, S., Van Obberghen, E., Bellot, F., Szapary, D., Schmidt, A., Ullrich, A., and Schlessinger, J.,** Point mutation at the ATP binding site of EGF receptor abolishes protein-tyrosine kinase activity and alters cellular routing, *Cell,* 51, 199, 1987.
377. **Felder, S., Miller, K., Moehren, G., Ullrich, A., Schlessinger, J., and Hopkins, C.R.,** Kinase activity controls the sorting of the epidermal growth factor receptor within the multivesicular body, *Cell,* 61, 623, 1990.
378. **Redemann, N., Holzmann, B., von Rüden, T., Wagner, E.F., Schlessinger, J., and Ullrich, A.,** Anti-oncogenic activity of signalling-defective epidermal growth factor receptor mutants, *Mol. Cell. Biol.,* 12, 491, 1992.
379. **Shiraishi, T., Domoto, T., Imai, N., Shimada, Y., and Watanabe, K.,** Specific inhibitors of tyrosine-specific protein kinase, synthetic 4-hydroxycinnamamide derivatives, *Biochem. Biophys. Res. Commun.,* 147, 322, 1987.
380. **Graf, T. and Beug, H.,** Avian leukemia viruses: interaction with their target cells *in vivo* and *in vitro, Biochim. Biophys. Acta,* 516, 269, 1978.
381. **Jurdic, P., Bouabdelli, M., Moscovici, M.G., and Moscovici, C.,** Embryonic erythroid cells transformed by avian erythroblastosis virus may proliferate and differentiate, *Virology,* 144, 73, 1985.
382. **Saule, S., Roussel, M., Lagrou, C., and Stehelin, D.,** Characterization of the oncogene (erb) of avian erythroblastosis virus and its cellular progenitor, *J. Virol.,* 38, 409, 1981.
383. **Graf, T. and Beug, H.,** Role of the v-*erb*A and v-*erb*B oncogenes of avian erythroblastosis virus in erythroid cell transformation, *Cell,* 34, 7, 1983.
384. **Jansson, M., Beug, H., Gray, C., Graf, T., and Vennström, B.,** Defective v-*erb*B genes can be complemented by v-*erb*A in erythroblast and fibroblast transformation, *Oncogene,* 1, 167, 1987.
385. **Yamamoto, T., Hihara, H., Nishida, T., Kawai, S., and Toyoshima, K.,** A new avian erythroblastosis virus, AEV-H, carries *erb*B gene responsible for the induction of both erythroblastosis and sarcomas, *Cell,* 34, 225, 1983.
386. **Miles, B.D. and Robinson, H.L.,** High-frequency transduction of c-*erb*B in avian leukosis virus-induced erythroblastosis, *J. Virol.,* 54, 295, 1985.
387. **Downward, J., Yarden, Y., Mayes, E., Scrace, G., Totty, N., Stockwell, P., Ullrich, A., Schlessinger, J., and Waterfield, M.D.,** Close similarity of epidermal growth factor receptor and v-*erb*-B oncogene protein sequences, *Nature (London),* 307, 521, 1984.
388. **Spurr, N.K., Solomon, E., Jansson, M., Sheer, D., Goodfellow, P.N., Bodmer, W.F., and Vennstrom, B.,** Chromosomal localisation of the human homologues to the oncogenes *erb*A and B, *EMBO J.,* 3, 159, 1984.
389. **Taglienti-Sian, C.A., Banner, B., Davis, R.J., and Robinson, H.L.,** Induction of renal adenocarcinoma by a nonmutated *erb*B oncogene, *J. Virol.,* 67, 1132, 1993.
390. **Hayman, M.J., Ramsey, G., Savin, K., Kitchener, G., Graf, T., and Beug, H.,** Identification and characterization of the avian erythroblastosis virus *erb*B gene product as a membrane glycoprotein, *Cell,* 32, 579, 1983.
391. **Hayman, M.J. and Beug, H.,** Identification of a form of the avian erythroblastosis virus *erb*-B gene product at the cell surface, *Nature (London),* 309, 460, 1984.

392. **Schatzman, R.C., Evan, G.I., Privalsky, M.L., and Bishop, J.M.,** Orientation of the v-*erb*-B gene product in the plasma membrane, *Mol. Cell. Biol.,* 6, 1329, 1986.
393. **Schmidt, J.A., Beug, H., and Hayman, M.J.,** Effects of inhibitors of glycoprotein processing on the synthesis and biological activity of the *erb* B oncogene, *EMBO J.,* 4, 105, 1985.
394. **Akiyama, T., Yamada, Y., Ogawara, H., Richert, N., Pastan, I., Yamamoto, T., and Kasuga, M.,** Site-specific antibodies to the *erb*B oncogene product immunoprecipitate epidermal growth factor receptor, *Biochem. Biophys. Res. Commun.,* 123, 797, 1984.
395. **Riedel, H., Schlessinger, J., and Ullrich, A.,** A chimeric, ligand-binding v-*erb*B/EGF receptor retains transforming potential, *Science,* 236, 197, 1987.
396. **Massoglia, S., Gray, A., Dull, T.J., Munemitsu, S., Kung, H.-J., Schlessinger, J., and Ullrich, A.,** Epidermal growth factor receptor cytoplasmic domain mutations trigger ligand-independent transformation, *Mol. Cell. Biol.,* 10, 3048, 1990.
397. **Kawai, S., Nishizawa, M., Yamamoto, T., and Toyoshima, K.,** Cell transformation by a virus containing a molecularly constructed *gag-erbB* fused gene, *J. Virol.,* 61, 1665, 1987.
398. **Haley, J.D., Hsuan, J.J., and Waterfield, M.D.,** Analysis of mammalian fibroblast transformation by normal and mutated human EGF receptors, *Oncogene,* 4, 273, 1989.
399. **Gamett, D.C., Tracy, S.E., and Robinson, H.L.,** Differences in sequences encoding the carboxyl-terminal domain of the epidermal growth factor receptor correlate with differences in the disease potential of viral *erbB* genes, *Proc. Natl. Acad. Sci. U.S.A.,* 83, 6053, 1986.
400. **Pelley, R.J., Maihle, N.J., Boerkel, C., Shu, H.-K., Carter, T.H., Moscovici, C., and Kung, H.-J.,** Disease tropism of c-*erbB:* effects of carboxyl-terminal tyrosine and internal mutations on tissue-specific transformation, *Proc. Natl. Acad. Sci. U.S.A.,* 86, 7164, 1989.
401. **Privalsky, M.L.,** Creation of a chimeric oncogene: analysis of the biochemical and biological properties of a v-*erbB/src* fusion polypeptide, *J. Virol.,* 61, 1938, 1987.
402. **Bassiri, M. and Privalsky, M.L.,** Mutagenesis of the avian erythroblastosis virus *erbB* coding region: an intact extracellular domain is not required for oncogenic transformation, *J. Virol.,* 59, 525, 1986.
403. **Bassiri, M. and Privalsky, M.L.,** Transmembrane domain of the AEV *erb* B oncogene protein is not required for partial manifestation of the transformed phenotype, *Virology,* 159, 20, 1987.
404. **Decker, S.J.,** Phosphorylation of the erbB gene product from an avian erythroblastosis virus-transformed chick fibroblast cell line, *J. Biol. Chem.,* 260, 2003, 1985.
405. **Gilmore, T., DeClue, J.E., and Martin, G.S.,** Protein phosphorylation at tyrosine is induced by the v-*erbB* gene product *in vivo* and *in vitro, Cell,* 40, 609, 1985.
406. **Lax, I., Bar-Eli, M., Yarden, Y., Liberman, T.A., and Schlessinger, J.,** Antibodies to two defined regions of the transforming protein pp60^{src} interact specifically with the epidermal growth factor receptor kinase system, *Proc. Natl. Acad. Sci. U.S.A.,* 81, 5911, 1984.
407. **Akiyama, T., Kadooka, T., Ogawara, H., and Sakakibara, S.,** Characterization of the epidermal growth factor receptor and the *erbB* oncogene product by site-specific antibodies, *Arch. Biochem. Biophys.,* 245, 531, 1986.
408. **Kris, R.M., Lax, I., Gullick, W., Waterfield, M.D., Ullrich, A., Fridkin, M., and Schlessinger, J.,** Antibodies against a synthetic peptide as a probe for the kinase activity of the avian EGF receptor and v-erbB protein, *Cell,* 40, 619, 1985.
409. **Gentry, L.E. and Lawton, A.,** Characterization of site-specific antibodies to the *erb*B gene product and EGF receptor: inhibition of tyrosine kinase activity, *Virology,* 152, 421, 1986.
410. **Kato, M., Kawai, S., and Takenawa, T.,** Altered signal transduction in erbB-transformed cells. Implication of enhanced inositol phospholipid metabolism in erbB-induced transformation, *J. Biol. Chem.,* 262, 5696, 1987.
411. **Ng, M. and Privalsky, M.L.,** Structural domains of the avian erythroblastosis virus *erbB* protein required for fibroblast transformation: dissection by in-frame insertional mutagenesis, *J. Virol.,* 58, 542, 1986.
412. **Hayman, M.J., Kitchener, G., Knight, J., McMahon, J., Watson, R., and Beug, H.,** Analysis of the autophosphorylation activity of transformation defective mutants of avian erythroblastosis virus, *Virology,* 150, 270, 1986.
413. **Choi, O.-R., Trainor, C., Graf, T., Beug, H., and Engel, J.D.,** A single amino acid substitution in v-*erbB* confers a thermolabile phenotype to *ts*167 avian erythroblastosis virus-transformed erythroid cells, *Mol. Cell. Biol.,* 6, 1751, 1986.

414. **Nair, N., Davis, R.J., and Robinson, H.L.,** Protein tyrosine kinase activities of the epidermal growth factor receptor and ErbB proteins: correlation of oncogenic activation with altered kinetics, *Mol. Cell. Biol.,* 12, 2010, 1992.
415. **Theroux, S.J., Taglienti-Sian, C., Nair, N., Counteway, J.L., Robinson, H.L., and Davis, R.J.,** Increased oncogenic potential of ErbB is associated with the loss of a COOH-terminal domain serine phosphorylation site, *J. Biol. Chem.,* 267, 7967, 1992.
416. **Hudson, L.G., Ertl, A.P., and Gill, G.N.,** Structure and inducible regulation of the human c-erbB2/*neu* promoter, *J. Biol. Chem.,* 265, 4389, 1990.
417. **White, M.R.A. and Hung, M.C.,** Cloning and characterization of the mouse *neu* promoter, *Oncogene,* 7, 677, 1992.
418. **Bargmann, C.I., Hung, M.-C., and Weinberg, R.A.,** The *neu* oncogene encodes an epidermal growth factor receptor-related protein, *Nature (London),* 319, 226, 1986.
419. **Maguire, H.C., Jr., Jaworsky, C., Cohen, J.A., Hellman, M., Weiner, D.B., and Greene, M.I.,** Distribution of neu (c-erbB-2) protein in human skin, *J. Invest. Dermatol.,* 92, 786, 1989.
420. **Mori, S., Akiyama, T., Yamada, Y., Morishita, Y., Sugawara, I., Toyoshima, K., and Yamamoto, T.,** c-*erbB*-2 gene product, a membrane protein commonly expressed on human fetal epithelial cells, *Lab. Invest.,* 61, 93, 1989.
421. **Quirke, P., Pickles, A., Tuzi, N.L., Mohamdee, O., and Gullick, W.J.,** Pattern of expression of c-*erb*B-2 oncoprotein in human fetuses, *Br. J. Cancer,* 60, 64, 1989.
422. **Wang, D., Fujii, S., Konishi, I., Nanbu, Y., Iwai, T., Nonogaki, H., and Mori, T.,** Expression of c-*erb*B-2 protein and epidermal growth factor receptor in normal tissues of the female genital tract and in the placenta, *Virchows Arch. A,* 420, 385, 1992.
423. **Cohen, J.A., Yachnis, A.T., Arai, M., Davis, J.G., and Scherer, S.S.,** Expression of the *neu* proto-oncogene by Schwann cells during peripheral nerve development and Wallerian degeneration, *J. Neurosci. Res.,* 31, 622, 1992.
424. **Akiyama, T., Sudo, C., Ogawara, H., Toyoshima, K., and Yamamoto, T.,** The product of the human c-*erb*B-2 gene: a 185-kilodalton glycoprotein with tyrosine kinase activity, *Science,* 232, 1644, 1986.
425. **Lupu, R., Colomer, R., Zugmaier, G., Sarup, J., Shepard, M., Slamon, D., and Lippman, M.E.,** Direct interaction of a ligand for the *erb*B2 oncogene product with the EGF receptor and $p185^{erbB2}$, *Science,* 249, 1552, 1990.
426. **Bacus, S.S., Huberman, E., Chin, D., Kiguchi, K., Simpson, S., Lippman, M., and Lupu, R.,** A ligand for the erbB-2 oncogene product (gp30) induces differentiation of human breast cancer cells, *Cell Growth Differ.,* 3, 401, 1992.
427. **Tarakhovsky, A., Zichuk, T., Prassolov, V., and Butenko, Z.A.,** A 25 kDa polypeptide is the ligand for p185neu and is secreted by activated macrophages, *Oncogene,* 6, 2187, 1991.
428. **Huang, S.S. and Huang, J.S.,** Purification and characterization of the neu/erb B2 ligand-growth factor from bovine kidney, *J. Biol. Chem.,* 267, 11508, 1992.
429. **Dobashi, K., Davis, J.G., Mikami, Y., Freeman, J.K., Hamuro, J., and Greene, M.I.,** Characterization of a neu/c-erbB-2 protein-specific activating factor, *Proc. Natl. Acad. Sci. U.S.A.,* 88, 8582, 1991.
430. **Lupu, R., Colomer, R., Kannan, B., and Lippman, M.E.,** Characterization of a growth factor that binds exclusively to the *erbB*-2 receptor and induces cellular responses, *Proc. Natl. Acad. Sci. U.S.A.,* 89, 2287, 1992.
431. **Holmes, W.E., Sliwkowski, M.X., Akita, R.W., Henzel, W.J., Lee, J., Park, J.W., Yansura, D., Abadi, N., Raab, H., Lewis, G.D., Shepard, H.M., Kuang, W.-J., Wood, W.I., Goeddel, D.V., and Vandlen, R.L.,** Identification of heregulin, a specific activator of $p185^{erbB2}$, *Science,* 256, 1205, 1992.
432. **Peles, E., Bacus, S.S., Koski, R.A., Lu, H.S., Wen, D., Ogden, S.G., Ben Levy, R., and Yarden, Y.,** Isolation of the Neu/HER-2 stimulatory ligand: a 44 kd glycoprotein that induces differentiation of mammary tumor cells, *Cell,* 69, 205, 1992.
433. **Wen, D., Peles, E., Cupples, R., Suggs, S.V., Bacus, S.S., Luo, Y., Trail, G., Hu, S., Silbiger, S.M., Ben Levy, R., Koski, R.A., Lu, H.S., and Yarden, Y.,** Neu differentiation factor: a transmembrane glycoprotein containing an EGF domain and an immunoglobulin homology unit, *Cell,* 69, 559, 1992.
434. **Marchionni, M.A., Goodearl, A.D.J., Chen M.S., Bermingham-McDonogh, O., Kirk, C., Hendricks, M., Danehi, F., Misumi, D., Sudhalter, J., Kobayashi, K., Wroblewski, D., Lynch, C., Baldassare, M., Hiles, I., Davis, J.B., Hsuan, J.J., Totty, N.F., Otsu, M., McBurney, R.N., Waterfield, M.D., Stroobant, P., and Gwynne, D.,** Glial growth factors are alternatively spliced erbB2 ligands expressed in the nervous system, *Nature (London),* 362, 312, 1993.

435. **Sistonen, L., Hölttä, E., Lehväslaiho, H., Lehtola, L., and Alitalo, K.,** Activation of the *neu* tyrosine kinase induces the *fos/jun* transcription factor complex, the glucose transporter, and ornithine decarboxylase, *J. Cell Biol.,* 109, 1911, 1989.
436. **Koskinen, P., Lehväslaiho, H., MacDonald-Bravo, H., Alitalo, K., and Bravo, R.,** Similar early gene responses to ligand-activated EGFR and *neu* tyrosine kinases in NIH/3T3 cells, *Oncogene,* 5, 615, 1990.
437. **Lehväslaiho, H., Sistonen, L., diRenzo, F., Partanen, J., Comoglio, P., Hölttä, E., and Alitalo, K.,** Regulation by EGF is maintained in an overexpressed chimeric EGFR/*neu* receptor tyrosine kinase, *J. Cell. Biochem.,* 42, 123, 1990.
438. **Segatto, O., Lonardo, F., Helin, K., Wexler, D., Fazioli, F., Rhee, S.G., and DiFiore, P.P.,** *erb*B-2 autophosphorylation is required for mitogenic action and high-affinity substrate coupling, *Oncogene,* 7, 1339, 1992.
439. **Hazan, R., Margolis, B., Dombalagian, M., Ullrich, A., Zilberstein, A., and Schlessinger, J.,** Identification of autophosphorylation sites of HER2/*neu, Cell Growth Differ.,* 1, 3, 1990.
440. **Akiyama, T., Matsuda, S., Namba, Y., Saito, T., Toyoshima, K., and Yamamoto, T.,** The transforming potential of the c-*erbB*-2 protein is regulated by its autophosphorylation at the carboxyl-terminal domain, *Mol. Cell. Biol.,* 11, 833, 1991.
441. **Akiyama, T., Saito, T., Ogawara, H., Toyoshima, K., and Yamamoto, T.,** Tumor promoter and epidermal growth factor stimulate phosphorylation of the c-*erbB*-2 gene product in MKN-7 human adenocarcinoma cells, *Mol. Cell. Biol.,* 8, 1019, 1988.
442. **King, C.R., Borrello, I., Bellot, F., Comoglio, P., and Schlessinger, J.,** EGF binding to its receptor triggers a rapid tyrosine phosphorylation of the *erb*B-2 protein in the mammary tumor cell line SK-BR-3, *EMBO J.,* 7, 1647, 1988.
443. **Lin, Y.J., Li, S., and Clinton, G.M.,** Insulin and epidermal growth factor stimulate phosphorylation of 185^{HER-2} in the breast carcinoma cell line, BT474, *Mol. Cell. Endocrinol.,* 69, 111, 1990.
444. **Wada, T., Qian, X., and Greene, M.I.,** Intermolecular association of the $p186^{neu}$ protein and EGF receptor modulates EGF receptor function, *Cell,* 61, 1339, 1990.
445. **Wada, T., Myers, J.N., Kokai, Y., Brown, V.I., Hamuro, J., LeVea, C.M., and Greene, M.I.,** Anti-receptor antibodies reverse the phenotype of cells transformed by two interacting proto-oncogene encoded receptor proteins, *Oncogene,* 5, 489, 1990.
446. **Lehtola, L., Sistonen, L., Koskinen, P., Lehväslaiho, H., Di Renzo, M.F., Comoglio, P.M., and Alitalo, K.,** Constitutively activated neu oncoprotein tyrosine kinase interferes with growth factor-induced signals for gene activation, *J. Cell. Biochem.,* 45, 69, 1991.
447. **Fujimoto, A., Kai, S., Akiyama, T., Toyoshima, K., Kaibuchi, K., Takai, Y., and Yamamoto, T.,** Transactivation of the TPA-responsive element by the oncogenic c-*erbB*-2 protein is partly mediated by protein kinase C, *Biochem. Biophys. Res. Commun.,* 178, 724, 1991.
448. **Wongsasant, B., Matuda, S., and Yamamoto, T.,** Active c-*erbB*-2 induces short-term growth of FDC-P2 cells after IL-3 depletion, *Biochem. Biophys. Res. Commun.,* 181, 981, 1991.
449. **Huang, S.S., Koh, H.A., Konish, Y., Bullock, L.D., and Huang, J.S.,** Differential processing and turnover of the oncogenically activated *neu/erb* B2 gene product and its normal cellular counterpart, *J. Biol. Chem.,* 265, 3340, 1990.
450. **Tsutsumi, Y., Naber, S.P., DeLellis, R.A., Wolfe, H.J., Marks, P.J., McKenzie, S.J., and Yin, S.,** *neu* oncogene protein and epidermal growth factor receptor are independently expressed in benign and malignant breast tissues, *Hum. Pathol.,* 21, 750, 1990.
451. **Perren, T.J.,** c-erbB-2 oncogene as a prognostic marker in breast cancer, *Br. J. Cancer,* 63, 328, 1991.
452. **Allred, D.C., Clark, G.M., Molina, R., Tandon, A.K., Schnitt, S.J., Gilchrist, K.W., Osborne, C.K., Tormey, D.C., and McGuire, W.L.,** Overexpression of HER-2/*neu* and its relationship with other prognostic factors change during the progression of *in situ* to invasive breast cancer, *Hum. Pathol.,* 23, 974, 1992.
453. **Umekita, Y., Enokizono, N., Sagara, Y., Kuriwaki, K., Takasaki, T., Yoshida, A., and Yoshida, H.,** Immunohistochemical studies on oncogene products (EGF-R, c-*erb*B-2) and growth factors (EGF, TGF-α) in human breast cancer: their relationship to oestrogen receptor status, histological grade, mitotic index and nodal status, *Virchows Arch. A,* 420, 345, 1992.
454. **Read, L.D., Keith, D., Jr., Slamon, D.J., and Katzenellenbogen, B.S.,** Hormonal modulation of HER-2/*neu* proto-oncogene messenger ribonucleic acid and p185 protein expression in human breast cancer cell lines, *Cancer Res.,* 50, 3947, 1990.

455. **Guy, C.T., Webster, M.A., Schaller, M., Parsons, T.J., Cardiff, R.D., and Muller, W.J.,** Expression of the *neu* proto-oncogene in the mammary epithelium of transgenic mice induces metastatic disease, *Proc. Natl. Acad. Sci. U.S.A.,* 89, 10578, 1992.
456. **Kraus, M.H., Issing, W., Miki, T., Popescu, N.C., and Aaronson, S.A.,** Isolation and characterization of *ERBB3,* a third member of the *ERBB*/epidermal growth factor receptor family: evidence for overexpression in a subset of human mammary tumors, *Proc. Natl. Acad. Sci. U.S.A.,* 86, 9193, 1989.
457. **Kraus, M.H., Fedi, P., Starks, V., Muraro, R., and Aaronson, S.A.,** Demonstration of ligand-depending signaling by the *erbB-3* tyrosine kinase and its constitutive activation in human breast tumor cells, *Proc. Natl. Acad. Sci. U.S.A.,* 90, 2900, 1993.
458. **Adamson, E.D. and Rees, A.R.,** Epidermal growth factor receptors, *Mol. Cell. Biochem.,* 34, 129, 1981.
459. **Mukku, V.R.,** Regulation of epidermal growth factor receptor levels by thyroid hormone, *J. Biol. Chem.,* 259, 6543, 1984.
460. **Edery, M., Pang, K., Larson, L., Colosi, T., and Nandi, S.,** Epidermal growth factor receptor levels in mouse mammary glands in various physiological states, *Endocrinology,* 117, 405, 1985.
461. **Mukku, V.R. and Stancel, G.M.,** Regulation of epidermal growth factor receptor by estrogen, *J. Biol. Chem.,* 260, 9820, 1985.
462. **Freidenberg, G.R., Klein, H.H., Kladde, M.P., Cordera, R., and Olefsky, J.M.,** Regulation of epidermal growth factor receptor number and phosphorylation by fasting in rat liver, *J. Biol. Chem.,* 261, 752, 1986.
463. **Johnson, A.C., Garfield, S.H., Merlino, G.T., and Pastan, I.,** Expression of epidermal growth factor receptor proto-oncogene mRNA in regenerating rat liver, *Biochem. Biophys. Res. Commun.,* 150, 412, 1988.
464. **Johansson, S. and Andersson, G.,** Similar induction of the hepatic EGF receptor *in vivo* by EGF and partial hepatectomy, *Biochem. Biophys. Res. Commun.,* 166, 661, 1990.
465. **Hanai, N., Nores, G., Torres-Méndez, C.-R., and Hakomori, S.,** Modified ganglioside as a possible modulator of transmembrane signaling mechanism through growth factor receptors: a preliminary note, *Biochem. Biophys. Res. Commun.,* 147, 127, 1987.
466. **Berthois, Y., Dong, X.F., and Martin, P.M.,** Regulation of epidermal growth factor-receptor by estrogen and antiestrogen in the human breast cancer cell line MCF-7, *Biochem. Biophys. Res. Commun.,* 159, 126, 1989.
467. **Murphy, L.J., Sutherland, R.L., and Lazarus, L.,** Regulation of growth hormone and epidermal growth factor receptors by progestins in breast cancer cells, *Biochem. Biophys. Res. Commun.,* 131, 767, 1985.
468. **Murphy, L.C., Murphy, L.J., and Shiu, R.P.C.,** Progestin regulation of EGF-receptor mRNA accumulation in T-47D human breast cancer cells, *Biochem. Biophys. Res. Commun.,* 150, 192, 1988.
469. **Snedeker, S.M., Brown, C.F., and DiAugustine, R.P.,** Expression and functional properties of transforming growth factor α and epidermal growth factor during mouse mammary gland ductal morphogenesis, *Proc. Natl. Acad. Sci. U.S.A.,* 88, 276, 1991.
470. **Traish, A.M. and Wotiz, H.H.,** Prostatic epidermal growth factor receptors and their regulation by androgens, *Endocrinology,* 121, 1461, 1987.
471. **Fanger, B.O., Austin, K.S., Earp, H.S., and Cidlowski, J.A.,** Cross-linking of epidermal growth factor receptors in intact cells: detection of initial stages of receptor clustering and determination of molecular weight of high-affinity receptors, *Biochemistry,* 25, 6414, 1986.
472. **Falette, N., Frappart, L., Lefebvre, M.F., and Saez, S.,** Increased epidermal growth factor receptor level in breast cancer cells treated by 1,25-dihydroxyvitamin D3, *Mol. Cell. Endocrinol.,* 63, 189, 1989.
473. **Lingham, R.B., Stancel, G.M., and Loose-Mitchell, D.S.,** Estrogen regulation of epidermal growth factor receptor messenger ribonucleic acid, *Mol. Endocrinol.,* 2, 230, 1988.
474. **Jansson, J.-O., Ekberg, S., Hoath, S.B., Beamer, W.G., and Frohman, L.A.,** Growth hormone enhances hepatic epidermal growth factor receptor concentration in mice, *J. Clin. Invest.,* 82, 1871, 1988.
475. **Westermark, K., Karlson, F.A., and Westermark, B.,** Thyrotropin modulates EGF receptor function in porcine thyroid follicle cells, *Mol. Cell. Endocrinol.,* 40, 17, 1985.
476. **Lev-Ran, A., Hwang, D.L., and Barseghian, G.,** Decreased expression of liver epidermal growth factor receptors in rats with alloxan and streptozotocin diabetes, *Biochem. Biophys. Res. Commun.,* 137, 258, 1986.

477. **Kasayama, S., Yoshimura, M., and Oka, T.,** Decreased expression of hepatic epidermal growth factor receptor gene in diabetic mice, *J. Mol. Endocrinol.,* 3, 49, 1989.
478. **Vonderhaar, B.K., Tang, E., Lyster, R.R., and Nascimento, M.C.S.,** Thyroid hormone regulation of epidermal growth factor receptor levels in mouse mammary glands, *Endocrinology,* 119, 580, 1986.
479. **Olashaw, N.E., O'Keefe, E.J., and Pledger, W.J.,** Platelet-derived growth factor modulates epidermal growth factor receptors by a mechanism distinct from that of phorbol esters, *Proc. Natl. Acad. Sci. U.S.A.,* 83, 3834, 1986.
480. **Olson, J.E. and Pledger, W.J.,** Transmodulation of epidermal growth factor binding by platelet-derived growth factor and 12-*O*-tetradecanoylphorbol-13-acetate is not sodium-dependent in Balb/c/3T3 cells, *J. Biol. Chem.,* 265, 1847, 1990.
481. **Decker, S.J. and Harris, P.,** Effects of platelet-derived growth factor on phosphorylation of the epidermal growth factor receptor in human skin fibroblasts, *J. Biol. Chem.,* 264, 9204, 1989.
482. **Lazarovici, P., Dickens, G., Kuzuya, H., and Guroff, G.,** Long-term, heterologous down-regulation of the epidermal growth factor receptor in PC12 cells by nerve growth factor, *J. Cell Biol.,* 104, 1611, 1987.
483. **Chang, E.H., Black, R., Zou, Z.-Q., Masnyk, T., Ridge, J., Noguchi, P., and Harford, J.B.,** γ-Interferon modulates growth of A431 cells and expression of EGF receptors, *UCLA Symp. Mol. Cell. Biol.,* 50, 335, 1986.
484. **Zoon, K.C., Karakaski, Y., zur Nedden, D.L., Hu, R., and Arnheiter, H.,** Modulation of epidermal growth factor receptors by human α interferon, *Proc. Natl. Acad. Sci. U.S.A.,* 83, 8226, 1986.
485. **Fernandez-Pol, J.A., Klos, D.J., and Hamilton, P.D.,** Modulation of transforming growth factor α-dependent expression of epidermal growth factor receptor gene by transforming growth factor β, triiodothyronine, and retinoic acid, *J. Cell. Biochem.,* 41, 159, 1989.
486. **Assoian, R.K., Frolik, C.A., Roberts, A.B., Miller, D., and Sporn, M.B.,** Transforming growth factor-β controls receptor levels for epidermal growth factor in NRK fibroblasts, *Cell,* 36, 35, 1984.
487. **Fernandez-Pol, J.A., Klos, D.J., Hamilton, P.D., and Talkad, V.D.,** Modulation of epidermal growth factor receptor gene expression by transforming growth factor-β in a human breast carcinoma cell line, *Cancer Res.,* 47, 4260, 1987.
488. **Palombella, V.J., Yamashiro, D.J., Maxfield, F.R., Decker, S.J., and Vilcek, J.,** Tumor necrosis factor increases the number of epidermal growth factor receptors on human fibroblasts, *J. Biol. Chem.,* 262, 1950, 1987.
489. **Zachary, I. and Rozengurt, E.,** Modulation of the epidermal growth factor receptor by mitogenic ligands: effects of bombesin and role of protein kinase C, *Cancer Surv.,* 4, 729, 1985.
490. **Mahadevan, L.C., Aitken, A., Heath, J., and Foulkes, J.G.,** Embryonal carcinoma-derived growth factor activates protein kinase C *in vivo* and *in vitro, EMBO J.,* 6, 921, 1987.
491. **Robinson, R.A., Vokenant, M.E., Ryan, R.J., and Moses, H.L.,** Decreased epidermal growth factor binding in cells growth arrested in G_1 by nutrient deficiency, *J. Cell. Physiol.,* 109, 517, 1981.
492. **Chua, C.C., Geiman, D.E., and Ladda, R.L.,** Receptor for epidermal growth factor retains normal structure and function in aging cells, *Mech. Ageing Dev.,* 34, 35, 1986.
493. **Rubin, R.A. and Earp, H.S.,** Dimethyl sulfoxide stimulates tyrosine residue phosphorylation of rat liver epidermal growth factor receptor, *Science,* 219, 60, 1983.
494. **Verdin, E.M., Maratos-Flier, E., Carpentier, J.-L., and Kahn, C.R.,** Persistent infection with a nontransforming RNA virus leads to impaired growth factor receptors and response, *J. Cell. Physiol.,* 128, 457, 1986.
495. **Harris, A.L.,** The epidermal growth factor receptor as a target for therapy, *Cancer Cells,* 2, 321, 1990.
496. **Lebeau, J. and Goubin, G.,** Amplification of the epidermal growth factor receptor gene in the BT20 breast carcinoma cell line, *Int. J. Cancer,* 40, 189, 1987.
497. **King, I.C.L. and Sartorelli, A.C.,** The relationship between epidermal growth factor receptors and the terminal differentiation of A431 carcinoma cells, *Biochem. Biophys. Res. Commun.,* 140, 837, 1986.
498. **Rizzino, A., Ruff, E., and Kazakoff, P.,** Isolation and characterization of A-431 cells that retain high epidermal growth factor binding capacity and respond to epidermal growth factor by stimulation, *Cancer Res.,* 48, 2377, 1988.
499. **Hunts, J., Ueda, M., Ozawa, S., Abe, O., Pastan, I., and Shimizu, N.,** Hypermethylation and gene amplification of the epidermal growth factor receptor in squamous cell carcinomas, *Jpn. J. Cancer Res.,* 76, 663, 1985.

500. **Yamamoto, T., Kamata, N., Kawano, H., Shimizu, S., Kuroki, T., Toyoshima, K., Rikimaru, K., Nomura, N., Ishizaki, R., Pastan, I., Gamou, S., and Shimizu, N.,** High incidence of amplification of the epidermal growth factor receptor gene in human squamous carcinoma cell lines, *Cancer Res.,* 46, 414, 1986.
501. **Kamata, N., Chida, K., Rikimaru, K., Horikoshi, M., Enomoto, S., and Kuroki, T.,** Growth-inhibitory effects of epidermal growth factor and overexpression of its receptors on human squamous cell carcinomas in culture, *Cancer Res.,* 46, 1648, 1986.
502. **Ozanne, B., Richards, C.S., Hendler, F., Burns, D., and Gusterson, B.,** Over-expression of the EGF receptor is a hallmark of squamous cell carcinomas, *J. Pathol.,* 149, 9, 1986.
503. **Knowles, A.F.,** Inhibition of growth and induction of enzyme activities in a clonal human hepatoma cell line (Li-7A): comparison of the effects of epidermal growth factor and an anti-epidermal growth factor receptor antibody, *J. Cell. Physiol.,* 134, 109, 1988.
504. **Ro, J., North, S.M., Gallick, G.E., Hortobagyi, G.N., Gutterman, J.U., and Blick, M.,** Amplified and overexpressed epidermal growth factor receptor gene in uncultured primary human breast carcinoma, *Cancer Res.,* 48, 161, 1988.
505. **Woloschak, G.E., Dewald, G.W., Bahn, R.S., Kyle, R.A., Greipp, P.R., and Ash, R.C.,** Amplification of RNA and DNA specific for erb B in unbalanced 1;7 chromosomal translocation associated with myelodysplastic syndrome, *J. Cell. Biochem.,* 32, 23, 1986.
506. **Eisbruch, A., Blick, M., Lee, J.S., Sacks, P.G., and Gutterman, J.,** Analysis of the epidermal growth factor receptor gene in fresh human head and neck tumors, *Cancer Res.,* 47, 3603, 1987.
507. **Hirata, Y., Uchihashi, M., Fujita, T., Matsukura, S., Motoyama, T., Kaku, M., and Koshimizu, K.,** Characteristics of specific binding of epidermal growth factor (EGF) on human tumor cell lines, *Endocrinol. Jpn.,* 30, 601, 1983.
508. **Hirata, Y., Uchihashi, M., Nakashima, H., Fujita, T., Matsukura, S., and Matsui, K.,** Specific receptors for epidermal growth factor in human bone tumour cells and its effect on synthesis of prostaglandin E_2 by cultured osteosarcoma cell line, *Acta Endocrinol.,* 107, 125, 1984.
509. **King, C.R., Kraus, M.H., Williams, L.T., Merlino, G.T., Pastan, I.H., and Aaronson, S.A.,** Human tumor cell lines with EGF receptor gene amplification in the absence of aberrant sized mRNAs, *Nucleic Acids Res.,* 13, 8477, 1985.
510. **Gullick, W.J., Marsden, J.J., Whittle, N., Ward, B., Bobrow, L., and Waterfield, M.D.,** Expression of epidermal growth factor receptors on human cervical, ovarian, and vulval carcinomas, *Cancer Res.,* 46, 285, 1986.
511. **Ishitoya, J., Toriyama, M., Oguchi, N., Kitamura, K., Ohshima, M., Asano, K., and Yamamoto, T.,** Gene amplification and overexpression of EGF receptor in squamous cell carcinomas of the head and neck, *Br. J. Cancer,* 59, 559, 1989.
512. **Barker, S. and Vinson, G.P.,** Epidermal growth factor in breast cancer, *Int. J. Biochem.,* 22, 939, 1990.
513. **Imai, Y., Leung, C.K.H., Friesen, H.G., and Shiu, R.P.C.,** Epidermal growth factor receptors and effect of epidermal growth factor on growth of human breast cancer cells in long-term tissue culture, *Cancer Res.,* 42, 4394, 1982.
514. **Fitzpatrick, S.L., LaChance, M.P., and Schultz, G.S.,** Characterization of epidermal growth factor receptor and action on human breast cancer cells in culture, *Cancer Res.,* 44, 3442, 1984.
515. **Roos, W., Fabbro, D., Küng, W., Costa, S.D., and Eppenberger, U.,** Correlation between hormone dependency and the regulation of epidermal growth factor receptor by tumor promoters in human mammary carcinoma cells, *Proc. Natl. Acad. Sci. U.S.A.,* 83, 991, 1986.
516. **Fabbro, D., Küng, W., Roos, W., Regazzi, R., and Eppenberger, U.,** Epidermal growth factor binding and protein kinase C activities in human breast cancer cell lines: possible quantitative relationship, *Cancer Res.,* 46, 2720, 1986.
517. **Lee, C.S.L., Koga, M., and Sutherland, R.L.,** Modulation of estrogen receptor and epidermal growth factor receptor mRNAs by phorbol ester in MCF 7 breast cancer cells, *Biochem. Biophys. Res. Commun.,* 162, 415, 1989.
518. **Filmus, J., Trent, J.M., Pollak, M.N., and Buick, R.N.,** Epidermal growth factor receptor gene-amplified MDA-468 breast cancer cell line and its nonamplified variants, *Mol. Cell. Biol.,* 7, 251, 1987.
519. **Koga, M., Eisman, J.A., and Sutherland, R.L.,** Regulation of epidermal growth factor receptor levels by 1,25-dihydroxyvitamin D_3 in human breast cancer cells, *Cancer Res.,* 48, 2734, 1988.
520. **Decker, S.J.,** Epidermal growth factor induces internalization but not degradation of the epidermal growth factor receptor in a human breast cancer cell line, *J. Receptor Res.,* 8, 853, 1988.

521. **Fitzpatrick, S.L., Brightwell, J., Wittliff, J.L., Barrows, G.H., and Schultz, G.S.,** Epidermal growth factor binding by breast tumor biopsies and relationship to estrogen receptor and progestin receptor levels, *Cancer Res.,* 44, 3448, 1984.
522. **Macias, A., Azavedo, E., Perez, R., Rutqvist, L.E., and Skoog, L.,** Receptors for epidermal growth factor in human mammary carcinomas and their metastases, *Anticancer Res.,* 6, 849, 1986.
523. **Cappelletti, V., Brivio, M., Miodini, P., Granata, G., Coradini, D., and Di Fronzo, G.,** Simultaneous estimation of epidermal growth factor receptors and steroid receptors in a series of 136 resectable primary breast tumors, *Tumor Biol.,* 9, 200, 1988.
524. **Spitzer, E., Grosse, R., Kunde, D., and Eberhard-Schmidt, H.,** Growth of mammary epithelial cells in breast-cancer biopsies correlates with EGF binding, *Int. J. Cancer,* 39, 279, 1987.
525. **Foekens, J.A., Portengen, H., van Putten, W.L.J., Trapman, A.M.A.C., Reubi, J.-C., Alexieva-Figusch, J., and Klijn, J.G.M.,** Prognostic value of receptors for insulin-like growth factor 1, somatostatin, and epidermal growth factor in human breast cancer, *Cancer Res.,* 49, 7002, 1989.
526. **Sainsbury, J.R.C., Malcolm, A.J., Appleton, D.R., Farndon, J.R., and Harris, A.L.,** Presence of epidermal growth factor receptor as an indicator of poor prognosis in patients with breast cancer, *J. Clin. Pathol.,* 38, 1225, 1985.
527. **Sainsbury, J.R., Farndon, J.R., Needham, G.K., Malcolm, A.J., and Harris, A.L.,** Epidermal-growth-factor receptor status as predictor of early recurrence of and death from breast cancer, *Lancet,* 1, 1398, 1987.
528. **Wrba, F., Reiner, A., Ritzinger, E., Holzner, J.H., and Reiner, G.,** Expression of epidermal growth factor receptors (EGFR) on breast carcinomas in relation to growth fractions, estrogen receptor status and morphological criteria, *Pathol. Res. Pract.,* 183, 25, 1988.
529. **Harris, A.J., Nicholson, S., Sainsbury, J.R.C., Farndon, J., and Wright, C.,** Epidermal growth factor receptors in breast cancer: association with early relapse and death, poor response to hormones and interactions with *neu, J. Steroid Biochem.,* 34, 123, 1989.
530. **Toi, M., Osaki, A., Yamada, H., and Toge, T.,** Epidermal growth factor receptor expression as a prognostic indicator in breast cancer, *Eur. J. Cancer,* 27, 977, 1991.
531. **Pérez, R., Pascual, M., Macías, A., and Lage, A.,** Epidermal growth factor receptors in human breast cancer, *Breast Cancer Res. Treat.,* 4, 189, 1984.
532. **Sainsbury, J.R.C., Farndon, J.R., Sherbet, G.V., and Harris, A.L.,** Epidermal growth factor receptors and oestrogen receptors in human breast cancer, *Lancet,* 1, 364, 1985.
533. **Wyss, R., Fabbro, D., Regazzi, R., Borner, C., Takahashi, A., and Eppenberger, U.,** Phorbol ester and epidermal growth factor receptors in human breast cancer, *Anticancer Res.,* 7, 721, 1987.
534. **Pekonen, F., Partanen, S., Mäkinen, T., and Rutanen, E.-M.,** Receptors for epidermal growth factor and insulin-like growth factor I and their relation to steroid receptors in human breast cancer, *Cancer Res.,* 48, 1343, 1988.
535. **Barker, S., Panahy, C., Puddefoot, J.R., Goode, A.W., and Vinson, G.P.,** Epidermal growth factor receptor and oestrogen receptors in the non-malignant part of the cancerous breast, *Br. J. Cancer,* 60, 673, 1989.
536. **Valverius, E.M., Velu, T., Shankar, V., Ciardiello, F., Kim, N., and Salomon, D.S.,** Overexpression of the epidermal growth factor receptor in human breast cancer cells fails to induce an estrogen-independent phenotype, *Int. J. Cancer,* 46, 712, 1990.
537. **Lacroix, H., Iglehart, J.D., Skinner, M.A., and Kraus, M.H.,** Overexpression of *erbB*-2 or EGF receptor proteins present in early stage mammary carcinoma is detected simultaneously in matched primary tumors and regional metastases, *Oncogene,* 4, 145, 1989.
538. **Banks-Schlegel, S.P. and Quintero, J.,** Human esophageal carcinoma cells have fewer, but higher affinity epidermal growth factor receptors, *J. Biol. Chem.,* 261, 4359, 1986.
539. **Yoshida, K., Kyo, E., Tsuda, T., Tsujino, T., Ito, M., Niimoto, M., and Tahara, E.,** EGF and TGF-α, the ligands of hyperproduced EGFR in human esophageal carcinoma cells, act as autocrine growth factors, *Int. J. Cancer,* 45, 131, 1990.
540. **Ozawa, S., Ueda, M., Ando, N., Abe, O., and Shimizu, N,** High evidence of EGF receptor hyperproduction in esophageal squamous-cell carcinomas, *Int. J. Cancer,* 39, 333, 1987.
541. **Lu, S.-H., Hsieh, L.-L., Luo, F.-C., and Weinstein, I.B.,** Amplification of the EGF receptor and c-*myc* genes in human esophageal cancers, *Int. J. Cancer,* 42, 502, 1988.
542. **Sakai, K., Mori, S., Kawamoto, T., Taniguchi, S., Kobori, O., Morioka, Y., Kuroki, T., and Kano, K.,** Expression of epidermal growth factor receptors on normal human gastric epithelia and gastric carcinomas, *J. Natl. Cancer Inst.,* 77, 1047, 1986.

543. **Yoshida, K., Kyo, E., Tsujino, T., Sano, T., Niimoto, M., and Tahara, E.,** Expression of epidermal growth factor, transforming growth factor-α and their receptor genes in human gastric carcinomas; implication for autocrine growth, *Jpn. J. Cancer Res.,* 81, 43, 1990.
544. **Yasui, W., Sumiyoshi, H., Hata, J., Kameda, T., Ochiai, A., Ito, H., and Tahara, E.,** Expression of epidermal growth factor receptor in human gastric and colonic carcinomas, *Cancer Res.,* 48, 137, 1988.
545. **Yasui, W., Hata, J., Yokozaki, H., Nakatani, H., Ochiai, A., Ito, H., and Tahara, E.,** Interaction between epidermal growth factor and its receptor in progression of human gastric carcinoma, *Int. J. Cancer,* 41, 211, 1988.
546. **Moorghen, M., Ince, P., Finney, K.J., Watson, A.J., and Harris, A.L.,** Epidermal growth factor receptors in colorectal carcinoma, *Anticancer Res.,* 10, 605, 1990.
547. **Dittadi, R., Gion, M., Brazzale, A., Nosadini, A., Meo, S., Tremolada, C., Sicari, U., Vinci, I., and Bruscagnin, G.,** Epidermal growth factor receptor and ERBB2 protein in colorectal tissue: comparison between cancer and normal mucosa, *Int. J. Oncol.,* 1, 587, 1992.
548. **Koenders, P.G., Peters, W.H.M., Wobbes, T., Beex, L.V.A.M., Nanengast, F.M., and Benraad, T.J.,** Epidermal growth factor receptor levels are lower in carcinomatous than in normal colorectal tissue, *Br. J. Cancer,* 65, 189, 1992.
549. **Murphy, L.D., Valverius, E.M., Tsokos, M., Mickley, L.A., Rosen, N., and Bates, S.E.,** Modulation of EGF receptor expression by differentiating agents in human colon carcinoma cell lines, *Cancer Commun.,* 2, 345, 1990.
550. **Costrini, N.V. and Beck, R.,** Epidermal growth factor-urogastrone receptors in normal human liver and primary hepatoma, *Cancer,* 51, 2191, 1983.
551. **Lev-Ran, A., Hwang, D., Josefsberg, Z., Barseghian, G., Kemeny, M., Meguid, M., and Beatty, D.,** Binding of epidermal growth factor (EGF) and insulin to human liver microsomes and Golgi fractions, *Biochem. Biophys. Res. Commun.,* 119, 1181, 1984.
552. **Lemoine, N.R., Hughes, C.M., Barton, C.M., Poulson, R., Jeffery, R.E., Klöppel, G., Hall, P.A., and Gullick, W.J.,** The epidermal growth factor receptor in human pancreatic cancer, *J. Pathol.,* 166, 7, 1992.
553. **Korc, M., Chandrasekar, B., Yamanaka, Y., Friess, H., Buchler, M., and Beger, H.G.,** Overexpression of the epidermal growth factor receptor in human pancreatic cancer is associated with concomitant increases in the levels of epidermal growth factor and transforming growth factor α, *J. Clin. Invest.,* 90, 1352, 1992.
554. **Korc, M., Meltzer, P., and Trent, J.,** Enhanced expression of epidermal growth factor receptor correlates with alterations of chromosome 7 in human pancreatic cancer, *Proc. Natl. Acad. Sci. U.S.A.,* 83, 5141, 1986.
555. **Smith, J.J., Derynck, R., and Korc, M.,** Production of transforming growth factor α in human pancreatic cancer cells: evidence for a superagonist autocrine cycle, *Proc. Natl. Acad. Sci. U.S.A.,* 84, 7567, 1987.
556. **Haeder, M., Rotsch, M., Bepler, G., Hennig, C., Havemann, K., Heimann, B., and Moelling, K.,** Epidermal growth factor receptor expression in human lung cancer cell lines, *Cancer Res.,* 48, 1132, 1988.
557. **Moody, T.W., Lee, M., Kris, R.M., Bellot, F., Bepler, G., Oie, H., and Gazdar, A.F.,** Lung carcinoid cell lines have bombesin-like peptides and EGF receptors, *J. Cell. Biochem.,* 43, 139, 1990.
558. **Veale, D., Kerr, N., Gibson, G.J., and Harris, A.L.,** Characterization of epidermal growth factor receptor in primary human non-small cell lung cancer, *Cancer Res.,* 49, 1313, 1989.
559. **Sakiyama, S., Nakamura, Y., and Yasuda, S.,** Expression of epidermal growth factor receptor gene in cultured human lung cancer cells, *Jpn. J. Cancer Res.,* 77, 965, 1986.
560. **Cerny, T., Barnes, D.M., Hasleton, P., Barber, P.V., Healy, K., Gullick, W., and Thatcher, N.,** Expression of epidermal growth factor receptor (EGF-R) in human lung tumours, *Br. J. Cancer,* 54, 265, 1986.
561. **Sobol, R.E., Astarita, R.W., Hofeditz, C., Masui, H., Fairshter, R., Royston, I., and Mendelsohn, J.,** Epidermal growth factor receptor expression in human lung carcinomas defined by a monoclonal antibody, *J. Natl. Cancer Inst.,* 79, 403, 1987.
562. **Hwang, D.L., Tay, Y.-C., Lin, S.S., and Lev-Ran, A.,** Expression of epidermal growth factor receptors in human lung tumors, *Cancer,* 58, 2260, 1986.
563. **Veale, D., Ashcroft, T., Marsh, C., Gibson, G.J., and Harris, A.L.,** Epidermal growth factor receptors in non-small cell lung cancer, *Br. J. Cancer,* 55, 513, 1987.

564. **Dazzi, H., Hasleton, P.S., Thatcher, N., Barnes, D.M., Wilkes, S., Swindel, R., and Lawson, R.A.M.,** Expression of epidermal growth factor receptor (EGF-R) in non-small cell lung cancer. Use of archival tissue and correlation of EGF-R with histology, tumour size, node status and survival, *Br. J. Cancer,* 59, 746, 1989.
565. **Dittadi, R., Gion, M., Pagan, V., Brazzale, A., Del Maschio, O., Bargossi, A., Busetto, A., and Bruscagnin, G.,** Epidermal growth factor receptor in lung malignancies. Comparison between cancer and normal tissue, *Br. J. Cancer,* 64, 741, 1991.
566. **Rusch, V., Baselga, J., Cordon-Cardo, C., Orazem, J., Zaman, M., Hoda, S., McIntosh, J., Kurie, J., and Dmitrovsky, E.,** Differential expression of the epidermal growth factor receptor and its ligands in primary non-small cell lung cancers and adjacent benign lung, *Cancer Res.,* 53, 2379, 1993.
567. **Chen, F.A., Repasky, E.A., Takita, H., Schepart, B.S., and Bankert, R.B.,** Cell surface glycoprotein associated with human lung tumors that is similar to but distinct from the epidermal growth factor receptor, *Cancer Res.,* 49, 3642, 1989.
568. **Libermann, T.A., Razon, N., Bartal, A.D., Yarden, Y., Schlessinger, J., and Soreq, H.,** Expression of epidermal growth factor receptors in human brain tumors, *Cancer Res.,* 44, 753, 1984.
569. **Steck, P.A., Gallick, G.E., Maxwell, S.A., Kloetzer, W.S., Arlinghaus, R.B., Moser, R.P., Gutterman, J.U., and Yung, W.K.A.,** Expression of epidermal growth factor receptor and associated glycoprotein on cultured human brain tumor cells, *J. Cell. Biochem.,* 32, 1, 1986.
570. **Chaffanet, M., Chauvin, C., Lainé, M., Berger, F., Chédin, M., Rost, N., Nissou, M.-F., and Benabid, A.L.,** EGF receptor amplification and expression in human brain tumors, *Eur. J. Cancer,* 28, 11, 1992.
571. **Baugnet-Mahieu, L., Lemaire, M., Brotchi, J., Levivier, M., Born, J., Gilles, J., Valkenaers-Michaux, A., and Vangheel, V.,** Epidermal growth factor receptors in human tumors of the central nervous system, *Anticancer Res.,* 10, 1275, 1990.
572. **Meyers, M.B., Shen, W.P., Spengler, B.A., Ciccarone, V., O'Brien, J.P., Donner, D.B., Furth, M.E., and Biedler, J.L.,** Increased epidermal growth factor receptor in multidrug-resistant human neuroblastoma cells, *J. Cell. Biochem.,* 38, 87, 1988.
573. **Nistér, M., Libermann, T.A., Betsholtz, C., Pettersson, M., Claesson-Welsh, L., Heldin, C.-H., Schlessinger, J., and Westermark, B.,** Expression of messenger RNAs for platelet-derived growth factor and transforming growth factor-α and their receptors in human malignant glioma cell lines, *Cancer Res.,* 48, 3910, 1988.
574. **Humphrey, P.A., Wong, A.J., Vogelstein, B., Friedman, H.S., Werner, M.H., Bigner, D.D., and Bigner, S.H.,** Amplification and expression of the epidermal growth factor receptor gene in human glioma xenografts, *Cancer Res.,* 48, 2231, 1988.
575. **Westphal, M., Harsh, G.R., IV, Rosenblum, M.L., and Hammonds, R.G., Jr.,** Epidermal growth factor receptors in the human glioblastoma cell line SF268 differ from those in epidermoid carcinoma cell line A431, *Biochem. Biophys. Res. Commun.,* 132, 284, 1985.
576. **Wells, A., Bishop, J.M., and Helmeste, D.,** Amplified gene for the epidermal growth factor receptor in a human glioblastoma cell line encodes an enzymatically inactive protein, *Mol. Cell. Biol.,* 8, 4561, 1988.
577. **Humphrey, P.A., Gangarosa, L.M., Wong, A.J., Archer, G.E., Lund-Johansen, M., Bjerkvig, R., Laerum, O.-D., Friedman, H.S., and Bigner, D.D.,** Deletion-mutant epidermal growth factor receptor in human gliomas: effect of type II mutation on receptor function, *Biochem. Biophys. Res. Commun.,* 178, 1413, 1991.
578. **Torp, S.H., Helseth, E., Ryan, L., Stolan, S., Dalen, A., and Unsgaard, G.,** Amplification of the epidermal growth factor receptor gene in human gliomas, *Anticancer Res.,* 11, 2095, 1991.
579. **Fleming, T.P., Saxena, A., Clark, W.C., Robertson, J.T., Oldfield, E.H., Aaronson, S.A., and Ali, I.U.,** Amplification and/or overexpression of platelet-derived growth factor receptors and epidermal growth factor receptor in human glial tumors, *Cancer Res.,* 52, 4550, 1992.
580. **Weisman, A.S., Raguet, S.S., and Kelly, P.A.,** Characterization of the epidermal growth factor receptor in human meningioma, *Cancer Res.,* 47, 2172, 1987.
581. **Yamazaki, H., Fukui, Y., Ueyama, Y., Tamaoki, N., Kawamoto, T., Taniguchi, S., and Shibuya, M.,** Amplification of the structurally and functionally altered epidermal growth factor receptor gene (c-*erbB*) in human brain tumors, *Mol. Cell. Biol.,* 8, 1816, 1988.
582. **Steck, P.A., Lee, P., Hung, M.-C., and Yung, W.K.A.,** Expression of an altered epidermal growth factor receptor by human glioblastoma cells, *Cancer Res.,* 48, 5433, 1988.

583. **Lund-Johansen, M., Bjerkvig, R., Humphrey, P.A., Bigner, S.H., Bigner, D.D., and Laerum, O.-D.,** Effect of epidermal growth factor on glioma cell growth, migration, and invasion *in vitro, Cancer Res.,* 50, 6039, 1990.
584. **Birman, P., Michard, M., Li, J.Y., Peillon, F., and Bression, D.,** Epidermal growth factor-binding sites, present in normal human and rat pituitaries, are absent in human pituitary adenomas, *J. Clin. Endocrinol. Metab.,* 65, 275, 1987.
585. **Hawkins, R.A., Killen, E., Whittle, I.R., Jack, W.J.L., Chetty, U., and Prescott, R.J.,** Epidermal growth factor receptors in intracranial and breast tumours: their clinical significance, *Br. J. Cancer,* 63, 553, 1991.
586. **Koprowski, H., Herlyn, M., Balaban, G., Parmiter, A., Ross, A., and Nowell, P.,** Expression of the receptor for epidermal growth factor correlates with increased dosage of chromosome 7 in malignant melanoma, *Somatic Cell Mol. Genet.,* 11, 297, 1985.
587. **Mueller, B.M., Romerdahl, C.A., Trent, J.M., and Reisfeld, R.A.,** Suppression of spontaneous melanoma metastasis in scid mice with an antibody to the epidermal growth factor receptor, *Cancer Res.,* 51, 2193, 1991.
588. **Gusterson, B., Cowley, G., Mc Ilhinney, J., Ozanne, B., Fisher, C., and Reeves, B.,** Evidence for increased epidermal growth factor receptors in human sarcomas, *Int. J. Cancer,* 36, 689, 1985.
589. **Ilekis, J. and Benveniste, R.,** Effects of epidermal growth factor, phorbol myristate acetate, and arachidonic acid on chroriogonadotropin secretion by cultured human choriocarcinoma cells, *Endocrinology,* 116, 2400, 1985.
590. **Werner, S., Viehweger, P., Hofschneider, P.H., and Roth, W.K.,** Low mitogenic response to EGF and TGF-α: a characteristic feature of cultured Kaposi's sarcoma derived cells, *Oncogene,* 6, 59, 1991.
591. **Noguchi, S., Yura, Y., Sherwood, E.R., Kakinuma, H., Kashikhara, N., and Oyasu, R.,** Stimulation of stromal cell growth by normal rat urothelial cell-derived epidermal growth factor, *Lab. Invest.,* 62, 538, 1990.
592. **Mydlo, J.H., Michaeli, J., Cordon-Cardo, C., Goldenberg, A.S., Heston, W.D.W., and Fair, W.R.,** Expression of transforming growth factor α and epidermal growth factor receptor messenger RNA in neoplastic and nonneoplastic human kidney tissue, *Cancer Res.,* 49, 3407, 1989.
593. **Pekonen, F., Partanen, S., and Rutanen, E.-M.,** Binding of epidermal growth factor and insulin-like growth-factor I in renal carcinoma and adjacent normal kidney tissue, *Int. J. Cancer,* 43, 1029, 1989.
594. **Ishikawa, J., Maeda, S., Umezu, K., Sugiyama, T., and Kamidono, S.,** Amplification and overexpression of the epidermal growth factor receptor gene in human renal-cell carcinoma, *Int. J. Cancer,* 45, 1018, 1990.
595. **Neal, D.E., Marsh, C., Bennett, M.K., Abel, P.D., Hall, R.R., Sainsbury, J.R.C., and Harris, A.L.,** Epidermal growth-factor receptors in human bladder cancer: comparison of invasive and superficial tumors, *Lancet,* 1, 366, 1985.
596. **Häder, M., Stach-Machado, D., Pflüger, K.-H., Rotsch, M., Heimann, B., Moelling, K., and Havemann, K.,** Epidermal growth factor receptor expression, proliferation, and colony stimulating activity production in the urinary bladder carcinoma cell line 5637, *J. Cancer Res. Clin. Oncol.,* 113, 579, 1987.
597. **Smith, K., Fennelly, J.A., Neal, D.E., Hall, R.R., and Harris, A.L.,** Characterization and quantitation of the epidermal growth factor receptor in invasive and superficial bladder tumors, *Cancer Res.,* 49, 5810, 1989.
598. **Maddy, S.Q., Chisholm, G.D., Busuttil, A., and Habib, F.K.,** Epidermal growth factor receptors in human prostate cancer: correlation with histological differentiation of the tumour, *Br. J. Cancer,* 60, 41, 1989.
599. **Reynolds, R.K., Talavera, F., Roberts, J.A., Hopkins, M.P., and Menon, K.M.J.,** Characterization of epidermal growth factor receptor in normal and neoplastic human endometrium, *Cancer,* 66, 1967, 1990.
600. **Korc, M., Padilla, J., and Grosso, D.,** Epidermal growth factor inhibits the proliferation of a human endometrial carcinoma cell line, *J. Clin. Endocrinol. Metab.,* 62, 874, 1986.
601. **Berns, E.M.J.J., Klijn, J.G.M., Henzen-Logmans, S.C., Rodenburg, C.J., van der Burg, M.E.L., and Foekens, J.A.,** Receptors for hormones and growth factors and (onco)-gene amplification in human ovarian cancer, *Int. J. Cancer,* 52, 218, 1992.
602. **Battaglia, F., Scambia, G., Benedetti Panici, P., Baiocchi, G., Perrone, L., Iacobelli, S., and Mancuso, S.,** Epidermal growth factor receptor expression in gynecological malignancies, *Gynecol. Obstet. Invest.,* 27, 42, 1989.

603. **Owens, O.J. and Leake, R.E.,** Epidermal growth factor receptor expression in malignant ovary, benign ovarian tumours and normal ovary: a comparison, *Int. J. Oncol.,* 2, 321, 1993.

604. **Miyamoto, M., Sugawa, H., Mori, T., Hase, K., Kuma, K., and Imura, H.,** Epidermal growth factor receptor on cultured neoplastic human thyroid cells and effects of epidermal growth factor and thyroid-stimulating hormone on their growth, *Cancer Res.,* 48, 3652, 1988.

605. **Kikuchi, A., Amagai, M., Hayakawa, K., Ueda, M., Hirohashi, S., Shimizu, N., and Nishikawa, T.,** Association of EGF receptor expression with proliferating cells and of *ras* p21 expression with differentiating cells in various skin tumours, *Br. J. Dermatol.,* 123, 49, 1990.

606. **Todaro, G.J., De Larco, J.E., and Cohen, S.,** Transformation by murine and feline sarcoma viruses specifically blocks binding of epidermal growth factor to cells, *Nature (London),* 264, 26, 1976.

607. **Wasilenko, W.J., Shawyer, L.K., and Weber, M.J.,** Down-modulation of EGF receptors in cells transformed by the *src* oncogene, *J. Cell. Physiol.,* 131, 450, 1987.

608. **Decker, S.,** Reduced binding of epidermal growth factor by avian sarcoma virus-transformed rat cells, *Biochem. Biophys. Res. Commun.,* 113, 678, 1983.

609. **Usui, T., Moriyama, N., Ishibe, T., and Nakatsu, H.,** Loss of epidermal growth factor receptor on the renal neoplasm induced *in vivo* with xenotropic pseudotype Kirsten murine sarcoma virus, *Biochem. Biophys. Res. Commun.,* 120, 879, 1984.

610. **Gebhardt, A., Bell, J.C., and Foulkes, J.G.,** Abelson transformed fibroblasts lacking the EGF receptor are not tumourigenic in nude mice, *EMBO J.,* 5, 2191, 1986.

611. **Berhanu, P. and Hollenberg, M.D.,** Epidermal growth factor-urogastrone receptor: selective alteration in simian virus 40 transformed mouse fibroblasts, *Arch. Biochem. Biophys.,* 203, 134, 1980.

612. **Chen, J.-K., Li, L., and Mioh, H.,** Differential responsiveness of normal and simian virus 40-transformed BALB/c 3T3 cells to retinoic acid: rapid enhancement of epidermal growth factor receptor binding in a simian virus 40-3T3 variant, *Cancer Res.,* 47, 4995, 1987.

613. **Guernsey, D.L., Duigou, G.J., Babiss, L.E., and Fisher, P.B.,** Regulation of thyroidal inducibility of Na,K-ATPase and binding of epidermal growth factor in wild-type and cold-sensitive E1a mutant type 5 adenovirus-transformed CREF cells, *J. Cell. Physiol.,* 133, 507, 1987.

614. **Sircar, S. and Weber, J.M.,** Normalization of epidermal growth factor receptor and transforming growth factor production in drug resistant variants derived from adenovirus transformed cells, *J. Cell. Physiol.,* 134, 467, 1988.

615. **Hwang, D.L., Roitman, A., Carr, B.I., Barseghian, G., and Lev-Ran, A.,** Insulin and epidermal growth factor receptors in rat liver after administration of the hepatocarcinogen 2-acetylaminofluorene: ligand binding and autophosphorylation, *Cancer Res.,* 46, 1955, 1986.

616. **Carr, B.I., Roitman, A., Hwang, D.L., Barseghian, G., and Lev-Ran, A.,** Effects of diethylnitrosamine on hepatic receptor binding and autophosphorylation of epidermal growth factor and insulin in rats, *J. Natl. Cancer Inst.,* 77, 219, 1986.

617. **Lev-Ran, A., Carr, B.I., Hwang, D.L., and Roitman, A.,** Binding of epidermal growth factor and insulin and the autophosphorylation of their receptors in experimental primary hepatocellular carcinomas, *Cancer Res.,* 46, 4656, 1986.

618. **Hwang, D.L., Lev-Ran, A., and Tay, Y.-C.,** Hepatocarcinogens induce decrease in mRNA transcripts of receptors for insulin and epidermal growth factor in the rat liver, *Biochem. Biophys. Res. Commun.,* 146, 87, 1987.

619. **Robinson, R.A., Branum, E.L., Volkenant, M.E., and Moses, H.L.,** Cell cycle variation in ^{125}I-labeled epidermal growth factor binding in chemically transformed cells, *Cancer Res.,* 42, 2633, 1982.

620. **Wakshull, E., Kraemer, P.M., and Wharton, W.,** Multistep change in epidermal growth factor receptors during spontaneous neoplastic progression in Chinese hamster embryo fibroblasts, *Cancer Res.,* 45, 2070, 1985.

621. **Santon, J.B., Cronin, M.T., MacLeod, C.L., Mendelsohn, J., Masui, H., and Gill, G.N.,** Effects of epidermal growth factor receptor concentration on tumorigenicity of A431 cells in nude mice, *Cancer Res.,* 46, 4701, 1986.

622. **Todaro, G.J., De Larco, J.E., and Fryling, C.M.,** Sarcoma growth factor and other transforming peptides produced by human cells: interactions with membrane receptors, *Fed. Proc.,* 41, 2996, 1982.

623. **Hunts, J., Gamou, S., Hirai, M., and Shimizu, N.,** Molecular mechanisms involved in increasing epidermal growth factor receptor levels on the cell surface, *Jpn. J. Cancer Res.,* 77, 423, 1986.

624. **Kaneko, Y., Shibuya, M., Nakayama, T., Hayashida, N., Toda, G., Endo, Y., Oka, H., and Oda, T.,** Hypomethylation of c-*myc* and epidermal growth factor receptor genes in human hepatocellular carcinoma and fetal liver, *Jpn. J. Cancer Res.,* 76, 1136, 1985.

625. **Lax, I., Kris, R., Sasson, I., Ullrich, A., Hayman, M.J., Beug, H., and Schlessinger, J.,** Activation of c-*erb*-B in avian leukosis virus-induced erythroblastosis leads to the expression of a truncated EGF receptor kinase, *EMBO J.,* 4, 3179, 1985.
626. **Weller, A., Meek, J., and Adamson, E.D.,** Preparation and properties of monoclonal antibodies to mouse epidermal growth factor (EGF) receptors: evidence for cryptic EGF receptors in embryonal carcinoma cells, *Development,* 100, 351, 1987.
627. **Chua, C.C., Geiman, D.E., Schreiber, A.B., and Ladda, R.L.,** Nonfunctional epidermal growth factor receptor in cells transformed by Kirsten sarcoma virus, *Biochem. Biophys. Res. Commun.,* 118, 538, 1984.
628. **Chua, C.C. and Ladda, R.L.,** Protein kinase C and non-functional EGF receptor in K-*ras* transformed cells, *Biochem. Biophys. Res. Commun.,* 135, 435, 1986.
629. **Bodine, P.V. and Tupper, J.T.,** Calmodulin antagonists decrease binding of epidermal growth factor to transformed, but not to normal human fibroblasts, *Biochem. J.,* 218, 629, 1984.
630. **Hollenberg, M.D., Barrett, J.C., Ts'o, P.O.P., and Berhanu, P.,** Selective reduction in receptors for epidermal growth factor-urogastrone in chemically transformed tumorigenic Syrian hamster embryo fibroblasts, *Cancer Res.,* 39, 4166, 1979.
631. **Masui, H., Kawamoto, T., Sato, J.D., Wolf, B., Sato, G., and Mendelsohn, J.,** Growth inhibition of human tumor cells in athymic mice by anti-epidermal growth factor receptor monoclonal antibodies, *Cancer Res.,* 44, 1002, 1984.
632. **Masui, H., Moroyama, T., and Mendelsohn, J.,** Mechanism of antitumor activity in mice for anti-epidermal growth factor receptor monoclonal antibodies with different isotypes, *Cancer Res.,* 46, 5592, 1986.
633. **Rodeck, U., Herlyn, M., Herlyn, D., Molthoff, C., Atkinson, B., Varello, M., Steplewski, Z., and Koprowski, H.,** Tumor growth modulation by a monoclonal antibody to the epidermal growth factor receptor: immunologically mediated and effector cell-independent effects, *Cancer Res.,* 47, 3692, 1987.
634. **Kudlow, J.E., Khosravi, M.J., Kobrin, M.S., and Mak, W.W.,** Inability of anti-epidermal growth factor receptor monoclonal antibody to block "autocrine" growth stimulation in transforming growth factor-secreting melanoma cells, *J. Biol. Chem.,* 259, 11895, 1984.
635. **Messing, E.M. and Reznikoff, C.A.,** Normal and malignant human urothelium: *in vitro* effects of epidermal growth factor, *Cancer Res.,* 47, 2230, 1987.
636. **Hollenberg, M.D., Nexo, E., Berhanu, P., and Hock, R.,** Phorbol ester and the selective modulation of receptors for epidermal growth factor-urogastrone, in *Receptor-Mediated Binding and Internalization of Toxins and Hormones,* Academic Press, New York, 1981, 181.
637. **Taketani, Y. and Oka, T.,** Tumor promoter 12-*O*-tetradecanoylphorbol 13-acetate, like epidermal growth factor, stimulates cell proliferation and inhibits differentiation of mouse mammary epithelial cells in culture, *Proc. Natl. Acad. Sci. U.S.A.,* 80, 1646, 1983.
638. **Kanter, P., Leister, K.J., Tomei, L.D., Wenner, P.A., and Wenner, C.E.,** Epidermal growth factor and tumor promoters prevent DNA fragmentation by different mechanisms, *Biochem. Biophys. Res. Commun.,* 118, 392, 1984.
639. **Shaw, G.P. and Hanson, P.J.,** Inhibitory effect of 12-*O*-tetradecanoylphorbol 13-acetate on acid secretion by rat stomach *in vivo, FEBS Lett.,* 201, 225, 1986.
640. **Friedman, B.A., Frackelton, A.R., Jr., Ross, A.H., Connors, J.M., Fujiki, H., Sugimura, T., and Rosner, M.R.,** Tumor promoters block tyrosine-specific phosphorylation of the epidermal growth factor receptor, *Proc. Natl. Acad. Sci. U.S.A.,* 81, 3034, 1984.
641. **Moon, S.O., Palfrey, H.C., and King, A.C.,** Phorbol esters potentiate tyrosine phosphorylation of epidermal growth factor receptors in A431 membranes by a calcium-independent mechanism, *Proc. Natl. Acad. Sci. U.S.A.,* 81, 2298, 1984.
642. **Davis, R.J. and Czech, M.P.,** Tumor-promoting phorbol diesters mediate phosphorylation of the epidermal growth factor receptor, *J. Biol. Chem.,* 259, 8545, 1984.
643. **Davis, R.J. and Czech, M.P.,** Tumor-promoting phorbol diesters cause the phosphorylation of epidermal growth factor receptors in normal human fibroblasts at threonine-654, *Proc. Natl. Acad. Sci. U.S.A.,* 82, 1974, 1985.
644. **Davis, R.J. and Czech, M.P.,** Inhibition of the apparent affinity of the epidermal growth factor receptor caused by phorbol diesters correlates with phosphorylation of threonine-654 but not other sites of the receptor, *Biochem. J.,* 233, 435, 1986.

645. **Davis, R.J., Ganong, B.R., Bell, R.M., and Czech, M.P.,** Structural requirements for diacylglycerols to mimic tumor-promoting phorbol diester action on the epidermal growth factor receptor, *J. Biol. Chem.,* 260, 5315, 1985.
646. **Northwood, I.C. and Davis, R.J.,** Protein kinase C inhibition of the epidermal growth factor receptor tyrosine protein kinase activity is independent of the oligomeric state of the receptor, *J. Biol. Chem.,* 264, 5746, 1989.
647. **Meyer, S.A. and Jirtle, S.A.,** Phenobarbital decreases hepatocyte EGF receptor expression independent of protein kinase C activation, *Biochem. Biophys. Res. Commun.,* 158, 652, 1989.
648. **Wattenberg, E.V., Fukiji, H., and Rosner, M.R.,** Heterologous regulation of the epidermal growth factor receptor by palytoxin, a non-12-*O*-tetradecanoylphorbol-13-acetate-type tumor promoter, *Cancer Res.,* 47, 4618, 1987.
649. **Selinfreund, R. and Wharton, W.,** Effects of 12-*O*-tetradecanoylphorbol-13-acetate on epidermal growth factor receptors in BALB/c-3T3 cells: relationship between receptor modulation and mitogenesis, *Cancer Res.,* 46, 4486, 1986.
650. **Wasilenko, W.J., Payne, D.M., Fitzgerald, D.L., and Weber, M.J.,** Phosphorylation and activation of epidermal growth factor receptors in cells transformed by the *src* oncogene, *Mol. Cell. Biol.,* 11, 309, 1991.
651. **Theodorescu, D., Cornil, I., Sheehan, C., Man, M.S., and Kerbel, R.S.,** Ha-*ras* induction of invasive phenotype results in up-regulation of epidermal growth factor receptors and altered responsiveness to epidermal growth factor in human papillary transitional cell carcinoma cells, *Cancer Res.,* 51, 4486, 1991.
652. **Madhukar, B.V., Oh, S.Y., Chang, C.C., Wade, M., and Trosko, J.E.,** Altered regulation of intercellular communication by epidermal growth factor, transforming growth factor-β and peptide hormones in normal human keratinocytes, *Carcinogenesis,* 10, 13, 1989.
653. **Fox, C.F., Linsley, P.S., and Wrann, M.,** Receptor remodeling and regulation in the action of epidermal growth factor, *Fed. Proc.,* 41, 2988, 1982.
654. **Pierce, J.H., Ruggiero, M., Fleming, T.P., Di Fiore, P.P., Greenberger, J.S., Varticovski, L., Schlessinger, J., Rovera, G., and Aaronson, S.A.,** Signal transduction through the EGF receptor transfected in IL-3-dependent hematopoietic cells, *Science,* 239, 628, 1988.
655. **Ball, R.L., Tanner, K.D., and Carpenter, G.,** Epidermal growth factor potentiates cyclic AMP accumulation in A-431 cells, *J. Biol. Chem.,* 265, 12836, 1990.
656. **Budnik, L.T. and Mukhopadhyay, A.K.,** Epidermal growth factor, a modulator of luteal adenylate cyclase. Characterization of epidermal growth factor receptors and its interaction with adenylate cyclase system in bovine luteal cell membranes, *J. Biol. Chem.,* 266, 13908, 1991.
657. **Yu, Y.M., Nair, B.G., and Patel, T.B.,** Epidermal growth factor stimulates cAMP accumulation in cultured rat cardiac myocytes, *J. Cell Physiol.,* 150, 559, 1992.
658. **Yazigi, R., Berchuck, A., Casey, M.L., and MacDonald, P.C.,** Effect of epidermal growth factor on ornithine decarboxylase activity in A431 cells, *Anticancer Res.,* 9, 209, 1989.
659. **Li, M., Morley, P., Asem, E.K., and Tsang, B.K.,** Epidermal growth factor elevates intracellular pH in chicken granulosa cells, *Endocrinology,* 129, 656, 1991.
660. **Bierman, A.J., Koenderman, L., Tool, A.J., and de Laat, S.W.,** Epidermal growth factor and bombesin differ strikingly in the induction of early responses in Swiss 3T3 cells, *J. Cell. Physiol.,* 142, 441, 1990.
661. **Hill, T.D., Kindmark, H., and Boynton, A.L.,** Epidermal growth factor-stimulated DNA synthesis requires an influx of extracellular calcium, *J. Cell. Biochem.,* 38, 137, 1988.
662. **Johnson, R.M. and Garrison, J.C.,** Epidermal growth factor and angiotensin II stimulate formation of inositol 1,4,5- and inositol 1,3,4-trisphosphate in hepatocytes. Differential inhibition by pertussis toxin and phorbol 12-myristate 13-acetate, *J. Biol. Chem.,* 262, 17285, 1987.
663. **Muldoon, L.L., Rodland, K.D., and Magun, B.E.,** Transforming growth factor β and epidermal growth factor alter calcium influx and phosphatidylinositol turnover in Rat-1 fibroblasts, *J. Biol. Chem.,* 263, 18834, 1988.
664. **Moolenaar, W.H., Aerts, R.J., Tertoolen, L.G.J., and de Laat, S.W.,** The epidermal growth factor-induced calcium signal in A431 cells, *J. Biol. Chem.,* 261, 279, 1986.
665. **Peppelenbosch, M.P., Tertoolen, L.G.J., and de Laat, S.W.,** Epidermal growth factor-activated calcium and potassium channels, *J. Biol. Chem.,* 266, 10938, 1991.

666. **Chapron, Y., Cochet, C., Crouzy, S., Jullien, T., Keramidas, M., and Verdetti, J.,** Tyrosine protein kinase activity of the EGF receptor is required to induce activation of receptor-operated calcium channels, *Biochem. Biophys. Res. Commun.,* 158, 527, 1989.
667. **Peppelenbosch, M.P., Tertoolen, L.G.J., den Hertog, J., and de Laat, S.W.,** Epidermal growth factor activates calcium channels by phospholipase A_2/5-lipoxigenase-mediated leukotriene C_4 production, *Cell,* 69, 295, 1992.
668. **Ross, A.H., Damsky, C., Phillips, P.D., Hwang, F., and Vance, P.,** Inhibition of epidermal growth factor-induced phosphorylation by trifluoperazine, *J. Cell. Physiol.,* 124, 499, 1985.
669. **Sardet, C., Counillon, L., Franchi, A., and Pouysségur, J.,** Growth factors induce phosphorylation of the Na^+/H^+ antiporter, a glycoprotein of 110 kD, *Science,* 247, 723, 1990.
670. **Ghishan, F.K., Kikuchi, K., and Riedel, B.,** Epidermal growth factor up-regulates intestinal Na^+/H^+ exchange activity, *Proc. Soc. Exp. Biol. Med.,* 201, 289, 1992.
671. **Church, J.G., Mills, G.B., and Buick, R.N.,** Activation of the Na^+/H^+ antiport is not required for epidermal growth factor-dependent gene expression, growth inhibition or proliferation in human breast cancer cells, *Biochem. J.,* 257, 151, 1989.
672. **Church, J.G. and Buick, R.N.,** G-protein-mediated epidermal growth factor signal transduction in a human breast cancer cell line. Evidence for two intracellular pathways distinguishable by pertussis toxin, *J. Biol. Chem.,* 263, 4242, 1988.
673. **Nair, B.G., Rashed, H.M., and Patel, T.B.,** Epidermal growth factor stimulates rat cardiac adenylate cyclase through a GTP-binding regulatory protein, *Biochem. J.,* 264, 563, 1989.
674. **Liu, X.G. and Pawson, T.,** The epidermal growth factor receptor phosphorylates GTPase-activating protein (GAP) at Tyr-460, adjacent to the GAP SH2 domains, *Mol. Cell. Biol.,* 11, 2511, 1991.
675. **Buday, L. and Downward, J.,** Epidermal growth factor regulates the exchange rate of guanine nucleotides on $p21^{ras}$ in fibroblasts, *Mol. Cell. Biol.,* 13, 1903, 1993.
676. **Wang, P.-L., Nakagawa, Y., and Humphreys-Beher, M.G.,** Epidermal growth factor and isoproterenol stimulation of the Ras-guanine nucleotide exchange factor in proliferating parotid gland acinar cells, *Biochem. Biophys. Res. Commun.,* 192, 860, 1993.
677. **Osterop, A.P.R.M., Medema, R.H., Zon, G.C.M.V.d., Bos, J.L., Möller, W., and Maasen, J.A.,** Epidermal-growth-factor receptors generate Ras-GTP more efficiently than insulin receptors, *Eur. J. Biochem.,* 212, 477, 1993.
678. **Lowenstein, E.J., Daly, R.J., Batzer, A.G., Li, W., Margolis, B., Lammers, R., Ullrich, A., Skolnik, E.Y., Bar-Sagi, D., and Schlessinger, J.,** The SH2 and SH3 domain-containing protein GRB2 links receptor tyrosine kinases to ras signaling, *Cell,* 70, 431, 1992.
679. **Rozakis-Adcock, M., McGlade, J., Mbamalu, G., Pelicci, G., Li, W., Batzer, A., Thomas, S., Brugge, J., Pelicci, P.G., Schlessinger, J., and Pawson, T.,** Association of the Shc and Grb2/Sem5 SH2-containing proteins is implicated in activation of the Ras pathway by tyrosine kinases, *Nature (London),* 360, 689, 1992.
680. **Egan, S.E., Giddings, B.W., Brooks, M.W., Buday, L., Sizeland, A.M., and Weinberg, R.A.,** Association of Sos Ras exchange protein with Grb2 is implicated in tyrosine kinase signal transduction and transformation, *Nature (London),* 363, 45, 1993.
681. **Rozakis-Adcock, M., Fernley, R., Wade, J., Pawson, T., and Botwell, D.,** The SH2 and SH3 domains of mammalian Grb2 couple the EGF receptor to the Ras activator mSos1, *Nature (London),* 363, 83, 1993.
682. **Li, N., Batzer, A., Daly, R., Yajnik, E., Scolnik, E., Chardin, P., Bar-Sagi, D., Margolis, B., and Schlessinger, J.,** Guanine-nucleotide-releasing factor hSos1 binds to Grb2 and links receptor tyrosine kinases to Ras signalling, *Nature (London),* 363, 85, 1993.
683. **Gale, N.W., Kaplan, S., Lowenstein, E.J., Schlessinger, J., and Bar-Sagi, D.,** Grb2 mediates the EGF-dependent activation of guanine nucleotide exchange on Ras, *Nature (London),* 363, 88, 1993.
684. **Chardin, P., Camonis, J.H., Gale, N.W., Van Aelst, L., Schlessinger, J., Wigler, M.H., and Bar-Sagi, D.,** Human Sos1: a guanine nucleotide exchange factor for Ras that binds to Grb2, *Science,* 260, 1338, 1993.
685. **Buday, L. and Downward, J.,** Epidermal growth factor regulates p21*ras* through the formation of a complex of receptor, Grb2 adapter protein, and Sos nucleotide exchange factor, *Cell,* 73, 611, 1993.
686. **Thompson, D.M., Cochet, C., Chambaz, E.M., and Gill, G.N.,** Separation and characterization of a phosphatidylinositol kinase activity that co-purifies with the epidermal growth factor receptor, *J. Biol. Chem.,* 260, 8824, 1985.

687. **Bjorge, J.D., Chan, T.-O., Antczak, M., Kung, H.-J., and Fujita, D.J.,** Activated type I phosphatidylinositol kinase is associated with the epidermal growth factor (EGF) receptor following EGF stimulation, *Proc. Natl. Acad. Sci. U.S.A.,* 87, 3816, 1990.
688. **Miller, E.S. and Ascoli, M.,** Anti-phosphotyrosine immunoprecipitation of phosphatidylinositol-3′ kinase activity in different cell types after exposure to epidermal growth factor, *Biochem. Biophys. Res. Commun.,* 173, 289, 1990.
689. **Cochet, C., Filhol, O., Payrastre, B., Hunter, T., and Gill, G.N.,** Interaction between the epidermal growth factor receptor and phosphoinositide kinases, *J. Biol. Chem.,* 266, 637, 1991.
690. **Pike, L.J. and Eakes, A.T.,** Epidermal growth factor stimulates the production of phosphatidylinositol monophosphate and the breakdown of polyphosphoinositides in A431 cells, *J. Biol. Chem.,* 262, 1644, 1987.
691. **Wahl, M.I., Sweatt, J.D., and Carpenter, G.,** Epidermal growth factor (EGF) stimulates inositol trisphosphate formation in cells which overexpress the EGF receptor, *Biochem. Biophys. Res. Commun.,* 142, 688, 1987.
692. **Wahl, M. and Carpenter, G.,** Regulation of epidermal growth factor-stimulated formation of inositol phosphates in A-431 cells by calcium and protein kinase C, *J. Biol. Chem.,* 263, 7581, 1988.
693. **Payrastre, B., Plantavid, M., Breton, M., Chambaz, E., and Chap, H.,** Relationship between phosphoinositide kinase activities and protein tyrosine phosphorylation in plasma membranes from A431 cells, *Biochem. J.,* 272, 665, 1990.
694. **Walker, D.H. and Pike, L.J.,** Phosphatidylinositol kinase is activated in membranes derived from cells treated with epidermal growth factor, *Proc. Natl. Acad. Sci. U.S.A.,* 84, 7513, 1987.
695. **Hepler, J.R., Nakahata, N., Lovenberg, T.W., DiGuiseppi, J., Herman, B., Earp, H.S., and Harden, T.K.,** Epidermal growth factor stimulates the rapid accumulation of inositol (1,4,5)-trisphosphate and a rise in cytosolic calcium mobilized from intracellular stores in A431 cells, *J. Biol. Chem.,* 262, 2951, 1987.
696. **Gilligan, A., Prentki, M., Glennon, M.C., and Knowles, B.B.,** Epidermal growth factor-induced increases in inositol trisphosphates, inositol tetrakisphosphates, and cytosolic Ca^{2+} in a human hepatocellular carcinoma-derived cell line, *FEBS Lett.,* 233, 41, 1988.
697. **Muldoon, L.L., Rodland, K.D., and Magun, B.E.,** Transforming growth factor β modulates epidermal growth factor-induced phosphoinositide metabolism and intracellular calcium levels, *J. Biol. Chem.,* 263, 5030, 1988.
698. **Vicentini, L.M., Cervini, R., Zippel, R., and Mantegazza, P.,** Epidermal growth factor-induced phosphoinositide hydrolysis. Modulation by protein kinase C, *FEBS Lett.,* 228, 346, 1988.
699. **Wahl, M.I., Nishibe, S., Suh, P.-G., Rhee, S.G., and Carpenter, G.,** Epidermal growth factor stimulates tyrosine phosphorylation of phospholipase C-II independently of receptor internalization and extracellular calcium, *Proc. Natl. Acad. Sci. U.S.A.,* 86, 1568, 1989.
700. **Imoto, M., Shimura, N., Ui, H., and Umezawa, K.,** Inhibition of EGF-induced phospholipase C activation in A431 cells by erbstatin, a tyrosine kinase inhibitor, *Biochem. Biophys. Res. Commun.,* 173, 208, 1990.
701. **Nishibe, S., Wahl, M.I., Rhee, S.G., and Carpenter, G.,** Tyrosine phosphorylation of phospholipase C-II *in vitro* by the epidermal growth factor receptor, *J. Biol. Chem.,* 264, 10335, 1989.
702. **Margolis, B., Rhee, S.G., Felder, S., Mervic, M., Lyall, R., Levitzki, A., Ullrich, A., Zilberstein, A., and Schlessinger, J.,** EGF induces tyrosine phosphorylation of phospholipase C-II: a potential mechanism for EGF receptor signaling, *Cell,* 57, 1101, 1989.
703. **Nishibe, S., Wahl, M.I., Wedegaertner, P.B., Kim, J.J., Rhee, S.G., and Carpenter, G.,** Selectivity of phospholipase C phosphorylation by the epidermal growth factor receptor, the insulin receptor, and their cytoplasmic domains, *Proc. Natl. Acad. Sci. U.S.A.,* 87, 424, 1990.
704. **Wahl, M.I., Nishibe, S., Kim, J.W., Kim, H., Rhee, S.G., and Carpenter, G.,** Identification of two epidermal growth factor-sensitive tyrosine phosphorylation sites of phospholipase C-γ in intact HSC-1 cells, *J. Biol. Chem.,* 265, 3944, 1990.
705. **Todderud, G., Wahl, M.I., Rhee, S.G., and Carpenter, G.,** Stimulation of phospholipase C-γ 1 membrane association by epidermal growth factor, *Science,* 249, 297, 1990.
706. **Kim, U.H., Kim, H.S., and Rhee, S.G.,** Epidermal growth factor and platelet-derived growth factor promote translocation of phospholipase-C-γ from cytosol to membrane, *FEBS Lett.,* 270, 33, 1990.
707. **Kato, M., Homma, Y., Nagai, Y., and Takenawa, T.,** Epidermal growth factor stimulates diacylglycerol kinase in isolated plasma membrane vesicles from A431 cells, *Biochem. Biophys. Res. Commun.,* 1985.

708. **Olashaw, N.E. and Pledger, W.J.,** Epidermal growth factor stimulates formation of inositol phosphates in BALB/c/3T3 cells pretreated with cholera toxin and isobutylmethylxanthine, *J. Biol. Chem.,* 263, 1111, 1988.
709. **Gomez, M.L., Medrano, E.E., Cafferatta, E.G.A., and Tellez-Iñon, M.T.,** Protein kinase C is differentially regulated by thrombin, insulin, and epidermal growth factor in human mammary tumor cells, *Exp. Cell Res.,* 175, 74, 1988.
710. **Moscat, J., Molloy, C.J., Fleming, T.P., and Aaronson, S.A.,** Epidermal growth factor activates phosphoinositide turnover and protein kinase C in BALB/MK keratinocytes, *Mol. Endocrinol.,* 2, 799, 1988.
711. **Brown, K.D., Blay, J., Irvine, R.F., Heslop, J.P., and Berridge, M.J.,** Reduction of epidermal growth factor receptor affinity by heterologous ligands: evidence for a mechanism involving the break down of phosphoinositides and the activation of protein kinase C, *Biochem. Biophys. Res. Commun.,* 123, 377, 1984.
712. **McCaffrey, P.G., Friedman, B.-A., and Rosner, M.R.,** Diacylglycerol modulates binding and phosphorylation of the epidermal growth factor receptor, *J. Biol. Chem.,* 259, 12501, 1984.
713. **Sinnett-Smith, J.W. and Rozengurt, E.,** Diacylglycerol treatment rapidly decreases the affinity of the epidermal growth factor receptors of Swiss 3T3 cells, *J. Cell. Physiol.,* 124, 81, 1985.
714. **Wahl, M.I., Daniel, T.O., and Carpenter, G.,** Antiphosphotyrosine recovery of phospholipase C activity after EGF treatment of A-431 cells, *Science,* 241, 968, 1988.
715. **Taylor, D., Uhing, R.J., Blackmore, P.F., Prpič, V., and Exton, J.H.,** Insulin and epidermal growth factor do not affect phosphoinositide metabolism in rat liver plasma membranes and hepatocytes, *J. Biol. Chem.,* 260, 2011, 1985.
716. **Raben, D.M. and Cunningham, D.D.,** Effects of EGF and thrombin on inositol-containing phospholipids of cultured fibroblasts: stimulation of phosphatidylinositol synthesis by thrombin but not EGF, *J. Cell. Physiol.,* 125, 582, 1985.
717. **L'Allemain, G. and Pouysségur, J.,** EGF and insulin action in fibroblasts. Evidence that phosphoinositide hydrolysis is not an essential mitogenic signalling pathway, *FEBS Lett.,* 197, 344, 1986.
718. **Pignataro, O.P. and Ascoli, M.,** Studies with insulin and insulin-like growth factor-I show that the increased labeling of phosphatidylinositol-3,4-bisphosphate is not sufficient to elicit the diverse actions of epidermal growth factor on MA-10 Leydig tumor cells, *Mol. Endocrinol.,* 4, 758, 1990.
719. **Besterman, J.M., Watson, S.P., and Cuatrecasas, P.,** Lack of association of epidermal growth factor-, insulin-, and serum-induced mitogenesis with stimulation of phosphoinositide degradation in BALB/c 3T3 fibroblasts, *J. Biol. Chem.,* 261, 723, 1986.
720. **Matuoka, K., Fukami, K., Nakanishi, O., Kawai, S., and Takenawa, T.,** Mitogenesis in response to PDGF and bombesin abolished by microinjection of antibody to PIP_2, *Science,* 239, 640, 1988.
721. **Gatalica, Z. and Banfic, H.,** Epidermal growth factor stimulates the incorporation of phosphate into phosphatidic acid and phosphoinositides but does not affect phosphoinositide breakdown by phospholipase C in renal cortical slices, *Biochim. Biophys. Acta,* 968, 379, 1988.
722. **Vega, Q.C., Cochet, C., Filhol, O., Chang, C.-P., Rhee, S.G., and Gill, G.N.,** A site of tyrosine phosphorylation in the C terminus of the epidermal growth factor receptor is required to activate phospholipase C, *Mol. Cell. Biol.,* 12, 128, 1992.
723. **Fisher, G.J., Henderson, P.A., Voorhees, J.J., and Baldassare, J.J.,** Epidermal growth factor-induced hydrolysis of phosphatidylcholine by phospholipase D and phospholipase C in human dermal fibroblasts, *J. Cell. Physiol.,* 146, 309, 1991.
724. **Cook, S.J. and Wakelam, M.J.O.,** Epidermal growth factor increases *sn*-1,2-diacylglycerol levels and activates phospholipase D-catalysed phosphatidylcholine breakdown in Swiss 3T3 cells in the absence of inositol-lipid hydrolysis, *Biochem. J.,* 285, 247, 1992.
725. **Ming, L., Morley, P., and Tsang, B.K.,** Epidermal growth factor elevates intracellular pH in chicken granulosa cells by activating protein kinase C, *Endocrinology,* 129, 2957, 1991.
726. **Handler, J.A., Danilowicz, R.M., and Eling, T.E.,** Mitogenic signaling by epidermal growth factor (EGF), but not platelet-derived growth factor, requires arachidonic acid metabolism in BALB/c 3T3 cells. Modulation of EGF-dependent c-*myc* expression by prostaglandins, *J. Biol. Chem.,* 265, 3669, 1990.
727. **Mitchell, M.D.,** Epidermal growth factor actions on arachidonic acid metabolism in human amnion cells, *Biochim. Biophys. Acta,* 928, 240, 1987.

728. **Blay, J.B. and Hollenberg, M.D.,** Epidermal growth factor stimulation of prostacyclin production by cultured aortic smooth muscle cells: requirement for increased cellular calcium levels, *J. Cell. Physiol.,* 139, 524, 1989.
729. **Margolis, B.L., Bonventre, J.V., Kremer, S.G., Kudlow, J.E., and Skorecki, K.L.,** Epidermal growth factor is synergistic with phorbol esters and vasopressin in stimulating arachidonate release and prostaglandin production in renal glomerular mesangial cells, *Biochem. J.,* 249, 587, 1988.
730. **Margolis, B.L., Holub, B.J., Troyer, D.A., and Skorecki, K.L.,** Epidermal growth factor stimulates phospholipase A_2 in vasopressin-treated rat glomerular mesangial cells, *Biochem. J.,* 256, 469, 1988.
731. **Bonventre, J.V., Gronich, J.H., and Nemenoff, R.A.,** Epidermal growth factor enhances glomerular mesangial cell soluble phospholipase A_2 activity, *J. Biol. Chem.,* 265, 4934, 1990.
732. **Goldberg, H.J., Viegas, M.M., Margolis, B.L., Schlessinger, J., and Skorecki, K.L.,** The tyrosine kinase activity of the epidermal growth factor receptor is necessary for phospholipase A_2 activation, *Biochem. J.,* 267, 461, 1990.
733. **Casey, M.L., Korte, K., and MacDonald, P.C.,** Epidermal growth factor stimulation of prostaglandin E_2 biosynthesis in amnion cells: induction of prostaglandin H_2 synthase, *J. Biol. Chem.,* 263, 7846, 1988.
734. **Yang, S.-G., Saifeddine, M., Chuang, M., Severson, D.L., and Hollenberg, M.D.,** Diacylglycerol lipase and the contractile action of epidermal growth factor-urogastrone: evidence for distinct signal pathways in a single strip of gastric smooth muscle, *Eur. J. Pharmacol.,* 207, 225, 1991.
735. **Glasgow, W.C., Afshari, C.A., Barrett, J.C., and Eling, T.E.,** Modulation of the epidermal growth factor mitogenic response by metabolites of linoleic and arachidonic acid in Syrian hamster embryo fibroblasts. Differential effects in tumor suppressor gene (+) and (–) phenotypes, *J. Biol. Chem.,* 267, 10771, 1992.
736. **Giugni, T.D., James, L.C., and Haigler, H.T.,** Epidermal growth factor stimulates tyrosine phosphorylation of specific proteins in permeabilized human fibroblast, *J. Biol. Chem.,* 260, 15081, 1985.
737. **Ahn, N.G., Weiel, J.E., Chan, C.P., and Krebs, E.G.,** Identification of multiple epidermal growth factor-stimulated protein serine/threonine kinases from Swiss 3T3 cells, *J. Biol. Chem.,* 265, 11487, 1990.
738. **Ahn, N.G. and Krebs, E.G.,** Evidence for an epidermal growth factor-stimulated protein kinase cascade in Swiss 3T3 cells. Activation of serine peptide kinase activity by myelin basic protein kinases *in vitro, J. Biol. Chem.,* 265, 11495, 1990.
739. **Ghosh-Dastidar, P. and Fox, C.F.,** c-AMP-dependent protein kinase stimulates epidermal growth factor-dependent phosphorylation of epidermal growth factor receptors, *J. Biol. Chem.,* 259, 3864, 1984.
740. **Connolly, J.L., Green, S.A., and Greene, L.A.,** Comparison of rapid changes in surface morphology and coated pit formation of PC12 cells in response to nerve growth factor, epidermal growth factor, and dibutyryl cyclic AMP, *J. Cell Biol.,* 98, 457, 1984.
741. **Shiba, T., Akiyama, T., Kadowaki, T., Fukami, Y., Tsuji, T., Osawa, T., Kasuga, M., and Takaku, F.,** Purified tyrosine kinases, the EGF receptor kinase and the src kinase, can catalyze the phosphorylation of the band 3 protein from human erythrocytes, *Biochem. Biophys. Res. Commun.,* 135, 720, 1986.
742. **David-Pfeuty, T. and Guesdon, F.,** Epidermal growth factor stimulates serine and tyrosine phosphorylation in a 59-kD protein in purified plasma membranes from rat liver, *Biochem. Biophys. Res. Commun.,* 145, 982, 1987.
743. **Greenberg, M.E. and Edelman, G.M.,** The 34 kd $pp60^{src}$ substrate is located at the inner face of the plasma membrane, *Cell,* 33, 767, 1983.
744. **Gould, K.L., Cooper, J.A., and Hunter, T.,** The 46,000-dalton tyrosine protein kinase substrate is widespread, whereas the 36,000-dalton substrate is only expressed at high levels in certain rodent tissues, *J. Cell Biol.,* 98, 487, 1984.
745. **Valentine-Braun, K.A., Northup, J.K., and Hollenberg, M.D.,** Epidermal growth factor (urogastrone)-mediated phosphorylation of a 35-kDa substrate in human placental membranes: relationship to the β subunit of the guanine nucleotide regulatory complex, *Proc. Natl. Acad. Sci. U.S.A.,* 83, 236, 1986.
746. **Valentine-Braun, K.A., Hollenberg, M.D., Fraser, E., and Northup, J.K.,** Isolation of a major human placental substrate for the epidermal growth factor (urogastrone) receptor kinase: immunological cross-reactivity with transducin and sequence homology with lipocortin, *Arch. Biochem. Biophys.,* 259, 262, 1987.

747. **Erikson, E. and Erikson, R.L.,** Identification of a cellular protein substrate phosphorylated by the avian sarcoma virus-transforming gene product, *Cell,* 21, 829, 1980.
748. **Ghosh-Dastidar, P. and Fox, C.F.,** Epidermal growth factor and epidermal growth factor receptor-dependent phosphorylation of a M_r = 34,000 protein substrate for $pp60^{src}$, *J. Biol. Chem.,* 258, 2041, 1983.
749. **Blay, J., Valentine-Braun, K.A., Northup, J.K., and Hollenberg, M.D.,** Epidermal-growth-factor-stimulated phosphorylation of calpactin II in membrane vesicles shed from cultured A-431 cells, *Biochem. J.,* 259, 577, 1989.
750. **Sawyer, S.T. and Cohen, S.,** Epidermal growth factor stimulates the phosphorylation of the calcium-dependent 35,000-dalton substrate in intact A431 cells, *J. Biol. Chem.,* 260, 8233, 1985.
751. **De, B.K., Misono, K.S., Lukas, T.J., Mrczkowski, B., and Cohen, S.,** A calcium-dependent 35-kilodalton substrate for epidermal growth factor receptor/kinase isolated from normal tissue, *J. Biol. Chem.,* 261, 13784, 1986.
752. **Sheets, E.E., Giugni, T.D., Coates, G.G., Schlaepfer, D.D., and Haigler, H.T.,** Epidermal growth factor dependent phosphorylation of a 35-kilodalton protein in placental membranes, *Biochemistry,* 26, 1164, 1987.
753. **Karasik, A., Pepinsky, R.B., and Kahn, C.R.,** Insulin and epidermal growth factor stimulate phosphorylation of a 170-kDa protein in intact hepatocytes immunologically related to lipocortin 1, *J. Biol. Chem.,* 263, 18558, 1988.
754. **Isacke, C.M., Lindberg, R.A., and Hunter, T.,** Synthesis of p36 and p35 is increased when U-937 cells differentiate in culture but expression is not inducible by glucocorticoids, *Mol. Cell. Biol.,* 9, 232, 1989.
755. **Thomas, G., Martin-Pérez, J., Siegmann, M., and Otto, A.M.,** The effect of serum, EGF, $PGF_{2\alpha}$ and insulin on S6 phosphorylation and the initiation of protein and DNA synthesis, *Cell,* 30, 235, 1982.
756. **Novak-Hofer, I. and Thomas, G.,** Epidermal growth factor-mediated activation of an S6 kinase in Swiss mouse 3T3 cells, *J. Biol. Chem.,* 260, 10314, 1985.
757. **Rosen, O.M., Rubin, C.S., Cobb, M.H., and Smith, C.J.,** Insulin stimulates the phosphorylation of ribosomal protein S6 in a cell-free system derived from 3T3-L1 adipocytes, *J. Biol. Chem.,* 256, 3630, 1981.
758. **Dazord, A., Genot, A., Langlois-Gallet, D., Mombrial, C., Haour, F., and Saez, J.M.,** hCG-increased phosphorylation of proteins in primary culture of Leydig cells: further characterization, *Biochem. Biophys. Res. Commun.,* 118, 8, 1984.
759. **Blenis, J. and Erikson, R.L.,** Phosphorylation of the ribosomal protein S6 is elevated in cells transformed by a variety of tumor viruses, *J. Virol.,* 50, 966, 1984.
760. **Jenö, P., Ballou, L.M., Novak-Hofer, I., and Thomas, G.,** Identification and characterization of a mitogen-activated S6 kinase, *Proc. Natl. Acad. Sci. U.S.A.,* 85, 406, 1988.
761. **Martin-Pérez, J., Siegmann, M., and Thomas, G.,** EGF, $PGF_{2\alpha}$ and insulin induce the phosphorylation of identical S6 peptides in Swiss mouse 3T3 cells: effect of cAMP on early sites of phosphorylation, *Cell,* 36, 287, 1984.
762. **Perisic, O. and Traugh, J.A.,** Protease-activated kinase II as the mediator of epidermal growth factor-stimulated phosphorylation of ribosomal protein S6, *FEBS Lett.,* 183, 215, 1985.
763. **Pelech, S.L. and Krebs, E.G.,** Mitogen-activated S6 kinase is stimulated via protein kinase C-dependent and independent pathways in Swiss 3T3 cells, *J. Biol. Chem.,* 262, 11598, 1987.
764. **Olivier, A.R., Ballou, L.M., and Thomas, G.,** Differential regulation of S6 phosphorylation by insulin and epidermal growth factor in Swiss mouse 3T3 cells: insulin activation of type 1 phosphatase, *Proc. Natl. Acad. Sci. U.S.A.,* 85, 4720, 1988.
765. **Kulkarni, R.K. and Straus, D.S.,** Insulin-mediated phosphorylation of ribosomal protein S6 in mouse melanoma cells and melanoma x fibroblast hybrid cells in relation to cell proliferation, *Biochim. Biophys. Acta,* 762, 542, 1983.
766. **Kadowaki, T., Koyasu, S., Nishida, E., Sakai, H., Takaku, F., Yahara, I., and Kasuga, M.,** Insulin-like growth factors, insulin, and epidermal growth factor cause rapid cytoskeletal reorganization in KB cells. Clarification of the roles of type I insulin-like growth factor receptors and insulin receptors, *J. Biol. Chem.,* 261, 16141, 1986.
767. **Akiyama, T., Kadowaki, T., Nishida, E., Kadooka, T., Ogawara, H., Fukami, Y., Sakai, H., Takaku, F., and Kasuga, M.,** Substrate specificities of tyrosine-specific protein kinases toward cytoskeletal proteins *in vitro, J. Biol. Chem.,* 261, 14797, 1986.

768. **Tsao, M.-S., Earp, H.S., and Grisham, J.W.,** The effects of epidermal growth factor and the state of confluence on enzymatic activities of cultured rat liver epithelial cells, *J. Cell. Physiol.,* 126, 167, 1986.
769. **Reiss, N., Kanety, H., and Schlessinger, J.,** Five enzymes of the glycolytic pathway serve as substrates for purified epidermal-growth-factor-receptor kinase, *Biochem. J.,* 239, 691, 1986.
770. **Warburg, O.,** On the origin of cancer cells, *Science,* 123, 309, 1956.
771. **Moule, S.K. and McGivan, J.D.,** Epidermal growth factor stimulates the phosphorylation of pyruvate kinase in freshly isolated rat hepatocytes, *FEBS Lett.,* 280, 37, 1991.
772. **Lamy, F., Wilkin, F., Baptist, M., Posada, J., Roger, P.P., and Dumont, J.E.,** Phosphorylation of mitogen-activated protein kinases is involved in the epidermal growth factor and phorbol ester, but not in the thyrotropin/cAMP, thyroid mitogenic pathway, *J. Biol. Chem.,* 268, 4398, 1993.
773. **Polet, H. and Fryxell, D.,** Effects of epidermal growth factor on protein degradation, the translocation of non-histone proteins to the nucleus and DNA synthesis, *Biochim. Biophys. Acta,* 1013, 279, 1989.
774. **Baldwin, G.S., Knesel, J., and Monckton, J.M.,** Phosphorylation of gastrin-17 by epidermal growth factor-stimulated tyrosine kinase, *Nature (London),* 301, 435, 1983.
775. **Baldwin, G.S., Grego, B., Hearn, M.T.W., Knesel, J.A., Morgan, F.J., and Simpson, R.J.,** Phosphorylation of human growth hormone by the epidermal growth factor-stimulated tyrosine kinase, *Proc. Natl. Acad. Sci. U.S.A.,* 80, 5276, 1983.
776. **Rao, K.V.S. and Fox, C.F.,** Epidermal growth factor stimulates tyrosine phosphorylation of human glucocorticoid receptor in cultured cells, *Biochem. Biophys. Res. Commun.,* 144, 512, 1987.
777. **Woo, D.D.L., Fay, S.P., Griest, R., Coty, W., Goldfine, I., and Fox, C.F.,** Differential phosphorylation of the progesterone receptor by insulin, epidermal growth factor, and platelet-derived growth factor receptor tyrosine protein kinases, *J. Biol. Chem.,* 261, 460, 1986.
778. **Wiley, H.S. and Kaplan, J.,** Epidermal growth factor rapidly induces a redistribution of transferrin receptor pools in human fibroblasts, *Proc. Natl. Acad. Sci. U.S.A.,* 81, 7456, 1984.
779. **Castagnola, J., MacLeod, C., Sunada, H., Mendelsohn, J., and Taetle, R.,** Effects of epidermal growth factor on transferrin receptor phosphorylation and surface expression in malignant epithelial cells, *J. Cell. Physiol.,* 132, 492, 1987.
780. **Sahai, A., Feuerstein, N., Cooper, H.L., and Salomon, D.S.,** Effect of epidermal growth factor and 12-*O*-tetradecanoylphorbol-13-acetate on the phosphorylation of soluble acidic proteins in A431 epidermoid carcinoma cells, *Cancer Res.,* 46, 4143, 1986.
781. **Nakamura, K.D., Martinez, R., and Weber, M.J.,** Tyrosine phosphorylation of specific proteins after mitogen stimulation of chicken embryo fibroblasts, *Mol. Cell. Biol.,* 3, 380, 1983.
782. **Vila, J. and Weber, M.J.,** Mitogen-stimulated tyrosine phosphorylation of a 42-kD cellular protein: evidence for a protein kinase-C requirement, *J. Cell. Physiol.,* 135, 285, 1988.
783. **Contor, L., Lamy, F., Lecocq, R., Roger, P.P., and Dumont, J.E.,** Differential protein phosphorylation in induction of thyroid cell proliferation by thyrotropin, epidermal growth factor, or phorbol ester, *Mol. Cell. Biol.,* 8, 2494, 1988.
784. **Baccarini, M., Gill, G.N., and Stanley, E.R.,** Epidermal growth factor stimulates phosphorylation of RAF-1 independently of receptor autophosphorylation and internalization, *J. Biol. Chem.,* 266, 10941, 1991.
785. **Baldwin, G.S., Stanley, I.J., and Nice, E.C.,** A synthetic peptide containing the autophosphorylation site of the transforming protein of Harvey sarcoma virus is phosphorylated by the EGF-stimulated tyrosine kinase, *FEBS Lett.,* 153, 257, 1983.
786. **Kamata, T. and Feramisco, J.R.,** Epidermal growth factor stimulates guanine nucleotide binding activity and phosphorylation of *ras* oncogene proteins, *Nature (London),* 310, 147, 1984.
787. **Matrisian, L.M., Rautmann, G., Magun, B.E., and Breatnach, R.,** Epidermal growth factor or serum stimulation of rat fibroblasts induces an elevation in mRNA levels for lactate dehydrogenese and other glycolytic enzymes, *Nucleic Acids Res.,* 13, 711, 1985.
788. **Binas, B. and Grosse, R.,** Demonstration of an epidermal growth factor-dependent 58 kDa phosphoprotein secreted by rat kidney fibroblasts, *FEBS Lett.,* 213, 164, 1987.
789. **Ghezzo, F., Lauret, E., Ferrari, S., and Baserga, R.,** Growth factor regulation of the promoter for calcyclin, a growth-regulated gene, *J. Biol. Chem.,* 263, 4758, 1988.
790. **Courtney, M.G., Schmidt, L.J., and Getz, M.J.,** Organization and expression of endogenous virus-like (VL-30) DNA sequences in nontransformed and chemically transformed mouse embryo cells in culture, *Cancer Res.,* 42, 569, 1982.

791. **Foster, D.N., Schmidt, L.J., Hodgson, C.P., Moses, H.L., and Getz, M.J.,** Polyadenylated RNA complementary to a mouse retrovirus-like multigene family is rapidly and specifically induced by epidermal growth factor stimulation of quiescent cells, *Proc. Natl. Acad. Sci. U.S.A.,* 79, 7317, 1982.
792. **Rodland, K.D., Muldoon, L.L., Dinh, T.-H., and Magun, B.E.,** Independent transcriptional regulation of a single VL30 element by epidermal growth factor and activators of protein kinase C, *Mol. Cell. Biol.,* 8, 2247, 1988.
793. **Gomperts, M., Pascall, J.C., and Brown, K.D.,** The nucleotide sequence of a cDNA encoding an EGF-inducible gene indicates the existence of a new family of mitogen-induced genes, *Oncogene,* 5, 1081, 1990.
794. **Matrisian, L.M., Leroy, P., Ruhlmann, C., Gesnel, M.-C., and Breathnach, R.,** Isolation of the oncogene and epidermal growth factor-induced transin gene: complex control in rat fibroblasts, *Mol. Cell. Biol.,* 6, 1679, 1986.
795. **Kerr, L.D., Holt, J.T., and Matrisian, L.M.,** Growth factors regulate transin gene expression by c-*fos*-dependent and c-*fos*-independent pathways, *Science,* 242, 1424, 1988.
796. **Machida, C.M., Muldoon, L.L., Rodland, K.D., and Magun, B.E.,** Transcriptional modulation of transin gene expression by epidermal growth factor and transforming growth factor β, *Mol. Cell. Biol.,* 8, 2479, 1988.
797. **Breathnach, R., Matrisian, L.M., Gesnel, M.-C., and Leroy, P.,** Sequences coding for part of oncogene-induced transin are highly conserved in a related rat gene, *Nucleic Acids Res.,* 15, 1139, 1987.
798. **Matrisian, L.M., Bowden, G.T., Krieg, P., Fürstenberger, G., Briand, J.-P., Leroy, P., and Breathnach, R.,** The mRNA coding for the secreted protease transin is expressed more abundantly in malignant than in benign tumors, *Proc. Natl. Acad. Sci. U.S.A.,* 83, 9413, 1986.
799. **Elder, P.K., Schmidt, L.J., Ono, T., and Getz, M.J.,** Specific stimulation of actin gene transcription by epidermal growth factor and cycloheximide, *Proc. Natl. Acad. Sci. U.S.A.,* 81, 7476, 1984.
800. **Blatti, S.P., Foster, D.N., Ranganathan, G., Moses, H.L., and Getz, M.J.,** Induction of fibronectin gene transcription and mRNA is a primary response to growth-factor stimulation of AKR-2B cells, *Proc. Natl. Acad. Sci. U.S.A.,* 85, 1119, 1988.
801. **Maness, P.F. and Walsh, R.C., Jr.,** Dihydrocytochalasin B disorganizes actin cytoarchitecture and inhibits initiation of DNA synthesis in 3T3 cells, *Cell,* 30, 253, 1982.
802. **Laverdure, G.R., Banerjee, D., Chackalaparampil, I., and Mukherjee, B.B.,** Epidermal and transforming growth factors modulate secretion of a 69 kDa phosphoprotein in normal rat kidney fibroblasts, *FEBS Lett.,* 222, 261, 1987.
803. **Lewis, E.J. and Chikaraishi, D.M.,** Regulated expression of the tyrosine hydroxylase gene by epidermal growth factor, *Mol. Cell. Biol.,* 7, 3332, 1987.
804. **Kido, H., Fukusen, N., and Katunuma, N.,** Epidermal growth factor as a new regulator of induction of tyrosine aminotransferase and tryptophan oxygenases by glucocorticoids, *FEBS Lett.,* 223, 223, 1987.
805. **Roger, P.P., Van Heuverswyn, B., Lambert, C., Reuse, S., Vassart, G., and Dumont, J.E.,** Antagonistic effects of thyrotropin and epidermal growth factor on thyroglobulin mRNA level in cultured thyroid cells, *Eur. J. Biochem.,* 152, 239, 1985.
806. **Vonderhaar, B.K. and Nakhasi, H.L.,** Bifunctional activity of epidermal growth factor on alpha- and kappa-casein gene expression in rodent mammary glands *in vitro, Endocrinology,* 119, 1178, 1986.
807. **Arakawa, M., Perry, J.W., Cossu, M.F., and Oka, T.,** Further characterization of the inhibition of casein production in a primary mouse mammary epithelial cell culture by epidermal growth factor, *Exp. Cell Res.,* 158, 111, 1985.
808. **Elsholtz, H.P., Mangalam, H.J., Potter, E., Albert, V.R., Supowit, S., Evans, R.M., and Rosenfeld, M.G.,** Two different *cis*-active elements transfer the transcriptional effects of both EGF and phorbol esters, *Science,* 234, 1552, 1986.
809. **Miskimins, R., Miskimins, W.K., Bernstein, H., and Shimizu, N.,** Epidermal growth factor-induced topoisomerase(s): intracellular translocation and relation to DNA synthesis, *Exp. Cell Res.,* 146, 53, 1983.
810. **Basu, M., Frick, K., Sen-Majumdar, A., Scher, C.D., and Das, M.,** EGF receptor-associated DNA-nicking activity is due to a M_r-100,000 dissociable protein, *Nature (London),* 316, 640, 1985.
811. **Ranganathan, G. and Getz, M.J.,** Cooperative stimulation of specific gene transcription by epidermal growth factor and transforming growth factor type β1, *J. Biol. Chem.,* 265, 3001, 1990.

812. **Heldin, N.-E. and Westermark, B.,** Epidermal growth factor, but not thyrotropin stimulates the expression of c-*fos* and c-*myc* messenger ribonucleic acid in porcine thyroid follicle cells in primary culture, *Endocrinology,* 122, 1042, 1988.
813. **Cutry, A.F., Kinniburgh, A.J., Twardzik, D.R., and Wenner, C.E.,** Transforming growth factor α (TGF α) induction of c-fos and c-myc expression in C3H 10T1/2 cells, *Biochem. Biophys. Res. Commun.,* 152, 216, 1988.
814. **Müller, R., Bravo, R., Burckhardt, J., and Curran, T.,** Induction of c-*fos* gene and protein by growth factors precedes activation of c-*myc, Nature (London),* 312, 716, 1984.
815. **Bravo, R., Macdonald-Bravo, H., Müller, R., Hübsch, D., and Almendral, J.M.,** Bombesin induces c-*fos* and c-*myc* expression in quiescent Swiss 3T3 cells: comparative study with other mitogens, *Exp. Cell Res.,* 170, 103, 1987.
816. **Lu, K., Levine, R.A., and Campisi, J.,** c-*ras*-Ha gene expression is regulated by insulin or insulinlike growth factor and epidermal growth factor in murine fibroblasts, *Mol. Cell. Biol.,* 9, 3411, 1989.
817. **Sagar, S.M., Edwards, R.H., and Sharp, F.R.,** Epidermal growth factor and transforming growth factor α induce c-*fos* gene expression in retinal Muller cells *in vivo, J. Neurosci. Res.,* 29, 549, 1991.
818. **Liboi, E., Pelosi, E., Testa, U., Peschle, C., and Rossi, G.B.,** Proliferative response and oncogene expression induced by epidermal growth factor in EL2 rat fibroblasts, *Mol. Cell. Biol.,* 6, 2275, 1986.
819. **Di Francesco, P. and Liboi, E.,** Role of the c-*fos* gene expression on the mitogenic response in EL2 rat fibroblasts, *J. Tissue React.,* 5, 311, 1988.
820. **Quantin, B. and Breathnach, R.,** Epidermal growth factor stimulates transcription of the c-*jun* proto-oncogene in rat fibroblasts, *Nature (London),* 334, 538, 1988.
821. **McDonnell, S.E., Kerr, L.D., and Matrisian, L.M.,** Epidermal growth factor stimulation of stromelysin mRNA in rat fibroblasts requires induction of proto-oncogenes c-*fos* and c-*jun* and activation of protein kinase C, *Mol. Cell. Biol.,* 10, 4284, 1990.
822. **den Hertog, J., de Groot, R.P., de Laat, S.W., and Kruijer, W.,** EGF-induced *jun B*-expression in transfected P19 embryonal carcinoma cells expressing EGF receptors is dependent on Jun D, *Nucleic Acids Res.,* 20, 125, 1992.
823. **Fernandez-Pol, J.A., Talkad, V.D., Klos, D.J., and Hamilton, P.D.,** Suppression of the EGF-dependent induction of c-myc proto-oncogene expression by transforming growth factor β in a human breast carcinoma cell line, *Biochem. Biophys. Res. Comm.,* 144, 1197, 1987.
824. **Fernandez-Pol, J.A., Hamilton, P.D., and Klos, D.J.,** Transcriptional regulation of proto-oncogene expression by epidermal growth factor, transforming growth factor β1, and triiodothyronine in MDA-468 cells, *J. Biol. Chem.,* 264, 4151, 1989.
825. **Greenberg, M.E., Greene, L.A., and Ziff, E.B.,** Nerve growth factor and epidermal growth factor induce rapid transient changes in proto-oncogene transcription in PC12 cells, *J. Biol. Chem.,* 260, 14101, 1985.
826. **Bravo, R., Burckhardt, J., Curran, T., and Müller, R.,** Stimulation and inhibition of growth by EGF in different A431 cell clones is accompanied by the rapid induction of c-*fos* and c-*myc* proto-oncogenes, *EMBO J.,* 4, 1193, 1985.
827. **Filmus, J., Benchimol, S., and Buick, R.N.,** Comparative analysis of the involvement of p53, c-*myc* and c-*fos* in epidermal growth factor-mediated signal transduction, *Exp. Cell Res.,* 169, 554, 1987.
828. **Black, R.J., Yu, Z.-P., Brown, D., and Chang, E.,** Modulation of oncogene expression by epidermal growth factor and γ-interferon in A431 squamous carcinoma cells, *J. Biol. Regul. Homeostat. Agents,* 2, 35, 1988.
829. **Hauguel-DeMouzon, S., Csermely, P., Zoppini, G., and Kahn, C.R.,** Quantitative dissociation between EGF effects on c-*myc* and c-*fos* gene expression, DNA synthesis, and epidermal growth factor receptor tyrosine kinase activity, *J. Cell. Physiol.,* 150, 180, 1992.
830. **McCaffrey, P., Ran, W., Campisi, J., and Rosner, M.R.,** Two independent growth factor-generated signals regulate c-*fos* and c-*myc* mRNA levels in Swiss 3T3 cells, *J. Biol. Chem.,* 262, 1442, 1987.
831. **Tsuda, T., Hakomori, Y., Fukumoto, Y., Kaibuchi, K., and Takai, Y.,** Epidermal growth factor increases c-*myc* mRNA without eliciting phosphoinositide turnover, protein kinase C activation, or calcium ion mobilization in Swiss 3T3 fibroblasts, *J. Biochem.,* 100, 1631, 1986.
832. **Franklin, C.C. and Kraft, A.S.,** Protein kinase C-independent activation of c-*jun* and c-*fos* transcription by epidermal growth factor, *Biochim. Biophys. Acta,* 1134, 137, 1992.
833. **Ran, W., Dean, M., Levine, R.A., Henkle, C., and Campisi, J.,** Induction of c-*fos* and c-*myc* mRNA by epidermal growth factor or calcium ionophore is cAMP dependent, *Proc. Natl. Acad. Sci. U.S.A.,* 83, 8216, 1986.

834. **Skouteris, G.G. and Kaser, M.R.,** Prostaglandins E_2 and F_{2a} mediate the increase in c-*myc* expression induced by EGF in primary rat hepatocyte cultures, *Biochem. Biophys. Res. Commun.*, 178, 1240, 1991.

835. **Prywes, R. and Roeder, R.G.,** Inducible binding of a factor to the c-*fos* enhancer, *Cell*, 47, 777, 1986.

836. **Mahadevan, L.C., Wills, A.J., Hirst, E.A., Rathjen, P.D., and Heath, J.K.,** 2-Aminopurine abolishes EGF- and TPA-stimulated pp33 phosphorylation and c-*fos* induction without affecting the activation of protein kinase C, *Oncogene*, 5, 327, 1990.

837. **Fisch, T.M., Prywes, R., and Roeder, R.G.,** An AP1-binding site in the c-*fos* gene can mediate induction by epidermal growth factor and 12-*O*-tetradecanoylphorbol-13-acetate, *Mol. Cell. Biol.*, 9, 1327, 1989.

838. **Taetle, R., Castagnola, J., and MacLeod, C.,** Induction of c-fos by an anti-epidermal growth factor receptor monoclonal antibody, *Biochem. Biophys. Res. Commun.*, 158, 129, 1989.

839. **Kost, D.P. and Michalopoulos, G.K.,** Effect of epidermal growth factor on the expression of proto-oncogenes c-myc and c-Ha-ras in short-term primary hepatocyte culture, *J. Cell. Physiol.*, 144, 122, 1990.

840. **Di Francesco, P., Favalli, C., and Liboi, E.,** Secreted proteins induced by epidermal growth factor and transforming growth factor β in EL2 rat fibroblasts. Role in the mitogenic response, *Cell Biol. Int. Rep.*, 12, 365, 1988.

841. **Brown, G.L., Curtsinger, L., III, Brightwell, J.R., Ackerman, D.M., Tobin, G.R., Polk, H.C., Jr., George-Nascimento, C., Valenzuela, P., and Schultz, G.S.,** Enhancement of epidermal regeneration by biosynthetic epidermal growth factor, *J. Exp. Med.*, 163, 1319, 1986.

842. **Reilly, C.F., Fritze, L.M.S., and Rosenberg, R.D.,** Antiproliferative effects of heparin on vascular smooth muscle cells are reversed by epidermal growth factor, *J. Cell. Physiol.*, 131, 149, 1987.

843. **Polet, H.,** Epidermal growth factor stimulates DNA synthesis while inhibiting cell multiplication of A-431 carcinoma cells, *Exp. Cell Res.*, 186, 390, 1990.

844. **Blay, J. and Brown, K.D.,** Contradistinctive growth responses of cultured rat intestinal epithelial cells to epidermal growth factor depending on cell population density, *J. Cell. Physiol.*, 129, 343, 1986.

845. **Matsuda, T., Okamura, K., Sato, Y., Morimoto, A., Ono, M., Kohno, K., and Kuwano, M.,** Decreased response to epidermal growth factor during cellular senescence in cultured human microvascular endothelial cells, *J. Cell. Physiol.*, 150, 510, 1992.

846. **Kinzel, V., Kaszkin, M., Blume, A., and Richards, J.,** Epidermal growth factor inhibits transiently the progression from G_2-phase to mitosis: a receptor-mediated phenomenon in various cells, *Cancer Res.*, 50, 7932, 1990.

847. **Eckl, P.M., Whitcomb, W.R., Michalopoulos, G., and Jirtle, R.L.,** Effects of EGF and calcium on adult parenchymal hepatocyte proliferation, *J. Cell. Physiol.*, 132, 363, 1987.

848. **Zendegui, J.G., Inman, W.H., and Carpenter, G.,** Modulation of the mitogenic response of an epidermal growth factor-dependent keratinocyte cell line by dexamethasone, insulin, and transforming growth factor-β, *J. Cell. Physiol.*, 136, 257, 1988.

849. **Ignar-Trowbridge, D.M., Nelson, K.G., Bidwell, M.C., Curtis, S.W., Washburn, T.F., McLachlan, J.A., and Korach, K.S.,** Coupling of dual signaling pathways: epidermal growth factor action involves the estrogen receptor, *Proc. Natl. Acad. Sci. U.S.A.*, 89, 4658, 1992.

850. **Dubeau, L. and Jones, P.A.,** Growth of normal and neoplastic urothelium and response to epidermal growth factor in a defined serum-free medium, *Cancer Res.*, 47, 2107, 1987.

851. **Serrero, G.,** EGF inhibits the differentiation of adipocyte precursors in primary culture, *Biochem. Biophys. Res. Commun.*, 146, 194, 1987.

852. **MacLeod, C.L., Luk, A., Castagnola, J., Cronin, M., and Mendelsohn, J.,** EGF induces cell cycle arrest of A431 human epidermoid carcinoma cells, *J. Cell. Physiol.*, 127, 175, 1986.

853. **Bishop, R., Martinez, R., Weber, M.J., Blackshear, P.J., Beatty, S., Lim, R., and Herschman, H.R.,** Protein phosphorylation in a tetradecanoylphorbol acetate-nonproliferative variant of 3T3 cells, *Mol. Cell. Biol.*, 5, 2231, 1985.

854. **McKeehan, W.L. and McKeehan, K.A.,** Epidermal growth factor modulates extracellular Ca^{2+} requirement for multiplication of normal human skin fibroblasts, *Exp. Cell Res.*, 123, 397, 1979.

855. **Lechner, J.F.,** Interdependent regulation of epithelial cell replication by nutrients, hormones, growth factors, and cell density, *Fed. Proc.*, 43, 116, 1984.

856. **Stastny, M. and Cohen, S.,** Epidermal growth factor. IV. The induction of ornithine decarboxylase, *Biochim. Biophys. Acta*, 204, 578, 1970.

857. **Bravo, R.,** Epidermal growth factor inhibits the synthesis of the nuclear protein cyclin in A431 human carcinoma cells, *Proc. Natl. Acad. Sci. U.S.A.,* 81, 4848, 1984.
858. **Nilsen-Hamilton, M. and Holley, R.W.,** Rapid selective effects by a growth inhibitor and epidermal growth factor on the incorporation of (^{35}S)methionine into protein secreted by African green monkey (BSC-1) cells, *Proc. Natl. Acad. Sci. U.S.A.,* 80, 5636, 1983.
859. **Hosokawa, M., Phillips, P.D., and Cristofalo, V.J.,** The effect of dexamethasone on epidermal growth factor binding and stimulation of proliferation in young and senescent WI38 cells, *Exp. Cell Res.,* 164, 408, 1986.
860. **Tomomura, A., Sawada, N., Sattler, G.L., Kleinman, H.K., and Pitot, H.C.,** The control of DNA synthesis in primary cultures from adult and young rats: interactions of extracellular matrix components, epidermal growth factor, and the cell cycle, *J. Cell. Physiol.,* 130, 221, 1987.
861. **Sand, T.-E. and Christoffersen, T.,** Temporal requirement for epidermal growth factor and insulin in the stimulation of hepatocyte DNA synthesis, *J. Cell. Physiol.,* 131, 141, 1987.
862. **Pietrzkowski, Z., Sell, C., Lammers, R., Ullrich, A., and Baserga, R.,** Roles of insulinlike growth factor 1 (IGF-1) and the IGF-1 receptor in epidermal growth factor-stimulated growth of 3T3 cells, *Mol. Cell. Biol.,* 12, 3883, 1992.
863. **Lin, S.L., Ts'o, P.O.P., and Hollenberg, M.D.,** The effects of interferon on epidermal growth factor action, *Biochem. Biophys. Res. Commun.,* 96, 168, 1980.
864. **Cruise, J.L. and Michalopoulos, G.,** Norepinephrine and epidermal growth factor: dynamics of their interaction in the stimulation of hepatocyte DNA synthesis, *J. Cell. Physiol.,* 125, 45, 1985.
865. **McKay, I.A., Marshall, C.J., Calès, C., and Hall, A.,** Transformation and stimulation of DNA synthesis in NIH-3T3 cells are titratable function of normal p21$^{N\text{-}ras}$ expression, *EMBO J.,* 5, 2617, 1986.
866. **Luttrell, D.K., Luttrell, L.M., and Parsons, S.J.,** Augmented mitogenic responsiveness to epidermal growth factor in murine fibroblasts that overexpress pp60$^{c\text{-}src}$, *Mol. Cell. Biol.,* 8, 497, 1988.
867. **Wilson, L.K., Luttrell, D.K., Parsons, J.T., and Parsons, S.J.,** pp60$^{c\text{-}src}$ tyrosine kinase, myristylation, and modulatory domains are required for enhanced mitogenic responsiveness to epidermal growth factor seen in cells overexpressing c-*src*, *Mol. Cell. Biol.,* 9, 1536, 1989.
868. **Weissman, B.E. and Aaronson, S.A.,** BALB and Kirsten murine sarcoma viruses alter growth and differentiation of EGF-dependent BALB/c mouse epidermal keratinocyte lines, *Cell,* 32, 599, 1983.
869. **Weissman, B. and Aaronson, S.A.,** Members of the *src* and *ras* oncogene families supplant the epidermal growth factor requirement of BALB/MK-2 keratinocytes and induce distinct alterations in their terminal differentiation process, *Mol. Cell. Biol.,* 5, 3386, 1985.
870. **Singletary, S.E., Frappaz, D., Tucker, S.L., Larry, L., Brock, W.A., and Spitzer, G.,** Epidermal growth factor effect on serum-free growth of primary and metastatic human tumors, *Int. J. Cell Cloning,* 7, 59, 1989.
871. **Heideran, M.A., Fleming, T.P., Bottaro, D.P., Bell, G.I., Di Fiore, P.P., and Aaronson, S.A.,** Transformation of NIH 3T3 fibroblasts by an expression vector for the human epidermal growth factor precursor, *Oncogene,* 5, 1265, 1990.
872. **Mori, K., Ibaragi, S., Kurobe, M., Furukawa, S., and Hayashi, K.,** Synthesis and secretion of an hEGF-like immunoreactive factor by human gastric cancer cells (MKN-45), *Biochem. Int.,* 14, 779, 1987.
873. **Mori, K., Ibaragi, S., Kurobe, M., Furukawa, S., and Hayashi, K.,** Production of an hEGF-like immunoreactive factor by human gastric cancer cells depends on differentiational state of the cells, *Biochem. Biophys. Res. Commun.,* 145, 1019, 1987.
874. **Tahara, E., Sumiyoshi, H., Hata, J., Yasui, W., Taniyama, K., Hayashi, T., Nagae, S., and Sakamoto, S.,** Human epidermal growth factor in gastric carcinoma as a biological marker of high malignancy, *Jpn. J. Cancer Res.,* 77, 145, 1986.
875. **Murphy, L.C., Murphy, L.J., Dubik, D., Bell, G.I., and Shiu, R.P.C.,** Epidermal growth factor gene expression in human breast cancer cells: regulation of expression by progestins, *Cancer Res.,* 48, 4555, 1988.
876. **Dickson, R.B., Huff, K.K., Spencer, E.M., and Lippman, M.E.,** Induction of epidermal growth factor-related polypeptides by 17 β-estradiol in MCF-7 human breast cancer cells, *Endocrinology,* 118, 138, 1986.
877. **Mori, K., Kurobe, M., Furukawa, S., Kubo, K., and Hayashi, K.,** Human breast cancer cells synthesize and secrete an EGF-like immunoreactive in culture, *Biochem. Biophys. Res. Commun.,* 136, 300, 1986.

878. **Mori, K., Yoshimura, T., Ibaragi, S., Kurobe, M., Furukawa, S., Kurosawa, K., Katoh, M., Tanaka, S., and Hayashi, K.,** Aberrant synthesis and secretion of a human epidermal growth factor-like immunoreactive factor by human breast cancer cells, *J. Clin. Biochem. Nutr.,* 4, 49, 1988.
879. **Dotzlaw, H., Miller, T., Karvelas, J., and Murphy, L.C.,** Epidermal growth factor gene expression in human breast cancer biopsy samples: relationship to estrogen and progesterone receptor gene expression, *Cancer Res.,* 50, 4204, 1990.
880. **Ethier, S.P. and Moorthy, R.,** Multiple growth factor independence in rat mammary carcinoma cells, *Breast Cancer Res. Treat.,* 18, 73, 1991.
881. **Ethier, S.P., Moorthy, R., and Dilts, C.A.,** Secretion of an epidermal growth factor-like growth factor by epidermal growth factor-independent rat mammary carcinoma cells, *Cell Growth Differ.,* 2, 593, 1991.
882. **Henrard, D.R., Thornley, A.T., Brown, M.L., and Rheinwald, J.G.,** Specific effects of *ras* oncogene expression on the growth and histogenesis of human epidermal keratinocytes, *Oncogene,* 5, 475, 1990.
883. **Leof, E.B., Lyons, R.M., Cunningham, M.R., and O'Sullivan, D.,** Mid-G_1 arrest and epidermal growth factor independence of *ras*-transfected mouse cells, *Cancer Res.,* 49, 2356, 1989.
884. **Tubo, R.A. and Rheinwald, J.G.,** Normal human mesothelial cells and fibroblasts transfected with *EJras* oncogene become EGF-independent, but are not malignantly transformed, *Oncogene Res.,* 1, 407, 1987.
885. **Reiss, M., Dibble, C.L., and Narayanan, R.,** Transcriptional activation of the c-*myc* proto-oncogene in murine keratinocytes enhances the response to epidermal growth factor, *J. Invest. Dermatol.,* 93, 136, 1989.
886. **Muakkassah-Kelly, S.F., Jans, D.A., Lydon, N., Bieri, F., Waechter, F., Bentley, P., and Stäubli, W.,** Electroporation of cultured rat hepatocytes with the c-*myc* gene potentiates DNA synthesis in response to epidermal growth factor, *Exp. Cell Res.,* 178, 296, 1988.
887. **Kelsey, K.T., Nagasawa, H., Umans, R.S., and Little, J.B.,** Epidermal growth factor induces cytogenetic damage in mammalian cells, *Carcinogenesis,* 8, 625, 1987.
888. **Messing, E.M., Hanson, P., Ulrich, P., and Erturk, E.,** Epidermal growth factor-interactions with normal and malignant urothelium: *in vivo* and *in situ* studies, *J. Urol.,* 138, 1329, 1987.
889. **Fisher, P.B., Mufson, R.A., Weinstein, I.B., and Little, J.B.,** Epidermal growth factor, like tumor promoters, enhances viral and radiation-induced cell transformation, *Carcinogenesis,* 2, 183, 1981.
890. **Herschman, H.R. and Brankow, D.,** Interaction of epidermal growth factor with initiators and complete carcinogens in the C3H10T1/2 cell culture system, *J. Cell. Biochem.,* 28, 1, 1985.
891. **Koyasu, S., Kadowaki, T., Nishida, E., Tobe, K., Abe, E., Kasuga, M., Sakai, H., and Yahara, I.,** Alteration in growth, cell morphology, and cytoskeletal structures of KB cells induced by epidermal growth factor and transforming growth factor-β, *Exp. Cell Res.,* 176, 107, 1988.
892. **Indo, K., Matsuoka, T., Nakasho, K., Funahashi, I., and Miyaji, H.,** Reversible induction of anchorage independent growth from normal mouse epidermal keratinocytes, MSK-C3H-NU, in soft agar medium by 12-*O*-tetradecanoylphorbol-13-acetate and epidermal growth factor, *Cancer Res.,* 48, 1566, 1988.
893. **Rose, S.P., Stahn, R., Passovoy, D.S., and Herschman, H.R.,** Epidermal growth factor enhancement of skin tumor induction in mice, *Experientia,* 32, 913, 1976.
894. **Aboud-Pirak, E., Hurwitz, E., Pirak, M.E., Bellot, F., Schlessinger, J., and Sela, M.,** Efficacy of antibodies to epidermal growth factor receptor against KB carcinoma *in vitro* and in nude mice, *J. Natl. Cancer Inst.,* 80, 1605, 1988.
895. **Kurachi, H., Okamoto, S., and Oka, T.,** Evidence for the involvement of the submandibular gland epidermal growth factor in mouse mammary tumorigenesis, *Proc. Natl. Acad. Sci. U.S.A.,* 82, 6940, 1985.
896. **Tsutsumi, O., Tsutsumi, A., and Oka, T.,** Importance of epidermal growth factor in implantation and growth of mouse mammary tumor in female nude mice, *Cancer Res.,* 47, 4651, 1987.
897. **Singletary, S.E., Baker, F.L., Spitzer, G., Tucker, S.L., Tomasovic, B., Brock, W.A., Ajani, J.A., and Kelly, A.M.,** Biological effect of epidermal growth factor on the *in vitro* growth of human tumors, *Cancer Res.,* 47, 403, 1987.
898. **den Hertog, J., de Laat, S.W., Schlessinger, J., and Kruijer, W.,** Neuronal differentiation in response to epidermal growth factor of transfected murine P19 embryonal carcinoma cells expressing human epidermal growth factor receptors, *Cell Growth Differ.,* 2, 155, 1991.

899. **zur Hausen, H.,** Papillomaviruses as carcinomaviruses, *Adv. Viral Oncol.,* 8, 1, 1989.
900. **Yasumoto, S., Taniguchi, A., and Sohma, K.,** Epidermal growth factor (EGF) elicits down-regulation of human papillomavirus type 16 (HPV-16) E6/E7 mRNA at the transcriptional level in an EGF-stimulated human keratinocyte cell line: functional role of EGF-responsive silencer in the HPV-16 long control region, *J. Virol.,* 65, 2000, 1991.

Chapter 4

Fibroblast Growth Factors

I. INTRODUCTION

The fibroblast growth factor (FGF), also called fibroblast-derived growth factor (FDGF) or heparin-binding growth factor (HBGF), was characterized as a 14- to 16-kDa polypeptide with mitogenic properties for fibroblasts as well as for a variety of neuroectoderm- and mesoderm-derived cell types.[1-4] In addition to its mitogenic activity, FGF has an important role in processes related to cellular differentiation during both prenatal and postnatal life. The effects of FGF on cell differentiation processes have been conveniently characterized in certain cellular systems *in vitro*. Differentiation of cultured muscle cells depends on an interaction between the inducing action of factors such as FGF and insulin-like growth factor (IGF) and the inhibitory action of other factors such as transforming growth factor (TGF).[5]

Factors with FGF activity have been characterized by their potent mitogenic capacities and heparin-binding properties. They have been isolated from a number of tissues, including bovine brain and pituitary and adrenal glands.[6-11] Two related forms of FGFs isolated from the bovine brain are the brain-derived growth factors (BDGFs) A and B (BDGF-A and BDGF-B).[12] Other FGF-related factors have been designated eye-derived growth factor (EDGF) and endothelial cell growth factor (ECGF). A factor detected in macrophages, designated macrophage-derived growth factor (MDGF), is a potent mitogen for nonlymphoid mesenchymal cells, including fibroblasts, smooth muscle cells, and endothelial cells; and is probably identical to FGF.[13] However, an MDGF of mol wt 40,000 isolated from mitogen-stimulated human peripheral blood lymphocyte conditioned medium stimulates proliferation of fibroblast and smooth muscle cells but is biochemically different from other known growth factors, including interleukin 1 (IL-1), platelet-derived growth factor (PDGF), and FGF.[14]

FGFs have been purified by using the selective method of heparin-sepharose affinity chromatography, and heparin has been shown to either inactivate or potentiate the biological activities of FGFs, according to the particular conditions *in vitro*.[15] Chick embryo fibroblasts (CEF cells) and adult human FGFs isolated by these methods share the same biochemical properties.[16] Two types of FGF have been distinguished, acidic FGF/HBGF-1 and basic FGF/HBGF-2. Although these mitogens resemble each other with respect to their *in vitro* mitogenic activity on cell types such as fibroblasts and endothelial cells and with respect to certain physicochemical characteristics, the acidic FGF and basic FGF are structurally different.[17] However, crystallographic analysis of acidic FGF and basic FGF indicates that both factors exhibit a three-dimensional folding pattern similar to that observed for IL-1α and IL-1β.[18] The genes coding for basic FGF and acidic FGF are located on different human chromosomes.[19] These genes encode multiple molecular forms of basic and acidic FGF, in which expression can be regulated by serum treatment of human dermal fibroblasts.[20]

II. THE FGF PROTEIN FAMILY

The purification and characterization of over 30 FGF-related factors have demonstrated that they constitute a family of proteins containing at least seven members: acidic FGF/ECGF/HBGF-1/FGF-1, basic FGF/HBGF-2/FGF-2, Int-2/FGF-3, k-FGF/Hst-1/FGF-4, FGF-5, Hst-2/FGF-6, and keratinocyte growth factor (KGF)/FGF-7 (Table 4.1). The members of this family are multipotential factors that display mitogenic, chemotactic, neurotrophic, and angiogenic activities.[21,22] They interact with multiple high- and low-affinity receptors and associate with both heparin and heparan sulfate proteoglycans on cell surfaces as well as in the extracellular matrix. The members of the FGF family play an important role in developmental processes. They function as mitogens and morphogens and are involved in angiogenesis and tissue regeneration. They may also have an important role in oncogenesis, and some of them exhibit transforming potential under specific conditions.

A. THE FGF-1/BASIC FGF PROTEIN

Basic FGF is a mitogen for many types of cells of mesodermal or neuroectodermal origin. Basic FGF is protein with angiogenic properties and is a potent mitogen and chemoattractant for capillary endothelial cells, stimulating angiogenesis *in vivo*.[23] Capillary endothelial cells express basic FGF, which acts as a

Table 4.1 **Fibroblast Growth Factors (FGFs) and Their Receptors (FGFRs)**

	Synonyms
FGF	
FGF-1	Acidic FGF
	Endothelial cell growth factor (ECGF)
	Heparin-binding growth factor 1 (HBGF-1)
FGF-2	Basic FGF
	Heparin-binding growth factor 2 (HBGF-2)
FGF-3	Int-2 protein
FGF-4	Hst-1 protein
	k-FGF
FGF-5	
FGF-6	Hst-2 protein
FGF-7	Keratinocyte growth factor (KGF)
FGFR	
FGFR-1	Flg protein
	N-Sam protein
FGFR-2	Bek protein
	K-Sam protein
FGFR-3	Sam-3 protein
FGFR-4	TKF protein

mitogen by promoting the growth of these cells under certain physiological conditions.[24] However, studies with monoclonal antibodies against basic FGF suggest that it is not essential as an autocrine or paracrine factor for angiogenesis *in vivo.*[25] Several other growth factors possess angiogenic properties, including angionenin, TGF-α, TGF-β, and tumor necrosis factor-α (TNF-α).

1. Production of Basic FGF

Basic FGF has been purified from a wide variety of normal tissues, including pituitary, brain, hypothalamus, retina, adrenal gland, thymus, ovary, kidney, and placenta; as well as from various types of tumors.[26] In cultured cells, the factor is detected only in cell extracts and is apparently not secreted into the culture medium. The lack of secretion of basic FGF by cells producing the factor is probably due to the absence of a secretory signal peptide sequence in the protein. However, basic FGF is secreted by some types of cells. Human keratinocytes produce basic FGF *in vitro,* and it seems to be responsible for maintaining the viability of melanocytes cultured in the presence of keratinocytes.[27] Ultraviolet irradiation of the skin may stimulate the production of basic FGF by the proliferating keratinocyte population *in vivo,* which would regulate the proliferation of melanocytes through a paracrine type of mechanism. There is evidence that basic FGF can be released by certain cells via a pathway independent of the endoplasmic reticulum-Golgi complex.[28] In certain types of cells, the synthesis of basic FGF transcripts may not result in the production of the protein. Expression of RNA transcripts for basic FGF in human melanocytes may not result in expression of basic FGF protein in the cells, suggesting a tight regulation of basic FGF mRNA translation in these cells.[29]

Other hormones and growth factors may contribute to regulation of the synthesis of basic FGF in different biological systems. Production of basic FGF by certain human endometrial carcinoma cell lines is stimulated by estradiol, and this stimulation is inhibited by progesterone.[30] Factors involved in regulating the production of basic FGF have been identified in the human astrocytoma cell line U87-MG.[31] These cells express two major basic FGF mRNA transcripts of 7.0 and 3.7 kb; and this expression is stimulated by treatment with phorbol ester, diacylglycerol analogue, or PDGF. The results suggest that protein kinase C is involved in the regulation of basic FGF mRNA expression and that some of the mitogenic effects of PDGF and phorbol esters may be mediated by a secondary increase in basic FGF expression.

2. Structure and Synthesis of Basic FGF

cDNA clones encoding human and bovine basic FGF have been isolated and expressed in *Escherichia coli* and the complete primary structure of the 146-amino acid basic FGF proteins have been determined.[27,32-35] The three-dimensional structure of human basic FGF was also determined.[36,37] Basic FGF is a structural homologous of IL-1β. The sequences of human and bovine basic FGF are remarkably homologous, differing in only two amino acids, which implicates a strong selection for maintenance of structure-associated functions. The gene encoding basic FGF is located on human chromosome region 4q26-q27.[38]

Basic FGF activity may correspond to several biologically active protein species generated by differential processing within the cells.[39] Amino-terminal extended forms of basic FGF have been detected in bovine brain and human placenta.[40,41] Recent evidence indicates that basic FGF activity produced by the human hepatoma cell line SK-HEP-1 is composed of four distinct molecules (17.8, 22.5, 23.1, and 24.2 kDa) which are generated by differential initiation of translation.[42] Whereas the synthesis of 17.8-kDa basic FGF protein is translationally initiated at the AUG codon of the mRNA, the synthesis of the other three molecular species of basic FGF unusually begins at non-AUG translation initiation codons. There may be tissue-specific variation in the biosynthesis of basic FGF. Normal fibroblasts cultured from human dermal foreskin synthesize four distinct mRNAs (7.0, 3.7, 2.2, and 1.5 kb) which hybridize to a specific basic FGF cDNA probe.[43] Whereas in bovine hypothalamus basic FGF is encoded by a single 5.0-kb mRNA, in a human hepatoma cell line the factor is encoded by two mRNAs of 4.6 and 2.2 kb.[33] Both the bovine and the human factors are synthesized with short amino-terminal extensions that are not found in the mature forms of the protein, and neither basic nor acidic FGF has a classical signal peptide. A factor isolated from benign hyperplastic tissue of the human prostate has an amino acid sequence which is identical to that of human basic FGF with the exception of an extension of eight amino acid residues located at the amino-terminal region of the molecule.[44] A 25-kDa form of basic FGF has been isolated from guinea pig brain along with the typical 18-kDa form.[45] The biological significance of different molecular forms of basic FGF is unknown.

Basic FGF may be identical to a factor present in several ocular tissues called eye-derived growth factor I (EDGF-I).[46] Basic FGF is identical with astroglial growth factor 2 (AGF-2).[47,48] A human pituitary-derived factor with mitogenic properties in a rabbit fetal chondrocyte bioassay, referred to as chondrocyte growth factor (CGF), is probably also identical to basic FGF.[49] A human hepatoma cell line (SK-HEP-1) synthesizes an endothelial cell mitogen with angiogenic properties which is apparently highly identical to bovine basic FGF.[50] The hepatoma-derived growth factor (HDGF) is a cationic polypeptide of 18,500 to 19,000 mol wt and contains amino acid sequences that are homologous to both amino-terminal and carboxyl-terminal sequences of basic FGF.

B. THE FGF-2/ACIDIC FGF PROTEIN

Acidic FGF was initially identified by its ability to cause the proliferation and delayed differentiation of myoblasts and was rediscovered on the basis of its ability to stimulate endothelial cell proliferation. Acidic FGF stimulates growth in various types of mesenchymal cells, including endothelial cells, chondrocytes, and fibroblasts. Synthesis of DNA in primary cultures of rat hepatocytes is stimulated by acidic FGF.[51] The factor may also have effects on bone tissue. Even at very low concentrations, it stimulates the proliferation of calvaria-derived osteoblastic cells and can maintain, in the presence of additional factors present in serum, the growth of these cells for multiple passages *in vitro*.[52] In the presence of heparin, acidic FGF induces blood vessel growth *in vivo,* and it may be similar or identical to the tumor angiogenic factor which is involved in carcinogenesis, including the transition from hyperplasia to neoplasia.[53]

1. Production of Acidic FGF

Acidic FGF has a distribution which is much more restricted than that of basic FGF. It has been found only in brain and retina. A form of acidic FGF, designated as acidic FGF-1, may be similar or identical to both a factor described as endothelial cell growth factor (ECGF) and another factor called eye-derived growth factor II (EDGF-II).[54] Acidic FGF is probably also identical with factors described under the names of astroglial growth factor-1 (AGF-1) and retinal-derived growth factor (RDGF).[48,55,56] Some established human tumor cell lines also produce heparin-binding growth factors related to bovine acidic FGF.[56] However, some of these factors may not be identical, depending in part on the source of isolation.

2. Structure and Synthesis of Acidic FGF

Acidic FGF exhibits sequence homology with IL-1 and is a potent mitogen for vascular endothelial cells in culture.[57] The complete amino acid sequences of bovine and human brain-derived acidic FGF have been determined.[55,58,59] The organization and structure of the human gene coding for the 155-amino acid human acidic FGF protein was deduced from the isolation of genomic clones.[60] A gene encoding bovine acidic FGF has been chemically synthesized, cloned, and expressed as a biologically active protein in *Escherichia coli*.[61] The primary structures of human and bovine brain-derived acidic FGF differ by 11 out of 140 residues, and all 11 amino acid substitutions between both proteins (92% homology) can be generated by single-base changes. Acidic FGF is a protein of 140 amino acid residues containing in its carboxyl-terminal portion a decapeptide (residues 102 to 111) with significant homology to various biologically active neuropeptides (neuromedin C, neuromedin K, bombesin, substance K, substance P, physalaemin, and eledoisin). At least some of these peptides are mitogenic, and the homology corresponds to a highly conserved region believed to participate in receptor binding at the cellular level.[62] Amino terminal sequences of both acidic and basic FGFs from human brain show extensive homology with each other and with their respective bovine counterparts.[63] The amino-terminal sequences of 25 amino acids of chick embryonic and adult acidic FGF are identical to human acidic FGF, indicating the structural conservation and biological importance of this growth factor in vertebrates.[64]

C. THE FGF-3/INT-2 PROTEIN

The *int*-1 and *int*-2 proto-oncogenes are frequently activated by proviral insertion in mouse mammary tumorigenesis induced by the MMTV retrovirus.[65,66] The *int*-1 gene exhibits no homology with *int*-2; and the protein product of the *int*-2 gene has significant homology to FGFs, especially basic FGF.[67] Multiple transcripts expressed from the *int*-2 gene in mouse embryonal carcinoma cell lines encode a protein with homology to FGFs.[68] The Int-2 protein is also known as FGF-3.

The Int-1 protein is involved in cell-signaling mechanisms in both insects and vertebrates.[69] The Int-1 homologous protein encoded by the *wingless* (*wnt*-1) gene of *Drosophila melanogaster* is involved in the establishment of early segmentation during development. During mouse development, expression of the *int*-1/*wnt*-1 gene is restricted to the developing neural plate and neural tube. This gene is not expressed in any adult mouse tissue with the exception of the testis, where its expression occurs only in postmeiotic round spermatids. The *int*-1/*wnt*-1 is a member of the *wnt* gene family, which encodes secreted proteins containing up to 23 or 24 cysteines in nearly parallel positions.[70] Wnt proteins are about 350 to 380 amino acids in length, and their structures show the characteristics of growth factors.

COS-1 monkey cells transfected with *int*-2 DNA express Int-2-like proteins.[71] These proteins have low mitogenic activity on cultured mammalian epithelial cells. The *int*-2 gene is differentially expressed during gastrulation and neurulation in the mouse, which suggests a role for *int*-2 in mammalian embryogenesis.[72] It may be responsible for induction of the inner ear.[73] The Int-2/FGF-3 protein can replace basic FGF in modulating the anchorage-independent growth of human SW13 adrenal cortical tumor cells.[74] Extracellular or surface-bound Int-2/FGF-3 protein may be instrumental for the morphological transformation of National Institutes of Health (NIH)/3T3 cells expressing *int*-2 cDNA.[75] These cells produce and secrete Int-2-related products representing discrete stages of processing and glycosylation. Autocrine expression of Int-2/FGF-3 in HC11 mouse mammary epithelial cells infected with a retroviral vector expressing the *int*-2 gene can stimulate anchorage-dependent and -independent growth in a fashion similar to that of basic FGF.[76] The Flg/FGF receptor is expressed in HC11 cells and may recognize Int-2/FGF-3 as a ligand. HC11 cells treated with basic FGF or expressing Int-2/FGF-3 are competent to respond to lactogenic hormones and synthesize β-casein in the absence of EGF.

D. THE FGF-4/HST-1 PROTEIN

The *hst* gene, also termed *hst*-1, *HSTF1, FGFK,* KS3, or K-*fgf,* was discovered through its transforming ability in the NIH/3T3 DNA transfection assay. The *hst* gene was first detected in 2 of 21 samples of primary or metastatic human stomach cancers and in a noncancerous portion of the stomach mucosa from one of the two positive patients.[77] A cDNA sequence for the human *hst* gene was isolated, and the coding sequence required for its oncogenic activity was identified.[78] A gene that is probably identical with *hst* was found to be active in a transfection assay with DNA from acquired immunodeficiency syndrome (AIDS)-associated Kaposi's sarcoma.[79,80] Secretion of the Hst/k-FGF/FGF-4 protein, or localization of the ligand-receptor complex on the cell surface, is required for *hst*-induced transformation.[81]

The human *hst* gene, as deduced from a cDNA sequence, encodes a protein of 206 amino acids which exhibits homology to FGF.[82] The Hst/FGF-4 protein is mitogenic for fibroblasts and endothelial cells, and promotes the growth of mouse NIH/3T3 cells in serum-free medium supplemented with insulin and transferrin.[83] A murine homolog of the human *hst* gene has been identified, and its complete nucleotide sequence was determined.[84] The *hst* gene is located not far from the *int*-2 gene on mouse chromosome 7, and its orientation and structure suggest that the two genes arose by duplication of a common ancestral gene. The mouse *hst* gene encodes a 202-amino acid protein which is 82% homologous to the human Hst product. Expression of the *hst* gene is extremely restricted in normal mouse tissues and is regulated, at least in part, by elements located in the 3′-noncoding region of exon 3.[85] The gene is expressed at the RNA and protein level in the mouse blastocyst, but not in adult mouse tissues.

The *hst* gene is expressed in mouse and human embryonal carcinoma cells and is downregulated upon induction of differentiation in these cells.[86] In contrast, the *int*-2 gene is transcriptionally induced in differentiated F9 cells. These results suggest that single members of the FGF family may perform distinct functions *in vivo,* and that the Hst/FGF-4 protein may be related to cell differentiation processes during early development.

E. THE FGF-5 AND FGF-6 PROTEINS

Two other genes of the FGF family are *fgf*-5 and *fgf*-6. The *fgf*-5 gene was detected in a human tumor-derived cell line by means of a transformation assay based on DNA transfection into NIH/3T3 mouse fibroblasts, and its protein product was characterized.[87,88] The *fgf*-5 gene is expressed at low levels in growing normal human fibroblasts *in vitro;* and its expression in quiescent fibroblasts is strongly induced by serum and growth factors such as PDGF, EGF, and TGF-α.[89] This induction is mediated by pathways involving protein kinase C or cyclic adenosine-3′,5′-monophosphate (cAMP)-dependent protein kinases. The *fgf*-5 gene is a primary response gene; and its product, the FGF-5 protein, is mitogenic for human fibroblasts. The *fgf*-5 gene has been assigned to human chromosome 4.[90]

The *fgf*-6 gene was first identified by screening a mouse cosmid library with a human *hst* probe, and a human homolog of *fgf*-6 was subsequently isolated and sequenced.[91] The cloned normal human *fgf*-6 gene was able to transform NIH/3T3 fibroblasts in focus and tumorigenicity assays. Nucleotide sequence of the human *fgf*-6 gene revealed the presence of an ORF capable of coding a protein of 208 amino acids which shares homology to other members of the FGF family within the carboxyl-terminal two thirds of the molecule.[92] The presence of a signal peptide sequence is essential for the *in vivo* transforming capacity of the *fgf*-6 gene. Cloning and genomic organization of the mouse *fgf*-6 gene have been determined.[93] The gene is located on mouse chromosome 6 and is capable of inducing transformation of mouse fibroblasts in culture. The purified FGF-6 protein displays a strong mitogenic activity on BALB/c 34T3 cells and is able to morphologically transform these cells. The levels of *fgf*-6 mRNA are developmentally regulated, with a peak of expression in the developing fetus at day 15 of gestation. Moderate levels of *fgf*-6 mRNA are present during late mouse gestation and in the mouse neonate. In the adult animal, *fgf*-6 transcripts can be detected in the testis, heart, and skeletal muscle.

F. KERATINOCYTE GROWTH FACTOR

The keratinocyte growth factor (KGF), a member of the FGF family, was derived from human embryonic lung fibroblasts.[96] A method that gives high yields of functional human recombinant KGF has been reported.[95] KGF is a single-chain polypeptide of 28 kDa with mitogenic activity specific for epithelial cells. This activity is strongly inhibited by heparin. KGF is normally expressed by stromal fibroblasts and acts on epithelial cells in a paracrine fashion. A large induction of KGF expression occurs in the dermis during wound healing.[96] In addition to KGF, other members of the FGF family are induced, although at lesser extents, in the skin after injury. Expression of KGF in basal keratinocytes of transgenic mice shows that KGF is a potent factor in eliciting global effects not only on growth but also on development and differentiation of skin and other tissues.[97] KGF is a mitogen for rat hepatocytes maintained in primary culture.[98] KGF may interfere with signaling of some mesenchymal-epithelial interactions.

III. FGF-RELATED GROWTH FACTORS

A number of FGF-related growth factors have been described in different biological systems. The precise relationship of some of these factors to FGF is not totally clear.

A. ENDOTHELIAL CELL GROWTH FACTORS

Blood vessels are protected from thrombogenesis by the presence of the endothelium. Endothelial cells form a single layer and do not undergo mitosis except after wounding or other types of damage. Several peptide factors are involved in mediating endothelial cell functions. These factors may contribute to the maintenance and regulation of blood vessel homeostasis and may be important in the pathobiology of tumor growth and development as well as in atherogenesis and the events related to the aging process.[99] Mitogenic activity produced by mesenchymal and epithelial cells may act in a paracrine manner to stimulate endothelial cell proliferation. A factor with ECGF activity was found to be released, for example, by the epithelial cells of cultured porcine thyroid follicles.[100] Production of this ECGF is regulated by thyroid-stimulating hormone (TSH), epidermal growth factor (EGF), and iodide in parallel with the effects of these agents on thyroid cell growth, suggesting a possible role of the thyroid ECGF in the pathogenesis of goiter. An ECGF isolated from bovine hypothalamus consists of two closely related molecular forms, α and β.[101,102]

A number of growth factors described as ECGF are closely related or identical to FGFs. cDNA clones encoding human ECGF have been constructed, and the primary structure of the polypeptide has been deduced from nucleotide sequence analysis.[99] The gene coding for ECGF has been localized to human chromosome region 5q31.2-q33.2, and is expressed as a 4.8-kb mRNA which is present in human brain stem cells. Human ECGF is a 155-amino acid polypeptide which shows very high homology with acidic FGF-1. Structural evidence shows that ECGF-β is the precursor of both ECGF-α and acidic FGF-1.[103] ECGF-α, ECGF-β, and acidic FGF are apparently coded by the same gene and represent various processing products of the same polypeptide precursors. Acidic FGF-1 would be generated by cleavage of ECGF-β involving serine proteases, which results in a polypeptide that is 15 amino acids smaller than human ECGF, the deletion being localized at the amino-terminal end of the molecule.[99] Human ECGF also shows structural homology, although at a much lesser degree (50%) with bovine basic FGF. Unlike many other growth factors, ECGF is not derived from proteolytic processing of a precursor and lacks a signal peptide sequence.

The biological activity of ECGF is exerted through its interaction with a high-affinity receptor which is present on the surface of endothelial cells.[54,104] Heparin binds ECGF and potentiates its biological action by structural interaction.[105] ECGF isolated from bovine brain stimulates *in vitro* growth of human umbilical vein endothelial cells and permits their serial propagation.[106] Both acidic and basic FGFs of human brain origin have equivalent specific mitogenic activities on human umbilical vein endothelial cells in culture in the presence of heparin.[55] The mechanism of action of FGFs on endothelial cells is apparently very complex and may involve the mediation of other growth factors. Addition of acidic FGF/HBGF-1 to quiescent cultures of human umbilical vein endothelial cells increases the level of PDGF-A mRNA but not PDGF-B/c-*sis* mRNA.[107] In contrast, factors that inhibit endothelial cell proliferation such as the phorbol ester phenylmercuric acetate (PMA) and the cytokines IL-2, IL-6, and TNF-α increase both PDGF-A and PDGF-B chain mRNA levels. The mitogenic effect of ECGF on endothelial cells in culture is inhibited by recombinant IL-1 in a concentration-dependent manner.[108]

In the presence of increasing concentration of EGF, confluent cultures of early and late passage human endothelial cells exhibit an increased incorporation of glucosamine and Na_2SO_4 into the glycosamineglycans, hyaluronic acid, chondroitin-4-sulfate, dermatan-4-sulfate, and chondroitin-6-sulfate.[106] Cells grown in the presence of ECGF are more adhesive to the substratum and more resistant to detachment by proteolytic enzymes than cells grown in medium lacking ECGF. Treatment of vascular endothelial cells with pituitary FGF induces an increase in angiotensin-converting enzyme, an enzyme involved in regulating the production of angiotensin II from renin substrate.[109] ECGF-α stimulates bone DNA synthesis, an effect that results in an increased number of cells capable of synthesizing bone matrix proteins, but ECGF-α has no effect on bone matrix degradation or bone resorption.[110] The effect of ECGF-α on the synthesis of bone matrix proteins is accompanied by increased production of prostaglandin (PGE_2); however, the possible significance of this production is unknown because indomethacin, which is an inhibitor of prostaglandin synthesis, does not abolish the ECGF-α effect on bone DNA synthesis.

B. FGF- AND ECGF-RELATED GROWTH FACTORS

A factor, termed endothelial cell stimulating angiogenesis factor (ESAF), was shown to induce angiogenesis *in vivo* and stimulated microvessel, but not aortic endothelial cells or fibroblast growth.[111] Basic FGF and ESAF displayed a positive synergistic effect in the stimulation of microvessel endothelial cells grown on a collagen matrix; and it was postulated that ESAF may represent, at least in part, the permissive effects of serum in the action of basic FGF on microvessel endothelial cells. It was also postulated that the factor described as tumor angiogenesis factor (TAF) may possess its potency toward microvessel endothelial cells due to the presence of both basic FGF and ESAF.

A growth factor activity, partially purified from normal and neoplastic tissue by heparin affinity chromatography, had angiogenic properties and was mitogenic for bovine capillary endothelial cells.[112] This type of activity was detected in mouse bladder tumors and urine from mice with bladder cancer. Two factors with mol wt 16,000 and 26,000 were responsible for this activity; however, their possible relation to other known growth factors, including EGF, TGF-α, and PDGF, could not be elucidated.

Human fibroblasts produce a factor which is a potent mitogen and chemoattractant for fetal bovine aortic endothelial cells and which is similar to basic FGF.[113] An ECGF isolated from serum-free culture supernatants of human embryo fibroblast cells was found to be an acidic glycoprotein of 30 kDa showing heterogeneity of the oligosaccharide moiety.[114] This factor was apparently different from ECGF isolated from bovine brain as well as from acidic FGF isolated from the same source.

The precise structural and physiological relationships between ECGF, or ECGF-related factors, and FGFs remain to be established; however, endothelial cells derived from either large vessels or capillaries are responsive to FGFs (more to basic FGF than to acidic FGF), and heparin strongly reduces the proliferative response of endothelial cells to basic and acidic FGF.[115] Bovine pulmonary artery endothelial cells produce basic FGF which induces the proliferation of the endothelial cells via an extracellular autocrine loop upon binding to specific receptors expressed on the cell surface.[116]

C. RETINA-DERIVED GROWTH FACTOR

An anionic endothelial cell mitogen, termed retina-derived growth factor (RDGF), with two molecular forms (mol wt 16,500 and 18,000), was purified from bovine retina by taking advantage of its high affinity to heparin.[117,118] RDGF may be closely related, if not identical to ECGF purified from hypothalamus. In addition to its action on cells from mesenchymal origin, RDGF has powerful effects on PC12 cells, which are cells of neuroectodermal origin. RDGF induces in PC12 cells neurite outgrowth, and this effect is markedly potentiated by heparin but is not blocked by antibodies to nerve growth factor (NGF).[119] The exact structural and functional relationship between RDGF and FGF remains to be established, but both basic FGF and acidic FGF present in retinal tissue can stimulate the proliferation of retinal capillary endothelial cells.[120]

A vascular endothelium growth factor (VEGF) found in retina is apparently different from other known growth factors, including bovine retina-derived ECGF.[121] The VEGF is extremely potent in stimulating the growth of fetal bovine aortic endothelial cells, and its activity is enhanced by the presence of Mg^{2+}.

D. BRAIN-DERIVED GROWTH FACTOR

Brain-derived growth factor (BDGF) is a 16- to 17-kDA protein that has been extracted and purified from the bovine brain.[12] BDGF is a potent mitogen with a broad spectrum of cell specificity including endothelial cells, fibroblasts, osteoblasts, astroglial cells, chondrocytes, smooth muscle cells, and epithelial cells. It may act as a chemoattractant for fibroblasts and astroglial cells and can induce anchorage-independent growth in NRK cells.[122,123] The precise physiological role of BDGF is unknown but it is localized in brain neurons, and a BDGF-like molecule is produced by human neuroblastoma cells.[124] There may be two forms of BDGF (BDGF-A and BDGF-B); BDGF downmodulates high-affinity EGF receptors in mouse 3T3 cells in a time-, temperature-, and concentration-dependent manner.[125] The functional relevance of this transmodulatory effect is not known. An 17- to 18-kDa HBGF isolated from developing rat brain and bovine uterus may be distinct from other members of the FGF family.[126,127] Further studies are required to characterize the structural and functional relationships of BDGFs with FGFs, EGF, neurotrophic factors, and other growth factors.

E. ASTROGLIAL GROWTH FACTORS (AGFS)

Astroglial growth factors (AGFs) 1 and 2 are closely related or identical to acidic and basic FGF, respectively.[47,48] The AGFs can induce proliferation and differentiation of neuroectodermal astroglial cells, oligodendrocytes, and PC12 rat pheochromocytoma cells. Both AGFs and FGFs can mimic the promotion of *in vitro* neuron survival and enhance the activity of a neurotransmitter-synthesizing enzyme, choline acetyltransferase, in neurons of the peripheral and central nervous system. Basic FGF/AGF-2 enhances the proliferation of neuronal precursors from embryonic rat brain.[128] Since in rat brain AGF-2 has been found to be localized essentially in neuronal cells,[129] it may function as an autocrine factor for these cells during ontogenesis.

F. OMENTUM-DERIVED ECGF

The omentum, a vascularized adipose tissue present in the abdominal cavity, can cause rapid formation of new blood vessels when transposed to normal or ischemic organs. The bovine omentum contains a factor capable of stimulating the proliferation of bovine aortic endothelial cells.[130] The omentum-derived factor (ODF) is represented by a polypeptide of 21 kDa which is not mitogenic to BALB/c 3T3 mouse fibroblasts and has no affinity for immobilized heparin. The ODF is apparently different from FGFs and other known growth factors.

G. FIBROBLAST-STIMULATING FACTOR (FSF)

Mature T lymphocytes elaborate a soluble factor that stimulates fibroblast growth.[131] This factor, referred to as fibroblast-stimulating factor (FSF), is a glycoprotein of apparent mol wt 17,000 and is different from FGF as well as from HGFs including colony-stimulating factor (CSF), IL-3, and IFN. Its normal function and possible role in fibrotic diseases are unknown.

H. EMBRYONAL CARCINOMA-DERIVED GROWTH FACTOR

PC13 embryonal carcinoma cells produce a mitogen called embryonal carcinoma-derived growth factor.[132,133] This factor is represented by a 17.5-kDa protein that binds heparin and modulates the activity of the EGF receptor. Embryonal carcinoma-derived growth factor induces phosphorylation of an 80-kDa cellular protein which has an unknown function.[134] Embryonal carcinoma-derived growth factor-induced phosphorylation of the 80-kDa protein seems to depend on the activation of protein kinase C through an enhanced rate of phosphoinositide metabolism. Embryonal carcinoma-derived growth factor derived from PC13 cells may be structurally related to basic FGF.

IV. FUNCTIONS OF FGFs

The members of the FGF family are mitogenic for a wide variety of cells. They are significantly involved in developmental processes and have angiogenic and neurotrophic properties. They participate in mesoderm induction during vertebrate embryogenesis.[135,136] Basic FGF may be involved in regulating the development of a basic body plan during early amphibian embryogenesis.[137] FGFs contained in amphibian eggs may be involved in the induction of mesoderm.[138] FGFs can stimulate blastoma cells which are involved in organ regeneration, for example, in the regeneration of newt limb after amputation.[139] The role of FGF in mammalian embryogenesis *in vivo* was shown by using an infusion system to study the direct effects of an antiserum to basic FGF and recombinant bovine FGF on growth and differentiation of embryos transplanted into syngeneic nonpregnant hosts.[140] FGF is important for the growth and differentiation of endodermal and some mesodermal tissues of mammalian embryos.

Acidic and basic FGF are able to stimulate the proliferation of a wide variety of cells including vascular endothelial cells as well as chondrocytes, myoblasts, and osteoblasts. Both types of FGFs act directly on vascular cells to induce endothelial cell growth and angiogenesis not only *in vitro* but also *in vivo*.[141] Acidic FGF promotes vascular repair, and basic FGF enhances the development of the collateral circulation after acute arterial occlusion.[142,143]

The proliferation of smooth muscle cells after vascular injury can be inhibited by an antibody against FGF.[144] Acidic FGF may have a role in the control of liver regeneration by autocrine and/or paracrine mechanisms.[145] Growth of calvaria-derived osteoblasts is stimulated *in vitro* by both acidic and basic FGF, and this effect is counteracted by parathyroid hormone.[146] Basic FGF may be a growth factor not only for cells of mesenchymal origin but also for neuroectodermal cells, and probably for ectodermal cells in general. Basic FGF, but not acidic FGF, is a mitogen for mouse keratinocytes.[147] IGF-I and insulin have mitogenic effects on keratinocytes similar to those of EGF or basic FGF. However, the role of basic FGF as a mitogen for epidermal keratinocytes *in vivo* is not clear since none of the cells located in the epidermis under normal conditions appears to be a source of FGF. Basic FGF and TGF-α are potent hepatotrophic mitogens *in vitro*.[148] Basic FGF has a permissive action in the initial stages of hematopoiesis.[149] It may influence early- and late-stage hematopoietic progenitor cells and exhibits synergistic activities with HGFs.[150] Basic FGF can counteract the suppressive effect of TGF-β on human myeloid progenitor cells.[151]

Growth factor-depleted mouse fibroblasts respond to basic FGF in culture with a burst of mitogenesis and with a rapid and marked increase in thrombospondin RNA levels.[152] Thrombospondin, the major constituent of platelet α granules, is secreted by a variety of cell types in culture and may display mitogenic activity. The kinetics of response of the thrombospondin gene to FGF resemble those of the responses of c-*fos* and c-*myc* genes in which expression contributes to progression through the cycle. The thrombospondin gene may be a member of the immediate-early response genes, and its expression is stimulated not only by basic FGF but also by PDGF. Fibroblasts from patients with Werner's syndrome (an autosomal recessive disorder characterized by an apparent acceleration of the processes associated with aging) show a markedly attenuated mitogenic response to both FGF and PDGF.[153] This impaired response is not due to a diminished number of receptors on the cell surface, but is localized at sites beyond the growth factor binding.

Intracellular protein accumulation is increased in cells treated with acidic or basic FGF, although the effects of FGFs are smaller than those of insulin and are not additive.[154] FGFs induce cell growth in responsive cells such as differentiated muscle cells in culture (the clonal mouse muscle cell line BC3H-1), and both pituitary-derived FGF or brain-derived FGF are as effective as serum in repressing the synthesis of creatine phosphokinase in these cells.[155] FGFs can induce, in a manner similar to NGF, neurite outgrowth in the rat pheochromocytoma cell line PC12.[156] On a molar basis, basic FGF is about 800-fold more potent than acidic FGF in inducing PC12 cell differentiation. While the neurotropic activities of acidic FGF and NGF are potentiated by heparin, those of basic FGF are both partially inhibited or stimulated depending on the concentration of basic FGF. Basic FGF has important actions at the level of the central nervous system, including inhibitory effects on gastric acid and pepsin secretion.[157]

FGF may be involved in the regulation of hormone synthesis and secretion. FGF may participate in an intrapituitary mechanism regulating prolactin and TSH secretion.[158] Basic FGF is present in the human thyroid, where it may function as a modulator of cell function and growth.[159] Acidic FGF functions as an autocrine factor in calcium-regulated parathyroid hormone (PTH) secretion.[160] The rat parathyroid cell line Pt-r expresses both acidic FGF and its receptor, and changes in extracellular calcium induce altered production of acidic FGF and rapid changes in FGF receptor expression on the cell surface. In the Jar human choriocarcinoma cell line, FGF enhances human chorionic gonadotropin (hCG) synthesis and secretion.[161] The stimulation of hCG production by FGF in Jar cells is independent of a mitogenic effect and is enhanced by the coadministration of insulin. Testis cells possess binding sites specific for FGF, and FGF inhibits steroidogenesis in primary cultures of neonatal rat testicular cells.[162] These results indicate that growth factors may participate in regulatory functions other than cell growth regulation, including the regulation of hormone secretion.

Arginine-dependent formation of nitric oxide by the inducible enzyme, nitric oxide synthase, is an important autocrine and paracrine-signaling pathway which has a role in the physiology of mammalian cells including an antiproliferative action.[163] Several growth factors are involved in the regulation of nitric oxide synthase activity, including TGFs, interferons (IFNs), and FGFs. In bovine retinal pigmented epithelial cells, which express the enzyme after activation with LPS and IFN-γ, TGF-β slightly increases the production of nitrite (an oxidation product of nitric oxide), whereas acidic and basic FGFs markedly inhibit the nitrite release due to LPS and IFN-γ activation.[164]

V. FGF RECEPTORS

The action of FGF is initiated by its binding to specific high-affinity receptors on the cell surface.[165] Receptors for FGF were isolated from rat brain.[166] Neuroectodermal and mesodermal cells from different mammalian species (man, mouse, rat, and hamster) express a 165-kDa receptor protein able to recognize both acidic FGF and basic FGF.[167] FGF receptors are found at relatively high levels in embryonic tissues, suggesting a role for FGF in embryonic development and a more restrictive role in the adult.

The actions of FGF are solely mediated by its high-affinity receptor. Bovine basic FGF binds to a receptor of 130 to 135 kDa that possesses tyrosine kinase activity.[168-170] A FGF receptor protein purified from the human hepatoma cell line HepG2 was identified as a transmembrane glycoprotein of 130 kDa.[171] Low-affinity FGF-binding sites are present in bovine capillary endothelial cells, but these sites are unable to mediate the physiological actions of basic FGF.[172] Thus, the low-affinity FGF-binding sites would only act as a reservoir of FGF for the cell. Binding of FGF to heparan sulfate proteoglycan on the basement membrane like matrix of certain normal and tumor cells may be a reservoir that could be involved in modulating the cellular actions of FGF.[173] However, there is evidence that the functional FGF receptor may be represented by a ternary complex of heparan sulfate proteoglycan, tyrosine kinase transmembrane glycoprotein, and ligand.[174]

FGF receptors have been characterized recently as transmembrane glycoproteins of 800 to 822 amino acids and 110 to 150 kDa which possess tyrosine kinase activity.[175] The extracellular portion of FGF receptors is composed of 346 to 356 amino acids and contains two or three Ig-like domains. There are five distinct molecular forms of FGF receptors: Flg/FGFR-1, Bek/FGFR-2, FGFR-3, FGFR-4, and Flg-2.[176] These forms of FGF receptors are encoded by separated genes and exhibit partially distinct ligand-binding specificities and cellular functions.[177]

A human FGF receptor gene, identical with the putative proto-oncogene *flg*, was isolated by low-stringency screening of an endothelial cDNA library by using human c-*fms* proto-oncogene cDNA as a probe and is located on human chromosome region 8p12.[178-180] The *flg* gene encodes the receptor protein Flg/FGFR-1, also called N-Sam. A second receptor, Bek/FGFR-2, is the product of a distinct but closely related gene, *bek*, which is located on chromosome region 17q23-qter.[181] A gene closely related or identical to *bek*, K-*sam*, was found to be amplified in the human stomach cancer cell line KATO-III.[182] The Bek and K-Sam proteins are identical. The receptor proteins encoded by the *flg* and *bek* genes bind both acidic FGF and basic FGF. A K-*sam*-related gene, N-*sam*, which is almost identical with *flg*, also encodes an FGF receptor.[183] The N-*sam* gene was found to be expressed in thymic T-cell leukemia cells, and its level of expression is enhanced by basic FGF and 12-*O*-tetradecanoylphorbol-13-acetate (TPA).

The complete amino acid sequence of a human receptor for acidic and basic FGF was deduced from a cDNA clone.[184] In addition to FGFR-1/Flg and FGFR-2/Bek, two other types of FGF receptors have been identified in human cells: FGFR-3/Sam-3 and TKF/FGFR-4.[185-187] The four types of FGF receptors are subjected to differential expression in human tissues. Transcripts of the FGFR genes are present in human megakaryocytic leukemia cell lines, and *bek* mRNA is found in human circulating platelets.[188]

Human cDNA clones encoding FGF receptor variants that contain two Ig-like domains in the extracellular region have been isolated.[189] Two of these clones encode membrane-spanning receptors, and the other two encode secreted forms of the receptor. These variants are generated by alternative splicing of RNAs transcribed from a single gene. In addition, FGFR-3 and FGFR-4, cloned from K562 leukemia cells, exhibit distinct patterns of expression in different tissues. FGFR-4 expression is very high in adrenal glands from human fetuses, where only minute amounts of Flg/FGFR-1, Bek/FGFR-2, and FGFR-3 mRNAs are detected. The FGFR-4 receptor is 55% identical with the Flg and Bek receptors and has three Ig-like domains in its extracellular part. FGFR-4 binds acidic FGF with high affinity, but does not bind basic FGF.

A murine FGF receptor expressed in Chinese hamster ovary (CHO) cells lacks 88 amino acids in the extracellular portion and may be produced by alternative RNA splicing.[190] Three distinct isoforms of the murine Flg/FGFR-1 receptor are generated by alternative RNA splicing, and expression of at least one of these forms is restricted to fibroblasts.[191] There is a cell type-specific expression of the Flg/FGFR-1 receptor gene and its isoform products in rat and human cells that result from combinations of splice variations.[192] Two distinct forms of this receptor are differentially expressed in mouse tissues: a larger form containing three Ig-like domains and a shorter form containing four Ig-like domains.[193] Both acidic and basic FGF enhance the synthesis of DNA in the mouse breast cancer cell line SC-3, and these cells express qualitatively altered FGF receptors with an insertion of 12 amino acids and a deletion of 2 amino acids.[194] An alternative form of the Flg/FGFR-1 receptor produced in normal and tumor rat tissues exhibits transforming activity when expressed at high levels in rat prostate fibroblasts.[195] The precise biological significance of altered forms of FGF receptors remains to be elucidated.

Chicken FGF receptors may contain two or three Ig-like domains in the extracellular region, and these forms mediate biological responses to both acidic and basic FGF.[196] The tyrosine kinase region of the chicken receptor shares 51 to 53% identity with the PDGF and CSF-1 receptors and contains a tyrosine residue (Tyr-651) at the position analogous to the major phosphorylation site of the v-Src oncoprotein. cDNA cloning and developmental expression of FGF receptors from *Xenopus laevis* has been reported.[197] A protein of 140 kDa with specific affinity for basic FGF, but not acidic FGF, was detected in *Drosophila melanogaster*.[198] However, the genome of *Drosophila* contains two FGF receptor homologous genes, *DFR1* and *DFR2*.[199] These two genes encode proteins containing two and five Ig-like domains, respectively, in the extracellular region and a split tyrosine kinase domain in the intracellular regions. The FGF receptor homologs of *Drosophila* are essential for generation of mesodermal and endodermal layers, invagination of various types of cells, and formation of the central nervous system.

Expression of high-affinity FGF receptors is regulated by local and general factors in the intact animal. Differential expression of FGF receptors in various tissues during the development of the rat indicates that FGF has a role in developmental events.[200] Changes in FGF receptor gene expression and receptor genotype may occur during liver regeneration and in hepatoma cells. Nucleotide sequence of cDNA predicts that three amino-terminal domain motifs may combine to form 6 to 12 homologous polypeptides that constitute the FGF receptor family in a single human liver cell population.[201] The three distinct structural domains that combine to form FGF receptor isoforms are likely to affect ligand binding, oligomerization, cellular location, and signal transduction. The physiological factors involved in the regulation of FGF receptor expression are little known. Binding of FGF to nontransformed cell lines *in vitro* decreases as cell density increases, and this is due to a reduction in the number of FGF receptors without change in receptor affinity.[201] Platelet factor-4 (PF-4) blocks the binding of basic FGF to the receptor and inhibits the spontaneous migration of vascular endothelial cells.[203]

Herpes simplex virus 1 (HSV-1), an ubiquitous virus responsible for considerable morbidity in the general population, uses the basic FGF receptor as a means of entry into the cells.[204] The availability of specific FGF receptor antagonists may prevent HSV-1 uptake and could be helpful in the development of strategies to control HSV-1-related diseases.

VI. POSTRECEPTOR MECHANISMS OF FGF ACTION

The postreceptor transductional mechanisms of action of FGF remain poorly characterized. The adenylyl cyclase system appears to have little, if any, importance in this action, although cAMP can potentiate basic FGF-induced neurite outgrowth in PC12 cells.[205] FGF may stimulate monovalent ion exchange in the membrane of its target cells. After binding to its specific receptor on the surface of BALB/c 3T3 mouse fibroblasts, FGFs rapidly stimulate the activity of the Na^+/K^+ pump and the entry of Na^+ into the cell in a concentration-dependent manner.[206] The FGF-stimulated increase of Na^+ influx into the cell is abolished by amiloride, but the molecular mechanisms responsible for the observed pump stimulation is unknown.

Phosphoinositide breakdown with the associated intracellular calcium mobilization and activation of protein kinase C may be involved in the mechanism of action of FGFs in at least some cellular systems. Generation of 1,2-diacylglycerol, mobilization of calcium, and activation of protein kinase C were observed in quiescent cultures of mouse 3T3 fibroblasts after addition of FGF to the medium.[207] Basic FGF was found to be up to tenfold more potent than acidic FGF in the stimulation of inositol phosphate accumulation in 3T3 cells.[208] Apoptosis in serum-depleted human vascular endothelial cells is suppressed by FGF by a mechanism involving protein kinase C.[209] However, the precise role of phosphoinositides in the mechanisms of FGF action in other systems is not understood. Protein kinase C is not a mediator of basic FGF action on rabbit chondrocytes.[210] Reinitiation of DNA synthesis induced by FGF in quiescent Chinese hamster fibroblasts can occur without increased polyphosphoinositide breakdown and activation of protein kinase C.[211] No significant changes in phosphoinositide breakdown occur in bovine epithelial lens cells in culture after addition of FGF to the medium.[212] The mitogenic effects of FGF in the epithelial lens cells are mediated by other mechanisms, including ionic changes (activation of Na^+/H^+ exchange and/or increased entry of Ca^{2+} into the cell). Point mutation at Tyr-766, a residue required for phosphorylation of phospholipase C-γ, eliminates FGF receptor-stimulated phosphatidylinositol hydrolysis without affecting the mitogenic effect of FGF.[213] Studies using a monoclonal antibody specific for phosphatidylinositol 4,5-bisphosphate suggest that breakdown of phosphoinositides is not crucially required for the mitogenic action of FGF in NIH/3T3 cells.[214] These results do not lend support to the hypothesis that inositol lipid breakdown is universally involved in the mechanisms of growth control associated with the action of hormones and growth factors on susceptible cells.

FGF, although unable to activate on its own the turnover of phospholipids in CCL39 hamster fibroblasts, potentiates the thrombin-induced formation of inositol phosphates mediated by stimulation of phospholipase C; this is a property shared with other growth factors known to activate receptor protein-tyrosine kinases, such as EGF and PDGF.[215] Prostaglandins and gangliosides may have a role in the transductional mechanisms of FGF action. Basic FGF stimulates prostaglandin ($PGF_{2\alpha}$) synthesis in rat corpus luteum.[216] Heparin and acidic FGF interact to decrease prostacyclin synthesis in human endothelial cells by affecting the activity of prostaglandin H synthase and prostacyclin synthase.[217] A synergism between gangliosides and basic FGF may favor the survival, growth, and motility of capillary endothelium.[218] The possible role of an $[Ca^{2+}]_i$ elevation in the mitogenic action of FGF is unknown. This elevation may be required for the mitogenic action of PDGF, but not basic or acidic FGF, in BALB/c 3T3 mouse cells.[219]

Treatment of 3T3 cells with either acidic or basic FGF activates a ribosomal S6 kinase that can phosphorylate a synthetic peptide substrate.[220] This result suggests that the mechanism of FGF action may involve the phosphorylation of certain cellular substrates on serine/threonine residues. Repression of muscle protein transcription by FGF is exerted through phosphorylation of the Thr-87 residue contained in the basic, DNA-binding region of the helix-loop-helix (HLH) protein myogenin; and this effect can be reproduced by activation of protein kinase C.[221] However, the possible role of protein kinase C, tyrosine kinases, and other protein kinases in the mechanism of action of FGF remains little understood. There is evidence that FGF-induced activation of S6 kinase may proceed via protein kinase C-independent mechanisms. Protein kinase C may not stimulate, but may inhibit, the mitogenic response of bovine capillary endothelial cells to the FGF-like factor, HDGF.[222]

Phosphorylation of cellular proteins on tyrosine residues depends on a balance between the activities of tyrosine-specific protein kinases and tyrosine-specific protein phosphatases. Vanadate, a powerful inhibitor of tyrosine-protein phosphatases, mimics the action of FGF on actin and creatine phosphokinase synthesis and causes BC3H-1 cells to exit the G_0 phase of the cycle, moving the cells into the G_1 phase and repressing the expression of a muscle phenotype.[223] Activation of tyrosine kinase may be an early event in the cellular mechanism of action of FGFs. Both acidic FGF and basic FGF activate a tyrosine kinase in mouse 3T3 cells, leading to phosphorylation on tyrosine of a 90-kDa protein substrate.[224] Acidic FGF stimulates phosphorylation of phospholipase C (PLC-γ) on tyrosine residues in intact cells with a potency similar to that of PDGF and EGF.[225] It is clear that FGF, in a manner similar to that of EGF, PDGF, insulin, and IGF-I, exerts its biological effects by the activation of tyrosine kinase associated with the FGF receptor.

The FGFs may have an important role in the regulation of gene transcription. Serum and FGF may inhibit the expression of muscle-specific gene products like the muscle isoenzyme of creatine kinase, as well as the mitogenic differentiation of the mouse cell line BC3H-1, through a mechanism dependent on protein synthesis and independent of cell proliferation.[226] FGF-inducible early gene products may be involved in the repression of myogenic differentiation. Interestingly, there is evidence that basic FGF may regulate transcription directly in the cell nucleus in a gene-specific manner after being transported into the nucleus by a targeting signal.[227] Nuclear localization of basic FGF has been reported in a variety of cultured cells and tissues.

The levels of expression of several proto-oncogenes may be altered by the action of FGFs on their target cells. Acidic FGF increases the expression of c-*myc*, c-*fos*, and c-*jun* mRNA in cultures of human umbilical vein endothelial cells.[107] In BC3H-1 cells, FGF induces c-*fos* gene expression and inhibits the synthesis of α-actin by a mechanism involving alteration in the transcriptional efficiency of α-actin mRNA.[223,228] In PC12 rat pheochromocytoma cells, basic FGF induces c-*fos* expression by a mechanism which is independent of protein kinase C activation.[229] The N-Ras protein selectively interacts with pathways involved in induction of c-*fos* expression initiated by the activated NGF and basic FGF receptors in PC12 cells.[230] In the FRTL5 thyroid cell line, basic FGF stimulates DNA synthesis, and this action is associated with its ability to increase the $[Ca^{2+}]_i$ and to induce a transient increase in c-*fos* mRNA levels.[231] Iodide uptake, which is regulated by TSH in FRTL5 cells through a cAMP signal, is inhibited by basic FGF despite its ability to stimulate DNA synthesis in these cells. An FGF-like factor contained in cultured Sertoli cells from immature rats induces an increase in c-*fos* mRNA levels in these cells through a mechanism that does not involve production of cAMP or activation of the Ca^{2+}/phosphoinositide pathway.[232] Both basic FGF and follicle-stimulating hormone (FSH) increase the levels of *jun*-B mRNA in Sertoli cells. Basic FGF decreases the levels of c-*sis*/PDGF-B mRNA and the amount of PDGF-like protein produced by human umbilical vein endothelial cells in culture.[233] The precise biological significance of FGF-induced proto-oncogene expression is unknown.

VII. ROLE OF FGFs IN ONCOGENESIS

Fibroblastic cells may have an important role in oncogenesis. Unidentified factors secreted by the mesenchyme may contribute to the growth of epithelial tumor cells *in vivo*. Transformed fibroblasts coinoculated with epithelial cells accelerate the growth and shorten the latency period of human epithelial tumors in athymic mice.[234] Even lethally irradiated transformed fibroblasts can shorten the latency period and increase the growth rate of human tumors, although with less efficiency than the live fibroblasts. The FGFs may contribute to the regulation of tumor growth *in vivo*.[235] The production and utilization of FGFs, ECGFs, and related factors by the tumor cells may contribute to stimulation of their own growth as well as to improvement of their nutrient supply through the formation of new blood vessels (neovascularization), which is crucially required for tumor progression. Basic FGF may function as an efficient angiogenic factor in tumorigenesis. An export of cell-associated basic FGF rather than new synthesis of the factor may represent a switch to the angiogenic phenotype, serving to activate the quiescent vasculature to the growth of new blood vessels.[236]

Basic FGF may have an important role in the growth of human melanoma cells.[237] It is the only known growth factor that can replace phorbol ester in culture medium of normal or neoplastic human melanocytes.[238] Melanocytes derived from dysplastic nevi and primary melanomas depend on exogenously supplied basic FGF for their growth *in vitro,* whereas melanocytes derived from metastatic melanomas do not require such supply. The mitogenic activity necessary for normal human melanocytes is constitutively present in cell lines derived from human metastatic melanomas, and this activity is inactivated by antibasic FGF antibodies.[239] Melanoma cells, but not normal melanocytes, express basic FGF mRNA. Exposure of primary human malignant melanoma cells and metastatic melanomas to antisense oligodeoxynucleotides targeted against different sequences of human basic FGF mRNA inhibits cell proliferation and colony formation in soft agar.[240] These results suggest that basic FGF may act as an autocrine and paracrine growth factor in human melanomas and that it may play an important role in the progression from melanocyte precursor lesions to malignant melanoma. Basic FGF is also required for the growth of other types of tumor cells, for example, primary rat fibrosarcoma cells cultured *in vitro* in a serum-free medium.[241] However, the simultaneous presence of albumin and transferrin in the medium is required for this action of basic FGF.

A. ONCOGENIC POTENTIAL OF FGFs

FGFs may exhibit oncogenic potential. Acidic FGF can promote, in a manner similar to TGF-β, anchorage-independent growth (colony formation of NRK cells in soft agar).[122] However, acidic FGF behaves more as a progression factor than as a transforming factor. It can enhance the tumorigenic potential of Rat-1 cells but is not capable of inducing the tumorigenic conversion of these cells.[242] Human cells may be susceptible to FGF-induced transformation. Basic FGF can induce anchorage-independent growth in cultured human fibroblasts.[243] The oncogenic potential of FGFs has been shown in transfection experiments. Transfection into mouse NR6 cells of a constructed plasmid vector expressing the acidic FGF gene results in loss of contact inhibition and induction of tumorigenic capability.[244] Expression of human basic FGF cDNA in mouse fibroblasts results in the expression of a transformed phenotype, associated with downregulation of FGF receptors.[245-247] Treatment of the transfected cells with suramin, which blocks the interaction of basic FGF with its receptor, results in reversion of the morphological transformation and restoration of FGF receptor expression. The phenotypic alteration of the transformed murine cells is also reversed by the addition of antihuman basic FGF antibodies to the medium, which suggests that basic FGF-mediated transformation may be associated with autocrine stimulation by secreted basic FGF protein. Normal murine melanocytes expressing a recombinant retrovirus carrying cDNA coding for basic FGF acquire properties similar to those of metastatic melanoma cells, but the constitutive production of basic FGF is insufficient to make melanocytes tumorigenic.[248]

Transformation induced by constitutive expression of basic FGF may require the secretion of high amounts of basic FGF and its interaction with the specific receptor on the cell surface. The adrenal carcinoma cell line SW-13 does not clone in soft agar even in the presence of 10% fetal calf serum unless supplemented with basic FGF. Transfection into SW-13 cells of a vector encoding the K-*fgf*/*hst* gene results in high levels of constitutive expression of basic FGF, but the growth factor requires to be released from the cells in order to function as an effective growth stimulator *in vitro* as well as *in vivo,* in athymic nude mice implanted with the manipulated cells.[249]

The K-*fgf*/*hst* gene was discovered through its oncogene-like properties. The k-FGF/Hst/FGF-4 protein can induce anchorage-independent growth of NRK-49F cells.[250] Transfection of the K-*fgf* gene into mouse NIH/3T3 cells results in the production of cell lines with increased motility.[251] Transformation of NIH/3T3 cells by a transfected K-*fgf* gene occurs by an autocrine mechanism and requires activation of the mitogenic pathway from signals arising at the cell surface.[252] A retrovirus coding for a fused Env/k-FGF protein can induce diffuse meningeal tumors and soft-tissue fibrosarcomas in mice.[253] Cells from Kaposi's sarcoma synthesize and release k-FGF/Hst and other active basic FGF-like molecules capable of inducing proliferation of themselves by an autocrine mechanism, as well as proliferation of endothelial cells by a paracrine type of mechanism.[254] Overexpression of the K-*fgf*/*hst* gene may contribute to the progression from a nonmetastatic to a metastatic phenotype in mouse mammary tumors.[255] Mice immunized with human k-FGF/Hst protein exhibit protection against tumors produced by injection of K-*fgf*-transformed cells.[256] Expression of the K-*fgf*/*hst* gene occurs in murine embryonal carcinoma cell lines, and the levels of this expression show a dramatic reduction when the cell lines were induced to differentiate.[257]

The mechanism of the FGF-associated potential oncogenicity is unknown but the genes of two members of the FGF family, *hst* and *int*-2, are considered as putative proto-oncogenes. Amplification of these two genes may occur in human mammary tumors; and it may be significantly associated with estrogen and progesterone receptor-positive, low-grade tumors.[258] Neoplastic transformation mediated by both the basic FGF and the Hst/FGF-4 proteins is mediated by interaction with the basic FGF receptor.

B. PRODUCTION OF FGFs BY TUMOR CELL LINES

Different types of tumor cell lines synthesize and secrete FGFs and express cell surface receptors for FGFs. Basic FGF was identified in eight of ten cell lines tested, and the amount of FGF activity produced by these cells had no apparent relation with their origin from normal or tumor tissue.[259] In the basic FGF-positive cell lines, the factor was detected only in cell extracts and is apparently not secreted into the culture medium. The lack of secretion of basic FGF by FGF-producing cultured cells is probably due to the absence of a secretory signal peptide sequence in the basic FGF protein. The artificial fusion of basic FGF with a signal secretory peptide sequence allows the entrance of the factor into the endoplasmic reticulum, and probably also its access and binding to the endogenous FGF receptor, which results in the acquisition of oncogenic potential.[260] NIH/3T3 fibroblasts transfected with a cDNA containing the coding sequence for a fused protein composed of basic FGF and a signal peptide undergo marked morphological alteration and are capable of producing rapidly growing tumors after inoculation into syngeneic mice.

Avian monocytic leukemia cells produce FGF, which may have implications to the leukemia-associated myelofibrosis.[261] A-204 human embryonal rhabdomyosarcoma cells release a mitogen which is indistinguishable from basic FGF.[262] Embryonal rhabdomyosarcomas are myoblast-derived tumors that affect primarily children and young adults and are constituted by immature muscle tissue arrested at an early stage of the normal muscle differentiation. A form of basic FGF is also produced by A431 human epidermoid carcinoma cells.[263] The human adrenal carcinoma cell line SW-13 produces a basic FGF-like protein that may act as an autocrine factor in these cells, enhancing their anchorage-independent growth.[264] The human squamous cell carcinoma cell line SCC-25 also expresses mRNA coding for basic FGF.[43] A polypeptide isolated from tissue extracts of human renal adenocarcinoma was mitogenic for mouse fibroblasts and human umbilical vein cells in culture and displayed angiogenic properties in the chorioallantoic membrane assay.[265] The biochemical and physiological properties of this heparin-binding factor suggested that it was similar to the family of basic FGFs. These results suggest that genes coding for FGFs or FGF-like factors may become activated in the cells of some tumors, including carcinomas, where the synthesized FGF may act as an autocrine growth factor.

FGF may act as an autocrine mediator of hormone action in tumor cells. Androgen stimulation of a Shionogi carcinoma 115-derived cell line (SC-3) results in secretion of heparin-binding growth factor which is represented by an FGF-like factor that interacts with the FGF receptor present in the same cells.[266] Human prostate carcinoma cell lines which differ in their dependence on androgens *in vitro* and in their metastatic potential in nude mice may also differ in their content of FGF receptor as well as in the expression of, and response to, exogenous basic FGF.[267]

Human glioma cell lines produce acidic FGF/ECGF and may possess receptors for the factor.[268] Acidic FGF/ECGF induces DNA synthesis in these glioma cells, and hence the expression of acidic FGF/ECGF in them may contribute to autocrine stimulation as well as to paracrine stimulation of other cell types including endothelial cells, which would result in angiogenetic effects favoring the vascularization of the tumor tissue. Basic FGF mRNA is abundantly expressed by human gliomas and meningiomas *in vivo,* whereas it is not present in metastatic brain tumors and normal human brain.[269] Acidic FGF mRNA is also expressed in human gliomas *in vivo.* Since human gliomas and meningiomas simultaneously express *in vivo* the *flg*- or *bek*-encoded FGF receptors, these results suggest that FGF may be an essential factor in the autocrine growth of these tumors.[270] Cell growth and tumorigenesis of human glioblastoma cells can be inhibited by a neutralizing antibody against basic FGF.[271] However, the role of FGFs in the growth of human brain tumors is not clear; and the level of expression of the two FGF receptor genes, *flg* and *bek,* is not different in tumor and normal brain tissues. In addition to FGFs, human gliomas express TGF-β mRNA as well as the genes for EGF/TGF-α receptors and the genes coding for PDGF chains and PDGF receptors. Thus, multiple growth factors may be involved in the progression of human gliomas.

The role of FGFs in the differentiation of tumor cells *in vitro* is not understood, but modulation of FGF receptor expression can be observed during the process of differentiation of certain tumor cell lines.

Differentiation of Tera-2 human embryonal carcinoma cells induced by treatment with retinoic acid is accompanied by marked downregulation of FGFR-2 and FGFR-3 mRNAs and loss of mRNA for FGFR-4, while the levels of FGFR-1 mRNA remain unchanged.[272] These changes are associated with altered FGF regulation of immediate-early gene expression and DNA synthesis.

C. ROLE OF FGFs IN HUMAN PRIMARY TUMORS

The possible role of FGFs in the growth of primary human tumors is poorly understood, but there is evidence that some FGFs may have a role in the development and/or progression of certain tumors. The *hst* and *int*-2 genes are located on the same human chromosome region, 11q13; and they may derive from a common ancestor by a process of duplication.[273] Coamplification of the two genes, *hst* and *int*-2, occurs in some primary human melanomas. A number of primary breast carcinomas may also contain amplified *hst* and *int*-2 gene sequences.[274] A correlation between RNA expression and gene amplification was found only in the case of *hst* but not of *int*-2.

Increased serum levels of basic FGF, but not acidic FGF, are found in renal cell carcinoma patients with an advanced stage in the extent of growth and spread and with an increased grade of malignancy.[275] Basic FGF may be produced by renal cell carcinoma cells and could have a role in the development of this tumor. Alternatively, basic FGF could be produced in endothelial cells of this hypervascular tumor as a consequence of its development. In any case, the serum basic FGF level could be a useful diagnostic and prognostic marker in patients with renal cell carcinoma.

Alterations in the receptors for FGF are found in different types of tumors. Particular subsets of primary human breast cancers contain amplified sequences of the *bek* and *flg* genes that encode receptors for members of the FGF family.[276] Analysis of amplified FGF receptor genes may be useful for the molecular phenotyping of breast cancer and other tumors.

D. FGFs AND ONCOGENE-INDUCED TRANSFORMATION

Oncogene-transformed cells may express high levels of FGFs. Normal Rat-1 cells transformed by the EJ c-H-*ras* oncogene synthesize elevated levels of basic FGF, as compared to nontransformed Rat-1 cells.[277] The basic FGF synthesized by the oncogene-transformed cells appeared in two biologically active molecular forms with molecular weights of 18 and 22 kDa. Cells transformed by other oncogenes *(sis, neu,* and *fms)* may also synthesize high amounts of basic FGF, while no synthesis of acidic FGF occurs in these cells. Mitogenic activity present in SSV-transformed NRK cells is due not only to the v-Sis protein but also (and to a higher grade) to the endogenous production of basic FGF and another growth factor protein of 18 kDa.[278] Kaposi's sarcoma-derived cells (KS cells) transfected with simian virus 40 (SV40) secrete an FGF-like mitogen and overexpress several proto-oncogenes (c-*myc,* c-H-*ras,* N-*ras*), as well as the p53 oncosuppressor gene.[279] In addition to FGF-related factors, Kaposi's sarcoma is associated with the expression of other types of growth factors, including PDGF, TGF-β, TNFs, and various cytokines.[280]

Oncogene-induced transformation may result in abrogation of the requirement for exogenous supply of FGF and other growth factors. Transfection of mouse fibroblasts (NIH/3T3 and BALB/c 3T3 cells) with expression vectors carrying the oncogenes v-*mos,* v-*src,* or v-*sis* or carrying a mutated oncogenic form of the c-H-*ras* proto-oncogene abolishes the requirement for supply of PDGF or pituitary-derived FGF.[281] However, not all oncogenes produce this abrogation. Transfection of the same cells with plasmids carrying the v-*fos* oncogene does not result in elimination of the requirement for exogenous supply of FGF or PDGF. Oncogene-induced abrogation of the requirement for exogenous supply of growth factors may occur by stimulation of the endogenous production of these factors or by other mechanisms.

In addition to their role in oncogenesis, FGFs may be involved in the pathogenesis of other human diseases, including hyperplastic lesions and benign tumors affecting different organs and tissues. Examination of human normal and benign hyperplastic prostates indicated that acidic FGF transcripts are generally undetectable in the prostate, but basic FGF mRNA is present in all of the prostates, either normal or hyperplastic.[282] However, the levels of basic FGF mRNA were higher in benign prostatic hyperplasia than in normal prostate. TGF-β_2, but not TGF-β_1, was expressed at significantly higher levels in the hyperplastic prostates, as compared to normal prostates. FGFs may also have an important role in the pathogenesis of arteriosclerosis. Although PDGF, as a potent smooth muscle cell mitogen, has been considered to play a central role in atherogenic processes, other growth factors are undoubtedly associated with these processes and one of these factors may be FGF.[283]

REFERENCES

1. **Gospodarowicz, D., Neufeld, G., and Schweigerer, L.,** Fibroblast growth factor, *Mol. Cell. Endocrinol.,* 46, 187, 1986.
2. **Gospodarowicz, D., Ferrara, N., Schweigerer, L., and Neufeld, G.,** Structural characterization and biological functions of fibroblast growth factor, *Endocr. Rev.,* 8, 1, 1987.
3. **Thomas, K.A.,** Fibroblast growth factors, *FASEB J.,* 1, 434, 1987.
4. **Gospodarowicz, D.,** Fibroblast growth factor, *Crit. Rev. Oncogenesis,* 1, 1, 1989.
5. **Florini, J.R. and Magri, K.A.,** Effects of growth factors on myogenic differentiation, *Am. J. Physiol.,* 256, C701, 1989.
6. **Holley, R.W. and Kiernan, J.A.,** "Contact inhibition" of cell division in 3T3 cells, *Proc. Natl. Acad. Sci. U.S.A.,* 60, 300, 1968.
7. **Böhlen, P., Baird, A., Esch, F., Ling, N., and Gospodarowicz, D.,** Isolation and partial molecular characterization of pituitary fibroblast growth factor, *Proc. Natl. Acad. Sci. U.S.A.,* 81, 5364, 1984.
8. **Gospodarowicz, D., Cheng, J., Lui, G.-M., Baird, A., and Böhlen, P.,** Isolation of brain fibroblast growth factor by heparin-sepharose affinity chromatography: identity with pituitary fibroblast growth factor, *Proc. Natl. Acad. Sci. U.S.A.,* 81, 6963, 1984.
9. **Gospodarowicz, D., Massoglia, S., Cheng, J., Lui, G.-M., and Böhlen, P.,** Isolation of pituitary fibroblast growth factor by fast protein liquid chromatography (FPLC): partial chemical and biological characterization, *J. Cell. Physiol.,* 122, 323, 1985.
10. **Böhlen, P., Esch, F., Baird, A., Jones, K.L., and Gospodarowicz, D.,** Human brain fibroblast growth factor: isolation and partial chemical characterization, *FEBS Lett.,* 185, 177, 1985.
11. **Gospodarowicz, D., Baird, A., Cheng, J., Lui, G.M., Esch, F., and Böhlen, P.,** Isolation of fibroblast growth factor from bovine adrenal gland: physicochemical and biological characterization, *Endocrinology,* 118, 82, 1986.
12. **Huang, J.S., Huang, S.S., and Kuo, M.-D.,** Bovine brain-derived growth factor. Purification and characterization of its interaction with responsive cells, *J. Biol. Chem.,* 261, 11600, 1986.
13. **Baird, A., Mormede, P, and Böhlen, P.,** Immunoreactive fibroblast growth factor in cells of peritoneal exudate suggests its identity with macrophage-derived growth factor, *Biochem. Biophys. Res. Commun.,* 126, 358, 1985.
14. **Singh, J.P. and Bonin, P.D.,** Purification and biochemical properties of a human monocyte-derived growth factor, *Proc. Natl. Acad. Sci. U.S.A.,* 85, 6374, 1988.
15. **Gospodarowicz, D. and Cheng, J.,** Heparin protects basic and acidic FGF from inactivation, *J. Cell. Physiol.,* 128, 475, 1986.
16. **Mascarelli, F., Raullais, D., Counis, M.F., and Courtois, Y.,** Characterization of acidic and basic fibroblast growth factors in brain, retina and vitreous chick embryo, *Biochem. Biophys. Res. Commun.,* 146, 478, 1987.
17. **Böhlen, P., Esch, F., Baird, A., and Gospodarowicz, D.,** Acidic fibroblast growth factor (FGF) from bovine brain: amino-terminal sequence and comparison with basic FGF, *EMBO J.,* 4, 1951, 1985.
18. **Zhu, X., Komiya, H., Chirino, A., Faham, S., Fox, G.M., Arakawa, T., Hsu, B.T., and Rees, D.C.,** Three-dimensional structures of acidic and basic fibroblast growth factors, *Science,* 251, 90, 1991.
19. **Mergia, A., Eddy, R., Abraham, J.A., Fiddes, J.C., and Shows, T.B.,** The genes for basic and acidic fibroblast growth factors are on different human chromosomes, *Biochem. Biophys. Res. Commun.,* 138, 644, 1986.
20. **Root, L.L. and Shipley, G.D.,** Human dermal fibroblasts express multiple bFGF and aFGF proteins, *In Vitro Cell. Dev. Biol.,* 27A, 815, 1991.
21. **Benharroch, D. and Birnbaum, D.,** Biology of the fibroblast growth factor gene family, *Israel J. Med. Sci.,* 26, 212, 1990.
22. **Baird, A. and Klagsbrun, M.,** The fibroblast growth factor family, *Cancer Cells,* 3, 239, 1991.
23. **Montesano, R., Vassalli, J.D., Baird, A., Guillemin, R., and Orci, L.,** Basic fibroblast growth factor induces angiogenesis *in vivo, Proc. Natl. Acad. Sci. U.S.A.,* 83, 7297, 1986.
24. **Schweigerer, L., Neufeld, G., Friedman, J., Abraham, J.A., Fiddes, J.C., and Gospodarowicz, D.,** Capillary endothelial cells express basic fibroblast growth factor, a mitogen that promotes their own growth, *Nature (London),* 325, 257, 1987.
25. **Matsuzaki, K., Yoshitake, Y., Matuo, Y., Sasaki, H., and Nishikawa, K.,** Monoclonal antibodies against heparin-binding growth factor II/basic fibroblast growth factor that block its biological activity: invalidity of the antibodies for tumor angiogenesis, *Proc. Natl. Acad. Sci. U.S.A.,* 86, 9911, 1989.

26. **Abraham, J.A., Whang, J.L., Tumolo, A., Mergia, A., Friedman, J., Gospodarowicz, D., and Fiddes, J.C.,** Human basic fibroblast growth factor: nucleotide sequence and genomic organization, *EMBO J.*, 5, 2523, 1986.
27. **Halaban, R., Langdon, R., Birchall, N., Cuono, C., Baird, A., Scott, G., Moellmann, G., and McGuire, J.,** Basic fibroblast growth factor from human keratinocytes is a natural mitogen for melanocytes, *J. Cell Biol.*, 107, 1611, 1988.
28. **Mignatti, P., Morimoto, T., and Rifkin, D.B.,** Basic fibroblast growth factor, a protein devoid of secretory signal sequence, is released by cells via a pathway independent of the endoplasmic reticulum-Golgi complex, *J. Cell. Physiol.*, 151, 81, 1992.
29. **Yamanishi, D.T., Graham, M.J., Florkiewicz, R.Z., Buckmeier, J.A., and Meyskens, F.L., Jr.,** Differences in basic fibroblast growth factor RNA and protein levels in human primary melanocytes and metastatic melanoma cells, *Cancer Res.*, 52, 5024, 1992.
30. **Presta, M.,** Sex hormones modulate the synthesis of basic fibroblast growth factor in human endometrial carcinoma cells: implications for the neovascularization of normal and neoplastic endometrium, *J. Cell. Physiol.*, 137, 593, 1988.
31. **Murphy, P.R., Sato, Y., Sato, R., and Friesen, H.G.,** Regulation of multiple basic fibroblast growth factor messenger ribonucleic acid transcripts by protein kinase C activators, *Mol. Endocrinol.*, 2, 1196, 1988.
32. **Abraham, J.A., Mergia, A., Whang, J.L., Tumolo, A., Friedman, J., Hjerrild, K.A., Gospodarowicz, D., and Fiddes, J.C.,** Nucleotide sequence of a bovine clone encoding the angiogenic protein, basic fibroblast growth factor, *Science*, 233, 545, 1986.
33. **Kurokawa, T., Sasada, R., Iwane, M., and Igarashi, K.,** Cloning and expression of cDNA encoding human basic fibroblast growth factor, *FEBS Lett.*, 213, 189, 1987.
34. **Iwane, M., Kurokawa, T., Sasada, R., Seno, M., Nakagawa, S., and Igarashi, K.,** Expression of cDNA encoding human basic fibroblast growth factor in *E. coli, Biochem. Biophys. Res. Commun.*, 146, 470, 1987.
35. **Wang, W.P., Quick, D., Baicerzak, S.P., Needleman, S.W., and Chiu, I.M.,** Cloning and sequence analysis of the human acidic fibroblast growth factor gene and its preservation in leukemic patients, *Oncogene*, 6, 1521, 1991.
36. **Eriksson, A.E., Cousens, L.S., Weaver, L.H., and Matthews, B.W.,** Three-dimensional structure of human basic fibroblast growth factor, *Proc. Natl. Acad. Sci. U.S.A.*, 88, 3441, 1991.
37. **Zhang, J.D., Cousens, L.S., Barr, P.J., and Sprang, S.R.,** Three-dimensional structure of human basic fibroblast growth factor, a structural homolog of interleukin-1β, *Proc. Natl. Acad. Sci. U.S.A.*, 88, 3446, 1991.
38. **Lafage-Pochilatoff, M., Galland, F., Simonetti, J., Pratts, H., Mattei, M.G., and Birnbaum, D.,** The human basic fibroblast growth factor gene is located on the long arm of chromosome 4, at bands q26-q27, *Oncogene Res.*, 5, 241, 1990.
39. **Klagsbrun, M., Smith, S., Sullivan, R., Shing, Y., Davidson, S., Smith, J.A., and Sasse, J.,** Multiple forms of basic fibroblast growth factor: amino-terminal cleavages by tumor cell-and brain cell-derived acid proteinases, *Proc. Natl. Acad. Sci. U.S.A.*, 84, 1839, 1987.
40. **Ueno, N., Baird, A., Esch, F., Ling, N., and Guillemin, R.,** Isolation of an amino terminal extended form of basic fibroblast growth factor, *Biochem. Biophys. Res. Commun.*, 138, 580, 1986.
41. **Sommer, A., Brewer, M.T., Thompson, R.C., Moscatelli, D., Presta, M., and Rifkin, D.B.,** A form of human basic fibroblast growth factor with an extended amino terminus, *Biochem. Biophys. Res. Commun.*, 144, 543, 1987.
42. **Florkiewicz, R.Z. and Sommer, A.,** Human fibroblast growth factor gene encodes four polypeptides: three initiate translation from non-AUG codons, *Proc. Natl. Acad. Sci. U.S.A.*, 86, 3978, 1989.
43. **Sternfeld, M.D., Hendrickson, J.E., Keeble, W.W., Rosenbaum, J.T., Robertson, J.E., Pittelkow, M.R., and Shipley, G.D.,** Differential expression of mRNA coding for heparin-binding growth factor type 2 in human cells, *J. Cell. Physiol.*, 136, 297, 1988.
44. **Story, M.T., Esch, F., Shimasaki, S., Sasse, J., Jacobs, S.C., and Lawson, R.K.,** Amino-terminal sequence of a large form of basic fibroblast growth factor isolated from human benign prostatic hyperplastic tissue, *Biochem. Biophys. Res. Commun.*, 142, 702, 1987.
45. **Moscatelli, D., Joseph-Silverstein, J., Manejias, R., and Rifkin, D.B.,** M_r 25,000 heparin-binding protein from guinea pig brain is a high molecular weight form of basic fibroblast growth factor, *Proc. Natl. Acad. Sci. U.S.A.*, 84, 5778, 1987.

46. **Courty, J., Chevalier, B., Moenner, M., Loret, C., Lagente, O., Böhlen, P., Courtois, Y., and Barritault, D.,** Evidence for FGF-like growth factor in adult bovine retina: analogies with EDGF I, *Biochem. Biophys. Res. Commun.,* 136, 102, 1986.
47. **Pettmann, B., Weibel, M., Sensenbrenner, M., and Labourdette, G.,** Purification of two astroglial growth factors from bovine brain, *FEBS Lett.,* 189, 102, 1985.
48. **Unsicker, K., Reichert-Preibsch, H., Schmidt, R., Pettmann, B., Labourdette, G., and Sensenbrenner, M.,** Astroglial and fibroblast growth factors have neurotrophic functions for cultured peripheral and central nervous system neurons, *Proc. Natl. Acad. Sci. U.S.A.,* 84, 5459, 1987.
49. **Too, C.K.L., Murphy, P.R., Hamel, A.-M., and Friesen, H.G.,** Further purification of human pituitary-derived chondrocyte growth factor: heparin-binding and cross-reactivity with antiserum to basic FGF, *Biochem. Biophys. Res. Commun.,* 144, 1128, 1987.
50. **Klagsbrun, M., Sasse, J., Sullivan, R., and Smith, J.A.,** Human tumor cells synthesize an endothelial cell growth factor that is structurally related to basic fibroblast growth factor, *Proc. Natl. Acad. Sci. U.S.A.,* 83, 2448, 1986.
51. **Houck, K.A., Zarnegar, R., Muga, S.J. and Michalopoulos, G.K.,** Acidic fibroblast growth factor (HBGF-1) stimulates DNA synthesis in primary rat hepatocyte cultures, *J. Cell. Physiol.,* 143, 129, 1990.
52. **Rodan, S.B., Wesolowski, G., Thomas, K., and Rodan, G.A.,** Growth stimulation of rat calvaria osteoblastic cells by acidic fibroblast growth factor, *Endocrinology,* 121, 1917, 1987.
53. **Folkman, J., Watson, K., Ingber, D., and Hanahan, D.,** Induction of angiogenesis during the transition from hyperplasia to neoplasia, *Nature (London),* 339, 58, 1989.
54. **Schreiber, A.B., Kenney, J., Kowalski, J., Thomas, K.A., Gimenez-Gallego, G., Rios-Candelore, M., DiSalvo, J., Barritault, D., Courty, J., Courtois, Y., Moenner, M., Loret, C., Burgess, W.H., Mehlman, T., Friesel, R., Johnson, W., and Maciag, T.,** A unique family of endothelial cell polypeptide mitogens: the antigenic and receptor cross-reactivity of bovine endothelial cell growth factor, brain-derived acidic fibroblast growth factor, and eye-derived growth factor-II, *J. Cell Biol.,* 101, 1623, 1985.
55. **Gimenez-Gallego, G., Conn, G., Hatcher, V.B., and Thomas, K.A.,** The complete amino acid sequence of human brain-derived acidic fibroblast growth factor, *Biochem. Biophys. Res. Commun.,* 138, 611, 1986.
56. **Lobb, R.R., Rybak, S.M., St.Clair, D.K., and Fett, J.W.,** Lysates of two established human tumor lines contain heparin-binding growth factors related to bovine acidic brain fibroblast growth factor, *Biochem. Biophys. Res. Commun.,* 139, 861, 1986.
57. **Thomas, K.A., Rios-Candelore, M., Gimenez-Gallego, G., DiSalvo, J., Bennett, C., Rodkey, J., and Fitzpatrick, S.,** Pure brain-derived acidic fibroblast growth factor is a potent angiogenic vascular endothelial cell mitogen with sequence homology to interleukin 1, *Proc. Natl. Acad. Sci. U.S.A.,* 82, 6409, 1985.
58. **Esch, F., Ueno, N., Baird, A., Hill, F., Denoroy, L., Ling, N., Gospodarowicz, D., and Guillemin, R.,** Primary structure of bovine brain acidic fibroblast growth factor (FGF), *Biochem. Biophys. Res. Commun.,* 133, 554, 1985.
59. **Gautschi-Sova, P., Müller, T., and Böhlen, P.,** Amino acid sequence of human acidic fibroblast growth factor, *Biochem. Biophys. Res. Commun.,* 140, 874, 1986.
60. **Mergia, A., Tischer, E., Graves, D., Tumolo, A., Miller, J., Gospodarowicz, D., Abraham, J.A., Shipley, G.D., and Fiddes, J.C.,** Structural analysis of the gene for human acidic fibroblast growth factor, *Biochem. Biophys. Res. Commun.,* 164, 1121, 1989.
61. **Linemeyer, D.L., Kelly, L.J., Menke, J.G., Gimenez-Gallego, G., DiSalvo, J., and Thomas, K.A.,** Expression in *Escherichia coli* of a chemically synthesized gene for biologically active bovine acidic fibroblast growth factor, *Biotechnology,* 5, 960, 1987.
62. **Gimenez-Gallego, G., Rodkey, J., Bennett, C., Rios-Candelore, M., DiSalvo, J., and Thomas, K.,** Brain-derived acidic fibroblast growth factor: complete amino acid sequence homologies, *Science,* 230, 1385, 1985.
63. **Gimenez-Gallego, G., Conn, G., Hatcher, V.B., and Thomas, K.A.,** Human brain-derived and basic fibroblast growth factors: amino terminal sequences and specific mitogenic activities, *Biochem. Biophys. Res. Commun.,* 135, 541, 1986.
64. **Risau, W., Gautschi-Sova, P., and Böhlen, P.,** Endothelial cell growth factors in embryonic and adult chick brain are related to human acidic fibroblast growth factor, *EMBO J.,* 7, 959, 1988.

65. **Mester, J., Wagenaar, E., Sluyser, M., and Nusse, R.,** Activation of *int*-1 and *int*-2 mammary oncogenes in hormone-dependent and -independent mammary tumors of GR mice, *J. Virol.,* 61, 1073, 1987.
66. **Nusse, R.,** The *int* genes in mammary tumorigenesis and in normal development, *Trends Genet.,* 4, 291, 1988.
67. **Dickson, C. and Peters, G.,** Potential oncogene product related to growth factors, *Nature (London),* 326, 833, 1987.
68. **Smith, R., Peters, G., and Dickson, C.,** Multiple RNAs expressed from the *int*-2 gene in mouse embryonal carcinoma cell lines encode a protein with homology to fibroblast growth factors, *EMBO J.,* 7, 1013, 1988.
69. **McMahon, A.P. and Moon, R.T.,** *int*-1 — a proto-oncogene involved in cell signalling, *Development,* Suppl. 1, 161, 1989.
70. **Nusse, R. and Varmus, H.E.,** *Wnt* genes, *Cell,* 69, 1073, 1992.
71. **Dixon, M., Deed, R., Acland, P., Moore, R., Whyte, A., Peters, G., and Dickson, C.,** Detection and characterization of the fibroblast growth factor-related oncoprotein INT-2, *Mol. Cell. Biol.,* 9, 4896, 1989.
72. **Wilkinson, D.G., Peters, G., Dickson, C., and McMahon, A.P.,** Expression of the FGF-related proto-oncogene *int*-2 during gastrulation and neurulation in the mouse, *EMBO J.,* 7, 691, 1988.
73. **Represa, J., Leon, Y., Miner, C., and Giraldez, F.,** The *int*-2 proto-oncogene is responsible for induction of the inner ear, *Nature (London),* 353, 561, 1991.
74. **Merlo, G.R., Blondel, B.J., Deed, R., MacAllan, D., Peters, G., Dickson, C., Liscia, D.S., Ciardiello, F., Valverius, E.M., Salomon, D.S., and Callahan, R.,** The mouse *int*-2 gene exhibits basic fibroblast growth factor activity in a basic fibroblast growth factor-responsive cell line, *Cell Growth Differ.,* 1, 463, 1990.
75. **Kiefer, P., Peters, G., and Dickson, C.,** The *Int-2/Fgf-3* oncogene product is secreted and associates with extracellular matrix: implications for cell transformation, *Mol. Cell. Biol.,* 11, 5929, 1991.
76. **Venesio, T., Taverna, D., Hynes, N.E., Deed, R., MacAllan, D., Ciardiello, F., Valverius, E.M., Salomon, D.S., Callahan, R., and Merlo, G.,** The *int*-2 gene product acts as a growth factor and substitutes for basic fibroblast growth factor in promoting the differentiation of a normal mouse mammary epithelial cell line, *Cell Growth Differ.,* 3, 63, 1992.
77. **Sakamoto, H., Mori, M., Taira, M., Yoshida, T., Matsukawa, S., Shimizu, K., Sekiguchi, M., Terada, M., and Sugimura, T.,** Transforming gene from human stomach cancers and a noncancerous portion of stomach mucosa, *Proc. Natl. Acad. Sci. U.S.A.,* 83, 3997, 1986.
78. **Taira, M., Yoshida, T., Miyagawa, K., Sakamoto, H., Terada, M., and Sugimura, T.,** cDNA sequence of human transforming gene *hst* and identification of the coding sequence required for transforming activity, *Proc. Natl. Acad. Sci. U.S.A.,* 84, 2980, 1987.
79. **Delli Bovi, P. and Basilico, C.,** Isolation of a rearranged human transforming gene following transfection of Kaposi sarcoma DNA, *Proc. Natl. Acad. Sci. U.S.A.,* 84, 5660, 1987.
80. **Delli Bovi, P., Curatola, A.M., Kern, F.G., Greco, A., Ittmann, M., and Basilico, C.,** An oncogene isolated by transfection of Kaposi's sarcoma DNA encodes a growth factor that is a member of the FGF family, *Cell,* 50, 729, 1987.
81. **Fuller-Pace, F., Peters, G., and Dickson, C.,** Cell transformation by *kFGF* requires secretion but not glycosylation, *J. Cell Biol.,* 115, 547, 1991.
82. **Yoshida, T., Miyagawa, K., Odagiri, H., Sakamoto, H., Little, P.F.R., Terada, M., and Sugimura, T.,** Genomic sequence of *hst,* a transforming gene encoding a protein homologous to fibroblast growth factors and the *int*-2-encoded protein, *Proc. Natl. Acad. Sci. U.S.A.,* 84, 7305, 1987.
83. **Delli-Bovi, P., Curatola, A.M., Newman, K.M., Sato, Y., Moscatelli, D., Hewick, R.M., Rifkin, D.B., and Basilico, C.,** Processing, secretion, and biological properties of a novel growth factor of the fibroblast family with oncogenic potential, *Mol. Cell. Biol.,* 8, 2933, 1988.
84. **Brookes, S., Smith, R., Thurlow, J., Dickson, C., and Peters, G.,** The mouse homologue of *hst/k*-FGF: sequence, genome organization and location relative to *int-2*, *Nucleic Acids Res.,* 17, 4037, 1989.
85. **Curatola, A.M. and Basilico, C.,** Expression of the K-*fgf* proto-oncogene is controlled by 3′ regulatory elements which are specific for embryonal carcinoma cells, *Mol. Cell. Biol.,* 10, 2475, 1990.
86. **Velcich, A., Delli-Bovi, P., Mansukhani, A., Ziff, E.B., and Basilico, C.,** Expression of the K-*fgf* proto-oncogene is repressed during differentiation of F9 cells, *Oncogene Res.,* 5, 31, 1989.
87. **Zhan, X., Culpepper, A., Reddy, M., Loveless, J., and Goldfarb, M.,** Human oncogenes detected by a defined medium culture assay, *Oncogene,* 1, 369, 1987.

88. **Zhan, X., Bates, B., Hu, X., and Goldfarb, M.,** The human FGF-5 oncogene encodes a novel protein related to fibroblast growth factors, *Mol. Cell. Biol.,* 8, 3487, 1988.
89. **Werner, S., Roth, W.K., Bates, B., Goldfarb, M., and Hofschneider, P.H.,** Fibroblast growth factor 5 proto-oncogene is expressed in normal human fibroblasts and induced by serum growth factors, *Oncogene,* 6, 2137, 1991.
90. **Dionne, C.A., Kaplan, R., Seuánez, H., O'Brien, S.J., and Jaye, M.,** Chromosome assignment by polymerase chain reaction techniques: assignment of the oncogene FGF-5 to human chromosome 4, *Biotechniques,* 8, 190, 1990.
91. **Marics, I., Adelaide, J., Raybaud, F., Mattei, M.-G., Coulier, F., Planche, J., de Lapeyriere, O., and Birnbaum, D.,** Characterization of the *HST*-related *FGF* 6 gene, a new member of the fibroblast growth factor gene family, *Oncogene,* 4, 335, 1989.
92. **Coulier, F., Batoz, M., Marics, I., de Lapeyrière, O., and Birnbaum, D.,** Putative structure of the *FGF6* gene product and role of the signal peptide, *Oncogene,* 6, 1437, 1991.
93. **de Lapeyriere, O., Rosnet, O., Benharroch, D., Raybaud, F., Marchetto, S., Planche, J., Galland, F., Mattei, M.-G., Copeland, N.G., Jenkins, N.A., Coulier, F., and Birnbaum, D.,** Structure, chromosome mapping and expression of the murine *Fgf*-6 gene, *Oncogene,* 5, 823, 1990.
94. **Rubin, J.S., Osada, H., Finch, P.W., Taylor, W.G., Rudikoff, S., and Aaronson, S.A.,** Purification and characterization of a newly identified growth factor specific for epithelial cells, *Proc. Natl. Acad. Sci. U.S.A.,* 86, 802, 1989.
95. **Ron, D., Bottaro, D.P., Finch, P.W., Morris, D., Rubin, J.S., and Aaronson, S.A.,** Expression of biologically active recombinant keratinocyte growth factor. Structure/function analysis of amino-terminal truncation mutants, *J. Biol. Chem.,* 268, 2984, 1993.
96. **Werner, S., Peters, K.G., Longaker, M.T., Fuller-Pace, F., Banda, M.J., and Williams, L.T.,** Large induction of keratinocyte growth factor expression in the dermis during wound healing, *Proc. Natl. Acad. Sci. U.S.A.,* 89, 6896, 1992.
97. **Guo, L., Yu, Q.-C., and Fuchs, E.,** Targeting expression of keratinocyte growth factor to keratinocytes elicits striking changes in epithelial differentiation in transgenic mice, *EMBO J.,* 12, 973, 1993.
98. **Itoh, T., Suzuki, M., and Mitsui, Y.,** Keratinocyte growth factor as a mitogen for primary culture of rat hepatocytes, *Biochem. Biophys. Res. Commun.,* 192, 1011, 1993.
99. **Jaye, M., Howk, R., Burgess, W., Ricca, G.A., Chiu, I.-M., Ravera, M.W., O'Brien, S.J., Modi, W.S., Maciag, T., and Drohan, W.N.,** Human endothelial cell growth factor: cloning, nucleotide sequence, and chromosome localization, *Science,* 233, 541, 1986.
100. **Greil, W., Rafferzeder, M., Bechner, G., and Gärtner, R.,** Release of an endothelial cell growth factor from cultured porcine thyroid follicles, *Mol. Endocrinol.,* 3, 858, 1989.
101. **Maciag, T., Cerundolo, S., Ilsley, S., Kelley, P.R., and Forand, R.,** An endothelial cell growth factor from bovine hypothalamus: identification and partial characterization, *Proc. Natl. Acad. Sci. U.S.A.,* 76, 5674, 1979.
102. **Maciag, T., Hoover, G.A., and Weinstein, R.,** High and low molecular weight forms of endothelial cell growth factor, *J. Biol. Chem.,* 257, 5333, 1982.
103. **Burgess, W.H., Mehlman, T., Marshak, D.R., Fraser, B.A., and Maciag, T.,** Structural evidence that endothelial cell growth factor β is the precursor of both endothelial cell growth factor α and acidic fibroblast growth factor, *Proc. Natl. Acad. Sci. U.S.A.,* 83, 7216, 1986.
104. **Friesel, R., Burgess, W.H., Mehlman, T., and Maciag, T.,** The characterization of the receptor for endothelial cell growth factor by covalent ligand attachment, *J. Biol. Chem.,* 261, 7581, 1986.
105. **Maciag, T., Mehlman, T., Friesel, R., and Schreiber, A.B.,** Heparin binds endothelial cell growth factor, the principal endothelial cell mitogen in bovine brain, *Science,* 225, 932, 1984.
106. **Gordon, P.B., Conn, G., and Hatcher, V.B.,** Glycosaminoglycan production in cultures of early and late passage human endothelial cells: the influence of an anionic endothelial cell growth factor and the extracellular matrix, *J. Cell. Physiol.,* 125, 596, 1985.
107. **Gay, C.G. and Winkles, J.A.,** Heparin-binding growth factor-1 stimulation of human endothelial cells induces platelet-derived growth factor A-chain gene expression, *J. Biol. Chem.,* 265, 3284, 1990.
108. **Norioka, K., Hara, M., Kitani, A., Hirose, T., Hirose, W., Harigai, M., Suzuki, K., Kawakami, M., Tabata, H., Kawagoe, M., and Nakamura, H.,** Inhibitory effect of human recombinant interleukin-1α and β on growth of human vascular endothelial cells, *Biochem. Biophys. Res. Commun.,* 145, 969, 1987.

109. **Okabe, T., Yamagata, K., Fujisawa, M., Takaku, F., Hidaka, H., and Umezawa, Y.,** Induction by fibroblast growth factor of angiotensin converting enzyme in vascular endothelial cells *in vitro, Biochem. Biophys. Res. Commun.,* 145, 1211, 1987.
110. **Canalis, E., Lorenzo, J., Burgess, W.H., and Maciag, T.,** Effects of endothelial cell growth factor on bone remodelling in vitro, *J. Clin. Invest.,* 79, 52, 1987.
111. **Odedra, R. and Weiss, J.B.,** A synergistic effect on microvessel cell proliferation between basic fibroblast growth factor (FGFb) and endothelial cell stimulating angiogenesis factor (ESAF), *Biochem. Biophys. Res. Commun.,* 143, 947, 1987.
112. **Chodak, G.W., Shing, Y., Borge, M., Judge, S.M., and Klagsbrun, M.,** Presence of heparin binding growth factor in mouse bladder tumors and urine from mice with bladder cancer, *Cancer Res.,* 46, 5507, 1986.
113. **Connolly, D.T., Stoddard, B.L., Harakas, N.K., and Feder, J.,** Human fibroblast-derived growth factor is a mitogen and chemoattractant for endothelial cells, *Biochem. Biophys. Res. Commun.,* 144, 705, 1987.
114. **Satoh, T., Kan, M., Kato, M., and Yamane, I.,** Purification and characterization of an endothelial cell growth factor from serum-free culture medium of human diploid fibroblast cells, *Biochim. Biophys. Acta,* 887, 86, 1986.
115. **Gospodarowicz, D., Massoglia, S., Cheng, J., and Fujii, D.K.,** Effect of fibroblast growth factor and lipoproteins on the proliferation of endothelial cells derived from bovine adrenal cortex, brain cortex, and corpus luteum capillaries, *J. Cell. Physiol.,* 127, 121, 1986.
116. **Sakaguchi, M., Kajio, T., Kawahara, K., and Kato, K.,** Antibodies against basic fibroblast growth factor inhibit the autocrine growth of pulmonary artery endothelial cells, *FEBS Lett.,* 233, 163, 1988.
117. **D'Amore, P.A. and Klagsbrun, M.,** Endothelial cell mitogens derived from retina and hypothalamus: biochemical and biological similarities, *J. Cell Biol.,* 99, 1545, 1984.
118. **Baird, A., Esch, F., Gospodarowicz, D., and Guillemin, R.,** Retina and eye derived endothelial cell growth factors: partial molecular characterization and identity with acidic and basic fibroblast growth factors, *Biochemistry,* 24, 7855, 1985.
119. **Wagner, J.A. and D'Amore, P.A.,** Neurite outgrowth induced by an endothelial cell mitogen isolated from retina, *J. Cell Biol.,* 103, 1363, 1986.
120. **Gospodarowicz, D., Massoglia, S., Cheng, J., and Fujii, D.K.,** Effect of retina-derived basic and acidic fibroblast growth factor and lipoproteins on the proliferation of retina-derived capillary endothelial cells, *Exp. Eye Res.,* 43, 459, 1986.
121. **Chen, C.-H. and Chen, S.C.,** Evidence of the presence of a specific vascular endothelial growth factor in fetal bovine retina, *Exp. Cell Res.,* 169, 287, 1987.
122. **Huang, S.S., Kuo, M.-D., and Huang, J.S.,** Transforming growth factor activity of bovine brain-derived growth factor, *Biochem. Biophys. Res. Commun.,* 139, 619, 1986.
123. **Senior, R.M., Huang, S.S., Griffin, G.L., and Huang, J.S.,** Brain-derived growth factor is a chemoattractant for fibroblasts and astroglial cells, *Biochem. Biophys. Res. Commun.,* 141, 67, 1986.
124. **Huang, S.S., Tsai, C.C., Adams, S.P., and Huang, J.S.,** Neuron localization and neuroblastoma cell expression of brain-derived growth factor, *Biochem. Biophys. Res. Commun.,* 144, 81, 1987.
125. **Huang, S.S., Lokeshwar, V.B., and Huang, J.S.,** Modulation of the epidermal growth factor receptor by brain-derived growth factor in Swiss mouse 3T3 cells, *J. Cell. Biochem.,* 36, 209, 1988.
126. **Rauvala, H.,** An 18-kd heparin-binding protein of developing brain that is distinct from fibroblast growth factors, *EMBO J.,* 8, 2933, 1989.
127. **Milner, P.G., Li, Y.-S., Hoffman, R.M., Kodner, C.M., Siegel, N.R., and Deuel, T.F.,** A novel 17 kD heparin-binding growth factor (HBGF-8) in bovine uterus: purification and N-terminal amino acid sequence, *Biochem. Biophys. Res. Commun.,* 165, 1096, 1989.
128. **Gensburger, C., Labourdette, G., and Sensenbrenner, M.,** Brain basic fibroblast growth factor stimulates the proliferation of rat neuronal precursor cells *in vitro, FEBS Lett.,* 217, 1, 1987.
129. **Pettmann, B., Labourdette, G., Weibel, M., and Sensenbrenner, M.,** The brain fibroblast growth factor (FGF) is localized in neurons, *Neurosci. Lett.,* 68, 175, 1986.
130. **Imaizumi, T., Hashi, K., and Kanoh, H.,** Non-heparin-binding endothelial cell growth factor from bovine omentum, *Exp. Cell Res.,* 187, 292, 1990.
131. **Lammie, P.J., Monroe, J.G., Michael, A.I., Johnson, G.D., Phillips, S.M., and Prystowsky, M.B.,** Partial characterization of a fibroblast-stimulating factor produced by cloned murine T lymphocytes, *Am. J. Pathol.,* 130, 189, 1988.

132. **Heath, J.K. and Isacke, C.M.,** PC13 embryonal carcinoma-derived growth factor, *EMBO J.,* 3, 2957, 1984.
133. **van Veggel, J.H., van Oostwaard, T.M.J., de Laat, S.W., and van Zoelen, E.J.J.,** PC13 embryonal carcinoma cells produce a heparin-binding growth factor, *Exp. Cell Res.,* 169, 280, 1987.
134. **Mahadevan, L.C., Aitken, A., Heath, J., and Foulkes, J.G.,** Embryonal carcinoma-derived growth factor activate protein kinase C *in vivo* and *in vitro, EMBO J.,* 6, 921, 1987.
135. **Kimelman, D. and Kirschner, M.,** Synergistic induction of mesoderm by FGF and TGF-β and the identification of an mRNA coding for FGF in the early *Xenopus* embryo, *Cell,* 51, 869, 1987.
136. **Paterno, G.D., Gillespie, L.L., Dixon, M.S., Slack, J.M.W., and Heath, J.K.,** Mesoderm-inducing properties of INT-2 and kFGF: two oncogene-encoded growth factors related to FGF, *Development,* 106, 79, 1989.
137. **Slack, J.M.W., Darlington, B.G., Heath, J.K., and Godsave, S.F.,** Mesoderm induction in early *Xenopus* embryos by heparin-binding growth factors, *Nature (London),* 326, 197, 1987.
138. **Kimelman, D., Abraham, J.A., Haaparanta, T., Palisi, T.M., and Kirschner, M.,** The presence of fibroblast growth factor in the frog egg: its role as a natural mesoderm inducer, *Science,* 242, 1053, 1988.
139. **Albert, P., Boilly, B., Courty, J., and Barritault, D.,** Stimulation in cell culture of mesenchymal cells of newt limb blastemas by EDGF I or II (basic of acidic FGF), *Cell Differ.,* 21, 63, 1987.
140. **Liu, L. and Nicoll, C.S.,** Evidence for a role of basic fibroblast growth factor in rat embryonic growth and differentiation, *Endocrinology,* 123, 2027, 1988.
141. **Nabel, E.G., Yang, Z., Plautz, G., Forough, R., Zhan, X., Haudenschäd, C.C., Maciag, T., and Nabel, G.J.,** Recombinant fibroblast growth factor-1 promotes intimal hyperplasia and angiogenesis in arteries *in vivo, Nature (London),* 362, 844, 1993.
142. **Björnsson, T.D., Dryjski, M., Tluzzek, J., Mennie, R., Ronan, J., Mellin, T.N., and Thomas, K.A.,** Acidic fibroblast growth factor promotes vascular repair, *Proc. Natl. Acad. Sci. U.S.A.,* 88, 8651, 1991.
143. **Chleboun, J.O., Martins, R.N., Mitchell, C.A., and Chirila, T.V.,** bFGF enhances the development of the collateral circulation after acute arterial occlusion, *Biochem. Biophys. Res. Commun.,* 185, 510, 1992.
144. **Lindner, V. and Reidy, M.A.,** Proliferation of smooth muscle cells after vascular injury is inhibited by an antibody against fibroblast growth factor, *Proc. Natl. Acad. Sci. U.S.A.,* 88, 3739, 1991.
145. **Kan, M., Huang, M., Mansson, P.-E., Yasumitsu, H., Carr, B., and McKeehan, W.L.,** Heparin-binding growth factor type 1 (acidic fibroblast growth factor): a potential biphasic autocrine and paracrine regulator of hepatocyte regeneration, *Proc. Natl. Acad. Sci. U.S.A.,* 86, 7432, 1989.
146. **Rodan, S.B., Wesolowski, G., Thomas, K.A., Yoon, K., and Rodan, G.A.,** Effects of acidic and basic fibroblast growth factors on osteoblastic cells, *Connect. Tissue Res.,* 20, 283, 1989.
147. **Ristow, H.-J. and Messmer, T.O.,** Basic fibroblast growth factor and insulin-like growth factor I are strong mitogens for cultured mouse keratinocytes, *J. Cell. Physiol.,* 137, 277, 1988.
148. **Hoffmann, B. and Paul, D.,** Basic fibroblast growth factor and transforming growth factor-α are hepatotrophic mitogens *in vitro, J. Cell. Physiol.,* 142, 149, 1990.
149. **Gabbianelli, M., Sargiacomo, M., Pelosi, E., Testa, U., Isacchi, G., and Peschle, C.,** "Pure" human hematopoietic progenitors: permissive action of basic fibroblast growth factor, *Science,* 249, 1561, 1990.
150. **Gallicchio, V.S., Hughes, N.K., Hulette, B.C., Della Puca, R., and Noblitt, L.,** Basic fibroblast growth factor (B-FGF) induces early- (CFU-s) and late-stage hematopoietic progenitor cell colony formation (CFU-gm, CFU-meg, and BFU-e) by synergizing with GM-CSF, Meg-CSF, and erythropoietin, and is a radioprotective agent *in vitro, Int. J. Cell Cloning,* 9, 220, 1991.
151. **Gabrilove, J.L., Wong, G., Bollenbacher, E., White, K., Kojima, S., and Wilson, E.L.,** Basic fibroblast growth factor counteracts the suppressive effect of transforming growth factor-β1 on human myeloid progenitor cells, *Blood,* 81, 909, 1993.
152. **Donoviel, D.B., Amacher, S.L., Judge, K.W., and Bornstein, P.,** Thrombosponding gene expression is associated with mitogenesis in 3T3 cells: induction by basic fibroblast growth factor, *J. Cell. Physiol.,* 145, 16, 1990.
153. **Bauer, E.A., Silverman, N., Busiek, D.F., Kronberger, A., and Deuel, T.F.,** Diminished response of Werner's syndrome fibroblasts to growth factors PDGF and FGF, *Science,* 234, 1240, 1986.
154. **Ross, M. and Ballard, F.J.,** Regulation of protein metabolism and DNA synthesis by fibroblast growth factor in BHK-21 cells, *Biochem. J.,* 249, 363, 1988.

155. **Lathrop, B., Olson, E., and Glaser, L.,** Control by fibroblast growth factor of differentiation in the BC_3H1 muscle cell line, *J. Cell Biol.,* 100, 1540, 1985.
156. **Neufeld, G., Gospodarowicz, D., Dodge, L., and Fujii, D.K.,** Heparin modulation of the neurotropic effects of acidic and basic fibroblast growth factors and nerve growth factor on PC12 cells, *J. Cell. Physiol.,* 131, 131, 1987.
157. **Okumura, T., Uehara, A., Isuji, K., Taniguchi, Y., Kitamori, S., Shibata, Y., Okamura, K., Takasugi, Y., and Namiki, M.,** Central nervous system action of basic fibroblast growth factor — Inhibition of gastric acid and pepsin secretion, *Biochem. Biophys. Res. Commun.,* 175, 527, 1991.
158. **Baird, A., Mormede, P., Ying, S.-Y., Wehrenberg, W.B., Ueno, N., Ling, N., and Guillemin, R.,** A nonmitogenic pituitary function of fibroblast growth factor: regulation of thyrotropin and prolactin secretion, *Proc. Natl. Acad. Sci. U.S.A.,* 82, 5545, 1985.
159. **Taylor, A.H., Millatt, L.J., Whitley, G.S., Johnstone, A.P., and Nussey, S.S.,** The effect of basic fibroblast growth factor on the growth and function of human thyrocytes, *J. Endocrinol.,* 136, 339, 1993.
160. **Sakaguchi, K.,** Acidic fibroblast growth factor autocrine system as a mediator of calcium-regulated parathyroid cell growth, *J. Biol. Chem.,* 267, 24554, 1992.
161. **Oberbauer, A.M., Linkhart, T.A., Mohan, S., and Longo, L.D.,** Fibroblast growth factor enhances human chorionic gonadotropin synthesis independent of mitogenic stimulation in Jar choriocarcinoma cells, *Endocrinology,* 123, 2696, 1988.
162. **Fauser, B.C.J.M., Baird, A., and Hsueh, A.J.W.,** Fibroblast growth factor inhibits luteinizing hormone-stimulated androgen production by cultured rat testicular cells, *Endocrinology,* 123, 2935, 1988.
163. **Moncada, S., Palmer, R.M.J., and Higgs, E.A.,** Nitric oxide — Physiology, pathophysiology, and pharmacology, *Pharmacol. Rev.,* 43, 109, 1991.
164. **Goureau, O., Lepoivre, M., Becquet, F., and Courtois, Y.,** Differential regulation of inducible nitric oxide synthase by fibroblast growth factors and transforming growth factor β in bovine retinal pigmented epithelial cells: inverse correlation with cellular proliferation, *Proc. Natl. Acad. Sci. U.S.A.,* 90, 4276, 1993.
165. **Olwin, B.B. and Hauschka, S.D.,** Identification of the fibroblast growth factor receptor of Swiss 3T3 cells and mouse skeletal muscle myoblasts, *Biochemistry,* 25, 3487, 1986.
166. **Imamura, T., Tokita, Y., and Mitsui, Y.,** Purification of basic FGF receptors from rat brain, *Biochem. Biophys. Res. Commun.,* 155, 583, 1988.
167. **Olwin, B.B. and Hauschka, S.D.,** Cell type and tissue distribution of the fibroblast growth factor receptor, *J. Cell. Biochem.,* 39, 443, 1989.
168. **Moenner, M., Chevalier, B., Badet, J., and Barritault, D.,** Evidence and characterization of the receptor to eye-derived growth factor I, the retinal form of basic fibroblast growth factor, on bovine epithelial lens cells, *Proc. Natl. Acad. Sci. U.S.A.,* 83, 5024, 1986.
169. **Huang, S.S. and Huang, J.S.,** Association of bovine brain-derived growth factor receptor with protein tyrosine kinase activity, *J. Biol. Chem.,* 261, 9568, 1986.
170. **Kuo, M.-D., Huang, S.S., and Huang, J.S.,** Acidic fibroblast growth factor receptor purified from bovine liver is a novel protein tyrosine kinase, *J. Biol. Chem.,* 265, 16455, 1990.
171. **DiSorbo, D., Shi, E., and McKeehan, W.L.,** Purification from human hepatoma cells of a 130-kDa membrane glycoprotein with properties of the heparin-binding growth factor receptor, *Biochem. Biophys. Res. Commun.,* 157, 1007, 1988.
172. **Moscatelli, D.,** High and low affinity binding sites for basic fibroblast growth factor on cultured cells: absence of a role for low affinity binding in the stimulation of plasminogen activator production by bovine capillary endothelial cells, *J. Cell. Physiol.,* 131, 123, 1987.
173. **Vigny, M., Ollier-Hartmann, M.P., Lavigne, M., Fayein, N., Jeanny, J.C., Laurent, M., and Courtois, Y.,** Specific binding of basic fibroblast growth factor to basement membrane-like structures and to purified heparan sulfate proteoglycan of the EHS tumor, *J. Cell. Physiol.,* 137, 321, 1988.
174. **Kan, M., Wang, F., Xu, J., Crabb, J.W., Hou, J., and McKeehan, W.L.,** An essential heparin-binding domain in the fibroblast growth factor receptor kinase, *Science,* 259, 1918, 1993.
175. **Jaye, M., Schlessinger, J., and Dionne, C.A.,** Fibroblast growth factor receptor tyrosine kinases — Molecular analysis and signal transduction, *Biochim. Biophys. Acta,* 1135, 185, 1992.
176. **Givol, D. and Yayon, A.,** Complexity of FGF receptors. Genetic basis for structural diversity and functional specificity, *FASEB J.,* 6, 3362, 1992.

177. **Vainikka, S., Partanen, J., Bellosta, P., Coulier, F., Basilico, C., Jaye, M., and Alitalo, K.,** Fibroblast growth factor receptor-4 shows novel features in genomic structure, ligand binding and signal transduction, *EMBO J.*, 11, 4273, 1992.
178. **Ruta, M., Howk, R., Ricca, G., Drohan, W., Zabelshansky, M., Laureys, G., Barton, D.E., Francke, U., Schlessinger, J., and Givol, D.,** A novel protein tyrosine kinase gene whose expression is modulated during endothelial cell differentiation, *Oncogene,* 3, 9, 1988.
179. **Ruta, M., Burgess, W., Givol, D., Epstein, J., Neiger, N., Kaplow, J., Crumley, G., Dionne, C., Jaye, M., and Schlessinger, J.,** Receptor for acidic fibroblast growth factor is related to the tyrosine kinase encoded by the *fms*-like gene (FLG), *Proc. Natl. Acad. Sci. U.S.A.,* 86, 8722, 1989.
180. **Safran, A., Avivi, A., Orr-Urtereger, A., Neufeld, G., Lonai, P., Givol, D., and Yarden, Y.,** The murine *flg* gene encodes a receptor for fibroblast growth factor, *Oncogene,* 5, 635, 1990.
181. **Dionne, C.A., Modi, W.S., Cruley, G., O'Brien, S.J., Schlessinger, J., and Jaye, M.,** *BEK,* a receptor for multiple members of the fibroblast growth factor (FGF) family, maps to human chromosome 17q23-qter, *Cytogenet. Cell Genet.,* 60, 34, 1992.
182. **Hattori, Y., Odagiri, H., Nakatani, H., Miyagawa, K., Naito, K., Sakamoto, H., Katoh, O., Yoshida, T., Sugimura, T., and Terada, M.,** K-*sam,* an amplified gene in stomach cancer, is a member of the heparin-binding growth factor receptor genes, *Proc. Natl. Acad. Sci. U.S.A.,* 87, 5983, 1990.
183. **Hattori, Y., Odagiri, H., Katoh, O., Sakamoto, H., Morita, T., Shimotohno, K., Tobinai, K., Sugimura, T., and Terada, M.,** K-*sam*-related gene, N-*sam,* encodes fibroblast growth factor receptor and is expressed in T-lymphocyte tumors, *Cancer Res.,* 52, 3367, 1992.
184. **Isacchi, A., Bergonzoni, L., and Sarmientos, P.,** Complete sequence of a human receptor for acidic and basic fibroblast growth factors, *Nucleic Acids Res.,* 18, 1906, 1990.
185. **Keegan, K., Johnson, D.E., Williams, L.T., and Hayman, M.J.,** Isolation of an additional member of the fibroblast growth factor receptor family, FGFR-3, *Proc. Natl. Acad. Sci. U.S.A.,* 88, 1095, 1991.
186. **Partanen, J., Mäkelä, T.P., Eerola, E., Kornohen, J., Hirvonen, H., Claesson-Welsh, L., and Alitalo, K.,** FGFR-4, a novel acidic fibroblast growth factor receptor with a distinct expression pattern, *EMBO J.,* 10, 1347, 1991.
187. **Holtrich, U., Bräuninger, A., Strebhardt, K., and Rübsamen-Waigmann, H.,** Two additional protein-tyrosine kinases expressed in human lung: fourth member of the fibroblast growth factor receptor family and an intracellular protein-tyrosine kinase, *Proc. Natl. Acad. Sci. U.S.A.,* 88, 10411, 1991.
188. **Katoh, O., Hattori, Y., Sato, T., Kimura, A., Kuramoto, A., Sugimura, T., and Terada, M.,** Expression of the heparin-binding growth factor receptor genes in human megakaryocytic leukemia cells, *Biochem. Biophys. Res. Commun.,* 183, 83, 1992.
189. **Johnson, D.E., Lee, P.L., Lu, J., and Williams, L.T.,** Diverse forms of a receptor for acidic and basic fibroblast growth factors, *Mol. Cell. Biol.,* 10, 4728, 1990.
190. **Mansukhani, A., Moscatelli, D., Talarico, D., Levytska, V., and Basilico, C.,** A murine fibroblast growth factor (FGF) receptor expressed in CHO cells is activated by basic and Kaposi FGF, *Proc. Natl. Acad. Sci. U.S.A.,* 87, 4378, 1990.
191. **Fasel, N.J., Bernard, M., Déglon, N., Rousseaux, M., Eisenberg, R.J., Bron, C., and Cohen, G.H.,** Isolation from mouse fibroblasts of a cDNA encoding a new form of the fibroblast growth factor receptor *(flg), Biochem. Biophys. Res. Commun.,* 178, 8, 1991.
192. **Xu, J., Nakahara, M., Crabb, J.W., Shi, E., Matuo, Y., Fraser, M., Kan, M., Hou, J., and McKeehan, W.L.,** Expression and immunochemical analysis of rat and human fibroblast growth factor receptor (flg) isoforms, *J. Biol. Chem.,* 267, 17792, 1992.
193. **Bernard, O., Li, M., and Reid, H.H.,** Expression of two different forms of fibroblast growth factor receptor 1 in different mouse tissues and cell lines, *Proc. Natl. Acad. Sci. U.S.A.,* 88, 7625, 1991.
194. **Kouhara, H., Kasayama, S., Saito, H., Matsumoto, K., and Sato, B.,** Expression cDNA cloning of fibroblast growth factor (FGF) receptor in mouse breast cancer cells: a variant form in FGF-responsive transformed cells, *Biochem. Biophys. Res. Commun.,* 176, 31, 1991.
195. **Yan, G., Wang, F., Fukabori, Y., Sussman, D., Hou, J., and McKeehan, W.L.,** Expression and transforming activity of a variant of the heparin-binding fibroblast growth factor receptor (*flg*) gene resulting from splicing of the α exon at an alternate 3′-acceptor site, *Biochem. Biophys. Res. Commun.,* 183, 423, 1992.
196. **Lee, P.L., Johnson, D.E., Cousens, L.S., Fried, V.A., and Williams, L.T.,** Purification and complementary DNA cloning of a receptor for basic fibroblast growth factor, *Science,* 245, 57, 1989.

197. **Friesel, R. and Dawid, I.B.,** cDNA cloning and developmental expression of fibroblast growth factor receptors from *Xenopus laevis, Mol. Cell. Biol.,* 11, 2481, 1991.
198. **Doctor, J.S., Hoffmann, F.M., and Olwin, B.B.,** Identification of a fibroblast growth factor-binding protein in *Drosophila melanogaster, Mol. Cell. Biol.,* 11, 2319, 1991.
199. **Shishido, E., Higashijima, S., Emori, Y., and Saigo, K.,** Two FGF-receptor homologues of *Drosophila:* one is expressed in mesodermal primordium in early embryos, *Development,* 117, 751, 1993.
200. **Wanaka, A., Milbrandt, J., and Johnson, E.M., Jr.,** Expression of FGF receptor gene in rat development, *Development,* 111, 455, 1991.
201. **Hou, J., Kan, M., McKeehan, K., McBride, G., Adams, P., and McKeehan, W.L.,** Fibroblast growth factor receptors from liver vary in three structural domains, *Science,* 251, 665, 1991.
202. **Veomett, G., Kuszynski, C., Kazakoff, P., and Rizzino, A.,** Cell density regulates the number of cell surface receptors for fibroblast growth factor, *Biochem. Biophys. Res. Commun.,* 159, 694, 1989.
203. **Sato, Y., Abe, M., and Takaki, R.,** Platelet factor-4 blocks the binding of fibroblast growth factor to the receptor and inhibits the spontaneous migration of vascular endothelial cells, *Biochem. Biophys. Res. Commun.,* 172, 595, 1990.
204. **Kaner, R.J., Baird, A., Mansukhani, A., Basilico, C., Summers, B.D., Florkiewicz, R.Z., and Hajjar, D.P.,** Fibroblast growth factor receptor is a portal of cellular entry for herpes simplex virus type 1, *Science,* 248, 1410, 1990.
205. **Ho, P.L. and Raw, I.,** Cyclic AMP potentiates bFGF-induced neurite outgrowth in PC12 cells, *J. Cell Physiol.,* 150, 647, 1992.
206. **Halperin, J.A. and Lobb, R.R.,** Effect of heparin-binding growth factors on monovalent cation transport in BALB/C 3T3 cells, *Biochem. Biophys. Res. Commun.,* 144, 115, 1987.
207. **Tsuda, T., Kaibuchi, K., Kawahara, Y., Fukuzaki, H., and Takai, Y.,** Induction of protein kinase C activation and Ca^{2+} mobilization by fibroblast growth factor in Swiss 3T3 cells, *FEBS Lett.,* 191, 205, 1985.
208. **Brown, K.D., Blakeley, D.M., and Brigstock, D.R.,** Stimulation of phosphoinositide hydrolysis in Swiss 3T3 cells by recombinant fibroblast growth factors, *FEBS Lett.,* 247, 227, 1989.
209. **Araki, S., Simada, Y., Kaji, K., and Hayashi, H.,** Role of protein kinase C in the inhibition by fibroblast growth factor of apoptosis in serum-depleted endothelial cells, *Biochem. Biophys. Res. Commun.,* 172, 1081, 1990.
210. **Hulkower, K.I., Georgescu, H.I., and Evans, C.H.,** Evidence that responses of articular chondrocytes to interleukin-1 and basic fibroblast growth factor are not mediated by protein kinase C, *Biochem. J.,* 276, 157, 1991.
211. **Magnaldo, I., L'Allemain, G., Chambard, J.C., Moenner, M., Barritault, D., and Pouysségur, J.,** The mitogenic signaling pathway of fibroblast growth factor is not mediated through polyphosphoinositide hydrolysis and protein kinase C activation in hamster fibroblasts, *J. Biol. Chem.,* 261, 16916, 1986.
212. **Moenner, M., Magnaldo, I., L'Allemain, G., Barritault, D., and Pouysségur, J.,** Early and late mitogenic events induced by FGF on bovine epithelial lens cells are not triggered by hydrolysis of polyphosphoinositides, *Biochem. Biophys. Res. Commun.,* 146, 32, 1987.
213. **Mohammadi, M., Dionne, C.A., Li, W., Li, N., Spivak, T., Honegger, A.M., Jaye, M., and Schlessinger, J.,** Point mutation in FGF receptor eliminates phosphatidylinositol hydrolysis without affecting mitogenesis, *Nature (London),* 358, 681, 1992.
214. **Matuoka, K., Fukami, K., Nakanishi, O., Kawai, S., and Takenawa, T.,** Mitogenesis in response to PDGF and bombesin abolished by microinjection of antibody to PIP_2, *Science,* 239, 640, 1988.
215. **Paris, S., Chambard, J.-C., and Pouysségur, J.,** Tyrosine kinase-activating growth factors potentiate thrombin-and AlF_4-induced phosphoinositide breakdown in hamster fibroblasts. Evidence for positive cross-talk between the two mitogenic signaling pathways, *J. Biol. Chem.,* 263, 12893, 1988.
216. **Tamura, K., Asakai, R., and Okamoto, R.,** Basic fibroblast growth factor in rat corpus luteum stimulates prostaglandin F2 α production, *Biochem. Biophys. Res. Commun.,* 178, 393, 1991.
217. **Weksler, B.B.,** Heparin and acidic fibroblast growth factor interact to decrease prostacyclin synthesis in human endothelial cells by affecting both prostaglandin H synthase and prostacyclin synthase, *J. Cell. Physiol.,* 142, 514, 1990.
218. **De Cristan, G., Morbidelli, L., Alessandri, G., Ziche, M., Cappa, A.P.M., and Gullino, P.M.,** Synergism between gangliosides and basic fibroblast growth factor in favouring survival, growth, and motility of capillary endothelium, *J. Cell. Physiol.,* 144, 505, 1990.

219. **Tucker, R.W., Chang, D.T., and Meade-Cobun, K.,** Effects of platelet-derived growth factor and fibroblast growth factor on free intracellular calcium and mitogenesis, *J. Cell. Biochem.,* 39, 139, 1989.
220. **Pelech, S.L., Olwin, B.B., and Krebs, E.G.,** Fibroblast growth factor treatment of Swiss 3T3 cells activates a subunit S6 kinase that phosphorylates a synthetic peptide substrate, *Proc. Natl. Acad. Sci. U.S.A.,* 83, 5968, 1986.
221. **Li, L., Zhou, J., James, G., Heller-Harrison, R., Czech, M.P., and Olson, E.N.,** FGF inactivates myogenic helix-loop-helix proteins through phosphorylation of a conserved protein kinase C site in their DNA-binding domains, *Cell,* 71, 1181, 1992.
222. **Doctrow, S.R. and Folkman, J.,** Protein kinase C activators suppress stimulation of capillary endothelial cell growth by angiogenic endothelial mitogens, *J. Cell Biol.,* 104, 679, 1987.
223. **Wice, B., Milbrandt, J., and Glaser, L.,** Control of muscle differentiation in BC_3H1 cells by fibroblast growth factor and vanadate, *J. Biol. Chem.,* 262, 1810, 1987.
224. **Coughlin, S.R., Barr, P.J., Cousens, L.S., Fretto, L.J., and Williams, L.T.,** Acidic and basic fibroblast growth factors stimulate tyrosine kinase activity *in vivo, J. Biol. Chem.,* 263, 988, 1988.
225. **Burgess, W.H., Dionne, C.A., Kaplow, J., Mudd, R., Friesel, R., Zilberstein, A., Schlessinger, J., and Jaye, M.,** Characterization and cDNA clonic of phospholipase C-γ, a major substrate for heparin-binding growth factor 1 (acidic fibroblast growth factor)-activated tyrosine kinase, *Mol. Cell. Biol.,* 10, 4770, 1990.
226. **Spizz, G., Roman, D., Strauss, A., and Olson, E.N.,** Serum and fibroblast growth factor inhibit myogenic differentiation through a mechanism dependent on protein synthesis and independent of cell proliferation, *J. Biol. Chem.,* 261, 9483, 1986.
227. **Nakanishi, Y., Kihara, K., Mizuno, K., Masamune, Y., Yoshitake, Y., and Nishikawa, K.,** Direct effect of basic fibroblast growth factor on gene transcription in a cell-free system, *Proc. Natl. Acad. Sci. U.S.A.,* 89, 5216, 1992.
228. **Spizz, G., Hu, J.-S., and Olson, E.N.,** Inhibition of myogenic differentiation by fibroblast growth factor or type β transforming growth factor does not require persistent c-*myc* expression, *Dev. Biol.,* 123, 500, 1987.
229. **Sigmund, O., Naor, Z., Anderson, D.J., and Stein, R.,** Effect of nerve growth factor and fibroblast growth factor on SCG10 and c-*fos* expression and neurite outgrowth in protein kinase C-depleted PC12 cells, *J. Biol. Chem.,* 265, 2257, 1990.
230. **Thomson, T.M., Green, S.H., Trotta, R.J., Burstein, D.E., and Pellicer, A.,** Oncogene N-*ras* mediates selective inhibition of c-*fos* induction by nerve growth factor and basic fibroblast growth factor in a PC12 cell line, *Mol. Cell. Biol.,* 10, 1556, 1990.
231. **Isozaki, O., Emoto, N., Tsushima, T., Sato, Y., Shizume, K., Demura, H., Akamizu, T., and Kohn, L.D.,** Opposite regulation of deoxyribonucleic acid synthesis and iodide uptake in rat thyroid cells by basic fibroblast growth factor: correlation with opposite regulation of c-*fos* and thyrotropin receptor gene expression, *Endocrinology,* 131, 2723, 1992.
232. **Smith, E.P., Hall, S.H., Monaco, L., French, F.S., Wilson, E.M., and Conti, M.,** A rat Sertoli cell factor similar to basic fibroblast growth factor increases c-*fos* messenger ribonucleic acid in cultured Sertoli cells, *Mol. Endocrinol.,* 3, 954, 1989.
233. **Kourembanas, S. and Faller, D.V.,** Platelet-derived growth factor production by human umbilical vein endothelial cells is regulated by basic fibroblast growth factor, *J. Biol. Chem.,* 264, 4456, 1989.
234. **Camps, J.L., Chang, S.-M., Hsu, T.C., Freeman, M.R., Hong, S.-J., Zhau, H.E., von Eschenbach, A.C., and Chung, L.W.K.,** Fibroblast-mediated acceleration of human epithelial tumor growth *in vivo, Proc. Natl. Acad. Sci. U.S.A.,* 87, 75, 1990.
235. **Gross, J.L., Herblin, W.F., Dusak, B.A., Czerniak, P., Diamond, M.D., Sun T., Eidsvoog, K., Dexter, D.L., and Yayon, A.,** Effects of modulation of basic fibroblast growth factor on tumor growth *in vivo, J. Natl. Cancer Inst.,* 85, 121, 1993.
236. **Kandel, J., Bossy-Wetzel, E., Radvanyi, F., Klagsbrun, M., Folkman, J., and Hanahan, D.,** Neovascularization is associated with a switch to the export of bFGF in the multistep development of fibrosarcoma, *Cell,* 66, 1095, 1991.
237. **Rodeck, U., Becker, D., and Herlyn, M.,** Basic fibroblast growth factor in human melanoma, *Cancer Cells,* 3, 308, 1991.
238. **Halaban, R.,** Responses of cultured melanocytes to defined growth factors, *Pigm. Cell Res.,* Suppl. 1, 18, 1988.
239. **Halaban, R., Kwon, B.S., Ghosh, S., Delli Bovi, P., and Baird, A.,** bFGF as an autocrine growth factor for human melanomas, *Oncogene Res.,* 3, 177, 1988.

240. **Becker, D., Meier, C.B., and Herlyn, M.,** Proliferation of human malignant melanomas is inhibited by antisense oligodeoxynucleotides targeted against basic fibroblast growth factor, *EMBO J.*, 8, 3685, 1989.
241. **Nagao, Y. and Nishikawa, K.,** Basic fibroblast growth factor, albumin, and transferrin purified from rat rhodamine fibrosarcoma tissue are all essential for growth of primary tumor cells from the same tissue in serum-free medium, *In Vitro Cell. Dev. Biol.*, 25, 873, 1989.
242. **Takahashi, J.B., Hoshimaru, M., Jaye, M., Kikuchi, H., and Hatanaka, M.,** Possible activity of acidic fibroblast growth factor as a progression factor rather than a transforming factor, *Biochem. Biophys. Res. Commun.*, 189, 398, 1992.
243. **Palmer, H., Maher, V.M., and McCormick, J.J.,** Platelet-derived growth factor or basic fibroblast growth factor induce anchorage-independent growth of human fibroblasts, *J. Cell. Physiol.*, 137, 588, 1988.
244. **Jaye, M., Lyall, R.M., Mudd, R., Schlessinger, J., and Sarver, N.,** Expression of acidic fibroblast growth factor cDNA confers growth advantage and tumorigenesis to Swiss 3T3 cells, *EMBO J.*, 7, 963, 1988.
245. **Sasada, R., Kurokawa, T., Iwane, M., and Igarashi, K.,** Transformation of mouse BALB/c 3T3 cells with human basic fibroblast growth factor cDNA, *Mol. Cell. Biol.*, 8, 588, 1988.
246. **Moscatelli, D. and Quarto, N.,** Transformation of NIH 3T3 cells with basic fibroblast growth factor or the hst/K-*fgf* oncogene causes downregulation of the fibroblast growth factor receptor: reversal of morphological transformation and restoration of receptor number by suramin, *J. Cell Biol.*, 109, 2519, 1989.
247. **Yayon, A. and Klagsbrun, M.,** Autocrine transformation by chimeric signal peptide-basic fibroblast growth factor: reversal by suramin, *Proc. Natl. Acad. Sci. U.S.A.*, 87, 5346, 1990.
248. **Dotto, G.P., Moellmann, G., Ghosh, S., Edwards, M., and Halaban, R.,** Transformation of murine melanocytes by basic fibroblast growth factor cDNA and oncogenes and selective suppression of the transformed phenotype in a reconstituted cutaneous environment, *J. Cell Biol.*, 109, 3115, 1989.
249. **Wellstein, A., Lupu, R., Zugmaier, G., Flamm, S.L., Cheville, A.L., Delli Bovi, P., Basilico, C., Lippman, M.E., and Kern, F.G.,** Autocrine growth stimulation by secreted Kaposi fibroblast growth factor but not by endogenous basic fibroblast growth factor, *Cell Growth Differ.*, 1, 63, 1990.
250. **Miyagawa, K., Kimura, S., Yoshida, T., Sakamoto, H., Takaku, F., Sugimura, T., and Terada, M.,** Structural analysis of a mature *hst*-1 protein with transforming growth factor activity, *Biochem. Biophys. Res. Commun.*, 174, 404, 1991.
251. **Taylor, W.R., Greenberg, A.H., Turley, E.A., and Wright, J.A.,** Cell motility, invasion, and malignancy induced by overexpression of K-FGF or bFGF, *Exp. Cell Res.*, 204, 295, 1993.
252. **Talarico, D. and Basilico, C.,** The K-*fgf*/*hst* oncogene induces transformation through an autocrine mechanism that requires extracellular stimulation of the mitogenic pathway, *Mol. Cell. Biol.*, 11, 1138, 1991.
253. **Talarico, D., Ittmann, M.M., Bronson, R., and Basilico, C.,** A retrovirus carrying the K-*fgf* oncogene induces diffuse meningeal tumors and soft-tissue fibrosarcomas, *Mol. Cell. Biol.*, 13, 1998, 1993.
254. **Ensoli, B., Nakamura, S., Salahuddin, S.Z., Biberfeld, P., Larsson, L., Beaver, B., Wong-Staal, F., and Gallo, R.C.,** AIDS-Kaposi's sarcoma-derived cells express cytokines with autocrine and paracrine growth effects, *Science*, 243, 223, 1989.
255. **Murakami, A., Tanaka, H., and Matsuzawa, A.,** Association of *hst* gene expression with metastatic phenotype in mouse mammary tumors, *Cell Growth Differ.*, 1, 225, 1990.
256. **Talarico, D., Ittmann, M., Balsari, A., Delli-Bovi, P., Basch, R.S., and Basilico, C.,** Protection of mice against tumor growth by immunization with an oncogene-encoded growth factor, *Proc. Natl. Acad. Sci. U.S.A.*, 87, 4222, 1990.
257. **Tiesman, J. and Rizzino, A.,** Expression and developmental regulation of the k-FGF oncogene in human and murine embryonal carcinoma cells, *In Vitro Cell. Dev. Biol.*, 25, 1193, 1989.
258. **Adnane, J., Gaudray, P., Simon, M.-P., Simony-Lafontaine, J., Jeanteur, P., and Theillet, C.,** Proto-oncogene amplification and human breast tumor phenotype, *Oncogene*, 4, 1389, 1989.
259. **Moscatelli, D., Presta, M., Joseph-Siverstein, J., and Rifkin, D.B.,** Both normal and tumor cells produce basic fibroblast growth factor, *J. Cell. Physiol.*, 129, 273, 1986.
260. **Rogelj, S., Weinberg, R.A., Fanning, P., and Klagsbrun, M.,** Basic fibroblast growth factor fused to a signal peptide transforms cells, *Nature (London)*, 331, 173, 1988.

261. **Dodge, W.H.,** Avian monocyte leukemia cells release fibroblast growth factor: implications to associated myelofibrosis, *Leukemia Res.,* 9, 1559, 1985.
262. **Schweigerer, L., Neufeld, G., Mergia, A., Abraham, J.A., Fiddes, J.C., and Gospodarowicz, D.,** Basic fibroblast growth factor in human rhabdomyosarcoma cells: implications for the proliferation and neovascularization of myoblast-derived tumors, *Proc. Natl. Acad. Sci. U.S.A.,* 84, 842, 1987.
263. **Masuda, Y., Yoshitake, Y., and Nishikawa, K.,** Secretion of DNA synthesis factor (DSF) by A431 cells that can grow in protein-free medium, *Cell Biol. Int. Rep.,* 11, 359, 1987.
264. **Corin, S.J., Chen, L.C., and Hamburger, A.W.,** Enhancement of anchorage-independent growth of a human adrenal carcinoma cell line by endogenously produced basic fibroblast growth factor, *Int. J. Cancer,* 46, 516, 1990.
265. **Mydlo, J.H., Heston, W.D.W., and Fair, W.R.,** Characterization of a heparin-binding growth factor from adenocarcinoma of the kidney, *J. Urol.,* 140, 1575, 1988.
266. **Nonomura, N., Lu, J., Tanaka, A., Yamanishi, H., Sato, B., Sonoda, T., and Matsumoto, K.,** Interaction of androgen-induced autocrine heparin-binding growth factor with fibroblast growth factor receptor on androgen-dependent Shionogi 115 cells, *Cancer Res.,* 50, 2316, 1990.
267. **Nakamoto, T., Chang, C., Li, A., and Chodak, G.W.,** Basic fibroblast growth factor in human prostate cancer cells, *Cancer Res.,* 52, 571, 1992.
268. **Libermann, T.A., Friesel, R., Jaye, M., Lyall, R.M., Westermark, B., Drohan, W., Schmidt, A., Maciag, T., and Schlessinger, J.,** An angiogenic growth factor is expressed in human glioma cells, *EMBO J.,* 6, 1627, 1987.
269. **Takahashi, J.A., Mori, H., Fukumoto, M., Igarashi, K., Jaye, M., Oda, Y., Kikuchi, H., and Hatanaka, M.,** Gene expression of fibroblast growth factors in human gliomas and meningiomas: demonstration of cellular source of basic fibroblast growth factor mRNA and peptide tumor tissues, *Proc. Natl. Acad. Sci. U.S.A.,* 87, 5710, 1990.
270. **Takahashi, J.A., Suzui, H., Yasuda, Y., Ito, N., Ohta, M., Jaye, M., Fukumoto, M., Oda, Y., Kikuchi, H., and Hatanaka, M.,** Gene expression of fibroblast growth factor receptors in the tissues of human gliomas and meningiomas, *Biochem. Biophys. Res. Commun.,* 177, 1, 1991.
271. **Takahashi, J.A., Fukumoto, M., Kozai, Y., Ito, N., Oda, Y., Kikuchi, H., and Hatanaka, M.,** Inhibition of cell growth and tumorigenesis of human glioblastoma cells by a neutralizing antibody against human basic fibroblast growth factor, *FEBS Lett.,* 288, 65, 1991.
272. **Pertovaara, L., Tienari, J., Vainikka, S., Partanen, J., Saksela, O., Lehtonen, E., and Alitalo, K.,** Modulation of fibroblast growth factor receptor expression and signalling during retinoic acid-induced differentiation of Tera-2 teratocarcinoma cells, *Biochem. Biophys. Res. Commun.,* 191, 149, 1993.
273. **Adelaide, J., Mattei, M.-G., Raybaud, F., Planche, J., De Lapeyriere, O., and Birnbaum, D.,** Chromosomal localization of the *hst* oncogene and its co-amplification with the *int*.2 oncogene in a human melanoma, *Oncogene,* 2, 413, 1988.
274. **Theillet, C., Le Roy, X., De Lapeyrière, O., Grosgeorges, J., Adnane, J., Raynaud, S.D., Simony-Lafontaine, J., Goldfarb, M., Escot, C., Birnbaum, D., and Gaudray, P.,** Amplification of *FGF*-related genes in human tumors: possible involvement of *HST* in breast carcinomas, *Oncogene,* 4, 915, 1989.
275. **Fujimoto, K., Ichimori, Y., Kakizoe, T., Okajima, E., Sakamoto, H., Sugimura, T., and Terada, M.,** Increased serum levels of basic fibroblast growth factor in patients with renal cell carcinoma, *Biochem. Biophys. Res. Commun.,* 180, 386, 1991.
276. **Adnane, J., Gaudray, P., Dionne, C.A., Crumley, G., Jaye, M., Schlessinger, J., Jeanteur, P., Birnbaum, D., and Theillet, C.,** *BEK* and *FLG,* two receptors to members of the FGF family, are amplified in subsets of human breast cancers, *Oncogene,* 6, 659, 1991.
277. **Iberg, N., Rogelj, S., Fanning, P., and Klagsbrun, M.,** Purification of 18- and 22-kDa forms of basic fibroblast growth factor from rat cells transformed by the *ras* oncogene, *J. Biol. Chem.,* 264, 19951, 1989.
278. **Milner, P.G.,** Simian sarcoma virus transformation of normal rat kidney fibroblasts is associated with markedly increased basic fibroblast growth factor expression, *Biochem. Biophys. Res. Commun.,* 180, 423, 1991.
279. **Werner, S., Hofschneider, P.H., Stürzl, M., Dick, I., and Roth, W.K.,** Cytochemical and molecular properties of simian virus 40 transformed Kaposi's sarcoma-derived cells: evidence for the secretion of a member of the fibroblast growth factor family, *J. Cell. Physiol.,* 141, 490, 1989.
280. **Roth, W.K., Brandstetter, H., and Stürzl, M.,** Cellular and molecular features of HIV-associated Kaposi's sarcoma, *AIDS,* 6, 895, 1992.

281. **Zhan, X. and Goldfarb, M.,** Growth factor requirements of oncogene-transformed NIH 3T3 and BALB/c 3T3 cells cultured in defined media, *Mol. Cell. Biol.,* 6, 3541, 1986.
282. **Mori, H., Maki, M., Oishi, K., Jaye, M., Igarashi, K., Yoshida, O., and Hatanaka, M.,** Increased expression of genes for basic fibroblast growth factor and transforming growth factor type β2 in human benign prostatic hyperplasia, *Prostate,* 16, 71, 1990.
283. **Klagsbrun, M. and Edelman, E.R.,** Biological and biochemical properties of fibroblast growth factors. Implications for the pathogenesis of atherosclerosis, *Arteriosclerosis,* 9, 269, 1989.

Chapter 5

Neurotrophic Growth Factors

I. INTRODUCTION

The development of the central and peripheral nervous system is a very complex process involving interactions of multiple cellular components. This process involves not only the proliferation of specific cells but also a widespread programmed death (apoptosis) of other cells in order to achieve a proper regulation of the number of neurons that project to a given target field. The development and function of neural tissue depends, in part, on the action of growth factors which may be transported in the blood or may be produced locally, acting in an autocrine or paracrine manner. Many proteins present in soluble tissue extracts and in the intercellular matrix influence the survival and development of cultured neurons.[1,2] Such proteins, called neurotrophic factors or neurotrophins, may have an important role in maintaining the structure and function of the nervous system. Identification and characterization of these factors is a difficult task due to their extremely low abundance. However, a number of them have been described recently. The best characterized neurotrophin is the nerve growth factor (NGF). Another neurotrophin may be represented by a 43-kDa protein, named neurite-promoting factor (NPF), which is released by brain cells and glioma cells in culture and can induce neurite outgrowth in neuroblastoma cells.[3] NPF was purified and characterized from chicken gizzard smooth muscle cells.[4] A factor, named ciliary neurotrophic factor (CNTF), detected in mouse and human neuroblastoma cells is able to support the *in vitro* survival of embryonic chicken and rat peripheral neurons. Other neurotrophins include the brain-derived neurotrophic factor (BDNF), neurotrophin 3 (NT-3), and neurotrophin 4 (NT-4). The receptors for neurotrophins are encoded by the Trk family of tyrosine kinases, which are related to the product of the c-*trk* proto-oncogene.[5,6] Cellular oncoproteins such as c-Fos, c-Jun, and c-Myc may be involved in intracellular signaling processes in neural tissues.[7]

II. NERVE GROWTH FACTOR

The nerve growth factor (NGF) is critically required for the differentiation and survival of sympathetic and sensory neurons in the peripheral nervous system and may also exert important effects in some regions of the central nervous system.[8-10] Neutralization of endogenous NGF by specific antibodies results in the death of sympathetic and sensory neurons. NGF does not attract sensory nerve fibers to their target fields during development but is crucially involved in the target-controlled survival of certain types of neurons.[11] The differentiation of cholinergic projection neurons is regulated by retrogradely transported NGF.[12] Administration of NGF to adult rats can inhibit retrograde degeneration of axotomized septal cholinergic neurons.[13]

The rat pheochromocytoma cell line PC12 has been widely used for studying the effects of NGF in cellular differentiation. In the presence of NGF, PC12 cells differentiate from adrenal chromaffin-like cells into sympathetic neuron-like cells.[14] This process of differentiation is associated with induction of gene expression and synthesis of a number of gene products.

A. ORIGIN AND FUNCTIONS OF NGF

NGF was discovered when mouse sarcoma tissue was transplanted into chick embryos, and it was observed that the transplants caused a marked increase in the size of spinal sensory and sympathetic ganglia. These effects were then attributed to the release from the sarcomatous tissue of a humoral factor into the chick's circulation. Subsequently, two particularly rich sources of NGF were discovered, namely, snake venoms and mouse submaxillary glands.[10] NGF is present in rat milk and is transported intact across the ileal epithelium of suckling rats.[15] Since the absorbed NGF is not cleared rapidly from the blood, it could potentially influence the development of distant neural tissues as well as enteric neurons. NGF is synthesized and secreted by several teratocarcinoma cell lines — although some of these lines, including the embryonal carcinoma PCC4, F9, and 1003 clones — contain only trace amounts of NGF mRNA.[16] Higher levels of NGF transcripts are contained in myocardial, myogenic, and adipogenic clones from these cell lines, which secrete mature NGF protein into the culture medium. Human fibroblast cells may synthesize and secrete NGF in culture.[17]

1. Functions of NGF in the Nervous System

NGF serves important functions in the peripheral nervous tissue: induction of neurite growth from peripheral ganglia, orientation of the neurite growth toward a source of NGF (neurotropism), and promotion of the survival of ganglionic cells (neurotrophism).[18] NGF induces the differentiation of the ganglionic cells into neurons by stimulating the production of molecules needed to build axons, synapses, and enzymes involved in neurotransmitter biosynthesis. The proliferation and differentiation of neuronal stem cells may be regulated by NGF.[19] Dorsal root ganglion neurons are destroyed by exposure *in utero* of rats and guinea pigs to maternal antibody to NGF.[20] The study of PC12 rat cells synchronized by serum starvation in culture indicates that the action of NGF is cell cycle-specific and that NGF-induced cell differentiation may be dissociated from cell proliferation.[21]

The actions of NGF are not limited to peripheral neurons but may also include the central nervous system. Studies on the expression of the NGF gene in the central nervous system indicate the existence of heterogeneity in the level and regional distribution of NGF mRNA and suggest that NGF functions as a factor for several distinct populations of neurons.[22] Whereas in the peripheral nervous system NGF is synthesized by nonneuronal cells such as Schwann cells, granulated tubular cells, and fibroblasts, in the central nervous system NGF is produced by neurons; this suggests that NGF may function not only as a trophic agent, but also as a modulator of neurotransmission.[23] Genes encoding NGF and its receptor are expressed in the developing rat brain, suggesting the possibility that NGF may contribute to the developmental regulation of neuronal systems involved in the neuroendocrine control of pituitary function.[24] There is also evidence that astrocytes produce NGF in the developing rat brain.[25] In tissues densely innervated by sensory or both sympathetic and sensory fibers, NGF is synthesized not only by Schwann cells ensheathing these fibers, but also by the target cells of the sensory and sympathetic neurons, i.e., epithelial cells and fibroblasts.[26] NGF may act as a trophic factor for the differentiation of cholinergic projection neurons in the basal forebrain and may be important for the maintenance and function of cholinergic neurons.[12] NGF may be involved in regulating the duration of the Ca^{2+}-mediated component of the somatic action potential in sensory ganglion neurons at mature stages, when they no longer require NGF for survival.[27]

The physiological effects of NGF in the nervous system may be exerted in cooperation with other growth factors. Interleukin 1 (IL-1) is induced by NGF in PC12 cells.[28] NGF is not the initial determinant of neuronal differentiation in the sympathoadrenal lineage, but instead fibroblast growth factor (FGF) or some other similar factor may play this role.[29] However, developing sympathetic neurons acquire a dependence on NGF as they differentiate; and this change may reflect, at least in part, the *de novo* induction of NGF receptor gene expression by FGF. The possible role of other growth factors in the physiological actions of NGF remains to be elucidated.

2. Functions of NGF in Nonneural Tissues

The physiological actions of NGF may not be limited to the nervous system but may include other organs and systems. NGF may have an important role in reproductive physiology. The factor is present in the semen, the seminal vesicle, and the prostate of mammalian species such as the bull, the rabbit, and the guinea pig. Male mouse germ cells synthesize both NGF mRNA and protein, and this RNA is also present in the epithelium of convoluted ducts of corpus epididymis.[30] These results, together with the demonstration of NGF receptor mRNA in mouse testis, suggest a role for NGF in maturation and/or motility of spermatozoa.

NGF may be involved in female sexual development. Treatment of neonatal rats with antibodies to NGF may result in ovarian dysfunction with alterations in follicular growth and androgen and estrogen production.[31] In the rats treated with NGF antibodies, the timing of first ovulation was delayed, estrous cyclicity was disrupted, and fertility was compromised. Plasma levels of the gonadotropin luteinizing hormone (LH) were found to be elevated, and LH pulsatility was enhanced, suggesting primary ovarian failure. These results show that development of the sympathetic innervation of the ovary is NGF-dependent and that NGF, by supporting the differentiation and survival of the innervating neurons, may contribute to the acquisition of mature ovarian function.

NGF may play a role in kidney morphogenesis. The NGF receptor is transiently synthesized by embryonic mammalian kidney and disappears from nephrons upon their terminal differentiation.[32] Expression of the NGF receptor may be required for the formation of epithelial kidney tubules.

The possibility that NGF may act as an immunoregulatory cytokine is suggested by the fact that human lymphocytes express functional NGF receptors.[33] Immunoprecipitation studies using monoclonal antibodies specific to the receptor found in neural cells allowed the detection of significant amounts of NGF receptors in human spleen and lymph node tissue. Both B and T lymphocytes possess NGF receptors, and addition of NGF to human B lymphocytes resulted in a dose-dependent increase in DNA synthesis, as determined by the incorporation of [^{3}H]thymidine. A proliferative response occurred in both B and T lymphocytes treated with NGF, and the B cells differentiated into antibody-secreting cells. NGF promoted immunoglobulin (Ig)M secretion, and to a lesser extent IgA secretion, but not IgG or IgE secretion. The NGF-induced B-cell growth-promoting activity was inhibited by a neutralizing anti-NGF monoclonal antibody.

B. THE NGF GENE AND THE BIOSYNTHESIS AND STRUCTURE OF NGF

NGF is a multimeric protein composed of three different types of subunits (α, β, and γ). The 7S NGF oligomer is represented in the mouse submaxillary gland by an $\alpha_2\beta\gamma_2$ complex. The biological activity of NGF is specifically associated with the β subunit, which is a 118-amino acid polypeptide synthesized from a 307-amino acid precursor. In the mouse submaxillary gland, the γ subunit (but not the α subunit) of the 7S NGF complex, possesses proteolytic activity capable of cleaving correctly the NGF precursor; this results in generation of functionally active β-NGF.[34] In other cells which synthesize NGF but not in the form of the $\alpha_2\beta\gamma_2$ complex, the proteolytic activation of the NGF precursor may depend on the action of other proteases. The Golgi-localized, membrane-associated endoprotease, furin, is capable of converting pro-β-NGF into mature NGF *in vivo*.[35] The gene coding for the β subunit of NGF is located on human chromosome region 1p21-p22.1; in the mouse, the gene is located on chromosome 3.[36] Two different pathways of RNA splicing can generate different transcripts of the NGF gene,[37] but the possible biological significance of these differential splicing mechanisms is unknown.

The production, purification, and characterization of biologically active recombinant human NGF has been reported.[38] Human NGF expressed in Chinese hamster ovary (CHO) cells is a protein of 120 amino acids with three disulfide linkages and has a molecular weight of 13,489. It is active in stimulating neurite outgrowth in PC12 cells. The nucleotide sequences of chicken and bovine cloned cDNAs encoding NGF have also been determined.[39,40] The deduced NGF amino acid sequences exhibit a high degree of homology among different species. The chicken genome contains a single-copy NGF gene; and the amino acid sequences deduced for the chicken NGF protein are highly homologous to the sequences of the murine and human NGF proteins, which suggests that the functional demands on the protein allowed very few changes during evolution.[41] The pattern of NGF transcript expression in chicken organs is similar to that of mammals, with high levels in brain and heart, intermediate levels in spleen and liver, and barely detectable levels in skeletal muscle. Both in mammals and chickens, the levels of NGF mRNA expression correlate with the density of peripheral adrenergic innervation, but the distribution of NGF within the brain differs markedly in the two taxa.

C. REGULATION OF NGF PRODUCTION

NGF is produced by a diversity of cells both *in vivo* and *in vitro*. Factors involved in regulating the production of NGF are only partially known but include other growth factors and hormones, as well as specific physiological conditions. Tumor necrosis factor-α (TNF-α), basic FGF, and epidermal growth factor (EGF) stimulate the synthesis and secretion of NGF in quiescent cultures of mouse fibroblasts.[42] The synthesis of NGF mRNA in nonneuronal cells or rat sciatic nerve after lesion of the nerve is apparently regulated by IL-1 produced by macrophages invading the site of the lesion.[43] In primary cultures of fibroblasts isolated from adult rat sciatic nerves, IL-1 not only enhances the transcriptional expression of the NGF gene but also increases the stability of NGF mRNA.[44] Expression of c-*fos* and c-*jun* is required for an increase in the levels of NGF mRNA which occurs after injury of the sciatic nerve in mice.[45] The c-Src protein could also have a role in the regulation of NGF synthesis.[46]

A number of hormones are involved in the regulation of NGF gene expression and NGF production. The active vitamin D derivative, calcitriol, is a potent inducer of NGF synthesis.[47] Testosterone, aldosterone, and cortisone, but not estradiol, progesterone, or thyroid hormone reduce the expression of β-NGF mRNA and protein in mouse fibroblastic cells.[48] The rapid decline of the pool of β-NGF mRNA elicited by glucocorticoids is consistent with a direct inhibitory effect on the rate of transcription of the β-NGF gene. Hormone-induced second messenger pathways including production of cyclic adenosine-3′,5′-monophosphate (cAMP), activation of protein kinase C, and induction of Fos and Jun proteins may converge in the regulation of NGF gene expression in murine L929 fibroblasts.[49] Retinoic acid, which exerts potent morphogenetic effects and induces the neuronal differentiation of teratocarcinoma and neuroblastoma cell lines, enhances the synthesis of NGF in cultures of L929 cells.[50] The factors involved in the regulation of NGF expression in L929 cells are partially different from those operating in other cell types, for example, in the mouse salivary glands. Thyroid hormone increases NGF gene expression in the submandibular gland of neonatal mice.[51] Further studies are required for a proper evaluation of the importance of different signaling mechanisms in the regulation of NGF synthesis in different types of cells.

D. NGF RECEPTORS

The actions of NGF on its target cells are initiated by the binding to a specific receptor located on the cell surface.[5,18] NGF receptors are present not only on NGF-dependent neurons but also in Schwann cells.[52] Interaction between neurons and Schwann cells is of crucial importance during the developmental processes of the nervous system and after nerve tissue injury. Two types of NGF receptors, I and II, are expressed in different cells types.

1. The Type II NGF Receptor

The gene encoding for a low-affinity NGF receptor of 75 kDa is located on human chromosome region 17q12-q22, closely distal to the AML-associated chromosome breakpoint at 17q21.[53,54] Gene transfer and molecular cloning of the human and rat receptor, as well as the structure and expression of the human NGF receptor, have been reported.[55-57] The receptor contains an extracellular ligand-binding region, a membrane-spanning domain, and an intracellular (cytoplasmic) region. Studies with constructed chimeric NGF-EGF receptors defined domains responsible for neuronal differentiation, indicating that the membrane-spanning and cytoplasmic domains of the NGF receptor are necessary for signal transduction.[58] The extracellular portion of the 75-kDa human receptor contains four negatively charged cysteine-rich repeats, and site-directed mutagenesis studies indicate that disruption of the third and fourth of these sequences leads to loss of ligand binding.[59] Insertions made outside the cysteine-rich region or in the cytoplasmic domain of the receptor do not inhibit its ability to bind NGF.

Two classes of NGF receptors have been identified by kinetic and molecular criteria: high-affinity or type I receptors and low-affinity or type II receptors. Both type I and type II receptors exist in PC12 rat pheochromocytoma cells, which represent a model system for the study of neuronal differentiation. The two classes of receptors share similar association constants with the ligand, but they are distinguished by their dissociation constants. The high-affinity, slowly dissociating NGF receptor complex has a molecular weight of 158 kDa and is preferentially internalized.[60] PC12 cell mutants that possess low- but not high-affinity NGF receptors neither respond to nor internalize NGF.[61] The low-affinity, type II receptor is represented by the 75-kDa protein (p75) which does not possess intrinsic tyrosine kinase activity but has a potential role in NGF-induced tyrosine phosphorylation.[62] NGF induces in PC12 cells the association of the 75-kDa, type II NGF receptor with a 130-kDa phosphoprotein of unknown function.[63] The type II NGF receptor is a substrate for protein kinases *in vitro* and is phosphorylated on serine, and to a lesser extent on threonine, by the cAMP-independent protein kinase F_A/GSK-3.[64] However, the possible role of such modifications for the biological activity of the receptor is unknown. The type II receptor binds NGF and other neurotrophic factors (BDNF, NT-3, and NT-4) with similar affinity.

A superfamily of cell surface proteins structurally related to the type II NGF receptor includes, in addition to this receptor, the TNF receptors I and II, the T-cell activation antigen OX40, the T-cell inducible antigen 4-1BB, the B-cell antigen CD40, the apoptosis-associated protein Fas, and the Shope sarcoma virus (SSV) protein SalF19R.[65] Another member of the NGF receptor superfamily is the transmembrane protein CD30, which is composed of 595 amino acids residues and contains six cysteinrich motifs.[66] The CD30 protein may be the receptor for an unidentified growth factor and is expressed in the Reed-Sternberg cells that are characteristic of Hodgkin's disease.

2. The Type I NGF Receptor

Recent evidence indicates that the ligand-activated high-affinity NGF receptor is tightly coupled to tyrosine kinase activity.[67] The *trk* proto-oncogene (*trk*-A gene), which is localized on human chromosome region 1q24-q26, encodes a glycoprotein of 140 kDa, Trk-A, which possesses intrinsic tyrosine kinase activity and is an essential component of the high-affinity NGF receptor.[68-70] The amino-terminus of NGF is involved in the interaction with the Trk-A tyrosine kinase.[71] The Tyr-785 residue is a major determinant of Trk-substrate interaction.[72] The low-affinity p75 NGF receptor (type II receptor) may be required for the normal activity of the high-affinity Trk-A/NGF receptor molecule (type I receptor) in relation to signal transduction.[73] However, it is still unclear whether Trk-A alone is the functional high-affinity NGF receptor or the p75 protein is a component of this receptor.

Kinetic studies of the association of NGF to its receptor on PC12 cells indicate the existence of complex interactions between the two molecules.[74] The α subunit of NGF (α-NGF) and the γ subunit of NGF (γ-NGF) block competitively all steady-state binding of the β subunit of NGF (β-NGF) to PC12 cells, whereas in intact dorsal root ganglionic cells only high-affinity binding is blocked.[75] αβ or βγ complexes do not bind to any form of the PC12 NGF receptor.

In addition to the 140-kDa Trk-A receptor, PC12 cells express two proteins of 220 and 300 kDa that can be immunoprecipitated by anti-Trk antisera raised against the intracellular domain of Trk-A.[76] The 220-kDa protein may be represented by a Trk-A molecule with alternate posttranslational modification, and the 300-kDa protein may be a Trk-A dimer. The possible biological significance of these two higher molecular weight forms of the NGF receptor is unknown, but they are phosphorylated on tyrosine when bound to NGF, suggesting that they may participate in signal transduction.

Truncated forms of the NGF receptor have been identified in the human melanoma cell line A875 as well as in normal human urine and amniotic fluid.[77] A truncated, soluble form of the NGF receptor (mol wt 50 to 52 kDa) detected in cultured rat Schwann cells is generated by posttranslational processing of the intact, surface-bound form of the receptor protein.[78] The possible biological significance of these forms of the NGF receptor is unknown but possible reasons for their appearance in extracellular fluids include NGF receptor metabolic turnover and secretion of soluble NGF receptor-related molecules. In general, the presence of soluble forms of growth factor receptors in the body fluids correlates with regulation of the action of growth factors on their target cells.

3. Regulation of NGF Receptor Expression

In contrast to other cell surface receptors, NGF receptors are preclustered and immobile on responsive cells, which suggests that immobilization of the receptor prior to ligand binding is required for signal transduction.[79] After binding, the NGF receptor complex is internalized and is transported to the lysosomes through the cell by both the microtubule and microfilament pathway.[80] These processes lead to degradation of the NGF protein. Internalization of NGF in rat glial cell lines occurs via the p75 receptor.[81] Internalization of the NGF receptor is probably a necessary step for the action of NGF on its target cells.

Homologous and heterologous regulation of NGF receptor expression can occur at both the transcriptional and posttranscriptional levels. Nonneural cells may express NGF receptors. A functional Trk protein is expressed in human monocytes.[82] Expression of β-NGF receptor RNA in Sertoli cells is downregulated by testosterone.[83] Basic FGF enhances NGF receptor gene promoter activity in the human neuroblastoma cell line CHP-100.[84] Retinoic acid promotes the survival of peripheral sensory and sympathetic neurons when the neurites reach their peripheral targets by inducing the expression of NGF receptors in these cells.[85] Retinoic acid regulates both the expression of NGF receptor and the sensitivity to NGF.[86]

Expression of NGF receptor mRNA is developmentally regulated in specific areas of the central and peripheral nervous system in a differential fashion.[87] The NGF receptor gene is selectively and specifically expressed in sympathetic (superior cervical) and sensory (dorsal root) ganglia of the periphery, and by the septum-basal forebrain centrally. A marked reduction in NGF receptor mRNA levels occurs in the sensory ganglia from adult rats. In contrast, an increase in this mRNA is observed in the basal forebrain and in the sympathetic ganglia during the transition from neonatal to adult life. These results suggest that NGF may serve a maintenance as well as a developmental function in the brain and periphery. Diminished NGF response observed in sympathetic ganglion neurons from aged mice is correlated to loss of both high- and low-affinity NGF receptors.[88]

E. POSTRECEPTOR MECHANISMS OF ACTION OF NGF

The postreceptor mechanisms of action of NGF are only partially understood. The NGF transduction signals may include alterations in monovalent and divalent ion fluxes and distribution, activation of the adenylyl cyclase system, ornithine decarboxylase (ODC) induction and generation of polyamines, changes in phosphoinositide turnover and activation of protein kinase C, changes in protein phosphorylation, regulation of growth factor receptors, and effects at the transcriptional level including alterations in protooncogene expression. In differentiating PC12 cells, NGF induces early changes in the synthesis of both nuclear and cytoplasmic proteins.[89] The synthesis, phosphorylation, and metabolic stability of neurofilament proteins is enhanced by NGF in PC12 cells.[90]

Other growth factors can elicit effects similar to those of NGF in neuronal cells. IL-6 can induce neuronal differentiation of PC12 cells,[91] suggesting the existence of common biochemical pathways shared in neural cells by NGF and IL-6. The possible role of IL-6 in neural physiology is unknown, but IL-6 mRNA has been detected in glioblastoma and astrocytoma cell lines. Acidic and basic FGFs promote the survival and differentiation of specific classes of neuronal cells. Studies performed on PC12 cells with deficiencies in particular second messenger systems suggest that NGF and FGFs induce neuronal differentiation (neurite outgrowth) and elevations in the transcription of early response genes via second messengers that are not dependent on cAMP and protein kinase C.[92] However, either protein kinase C or cAMP responsiveness is required for NGF and FGFs to maximally activate early response gene transcription. The results of these studies suggest that NGF and FGFs act via at least three second messenger systems including those that involve cAMP and protein kinase C.

1. Monovalent and Divalent Ion Fluxes and Distribution

All growth factors and mitogenic polypeptides that are active on fibroblasts and epithelial cells examined until now activate an Na^+/H^+ antiport but, as an apparent exception to this rule, NGF would not activate Na^+/H^+ exchange in PC12 cells.[93] However, treatment of PC12 cells with NGF may lead to an increase in Na^+ channels, although the newly synthesized channels are apparently identical to those existing in undifferentiated PC12 cells.[94] In any case, the mechanism by which NGF treatment of PC12 cells leads to an increase in the number of Na^+ channels is not known.

Cytosolic Ca^{2+} may serve as a second messenger in the cellular mechanisms of action of NGF.[95] An increase in cytosolic Ca^{2+} is induced within a few tens of seconds by NGF in PC12 and bovine chromaffin cells.[96] This rise in cytosolic Ca^{2+} represents the earliest biochemical event occurring in target cells treated with NGF and may be a direct consequence of NGF receptor activation.

2. Adenylyl Cyclase System

The possible role of an activation of the adenylyl cyclase system, with generation of elevated intracellular levels of cAMP, in the mechanism of action of NGF is not clear. The levels of an RNA coding for the regulatory α subunit of the stimulatory G protein (G_s-α), which is an adenylyl cyclase-activating protein, are altered in a density-dependent manner by treatment of PC12 cells with NGF.[97] Exposure of PC12 cells to agents such as forskolin and cholera toxin, which selectively activate adenylyl cyclase, may result in a reversible, dose-dependent suppression of NGF-promoted regeneration of neurites.[98] However, when tested on NGF-stimulated initiation of process outgrowth, cholera toxin and forskolin exert a dual effect on the morphological characteristics of PC12 cells. Other studies suggest that NGF may induce phosphorylation-activation of the ribosomal protein kinase S6 and that this phosphorylation is mediated, directly or indirectly, by an activation of cAMP-dependent protein kinase.[99] There is more than one S6 kinase, and NGF and EGF increase the activities of different S6 protein kinases in PC12 cells.[100] The physiological effects of cAMP are exerted through the activation of protein kinase A, and NGF may regulate neuronal differentiation of PC12 cells by pathways that are independent of this enzyme.[101]

3. ODC Activity and Generation of Polyamines

NGF modulates the differentiation of several neuronal cell types *in vivo* and *in vitro*.[102] In the PC12 cell line, NGF induces a diversity of morphological and physiological changes that include neurite growth, modification of the cytoskeleton, and changes in neurotransmitter synthesis. PC12 cells respond to NGF by shifting a chromaffin cell-like phenotype to a neurite-bearing sympathetic neurone-like phenotype. The enzyme ODC, which is crucially involved in polyamine biosynthesis, is induced by NGF in PC12 cells. Benzodiazepines can interact with NGF to modify NGF action on both neurite outgrowth and induction of ODC.[103]

4. Phosphoinositide Turnover and Protein Kinase C Activity

Protein kinase C activation is a component of the NGF-sensitive phosphorylation system in PC12 cells.[104,105] Phosphoinositide metabolism and mobilization of intracellular calcium stores may be involved in the activation of protein kinase C and phosphorylation of cellular proteins induced by NGF. Neurite outgrowth induced by NGF in PC12 cells is governed, at least in part, by the activation of protein kinase C.[106] A 48-kDa protein, termed B-50, which is a substrate for protein kinase C, is phosphorylated in PC12 cells exposed to NGF.[107] B-50 is located mainly on the cell membrane and growth cones, suggesting that it may be associated with neurite outgrowth, but its precise role is unknown. Phosphorylation of a soluble 100-kDa protein, Nsp 100, is decreased in PC12 cells after treatment with either NGF or phorbol 12-myristate 13-acetate (PMA).[104] Nsp 100 phosphorylation would depend on a protein kinase with an activity that appears to be decreased upon its phosphorylation by protein kinase C.

The mechanism by which NGF increases protein kinase C activity is unknown. In PC12 cells NGF rapidly stimulates the phosphorylation on tyrosine of phospholipase C-γ1, and this effect is mediated by the specific kinase activity associated with the Trk-A receptor protein.[108-110] At least some of the effects of NGF in PC12 cells may be mediated by phospholipase C-catalyzed phosphatidylinositol hydrolysis.[111] NGF stimulates tyrosine phosphorylation of a 38-kDa protein that specifically associates with the SH2 domain of phospholipase C-γ.[112] The enzyme is phosphorylated on tyrosine in *trk*-transformed cells, and a proportion of the hybrid Trk oncoprotein is present in the form of a complex with phospholipase C-γ.[113] These results suggest that phospholipase C-γ may play a role in signal transduction by the Trk oncoprotein, as well as by the Trk-A protein and its physiological ligand, NGF.

In addition to its effects on phospholipase C-γ, NGF may have a role in the activation of phosphatidylinositol 3-kinase (PI 3-kinase), an enzyme involved in the generation of a family of phosphoinositides with phosphate at position D-3 of the inositol ring. NGF stimulates the phosphorylation of the 85-kDa subunit of the PI 3-kinase on tyrosine, which may result in direct activation of the enzyme.[114]

5. Protein Phosphorylation

Phosphorylation of cellular proteins may have an important role in the mechanism of action of NGF.[115] Treatment of PC12 cells with NGF induces phosphorylation of various proteins including tyrosine hydroxylase, ribosomal protein S6, and histone and nonhistone nuclear proteins.[116] Microtubule-associated protein-2 (MAP-2) is phosphorylated in PC12 cells by the action of NGF through the stimulation of a protein kinase.[117,118] MAP-2 is protein kinase with specificity for serine/threonine residues.[119] Activation of MAP kinases by NGF in PC12 cells involves the action of MAP kinases that promote the phosphorylation of MAP kinases on tyrosine and serine/threonine residues, thus establishing a cascade of protein phosphorylations implicated in the action of NGF on its target cells.[120] MAP kinases may mediate NGF-induced phosphorylation of the B-Raf protein on serine residues.[121] B-Raf is a serine-specific protein kinase which may be involved in the cascade of phosphorylations initiated by NGF in its target cells.

The enzymes involved in NGF-induced phosphorylation of cellular proteins on serine and threonine residues are known only in part. Exposure of PC12 cells to NGF results in a rapid increase in the phosphorylation of the translation initiation factor eIF-4E (a factor which binds to the 5′ cap structure of eukaryotic mRNAs), and this modification is mediated by Ras-dependent and protein kinase C-independent signal transduction pathways.[122] Some of the NGF-induced protein phosphorylations are mediated by activation of cAMP-dependent protein kinases. NGF-induced phosphorylation of tyrosine-hydroxylase (the rate-limiting enzyme for catecholamine neurotransmitter biosynthesis) is exerted at two distinct sites of the enzyme molecule through activation of both protein kinase C and cAMP-dependent protein kinase.[123] Thus, NGF can independently stimulate at least two different kinase pathways. NGF may be involved in activation of other kinases, including a novel kinase of 22 to 25 kDa that phosphorylates proteins on serine residues.[124] The activity of other kinases may be reduced by NGF. Treatment of PC12 cells with NGF results in reduction in the activity of Ca^{2+}/calmodulin-dependent kinase III, an enzyme for which the major substrate is a 100-kDa protein.[125] An even greater effect on this kinase is produced when cells are treated with dibutyryl cAMP, forskolin, or other agents that raise the cAMP content of the cells. The mechanisms of regulation of serine/threonine kinases by NGF are little known as yet, but there is evidence that the Ras protein mediates in PC12 cells modulation of the signal-transducing protein kinases MAP kinase, Raf-1, and RSK by the activated NGF receptor.[126] Treatment of PC12 cells with NGF also results in the activation of a distinct protein kinase that phosphorylates c-Fos on Ser-362, a residue located on the carboxyl-terminal part of the protein, at a site critically implicated in the capacity of c-Fos to exhibit *trans*-repressive activity on gene expression.[127] The c-Fos-specific protein kinase activated by NGF in PC12 cells may be regulated also by EGF but not by depolarization or other agents.

Protein tyrosine phosphorylation is an early step in the pathway of NGF-induced differentiation of PC12 cells and may be required for all the subsequent effects of NGF on these tumor cells. NGF stimulates the rapid phosphorylation of tyrosine residues in a limited number of proteins of PC12 cells.[128,129] Tyrosine kinase activity is intrinsic to Trk-A, which is an essential component of the high-affinity NGF receptor.[5,73]

6. Heterologous Regulation of Growth Factor Receptors

Heterologous receptor regulation may be induced by NGF. Treatment of PC12 cells with NGF induces a progressive and nearly total loss in the specific binding of EGF, which is due to decreased EGF receptor biosynthesis.[130] In addition rapid EGF responses are completely lost in NGF-treated PC12 cells.[131]

NGF-induced downregulation of EGF receptors and lack of response to EGF in PC12 cells may be one of the mechanisms responsible for the cessation of DNA synthesis and cell division when the cells are treated with NGF.

7. Transcriptional Effects of NGF

Transcription of a specific set of cellular genes is activated in PC12 rat pheochromocytoma cells induced to differentiate from adrenal chromaffin-like cells into sympathetic neuronlike cells by treatment with NGF.[132,133] At least eight NGF-regulated genes were identified in a study with differential screening of cDNA libraries derived from mRNA from PC12 cells treated for 2 weeks with NGF.[134] Five of these mRNAs increased up to tenfold and three decreased down to tenfold after long-term exposure to NGF. Agents that stimulate PC12 cells but that are not capable of promoting differentiation of these cells do not induce the same patterns of mRNA expression as NGF.

NGF-induced differentiation of PC12 cells is associated with the induction of genes including those coding for a nerve growth factor-inducible large external glycoprotein (NILE),[135] ODC,[136] and the protease transin.[137] Treatment of PC12 cells with NGF also results in increased levels of the mRNAs coding for the 68-kDa neurofilament protein NF68 and the cell surface glycoprotein Thy-1.[138] Two genes induced by NGF have been termed NGFI-A and NGFI-B. NGFI-A encodes a 54-kDa protein of unknown function.[139] The NGFI-A gene is expressed at relatively high levels in rat brain, lung, and superior cervical ganglia. NGFI-B is an early response gene that is induced by NGF in PC12 cells and may have a function in NGF-induced differentiation. Sequence analysis indicated that NGFI-B shares similarity with the steroid-thyroid hormone receptor superfamily.[140] An NGFI-B response element, NBRE, is structurally similar to but functionally distinct from DNA elements recognized by the estrogen and thyroid hormone receptors. Cotransfection experiments showed that NGFI-B can activate transcription from the NBRE with or without its putative ligand-binding domain.

One of the genes in which expression is stimulated by NGF in PC12 cells encodes a protein, GTPase-activating protein-43 (GAP-43), which is expressed exclusively in neurons and which may be integral to neurite growth and to the plasticity of neuronal structure.[141,142] Glucocorticoids exert an inhibitory effect on NGF-stimulated GAP-43 gene expression. Another gene, termed *d5*, in which transcription is stimulated by NGF in PC12 cells is a member of a gene family that contains several hundred members of elements, called nerve growth factor-inducible cAMP-extinguishable retrovirus-like (NICER), which are closely related to endogenous retroviruses and transposons.[143] Induction of NICER element expression by NGF is repressed by cAMP. The physiological role of NICER elements is unknown. A nervous system-specific mRNA, NGF33.1, that is rapidly and selectively induced by treatment of PC12 cells with NGF and basic FGF is identical to the NGF-inducible mRNA called VGF.[144,145] The VGF genomic sequence predicts a polypeptide of 617 amino acids with a molecular weight of 68 kDa. The VGF protein shows no strong sequence homology to any known protein and its function is unknown. The fact that VGF mRNA is expressed exclusively in the nervous system and that peak levels occur during a crucial phase of neuronal development suggests that the VGF protein may be involved in the regulation of cell-cell interactions or in synaptogenesis during the maturation of the nervous system.

The mechanisms involved in the regulation of gene expression by NGF have not been characterized. In different cell lines, NGF is tightly bound to chromatin after 1 h of incubation, and the binding of NGF to isolated chromatin is inhibited by monoclonal antibodies specific for NGF receptor.[146] NGF translocated to the nucleus and bound to the chromatin inhibits PDGF-stimulated rRNA synthesis in human colon carcinoma cell lines.[147] Phosphorylation of a specific nonhistone nuclear protein of 30 kDa is stimulated by NGF through activation of a kinase which is not cAMP-dependent and is not similar to either protein kinase C or casein kinase.[148] The increased phosphorylation of nonhistone nuclear protein produced by NGF is not transient, the stimulation being persistent for at least 3 d in the continuous presence of NGF.

Regulation of gene expression by NGF may take place by more than one mechanism.[133] NGF regulates the expression of the tyrosine hydroxylase gene in PC12 cells through a cascade of alterations in gene expression.[149] The series of events includes the activation of an NGF-inducible transcription factor which is represented by the c-Fos protein, and this product contributes to generating a sequence-specific DNA-nucleoprotein complex within the tyrosine hydroxylase promoter during the growth factor-induced gene transcription. This DNA-nucleoprotein-binding complex thereby regulates the tyrosine hydroxylase gene, which has a protein product that is crucial for neuronal differentiation.

8. NGF-Induced Changes in Proto-oncogene Expression

NGF and EGF induce rapid changes in proto-oncogene transcription in PC12 cells. In contrast to most types of tumor cells in which the induction of differentiation is associated with decreased expression of the c-*myc* gene, NGF-induced differentiation of PC12 cells is not accompanied by reduced transcription and steady-state levels of c-*myc* mRNA.[150] Induction of c-*fos* expression occurs very rapidly in NGF-stimulated PC12 cells, being detected 30 min after treatment.[103,132,151-153] NGF-induced expression of c-*fos* in PC12 cells occurs only during the G_1 phase of the cycle.[20] The c-*fos* induction is mediated by both protein kinase C-dependent and -independent pathways.[154] It is enhanced more than 100-fold by peripherally active benzodiazepines. The mechanisms involved in the control of c-*fos* and c-*myc* gene expression by NGF in PC12 cells are distinct. Whereas the expression of c-*fos* is activated by NGF via a mechanism independent of protein synthesis, the activation of c-*myc* expression requires new protein synthesis. Constitutive c-*fos* expression inhibits NGF-induced differentiation in PC12 cells, indicating that deregulated expression of c-*fos* can interfere with the role of NGF in neuronal differentiation.[155] In addition to c-*myc* and c-*fos*, NGF induces the expression of c-*jun* in PC12 cells.[156] Other proto-oncogenes may also be involved in the mechanisms of NGF action. NGF stimulates the activities of the Raf-1 and MAP kinases via the Trk proto-oncogene product.[157]

The effect of NGF on c-*fos* gene expression is specific, as very little change is observed in the levels of transcripts of other proto-oncogenes, including c-H-*ras*, c-K-*ras*, and N-*myc*. The effect of NGF may be mediated by a transcriptional enhancer element which is located upstream from the c-*fos* gene.[158] Under the conditions of these experiments, NGF treatment ultimately results in neurite growth, with a reduction or cessation of cell division. Thus, NGF-induced expression of c-*fos* in PC12 cells is associated with cell differentiation and not cell proliferation. However, expression of the c-*fos* gene alone would not be sufficient to induce differentiation in PC12 cells since agents that do not promote neurite formation in these cells — such as EGF, PMA, and the calcium ionophore A23187 — can also stimulate c-*fos* expression in PC12 cells.[153] Induction of c-*fos* and c-*myc* gene expression is apparently unnecessary for NGF to trigger neurite outgrowth in PC12 cells.[159] Thus, the role of c-*myc* and c-*fos* gene expression in NGF-induced differentiation of neuronal cells is uncertain. Constitutive expression of the c-*myc* gene in PC12 cells may allow NGF to act not as a differentiation inducer but as a mitogenic agent.[160] However, the c-*fos* gene is importantly involved in the mechanism of action of NGF in at least certain types of cells, and the effects of NGF on c-*fos* expression have been confirmed *in vivo*. Stereotaxic injections of NGF into rat forebrain neocortex induce expression c-Fos, as assessed by immunofluorescence staining using a purified antiserum specific to an oligopeptide contained in the c-Fos protein.[161] Adjacent structures including hippocampus, septal nuclei, globus pallidus, and thalamus are unaffected by the treatment. Induction of c-Fos expression is also observed in animals with cortical lesions.

Ras proteins probably have an important role in the mechanisms of action of NGF. Differentiation of PC12 cells induced by NGF, FGF, IL-6, and other factors may depend on the accumulation of active Ras-guanosine triphosphate (GTP) complexes.[162] The ligand-activated NGF receptor system may enhance the activities of factors involved in the regulation of Ras protein activity such as the GTPase-activating proteins p120 GAP and neurofibromin (the product of the neurofibromatosis type I gene) and the guanine nucleotide exchange factor (GNEF).[163] As is true for other receptors with tyrosine kinase activity such as the EGF receptor, the NGF receptor may transmit its signal to Ras, at least in part, through the formation of a complex including the Grb1 adapter protein and the Sos nucleotide exchange factor.[164] However, the differentiation of PC12 cells induced by NGF can be effected through both Ras-dependent and -independent pathways and may take place in a way which is independent of cAMP, calcium, and protein kinase C second messenger systems.[165] Ras proteins may be involved in the regulation of gene expression by NGF. The product of the N-*ras* gene selectively inhibits NGF-induced c-*fos* gene expression in PC12 cells.[166]

9. Abrogation of NGF Requirement by Oncoproteins

Oncoproteins can mimic the physiological effects of NGF and/or other neurotrophic factors. Microinjection of the T24 Ras protein into the cytoplasm of dissociated chick embryonic neurons promotes the *in vitro* survival and neurite outgrowth of NGF-responsive dorsal root ganglion neurons.[167] Similar results were obtained with introduction of the oncoprotein into other types of chick embryonic neurons such as nodose ganglion neurons responsive to brain-derived neurotrophic factor and ciliary ganglion neurons responsive to ciliary neuronotrophic factor.

Microinjection of the T24 c-Ras oncoprotein, but not the normal c-Ras protein, into PC12 cells results in the induction of a morphologically differentiated phenotype with outgrowth of neuronlike processes.[168] The differentiation occurs in the absence of NGF, which suggests that the mutant Ras protein has an ability to eliminate the exogenous signal represented by the specific growth factor. Similar results have been obtained with transfection of an activated N-*ras* proto-oncogene.[159,169] Moreover, cultures of the NGF-resistant variant subline U7 of PC12 cells also show outgrowth of neurites and cessation of cell division following transfection with the mutated N-*ras* gene. However, the N-*ras*-induced phenotype changes in PC12 cells, as well as in its U7 variant, do not precisely duplicate the changes induced by NGF since differences in size and shape and degenerative changes are seen in the transfectants that are not characteristic of NGF effects.

Growth factor-induced differentiation of PC12 cells can be blocked by microinjection of an antibody to Ras.[170] This antibody inhibits neurite formation and produces temporary regression of partially extended neurites, an effect which is observed up to 36 h after initiation of NGF treatment. Neurite formation induced by cAMP is unaffected by injection of the anti-Ras antibody, indicating that Ras is involved in the initiation phase of NGF-induced neurite formation in PC12 cells and does not interfere with neurite formation per se, but specifically inhibits NGF-induced neurite formation. Expression of a v-*src* oncogene has an inductive effect that resembles the physiological action of NGF.[171] However, it is not known whether the c-*src* proto-oncogene is able to display a similar effect, and in the case of a positive answer whether the c-Src protein and NGF share a common pathway for their action on neural tissue.

F. ROLE OF NGF IN NEOPLASTIC PROCESSES

The role of NGF in neoplasia is little known. NGF receptors are expressed in different types of tumor cells, including cultured human melanoma cells,[172] PC12 rat pheochromocytoma cells,[173] human neuroblastoma cell lines,[174] and human neurofibroma Schwann-like cells.[175] A high proportion of primary tumors of the peripheral and central nervous system may exhibit NGF receptors, including peripheral nervous system neuroectodermal tumors (neuroblastomas, ganglioneuroblastomas, ganglioneuromas), pediatric primitive neuroectodermal tumors, astrocytomas, ependymomas, and gliomas.[176]

Less than half of human neuroblastoma cell lines express NGF receptors, and only a small proportion of the cells from these lines are positive for the presence of the receptors. In contrast, all human neuroepithelioma cell lines express NGF receptors, and up to 69% of the cells from these cell lines are positive for NGF receptors.[177] Thus, the NGF receptor may represent a reliable marker for neuroepithelioma. Although both neuroblastomas and neuroepitheliomas are tumors derived from the neural crest, their biological behavior is different. Neuroblastoma, one of the most common pediatric tumors, occurs usually in early infancy; and the primary lesion is localized in the adrenals and sympathetic chain.[178] The clinical behavior of neuroblastomas varies widely among different patients, with some tumors showing spontaneous regression and others growing relentlessly even while being treated. Oncoproteins and growth factors may be responsible, at least in part, for the variable clinical course of neuroblastomas. Neuroblastoma cells express biosynthetic enzymes for adrenergic neurotransmitters; and about half of the clinically advanced cases of neuroblastoma have amplified N-*myc* gene sequences and express high levels of N-Myc mRNA and protein, which correlates with an adverse prognosis.[179] An inverse relationship may exist between the amplification and overexpression of N-*myc* and the expression of the NGF receptor in neuroblastomas.[180] High levels of Trk/NGF receptor gene expression in neuroblastoma are predictive of a favorable outcome.[181,182] The most immature, undifferentiated neuroblastomas, which are often metastatic, frequently contain N-*myc* gene amplification and express high levels of N-Myc mRNA and protein, but express low or undetectable levels of both high- and low-affinity NGF receptors. On the other hand, patients with differentiated neuroblastoma tumors usually express high levels of Trk/NGF receptor mRNA and protein, and they have good or excellent survival rates. Neuroblastomas coexpressing mRNA for both TRK and low-affinity NGF receptor subtypes are favorable tumors likely to differentiate or regress spontaneously or respond to conventional therapy.[183] These results suggest that NGF may have an important role in neuroblastoma differentiation and regression. In contrast to neuroblastomas, the neuroepitheliomas usually occur in older children and adolescents, and their typical presentation consists in a localized chest wall mass.

Tumors of nonneural origin may occasionally possess NGF receptors. In a series of 144 human soft tissue tumors, NGF receptors were detected in 74% of tumors of neural origin including granular cell tumors, schwannoma/neurofibroma, malignant schwannoma, and paraganglioma; whereas less than 15% of tumors of smooth muscle, fibrous, or fibrohistiocytic origin exhibited NGF receptors.[184] In this series, NGF receptors were not found in rhabdomyosarcoma, angiosarcoma, liposarcoma, Ewing's sarcoma, and alveolar soft part sarcoma; however, they were found in synovial sarcomas, undifferentiated sarcomas, and hemangioperycytomas, suggesting a relationship to the neural phenotype. However, functional NGF receptors were detected in the International Agency for Research on Cancer (IARC)-EW1 Ewing's sarcoma cell line, which responded to NGF with a marked increase in c-*fos* mRNA levels.[185] Three other Ewing's sarcoma cell lines and one Wilms' tumor cell line possessed NGF receptors but responded to NGF exposure with only a small increase in c-*fos* mRNA levels.

G. NGF-INDUCED DIFFERENTIATION OF TUMOR CELLS

NGF is capable of inducing differentiation in some tumor cells. Treatment of the PC12 rat pheochromocytoma cell line with NGF results in cessation of mitosis and induction of differentiated characteristics such as the appearance of branching neuronal-like processes. Differentiation of PC12 cells is also induced by the v-Src protein.[186] However, some differences in the characteristics of PC12 cell differentiation induced by NGF and the v-Src protein indicate that partially different pathways may be used for the induction of a differentiated phenotype in PC12 cells. The *ras* oncogene can also induce PC12 cell differentiation. Expression of c-*fos* and c-*jun* is involved in oncogene- and NGF-induced PC12 cell differentiation.[187] Both the SH2 and SH3 domains of human Crk are required for neuronal differentiation of PC12 cells.[188]

An interesting aspect of neuroblastomas is the potential of the tumor cells to differentiate either with spontaneity or by the action of some inducers, including NGF.[189] However, NGF neither reduces the growth rate nor enhances survival of neuroblastoma cells *in vitro*.[174] Neuroblastoma cells containing an amplified N-*myc* gene, as well as neuroepithelioma cell lines, do not usually respond to NGF.[190] In contrast, treatment of anaplastic glioma cells T9 with NGF results in retarded growth rate as well as in the irreversible appearance of a flattened extended cytoplasm with numerous protruding processes.[191] The mechanisms responsible for NGF-induced differentiation of tumor cells are unknown.

III. BRAIN-DERIVED NEUROTROPHIC FACTOR

The brain-derived neurotrophic factor (BDNF) was initially characterized as a basic protein of 12 kDa present in brain extracts and capable of increasing the survival of dorsal root ganglia explants at concentrations around 0.4 n*M*.[192,193] BDNF is structurally related to NGF, as demonstrated by analysis of the cloned BDNF gene.[194] Both NGF and BDNF act on populations of sensory neurons from dorsal root ganglia, derived from neural crest, but only BDNF supports the sensory neurons of the placode-derived nodose ganglion. BDNF may be one of the factors that regulate motor neuron survival during development.[196-198] It is retrogradely transported by motor neurons, and local application of BDNF to transected sciatic nerve can prevent the massive death of motor neurons that follows axotomy in the neonatal period of rats. BDNF increases the survival of dopaminergic neurons in cultures of embryonic rat ventral mesencephalon and may also be a neurotrophic factor for dopaminergic neurons of the substantia nigra.[199]

The highest concentrations of BDNF are present in the hippocampus, followed by the cerebral cortex, where BDNF mRNA is more abundant than NGF mRNA.[200] A wide variety of neurons contacting those producing BDNF might be responsive to this neurotrophic factor. Significant levels of BDNF are found in heart, lung, and muscle. The surface receptor for both BDNF and the related neurotrophic factor, neurotrophin-3, (but not for NGF) is represented by the product of the *trk*-B gene, which is a glycoprotein of 145 kDa with tyrosine-specific kinase activity.[201-203]

IV. NEUROTROPHIN-3

Cloning of mouse, rat, and human genes encoding a neurotrophic factor, nerve growth factor 2 (NGF-2) or neurotrophin-3 (NT-3), has been reported.[204-207] Hippocampus-derived neurotrophic factor (HDNF) is identical with NGF-2/NT-3. The predicted size of the mature NT-3 protein in the rat is 119 amino acids and its computed isoelectric point is 9.5, thus resembling in both size and charge NGF and BDNF. Comparison of the NT-3 amino acid sequence with that of NGF and BDNF precursors revealed the existence of two regions of homology upstream of the mature NT-3 sequence. The conservation of these upstream regions suggests that they may play important and specific roles in the folding, processing, or transport of the three neurotrophic factors. The neurotrophic activity of NT-3 was confirmed by its ability to induce profuse neurite outgrowth on populations of sensory neurons from dorsal root ganglia explants. NT-3 is a mitogen for cultured neural crest cells.[208] In contrast to NGF and BDNF, NT-3 can promote the outgrowth of neurites derived from both neural placode-derived nodose ganglion and neurons of the paravertebral chain sympathetic ganglia. A mouse NT-3 genomic clone encodes a protein of 258 amino acids. Alignment of the sequences of NGF, BDGF, and NT-3 revealed a remarkable number of amino acid identities, including all cysteine residues. The alignment delineated four variable domains, each of 7 to 11 amino acids, indicating structural elements presumably involved in the neuronal specificity of these neurotrophic factors. The cell surface receptor for both BDGF and NT-3 (but not NGF) is represented by the product of the *trk*-B gene, which is a glycoprotein of 145 kDa with tyrosine kinase activity.[201,202] However, BDGF is a more effective ligand for the Trk-B protein than is NT-3; and the identification of a cell line that responds to BDNF (but not NT-3) and does not express Trk-B suggested the existence of receptors that allow for distinct responses to BDNF and NT-3. A novel member of the tyrosine kinase gene family, *trk*-C, was found to code for an NT-3 receptor.[209] The Trk-C/NT-3 receptor is a cell surface glycoprotein of 145 kDa which is preferentially expressed in the brain. NT-3 binds also to low-affinity receptors that are abundantly expressed in embryonic chick sensory neurons.[210] NT-3 is capable of supporting the survival of developing muscle sensory neurons in culture.[211]

The developmental expression and regional distribution of three members of the NGF family (NGF, NGF-2/NT-3, and BDNF) are different, suggesting that each factor has a distinct function in the nervous system during neural development.[212] The distribution of NT-3 mRNA in adult mouse tissue is different from that of BDNF and NGF. NT-3 transcripts were detected in all rat tissues examined, although with considerable variation in their levels of expression.[205] In the adult mouse brain, the highest NT-3 mRNA levels are present in cerebellum and hippocampus. The NT-3 protein, expressed in monkey kidney COS cells from a plasmid vector, is able to support the survival of neurons isolated from the nodose ganglion; and its effect is additive to that of BDNF. NT-3 mRNA is present in the visceral targets of the nodose ganglion, including heart, gut, liver, and lung. Autocrine expression of NT-3 and its *trk*-C-encoded receptor results in morphological transformation of National Institute of Health (NIH)/3T3 cells.[209]

V. NEUROTROPHIN-4

The gene for another member of the NGF family, neurotrophin 4 (NT-4), has been isolated from rat, frog, and viper.[213] The NT-4 gene encodes a precursor protein of 236 amino acids which is processed into a 123-residue mature NT-4 protein with 50 to 60% amino acid identity to NGF, BDNF, and NT-3. The NT-4 protein interacts with the low-affinity NGF receptor and elicits neurite outgrowth from explanted dorsal root ganglia with no and lower activity in sympathetic and nodose ganglia, respectively. Analysis of different tissues from *Xenopus* showed NT-4 mRNA in ovary, with levels over 100-fold higher than those of NGF mRNA in heart.

VI. CILIARY NEUROTROPHIC FACTOR

The ciliary neurotrophic factor (CNTF) was isolated on the basis of its ability to support the survival of embryonic chick ciliary ganglion neurons *in vitro.* It can support the *in vitro* survival of a spectrum of embryonic chicken and rat peripheral neurons.[214] CNTF may not be a target-derived factor in adult animals, but instead mRNA and protein are concentrated in peripheral nervous tissue. It is increased at the site of peripheral nerve injury and can promote the survival of all classes of peripheral neurons: motor, sensory, sympathetic, and parasympathetic. Whereas NGF, BDNF, and NT-3 fail to promote survival or neurite outgrowth from neuron-enriched cultures of the chick ciliary ganglion, the parasympathetic neurons that comprise this ganglion respond to CNTF. The factor induces cholinergic differentiation of cultured rat neurons and promotes the differentiation of glial progenitor cells into a specific type of astrocyte. CNTF is unrelated to the family of neurotrophic factors exemplified by NGF, and instead shares several features with cytokines such as IL-6, G-colony-stimulating factor (CSF)/CSF-3, and the leukemia inhibitory factor (LIF). A distinct set of proteins called CLIPs (CNTF and LIF-induced phosphoproteins) become rapidly phosphorylated on tyrosine after stimulation by CNTF and LIF. Two of these proteins, CLIP-1 and CLIP-2, are transmembrane proteins that may represent receptor components. CLIP-1 is identical with the LIF-binding protein LIF-Rβ, and CLIP-2 is also called gp130.

Expression and characterization of recombinant human CNTF has been reported.[215] The CNTF protein was purified to homogeneity and was found to be biologically active. Recombinant human CNTF is able to promote the survival in culture of chick embryo parasympathetic (ciliary), sympathetic chain, and sensory (dorsal root) neurons. A neurotrophic factor apparently identical with CNTF has been found in some mouse and human neuroblastoma cell lines.[216]

The physiological actions of CNTF are exerted through its binding to a specific receptor located on the surface of its target cells. The CNTF receptor is expressed exclusively in the nervous system and skeletal muscle.[217] Cloning of the CNTF receptor gene from a human neuroblastoma cell line showed that the CNTF receptor is unrelated to the receptors of the NGF family of neurotrophic molecules, but instead is similar to the IL-6 receptor. The CNTF receptor is represented by a 72-kDa protein composed of 372 amino acids and is unique in that it completely lacks a cytoplasmic domain and is bound to the cell surface by means of a glycosyl-phosphatidylinositol linkage. The mechanism of CNTF signal transduction is unknown but it may require the cooperation of another protein. In addition to the 72-kDa receptor, CNTF utilizes the IL-6 and LIF signal-transducing 130-kDa component, gp130 or CLIP-2, which contains a large intracytoplasmic domain.[218] The two LIF receptor components are recruited by CNTF. The CNTF has been recently characterized as a tripartite complex containing a CNTF-binding protein (CNTF-Rα) as well as the components of the LIF receptor, LIF-Rβ and gp130/CLIP-2.[219] The CLIP-2 protein is rapidly phosphorylated in response to ligand binding.[220] CNTF can upregulate fibrinogen gene expression in rat hepatocytes by binding to the IL-6 receptor.[221] At least some cellular actions of CNTF may be mediated by its interaction with the IL-6 receptor.

VII. GLIAL GROWTH FACTOR

Schwann cells are importantly involved in the development, function, and regeneration of peripheral nerves. Development of peripheral axons rely on the presence of Schwann cells. During embryonesis, progenitors of these cells migrate from the neural crest and proliferate along tracts to be occupied by peripheral axons, thus supporting axon extension. Adult Schwann cells are normally quiescent but retain the capacity for cell division and can dedifferentiate to a proliferative state after nerve injury. Several growth factors appear to be capable of regulating the proliferation, migration, and differentiation of Schwann cells. In particular, a protein of 31 kDa called glial growth factor (GGF), purified from bovine pituitary glands, exhibits mitogenic activity for Schwann cells. In fact, GGF activity purified from bovine pituitary extracts is represented by three activities of different molecular weights: GGF-I (34 kDa), GGF-II (59 kDa), and GGF-III (45 kDa).[222] These proteins are encoded by a single human GGF gene and are generated by alternative splicing. The GGF proteins are homologous to the previously described proteins heregulin and Neu differentiation factor, and are specific activators of the c-Neu/Erb-B2 receptor tyrosine kinase. The term neuregulins has been proposed for the designation of these homologous proteins.[222] GGFs and other neuregulins may have important functions in paracrine and autocrine mechanisms associated with the development and regeneration of the nervous system.

VIII. GLIA MATURATION FACTOR

The glia maturation factor (GMF) was detected by its ability to promote morphological and biochemical differentiation of glioblasts. It can induce the histiotypic differentiation of cultured astroblasts.[223,224] Brain GMF is an acidic protein of 17 kDa which was purified to homogeneity.[225] A monoclonal antibody against GMF binds the purified GMF protein and abolishes its activity. The complete amino acid sequence of a specific form of GMF (GMF-β), isolated from bovine brain, has been determined.[226] The GMF-β protein is composed of 141 amino acid residues and possesses no potential glycosylation sites. Its primary sequence does not exhibit significant homology with other proteins.

GMF may have dual functions in stimulating or inhibiting the proliferation or differentiation of normal and tumor cells, according to the specific type of cell and the predominant environmental conditions. It restores contact inhibition to the glioma (schwannoma) cell line 354A, derived from rat trigeminal nerve.[227] GMF promotes the initial growth of rodent and human glioma cell lines when the cells are sparse but limits growth by restoring contact inhibition when the cells are confluent.[228] Treatment of anaplastic glioma T9 cells with GMF *in vitro* results in the production of slender cells with long branching processes forming an interconnecting cell net.[191] These effects, in contrast to those induced by NGF, are readily reversible in the absence of GMF. An autocrine growth factor produced by both normal and immortalized Schwann cells may be represented by GMF or a similar factor.[228] Subcutaneous inoculation of the cloned C6 glioma cell line (derived from a DMNU-induced rat brain tumor) into nude mice results in 100% formation of single solid tumors at the site of injection, but the tumor growth is slowed down in animals treated with GMF.[229]

IX. GLIAL-DERIVED NEUROTROPHIC FACTOR

A specific dopaminergic neurotrophic factor, the glial-derived neurotrophic factor (GDNF), was isolated and characterized from the rat glial cell line B49.[230] Analysis of cDNA clones coding for human and rat GDNF showed that they are 93% identical and are synthesized as a precursor; this precursor is processed and secreted as a protein of 134 amino acids that is a distantly related member of the TGF-β superfamily. Mature GDNF is a glycosylated disulfide-bonded homodimer of 32 to 42 kDa. Recombinant human GDNF enhances survival of dopaminergic neurons and increases their high-affinity dopamine uptake in dissociated rat embryo midbrain cultures. GDNF also increases morphological differentiation of neurons positive for tyrosine hydroxylase, a marker for dopaminergic survival and differentiation. GDNF is potentially useful for the treatment of Parkinson's disease, which is characterized by a marked and progressive degeneration of midbrain dopaminergic neurons.

REFERENCES

1. **Thoenen, H. and Edgar, D.,** Neurotrophic factors, *Science,* 229, 238, 1985.
2. **Walicke, P.A.,** Novel neurotrophic factors, receptors, and oncogenes, *Annu. Rev. Neurosci.,* 12, 103, 1989.
3. **Guenther, J., Nick, H., and Monard, D.,** A glia-derived neurite-promoting factor with protease inhibitory activity, *EMBO J.,* 4, 1963, 1985.
4. **Hayashi, Y. and Miki, N.,** Purification and characterization of a neurite outgrowth factor from chicken gizzard smooth muscle, *J. Biol. Chem.,* 260, 14269, 1985.
5. **Barbacid, M., Lamballe, F., Pulido, D., and Klein, R.,** The *trk* family of tyrosine protein kinase receptors, *Biochim. Biophys. Acta,* 1072, 115, 1991.
6. **Raffioni, S., Bradshaw, R.A., and Buxser, S.E.,** The receptors for nerve growth factor and other neurotrophins, *Annu. Rev. Biochem.,* 62, 823, 1993.
7. **Sudol, M., Grant, S.G.N., and Maisonpierre, P.C.,** Proto-oncogenes and signaling processes in neural tissues, *Neurochem. Int.,* 22, 369, 1993.
8. **Levi-Montalcini, R. and Angeletti, P.U.,** Nerve growth factor, *Physiol. Rev.,* 48, 534, 1968.
9. **Greene, L.A. and Shooter, E.M.,** The nerve growth factor: biochemistry, synthesis, and mechanism of action, *Annu. Rev. Neurosci.,* 3, 353, 1980.
10. **Levi-Montalcini, R.,** The nerve growth factor: thirty-five years later, *Science,* 237, 1154, 1987.
11. **Davies, A.M., Bandtlow, C., Heumann, R., Korsching, S., Rohrer, H., and Thoenen, H.,** Timing and site of nerve growth factor synthesis in developing skin in relation to innervation and expression of the receptor, *Nature (London),* 326, 353, 1987.
12. **Large, T.H., Bodary, S.C., Clegg, D.O., Weskamp, G., Otten, U., and Reichardt, L.F.,** Nerve growth factor gene expression in the developing rat brain, *Science,* 234, 352, 1986.
13. **Kromer, L.F.,** Nerve growth factor treatment after brain injury prevents neuronal death, *Science,* 235, 214, 1987.
14. **Greene, L.A. and Tischler, A.S.,** Establishment of a nonadreneric clonal line of rat adrenal pheochromocytoma cells which respond to nerve growth factor, *Proc. Natl. Acad. Sci. U.S.A.,* 73, 2424, 1976.
15. **Siminoski, K., Gonnella, P., Bernanke, J., Owen, L., Neutra, M., and Murphy, R.A.,** Uptake and transepithelial transport of nerve growth factor in suckling rat ileum, *J. Cell Biol.,* 103, 1979, 1986.
16. **Dicou, E., Houlgatte, R., and Brachet, P.,** Synthesis and secretion of β-nerve growth factor by mouse teratocarcinoma cell lines, *Exp. Cell Res.,* 167, 287, 1986.
17. **Murase, K., Murakami, Y., Takayanagi, K., Furukawa, Y., and Hayashi, K.,** Human fibroblast cells synthesize and secrete nerve growth factor in culture, *Biochem. Biophys. Res. Commun.,* 184, 373, 1992.
18. **Eveleth, D.D.,** Nerve growth factor receptors: structure and function, *In Vitro Cell. Dev. Biol.,* 24, 1148, 1988.
19. **Cattaneo, E. and McKay, R.,** Proliferation and differentiation of neuronal stem cells regulated by nerve growth factor, *Nature (London),* 347, 762, 1990.
20. **Johnson, E.M., Jr., Gorin, P.D., Brandeis, L.D., and Pearson, J.,** Dorsal root ganglion neurons are destroyed by exposure in utero to maternal antibody to nerve growth factor, *Science,* 210, 916, 1980.

21. **Rudkin, B.B., Lazarovici, P., Levi, B.-Z., Abe, Y., Fujita, K., and Guroff, G.,** Cell cycle-specific action of nerve growth factor in PC12 cells: differentiation without proliferation, *EMBO J.,* 8, 3319, 1989.
22. **Shelton, D.L. and Reichardt, L.F.,** Studies on the expression of the nerve growth factor (NGF) gene in the central nervous system level and regional distribution of NGF mRNA suggest that NGF functions as a trophic factor for several distinct populations of neurons, *Proc. Natl. Acad. Sci. U.S.A.,* 83, 2714, 1986.
23. **Rennert, P.D. and Heinrich, G.,** Nerve growth factor mRNA in brain: localization by *in situ* hybridization, *Biochem. Biophys. Res. Commun.,* 138, 813, 1986.
24. **Ojeda, S.R., Hill, D.F., and Katz, K.H.,** The genes encoding nerve growth factor and its receptor are expressed in the developing female rat hypothalamus, *Mol. Brain Res.,* 9, 47, 1991.
25. **Yamakuni, T., Ozawa, F., Hishinuma, F., Kuwano, R., Takahashi, Y., and Amano, T.,** Expression of β-nerve growth factor mRNA in rat glioma cells and astrocytes from rat brain, *FEBS Lett.,* 223, 117, 1987.
26. **Bandtlow, C.E., Heumann, R., Schwab, M.E., and Thoenen, H.,** Cellular localization of nerve growth factor synthesis by *in situ* hybridization, *EMBO J.,* 6, 891, 1987.
27. **Chalazonitis, A., Peterson, E.R., and Crain, S.M.,** Nerve growth factor regulates the action potential duration of mature sensory neurons, *Proc. Natl. Acad. Sci. U.S.A.,* 84, 289, 1987.
28. **Alheim, K., Andersson, C., Tinsborg, S., Ziolkowska, M., Shultzberg, M., and Bartfai, T.,** Interleukin-1 expression is inducible by nerve growth factor in PC12 pheochromocytoma cells, *Proc. Natl. Acad. Sci. U.S.A.,* 88, 9302, 1991.
29. **Birren, S.J. and Anderson, D.J.,** A v-*myc*-immortalized sympathoadrenal progenitor cell line in which neuronal differentiation is initiated by FGF but not NGF, *Neuron,* 4, 189, 1990.
30. **Ayer-LeLievre, C., Olson, L., Ebendal, T., Hallböök, F., and Persson, H.,** Nerve growth factor mRNA and protein in the testis and epididimis of mouse and rat, *Proc. Natl. Acad. Sci. U.S.A.,* 85, 2628, 1988.
31. **Lara, H.E., McDonald, J.K., and Ojeda, S.R.,** Involvement of nerve growth factor in female sexual development, *Endocrinology,* 126, 364, 1990.
32. **Sariola, H., Saarma, M., Sainio, K., Arumäe, U., Palgi, J., Vaahtokari, A., Thesleff, I., and Karavanov, A.,** Dependence of kidney morphogenesis on the expression of nerve growth factor receptor, *Science,* 254, 571, 1991.
33. **Otten, U., Ehrhard, P., and Peck, R.,** Nerve growth factor induces growth and differentiation of human B lymphocytes, *Proc. Natl. Acad. Sci. U.S.A.,* 86, 10059, 1989.
34. **Edwards, R.H., Selby, M.J., Garcia, P.D., and Rutter, W.J.,** Processing of the native nerve growth factor precursor to form biologically active nerve growth factor, *J. Biol. Chem.,* 263, 6810, 1988.
35. **Bresnahan, P.A., Leduc, R., Thomas, L., Thorner, J., Gibson, H.L., Brake, A.J., Barr, P.J., and Thomas, G.,** Human *fur* gene encodes a yeast KEX2-like endoprotease that cleaves pro-β-NGF *in vivo, J. Cell Biol.,* 111, 2851, 1990.
36. **Zabel, B.U., Eddy, R.L., Lalley, P.A., Scott, J., Bell, G.I., and Shows, T.B.,** Chromosomal locations of the human and mouse genes for precursors of epidermal growth factor and the β subunit of nerve growth factor, *Proc. Natl. Acad. Sci. U.S.A.,* 82, 469, 1985.
37. **Edwards, R.H., Selby, M.J., and Rutter, W.J.,** Differential RNA splicing predicts two distinct nerve growth factor precursors, *Nature (London),* 319, 784, 1986.
38. **Iwane, M., Kitamura, Y., Kaisho, Y., Yoshimura, K., Shintani, A., Sasada, R., Nakagawa, S., Kawahara, K., Nakahama, K., and Kakinuma, A.,** Production, purification, and characterization of biologically active recombinant human nerve growth factor, *Biochem. Biophys. Res. Commun.,* 171, 116, 1990.
39. **Ebendal, T., Larhammar, D., and Persson, H.,** Structure and expression of the chicken β nerve growth factor gene, *EMBO J.,* 5, 1483, 1986.
40. **Meier, R., Becker-André, M., Götz, R., Heumann, R., Shaw, A., and Thoenen, H.,** Molecular cloning of bovine and chick nerve growth factor (NGF): delineation of conserved and unconserved domains and their relationship to the biological activity and antigenicity of NGF, *EMBO J.,* 5, 1489, 1986.
41. **Goeddert, M.,** Molecular cloning of the chicken nerve growth factor gene: mRNA distribution in developing and adult tissues, *Biochem. Biophys. Res. Commun.,* 141, 1116, 1986.

42. **Hattori, A., Tanaka, E., Murase, K., Ishida, N., Chatani, Y., Tsujimoto, M., Hayashi, K., and Kohno, M.,** Tumor necrosis factor stimulates the synthesis and secretion of biologically active nerve growth factor in non-neuronal cells, *J. Biol. Chem.*, 268, 2577, 1993.
43. **Lindholm, D., Heumann, R., Meyer, M., and Thoenen, H.,** Interleukin-1 regulates synthesis of nerve growth factor in non-neuronal cells of rat sciatic nerve, *Nature (London),* 330, 658, 1987.
44. **Lindholm, D., Heumann, R., Hengerer, B., and Thoenen, H.,** Interleukin 1 increases stability and transcription of mRNA encoding nerve growth factor in cultured rat fibroblasts, *J. Biol. Chem.*, 263, 16348, 1988.
45. **Hengerer, B., Lindholm, D., Heumann, R., Rüther, U., Wagner, E.F., and Thoenen, H.,** Lesion-induced increase in nerve growth factor mRNA is mediated by c-*fos*, *Proc. Natl. Acad. Sci. U.S.A.*, 87, 3899, 1990.
46. **Nemoto, K., Tashiro, F., Hagiwara, T., Kitamura, A., Habano, W., Hirano, N., Ishii, K., Ueno, Y., Omae, F., Shiroki, K., Kuchino, Y., and Furukawa, S.,** Elevation of nerve growth factor synthesis by constitutive expression of v-*src* oncogene in cultured rat fibroblasts, *Neurosci. Lett.*, 129, 281, 1991.
47. **Wion, D., MacGrogan, D., Neveu, I., Jehan, F., Houlgatte, R., and Brachet, P.,** 1,25-dihydroxyvitamin D_3 is a potent inducer of nerve growth factor synthesis, *J. Neurosci. Res.*, 28, 110, 1991.
48. **Siminoski, K., Murphy, R.A., Rennert, P., and Heinrich, G.,** Cortisone, testosterone, and aldosterone reduce levels of nerve growth factor messenger ribonucleic acid in L-929 fibroblasts, *Endocrinology,* 121, 1432, 1987.
49. **Jehan, F., Neveu, I., Naveilhan, P., Brachet, P., and Wion, D.,** Complex interactions among second messenger pathways, steroid hormones, and proto-oncogenes of the Fos and Jun families converge in the regulation of the nerve growth factor gene, *J. Neurochem.*, 60, 1843, 1993.
50. **Wion, D., Houlgatte, R., Barbot, N., Barrand, P., Dicou, E., and Brachet, P.,** Retinoic acid increases the expression of NGF gene in mouse L cells, *Biochem. Biophys. Res. Commun.*, 149, 510, 1987.
51. **Black, M.A., Pope, L., Lefebvre, Y.A., and Walker, P.,** Thyroid hormone precociously increase nerve growth factor gene expression in the submandibular gland of neonatal mice, *Endocrinology,* 130, 2083, 1992.
52. **DiStefano, P.S. and Johnson, E.M., Jr.,** Nerve growth factor receptors on cultured rat Schwann cells, *J. Neurosci.*, 8, 231, 1988.
53. **Huebner, K., Isobe, M., Chao, M., Bothwell, M., Ross, A.H., Finan, J., Hoxie, J.A., Sehgal, A., Buck, C.R., Lanahan, A., Nowell, P.C., Koprowski, H., and Croce, C.M.,** The nerve growth factor receptor gene is at human chromosome region 17q12-17q22, distal to the chromosome 17 breakpoint in acute leukemias, *Proc. Natl. Acad. Sci. U.S.A.*, 83, 1403, 1986.
54. **Rettig, W.J., Thomson, T.M., Spengler, B.A., Biedler, J.L., and Old, L.J.,** Assignment of human nerve growth factor receptor gene to chromosome 17 and regulation of receptor expression in somatic cell hybrids, *Somatic Cell Mol. Genet.*, 12, 441, 1986.
55. **Chao, M.V., Bothwell, M.A., Ross, A.H., Koprowski, H., Lanahan, A.A., Buck, C.R., and Sehgal, A.,** Gene transfer and molecular cloning of human NGF receptor, *Science,* 232, 518, 1986.
56. **Johnson, D., Lahanan, A., Buck, C.R., Sehgal, A., Morgan, C., Mercer, E., Bothwell, M., and Chao, M.,** Expression and structure of the human NGF receptor, *Cell,* 47, 545, 1986.
57. **Radeke, M.J., Misko, T.P., Hsu, C., Herzenberg, L.A., and Shooter, E.M.,** Gene transfer and molecular cloning of the rat nerve growth factor receptor, *Nature (London),* 325, 593, 1987.
58. **Yan, H., Schlessinger, J., and Chao, M.V.,** Chimeric NGF-EGF receptors define domains responsible for neuronal differentiation, *Science,* 252, 561, 1991.
59. **Yan, H. and Chao, M.V.,** Disruption of cysteine-rich repeats of the p75 nerve growth factor receptor leads to loss of ligand binding, *J. Biol. Chem.*, 266, 12099, 1991.
60. **Hosang, M. and Shooter, E.M.,** The internalization of nerve growth factor by high-affinity receptors on pheochromocytoma PC12 cells, *EMBO J.*, 6, 1197, 1987.
61. **Green, S.H., Rydel, R.E., Connolly, J.L., and Greene, L.A.,** PC12 cell mutants that possess low- but not high-affinity nerve growth factor receptors neither respond to nor internalize nerve growth factor, *J. Cell Biol.*, 102, 830, 1986.
62. **Berg, M.M., Sternberg, D.W., Hempstead, B.L., and Chao, M.V.,** The low-affinity p75 nerve growth factor (NGF) receptor mediates NGF-induced tyrosine phosphorylation, *Proc. Natl. Acad. Sci. U.S.A.*, 88, 7106, 1991.

63. **Ohmichi, M., Decker, S.J., and Saltiel, A.R.,** Nerve growth factor induces the association of a 130-Kd phosphoprotein with its receptor in PC-12 pheochromocytoma cells, *Cell Regul.,* 2, 691, 1991.
64. **Taniuchi, M., Johnson, E.M., Jr., Roach, P.J., and Lawrence, J.C., Jr.,** Phosphorylation of nerve growth factor receptor proteins in sympathetic neurons and PC12 cells. *In vitro* phosphorylation by the cAMP-independent protein kinase F_A/GSK-3, *J. Biol. Chem.,* 261, 13342, 1986.
65. **Mallett, S. and Barclay, A.N.,** A new superfamily of cell surface proteins related to the nerve growth factor receptor, *Immunol. Today,* 12, 220, 1991.
66. **Dürkop, H., Latza, U., Hummel, M., Eitelbach, F., Seed, B., and Stein, H.,** Molecular cloning and expression of a new member of the nerve growth factor receptor family that is characteristic for Hodgkin's disease, *Cell,* 68, 421, 1992.
67. **Meakin, S.O. and Shooter, E.M.,** Tyrosine kinase activity coupled to the high-affinity nerve growth factor receptor complex, *Proc. Natl. Acad. Sci. U.S.A.,* 88, 5862, 1991.
68. **Klein, R., Jing, S.Q., Nanduri, V., O'Rourke, E., and Barbacid, M.,** The *trk* proto-oncogene encodes a receptor for nerve growth factor, *Cell,* 65, 189, 1991.
69. **Hempstead, B.L., Martin-Zanca, D., Kaplan, D.R., Parada, L.F., and Chao, M.V.,** High-affinity NGF binding requires coexpression of the *trk* proto-oncogene and the low-affinity NGF receptor, *Nature (London),* 350, 678, 1991.
70. **Kaplan, D.R., Hempstead, B.L., Martin-Zanca, D., Chao, M.V., and Parada, L.F.,** The *trk* proto-oncogene product — a signal transducing receptor for nerve growth factor, *Science,* 252, 554, 1991.
71. **Kahle, P., Burton, L.E., Schmelzer, C.H., and Hertel, C.,** The amino terminus of nerve growth factor is involved in the interaction with the receptor tyrosine kinase $p140^{trkA}$, *J. Biol. Chem.,* 267, 22707, 1992.
72. **Obermeier, A., Halfter, H., Wiesmüller, K.H., Jung, G., Schlessinger, J., and Ullrich, A.,** Tyrosine-785 is a major determinant of Trk-substrate interaction, *EMBO J.,* 12, 943, 1993.
73. **Ross, A.H.,** Identification of tyrosine kinase Trk as a nerve growth factor receptor, *Cell Regul.,* 2, 685, 1991.
74. **Woodruff, N.R. and Neet, K.E.,** β Nerve growth factor binding to PC12 cells. Association kinetics and cooperative interactions, *Biochemistry,* 25, 7956, 1986.
75. **Woodruff, N.R. and Neet, K.E.,** Inhibition of β nerve growth factor binding to PC12 cells by α nerve growth factor and γ nerve growth factor, *Biochemistry,* 25, 7967, 1986.
76. **Hartman, D.S., McCormick, M., Schubenel, R., and Hertel, C.,** Multiple trkA proteins in PC12 cells bind NGF with a slow association rate, *J. Biol. Chem.,* 267, 24516, 1992.
77. **Zupan, A.A., Osborne, P.A., Smith, C.E., Siegel, N.R., Leimgruber, R.M., and Johnson, E.M., Jr.,** Identification, purification, and characterization of truncated forms of the human nerve growth factor receptor, *J. Biol. Chem.,* 264, 11714, 1989.
78. **Barker, P.A., Miller, F.D., Large, T.H., and Murphy, R.A.,** Generation of the truncated form of the nerve growth factor receptor by rat Schwann cells. Evidence for post-translational processing, *J. Biol. Chem.,* 266, 19113, 1991.
79. **Venkatakrishnan, G., McKinnon, C.A., Pilapil, C.G., Wolf, D.E., and Ross, A.H.,** Nerve growth factor receptors are preaggregated and immobile on responsive cells, *Biochemistry,* 30, 2748, 1991.
80. **Kasaian, M.T. and Neet, K.E.,** Internalization of nerve growth factor by PC12 cells. A description of cellular pools, *J. Biol. Chem.,* 263, 5083, 1988.
81. **Kahle, P. and Hertel, C.,** Nerve growth factor (NGF) receptor on glial cell lines — Evidence for NGF internalization via p75 (NGFR), *J. Biol. Chem.,* 267, 13917, 1992.
82. **Ehrhard, P.B., Ganter, U., Bauer, J., and Otten, U.,** Expression of functional *trk* proto-oncogene in human monocytes, *Proc. Natl. Acad. Sci. U.S.A.,* 90, 5423, 1993.
83. **Persson, H., Ayer-Le Lievre, C., Söder, O., Villar, M.J., Metsis, M., Olson, L., Ritzen, M., and Hökfelt, T.,** Expression of β-nerve growth factor receptor mRNA in Sertoli cells is downregulated by testosterone, *Science,* 247, 704, 1990.
84. **Taiji, M., Taiji, K., Deyerle, K., and Bothwell, M.,** Basic fibroblast growth factor enhances nerve growth factor receptor gene promoter activity in human neuroblastoma cell line CHP100, *Mol. Cell. Biol.,* 12, 2193, 1992.
85. **Rodriguez-Tébar, A. and Rohrer, H.,** Retinoic acid induces NGF-dependent survival response and high-affinity NGF receptors in immature chick sympathetic neurons, *Development,* 112, 813, 1991.
86. **Scheibe, R.J. and Wagner, J.A.,** Retinoic acid regulates both expression of the nerve growth factor receptor and sensitivity to nerve growth factor, *J. Biol. Chem.,* 267, 17611, 1992.

87. **Buck, C.R., Martinez, H.J., Black, I.B., and Chao, M.V.,** Developmentally regulated expression of the nerve growth factor receptor gene in the periphery and brain, *Proc. Natl. Acad. Sci. U.S.A.,* 84, 3060, 1987.
88. **Uchida, Y. and Tomonaga, M.,** Loss of nerve growth factor receptors in sympathetic ganglia from aged mice, *Biochem. Biophys. Res. Commun.,* 146, 797, 1987.
89. **Tiercy, J.-M. and Shooter, E.M.,** Early changes in the synthesis of nuclear and cytoplasmic proteins are induced by nerve growth factor in differentiating rat PC12 cells, *J. Cell Biol.,* 103, 2367, 1986.
90. **Lindenbaum, M.H., Carbonetto, S., and Mushynski, W.E.,** Nerve growth factor enhances the synthesis, phosphorylation, and metabolic stability of neurofilament proteins in PC12 cells, *J. Biol. Chem.,* 262, 605, 1987.
91. **Satoh, T., Nakamura, S., Taga, T., Matsuda, T., Hirano, T., Kishimoto, T., and Kaziro, Y.,** Induction of neuronal differentiation in PC12 cells by B-cell stimulatory factor 2/interleukin 6, *Mol. Cell. Biol.,* 8, 3546, 1988.
92. **Damon, D.H., D'Amore, P.A., and Wagner, J.A.,** Nerve growth factor and fibroblast growth factor regulate neurite outgrowth and gene expression in PC12 cells via both protein kinase C- and cAMP-independent mechanisms, *J. Cell Biol.,* 110, 1333, 1990.
93. **Chandler, C.E., Cragoe, E.J., Jr., and Glaser, L.,** Nerve growth factor does not activate Na^+/H^+ exchange in PC12 pheochromocytoma cells, *J. Cell. Physiol.,* 125, 367, 1985.
94. **Reed, J.K. and England, D.,** The effect of nerve growth factor on the development of sodium channels in PC12 cells, *Biochem. Cell Biol.,* 64, 1153, 1986.
95. **Davis, L.H. and Kauffman, F.C.,** Calcium-dependent activation of glycogen phosphorylase in rat pheochromocytoma PC12 cells by nerve growth factor, *Biochem. Biophys. Res. Commun.,* 138, 917, 1986.
96. **Pandiella-Alonso, A., Malgaroli, A., Vicentini, L.M., and Meldolesi, J.,** Early rise of cytosolic Ca^{2+} induced by NGF in PC12 and chromaffin cells, *FEBS Lett.,* 208, 48, 1986.
97. **Tjaden, G., Aguanno, A., Kumar, R., Benincasa, D., Gubits, R.M., Yu, H., and Dolan, K.P.,** Density-dependent nerve growth factor regulation of G_s-α RNA in pheochromocytoma 12 cells, *Mol. Cell. Biol.,* 10, 3277, 1990.
98. **Greene, L.A., Drexler, S.A., Connolly, J.L., Rukenstein, A., and Green, S.H.,** Selective inhibition of responses to nerve growth factor and of microtubule-associated protein phosphorylation by activators of adenylate cyclase, *J. Cell Biol.,* 103, 1967, 1986.
99. **Matsuda, Y. and Guroff, G.,** Purification and mechanism of activation of a nerve growth factor-sensitive S6 kinase from PC12 cells, *J. Biol. Chem.,* 262, 2832, 1987.
100. **Mutoh, T., Rudkin, B.B., Koizumi, S., and Guroff, G.,** Nerve growth factor, a differentiating agent, and epidermal growth factor, a mitogen, increase the activities of different S6 kinases in PC12 cells, *J. Biol. Chem.,* 263, 15853, 1988.
101. **Ginty, D.D., Glowacka, D., DeFranco, C., and Wagner, J.A.,** Nerve growth factor-induced neuronal differentiation after dominant repression of both type I and type II cAMP-dependent protein kinase activities, *J. Biol. Chem.,* 266, 15235, 1991.
102. **Feinstein, S.C., Dana, S.L., McConlogue, L., Shooter, E.M., and Coffino, P.,** Nerve growth factor rapidly induces ornithine decarboxylase mRNA in PC12 rat pheochromocytoma cells, *Proc. Natl. Acad. Sci. U.S.A.,* 5761, 1985.
103. **Curran, T. and Morgan, J.I.,** Superinduction of c-*fos* by nerve growth factor in the presence of peripherally active benzodiazepines, *Science,* 229, 1265, 1985.
104. **Hama, T., Huang, K.-P., and Guroff, G.,** Protein kinase C as a component of a nerve growth factor-sensitive phosphorylation system in PC12 cells, *Proc. Natl. Acad. Sci. U.S.A.,* 83, 2353, 1986.
105. **Blenis, J. and Erikson, R.L.,** Regulation of protein kinase activities in PC12 pheochromocytoma cells, *EMBO J.,* 5, 3441, 1986.
106. **Hall, F.L., Fernyhough, P., Ishii, D.N., and Vulliet, P.R.,** Suppression of nerve growth factor-directed neurite outgrowth in PC12 cells by sphingosine, an inhibitor of protein kinase C, *J. Biol. Chem.,* 263, 4460, 1988.
107. **Van Hooff, C.O.M., De Graan, P.N.E., Boonstra, J., Oestreicher, A.B., Schmidt-Michels, M.H., and Gispen, W.H.,** Nerve growth factor enhances the level of the protein kinase C substrate B-50 in pheochromocytoma PC12 cells, *Biochem. Biophys. Res. Commun.,* 139, 644, 1986.
108. **Kim, U.H., Fink, D., Park, D.J., Contreras, M.L., Guroff, G., and Rhee, S.G.,** Nerve growth factor stimulates phosphorylation of phospholipase C-γ in PC12 cells, *J. Biol. Chem.,* 266, 1359, 1991.

109. **Vetter, M.L., Martin-Zanca, D., Parada, L.F., Bishop, J.M., and Kaplan, D.R.,** Nerve growth factor rapidly stimulates tyrosine phosphorylation of phospholipase C-γ 1 by a kinase activity associated with the product of the *trk* proto-oncogene, *Proc. Natl. Acad. Sci. U.S.A.,* 88, 5650, 1991.
110. **Ohmichi, M., Decker, S.J., Pang, L., and Saltiel, A.R.,** Nerve growth factor binds to the 140 kd *trk* proto-oncogene product and stimulates its association with the *src* homology domain of phospholipase C γ1, *Biochem. Biophys. Res. Commun.,* 179, 217, 1991.
111. **Chan, B.L., Chao, M.V., and Saltiel, A.R.,** Nerve growth factor stimulates the hydrolysis of glycosylphosphatidylinositol in PC-12 cells: a mechanism of protein kinase C regulation, *Proc. Natl. Acad. Sci. U.S.A.,* 86, 1756, 1989.
112. **Ohmichi, M., Decker, S.J., Pang, L., and Saltiel, A.R.,** Phospholipase C-γ1 directly associates with the p70 *trk* oncogene product through its *src* homology domains, *J. Biol. Chem.,* 266, 14858, 1991.
113. **Ohmichi, M., Decker, S.J., and Saltiel, A.R.,** Nerve growth factor stimulates the tyrosine phosphorylation of a 38-kDa protein that specifically associates with the *src* homology domain of phospholipase C-γ1, *J. Biol. Chem.,* 267, 21601, 1992.
114. **Soltoff, S.P., Rabin, S.L., Cantley, L.C., and Kaplan, D.R.,** Nerve growth factor promotes the activation of phosphatidylinositol 3-kinase and its association with the *trk* tyrosine kinase, *J. Biol. Chem.,* 267, 17472, 1992.
115. **Miyasaka, T., Sternberg, D.W., Miyasaka, J., Sherline, P., and Saltiel, A.R.,** Nerve growth factor stimulates protein tyrosine phosphorylation in PC-12 pheochromocytoma cells, *Proc. Natl. Acad. Sci. U.S.A.,* 88, 2653, 1991.
116. **Halegoua, S. and Patrick, J.,** Nerve growth factor mediates phosphorylation of specific proteins, *Cell,* 22, 571, 1980.
117. **Aletta, J.M., Lewis, S.A., Cowan, N.J., and Greene, L.A.,** Nerve growth factor regulates both the phosphorylation and steady-state levels of microtubule-associated protein 1.2 (MAP1.2), *J. Cell Biol.,* 106, 1573, 1988.
118. **Miyasaka, T., Chao, M.V., Sherline, P., and Saltiel, A.R.,** Nerve growth factor stimulates a protein kinase in PC-12 cells that phosphorylates microtubule-associated protein-2, *J. Biol. Chem.,* 265, 4630, 1990.
119. **Rossomando, A.J., Payne, D.M., Weber, M.J., and Sturgill, T.W.,** Evidence that pp42, a major tyrosine kinase target protein, is a mitogen-activated serine/threonine protein kinase, *Proc. Natl. Acad. Sci. U.S.A.,* 86, 6840, 1989.
120. **Gómez, N. and Cohen, P.,** Dissection of the protein kinase cascade by which nerve growth factor activates MAP kinases, *Nature (London),* 353, 170, 1991.
121. **Oshima, M., Sithanandam, G., Rapp, U.R., and Guroff, G.,** The phosphorylation and activation of B-raf in PC12 cells stimulated by nerve growth factor, *J. Biol. Chem.,* 266, 23753, 1991.
122. **Frederickson, R.M., Mushynski, W.E., and Sonenberg, N.,** Phosphorylation of translation initiation factor eIF-4E is induced in a *ras*-dependent manner during nerve growth factor-mediated PC12 cell differentiation, *Mol. Cell. Biol.,* 12, 1239, 1992.
123. **Cremins, J., Wagner, J.A., and Halegoua, S.,** Nerve growth factor action is mediated by cyclic AMP- and Ca^{2+}/phospholipid-dependent protein kinases, *J. Biol. Chem.,* 103, 887, 1986.
124. **Rowland, E.A., Müller, T.H., Goldstein, M., and Greene, L.A.,** Cell-free detection and characterization of a novel nerve growth factor-activated protein kinase in PC12 cells, *J. Biol. Chem.,* 262, 7504, 1987.
125. **Nairn, A.C., Nichols, R.A., Brady, M.J., and Palfrey, H.C.,** Nerve growth factor treatment of cAMP elevation reduces Ca^{2+}/calmodulin-dependent protein kinase III activity in PC12 cells, *J. Biol. Chem.,* 262, 14265, 1987.
126. **Wood, K.W., Sarnecki, C., Roberts, T.M., and Blenis, J.,** *Ras* mediates nerve growth factor receptor modulation of three signal-transducing protein kinases: MAP kinase, Raf-1, and RSK, *Cell,* 68, 1041, 1992.
127. **Taylor, L.K., Marshak, D.R., and Landreth, G.E.,** Identification of a nerve growth factor- and epidermal growth factor-regulated protein kinase that phosphorylates the proto-oncogene product c-Fos, *Proc. Natl. Acad. Sci. U.S.A.,* 90, 368, 1993.
128. **Maher, P.A.,** Nerve growth factor induces protein-tyrosine phosphorylation, *Proc. Natl. Acad. Sci. U.S.A.,* 85, 6788, 1988.
129. **Maher, P.A.,** Role of protein tyrosine phosphorylation in the NGF response, *J. Neurosci. Res.,* 24, 29, 1989.

130. **Lazarovici, P., Dickens, G., Kuzuya, H., and Guroff, G.,** Long-term, heterologous regulation of the epidermal growth factor receptor in PC12 cells by nerve growth factor, *J. Cell Biol.,* 104, 1611, 1987.
131. **Boonstra, J., Mummery, C.L., Feyen, A., de Hoog, W.J., van der Saag, P.T., and de Laat, S.W.,** Epidermal growth factor receptor expression during morphological differentiation of pheochromocytoma cells, induced by nerve growth factor or dibutyryl cyclic AMP, *J. Cell. Physiol.,* 131, 409, 1987.
132. **Kujubu, D.A., Lim, R.W., Varnum, B.C., and Herschman, H.R.,** Induction of transiently expressed genes in PC-12 pheochromocytoma cells, *Oncogene,* 1, 257, 1987.
133. **Cho, K.-O., Skarnes, W.C., Minsk, B., Palmieri, S., Jackson-Grusby, L., and Wagner, J.A.,** Nerve growth factor regulates gene expression by several distinct mechanisms, *Mol. Cell. Biol.,* 9, 135, 1989.
134. **Leonard, D.G.B., Ziff, E.B., and Greene, L.A.,** Identification and characterization of mRNAs regulated by nerve growth factor in PC12 cells, *Mol. Cell. Biol.,* 7, 3156, 1987.
135. **Salton, S.R., Shelanski, M.L., and Greene, L.A.,** Biochemical properties of the nerve growth factor-inducible large external (NILE) glycoprotein, *J. Neurosci.,* 3, 2420, 1983.
136. **Masiakowski, P. and Shooter, E.M.,** Nerve growth factor induces the genes for two proteins related to a family of calcium-binding proteins in PC12 cells, *Proc. Natl. Acad. Sci. U.S.A.,* 85, 1277, 1988.
137. **Machida, C.M., Rodland, K.D., Matrisian, L., Magun, B.E., and Ciment, G.,** NGF induction of the gene encoding the protease transin accompanies neuronal differentiation of PC12 cells, *Neuron,* 2, 1587, 1989.
138. **Dickson, G., Prentice, H., Julien, J.-P., Ferrari, G., Leon, A., and Walsh, F.S.,** Nerve growth factor activates Thy-1 and neurofilament gene transcription in rat PC12 cells, *EMBO J.,* 5, 3449, 1986.
139. **Milbrandt, J.,** A nerve growth factor-induced gene encodes a possible transcriptional regulatory factor, *Science,* 238, 797, 1987.
140. **Wilson, T.E., Fahrner, T.J., Johnston, M., and Milbrandt, J.,** Identification of the DNA binding site for NGFI-B by genetic selection in yeast, *Science,* 252, 1296, 1991.
141. **Basi, G.S., Jacobson, R.D., Virag, I., Schilling, J., and Skene, J.H.P.,** Primary structure and transcriptional regulation of GAP-43, a protein associated with nerve growth factor, *Cell,* 49, 785, 1987.
142. **Federoff, H.J., Grabczyk, E., and Fishman, M.C.,** Dual regulation of GAP-43 gene expression by nerve growth factor and glucocorticoids, *J. Biol. Chem.,* 263, 19290, 1988.
143. **Cho, K.-O., Minsk, B., and Wagner, J.A.,** NICER elements: a family of nerve growth factor-inducible cAMP-extinguishable retrovirus-like elements, *Proc. Natl. Acad. Sci. U.S.A.,* 87, 3778, 1990.
144. **Levi, A., Eldridge, J.D., and Paterson, B.M.,** Molecular cloning of a gene sequence regulated by nerve growth factor, *Science,* 229, 393, 1985.
145. **Salton, S.R.J., Fischberg, D.J., and Dong, K-W.,** Structure of the gene encoding VGF, a nervous system-specific mRNA that is rapidly and selectively induced by nerve growth factor in PC12 cells, *Mol. Cell. Biol.,* 11, 2335, 1991.
146. **Rakowicz-Szulczynska, E.M., Rodeck, U., Herlyn, M., and Koprowski, H.,** Chromatin binding of epidermal growth factor, nerve growth factor, and platelet-derived growth factor in cells bearing the appropriate surface receptors, *Proc. Natl. Acad. Sci. U.S.A.,* 83, 3728, 1986.
147. **Rakowicz-Szulczynska, E.M. and Koprowski, H.,** Antagonistic effect of PDGF and NGF on transcription of ribosomal DNA and tumor cell proliferation, *Biochem. Biophys. Res. Commun.,* 163, 649, 1989.
148. **Nakanishi, N. and Guroff, G.,** Nerve growth factor-induced increase in the cell-free phosphorylation of a nuclear protein in PC12 cells, *J. Biol. Chem.,* 260, 7791, 1985.
149. **Gizang-Ginsberg, E. and Ziff, E.B.,** Nerve growth factor regulates tyrosine hydroxylase gene transcription through a nucleoprotein complex that contains c-Fos, *Genes Dev.,* 4, 477, 1990.
150. **Greenberg, M.E., Greene, L.A., and Ziff, E.B.,** Nerve growth factor and epidermal growth factor induce rapid transient changes in proto-oncogene transcription in PC12 cells, *J. Biol. Chem.,* 260, 14101, 1985.
151. **Kruijer, W., Schubert, D., and Verma, I.M.,** Induction of the proto-oncogene *fos* by nerve growth factor, *Proc. Natl. Acad. Sci. U.S.A.,* 82, 7330, 1985.
152. **Greenberg, M.E., Hermanowski, A.L., and Ziff, E.B.,** Effect of protein synthesis inhibitors on growth factor activation of c-*fos,* c-*myc,* and actin gene transcription, *Mol. Cell. Biol.,* 6, 1050, 1986.
153. **Milbrandt, J.,** Nerve growth factor rapidly induces c-*fos* mRNA in PC12 rat pheochromocytoma cells, *Proc. Natl. Acad. Sci. U.S.A.,* 83, 4789, 1986.

154. **Sigmund, O., Naor, Z., Anderson, D.J., and Stein, R.,** Effect of nerve growth factor and fibroblast growth factor on SCG10 and c-*fos* expression and neurite outgrowth in protein kinase C-depleted PC12 cells, *J. Biol. Chem.,* 265, 2257, 1990.
155. **Ito, E., Sonnenberg, J.L., and Narayanan, R.,** Nerve growth factor-induced differentiation in PC-12 cells is blocked by *fos* oncogene, *Oncogene,* 4, 1193, 1989.
156. **Wu, B., Fodor, E.J.B., Edwards, R.H., Rutter, W.J.,** Nerve growth factor induces the proto-oncogene c-*jun* in PC12 cells, *J. Biol. Chem.,* 264, 9000, 1989.
157. **Ohmichi, M., Pang, L., Decker, S.J., and Saltiel, A.R.,** Nerve growth factor stimulates the activities of the Raf-1 and the mitogen-activated protein kinases via the *trk* proto-oncogene, *J. Biol. Chem.,* 267, 14604, 1992.
158. **Deschamps, J., Meijlink, F., and Verma, I.M.,** Identification of a transcriptional enhancer element upstream from proto-oncogene *fos, Science,* 230, 1174, 1985.
159. **Guerrero, I., Pellicer, A., and Burstein, D.E.,** Dissociation of c-fos from ODC expression and neuronal differentiation in a PC12 subline stably transfected with an inducible N-ras oncogene, *Biochem. Biophys. Res. Commun.,* 150, 1185, 1988.
160. **Maruyama, K., Schiavi, S.C., Huse, W., Johnson, G.L., and Ruley, H.E.,** *myc* and E1A oncogenes alter the responses of PC12 cells to nerve growth factor and block differentiation, *Oncogene,* 1, 361, 1987.
161. **Sharp, F.R., Gonzalez, M.F., Hisanaga, K., Mobley, W.C., and Sagar, S.M.,** Induction of the c-*fos* gene product in rat forebrain following cortical lesions and NGF injections, *Neurosci. Lett.,* 100, 117, 1989.
162. **Nakafuku, M., Satoh, T., and Kaziro, Y.,** Differentiation factors, including nerve growth factor, fibroblast growth factor, and interleukin-6, induce an accumulation of an active Ras-GTP complex in rat pheochromocytoma PC12 cells, *J. Biol. Chem.,* 267, 19448, 1992.
163. **Li, B.-Q., Kaplan, D., Kung, H., and Kamata, T.,** Nerve growth factor stimulation of the Ras-guanine nucleotide exchange factor and GAP activities, *Science,* 256, 1456, 1992.
164. **Buday, L. and Downward, J.,** Epidermal growth factor regulates $p21^{ras}$ through the formation of a complex of receptor, Grb2 adapter protein, and Sos nucleotide exchange factor, *Cell,* 73, 611, 1993.
165. **Szeberényi, J., Erhardt, P., Cai, H., and Cooper, G.M.,** Role of Ras in signal transduction from the nerve growth factor receptor: relationship to protein kinase c, calcium and cyclic AMP, *Oncogene,* 7, 2105, 1992.
166. **Thomson, T.M., Green, S.H., Trotta, R.J., Burstein, D.E., and Pellicer, A.,** Oncogene N-ras mediates selective inhibition of c-*fos* induction by nerve growth factor and basic fibroblast growth factor in a PC12 cell line, *Mol. Cell. Biol.,* 10, 1556, 1990.
167. **Borasio, G.D., John, J., Wittinghofer, A., Barde, Y.-A., Sendtner, M., and Heumann, R.,** *ras* p21 protein promotes survival and fiber outgrowth of cultured embryonic neurons, *Neuron,* 2, 1087, 1989.
168. **Bar-Sagi, D. and Feramisco, J.R.,** Microinjection of the *ras* oncogene protein into PC12 cells induces morphological differentiation, *Cell,* 42, 841, 1985.
169. **Guerrero, I., Wong, H., Pellicer, A., and Burstein, D.E.,** Activated N-ras gene induces neuronal differentiation of PC12 rat pheochromocytoma cells, *J. Cell. Physiol.,* 129, 71, 1986.
170. **Hagag, N., Halegoua, S., and Viola, M.,** Inhibition of growth factor-induced differentiation of PC12 cells by microinjection of antibody to *ras* p21, *Nature (London),* 319, 680, 1986.
171. **Alemà, S., Casalbore, P., Agostini, E., and Tato, F.,** Differentiation of PC12 phaeochromocytoma cells induced by v-*src* oncogene, *Nature (London),* 316, 557, 1985.
172. **Fabricant, R.N., DeLarco, J.E., and Todaro, G.J.,** Nerve growth factor receptors on human melanoma cells in culture, *Proc. Natl. Acad. Sci. U.S.A.,* 74, 565, 1977.
173. **Schechter, A.L. and Bothwell, M.A.,** Nerve growth factor receptors on PC12 cells: evidence for two receptor classes with differing cytoskeletal association, *Cell,* 24, 867, 1981.
174. **Sonnenfeld, K.H. and Ishii, D.N.,** Nerve growth factor effects and receptors in cultured human neuroblastoma cell lines, *J. Neurosci. Res.,* 8, 375, 1982.
175. **Sonnenfeld, K.H., Bernd, P., Sobue, G., Lebwohl, M., and Rubenstein, A.E.,** Nerve growth factor receptors on dissociated neurofibroma Schwann-like cells, *Cancer Res.,* 46, 1446, 1986.
176. **Baker, D.L., Molenaar, W.M., Trojanowski, J.Q., Evans, A.E., Ross, A.H., Rorke, L.B., Packer, R.J., Lee, V.M.-Y., and Pleasure,** Nerve growth factor receptor expression in peripheral and central neuroectodermal tumors, other pediatric brain tumors, and during development of the adrenal gland, *Am. J. Pathol.,* 139, 115, 1991.

177. **Baker, D.L., Reddy, U.R., Pleasure, D., Thorpe, C.L., Evans, A.E., Cohen, P.S., and Ross, A.H.,** Analysis of nerve growth factor receptor expression in human neuroblastoma and neuroepithelioma cell lines, *Cancer Res.,* 49, 4142, 1989.
178. **Berthold, F.,** Current concepts on the biology of neuroblastoma, *Blut,* 50, 65, 1985.
179. **Schwab, M.,** Amplification of N-*myc* in human neuroblastomas, *Trends Genet.,* 1, 271, 1985.
180. **Christiansen, H., Christiansen, N.M., Wagner, F., Altmannsberger, M., and Lampert, F.,** Neuroblastoma: inverse relationship between expression of N-*myc* and NGF-r, *Oncogene,* 5, 437, 1990.
181. **Nakagawara, A., Arima-Nakagawara, M., Scavarda, N.J., Azar, C.G., Cantor, A.B., and Brodeur, G.M.,** Association between high levels of expression of the TRK gene and favorable outcome in human neuroblastoma, *N. Engl. J. Med.,* 328, 847, 1993.
182. **Suzuki, T., Bogenmann, E., Shimada, H., Stram, D., and Seeger, R.C.,** Lack of high-affinity nerve growth factor receptors in aggressive neuroblastomas, *J. Natl. Cancer Inst.,* 85, 377, 1993.
183. **Kogner, P., Barbany, G., Dominici, C., Castello, M.A., Raschallá, G., and Persson, H.,** Coexpression of messenger RNA for *TRK* proto-oncogene and low affinity nerve growth factor receptor in neuroblastoma with favorable prognosis, *Cancer Res.,* 53, 2044, 1993.
184. **Perosio, P.M. and Brooks, J.J.,** Expression of nerve growth factor receptor in paraffin-embedded soft tissue tumors, *Am. J. Pathol.,* 132, 152, 1988.
185. **Thomson, T.M., Pellicer, A., and Greene, L.A.,** Functional receptors for nerve growth factor on Ewing's sarcoma and Wilm's tumor cells, *J. Cell. Physiol.,* 141, 60, 1989.
186. **Rausch, D.M., Dickens, G., Doll, S., Fujita, K., Koizumi, S., Rudkin, B.B., Tocco, M., Eiden, L.E., and Guroff, G.,** Differentiation of PC12 cells with v-src: comparison with nerve growth factor, *J. Neurosci. Res.,* 24, 49, 1989.
187. **Sassone-Corsi, P., Der, C.J., and Verma, I.M.,** *ras*-Induced neuronal differentiation of PC12-cells: possible involvement of *fos* and *jun, Mol. Cell. Biol.,* 9, 3174, 1989.
188. **Tanaka, S., Hattori, S., Kurata, T., Nagashima, K., Fukui, Y., Nakamura, S., and Matsuda, M.,** Both the SH2 and SH3 domains of human CRK protein are required for neuronal differentiation of PC12 cells, *Mol. Cell. Biol.,* 13, 4409, 1993.
189. **Pavelic, K. and Spaventi, S.,** Nerve growth factor (NGF) induced differentiation of human neuroblastoma cells, *Int. J. Biochem.,* 19, 1237, 1987.
190. **Chen, J., Chattopadhyay, B., Venkatakrishnan, G., and Ross, A.H.,** Nerve growth factor-induced differentiation of human neuroblastoma and neuroepithelioma cell lines, *Cell Growth Differ.,* 1, 79, 1990.
191. **Marushige, Y., Raju, N.R., Marushige, K., and Koestner, A.,** Modulation of growth and of morphological characteristics in glioma cells by nerve growth factor and glia maturation factor, *Cancer Res.,* 47, 4109, 1987.
192. **Lindsay, R.M., Thoenen, H., and Barde, Y.A.,** Placode-and neural crest-derived sensory neurons are responsive at early developmental stages to brain-derived neurotrophic factor, *Dev. Biol.,* 112, 319, 1985.
193. **Barde, Y.A., Davies, A.M., Johnson, J.E., Lindsay, R.M., and Thoenen, H.,** Brain-derived neurotrophic factor, *Progr. Brain Res.,* 71, 185, 1987.
194. **Leibrock, J., Lottspeich, F., Hohn, A., Hofer, M., Hengerer, B., Masiakowski, P., Thoenen, H., and Barde, Y.-A.,** Molecular cloning and expression of brain-derived neurotrophic factor, *Nature (London),* 341, 149, 1989.
195. **Hofer, M.M. and Barde, Y.A.,** Brain-derived neurotrophic factor prevents neuronal death *in vivo, Nature (London),* 331, 261, 1988.
196. **Yan, Q., Elliott, J., and Snyder, W.D.,** Brain-derived neurotrophic factor rescues spinal motor neurons from axotomy-induced cell death, *Nature (London),* 360, 753, 1992.
197. **Oppenheim, R.W., Yin, Q.W., Prevette, D., and Yan, Q.,** Brain-derived neurotrophic factor rescues developing avian motoneurons from cell death, *Nature (London),* 360, 755, 1992.
198. **Sendtner, M., Holtmann, B., Kolbeck, R., Thoenen, H., and Barde, Y.-A.,** Brain-derived neurotrophic factor prevents the death of motoneurons in newborn rats after nerve section, *Nature (London),* 360, 757, 1992.
199. **Hyman, C., Hofer, M., Barde, Y.-A., Juhasz, M., Yancopoulos, G.D., Squinto, S.P., and Lindsay, R.M.,** BDNF is a neurotrophic factor for dopaminergic neurons of the substantia nigra, *Nature (London),* 350, 230, 1991.

200. **Hofer, M., Pagliusi, S.R., Hohn, A., Leibrock, J., and Barde, Y.-A.,** Regional distribution of brain-derived neurotrophic factor mRNA in the adult mouse brain, *EMBO J.*, 9, 2459, 1990.
201. **Squinto, S.P., Stitt, T.N., Aldrich, T.H., Davis, S., Bianco, S.M., Radziewski, C., Glass, D.J., Masiakowski, P., Furth, M.E., Valenzuela, D.M., DiStefano, P.S., and Yancopoulos, G.D.,** *trkB* encodes a functional receptor for brain-derived neurotrophic factor and neurotrophin-3 but not nerve growth factor, *Cell,* 65, 885, 1991.
202. **Soppet, D., Escandon, E., Maragos, J., Middlemas, D.S., Reid, S.W., Blair, J., Burton, L.E., Stanton, B.R., Kaplan, D.R., Hunter, T., et al.,** The neurotrophic factors brain-derived neurotrophic factor and neurotrophin-3 are ligands for the *trkB* tyrosine kinase receptor, *Cell,* 65, 895, 1991.
203. **Klein, R., Nanduri, V., Jing, S., Lamballe, F., Tapley, P., Bryant, S., Cordon-Cardo, C., Jones, K.R., Reichardt, L.F., and Barbacid, M.,** The *trkB* tyrosine kinase is a receptor for brain-derived neurotrophic factor and neurotrophin-3, *Cell,* 66, 395, 1991.
204. **Hohn, A., Leibrock, J., Bailey, K., and Barde, Y.-A.,** Identification and characterization of a novel member of the nerve growth factor/brain-derived neurotrophic factor family, *Nature (London),* 334, 339, 1990.
205. **Maisonpierre, P.C., Belluscio, L., Squinto, S., Ip, N.Y., Furth, M.E., Lindsay, R.M., and Yancopoulos, G.D.,** Neurotrophin-3: a neurotrophic factor related to NGF and BDNF, *Science,* 247, 1446, 1990.
206. **Ernfors, P., Ibañez, C.F., Ebendal, T., Olson, L., and Persson, H.,** Molecular cloning and neurotrophic activities of a protein with structural similarities to nerve growth factor: developmental and topographical expression in the brain, *Proc. Natl. Acad. Sci. U.S.A.,* 87, 5454, 1990.
207. **Jones, K.R. and Reichardt, L.F.,** Molecular cloning of a human gene that is a member of the nerve growth factor family, *Proc. Natl. Acad. Sci. U.S.A.,* 87, 8060, 1990.
208. **Kalcheim, C., Carmelli, C., and Rosenthal, A.,** Neurotrophin-3 is a mitogen for cultured neural crest cells, *Proc. Natl. Acad. Sci. U.S.A.,* 89, 1661, 1992.
209. **Lamballe, F., Klein, R., and Barbacid, M.,** *trk*C, a new member of the *trk* family of tyrosine kinases, is a receptor for neurotrophin-3, *Cell,* 66, 967, 1991.
210. **Rodríguez-Tébar, A., Dechant, G., Götz, R., and Barde, Y.-A.,** Binding of neurotrophin-3 to its neuronal receptors and interactions with nerve growth factor and brain-derived neurotrophic factor, *EMBO J.,* 11, 917, 1992.
211. **Horylee, F., Russell, M., Lindsay, R.M., and Franck, E.,** Neurotrophin-3 supports the survival of developing muscle sensory neurons in culture, *Proc. Natl. Acad. Sci. U.S.A.,* 90, 2613, 1993.
212. **Kaisho, Y., Shintani, A., Ono, Y., Kato, K., and Igarashi, K.,** Regional expression of the nerve growth factor gene family in rat brain during development, *Biochem. Biophys. Res. Commun.,* 174, 379, 1991.
213. **Hallböök, F., Ibañez, C.F., and Persson, H.,** Evolutionary studies of the nerve growth factor family reveal a novel member abundantly expressed in *Xenopus* ovary, *Neuron,* 6, 845, 1991.
214. **Barbin, G., Manthorpe, M., and Varon, S.,** Purification of the chick eye ciliary neuronotrophic factor, *J. Neurochem.,* 430, 1468, 1984.
215. **McDonald, J.R., Ko, C., Mismer, D., Smith, D.J., and Collins, F.,** Expression and characterization of recombinant human ciliary neurotrophic factor from *Escherichia coli, Biochim. Biophys. Acta,* 1090, 70, 1991.
216. **Heymanns, J. and Unsicker, K.,** Neuroblastoma cells contain a trophic factor sharing biological and molecular properties with ciliary neurotrophic factor, *Proc. Natl. Acad. Sci. U.S.A.,* 84, 7758, 1987.
217. **Davis, S., Aldrich, T.H., Valenzuela, D.M., Wong, V., Furth, M.E., Squinto, S.P., and Yacopoulos, G.D.,** The receptor for ciliary neurotrophic factor, *Science,* 253, 59, 1991.
218. **Ip, N.Y., Nye, S.H., Boulton, T.G., Davis, S., Taga, T., Li, Y., Birren, S.J., Yasukawa, K., Kishimoto, T., Anderson, D.J., Stahl, N., and Yancopoulos, G.D.,** *Cell,* 69, 1121, 1992.
219. **Davis, S., Aldrich, T.H., Stahl, N., Pan, L., Taga, T., Kishimoto, T., Ip, N.Y., and Yancopoulos, G.D.,** LIFRβ and gp130 as heterodimerizing signal transducers of the tripartite CNTF receptor, *Science,* 260, 1805, 1993.
220. **Stahl, N., Davis, S., Wong, V., Taga, T., Kishimoto, T., Ip, N.Y., and Yancopoulos, G.D.,** Cross-linking identifies leukemia inhibitory factor-binding protein as a ciliary neurotrophic factor receptor component, *J. Biol. Chem.,* 268, 7628, 1993.
221. **Nesbitt, J.E., Fuentes, N.L., and Fuller, G.M.,** Ciliary neurotrophic factor regulates fibrinogen gene expression in hepatocytes by binding to the interleukin-6 receptor, *Biochem. Biophys. Res. Commun.,* 190, 544, 1993.

222. **Marchionni, M.A., Goodearl, A.D.J., Chen, M.S., Bermingham-McDonogh, O., Kirk, C., Hendricks, M., Danehi, F., Misumi, D., Sudhalter, J., Kobayashi, K., Wroblewski, D., Lynch, C., Baldassare, M., Hiles, I., Davis, J.B., Hsuan, J.J., Totty, N.F., Otsu, M., McBurney, R.N., Waterfield, M.D., Stroobant, P., and Gwynne, D.,** Glial growth factors are alternatively spliced erbB2 ligands expressed in the nervous system, *Nature (London),* 362, 312, 1993.
223. **Lim, R., Mitsonobu, K., and Li, W.K.P.,** Maturation-stimulating effect of brain extract and dibutyryl cyclic AMP on dissociated embryonic brain cells in culture, *Exp. Cell Res.,* 79, 243, 1973.
224. **Lim, R. and Mitsonobu, K.,** Brain cells in culture: morphological transformation by a protein, *Science,* 185, 63, 1974.
225. **Lim, R., Miller, J.F., Hicklin, D.J., and Andresen, A.A.,** Purification of bovine glia maturation factor and characterization with monoclonal antibody, *Biochemistry,* 24, 8070, 1985.
226. **Lim, R., Zaheer, A., and Lane, W.S.,** Complete amino acid sequence of bovine glia maturation factor β, *Proc. Natl. Acad. Sci. U.S.A.,* 87, 5233, 1990.
227. **Lim, R., Nakagawa, S., Arnason, B.G.W., and Turriff, D.E.,** Glia maturation factor promotes contact inhibition in cancer cells, *Proc. Natl. Acad. Sci. U.S.A.,* 78, 4373, 1981.
228. **Lim, R., Hicklin, D.J., Ryken, T.C., Han, X.-M., Liu, K.-N., Miller, J.F., and Baggenstoss, B.A.,** Suppression of glioma growth *in vitro* and *in vivo* by glia maturation factor, *Cancer Res.,* 46, 5241, 1987.
229. **Porter, S., Glaser, L., and Bunge, R.P.,** Release of autocrine growth factor by primary and immortalized Schwann cells, *Proc. Natl. Acad. Sci. U.S.A.,* 84, 7768, 1987.
230. **Lin, L.-F.H., Doherty, D.H., Lile, J.D., Bektesh, S., and Collins, F.,** GDNF: a glial cell line-derived neurotrophic factor for midbrain dopaminergic neurons, *Science,* 260, 1130, 1993.

Chapter 6

Organ-Specific Growth Factors

I. INTRODUCTION

The growth and survival of the specific organs of the body depend on complex interactions between hormones and growth factors which may act in a rather ubiquitous and nonspecific fashion. However, there is evidence that, for their growth and survival, certain organs require the action of factors which display local effects. Unfortunately, little is known about the structure and function of organ-specific growth factors. Liver- and prostate-specific growth factors have been preliminarily characterized, and they would be involved in regulating the growth and function of liver and prostate cells, respectively. The relationships between these factors and the structure and function of specific gene products could give clues for a better understanding of their mechanisms of action and for their possible roles in oncogenic processes affecting these tissues.

II. LIVER GROWTH FACTORS

Parenchymal cells of the vertebrate liver (hepatocytes) in adult animals are usually arrested on the G_0 phase of the cell cycle. Only a very small proportion (0.1 to 0.5%) of the hepatocytes is engaged in replicative DNA synthesis. However, in the case of surgical resection of the organ (partial hepatectomy), or after toxic or infectious injury, most hepatocytes enter the cell cycle and proliferate in order to regenerate the lost liver tissue. Regulation of such regenerative liver growth has to fulfill several requirements.[1] First, cell growth has to proceed in such a manner that the functions the organ supplies to the body are not impaired significantly, which implies that not all hepatocytes should replicate simultaneously. Second, growth has to be restricted to the liver, which means that growth factors with relative specificity for the liver should be involved in the regenerative process. Third, the process of regeneration should stop once the organ has achieved its previous normal size, which indicates that factors with growth inhibiting properties should act in the final stage of the process.

Evidence for the existence of liver-specific growth factors has been accumulated over many years from experiments with parabiotic rats as well as from the presence of growth-promoting activity in the cytosol of the remnant liver after partial hepatectomy. Unfortunately, it has been difficult to characterize the structure and functional properties of the putative liver-specific growth factors.[1] The mechanisms responsible for the control of hepatocyte proliferation during liver development or regeneration are very complex and involve, in addition to signals from factors circulating in the blood, signals from neighboring hepatocytes and nonparenchymal liver cells (sinusoidal endothelial cells, Kupffer cells, fat-storing cells, and fibroblasts) through membrane-bound factors (cell-cell contact) or secreted factors.[2] The different factors may have additive or differential effects on liver regeneration and repair. Many of the factors involved in the control of liver regeneration act at the level of regulation of gene expression, including proto-oncogene expression.[3]

Well-characterized growth factors such as basic fibroblast growth factor (FGF) and transforming growth factor (TGF)-α act as potent hepatotrophic mitogens in systems such as primary cultures of rat hepatocytes.[4] Insulin and epidermal growth factor (EGF) can also stimulate growth of hepatocytes in primary culture, and it has been thought that these two hormones may be hepatotrophic factors capable of inducing the proliferation of hepatocytes during liver regeneration. However, insulin and EGF do not appear to augment in the circulation after partial hepatectomy; on the contrary, several reports indicate that the levels of insulin in peripheral and portal blood fall rapidly after the operation.[5] Moreover, the concentrations of EGF in normal rat blood are below 0.5 ng/ml, which are less than one tenth of the effective dose for DNA synthesis in cultured hepatocytes. Thus, serum may contain other growth factors for mature hepatocytes besides insulin and EGF.

A. HEPATOTROPINS AND HEPATOPOIETINS

Growth-promoting activities with specificity for the liver (hepatotropins) have been detected in the blood as well as in liver cells.[6] Circulating hepatotropins may derive directly from plasma or may correspond to factors released by platelets. Factors capable of specifically stimulating DNA synthesis in liver cells have been isolated from several sources, including the serum of normal and partially hepatectomized rats,[7,8] the liver of weanling rats,[9] rat hepatocellular carcinoma cells,[10] and the conditioned medium from H-35 Reuber hepatoma cells.[11] The factor isolated from Reuber hepatoma cells appears to have a glycan structure and to act as an autocrine factor for these cells. A factor purified from the liver of rats treated with D-galactosamine is a heat-labile, acid-stable protein and is distinct from insulin, multiplication-stimulating activity (MSA), EGF, and other known hepatocyte growth factors.[12] Two fractions of different molecular weight, termed hepatopoietin A and B, exhibiting liver-specific growth-stimulatory activity, were found in rat and human serum.[13] The purification of hepatopoietin A from human plasma and rabbit serum was reported.[14] The purified product is a heterodimeric heparin-bind protein of high molecular weight (70 to 100 kDa), consisting of a heavy and a light polypeptide chain of 50 to 70 kDa and 30 to 35 kDa, respectively. Hepatopoietin A may exist in the form of aggregates by itself or with other substances. The precise origin and physiologic role of hepatopoietins remain uncertain, but hepatopoietin A appears to be identical with the hepatocyte growth factor.

B. HEPATOCYTE GROWTH FACTOR (HCGF OR HGF)

The isolation of a hepatocyte growth factor (HCGF or HGF), also called scatter factor, represents an important advance in the field of liver physiology.[15-17] HCGF was initially purified from the serum of hepatectomized rats and was identified as a heat- and acid-labile protein.[18,19] HCGF activity in rat serum increases progressively after partial hepatectomy, reaching a maximum about 24 h after the operation. A marked increase in HCGF activity occurs in rat plasma after liver injury provoked by administration of CCl_4 to rats. Experimental hepatitis induced in rats by administration of CCl_4 or D-galactosamine results in marked increases in HCGF mRNA in the liver.[20] However, no significant increase in HCGF gene transcription occurs in the rat liver during regeneration of the organ after partial hepatectomy.[21] Stimulation of liver tissue repair by HCGF appears to occur by a paracrine type of mechanism. *In situ* hybridization studies indicate that HCGF-producing cells are mesenchymal cells such as fibroblasts, Kupffer cells, and endothelial cells, but not epithelial cells. HCGF may bind ionically to heparan sulfate proteoglycans in the extracellular matrix in liver Disse's space, which may play a role in stabilization and maintaining the pool of HCGF in the organ.[22] Human intrahepatic biliary epithelial cells proliferate *in vitro* in response to HCGF.[23] HCGF is produced by tissues other than liver, including kidney tubule cells, skin keratinocytes, melanocytes, mammary duct cells, and bronchial epithelial cells. Interestingly HCGF transcripts are expressed at high levels in the rat lung following injuries of distant organs, suggesting a possible function of HCGF as a mediator acting via the circulatory system.[24] The lung may thus perform an endocrine function by producing HCGF that may be utilized for the regeneration of injured tissues and organs. HCGF is produced in glucagon-positive A-cells of the rat and human pancreatic islets, suggesting a potential role for HCGF in the paracrine regulation of insulin secretion.[25]

The mechanisms involved in regulation of HCGF synthesis and secretion remain little characterized. Several cytokines, including IL-1α, IL-1β, and TNF-α, enhance the production of HCGF by cultured fibroblasts.[26,27] Cultures of human adult primary fibroblasts from breast and prostate produce HCGF. TGF-α, TGF-β, and EGF can act as inhibitors of HCGF secretion by MRC-5 human lung fibroblasts.[28,29] The tumor cells can exert either positive or negative paracrine effects, depending on the culture conditions, suggesting that a delicate balance could exist between epithelial and mesenchymal compartments *in vivo*.

HCGF binds to a cell surface protein that is represented by the c-*met* proto-oncogene product. The activated HCGF/Met protein complex is involved in a diversity of functions, including mitogenic responses, cell motility, and promotion of an ordered spatial arrangement of many tissues, including lumen formation.[30]

1. Structure and Function of HCGF

The HCGF gene resides on human chromosome 7,[31] at region 7q21.1.[32] DNA sequencing shows that HCGF is derived from a precursor of 728 amino acids which is proteolitically processed to form a mature HCGF protein which is composed of two chains. Processing of HCGF to the heterodimeric form is required for biological activity.[33] Rat HCGF is a protein composed of the 62-kDa α subunit and the 34-kDa β subunit, linked covalently by a disulfide bond.[34] Purified HCGF stimulated DNA synthesis of adult rat hepatocytes, but not of fibroblastic cells, and was found to be different from other known growth factors. Cloning and sequence analysis of cDNAs for the rat HCGF gene and its promoter region have been reported.[35,36] Both HCGF α and β subunits are specified by a single ORF with a capacity for coding a protein of 728 amino acids. The α-subunit contains four "kringle" structures and the β-subunit has a

37% amino acid identity with the serine protease domain of plasmin. Northern blot analysis revealed that HCGF mRNA is expressed in various rat tissues, including liver, kidney, lung, and brain.

Complex mechanisms may be involved in the regulation of HCGF synthesis in different types of cells and tissues. The human promyelocytic leukemia cell line HL-60 produces HCGF,[37] and expression of the HCGF in both HL-60 cells and the human embryonic lung fibroblast cell line MRC-5 is negatively regulated by glucocorticoids and TGF-β.[38]

The human placenta is a rich source of HCGF.[39] Placental HCGF is expressed in the villous syncytium, extravillous trophoblast, and amnionic epithelium. HCGF was identified in the trophoblast of hydatidiform moles and choriocarcinomas. cDNA sequences coding for a tumor cytotoxic factor (TCF), isolated from a library prepared with mRNA from the human embryonic lung fibroblast cell line IMR-90, were identical or very similar to placental type HCGF.[40] TCF has cytotoxic activity against several tumor cell lines. Binding of HCGF to the c-Met receptor on the surface of normal and malignant human melanocytes may result in the stimulation of cellular proliferation and motility as well as in increased tyrosine kinase activity and melanin synthesis.[41] Such findings suggest that the growth-regulating effects of HCGF are not limited to the liver and that HCGF or similar factors may be involved in inflammation and repair processes as well as in tumorigenesis.

HCGF activity was also detected in the sera of patients with fulminant hepatic failure and after partial hepatectomy for removal of a hepatoma.[5] Human HCGF purified from plasma of patients with fulminant hepatic failure appears to be represented by a heterodimeric protein of about 83 kDa.[42] The immunological and biological characteristics of human HCGF are different from those of EGF and TGF-α.[43] The amino acid sequence of human HCGF, determined by cDNA cloning and the expression of biologically active human HCGF in COS-1 cells transfected with the cloned cDNA, revealed that both α- and β-chains are contained in a single ORF coding for a precursor protein of 728 amino acids.[44,45] The human HCGF gene is expressed not only in the liver but also in leukocytes and placenta.

2. The HCGF Receptor

The physiologic actions of HCGF are initiated by its binding to a specific receptor on the cell surface.[46] Binding of HCGF to the receptor stimulates the tyrosine kinase activity associated with the receptor.[47] The HCGF receptor is composed of two distinct polypeptide chains, and the protein product of the c-*met* proto-oncogene was identified with the 145-kDa β chain of the HCGF receptor.[48] The Met/β chain of the HCGF receptor possesses intrinsic tyrosine kinase activity and is associated through disulphide bonds to an α chain of 50 kDa. The kinase activity of the HCGF receptor is activated by autophosphorylation on a single tryptic peptide.[49] On the other hand, the kinase activity of the receptor is negatively regulated by protein kinase C through phosphorylation of serine residues as well as by an increase in $[Ca^{2+}]_i$.[50] The mechanism of the latter effect is unknown but may involve the action Ca^{2+}-dependent protein kinase(s).

3. Postreceptor Mechanisms of HCGF Action

The postreceptor mechanisms of HCGF action are little known. In rat hepatocytes, HCGF induces mobilization of intracellular Ca^{2+} and production of inositol 1,4,5-trisphosphate, stimulating the generation of 1,2-diacylglycerol via both phosphatidylcholine and phosphoinositide-specific phospholipase C.[51,52] HCGF stimulates tyrosine phosphorylation of phospholipase C-γ by the c-Met/HGCF receptor kinase and leads to enhanced generation of inositol 1,4,5-trisphosphate in primary cultures of rat hepatocytes.[53] Phosphorylation of cellular proteins may represent a central mechanism of action of the activated HCG receptor kinase. HCGF rapidly induces the tyrosine phosphorylation of 41- and 43-kDa proteins in mouse keratinocytes.[54] The action of the activated HCGF receptor, as those of other receptor tyrosine kinases, may be associated with the formation of activated GTP-bound Ras proteins. HCGF can activate Ras protein through the stimulation of Ras guanine nucleotide exchanger.[55]

HCGF may be involved in regulation of gene expression in the liver. Expression of the α-fetoprotein (AFP) gene, which is located on human chromosome 4, is downregulated by HCGF through the suppression of its promoter activity in human hepatoma cells.[56] The AFP gene is expressed at very high levels in the embryo, but decreases to almost undetectable levels after birth. Expression of the AFP gene is activated during liver regeneration and hepatocellular carcinoma. The possible effects of HCGF on proto-oncogene expression remain little known, but fetal rat hepatocytes in primary culture respond to HCGF with increased DNA synthesis and entering into S phase and mitosis, and these effects are associated with induction of c-*myc* and c-*fos* gene expression in the stimulated cells.[57]

4. Effects of HCGF on Normal and Neoplastic Cells

HCGF exerts a potent growth stimulatory effect on normal liver cells, but it acts as a negative regulator for the growth of hepatocellular carcinoma cells.[58] The reason(s) for the different effects of HCGF in normal and transformed hepatocytes is not understood but it is clear that HCGF may not function as an autocrine growth factor for transformed liver cells.

The mitogenic action of HCGF is not limited to hepatocytes but may include other types of cells. HCGF can stimulate the growth of rat kidney proximal tubule epithelial (RPTE) cells, rat nonparenchymal liver cells, human melanoma cells, and mouse keratinocytes.[59] A fibroblast-derived morphogen of kidney epithelial cells is represented by HCGF.[60] There is evidence that HCGF is a potent growth factor for human melanocytes, since these cells have high-affinity HCGF receptors and respond to HCGF *in vitro* with increased DNA synthesis and cell proliferation.[61] In contrast, human foreskin fibroblasts, human endothelial cells, and HEP3B cells do not respond to HCGF. HCGF can substitute for basic FGF/HBGF-2 in stimulating the anchorage-independent growth of simian virus 40- (SV40)-transformed RPTE cells. Fenestrated endothelial cells in the kidney produce HCGF, which may act as a renotropic factor in compensatory renal growth and renal regeneration *in vivo*.[62] Migration and proliferation of human microvascular endothelial cells is modulated in culture by HCGF.[63] HCGF is a synergistic factor for the growth of hematopoietic progenitor cells.[64] Available evidence indicates that HCGF is a multifunctional factor capable of influencing the growth of many different types of epithelial cells through autocrine and/or paracrine mechanisms.

Low concentrations of HCGF have cytostatic, antiproliferative effects on some tumor cell lines, including hepatocellular carcinoma and epidermoid carcinoma cells as well as mouse melanoma cells.[65] A tumor cytotoxic factor purified from the culture broth of IMR-90 human embryonic lung diploid fibroblasts is similar or identical to HCGF.[66,67] This factor is moderately cytotoxic for tumor cell lines of human origin but exhibits strong cytotoxicity for sarcoma cell lines of murine origin. The human HCGF-like factor is a potent mitogen not only for rat hepatocytes but also for human endothelial cells and melanocytes. Moreover, the factor induces differentiation of HL-60 human leukemia cells into granulocyte-like cells.

The possible role of HCGF in oncogenic processes is little known. Complex interactions between HCGF and other growth factors such as EGF, acidic FGF, and TGF-β may determine the final effect of HCGF on its target cells. Long-term treatment of rats with hepatic tumor promoters such as phenobarbital and hexachlorocyclohexane can enhance mitogenic responses of hepatocytes and EGF and inhibit the responses to acidic FGF and HCGF.[68] The Met/HCGF receptor is expressed in various types of solid tumors.[69] Overexpression of the HCGF receptor has been found in approximately one half of follicular carcinomas of the human thyroid gland.[70] HCGF and its receptors are co-expressed in small-cell lung cancer cell (SCLC) lines, suggesting an autocrine regulatory role in these cells.[71]

C. HEPATOCYTE GROWTH INHIBITORY FACTORS

The action of hepatocyte growth factors must be counteracted by that of factors with growth inhibitory properties in order to stop the hepatocellular proliferation associated, for example, with liver regeneration when the organ has reached its normal size. A factor previously classified as a chalone, with powerful growth inhibitory properties on hepatocellular proliferation, was purified to homogeneity and was identified as TGF-β.[72] This factor may play an important role in hepatocyte differentiation during liver regeneration as well as in hepatocarcinogenesis. Increased levels of TGF-β mRNA and protein are expressed in precursor cells during early stages of hepatocytic lineage differentiation during liver regeneration in the intact rat, and TGF-β can induce differentiation of rat liver epithelial cells *in vitro* along the hepatocytic lineage.[73] The pattern of TGF-β expression changes from endothelial and periductal cells observed in normal liver to predominantly mesenchymal cells during development of hepatic neoplasia. TGF-β not only is a potent growth inhibitor for both adult and immature hepatocytes but is also capable of inducing the differentiation of liver cells along the hepatocytic lineage *in vitro* and is associated *in vivo* with hepatocytic differentiation of oval cells, a particular type of liver cell progenitors.

The growth modulatory effects of a rat liver-derived growth inhibitor (LDGI), isolated from normal rat liver cells, have been compared to other known factors with growth inhibitory properties such as TGF-β and TNF-α.[74] LDGI is a 20-kDa polypeptide which produces a reversible inhibition in the growth of normal rat liver cells maintained in primary culture. It can also exert a growth inhibitory effect on aflatoxin-B1-transformed rat liver cells, MCF-7 human breast carcinoma cells, and Reuber rat hepatoma cells. In contrast, the rat hepatoma cell line UVM 777 and the human hepatoma cell line HepG2 are resistant to the antiproliferative effect of LDGI. Moreover, the growth of rat and human mesenchymal cell lines is stimulated in response to LDGI. These results indicate that LDGI, in a manner similar to TGF-β and TNF-α, is a bifunctional modulator of cell proliferation.

III. PROSTATE GROWTH FACTORS

Prostate cancer (prostatic adenocarcinoma) is a tumor that afflicts over one half of male individuals after the age of 70 years. It is usually a slow-growing tumor, but some of these tumors reach a relatively large size and in some cases the cells of the tumor escape from the primary site, invade neighboring tissues, and metastasize. Such cases are associated with a high mortality rate and, in fact, prostate cancer represents one of the most frequent causes of cancer death among males. The metastases of prostatic adenocarcinoma are found most commonly in bone marrow, where their growth rate appears to be considerably more rapid than that observed for primary prostatic tumors.[75] The rapidity of this growth and its poor response to therapeutic manipulations suggest that growth factors contained in the marrow microenvironment may stimulate the growth of metastatic prostate cells.

Regulation of the growth of normal and neoplastic prostatic cells is a complex process.[76] Androgens are considered to be the main hormones involved in the regulation of prostate cell growth and differentiation. Treatment of prostate cancer patients with LHRH analogs may result in suppression of the pituitary-testicular axis and improvement of the disease.[77] However, as other steroid hormones, androgens may not act in a direct fashion but through the local production of growth factor(s). Moreover, nonandrogenic hormones as well as some growth factors may have independent effects in regulating prostate morphology and function. The normal human prostate, as well as tissue samples from human prostate cancers and benign prostate hyperplasias, may contain membrane receptors for a wide variety of well characterized hormones and growth factors, including LHRH, somatostatin, prolactin, and EGF receptors.[78] Bombesin receptors are present in human prostatic carcinoma cell lines, and bombesin stimulates the growth rate of these cell lines in a dose-dependent manner.[79] The role of hormones and growth factors in prostatic benign or malignant diseases is not well characterized. Other factors with apparent specificity to prostate cells may be present in the prostate gland or the bone marrow.

A. PROSTATE-DERIVED GROWTH FACTORS (PGF)

A prostate growth factor (PGF) was purified from the cytosol of human benign hypertrophic prostates by heparin-Sepharose chromatography.[80] A similar or identical factor was purified from acid extracts of rat prostatic tissue.[81] Two types of PGFs were subsequently separated from the rat prostate according to their different affinity for heparin.[82] One PGF was present in normal prostate tissue and the other was isolated from a rat prostatic adenocarcinoma, the Dunning tumor R3327, which is believed to have originated in the dorsolateral prostate. The two types of PGFs are separable by a different affinity to heparin-Sepharose and show similarities to EGF and acidic FGF.[83]

PGF derived from human benign hypertrophic prostates has a molecular weight between 11,000 and 13,000 and is able to stimulate DNA synthesis at concentrations as low as 10 ng/ml.[80] PGF derived from the Dunning rat tumor has an approximate molecular weight of 19,000 and can stimulate DNA synthesis at a concentration of about 0.25 n*M*.[83] The chemical composition of human and rat PGFs and their relationships to other growth factors described previously, especially to FGFs and TGFs, remains to be established. It is also interesting to consider the possibility that PGFs mediate the growth-promoting action of other hormones, especially of testosterone, in the prostate gland. PGFs of 10 to 13 kDa, extracted from benign prostatic hyperplasia and prostatic adenocarcinoma, can increase cellular proliferation and alkaline phosphatase activity in osteoblast-like cells.[84] These activities are interesting since prostatic adenocarcinoma is characterized by a very high association with bone metastases.

B. BONE MARROW-DERIVED PROSTATE GROWTH FACTOR

Bone marrow has two main components: hematopoietic cells and stromal cells. Whereas hematopoietic cells are transient in the marrow and move to the bloodstream upon maturation, the stroma remains there and serves as a scaffold for the normal processes of hematopoiesis. In addition, the stromal cells produce CSFs and other types of HGFs which regulate the growth and differentiation of blood cells.

Stromal cells of the bone marrow may produce a factor which is distinct from other known growth factors, including HGFs, and which can exert a distinct stimulus on the growth of human prostatic carcinoma cells in culture.[85] The factor is present in the conditioned media from unstimulated human, rat, and bovine bone marrow cultures as well as in stromal cell conditioned medium, and its growth stimulatory effect is apparently specific for prostatic carcinoma cells. Other tumor cell types do not respond to the putative growth factor. The identification of the agent responsible for this activity remains to be determined.

IV. MAMMARY GROWTH FACTORS

The development and function of the mammary gland depends on a complex interaction between different hormones, growth factors, and regulatory peptides. Protein, steroid, and thyroid hormones contribute to the regulation of mammary gland structure and function, and there is evidence that some of their actions are mediated by the local production of growth factors. In addition to well characterized growth factors, the possible existence of growth factors specifically involved in the regulation of the mammary gland is suggested by their identification in either normal or neoplastic mammary tissues.

A. MAMMARY-DERIVED GROWTH FACTOR 1 (MDGF-1)

Human milk and mammary tumors may contain a growth factor that selectively enhances the production of type IV collagen in cultures of mammary ducts and alveoli. Purification of the factor yielded an acidic protein of 62 kDa, which was termed mammary-derived growth factor 1 (MDGF-1) and was shown to be different from other known proteins, including EGF. Various normal and malignant human epithelial mammary cells secrete MDGF-1 in culture. MDGF-1 recognizes a binding site of 120- to 140-kDa on the cell surface and induces phosphorylation of a 180- to 185-kDa protein on tyrosine.[86] The precise nature of this protein and its possible relationship to the putative MDGF-1 receptor are unknown.

B. MAMMARY TUMOR-DERIVED GROWTH FACTOR (MTGF)

A potent mitogen, called human mammary tumor-derived growth factor (MTGF), was isolated from biopsy samples of human mammary tumors.[87] Human MTGF is a cationic and acid-labile polypeptide apparently distinct from other known major peptide growth factors. MTGF shows an affinity for heparin, suggesting that it may play a role in the process of neovascularization in solid human mammary tumors. Human MTGF is mitogenic not only for human breast cancer cells (T-47D cells) but also for human fibroblasts and bovine corneal endothelial cells. It may promote mammary tumor growth by stimulating the neoplastic cell population in addition to the fibrovascular elements in the mammary tumor stroma. The site of synthesis, the type of secretory system (autocrine or paracrine), and the precise physiological role of MTGF are unknown.

V. HEART GROWTH FACTORS

The heart consists of a syncytium of muscle cells (myocytes) that are surrounded by other types of cells, collectively called nonmyocytes, which include fibroblasts, endothelial cells, and smooth muscle cells. Although myocytes make up most of the adult myocardial mass, they comprise only about 30% of the total number of cells present in the heart. Cardiac nonmyocytes are intimately associated with myocytes and may influence myocyte growth and/or development through cell-cell contact or indirectly via the production of one or more paracrine factors. Soon after birth, cardiac myocytes lose the ability to undergo cell division, and further growth occurs through hypertrophy of the individual cells. Both nonmyocytes and the extracellular matrix are increased, however, in myocardial hypertrophy that occurs in response to injury or infarction. Myocardial hypertrophy is an adaptative response to pressure overload of the heart and in its initial stages can normalize the acute elevation of systolic wall stress, whereas in later stages the capacity to hypertrophy may diminish. Hypertrophy is accompanied by alterations in protein synthesis and isoenzyme expression as well as by cell growth and remodeling.

Growth factors responsible for the structural and functional properties of the heart in normal or abnormal conditions remain little characterized. Several growth factors produced by heart cell nonmyocytes can influence the structure and function of the organ through autocrine and/or paracrine mechanisms. These factors include acidic FGF, basic FGF, PDGF, TGF-β, and TNF-α. In addition to these factors, cardiac nonmyocytes may produce a paracrine heparin-binding growth factor (HBGF) for cardiac myocytes.[88] This factor, termed nonmyocyte-derived growth factor (NMDGF), may also act in an autocrine fashion to stimulate fibroblast DNA synthesis and proliferation. NMDGF appears to be a polypeptide of 45 to 50 kDa, but its precise structure and function as well as its mechanism of action remain to be elucidated.

Proto-oncogenes may play an important role in heart function and in the adaptative responses of the heart to mechanical stimuli.[88-92] Stretching of cardiac myocytes stimulates c-*fos* expression in a stretch-, length-dependent manner, and this effect is enhanced by the protein synthesis inhibitor, cycloheximide. Increased c-*myc* and c-*fos* gene expression is observed in the isolated beating rat heart in comparison to the arrested isolated heart. The c-*fos* and c-*myc* genes, as well as the *HSP*-70 gene, are induced in the ventricular myocardium of adult rats within 1 h after imposition of blood pressure overload. Induction of c-*fos* and c-*jun* gene expression occurs in heart cells as a response to an acute increase in left ventricle systolic wall stresses distinct from passive diastolic wall stretch.[93]

REFERENCES

1. **Fleig, W.E.,** Liver-specific growth factors, *Scand. J. Gastroenterol.,* 23(Suppl. 151), 31, 1988.
2. **Shimaoka, S., Nakamura, T., and Ichihara, A.,** Stimulation of growth of primary cultured adult rat hepatocytes without growth factors by coculture with nonparenchymal liver cells, *Exp. Cell Res.,* 172, 228, 1987.
3. **Koch, K.S., Lu, X.P., Brenner, D.A., Fey, G.H., Martinez-Conde, A., and Leffert, H.L.,** Mitogens and hepatocyte growth control *in vivo* and *in vitro, In Vitro Cell. Dev. Biol.,* 26, 26, 1011, 1990.
4. **Hoffmann, B. and Paul, D.,** Basic growth factor and transforming growth factor-α are hepatotrophic mitogens *in vitro, J. Cell. Physiol.,* 142, 149, 1990.
5. **Nakamura, T.,** Growth control of mature hepatocytes by growth factor and growth inhibitor from platelets, *Gunma Symp. Endocrinol. (Tokyo),* 25, 179, 1988.
6. **Michalopoulos, G.K.,** Liver regeneration: molecular mechanisms of growth control, *FASEB J.,* 4, 176, 1990.
7. **Morioka, K., Shimada, H., and Terayama, H.,** Serum factor stimulating DNA synthesis in the isolated nuclear system from rat liver, *Biochem. Biophys. Res. Commun.,* 51, 451, 1973.
8. **Thaler, F.J. and Michalopoulos, G.K.,** Hepatopoietin A: partial characterization and trypsin activation of a hepatocyte growth factor, *Cancer Res.,* 45, 2545, 1985.
9. **Labrecque, D.R., Wilson, M., and Fogerty, S.,** Stimulation of HTC hepatoma cell growth *in vitro* by hepatic stimulator substance (HSS). Interactions with serum, insulin, glucagon, epidermal growth factor and platelet derived growth factor, *Exp. Cell Res.,* 150, 419, 1984.
10. **Luetteke, N.C. and Michalopoulos, G.K.,** Partial purification and characterization of a hepatocyte growth factor produced by rat hepatocellular carcinoma cells, *Cancer Res.,* 45, 6331, 1985.
11. **Witters, L.A. and Watts, T.D.,** An autocrine factor from Reuber hepatoma cells that stimulates DNA synthesis and acetyl-CoA carboxylase. Characterization of biologic activity and evidence for a glycan structure, *J. Biol. Chem.,* 263, 8027, 1988.
12. **Suemori, S., Eto, T., Yamada, K., Nakamura, T., Nakanishi, T., and Kajiyama, G.,** Partial purification and characterization of hepatocyte proliferation stimulatory factor from liver of rats treated with D-galactosamine, *Biochem. Biophys. Res. Commun.,* 150, 133, 1988.
13. **Michalopoulos, G., Houck, K.A., Dolan, M.L., and Luetteke, N.C.,** Control of hepatocyte replication by two serum factors, *Cancer Res.,* 44, 4414, 1984.
14. **Zarnegar, R. and Michalopoulos, G.,** Purification and biological characterization of human hepatopoietin A, a polypeptide growth factor for hepatocytes, *Cancer Res.,* 49, 3314, 1989.
15. **Gherardi, E. and Stoker, M.,** Hepatocyte growth factor — scatter factor: mitogen, motogen, and *Met, Cancer Cells,* 3, 227, 1991.
16. **Matsumoto, K. and Nakamura, T.,** Hepatocyte growth factor: molecular structure, roles in liver regeneration, and other biological functions, *Crit. Rev. Oncogenesis,* 3, 27, 1992.
17. **Strain, A.J.,** Hepatocyte growth factor — Another ubiquitous cytokine, *J. Endocrinol.,* 137(Suppl.), 1, 1993.
18. **Nakamura, T., Nawa, K., and Ichihara, A.,** Partial purification and characterization of hepatocyte growth factor from serum of hepatectomized rats, *Biochem. Biophys. Res. Commun.,* 122, 1450, 1984.
19. **Nakamura, T., Teramoto, H., and Ichihara, A.,** Purification and characterization of a growth factor from rat platelets for mature parenchymal hepatocytes in primary cultures, *Proc. Natl. Acad. Sci. U.S.A.,* 83, 6489, 1986.
20. **Kinoshita, T., Tashiro, K., and Nakamura, T.,** Marked increase of HGF mRNA in non-parenchymal liver cells of rats treated with hepatotoxins, *Biochem. Biophys. Res. Commun.,* 165, 1229, 1989.
21. **Yanagita, K., Nagaike, M., Ishibashi, H., Niho, Y., Matsumoto, K., and Nakamura, T.,** Lung may have an endocrine function producing hepatocyte growth factor in response to injury of distal organs, *Biochem. Biophys. Res. Commun.,* 182, 802, 1992.

22. **Masumoto, A. and Yamamoto, N.,** Stimulation of DNA synthesis in hepatocytes by hepatocyte growth factor bound to extracellular matrix, *Biochem. Biophys. Res. Commun.,* 191, 1218, 1993.
23. **Joplin, R., Hishida, T., Tsubouchi, Y., Ayres, R., Neuberger, J.M., and Strain, A.J.,** Human intrahepatic biliary epithelial cells proliferate *in vitro* in response to human hepatocyte growth factor, *J. Clin. Invest.,* 90, 1284, 1992.
24. **Zarnegar, R., Petersen, B., DeFrances, M.C., and Michalopoulos, G.,** Localization of hepatocyte growth factor (HGF) gene on human chromosome 7, *Genomics,* 12, 147, 1992.
25. **Tsuda, H., Iwase, T., Matsumoto, K., Ito, M., Hirono, I., Nishida, Y., Yamamoto, M., Tatematsu, M., Matsumoto, K., and Nakamura, T.,** Immunohistochemical localization of hepatocyte growth factor protein in pancreas islet A-cells of man and rats, *Jpn. J. Cancer Res.,* 83, 1262, 1992.
26. **Matsumoto, K., Okazaki, H., and Nakamura, T.,** Up-regulation of hepatocyte growth factor gene expression by interleukin-1 in human skin fibroblasts, *Biochem. Biophys. Res. Commun.,* 188, 235, 1992.
27. **Tamura, M., Arakaki, N., Tsubouchi, H., Takada, H., and Daikuhara, Y.,** Enhancement of human hepatocyte growth factor production by interleukin-1α and interleukin-1β and tumor necrosis factor-α by fibroblasts in culture, *J. Biol. Chem.,* 268, 8140, 1993.
28. **Gohda, E., Matsunaga, T., Kataoka, H., Yamamoto, I.,** TGF-β is a potent inhibitor of hepatocyte growth factor secretion by human fibroblasts, *Cell Biol. Int. Rep.,* 16, 917, 1992.
29. **Seslar, S.P., Nakamura, T., and Byers, S.W.,** Regulation of fibroblast hepatocyte growth factor/scatter factor expression by human breast carcinoma cell lines and peptide growth factors, *Cancer Res.,* 53, 1233, 1993.
30. **Tsarfaty, I., Resau, J.H., Rulong, S., Keydar, I., Faletto, D.L., and Vande Woude, G.F.,** The *met* proto-oncogene receptor and lumen formation, *Science,* 257, 1258, 1992.
31. **Noji, S., Tashiro, K., Koyama, E., Nohno, T., Ohyama, K., Taniguchi, S., and Nakamura, T.,** Expression of hepatocyte growth factor gene in endothelial and Kupffer cells of damaged rat livers, as revealed by in situ hybridization, *Biochem. Biophys. Res. Commun.,* 173, 42, 1990.
32. **Saccone, S., Narsimhan, R.P., Gaudino, G., Dalpra, L., Comoglio, P.M., and Delavalle, G.,** Regional mapping of the human hepatocyte growth factor (HGF)-scatter factor gene to chromosome 7q21.1, *Genomics,* 13, 912, 1992.
33. **Gak, E., Taylor, W.G., Chan, A.M.L., and Rubin, J.S.,** Processing of hepatocyte growth factor to the heterodimeric form is required for biological activity, *FEBS Lett.,* 311, 17, 1992.
34. **Nakamura, T., Nawa, K., Ichihara, A., Kaise, N., and Nishino, T.,** Purification and subunit structure of hepatocyte growth factor from rat platelets, *FEBS Lett.,* 224, 311, 1987.
35. **Nakamura, T., Nishizawa, T., Hagiya, M., Seki, T., Shimonishi, M., Sugimura, A., Tashiro, K., and Shimizu, S.,** Molecular cloning and expression of human hepatocyte growth factor, *Nature,* 342, 440, 1989.
36. **Okajima, A., Miyazawa, K., and Kitamura, N.,** Characterization of the promoter region of the rat hepatocyte growth factor/scatter factor gene, *Eur. J. Biochem.,* 213, 113, 1993.
37. **Nishino, T., Kaise, N., Sindo, Y., Nishino, N., Nishida, T., Yasuda, S., and Masui, Y.,** Promyelocytic leukemia cell line, HL-60, produces human hepatocyte growth factor, *Biochem. Biophys. Res. Commun.,* 181, 323, 1991.
38. **Matsumoto, K., Tajima, H., Okazaki, H., and Nakamura, T.,** Negative regulation of hepatocyte growth factor gene expression in human lung fibroblasts and leukemic cells by transforming growth factor-β1 and glucocorticoids, *J. Biol. Chem.,* 267, 24917, 1992.
39. **Wolf, H.K., Zarnegar, R., Oliver, L., and Michalopoulos, G.K.,** Hepatocyte growth factor in human placenta and trophoblastic disease, *Am. J. Pathol.,* 138, 1035, 1991.
40. **Shima, N., Nagao, M., Ogaki, F., Tsuda, E., Murakami, A., and Higashio, K.,** Tumor cytotoxic factor/hepatocyte growth factor from human fibroblasts: cloning of its cDNA, purification and characterization of recombinant protein, *Biochem. Biophys. Res. Commun.,* 180, 1151, 1991.
41. **Halaban, R., Rubin, J.S., Funasaka, Y., Cobb, M., Boulton, T., Faletto, D., Rosen, E., Chan, A., Yoko, K., White, W., Cook, C., and Moellmann, G.,** Met and hepatocyte growth factor/scatter factor signal transduction in normal melanocytes and melanoma cells, *Oncogene,* 7, 2195, 1992.
42. **Gohda, E., Tsubouchi, H., Nakayama, H., Hirono, S., Sakiyama, O., Takahashi, K., Miyazaki, H., Hashimoto, S., and Dakuhara, Y.,** Purification and partial characterization of hepatocyte growth factor from plasma of a patient with fulminant hepatic failure, *J. Clin. Invest.,* 81, 414, 1988.

43. **Gohda, E., Yamasaki, T., Tsobouchi, H., Kurobe, M., Sakiyama, O., Aoki, H., Niidani, N., Shin, S., Hayashi, K., Hashimoto, S., Daikuhara, Y., and Yamamoto, I.,** Biological and immunological properties of human hepatocyte growth factor from plasma of patients with fulminant hepatic failure, *Biochim. Biophys. Acta,* 1053, 21, 1990.
44. **Tashiro, K., Hagiya, M., Nishizawa, T., Seki, T., Shimonishi, M., Shimizu, S., and Nakamura, T.,** Deduced primary structure of rat hepatocyte growth factor and expression of the mRNA in rat tissues, *Proc. Natl. Acad. Sci. U.S.A.,* 87, 3200, 1990.
45. **Seki, T., Ihara, I., Sugimura, A., Shimonishi, M., Nishizawa, T., Asami, O., Hagiya, M., Nakamura, T., and Shimizu, S.,** Isolation and expression of cDNA for different forms of hepatocyte growth factor from human leukocyte, *Biochem. Biophys. Res. Commun.,* 172, 321, 1990.
46. **Zarnegar, R., DeFrances, M.C., Oliver, L., and Michalopoulos, G.,** Identification and partial characterization of receptor binding sites for HGF on rat hepatocytes, *Biochem. Biophys. Res. Commun.,* 173, 1179, 1990.
47. **Naldini, L., Vigna, E., Nasimhan, R.P., Gaudino, G., Zarnegar, R., Michalopoulos, G.K., and Comoglio, P.M.,** Hepatocyte growth factor (HGF) stimulates the tyrosine kinase activity of the receptor encoded by the proto-oncogene c-MET, *Oncogene,* 6, 501, 1991.
48. **Bottaro, D.P., Rubin, J.S., Faletto, D.L., Chan, A.M.L., Kmiecik, T.E., Vande Woude, G.F., and Aaronson, S.A.,** Identification of the hepatocyte growth factor receptor as the c-*met* proto-oncogene product, *Science,* 251, 802, 1991.
49. **Naldini, L., Vigna, E., Ferracini, R., Longati, P., Gandino, L., Prat, M., and Comoglio, P.M.,** The tyrosine kinase encoded by the MET proto-oncogene is activated by autophosphorylation, *Mol. Cell. Biol.,* 11, 1793, 1991.
50. **Gandino, L., Munaron, L., Naldini, L., Ferracini, R., Magni, M., and Comoglio, P.M.,** Intracellular calcium regulates the tyrosine kinase receptor encoded by the MET oncogene, *J. Biol. Chem.,* 266, 16098, 1991.
51. **Osada, S., Nakashima, S., Saji, S., Nakamura, T., and Nozawa, Y.,** Hepatocyte growth factor (HGF) mediates the sustained formation of 1,2-diacylglycerol via phosphatidylcholine phospholipase C in cultures of rat hepatocytes, *FEBS Lett.,* 297, 271, 1992.
52. **Baffy, G., Yang, L.J., Michalopoulos, G.K., and Williamson, J.R.,** Hepatocyte growth factor induces calcium mobilization and inositol phosphate production in rat hepatocytes, *J. Cell. Physiol.,* 153, 332, 1992.
53. **Okano, Y., Mizuno, K., Osada, S., Nakamura, T., and Nozawa, Y.,** Tyrosine phosphorylation of phospholipase Cγ in v-*met*/HGF receptor-stimulated hepatocytes: comparison with HepG2 hepatocarcinoma cells, *Biochem. Biophys. Res. Commun.,* 190, 842, 1993.
54. **Chatani, Y., Itoh, A., Tanaka, E., Hattori, A., Nakamura, T., and Kohno, M.,** Hepatocyte growth factor rapidly induces the tyrosine phosphorylation of 41-kDa and 43-kDa protein in mouse keratinocytes, *Biochem. Biophys. Res. Commun.,* 185, 860, 1992.
55. **Graziani, A., Gramaglia, D., Dalla Zonca, P., and Comoglio, P.,** Hepatocyte growth factor/scatter factor stimulates the Ras guanine nucleotide exchanger, *J. Biol. Chem.,* 268, 9165, 1993.
56. **Hatano, M., Nakata, K., Nakao, K., Tsutsumi, T., Ohtsuru, A., Nakamura, T., Tamaoki, T., and Nagataki, S.,** Hepatocyte growth factor down-regulates the α-fetoprotein gene expression in PLC/PRF/5 human hepatoma cells, *Biochem. Biophys. Res. Commun.,* 189, 385, 1992.
57. **Fabregat, I., de Juan, C., Nakamura, T., and Benito, M.,** Growth stimulation of rat fetal hepatocytes in response to hepatocyte growth factor: modulation of c-*myc* and c-*fos* expression, *Biochem. Biophys. Res. Commun.,* 189, 684, 1992.
58. **Shiota, G., Rhoads, D.B., Wang, T.C., Nakamura, T., and Schmidt, E.V.,** Hepatocyte growth factor inhibits growth of hepatocellular carcinoma cells, *Proc. Natl. Acad. Sci. U.S.A.,* 89, 373, 1992.
59. **Kan, M., Zhang, G., Zarnegar, R., Michalopoulos, G., Myoken, Y., McKeehan, W.L., and Stevens, J.I.,** Hepatocyte growth factor/hepatopoietin A stimulates the growth of rat kidney proximal tubule epithelial cells (RPTE), rat nonparenchymal liver cells, human melanoma cells, mouse keratinocytes and stimulates anchorage-independent growth of SV-40 transformed RPTE, *Biochem. Biophys. Res. Commun.,* 174, 331, 1991.
60. **Montesano, R., Matsumoto, K., Nakamura, T., and Orci, L.,** Identification of a fibroblast-derived epithelial morphogen as hepatocyte growth factor, *Cell,* 67, 901, 1991.
61. **Matsumoto, K., Tajima, H., and Nakamura, T.,** Hepatocyte growth factor is a potent stimulator of human melanocyte DNA synthesis and growth, *Biochem. Biophys. Res. Commun.,* 176, 45, 1991.

62. **Igawa, T., Kanda, S., Kanetake, H., Saitoh, Y., Ichihara, A., Tomita, Y., and Nakamura, T.,** Hepatocyte growth factor is a potent mitogen for cultured rabbit renal tubular epithelial cells, *Biochem. Biophys. Res. Commun.,* 174, 831, 1991.
63. **Morimoto, A., Okamura, K., Hamanaka, R., Sato, Y., Shima, N., Higashio, K., and Kuwano, M.,** Hepatocyte growth factor modulates migration and proliferation of human microvascular endothelial cells in culture, *Biochem. Biophys. Res. Commun.,* 179, 1042, 1991.
64. **Kmiecik, T.E., Keller, J.R., Rosen, E., and Vande Woude, G.F.,** Hepatocyte growth factor is a synergistic factor for the growth of hematopoietic progenitor cells, *Blood,* 80, 2454, 1992.
65. **Tajima, H., Matsumoto, K., and Nakamura, T.,** Hepatocyte growth factor has potent anti-proliferative activity in various tumor cell lines, *FEBS Lett.,* 291, 229, 1991.
66. **Higashio, K., Shima, N., Goto, M., Itagaki, Y., Nagao, M., Yasuda, H., and Morinaga, T.,** Identity of a tumor cytotoxic factor from human fibroblasts and hepatocyte growth factor, *Biochem. Biophys. Res. Commun.,* 170, 397, 1990.
67. **Shima, N., Itagaki, Y., Nagao, M., Yasuda, H., Morinaga, T., and Higashio, K.,** A fibroblast-derived tumor cytotoxic factor/F-TCF (hepatocyte growth factor/HGF) has multiple functions *in vitro, Cell Biol. Int. Rep.,* 15, 397, 1991.
68. **Tsai, W.-H., Zarnegar, R., and Michalopoulos, G.K.,** Long-term treatment with hepatic tumor promoters inhibits mitogenic responses of hepatocytes to acidic fibroblast growth factor and hepatocyte growth factor, *Cancer Lett.,* 59, 103, 1991.
69. **Prat, M., Narsimhan, R.P., Crepaldi, T., Nicostra, M.R., Natali, P.G., and Comoglio, P.M.,** The receptor encoded by the human c-MET oncogene is expressed in hepatocytes, epithelial cells and solid tumors, *Int. J. Cancer,* 49, 323, 1991.
70. **Di Renzo, M.F., Olivero, M., Ferro, S., Prat, M., Bongarzone, I., Pilotti, S., Belfiore, A., Cosantino, A., Vigneri, R., Pierotti, M.A., and Comoglio, P.M.,** Overexpression of the c-*MET*/HGF receptor gene in human thyroid carcinomas, *Oncogene,* 7, 2549, 1992.
71. **Rygaard, K., Nakamura, T., and Spang-Thomsen, M.,** Expression of the proto-oncogenes c-*met* and c-*kit* and their ligands, hepatocyte growth factor/scatter factor and stem cell factor, in SCLC cell lines and xenografts, *Br. J. Cancer,* 67, 37, 1993.
72. **Strain, A.J.,** Transforming growth factor β and inhibition of hepatocellular proliferation, *Scand. J. Gastroenterol.,* 23(Suppl. 151), 37, 1988.
73. **Nagy, P., Evarts, R.P., McMahon, J.B., and Thorgeirsson, S.S.,** Role of TGF-β in normal differentiation and oncogenesis in rat liver, *Mol. Carcinogenesis,* 2, 345, 1989.
74. **Chapekar, M.S., Huggett, A.C., and Thorgeirsson, S.S.,** Growth modulatory effects of a liver-derived growth inhibitor, transforming growth factor β_1, and recombinant tumor necrosis factor α, in normal and neoplastic cells, *Exp. Cell Res.,* 185, 247, 1989.
75. **Jacobs, S.C.,** Spread of prostatic cancer to bone, *Urology,* 21, 337, 1983.
76. **Trapman, J., Jenster, G., Riegman, P., Klaassen, P., van der Korput, J.A.G.M., van Steenbrugge, G.J., and Romijn, J.C.,** Expression of (proto)oncogenes and genes encoding growth factors, growth factor receptors and prostate antigen in human prostate carcinoma cell lines, *Prog. Cancer Res. Ther.,* 3, 102, 1988.
77. **Gonzalez-Barcena, D., Perez-Sanchez, P.L., Graef, A., Gomez, A.M., Berea, H., Comaru-Schally, A.M., and Schally, A.V.,** Inhibition of the pituitary-gonadal axis by a single intramuscular administration of D-Trp-6-LH-RH (Decapeptyl) in a sustained-release formulation in patients with prostatic carcinoma, *Prostate,* 14, 291, 1989.
78. **Fekete, M., Redding, T.W., Comaru-Schally, A.M., Pontes, J.E., Connelly, R.W., Srkalovic, G., and Schally, A.V.,** Receptors for luteinizing hormone-releasing hormone, somatostatin, prolactin, and epidermal growth factor in rat and human prostate cancers and in benign prostate hyperplasia, *Prostate,* 14, 191, 1989.
79. **Bologna, M., Festuccia, C., Muzi, P., Biordi, L., and Ciomei, M.,** Bombesin stimulates growth of human prostatic cancer cells *in vitro, Cancer,* 63, 1714, 1989.
80. **Nishi, N., Matuo, Y., Muguruma, Y., Yoshitake, Y., Nishikawa, K., and Wada, F.,** A human prostatic growth factor (hPGF): partial purification and characterization, *Biochem. Biophys. Res. Commun.,* 132, 1103, 1985.
81. **Maehama, S., Li, D., Nanri, H., Leykam, J.F., and Deuel, T.F.,** Purification and partial characterization of prostate-derived growth factor, *Proc. Natl. Acad. Sci. U.S.A.,* 83, 8162, 1986.

82. **Matuo, Y., Nishi, N., Matsui, S., Sandberg, A.A., Isaacs, J.T., and Wada, F.,** Heparin binding affinity of rat prostatic growth factor in normal and cancerous prostates: partial purification and characterization of rat prostatic growth factor in the Dunning tumor, *Cancer Res.,* 47, 188, 1987.
83. **Matuo, Y., Nishi, N., and Wada, F.,** Growth factors in the prostate, *Arch. Androl.,* 19, 193, 1987.
84. **Koutsilieris, M., Rabbani, S.A., Bennett, H.P.J., and Goitzman, D.,** Characteristics of prostate-derived growth factors for cells of the osteoblast phenotype, *J. Clin. Invest.,* 80, 941, 1987.
85. **Chackal-Roy, M., Niemeyer, C., Moore, M., and Zetter, B.R.,** Stimulation of human prostatic carcinoma cell growth by factors present in human bone marrow, *J. Clin. Invest.,* 84, 43, 1989.
86. **Bano, M., Worland, P., Kidwell, W.R., Lippman, M.E., and Dickson, R.B.,** Receptor-induced phosphorylation by mammary-derived growth factor 1 in mammary epithelial cell lines, *J. Biol. Chem.,* 267, 10389, 1992.
87. **Rowe, J.M., Kasper, S., Shiu, R.P.C., and Friesen, H.G.,** Purification and characterization of a human mammary tumor-derived growth factor, *Cancer Res.,* 46, 1408, 1986.
88. **Long, C.S., Henrich, C.J., and Simpson, P.C.,** A growth factor for cardiac myocytes is produced by cardiac nonmyocytes, *Cell Regul.,* 2, 1081, 1991.
89. **Barka, T., van der Noen, H., and Shaw, P.A.,** Proto-oncogene *fos* (c-*fos*) expression in the heart, *Oncogene,* 1, 439, 1987.
90. **Komuro, I., Kaida, T., Shibazaki, Y., Kurabayashi, M., Katoh, Y., Hoh, E., Takaku, F., and Yazaki, Y.,** Stretching cardiac myocytes stimulates proto-oncogene expression, *J. Biol. Chem.,* 265, 3595, 1990.
91. **Bauters, C., Moalic, J.-M., Bercovici, J., Mouas, C., Emanoil-Ravier, R., Schiaffino, S., and Swynghedauw, B.,** Augmentation de l'expression des oncogènes c-myc et c-fos en fonction de l'activité mécanique due coeur isolé de rat adulte, *C. R. Acad. Sci. Paris,* 306, 597, 1988.
92. **Izumo, S., Nadal-Ginard, B., and Mahdavi, V.,** Proto-oncogene induction and reprogramming of cardiac gene expression produced by pressure overload, *Proc. Natl. Acad. Sci. U.S.A.,* 85, 339, 1988.
93. **Schunkert, H., Jahn, L., Izumo, S., Apstein, C.S., and Lorell, B.H.,** Localization and regulation of c-*fos* and c-*jun* proto-oncogene induction by systolic wall stress in normal and hypertrophied rat hearts, *Proc. Natl. Acad. Sci. U.S.A.,* 88, 11480, 1991.

Chapter 7

Cell-Specific Growth Factors

I. MELANOCYTE GROWTH FACTORS

Melanocytes reside on the basal layer of the epidermis and are closely associated with keratinocytes and Langerhans cells, which are cells that synthesize and secrete a number of growth factors and cytokines, including transforming growth factor (TGF)-α, TGF-β, interleukin (IL)-1, IL-6, IL-8, and transforming necrosis factor (TNF)-α. The growth of melanocytes *in vivo* may be controlled, at least in part, by paracrine mechanisms which may depend on the activity of keratinocytes and/or Langerhans cells. Normal melanocytes do not express growth factors constitutively, but normal neonatal human melanocytes cultured in the presence of 12-*O*-tetradecanoylphorbal-13-acetate (TPA) respond to IL-1α, IL-6, and TNF-α.[1] *In vitro,* the three cytokines inhibit the proliferation and melanogenesis of human melanocytes. Hepatocyte growth factor (HCGF) may also be a paracrine growth factor for human melanocytes, since these cells express HCGF receptors and respond to HCGF with increased DNA synthesis and cell proliferation.[2] However, melanocytes only rarely undergo mitosis *in vivo* and the precise roles of different growth factors and cytokines in melanocyte physiology is unknown.

A complex interplay between hormones, growth factors, and proto-oncogene protein products is implicated in melanomagenesis.[3] In contrast to normal melanocytes, human melanoma cell lines may constitutively and simultaneously produce growth factors including TGF-α, TGF-β, nerve growth factor (NGF), fibroblast growth factor (FGF), platelet-derived growth factor (PDGF)-A, PDGF-B, IL-1α, and IL-1β, but not IGF-I or IGF-II.[4] Basic FGF, but not PDGF and TGF-α, may be involved in the autocrine growth of human melanoma cells. In a comparative study between human metastatic melanoma cell lines and normal human foreskin melanocyte cell lines it was concluded that TGF-β_2, TGF-α, and basic FGF may be important in melanomagenesis, while acidic FGF and FGF-5 are probably involved in melanoma progression.[5] Basic FGF can stimulate the proliferation of mouse epidermal melanoblasts in a serum-free medium supplemented with dibutyryl cAMP.[6] However, the exact role of endogenous growth factors in the proliferation of melanoma cells *in vivo* is unknown.

Fibroblasts may participate in regulating melanocyte growth. A growth factor of 40 kDa extracted from the human embryo fibroblast cell line WI-38 supports continued proliferation of cultured melanocytes in the absence of the tumor promoter phorbol ester PMA.[7] Moreover, exposure of melanocytes to the combined action of TPA and active WI-38 extracts results in a synergistic effect on cell proliferation. In contrast to normal melanocytes, human melanoma cells generally grow vigorously in the absence of TPA, suggesting that their independent growth *in vivo* might be associated with the endogenous production of autostimulatory growth factors which may include TGFs and/or a melanocyte growth factor(s).[7,8] Basic FGF produced by the melanocytes themselves or by proliferating skin keratinocytes may be an important stimulator of melanocyte proliferation through an autocrine or paracrine type of mechanism, respectively.[9-12]

A. MELANOCYTE-STIMULATING HORMONE (MSH)

The melanocyte-stimulating hormone (MSH) is synthesized from the pro-opiomelanocortin (POMC) precursor and is secreted by the pars intermedia of the pituitary. MSH has a stimulating effect on melanogenesis in mammalian skin. In addition to this effect, MSH induces the differentiation of melanocytes and stimulates the initiation of melanin synthesis in the undifferentiated melanocyte precursors (melanoblasts). Keratinocyte-derived factors are required for the induction of melanocyte differentiation by MSH.[13]

Processing of the POMC molecule in the pituitary gives rise to adrenocorticotropin (ACTH), MSH, and lipotropin (LPH).[14,15] Three forms of MSH (α-, β-, and γ-MSH) are synthesized in the pituitary during POMC processing. α-MSH is a tridecapeptide with a molecular weight of 1663 and its amino acid sequence is identical to the 1 to 13 residues of ACTH. In addition to α-MSH and β-MSH, ACTH and β-LPH may also be involved in melanocyte growth and melanin production.

The action of MSH on its target cells is initiated by its binding to G protein-coupled cell surface receptors.[16] Cloning of a cDNA encoding the human MSH receptor has been reported.[17] The receptor is a transmembrane protein of 317 amino acids which exhibits no striking similarities to other characterized

G protein-coupled hormone receptors. MSH upregulates the expression of its receptor in Cloudman melanoma cells.[18]

B. MELANOMA GROWTH-STIMULATORY ACTIVITY (MGSA)

Endogenous growth factors produced by malignant melanoma cell lines are referred to as melanoma growth-stimulatory activity (MGSA), and this activity is separable from that of TGFs produced by the same cell lines.[7] The Hs0294 human malignant melanoma cell line produces a 16-kDa MGSA which is active in the picomolar concentration range and is different from other known growth factors, including TGFs.[19-21] MGSA is structurally related to β-thromboglobulin and analysis of an MGSA cDNA clone demonstrated that the oligonucleotide sequence is identical with the human *gro* gene. MGSA/GRO mRNA expression is induced in Hs294T melanoma cells by the addition of either exogenous MGSA or PDGF and this induction correlates with cell growth.[22] MGSA may act as a local tissue factor involved in the growth regulation of benign and malignant nevocytes.

A MGSA, isolated from the human melanoma cell line SK-MEL-178, has been purified and identified as a polypeptide of 14 kDa from the family of heparin-binding growth factors.[23] The purified MGSA stimulates DNA synthesis of normal human melanocytes in a dose-dependent manner and sustains them in culture without the aid of phorbol ester and cholera toxin. However, production of MGSA is not restricted to melanoma tissue but is found in various growing tissues with pathologic alterations as well as in keratinocytes and other cells derived from the basal epithelium of normal skin.[24] MGSA may be a growth regulator, and the microenvironment of the cell may regulate both the production of MGSA and the response to MGSA.

As MGSA/GRO, IL-8 is a member of the β-thyroglobulin (β-TG) family, and the genes coding for MGSA/GRO and IL-8 are both located on human chromosome region 4q12-q21. The genes encoding other members of the β-TG gene family, the platelet factor-4 (PF4) and the IFN-γ-inducible protein IP-10, are also located on human chromosome 4q12-q21. Melanoma cells, monocytes/macrophages, fibroblasts, endothelial cells, and keratinocytes have been shown to produce both IL-8 and MGSA/GRO. Interestingly, MGSA/GRO partially competes with IL-8 for binding to the IL-8 receptor in human placental membranes, but there is more than one class of receptor for MGSA/GRO.[25] A specific type of IL-8 receptor, IL-8R-B, binds both IL-8 and MGSA with high affinity.[26] The signal transduction mechanism of MGSA/GRO is unknown, but binding of MGSA/GRO to placental membranes stimulates the phosphorylation of several proteins on tyrosine residues.

C. MELANOMA GROWTH-INHIBITORY ACTIVITY (MIA)

A factor, called melanoma growth-inhibitory activity (MIA), was isolated from the human malignant melanoma cell line HTZ-19 dM, which was established from a central nervous system metastasis of melanoma.[27] The MIA is represented by a polypeptide of about 8 kDa and has potent growth inhibitory function on autologous tumor cells as well as in a number of neuroectoderm-derived tumors of different histology and grades of malignancy. MIA inhibits cell proliferation by prolongation of the S phase of the cycle and arrest of cells in the G_2 compartment.[28]

II. ANGIOGENIC FACTORS

The physiologic process of neomorphogenesis of blood vessels (neovascularization) is called angiogenesis. This process is a normal component of embryogenesis and is also observed in adult animals in conditions such as organ regeneration and wound healing or during the changes that occur in the uterus in association with the menstrual cycle. Progressive growth of solid tumors, even in the initial stages, requires the formation of new blood vessels, which is associated with the production of angiogenic factors. These factors are locally produced by stromal or parenchymal cells and act directly on vascular endothelial cells to stimulate their locomotion and/or mitosis.[29-31] They can also have an indirect action by mobilizing cells such as macrophages to release growth factors with specificity for endothelial cells. Extracellular matrix molecules and soluble growth factors may interact to control capillary endothelial cell growth.[32] Endothelial cells have a very important role in angiogenesis. The survival of these cells depends on a balance between growth-stimulatory and growth-inhibitory factors.[33] Factors with growth-stimulatory activity on endothelial cells are generically called endothelial cell growth factors (ECGFs). One of these factors is the platelet-derived ECGF (PD-ECGF).

In general, tumor growth comprises two stages, a prevascular phase, which is associated with limited growth potential, and a vascular phase, which is followed by rapid tumor growth, bleeding, and the potential for metastasis. There is a clear correlation between the density of microvessels in histologic sections of invasive tumors such as human breast carcinoma and the occurrence of metastases.[34] Hormones and growth factors can support tumor growth through either a direct stimulation of tumor cell growth or by promoting angiogenic processes that are required for nutrient supply. Angiogenic factors act through autocrine, paracrine, or endocrine mechanisms. Factors with this type of activity are produced by a diversity of normal and neoplastic tissues.

Factors with angiogenic activity include EGF, acidic FGF, basic EGF, TGF-α, TGF-β, IL-1α, IL-11, TNF-α, and PD-ECGF, as well as angiotensin II, polyamines, PGE_2, heparin, plasminogen activator, ceruloplasmin, and certain adipocyte lipids. Salt extracts of the extracellular matrix that is produced by endothelial cells contain mitogenic factors which are indistinguishable from basic and acid FGFs.[35,36] The FGFs bind tightly to immobilized heparin, and heparinase-like enzymes produced by certain tissues, including tumor tissues, may contribute to decrease the adsorption of FGFs to the matrix, which may result in the release of FGFs, thus favoring neovascularization. Both the FGFs and PD-ECGF lack a signal peptide required for extracellular transport, which suggests that they may act either intracellularly or after cell death. An export of cell-associated basic FGF, rather than new synthesis of the factor, may represent a switch mechanism favoring the neovascularization that is required for tumor progression.[37] Macrophage-derived IL-8 may act as a mediator of angiogenesis.[38] A specific angiogenic factor, angiogenin, is represented by a single chain polypeptide of 14,400 mol wt which was isolated from a human adenocarcinoma cell line. Another angiogenic factor, angiotropin, was isolated and purified from activated monocytes.[39] An ECGF activity isolated from adult bovine brain is identical to pleiotropin, a molecule purified from bovine uterus.[40]

Angiogenic activities have been isolated from a diversity of tumor tissues. Walker carcinoma 256, a rat transplantable tumor of mammary origin, is a reach source of angiogenic factors. One of the factors contained in Walker carcinoma is represented by the vitamin nicotinamide, and other angiogenic factors present in the same tumor contain nicotinamide as part of a more complex molecular arrangement.[41] An angiogenic activity contained in the human melanoma cell line A-375/2 was identified as a protein of 67 kDa, but its structure and function were not determined.[42] Nicotinamide and a complex containing nicotinamide were identified as angiogenic factors contained in Walker rat carcinoma cells.[43] The polyamines may have angiogenic properties, as suggested by the antiangiogenic action of α-difluoromethylornithine (DFMO), an inhibitor of ODC activity.[44] Other substances with angiogenic properties have been detected in different cellular systems, but they are less well characterized, and some of them are represented by lipids. The action of factors with angiogenic activity may be counteracted by that of antiangiogenic factors. Some factors with relatively specific angiogenic or antiangiogenic activity are discussed next.

A. ANGIOGENIN

Angiogenin, a 14,400-mol wt polypeptide first isolated from the conditioned medium of a human adenocarcinoma cell line, is a potent stimulator of angiogenesis.[45] The angiogenin gene has been cloned from a human liver cDNA library and the deduced amino acid sequence indicates that the polypeptide has 35% homology to a family of pancreatic ribonucleases. The cellular mechanism of angiogenin action remains unknown.

B. ANGIOTROPIN

Angiotropin, a substance isolated from conditioned media of cultures of porcine peripheral blood monocytes, exhibits potent angiogenic activity in rabbit cornea and chicken chorioallantois membrane tests.[46] In the rabbit skin, angiogenesis induced by angiotropin is associated with other proliferative tissue reactions without tissue damage.[47] Angiotropin stimulates the migration of capillary endothelial cells *in vitro* and the spatial organization of these cells to form tubular structures.[48] Angiotropin may be involved in monocyte-induced angiogenesis.

C. PLATELET-DERIVED ECGF (PD-ECGF)

Platelet-derived ECGF (PD-ECGF) is a 45-kDa endothelial cell mitogen which has angiogenic properties *in vivo*.[49] In contrast to FGFs, PD-ECGF does not bind heparin and does not stimulate the growth of fibroblastic cells. The cDNA cloning and heterologous expression of biologically active PD-ECGF has been reported.[50,51] The PD-ECGF gene is located on human chromosome 22q13.[52] The deduced primary structure of PD-ECGF shows no similarity with other known proteins and is conserved phylogenetically among vertebrates. PD-ECGF is the only ECGF activity present in human platelets. It stimulates endothelial cell growth as well as chemotaxis *in vitro* and angiogenesis *in vivo*. PD-ECGF is produced by human skin fibroblasts as well as by human squamous cell carcinoma and thyroid carcinoma cell lines.[53] Aberrant expression of PD-ECGF could play a role in tumor angiogenesis.

D. VASCULAR ENDOTHELIAL GROWTH FACTOR (VEGF)

A family of endothelial cell mitogens exhibiting homology to PDGF has been identified recently. Unlike other factors with similar properties, some of these vascular endothelial cell growth factors (VEGFs) are secreted and may act specifically on endothelial cells.[54] A VEGF secreted by bovine pituitary folliculo-stellate cells maintained in culture is, as predicted from the cDNA sequences, a 23-kDa protein composed of 164 amino acid residues.[55] A glioma-derived VEGF was characterized as a homodimeric glycoprotein of 46 kDa whose subunits show distant homology to the PDGF A and B chains.[56] A vascular permeability factor (VPF) identified in rodent tumor cell lines and purified from the serum-free conditioned medium of the human histiocytic lymphoma cell line U937 is similar or identical to VEGF. VPF is a 40-kDa polypeptide composed of 189 amino acids. It is active in increasing blood vessel permeability, endothelial cell growth, and angiogenesis.[57] VEGF/VPF, also called vasculotropin, can stimulate migration of monocytes across an endothelial cell monolayer. There is a potent synergism between VEGF and basic FGF in the induction of angiogenesis *in vitro* and probably also *in vivo*.[58]

1. Production of VEGF

The VEGF gene is expressed differentially in a diversity of normal adult organs and tissues as well as in several human tumors.[59] The mechanisms involved in the regulation of VEGF expression are little known. VEGF induced by hypoxia may mediate hypoxia-initiated angiogenesis.[60] PDGF may be involved in inducing VEGF gene transcription, and this effect appears to be mediated by protein kinase C.[61]

2. Molecular Forms of VEGF

VEGF activity (vasculotropin) is represented by a number of distinct, but structurally related, polypeptides. Analysis of VEGF transcripts contained in human fetal VSM cells by PCR and cDNA cloning revealed the existence of three different forms of the VEGF coding region and predicted the existence of three forms of the human VEGF protein which are generated by alternative exon splicing.[62] A fourth molecular species of VEGF was characterized more recently.[63] These forms are polypeptides of 121, 165, 189, and 206 amino acids, respectively, which contain identical amino-terminal residues. The VEGF forms may have different biological activities. Whereas the VEGF forms with 121 and 165 amino acids are secreted, the forms with 189 and 206 amino acids are mainly cell-associated. Endothelial cell mitogenic activity appears to be restricted to the secreted forms of VEGF.

3. Mechanism of VEGF Action

The highest density of VEGF binding sites in rats are found in brain, spinal cord, lung, adrenal cortex, glandular stomach, spleen, and pancreas. A VEGF purified from the conditioned medium of A431 human epidermoid carcinoma cells is recognized in human umbilical vein endothelial cells by a 190-kDa receptor protein with an associated tyrosine kinase activity.[64] A receptor for VEGF has been identified as the tyrosine kinase Flt (Flt-1), which is encoded by the putative proto-oncogene *flt*.[65] The Flt-1/VEGF receptor shows a singular structure. It has seven Ig-like domains in its extracellular region, a single transmembrane region, and a tyrosine kinase sequence that is interrupted by a kinase insert domain. Two Flt-related proteins, Flk-1 and Kdr, may also function as receptors for VEGF.[66,67] As Flt-1, these two proteins have seven Ig-like domains in their extracellular regions and function as tyrosine kinases. The precise roles of Flt-1, Flk-1, and Kdr in VEGF signal transduction remain to be evaluated, but present evidence suggests that the authentic high-affinity, endothelial cell-specific VEGF receptor is represented by the Flk-1 protein.[68]

A *flt*-related murine gene, *flt*-3 or *flk*-2, encodes a transmembrane receptor protein of 158 kDa with tyrosine kinase activity that is expressed in placenta and various adult tissues, including liver, gonads, and brain as well as hematopoietic cells.[69,70] The normal ligand of the Flt-3/Flk-2 receptor protein has not been identified but the receptor is phosphorylated to a high degree in the absence of the ligand. A chimeric molecule containing the ligand-binding domain of the CSF-1 receptor and the kinase domain of Flt-3 is capable of inducing transformation of Rat-2 cells.[71]

A human *flt*-related gene, FLT4 or *flt*-4, maps to chromosome region 5q35 and encodes a receptor tyrosine kinase.[72] The translational product of this gene, the Flt-4 of FLT4 protein, is 1298 amino acids in length and has an apparent molecular weight of 170 kDa. Flt-4 shows a high degree of homology with the Flt-1 and Flk-1 proteins. The normal ligand of the Flt-4 receptor is unknown but may be a growth factor related to VEGF. Further studies are required for a better characterization of VEGF-related receptors and their ligands.

The postreceptor mechanisms of VEGF action are little known but may include autophosphorylation of the VEGF receptor on tyrosine residues and phosphorylation of cellular proteins. The action of VEGF may also be associated with Ca^{2+} entry into the cells as well as with activation of phospholipase C and hydrolysis of phosphoinositides. Further studies are required for a better characterization of the action mechanisms of VEGF.

4. Role of VEGF in Neoplastic Processes

The development of new blood vessels (angiogenesis) is required for the growth of solid tumors, and VEGF may have an important role in tumor growth by acting through paracrine mechanisms as an angiogenesis factor for different types of tumors, including human gliomas *in vivo*.[73] AIDS-associated Kaposi's sarcoma cells in culture express VEGF.[74] Inhibition of VEGF-induced angiogenesis by a VEGF-specific monoclonal antibody results in suppression of the growth of various human tumor cell lines in nude mice but has no effect on the rate of tumor cell growth *in vitro*.[75] VEGF may thus be an important mediator of tumor angiogenesis and blocking of its action may have potential therapeutic application for some highly vascularized and aggressive human malignancies. The possible role of VEGF receptor expression in oncogenesis is not understood. Human melanoma cells, but not normal melanocytes, express VEGF receptors.[76]

E. PLEIOTROPHIN

Pleiotrophin, an 18-kDa heparin-binding growth factor, was first purified from the human breast cancer cell line MDA-MB-231.[77] The amino-terminal sequence of this factor does not exhibit homology to heparin-binding FGFs but is homologous to a developmentally regulated protein which has been described under different names such as heparin-binding regulatory protein (HBRP), heparin-binding growth-associated molecule, heparin-binding neurotrophic factor, heparin-binding neurite promoting factor, and osteoblast-specific factor. The cloning, expression, and structural analysis of the rat and human *Ptn* genes, which code for pleiotrophin, have been reported.[78,79] Pleiotrophin displays mitogenic activity on fibroblasts and endothelial cells and functions as a tumor angiogenic factor.[80] Moreover, overexpression of the *Ptn* gene in NIH/3T3 cells results in phenotypic transformation and tumorigenic conversion of the cells.[81] Both carcinogen-induced rat mammary carcinomas and primary human breast cancers frequently express the *Ptn* gene. Inhibition of heparin-binding growth factors involved in tumor neoangiogenesis could represent an approach to suppress local malignant growth and to prevent metastasis.[82]

F. ANTIANGIOGENIC FACTORS

The action of angiogenic factors may be counteracted by that of factors with antiangiogenic activity. An antiangiogenic activity was isolated from the clonal human chondrosarcoma cell line HCS-2/8.[83] The inhibitory aspect of this activity is lost by treatment with protease, indicating that it is represented by a protein, but further work is required for its characterization. A tumor suppressor-dependent inhibitor of angiogenesis is a 140-kDa protein which was found to be similar to a fragment of the ubiquitous adhesive glycoprotein thrombospondin.[84] In general, the protein products of proto-oncogenes and tumor suppressor genes (antioncogenes) may be involved in the stimulation or inhibition of angiogenic processes, respectively, which may be associated with tumor progression or tumor regression.[85] Neutralization of the activity of angiogenic factors by specific antibodies, or the use of factors with antiangiogenic activity, could be helpful in the treatment of malignant diseases.

REFERENCES

1. **Swope, V.P., Abdel-Malek, Z., Kassem, L.M., and Nordlund, J.J.,** Interleukins 1α and 6 and tumor necrosis factor-α are paracrine inhibitors of human melanocyte proliferation and melanogenesis, *J. Invest. Dermatol.*, 96, 180, 1991.
2. **Matsumoto, K., Tajima, H., and Nakamura, T.,** Hepatocyte growth factor is a potent stimulator of human melanocyte DNA synthesis and growth, *Biochem. Biophys. Res. Commun.*, 176, 45, 1991.
3. **Albino, A.P.,** The role of oncogenes and growth factors in progressive melanoma-genesis, *Pigment Cell Res.*, Suppl. 2, 199, 1992.

4. **Rodeck, U., Melber, K., Kath, R., Menssen, H.-D., Varello, M., Atkinson, B., and Herlyn, M.,** Constitutive expression of multiple growth factor genes by melanoma cells but not normal melanocytes, *J. Invest. Dermatol.,* 97, 20, 1991.
5. **Albino, A.P., Davis, B.M., and Nanus, D.M.,** Induction of growth factor RNA expression in human malignant melanoma: markers of transformation, *Cancer Res.,* 51, 4815, 1991.
6. **Hirobe, T.,** Basic fibroblast growth factor stimulates the sustained proliferation of mouse epidermal melanoblasts in a serum-free medium in the presence of dibutyryl cyclic AMP and keratinocytes, *Development,* 114, 435, 1992.
7. **Eisinger, M., Marko, O., Ogata, S.-I., and Old, L.J.,** Growth regulation of human melanocytes: mitogenic factors in extracts of melanoma, astrocytoma, and fibroblast cell lines, *Science,* 229, 984, 1985.
8. **Richmond, A., Lawson, D.H., Nixon, D.W., and Chawla, R.K.,** Characterization of autostimulatory and transforming growth factors from human melanoma cells, *Cancer Res.,* 45, 6390, 1985.
9. **Halaban, R., Langdon, R., Birchall, N., Cuono, C., Baird, A., Scott, G., Moellmann, G., and McGuire, J.,** Basic fibroblast growth factor from human keratinocytes is a natural mitogen for melanocytes, *J. Cell Biol.,* 107, 1611, 1988.
10. **Halaban, R., Kwon, B.S., Ghosh, S., Delli Bovi, P., and Baird, A.,** bFGF as an autocrine growth factor for human melanomas, *Oncogene Res.,* 3, 177, 1988.
11. **Rodeck, U., Becker, D., and Herlyn, M.,** Basic fibroblast growth factor in human melanoma, *Cancer Cells,* 3, 308, 1991.
12. **Rodeck, U. and Herlyn, M.,** Growth factors in melanoma, *Cancer Metast. Rev.,* 10, 89, 1991.
13. **Hirobe, T.,** Melanocyte stimulating hormone induces the differentiation of mouse epidermal melanocytes in serum-free culture, *J. Cell. Physiol.,* 152, 337, 1992.
14. **Whitfeld, P.L., Seeburg, P.H., and Shine, J.,** The human pro-opiomelanocortin gene: organization, sequence, and interspersion with repetitive DNA, *DNA,* 1, 133, 1982.
15. **Lowry, P.J.,** Pro-opiocortin: the multiple adrenal hormone precursor, *Biosci. Rep.,* 4, 467, 1984.
16. **Mountjoy, K.G., Robbins, L.S., Mortrud, M.T., and Cone, R.D.,** The cloning of a family of genes that encode the melanocortin receptors, *Science,* 257, 1248, 1992.
17. **Chhajlani, V. and Wikberg, J.E.S.,** Molecular cloning and expression of the human melanocyte stimulating hormone receptor cDNA, *FEBS Lett.,* 309, 417, 1992.
18. **Chakraborty, A.K. and Pawelek, J.M.,** Up-regulation of MSH receptors by MSH in Cloudman melanoma cells, *Biochem. Biophys. Res. Commun.,* 188, 1325, 1992.
19. **Richmond, A. and Thomas, H.G.,** Purification of melanoma growth stimulatory activity, *J. Cell. Physiol.,* 129, 375, 1986.
20. **Thomas, H.G. and Richmond, A.,** Immunoaffinity purification of melanoma growth stimulatory activity, *Arch. Biochem. Biophys.,* 260, 719, 1988.
21. **Richmond, A., Balenstien, E., Thomas, H.G., Flaggs, G., Barton, D.E., Spiess, J., Bordoni, R., Francke, U., and Derynck, R.,** Molecular characteristics and chromosomal mapping of melanoma growth stimulatory activity, a growth factor structurally related to β-thromboglobulin, *EMBO J.,* 7, 2025, 1988.
22. **Bordoni, R., Thomas, G., and Richmond, A.,** Growth factor modulation of melanoma growth stimulatory activity mRNA expression in human malignant melanoma cells correlates with cell growth, *J. Cell. Biochem.,* 39, 421, 1989.
23. **Ogata, S., Furuhashi, Y., and Eisinger, M.,** Growth stimulation of human melanocytes: identification and characterization of melanoma-derived growth factor (M-McGF), *Biochem. Biophys. Res. Commun.,* 1204, 1987.
24. **Richmond, A. and Thomas, H.G.,** Melanoma growth stimulatory activity: isolation from human melanoma tumors and characterization of tissue distribution, *J. Cell. Biochem.,* 36, 185, 1988.
25. **Cheng, Q.C., Han, J.H., Thomas, H.G., Balentien, E., and Richmond, A.,** The melanoma growth stimulatory activity receptor consists of two proteins. Ligand binding results in enhanced tyrosine phosphorylation, *J. Immunol.,* 148, 451, 1992.
26. **Lee, J., Horuk, R., Rice, G.C., Bennett, G.L., Camerato, T., and Wood, W.I.,** Characterization of two high affinity human interleukin-8 receptors, *J. Biol. Chem.,* 267, 16283, 1992.
27. **Bogdahn, U., Apfel, R., Hahn, M., Gerlach, M., Behl, C., Müller, F., Hoppe, J., and Martin, R.,** Autocrine tumor cell growth inhibiting activities from human malignant melanoma, *Cancer Res.,* 49, 5358, 1989.

28. **Weilbach, F.X., Bogdahn, U., Poot, M., Apfel, R., Behl, C., Drenkhard, D., Martin, R., and Hoehn, H.,** Melanoma-inhibiting activity inhibits cell proliferation by prolongation of the S-phase and arrest of cells in the G_2 compartment, *Cancer Res.,* 50, 6981, 1990.
29. **Folkman, J.,** Tumor angiogenesis, *Adv. Cancer Res.,* 43, 175, 1985.
30. **Klagsbrun, M.,** Regulators of angiogenesis, *Annu. Rev. Physiol.,* 53, 217, 1991.
31. **Folkman, J. and Shing, Y.,** Angiogenesis, *J. Biol. Chem.,* 267, 10931, 1992.
32. **Ingber, D.F., Madri, J.A., and Folkman, J.,** Endothelial growth factors and extracellular matrix regulate DNA synthesis through modulation of cell and nuclear expansion, *In Vitro Cell. Devel. Biol.,* 23, 387, 1987.
33. **Etoh, T., Takehara, K., Igarashi, A., and Ishibashi, Y.,** The effects of various growth factors on endothelial cell survival *in vitro, Biochem. Biophys. Res. Commun.,* 162, 1010, 1989.
34. **Weidner, N., Semple, J.P., Welch, W.R., and Folkman, J.,** Tumor angiogenesis and metastasis — correlation in invasive breast carcinoma, *N. Engl. J. Med.,* 324, 1, 1991.
35. **Baird, A. and Ling, N.,** Fibroblast growth factors are present in the extracellular matrix produced by endothelial cells *in vitro:* implications for a role of heparinase-like enzymes in the neovascular response, *Biochem. Biophys. Res. Commun.,* 142, 428, 1987.
36. **Vlodavsky, I., Folkman, J., Sullivan, R., Fridman, R., Ishai-Michaeli, R., Sasse, J., and Klagsbrun, M.,** Endothelial cell-derived basic fibroblast growth factor: synthesis and deposition into subendothelial extracellular matrix, *Proc. Natl. Acad. Sci. U.S.A.,* 84, 2292, 1987.
37. **Kandel, J., Bossy-Wetzel, E., Radvanyi, F., Klagsbrun, M., Folkman, J., and Hanahan, D.,** Neovascularization is associated with a switch to the export of bFGF in the multistep development of fibrosarcoma, *Cell,* 66, 1095, 1991.
38. **Koch, A.E., Polverini, P.J., Kunkel, S.L., Harlow, L.A., DiPietro, L.A., Elner, V.M., Elner, S.G., and Strieter, R.M.,** Interleukin-8 as a macrophage-derived mediator of angiogenesis, *Science,* 258, 1798, 1992.
39. **Höckel, M., Sasse, J., and Wissler, J.H.,** Purified monocyte-derived angiogenic substance (angiotropin) stimulates migration, phenotypic changes, and "tube formation" but not proliferation of capillary endothelial cells *in vitro, J. Cell. Physiol.,* 133, 1, 1987.
40. **Courty, J., Dauchel, M.C., Caruelle, C., Perderiset, M., and Barritault, D.,** Mitogenic properties of a new endothelial cell growth factor related to pleiotropin, *Biochem. Biophys. Res. Commun.,* 180, 145, 1991.
41. **Kull, F.C., Jr., Brent, D.A., Parikh, I., and Cuatrecasas, P.,** Chemical identification of a tumor-derived angiogenic factor, *Science,* 236, 843, 1987.
42. **Osthoff, K.S., Frühbeis, B., Overwien, B., Hilbig, B., and Sorg, C.,** Purification and characterization of a novel human angiogenic factor (h-AF), *Biochem. Biophys. Res. Commun.,* 146, 945, 1987.
43. **Kull, F.C., Jr., Brent, D.A., Parikh, I., and Cuatrecasas, P.,** Chemical identification of a tumor-derived angiogenic factor, *Science,* 236, 843, 1987.
44. **Takigawa, M., Enomoto, M., Nishida, Y., Pan, H-O., Kinoshita, A., and Suzuki, F.,** Tumor angiogenesis and polyamines: α-difluoromethylornithine, an irreversible inhibitor of ornithine decarboxylase, inhibits B16 melanoma-induced angiogenesis *in ovo* and the proliferation of vascular endothelial cells *in vitro, Cancer Res.,* 50, 4131, 1990.
45. **Takigawa, M., Pan, H.-O., Enomoto, M., Kinoshita, A., Nishida, Y., Suzuki, F., and Tajima, K.,** A clonal human chondrosarcoma cell line produces an anti-angiogenic antitumor factor, *Anticancer Res.,* 10, 311, 1990.
46. **Wissler, J.H. and Renner, H.,** Inflammation, chemotropism and morphogenesis: novel leukocyte-derived mediators for directional growth of blood vessels and regulation of tissue neovascularization, *Z. Physiol. Chem.,* 362, 244, 1981.
47. **Höckel, M., Beck, T., and Wissler, J.H.,** Neomorphogenesis of blood vessels in rabbit skin induced by a highly purified monocyte-derived polypeptide (monocyte-angiotropin) and associated tissue reactions, *Int. J. Tissue React.,* 6, 323, 1984.
48. **Höckel, M., Sasse, J., and Wissler, J.H.,** Purified monocyte-derived angiogenic substance (angiotropin) stimulates migration, phenotypic changes, and "tube formation" but not proliferation of capillary endothelial cells *in vitro, J. Cell. Physiol.,* 133, 1, 1987.
49. **Miyazono, K. and Heldin, C.-H.,** High-yield purification of platelet-derived endothelial cell growth factor: structural characterization and establishment of a specific antiserum, *Biochemistry,* 28, 1704, 1989.

50. **Ishikawa, F., Miyazono, K., Hellman, U., Drexler, H., Wernstedt, C., Hagiwara, K., Usuki, K., Takaku, F., Risau, W., and Heldin, C.-H.,** Identification of angiogenic activity and cloning and expression of platelet-derived endothelial cell growth factor, *Nature,* 338, 557, 1989.
51. **Hagiwara, K., Stenman, G., Honda, H., Sahlin, P., Anderson, A., Miyazono, K., Heldin, C.-H., Ishikawa, F., and Takaku, F.,** Organization and chromosomal localization of the human platelet-derived endothelial cell growth factor gene, *Mol. Cell. Biol.,* 11, 2125, 1991.
52. **Stenman, G., Sahlin, P., Dumanski, J.P., Hagiwara, K., Ishikawa, F., Miyazono, K., Collins, V.P., and Heldin, C.-H.,** Regional localization of the human platelet-derived endothelial cell growth factor (ECGF1) gene to chromosome 22q13, *Cytogenet. Cell Genet.,* 59, 22, 1992.
53. **Usuki, K., Heldin, N.-E., Miyazono, K., Ishikawa, F., Takaku, F., Westermark, B., and Heldin, C.-H.,** Production of platelet-derived endothelial cell growth factor by normal and transformed human cells in culture, *Proc. Natl. Acad. Sci. U.S.A.,* 86, 7427, 1989.
54. **Ferrara, N., Houck, K., Jakeman, L., and Leung, D.W.,** Molecular and biological properties of the vascular endothelial growth factor of proteins, *Endocrine Rev.,* 13, 18, 1992.
55. **Tischer, E., Gospodarowicz, D., Mitchell, R., Silva, M., Schilling, J., Lau, K., Crisp, T., Fiddes, J.C., and Abraham, J.A.,** Vascular endothelial growth factor: a new member of the platelet-derived growth factor gene family, *Biochem. Biophys. Res. Commun.,* 165, 1198, 1989.
56. **Conn, G., Bayne, M.L., Soderman, D.D., Kwok, P.W., Sullivan, K.A., Palisi, T.M., Hope, D.A., and Thomas, K.A.,** Amino acid and cDNA sequences of a vascular endothelial cell mitogen that is homologous to platelet-derived growth factor, *Proc. Natl. Acad. Sci. U.S.A.,* 87, 2628, 1990.
57. **Keck, P.J., Hauser, S.D., Krivi, G., Sanzo, K., Warren, T., Feder, J., and Connolly, D.T.,** Vascular permeability factor, an endothelial cell mitogen related to PDGF, *Science,* 246, 1309, 1989.
58. **Pepper, M.S., Ferrara, N., Orci, L., and Montesano, R.,** Potent synergism between vascular endothelial growth factor and basic fibroblast growth factor in the induction of angiogenesis *in vitro, Biochem. Biophys. Res. Commun.,* 189, 824, 1992.
59. **Berse, B., Brown, L.F., Van de Water, L., Dvorak, H.F., and Senger, D.R.,** Vascular permeability factor (vascular endothelial growth factor) gene is expressed differentially in normal tissues, macrophages, and tumors, *Mol. Biol. Cell,* 3, 211, 1992.
60. **Shweiki, D., Itin, A., Soffer, D., and Keshet, E.,** Vascular endothelial growth factor induced by hypoxia may mediate hypoxia-initiated angiogenesis, *Nature,* 359, 843, 1992.
61. **Finkenzeller, G., Marmé, D., Weich, H.A., and Hug, H.,** Platelet-derived growth factor-induced transcription of the vascular endothelial growth factor gene is mediated by protein kinase C, *Cancer Res.,* 52, 4821, 1992.
62. **Tischer, E., Mitchell, R., Hartman, T., Silva, M., Gospodarowicz, D., Fiddes, J.C., and Abraham, J.A.,** The human gene for vascular endothelial growth factor. Multiple protein forms are encoded through alternative exon splicing, *J. Biol. Chem.,* 266, 11947, 1991.
63. **Houck, K.A., Ferrara, N., Winer, J., Cachianes, G., Li, B., and Leung, D.W.,** The vascular endothelial growth factor family: identification of a fourth molecular species and characterization of alternative splicing of RNA, *Mol. Endocrinol.,* 5, 1986, 1991.
64. **Myoken, Y., Kayada, Y., Okamoto, T., Kan, M., Sato, G.H., and Sato, J.D.,** Vascular endothelial cell growth factor (VEGF) produced by A-431 human epidermoid carcinoma cells and identification of VEGF membrane binding sites, *Proc. Natl. Acad. Sci. U.S.A.,* 88, 5819, 1991.
65. **de Vries, C., Escobedo, J.A., Ueno, H., Houck, K., Ferrara, N., and Williams, L.T.,** The *fms*-like tyrosine kinase, a receptor for vascular endothelial growth factor, *Science,* 255, 989, 1992.
66. **Matthews, W., Jordan, C.T., Gavin, M., Jenkins, N.A., Copeland, N.G., and Lemischka, I.R.,** A receptor tyrosine kinase cDNA isolated from a population of enriched primitive hematopoietic cells and exhibiting close genetic linkage to c-*kit, Proc. Natl. Acad. Sci. U.S.A.,* 88, 9026, 1991.
67. **Terman, B.I., Dougher-Vermazen, M., Carrion, M.E., Dimitrov, D., Armelino, D.C., Gospodarowicz, D., and Böhlen, P.,** Identification of the KDR tyrosine kinase as a receptor for vascular endothelial cell growth factor, *Biochem. Biophys. Res. Commun.,* 187, 1579, 1992.
68. **Millauer, B., Wizigmann-Voos, S., Schnürch, H., Martinez, R., Moller, N.P.H., Risau, W., and Ullrich, A.,** High affinity VEGF binding and developmental expression suggest Flk-1 as a major regulator of vasculogenesis and angiogenesis, *Cell,* 72, 835, 1993.
69. **Rosnet, O., Marchetto, S., de Lapeyriere, O., and Birnbaum, D.,** Murine Flt3, a gene encoding a novel tyrosine kinase receptor of the PDGFR/CSF1R family, *Oncogene,* 6, 1641, 1991.

70. **Lyman, S.D., James, L., Zappone, J., Sleath, P.R., Beckmann, M.P., and Bird, T.,** Characterization of the protein encoded by the *flt3* (*flk2*) receptor-like tyrosine kinase gene, *Oncogene,* 8, 815, 1993.
71. **Maroc, N., Rottapel, R., Rosnet, O., Marchetto, S., Lavezzi, C., Mannoni, P., Birnbaum, D., and Dubreuil, P.,** Biochemical characterization and analysis of the transforming potential of the FLT3/ FLK2 receptor tyrosine kinase, *Oncogene,* 8, 909, 1993.
72. **Galland, F., Karamysheva, A., Pebusque, M.-J., Borg, J.-P., Rottapel, R., Dubreuil, P., Rosnet, O., and Birnbaum, D.,** The FLT4 gene encodes a transmembrane tyrosine kinase related to the vascular endothelial growth factor receptor, *Oncogene,* 8, 1233, 1993.
73. **Plate, K.H., Breier, G., Weich, H.A., and Risau, W.,** Vascular endothelial cell growth factor is a potential tumour angiogenesis factor in human gliomas *in vivo, Nature,* 359, 845, 1992.
74. **Windel, K., Marme, D., and Weich, H.A.,** AIDS-associated Kaposi's sarcoma cells in culture express vascular endothelial growth factor, *Biochem. Biophys. Res. Commun.,* 183, 1167, 1992.
75. **Kim, K.J., Li, B., Winer, J., Armanini, M., Gillett, N., Phillips, H.S., and Ferrara, N.,** Inhibition of vascular endothelial growth factor-induced angiogenesis suppresses tumour growth *in vivo, Nature,* 362, 841, 1993.
76. **Gitaygoren, H., Halaban, R., and Neufeld, G.,** Human melanoma cells but not normal melanocytes express vascular endothelial growth factor receptors, *Biochem. Biophys. Res. Commun.,* 190, 702, 1993.
77. **Wellstein, A., Fang, W.J., Khatri, A., Lu, Y., Swain, S.S., Dickson, R.B., Sasse, J., Riegel, A.T., Lippman, M.E.,** A heparin-binding growth factor secreted from breast cancer cells homologous to a developmentally regulated cytokine, *J. Biol. Chem.,* 267, 2582, 1992.
78. **Li, Y.S., Milner, P.G., Chauhan, A.K., Watson, M.A., Hoffman, R.M., Kodner, C.M., Milbrandt, J., and Deuel, T.F.,** Cloning and expression of a developmentally regulated protein that induces mitogenic and neurite outgrowth activity, *Science,* 250, 1690, 1990.
79. **Lai, S., Czubayko, F., Riegel, A.T., and Wellstein, A.,** Structure of the human heparin-binding growth factor gene pleiotrophin, *Biochem. Biophys. Res. Commun.,* 187, 1113, 1992.
80. **Fang, W., Hartmann, N., Chow, D.T., Riegel, A.T., and Wellstein, A.,** Pleiotrophin stimulates fibroblasts and endothelial and epithelial cells and is expressed in human cancer, *J. Biol. Chem.,* 267, 25889, 1992.
81. **Chauhan, A.K., Li, Y.-S., and Deuel, T.F.,** Pleiotrophin transforms NIH 3T3 cells and induces tumors in nude mice, *Proc. Natl. Acad. Sci. U.S.A.,* 90, 679, 1993.
82. **Zugmaier, G., Lippman, M.E., and Wellstein, A.,** Inhibition by pentosan polysulfate (PPS) of heparin-binding growth factors released from tumor cells and blockage by PPS of tumor growth in animals, *J. Natl. Cancer Inst.,* 84, 1716, 1992.
83. **Takigawa, M., Pan, H.-O., Enomoto, M., Kinoshita, A., Nishida, Y., Suzuki, F, and Tajima, K.,** A clonal human chondrosarcoma cell line produces an anti-angiogenic antitumor factor, *Anticancer Res.,* 10, 311, 1990.
84. **Good, D.J., Polverini, P.J., Rastinejad, F., Le Beau, M.M., Lemons, R.S., Frazier, W.A, and Bouck, N.P.,** A tumor suppressor-dependent inhibitor of angiogenesis is immunologically and functionally indistinguishable from a fragment of thrombospodin, *Proc. Natl. Acad. Sci. U.S.A.,* 87, 6624, 1990.
85. **Bouck, N.,** Tumor angiogenesis: the role of oncogenes and tumor suppressor genes, *Cancer Cells,* 2, 179, 1990.

Chapter 8

Transforming Growth Factors

I. INTRODUCTION

The transforming growth factors (TGFs) are peptides involved in the regulation of cell growth and differentiation. TGFs were first detected in cells transformed *in vitro* by murine and feline sarcoma viruses and were named sarcoma growth factor (SGF).[1,2] It was later shown that SGF activity is composed of a mixture of two distinct factors, TGF-α (type-1 TGF) and TGF-β (type-2 TGF).[3] TGF-α and TGF-β may be produced in an endocrine, paracrine, or autocrine fashion by neoplastic cells. They are also synthesized and secreted by a diversity of normal cells. The growth of normal human keratinocytes may be regulated in an autocrine fashion by TGFs.[4] The TGFs are involved in regulating the expression of differentiated functions and have an important role in many physiological processes, including wound healing.[5] Regeneration of the liver after injury caused by chemical or infectious agents or after partial hepatectomy is regulated by a complex interaction of proto-oncogene products and growth factors, including TGFs.[6]

TGFs reversibly induce the expression of a transformed phenotype in certain types of cultured cells; in particular, they are able to stimulate anchorage-independent growth of certain types of nontransformed cells in soft agar.[7-13] A concerted action of TGF-α and TGF-β is required for inducing the expression of a transformed phenotype in some types of cultured cells.[3] TGFs can be detected by a microassay where colony-forming activity in soft agar medium is examined using BALB/3T3 or NRK cells.[14] TGFs may not be able to stimulate soft agar colony formation of other cells such as C3H/10T1/2 cells and human foreskin diploid fibroblasts.

TGFs have been isolated and purified from a variety of tumor cell lines or from cells transformed by acute retroviruses or chemical carcinogens.[15-36] TGFs may exert autocrine or paracrine effects on tumor cell growth.[37] They may be, at least in part, responsible for the continuous ruffling observed in cultured cells transformed by Kirsten murine sarcoma virus (K-MuSV) and other oncogenic agents.[38,39] TGFs are also produced by flat revertants from K-MuSV-transformed cells.[40,41] These revertants express TGFs and the v-Ras oncoprotein but fail to exhibit a fully transformed phenotype, suggesting that v-Ras action may be blocked at a point distal to its transforming activity and that expression of v-Ras and TGFs may be necessary but not sufficient for maintaining the transformed state.

The TGFs constitute a family of structurally and functionally related polypeptides. These growth factors are present in almost all human benign and malignant neoplasms, including leukemic and carcinoma cells.[42-45] However, the two main types of TGFs, TGF-α and TGF-β, may be differentially expressed in tumors. Whereas TGF-β mRNA is universally present in human tumors, TGF-α mRNA is found in a variety of solid human tumors, particularly carcinomas, but is not present in hematopoietic tumors.[46] TGF-like activity was detected in human premalignant adenoma tissue from familial polyposis coli colectomy specimens as well as in serum-free culture supernatant of an adenoma cell line derived from one of these lesions, which are characterized by their malignant potential.[47] Although no TGF-like activity was detected in normal colon mucosa, high levels of this activity, comparable to that of the adenoma tissue, were found in extracts of separated muscle and submucosa layers, deep to the muscularis mucosae. TGFs are present in pleural and peritoneal effusions not only from cancer patients, but also from patients with nonmalignant diseases.[48] It is thus clear that TGFs cannot represent, at least in qualitative terms, a reliable marker for the detection of cancer in general. The secretion of TGFs, or substances with TGF-like activity, may be a property of many types of both neoplastic and nonneoplastic primary human cells.[45] TGFs are present in serum and tissues of embryos and fetuses, as well as in human term placenta, human colostrum and milk, and some adult tissues, including lung and kidney.[49-56] Both TGF-α- and TGF-β-related molecules are present in the urine of normal individuals as well as in the urine of cancer patients.[57-59] TGF activity is present in effusions from cancer patients, but the same activity is found, although less frequently, in effusions from patients with benign diseases.

II. TRANSFORMING GROWTH FACTOR α (TGF-α)

The transforming growth factor α (TGF-α) is a functional and structural analog of EGF.[60,61] TGF-α recognizes the EGF receptor and the cellular actions of TGF-α are elicited through its binding to the epidermal growth factor (EGF) receptor on the cell surface. TGF-α was first detected in murine cells transformed by acute retroviruses such as MuSV.[1,2,9] A biologically active precursor for TGF-α may be released by retrovirally transformed cells.[62] It was initially believed that TGF-α is produced only by neoplastically transformed cells and it was suggested that autostimulation by endogenous TGFs may contribute to tumor progression.[7,9] However, the TGF-α gene is expressed in many normal animal tissues, including brain, liver, and kidney.[63] TGF-α protein is present in normal human saliva.[64]

A. PRODUCTION AND PHYSIOLOGICAL EFFECTS OF TGF-α

TGF-α is produced in various types of normal human cells. Primary cultures of human keratinocytes synthesize TGF-α, and the addition of EGF or TGF-α to these cultures induces TGF-α gene expression, suggesting an autoinduction mechanism.[65] Microvascular endothelial tubular morphogenesis is induced by human keratinocytes in culture by a paracrine mechanism involving TGF-α secretion.[66] Apparently, TGF-α secreted by keratinocytes stimulates the growth of these cells by an autocrine mechanism and at the same time induces skin angiogenesis by a paracrine mechanism. TGF-α can enhance the locomotion of human keratinocytes in culture.[67] High amounts of TGF-α mRNA and protein are produced in the lesional cells of psoriasis, a common hyperproliferative skin disease.[68] Mice with null mutation of the TGF-α gene show abnormal skin architecture, wavy hair, and curly whiskers and develop eye abnormalities including corneal inflammation.[69] Homozygous TGF-α mutant mice, obtained by disruption of the mouse TGF-α gene by homologous recombination in embryonic stem cells, do not produce significant amounts of TGF-α.[70] Development and fertility of the mutant mice are essentially normal. The most obvious defects of these animals are limited to hair follicles and eyes, with expression of a phenotype identical to that observed in the recessive mutation waved-1 (*wa*-1).

Human peripheral blood white cells, including macrophages and eosinophils, are a source of TGF-α, which may have a role in epithelial cell proliferation, as well as in inflammation and wound repair.[71,72] TGF-α is expressed throughout the normal human gastrointestinal mucosa in higher amounts than EGF.[73] Secretion of gastric acid is inhibited by TGF-α.[93] Expression of TGF-α is associated with the proliferation and differentiation of liver cells.[74] High levels of TGF-α mRNA are found in prenatal and postnatal rat livers, in particular in replicating oval cells and perisinusoidal stellate cells, as well as in basophilic small hepatocytes, after partial hepatectomy. These results suggest that TGF-α may be involved in growth regulation of both immature and fully differentiated hepatocytes. Both TGF-α and basic FGF have been identified as potent hepatotrophic mitogens *in vitro*.[75] Low levels of constitutive TGF-α mRNA expression are found in the normal human kidney.[76]

TGF-α is produced by a variety of human solid tumors and cell lines derived from these tumors, including the majority of human colon cancer cell lines.[77,78] A factor closely related to TGF-α is produced by a variant of PC12 rat pheochromocytoma cells that express an activated c-*ras* oncogene.[79] This factor induces neurites in PC12 cells and enhances the survival of embryonic brain neurons in primary culture. However, some cell lines that produce TGF-α/EGF-like activity may not express TGF-α receptor mRNA or protein, which indicates that TGF-α is not an autocrine factor for the growth of these cells.[35] TGF-α mRNA has not been detected in human tumor cell lines of hematopoietic origin.

TGF-α may function as a "juxtacrine stimulation factor" in developmental processes through the binding of membrane-anchored pro-TGF-α on one cell to the EGF/TGF-α receptor on the adjacent cell.[80] TGF-α has important physiological effects in the fetus and in newborn animals. Chicken erythrocytic progenitors express the c-*erb*-B/EGF receptor, which transmit mitogenic signals induced by TGF-α.[81] Differential expression of TGF-α can occur during prenatal development in rodents.[82,83] Expression of the TGF-α receptor is developmentally regulated during late gestation in the fetal rat.[84] These results indicate that TGF-α plays a role in development, possibly as a fetal growth factor. TGF-α mRNA is present at relatively high levels in 8- to 10-day-old rat embryos and then declines to the low levels which are characteristic of adult tissues. The level of TGF-α mRNA present during early gestation is similar to that of retrovirus-transformed cells in culture. High levels of TGF-α observed during early rodent development may be the result of expression in the maternal decidua and not in the embryo.[85] The growth factor produced in the decidua may act not only locally through an autocrine mechanism but also through paracrine and/or endocrine mechanisms. TGF-α, as EGF, is active in promoting eyelid opening in newborn mice.[86] Treatment of newborn mice with TGF-α may cause precaucious eyelid opening and accelerated incisor eruption. Such facts suggest that TGF-α and EGF may have important functions in the development of immature animals.

TGF-α may be involved in the control of gonadal functions. Ovarian theca cells synthesize and secrete TGF-α as the primary EGF-like activity in the follicle.[87] In contrast, granulosa cells do not express TGF-α mRNA transcripts or protein. The production of TGF-α by theca cells provides a possible autocrine role for TGF-α in the regulation of theca cell growth and a paracrine role in the regulation of granulosa cell growth. Both TGF-α and EGF attenuate the acquisition of aromatase activity and the follicle-stimulating hormone (FSH)-induced accumulation of estrogen in cultured rat granulosa cells.[88] The two growth factors neutralize FSH action at sites both proximal and distal to cAMP generation. TGF-α is produced and secreted by Sertoli and peritubular cells in the testis and it may act as an autocrine/paracrine factor involved in the control and maintenance of testicular function, including spermatogenesis.[89,90] However, TGF-α does seem to affect the testicular germ cells in a direct manner. It is not a mitogen for prepubertal Sertoli cells, and other locally produced growth factors may be responsible for the proliferation of prepubertal Sertoli cells. Purified TGF-α is a stimulator of testicular ornithine decarboxylase (ODC) activity, which is involved in the generation of polyamines, and this stimulation is coupled to new protein synthesis and testicular weight gain.[91] Thus, TGF-α may have a role in testicular maturation and function. Both TGF-α and EGF are involved in the development of the mammary gland, as shown *in vivo* with the use of slow-releasing cholesterol-based pellets inserted directly into the gland.[92] The presence of estrogen and progesterone is required for the effect of TGF-α and EGF on the mammary gland.

Both TGF-α and EGF stimulate bone resorption by increasing the proliferation of osteoclast precursors, which leads to increased numbers of osteoclasts.[94] TGF-α and EGF have important vascular effects. They are potent stimulators of femoral arterial blood flow in the anesthetized dog *in vivo* and inhibit the contractile response of helical coronary arterial strips to various smooth muscle agonists *in vitro,* including norepinephrine.[95,96] TGF-α is a more potent angiogenic mediator than EGF and may be involved in tumor-associated neovascularization, contributing to the generation of a local microenvironment that is favorable for the growth of solid tumors.[97]

B. THE TGF-α GENE

The human and rat gene for TGF-α have been cloned, and the human gene has been expressed in *Escherichia coli* and the structure of the TGF-α precursor peptide has been determined.[63,98] The human TGF-α gene, cloned from a cDNA library prepared using RNA from a human renal carcinoma cell line, encodes a precursor polypeptide of 160 amino acids. The 50 amino acid mature human TGF-α produced by expression of the coding sequences in *Escherichia coli* binds to the EGF receptor and induces anchorage independent cell growth in a soft agar assay, two biological characteristics of EGF and natural TGF-α.[48] On its 5′-flanking region, the human TGF-α gene contains an estrogen-responsive element which may mediate the regulatory action of estrogen on TGF-α protein expression.[99] The TGF-α gene has been mapped to human chromosome region 2p11-p13, close to the breakpoint of the Burkitt's lymphoma t(2;8) variant translocation.[100] The biological significance of this association, if any, is not understood.

The rat TGF-α gene contains six exons which span 85kb of DNA, and the complete nucleotide sequence of the respective mRNA was determined.[101] Exons 1 to 6 are separated by large introns and, in contrast to its human counterpart, the rat TGF-α gene promoter directs transcription from numerous sites spanning almost 250 bp of DNA. A cDNA clone encoding rat TGF-α hybridized to a 4.5-kb mRNA that is 30 times larger than necessary to code for a 50-amino acid TGF-α polypeptide.[63] TGF-α mRNA is present not only in transformed cells but also, although at lower levels, in several normal rat tissues, and its nucleotide sequence predicts that TGF-α is synthesized as a larger product and that the larger form may exist as a transmembrane protein. The translation product of rat TGF-α mRNA is composed of 159 amino acids. Expression of a transfected cDNA, encoding the complete rat TGF-α precursor protein in nontransformed BHK fibroblasts, resulted in the synthesis of proteins with molecular weights between 13 and 17 kDa, which are proteolytically processed to the mature 6-kDa TGF-α species.[102] Thus, the nontransformed BHK fibroblasts possess the ability to process the TGF-α precursor molecule into its native form.

C. STRUCTURE AND BIOSYNTHESIS OF TGF-α

TGF-α purified from human melanoma cell lines and from rat and MuSV-transformed mouse cells show a high degree of structural homology, differing from each other by only a few amino acid substitutions.[19] TGF-α isolated from the conditioned medium of human melanoma cells or obtained by recombinant DNA technology is a single-chain polypeptide of 50 amino acids, with a molecular weight of 7400, and contains three disulfide bridges in positions homologous to those of EGF.[103,104] Mutational analysis indicates that disruption of disulfide bonds 8 to 21 and 34 to 43 results in loss of biological activity of the TNF-α molecule.[105] The presence of an aromatic side chain at position 38 of TGF-α seems to be essential for its activity. The carboxyl-terminal part of the TNF-α molecule, in particular the Leu-49 and Ala-50 residues, are important in terms of the binding affinity and biological activity of the polypeptide.[106] Human TGF-α identical to the recombinant molecule was chemically synthesized by a stepwise solid-phase method with an overall yield of 26%.[107]

Larger polypeptides, structurally and functionally related to TGF-α, were detected by gel filtration of conditioned medium from cultured retrovirus-transformed rat cells.[108] An antiserum raised against an oligopeptide corresponding to the carboxyl-terminal 17 amino acids of rat TGF-α was used to develop a competitive radioimmunoassay for the immunizing peptide. Immunoblotting analysis revealed three TGF-α-related polypeptides with molecular weights of 24, 40, and 42 kDa, respectively, which could represent the products of a cleaved TGF-α precursor molecule. The precursor may be, at least partially, processed extracellularly. Different TGF-α species are derived from a glycosylated and palmitoylated transmembrane precursor molecule.[109]

Studies with cDNA cloning revealed that human and rat TGF-α are synthesized as highly conserved precursor proteins of 160 and 159 amino acids, respectively, termed pro-TGF-α.[63,98] Both precursors contain an amino-terminal sequence of 23 uncharged and apolar amino acid residues that may function as a signal peptide. This hydrophobic sequence is then followed, in order, by a domain of 74 or 75 residues that includes potential glycosylation sites as well as the mature TGF-α sequence, an extremely hydrophobic domain of 23 residues, and finally, a cysteine-rich sequence of 39 residues at the carboxyl terminus. The presence of the second hydrophobic, bordered by pairs of basic residues, suggested that pro-TGF-α is an integral membrane protein. In fact, results from recent studies indicate that the integral membrane precursor of TGF-α is biologically active and can interact with EGF receptor molecules on adjacent cells, leading to signal transduction.[110]

The complete amino acid sequence of rat TGF-α has been determined and a high degree of structural and functional homology between this factor and EGF was recognized.[111] Rat TGF-α is a single peptide chain containing 50 amino acid residues and 3 disulphide linkages. A synthetic peptide corresponding to rat TGF-α was produced by a stepwise solid-phase approach, following the general principles of the Merrifield method. The synthetic product has chemical and biological properties that are indistinguishable from those of natural rat TGF-α purified from rat embryo fibroblasts transformed by FeSV.[112] A synthetic fragment of TGF-α comprising a decapeptide of the third disulfide loop (residues 34 to 43) has no mitogenic activity but prevents the mitogenic effects of EGF and TGF-α on fibroblasts.[113] This fragment acts as an antagonist of the induction of cellular proliferation by EGF and contains a receptor binding sequence of TGF-α. The results obtained with synthetic TGF-α fragments suggest that the receptor binding domain of TGF-α is composed of separate regions of the molecule that fold into the correct alignment when the hormone assumes its native conformation.[114]

Site-directed mutagenesis studies showed that, whereas mutation of aspartic acid-47 to alanine or asparagine in the TGF-α molecule is compatible with retention of biological activity, mutation of leucine-48 to alanine results in a complete loss of binding and colony-formation abilities (i.e., induction of anchorage-independent growth).[115] These results suggest that two adjacent conserved amino acids in positions 47 and 48 of the TGF-α molecule play different roles in defining the structure and biological activity of the growth factor and that the carboxyl terminus of TGF-α is involved in interactions with cellular TGF-α/EGF receptors. Preparation of synthetic peptide fragments and recombinant mutant proteins of TGF-α and the testing of these TGF-α derivatives in receptor binding and mitogenesis assays indicated that at least three distinct regions of the TGF-α molecule contribute to biological activity.[116] These studies suggest that simple substitution mutants cannot be used to generate TGF-α antagonists.

D. CELLULAR MECHANISMS OF ACTION OF TGF-α

The mitogenic action of TGF-α is mediated by its interaction with the EGF receptor. This idea was confirmed by studies using a human EGF receptor with an amino acid residue covalently cross-linked to mouse EGF or TGF-α, which demonstrated that both factors recognize identical or overlapping binding sites on the EGF receptor.[117] TGF-α efficiently competes with EGF for binding to EGF receptors, while TGF-β possesses a distinct type of surface receptor and does not compete with the EGF receptor. Recombinant human TGF-α binds to the EGF receptor with an affinity about 55% that of EGF.[104] Occupation of the EGF receptor by TGF-α is required for both the mitogenic and colony-forming activity of TGF-α.[118] Expression of the EGF receptors on the cell surface is downregulated after interaction of TGF-α with the receptor.[119] In addition, TGF-α modulates expression of EGF receptor mRNA, and this effect involves synergistic interactions with other growth factors and hormones, including TGF-β, thyroid hormone, and retinoic acid.[120]

It is unclear how both TGF-α and EGF, which bind to the same receptor, can trigger differential cellular responses. Comparative studies on the effects of TGF-α and EGF have shown that TGF-α is generally more potent than EGF in a variety of biological systems. This is not explained by differences in affinities of the ligands for the receptor, which are similar, but there is evidence that EGF and TGF-α may be differentially processed after binding to the receptor.[121,122] Whereas EGF routes the EGF receptor directly to a degradative pathway, TGF-α allows receptor recycling prior to degradation, and tyrosine phosphorylation may play a role in this differential receptor processing.[123] A monoclonal antibody to the EGF receptor (13A9) has only small effects on the binding of EGF to its receptor, but has very large effects on the binding of TGF-α to the EGF receptor.[124] The different biological responses to TGF-α and EGF could be mediated through their binding to different sites of the EGF receptor molecule or the two growth factors would cause different conformational changes in the receptor.

Human TGF-α binds to the chicken EGF receptor, expressed in murine National Institutes of Health (NIH)/3T3 cells devoid of endogenous EGF receptors and transfected with an appropriate chicken cDNA clone, with an affinity which is even higher than its affinity for the human receptor.[125] Moreover, human TGF-α stimulated the kinase activities of the chicken and human EGF receptors to a similar extent and, surprisingly, it was 100-fold more efficient than EGF in inducing DNA synthesis in the murine cells expressing the chicken EGF receptor.

Tyrosine-specific protein kinase activity intrinsic to the EGF receptor is stimulated upon ligand binding.[126] TGF-α stimulates tyrosine phosphorylation at specific sites of the EGF receptor and, as a consequence, a number of cellular proteins may be phosphorylated by the activated TGF-α/EGF receptor kinase.[127,128] TGF-α can also stimulate the phosphorylation on tyrosine of a synthetic oligopeptide whose sequence is related to that of the site of tyrosine phosphorylation in the v-Src oncoprotein.[129]

The use of polyclonal or monoclonal antibodies to TGF-α or the EGF receptor may contribute to a better knowledge of the physiological properties of TGF-α. The monoclonal antibody 425 binds to a protein epitope of the human EGF receptor and blocks TGF-α-induced second messengers including the formation of inositol 1,4,5-trisphosphate and the increase of $[Ca^{2+}]_i$, two signals that are normally elicited by TGF-α.[130] A synthetic linear TGF-α was produced to study the biological activity of polyclonal sheep antibodies against the carboxyl-terminal part (17 amino acids) of rat TGF-α.[131] Although the antibodies recognized the linear TGF-α molecule, they failed to inhibit the binding to EGF receptors and to inhibit TGF-α-stimulated colony formation of normal rat fibroblasts. No cytotoxic activity of these antibodies was found against fresh human tumor specimens in a human tumor cloning assay. Thus, antibodies against the complete TGF-α molecule may be necessary in order to interfere with autocrine secretion of TGFs and to evaluate their potential biological effects.

E. REGULATION OF PROTO-ONCOGENE EXPRESSION BY TGF-α

In a manner similar to EGF, TGF-α rapidly induces c-*myc* and c-*fos* gene expression in CH3/10T1/2 mouse fibroblasts.[132] The induction of c-*myc* mRNA by TGF-α exhibits slower kinetics than by EGF. In primary cultures of adult rat hepatocytes, TGF-α initiates a mitogenic program which is associated with an increase in the steady-state levels of c-*jun* mRNA.[133] The biological significance of TGF-α-induced proto-oncogene expression is unknown.

F. ROLE OF TGF-α IN NEOPLASTIC PROCESSES

TGF-α may be involved in regulating the growth of tumor cells. Human and nonhuman tumor cell lines may produce TGF-α. Production of TGF-α has been detected in approximately one fourth of human tumor cell lines of diverse origins.[134] Some of these cell lines may produce TGF-α in large quantities. Human non-small cell lung carcinoma cell lines produce TGF-α and may express high levels of EGF receptors.[135] Human squamous cell carcinoma cell lines may frequently exhibit a combination of constitutive secretion of TGF-α and overexpression of EGF receptors.[136] Poorly differentiated human colon carcinoma cells in culture exhibit an upregulation of both the TGF-α and the c-*myc* genes during the establishment of quiescence, which may be related to their independence of growth factors.[137] The human colon cancer cell line GEO expresses low levels of TGF-α and EGF receptor and exhibits relatively indolent transformed properties, as indicated by their low cloning efficiency in anchorage-independent growth assays and poor tumorigenicity in athymic nude mice, but the malignant properties of GEO cells are greatly enhanced by transfection of an expression vector encoding the human TGF-α gene.[138] The increased malignant behavior of GEO cells is associated with the operation of an autocrine mechanism involving TGF-α and the EGF receptor.

TGF-α may act as an autocrine or paracrine factor in tumor promotion. Exposure of human epidermal keratinocytes to TPA results in enhanced accumulation of TGF-α mRNA and secretion of TGF-α protein.[139] Thus, a possible mechanism for the hyperplastic response and tumor promoting activity of TPA in epidermic cells may consist in the production of TGF-α. Repeated topical application of a tumor promoter such as the phorbol ester TPA to mouse skin may lead to sustained loss of a negative feedback mechanism involving phosphorylation of Thr-654 of the EGF receptor by protein kinase C, and the concomitant elevation of the EGF receptor ligand TGF-α would provide a mechanism for sustained cellular proliferation essential for skin tumor promotion.[140] 12-*O*-Tetradecanoylphorbal-13-acetate (TPA) may increase the expression of TGF-α in neoplastic cells such as the chemically transformed rat hepatic epithelial cell line GP6ac.[141] Treatment of GP6ac cells with hormones (e.g., EGF, angiotensin II, epinephrine, or bradykinin) results in increased levels of TGF-α expression. TPA-induced increase of TGF-α expression in GP6ac cells occurs, at least in part, at the transcriptional level and is blocked by concurrent incubation with agents that inhibit protein synthesis. The human pancreatic cancer cell line MIA-PaCa 2 produces both TGF-α and insulin-like growth factor (IGF)-I and possesses receptors for both growth factors, which may be involved in autocrine loops that stimulate the growth of these cells.[142] Phorbol ester markedly inhibits the growth of MIA-PaCa 2 cells due, at least in part, to decreased binding of TGF-α to the cells associated with reduced affinity of the TGF-α receptors for the ligand.

Primary human tumors may produce TGF-α. As shown by *in situ* hybridization analysis, the tumor cells of primary gastric carcinomas, but not the adjacent nonmalignant cells, express TGF-α mRNA.[143] Immunoreactive TGF-α is commonly present in colorectal cancers.[144] The normal human kidney expresses low levels of TGF-α mRNA, and increased expression of this mRNA is found in renal cell carcinomas.[63] Elevated levels of TGF-α mRNA and protein, as well as increased expression of EGF receptor mRNA and protein, have been detected in malignant kidney tissue specimens from patients with localized renal cell carcinomas but not in the autologous normal homologues.[145] Immunohistochemical analyses of lung adenocarcinomas showed that the cells of these tumors frequently exhibit intense immunostaining for TGF-α and that the amount of TGF-α may correlate with the prognosis of the disease.[146] Primary papillary carcinomas of the thyroid may coexpress TGF-α and the EGF receptor.[147] Immunoreactive EGF and TGF-α are present in tissue extracts from benign prostate hyperplasia and cancer of the prostate at similar concentrations.[148]

Malignant tumors of the ovary may express TGF-α,[149] and it has been suggested that an autocrine mechanism involving TGF-α and its receptor may frequently operate in ovarian adenocarcinomas.[150] However, expression of the TGF-α and the EGF receptor in primary ovarian carcinomas shows wide variation, and many of these tumors are negative for TGF-α, the EGF receptor, or both.[151] TGF-α is widely distributed in both normal and neoplastic human endocrine tissues and it is unlikely that it may be useful as a tumor marker for these tissues.[152] The autocrine/paracrine effects of TGF/α/EGF on human ovarian cancer cell lines is either growth-stimulatory or growth-inhibitory, depending on the specific type of cell line.[153] In any case, an autocrine function of TGF-α is not restricted to tumor tissues. Coexpression of TGF-α and EGF/TGF-α receptor mRNA occurs, for example, in normal and adenomatous human colonic epithelium.[154] TGF-α may be an important stimulating factor for the proliferation of the epithelium of the normal human colon by either autocrine or paracrine mechanisms.

A TGF-α-like factor was detected in urine from women with disseminated adenocarcinoma of the breast but not in urine from normal individuals.[155] It was suggested that the presence of TGF-α in the urine could represent a tumor marker in patients that are being followed with a diagnosis of melanoma.[156] Further studies are required, however, for a proper evaluation of the possible use of urinary TGF-α and TGF-α-related peptides as markers in patients with different types of tumors.

In certain tumors there may be a dissociation between TGF-α mRNA and TGF-α protein expression. Primary human colon carcinomas may express high amounts of immunoreactive TGF-α in comparison with the normal colon mucosa, but expression of TGF-α mRNA in these tumors, as assessed by nucleic acid hybridization, is rare.[157] The discrepancy between the levels of TGF-α mRNA and TGF-α protein expression may be caused by different posttranscriptional controls that exist in normal and tumor cells and/or by some other mechanism(s). In any case, the marked heterogeneity of TGF-α expression, even among cancers of the same organ or histological type, suggests that TGF-α is unlikely to play a consistent and direct role in the pathogenesis of all cancers. Moreover, response of tumor cells to TGF-α is heterogeneous not only between different types of tumors but also within a given type of tumor. The effects of human TGF-α on the *in vitro* growth of fresh human tumors can be determined by using a capillary cloning system.[158] Growth of most human tumors capable of colony formation under regular assay conditions is further stimulated in a concentration-dependent manner by TGF-α. Individual tumors, even of the same type, display different sensitivities to TGF-α. A number of tumor specimens incapable of colony formation under standard assay conditions are stimulated by TGF-α, but the sensitivity to the factor varies widely among tumor types, and some tumor specimens do not display colony formation even in the presence of TGF-α.

TGF-α may function as an autocrine or paracrine growth factor for tumor cells. It is present in hamster oral cells transformed by 7,12-dimethylbenz(a)anthracene (DMBA) but not in normal oral cells of adult hamsters; moreover, oral tumor cells express EGF receptors, suggesting that an autocrine growth mechanism may operate in these tumors.[159] The human lung adenocarcinoma cell lines A-549 and PC-9 produce TGF-α and express TGF-α receptors, and exposure of these cells to a monoclonal antibody against human TGF-α results in inhibition of cell growth.[160] Interaction between TGF-α and its receptor may be responsible for the autonomous growth of the androgen-independent prostatic carcinoma cell line PC3 and may be a mechanism of escape from androgen-dependent growth in human prostatic carcinoma.[161] TNF-α acts as an autocrine growth factor in ovarian carcinoma cell lines.[162] Tumor-derived cell lines derived from neoplastic tissue from rats with hereditary renal cell carcinoma express abundant TGF-α and EGF receptor mRNA, suggesting the operation of an autocrine loop in the tumor cells.[163] The role of TGF-α in tumor development is not clear, however. Studies with introduction of human TGF-α cDNA into primary mouse epidermal cells or papilloma cells using a replication-defective retroviral vector showed that expression of TGF-α, in either an autocrine or paracrine fashion, can stimulate tumor growth of skin grafts made with these cells, but expression of TGF-α does not appear to influence tumor progression directly.[164] Human EGF-dependent cell lines of melanoma and cervical carcinoma origin secrete TGF-α and this secretion is dependent on the continuous presence of EGF; these cells no longer grow when deprived of EGF, which indicates that an effective autocrine capacity from TGF-α secretion does not exist in these cells.[165]

Expression of TGF-α at high levels may be sufficient, in defined conditions, to induce neoplastic transformation. Rat fibroblasts expressing a vector coding for human TGF-α may become transformed and acquire tumorigenic properties in nude mice.[166] Expression in NRK cells of a rat TGF-α gene allows the cells to grow in soft agar in the presence of TGF-β.[167] Overexpression of a human TGF-α cDNA in immortalized, nontransformed mouse mammary epithelial cells can result in the induction of a transformed phenotype.[168] HC11 mouse mammary gland cells expressing a TGF-α gene or an activated c-H-*ras* oncogene are no longer able to respond to lactogenic hormones and express a transformed phenotype due, at least in part, to an autocrine mechanism involving the EGF receptor.[169] In contrast, HC11 cells expressing a transforming variant of the rat c-*neu* gene or a mutated copy of the human c-*erb*-B-2 gene can respond to lactogenic hormones, as assessed by the expression of the β-casein gene.

TGF-α can display oncogene-like activity, but the type of cell and the species of origin are critically important for this activity. Infection of normal mouse mammary epithelial cells (NOG-8 cells), but not normal rat fibroblasts (NRK-49F cells), with a TGF-α-expressing vector resulted in the induction of transformation.[170] Similar levels of TGF-α were expressed in both types of cells, but the levels of EGF receptor expression in NOG-8 cells were tenfold higher than in NRK-49F cells. Expression of a noncleavable, membrane-anchored form of pro-TGF-α in NRK cells resulted in transformation, as measured by anchorage-independent growth in soft agar and tumorigenicity in nude mice.[171] These results suggest that secreted TGF-α may not be the only physiologically active form of this factor. Human cells are generally resistant to growth factor-induced transformation. Expression of TGF-α in diploid human skin fibroblasts from a transfected vector containing a cDNA encoding human TGF-α resulted in the expression of a refractile morphology and a slight increase in saturation density, but the manipulated cells did not grow in soft agarose or show an extended lifespan, and the cells did not exhibit tumorigenic potential when injected into nude mice.[172]

Expression of high levels of TGF-α by certain neoplastic cells may be associated with an enhanced aggressive behavior of the cells. Transfection of the NBTII rat bladder carcinoma cell line with a vector expressing the human TGF-α gene resulted in the development of a motile, fibroblast-like phenotype with matrix-degrading potential.[173] The clones expressed a biologically active 18-kDa form of TGF-α and secreted a gelatinolytic metalloproteinase.

TGF-α and other growth factors may function as "second messengers" for the induction and/or maintenance of a transformed phenotype in hormone-dependent cells. Several human esophageal carcinoma cell lines produce EGF and TGF-α and express high levels of EGF receptors.[174] Anti-EGF and anti-TGF-α monoclonal antibodies can inhibit DNA synthesis in these lines, suggesting that EGF and TGF-α function as autocrine factors for the growth of human esophageal carcinomas. Human pancreatic carcinoma cells overexpressing EGF receptors in culture may utilize endogenously produced TGF-α as a superagonist in stimulating anchorage-independent growth.[175]

In the estrogen-responsive human breast cancer cell line MCF-7, estrogen induces growth factor activities, including TGF-α and IGF-I, and these activities are sufficient to stimulate MCF-7 cells to form tumors in ovariectomized mice, thus partially replacing for estradiol.[176] TGF-α mRNA was detected in 70% of estrogen receptor positive and negative primary human breast tumors from 40 patients.[177] However, the role of TGF-α in estradiol-dependent and estradiol-independent mammary tumors is not totally clear. Blockade of EGF receptors in MCF-7 cells with monoclonal or polyclonal antibodies specific to the receptor does not alter estrogen-regulated growth, suggesting that secreted TGF-α may not be a primary mediator of the growth effects of estrogen in these cells.[178] TGF-α is produced by well-differentiated, estrogen-dependent rat mammary adenocarcinomas, but the same factor is present in nonpregnant normal rat mammary glands.[179] Moreover, only low levels of TGF-α expression are observed in the more aggressive, metastatic transplantable rat mammary tumors that are estrogen-independent and lack estrogen receptors, indicating that TGF-α may be unnecessary for the growth of all types of rat tumors and that escape from hormone dependency may be unrelated to the constitutive production of TGF-α or to an autocrine response to this growth factor. The results of studies on growth inhibition caused by progestins and nonsteroidal antiestrogens in T-47D human breast cancer cells are not consistent with the hypothesis that TGF-α and TGF-β function directly as autocrine modulators.[180]

The effects of TGF-α overexpression in transgenic mice bearing a fusion gene consisting of the metallothionein 1 promoter and a human TGF-α cDNA are pleiotropic and tissue specific.[181,182] A spectrum of changes in the growth and differentiation of certain adult tissues is seen in the transgenic mice. The liver, breast, and pancreas are targets of overproduced TGF-α and the response of each of these tissues to the factor are phenotypically distinct. The liver frequently develops multifocal, well-differentiated hepatocellular carcinomas that express high levels of human TGF-α. In contrast, the pancreas exhibits progressive interstitial fibrosis and a florid acinoductular metaplasia. Secretory adenocarcinomas may appear in the postlactational mammary gland of these animals. Transgenic mice, in which a human TGF-α cDNA is expressed under the control of the mouse mammary tumor virus (MMTV) enhancer/promoter, exhibit a variety of histologic abnormalities in the mammary glands, ranging from simple hyperplasia to adenocarcinoma.[183] Elevated expression of TGF-α under the control of heterologous promoters in transgenic mice markedly accelerates oncogene-induced tumorigenesis in the pancreas and liver.[184] It is thus clear that TGF-α may display, under certain experimental conditions, oncogene-like activity. Further studies are required to elucidate the role of TGF-α in natural oncogenesis but available evidence strongly suggests that TGF-α may contribute to multistage carcinogenesis *in vivo.*

G. TGF-α AND ONCOPROTEIN EXPRESSION

Cells transformed by acute retroviruses or activated proto-oncogenes may produce high amounts of TGFs, especially TGF-α. SSV-transformed NRK cells produce and secrete high amounts of both TGF-α and TGF-β.[32] As a result, the cells have a reduced ability to bind external platelet-derived growth factor (PDGF) and EGF. The expression of cellular genes coding for TGFs may be essential for full cellular transformation induced by SSV. Mouse mammary epithelial cells transformed by the v-H-*ras* oncogene also produce increased amounts of TGF-α.[185] The level of expression of v-H-Ras oncoprotein in these cells correlates with the amount of TGF-α produced as well as with their cloning efficiency in soft agar.

A mutant c-*ras* proto-oncogene is contained in several human tumor cell lines that produce TGFs. Transfection of mutant human c-*ras* genes into rodent cells may result in the induction of TGFs, especially TGF-α, which is associated with anchorage-independent growth of the cells, followed by morphological transformation and *in vivo* tumorigenicity of the recipients.[186-188] The induction of TGF-α occurs at the level of transcription. Reversion to a normal phenotype by deletion of exogenous c-*ras* genes is accompanied by loss of the ability of these cells to produce TGFs. At least one of the pathways by which mutant c-*ras* genes induce the malignant transformation of rodent fibroblasts could consist in the activation of TGF production by these cells, especially the production of TGF-α. Immunohistochemical analysis of 174 human gastric carcinomas showed the existence of a correlation between the simultaneous expression of TGF-α and c-H-Ras product and the invasive properties of the tumor.[189]

Coexpression of the c-*myc* proto-oncogene and TGF-α in a transgenic mouse model results in a tremendous acceleration of neoplastic development in the liver.[190] Overexpression of TGF-α in the basal epidermal layer of transgenic mice can bypass the need for c-H-*ras* gene mutations in the formation of mouse skin papillomas.[191] TGF-α overexpression on its own is not sufficient for papilloma formation, but it can act in a synergistic way with the tumor promoter TPA to enhance the proliferation of epidermal cells. However, papillomas are benign lesions that can regress and heal after cessation of the promoting agent. The genetic program for terminal cell differentiation may remain intact in papilloma cells.

III. TRANSFORMING GROWTH FACTOR β

TGF-β was first identified by its ability to elicit anchorage-independent growth (i.e., formation of colonies in soft agar medium) and a transformed phenotype in nontumorigenic mesenchymal cells, including fibroblasts of rodent origin such as NRK or AKR-2B cells. Substances with this type of activity may be released into the media by normal chicken, mouse and human embryo fibroblasts.[54] TGF-β-like activity is found in fetal rat tissues.[56] Subsequently, it was shown that TGF-β is a potent growth inhibitor for cells of epithelial origin, and that it can influence a wide range of processes associated with cell proliferation and differentiation, including hematopoiesis, adipogenesis, myogenesis, and chondrogenesis. TGF-β also has an important influence on the differentiation of epithelial cells. It is secreted by blood cells such as platelets, monocytes, and lymphocytes. It plays a crucial role in embryogenesis and tissue repair, as well as in host response to tumors.[192-200] Production of TGF-β may correlate with the mitotic activity of normal and tumor cell populations. TGF-β may also have an important role in some nontumoral pathologic processes. Experimental glomerulonephritis can be suppressed by an antiserum against TGF-β.[201]

TGF-β activity is represented by a family of closely related peptides that are encoded by duplicated genes.[202,203] Four distinct subtypes of TGF-β have been identified: TGF-β_1, TGF-β_2, TGF-β_3, and TGF-β_5.[204] An activity present in the chicken, termed TGF-β_4,[205] is encoded by a gene homologous to mammalian TGF-β_1. Expression of the genes encoding the four types of TGF-β is subjected to a complex pattern of autologous and heterologous regulation.[206] The different forms of TGF-β exhibit differential expression during mammalian embryogenesis.[207] TGF-β_1 corresponds to the classical form of TGF-β activity, and the discussion that follows, unless indicated otherwise, refers to TGF-β/TGF-β_1. The other isoforms of TGF-β are discussed at the end of this chapter.

A. PRODUCTION AND METABOLISM OF TGF-β

TGF-β is an ubiquitous protein that is present in a diversity of normal and transformed cells and has been purified to homogeneity from human placenta, human platelets, bovine kidney, and other sources.[27,51,52,208,209] Human peripheral blood monocytes and neutrophils secrete a form of TGF-β which does not require acid treatment for activation.[210] TGF-β is also produced by human T lymphocytes and may be involved in the regulation of T-cell proliferation.[211] Megakaryocytes are a major site of synthesis of TGF-β, which is stored in the α-granules of these cells and mature platelets and may be used in the wound healing response.[212]

The amount of TGF-β secreted by different types of cells may not be correlated with transformation. For example, cultured human mesothelial cells secrete higher amounts of TGF-β into the medium than do mesothelioma cell lines.[213] This apparent paradox could be accounted for by the fact that TGF-β produced by normal cells is secreted in a latent or inactive form and, hence, exerts no significant biological effect.[37] TGF-β is stored in the form of a high molecular weight complex in which its activity is masked.[214] The cDNA cloning and the structural and functional characterization of a component of this complex, the TGF-β_1-binding protein, has been reported.[215] This protein has an unusual sequence consisting of 16 EGF-like, cysteine-rich repeats and, in addition, three copies of a novel motif. However, the binding protein is not necessary for the latency of TGF-β_1 and its exact function remains to be determined. Activation of TGF-β secreted by cells in a latent form may take place by proteolytic processes and may play a role in the regulation of TGF-β biological effects.[216] In contrast to normal cells, tumor cells often secrete active TGF-β and, in some instances, this factor may exert an autocrine growth stimulatory effect. However, an activated form of TGF-β is also secreted by cocultures of bovine endothelial cells and pericytes, suggesting that contacts between two specific types of normal cells may result in the secretion of unmasked TGF-β.[217]

TGF-β present in conditioned cell culture medium independent of cell transformation may derive from serum.[218] TGF-β activity can be detected in the normal human serum and is augmented in the sera of patients with CML during the chronic phase of the disease.[219] TGF-β is synthesized and secreted in a latent form from a wide variety of normal cells. A carrier protein of 440 kDa contained in platelets is associated with TGF-β and may function as a masking protein to regulate the local activity of the factor in processes such as wound healing and liver regeneration.[220] The TGF-β masking protein is composed of a dimeric amino-terminal part of a TGF-β precursor of 39 kDa and a large subunit of 105 to 120 kDa whose structure has been elucidated by determining the nucleotide sequence of its cDNA.[221] A precursor of the masking protein has seven N-glycosylation sites and an unusual structure containing 18 EGF-like domains and 4 cysteine-rich internal repeats. mRNA for the large subunit of the masking protein is synthesized in parallel with the expression of TGF-β mRNA in various rat tissues. α_2-Macroglobulin is a major TGF-β-binding protein present in serum and may be responsible, at least in part, for the latent form of TGF-β activity present in serum.[222,223] The liver plays a major role in the clearance and metabolism of TGF-β.[224] Metabolites of TGF-β are present in bile and urine. Biologically active TGF-β has been detected in normal human urine as well as in the urine obtained from cancer patients.[57,225]

B. THE TGF-β GENE

The TGF-β_1 gene is located on human chromosome 19, at region 19q13.1-q13.3.[226] Analysis of overlapping cDNAs and gene fragments led to the determination of a continuous sequence of 2439 bp corresponding to the TGF-β precursor mRNA.[227] TGF-β_1 cDNA encodes a precursor polypeptide of 391 amino acids, the carboxyl-terminal 112 amino acids which correspond to the sequence of the mature factor protein. Transcripts of the TGF-β_1 gene are found in various types of normal and transformed cells. Promoter DNA sequences contained in the 5′-flanking region of the gene are responsive to autoregulation of transcription.[228] Seven distinct factors present in nuclear extracts from human lung adenocarcinoma cells interact with sequences between –454 and –323 bp upstream of the 5′-most start site of the gene, supporting the involvement of sequence-specific transcription factors in the transcriptional autoactivation of the TGF-β_1 gene. The murine TGF-β_1 gene is located on chromosome 7 and its sequences have been characterized, including the 5′ untranslated and regulatory region.[229]

High levels of TGF-β_1 are present in cells transformed by the *ras* oncogene, which is due to transcriptional activation of the TGF-β_1 promoter by the Ras oncoprotein. The RB tumor suppressor protein regulates TGF-β_1 gene expression either positively or negatively, depending on the cell type.[230]

C. BIOSYNTHESIS AND STRUCTURE OF TGF-β

TGF-β purified from human placenta was characterized as a 23-to 25-kDa protein composed of two polypeptide chains held together by interchain disulfide linkages.[51,192] The protein purified from rat embryo fibroblasts transformed by Snyder-Theilen FeSV or from the mouse transformed cell line L-929 has similar physicochemical characteristics.[24,231] The primary structure of the human TGF-β monomer has been determined from a TGF-β cDNA clone.[227] The 112-amino acid monomeric form of the natural TGF-β homodimer is derived proteolytically from a much larger precursor polypeptide which may be secreted. The monomeric form of the simian TGF-β precursor molecule (pro-TGF-β_1) expressed in COS cells transfected with a plasmid vector consists of 390 amino acids and has a molecular weight of 45,000.[202,232] The mature growth factor sequences are represented by the carboxyl-terminal 112 amino acids (residues 279 to 390) of the TGF-β precursor. Disulfide bond formation of TGF-β would occur co- or posttranslationally and results in the dimeric factor molecule. The recombinant simian TGF-β precursor expressed in Chinese hamster ovary (CHO) cells is glycosylated by the addition of sialic acid residues and *N*-linked carbohydrates and is phosphorylated in the amino-terminal portion of the molecule.[233] Glycosylation and early stage remodeling of oligosaccharide side chains are necessary for secretion of TGF-β_1 by transfected CHO cells.[234] The structural characteristics of murine TGF-β precursor polypeptide have been determined as well.[235]

Normal cells and transformed cells may secrete a biologically inactive, high molecular weight form of TGF-β, which is converted into the active form by the action of unknown mechanisms. The latency of this form of TGF-β is due to the formation of a complex between TGF-β and a high molecular weight protein that prevents its binding to the receptor. This "masking protein" can be dissociated from the complex, with the appearance of TGF-β activity, by treatment with 1 *N* acetic acid or 6 *M* urea. Characterization of the masking protein has shown that it consists of two subunits with molecular weight of 39 and of 105 to 120 kDa, respectively.[236] These two subunits form a complex of 180 to 210 kDa in which the 39-kDa subunits are linked with the 105- to 120-kDa subunits by disulfide bonds. Since the masking protein has an estimated molecular weight of 400 to 500 kDa, the 180 to 210 components of the protein would form a dimer with noncovalent bonds. Amino terminal sequencing showed that the 39-kDa subunit of the complex is identical with the amino-terminal part of the TGF-β precursor molecule. The 39-kDa subunit (amino-terminal part) and TGF-β monomer (carboxyl-terminal part) are probably bound in a 1:1 molar ratio in the latent form of TGF-β.

D. STRUCTURAL HOMOLOGIES OF TGF-β PROTEINS

The TGF-β proteins belong to a superfamily of homologous proteins involved in the control of tissue development in organisms ranging from vertebrates to arthropods. The TGF-β homologous proteins include inhibins, Müllerian inhibiting substances, and a decapentaplegic gene complex protein of *Drosophila*.[237]

Analysis of cDNA sequences and primary protein sequences of TGF-β proteins shows the existence of significant structural homology between TGF-β and the β-chains of inhibin, a regulatory agent contained in the ovarian follicular fluid.[237,238] Inhibin is a hormone involved in the regulation of adenohypophysial function, acting as a potent inhibitor of FSH secretion. It is interesting that inhibin and TGF-β have opposite modulating effects on the FSH-induced aromatase activity of cultured granulosa cells from the rat ovarian follicle.[239] Inhibin concurrently added with FSH for 48 h on rat ovary granulosa cells prevents the FSH-stimulated conversion of androstenedione to estrogen, which demonstrates that the granulosa cell not only can modulate the secretion of FSH at the level of the pituitary gland by secreting inhibin as a classical hormone, but can also modulate the effects of FSH as the primary inducer of granulosa aromatase activity by utilizing its own inhibin in some paracrine or autocrine fashion. Similar to other polypeptide molecules, FSH may have totally unrelated biological activities depending on the nature of the target cell they reach and with which they interact.

Two cartilage-inducing factors, termed CIF-A and CIF-B, have been isolated from bovine demineralized bone.[240] The CIFs are dimeric proteins of 2 kDa capable of inducing rat muscle mesenchymal cells to undergo differentiation and synthesize cartilage-specific macromolecules. The CIF-A is probably identical to TGF-β.[241,242] On the other hand, CIF-B is a distinct molecule, although it exhibits extensive structural homology to CIF-A/TGF-β and may compete with this factor for the same membrane receptors in NRK-49F cells.[243] CIF-B induces anchorage-independent proliferation of NRK-49F cells when these cells are simultaneously treated with EGF. CIF-A and CIF-B are potent inhibitors of DNA synthesis and anchorage-independent growth of a variety of human and nonhuman tumor cells.[244] However, there is some evidence of differential binding properties of TGF-β and CIF-B to cell surface components, suggesting the existence of biological activities that may be unique to each of the proteins.[245]

The decapentaplegic complex encoded by the *dpp* gene of *Drosophila melanogaster* has been implicated in several morphogenetic events occurring in the insect. The carboxyl terminus of the *dpp*-encoded protein exhibits strong sequence homology to the carboxyl termini of mammalian TGF-β, inhibin, and Müllerian inhibiting substance (MIS).[246] Similar to other proteins of the TGF-β family, the *dpp* protein of *Drosophila* is proteolytically cleaved, and both portions of the protein are secreted from the cells.[247] These findings indicate that members of the TGF-β family of proteins may exist even in insects and that an ancestral gene was present before arthropods and vertebrates diverged millions of years ago.

E. TGF-β RECEPTORS

The complex cellular effects of TGF-β are mediated by the initial interaction of the growth factor with specific receptors located on the cell surface. TGF-β binding sites are present in a variety of cultured cells of both mesenchymal and epithelial origin, including human fibroblasts and keratinocytes.[248] Lower levels of TGF-β receptors are detected in some mouse and human tumor cell lines, which could be associated with a downregulation phenomenon caused by the production of TGF-β by these cells. The polyanionic compound suramin can inhibit TGF-β-induced mitogenicity in certain types of cells by altering the interaction between TGF-β and its receptor on the cell surface.[249] After ligand binding, the TGF-β receptor is internalized and partially degraded in the lysosomes, and the receptor is subjected to downregulatory phenomena in a manner similar to that of other growth factors.[250]

1. Molecular Forms of the TGF-β Receptor

The high-affinity TGF-β receptor was initially characterized as a large glycoprotein molecule with a molecular weight of 565 kDa in the mouse and 615 kDa in the human.[251] This big molecule would be composed of a disulfide-linked complex in which a 250-to 330-kDa subunit would contain the ligand site. More recent evidence indicates that the TGF-β receptor is represented by different protein complexes of 55 to 350 kDa. At least three distinct types of surface TGF-β receptors, (I, II, and III) have been identified in different types of cells.[252,253] The type I, II, and III receptors can be distinguished by their structural and functional properties. TGF-β receptors types I and II have high affinity for TGF-β_1 and lower affinity for TGF-β_2. Type I TGF-β receptors are represented by 55- to 65-kDa complexes in all animal species tested. Type II TGF-β receptors are represented by protein complexes of 85 to 110 kDa, depending on the species of origin.[254] The type III TGF-β receptor has an apparent molecular weight of 250 to 350 kDa, forms disulfide-linked complexes in mouse 3T3 fibroblasts, and has high affinity for TGF-β_1 and TGF-β_2. The type III receptor is of low abundance on the cell surface of many mammalian and avian species and its deglycosylation by the action of trifluoromethanesulfonic acid yields receptor cores of 110 to 130 kDa.[255] The type III TGF-β receptor has the properties of a membrane proteoglycan that carries heparan and chondroitin sulfate glycosaminoglycan chains.[256,257] The proteoglycan nature of the type III TGF-β receptor is unusual among cell surface receptors for peptide growth factors. The TGF-β binding site in the type III receptor appears to reside in its lower molecular weight (100 to 130 kDa) core. The role for the proteoglycan component of the type III TGF-β receptor remains undefined. The different forms of TGF-β receptors may have specific functional properties. Differential binding of TGF-β_1 to the three molecular forms of TGF-β receptor is observed in resting and mitogen-stimulated mouse T cells.[258] In addition to the type I, type II, and type III TGF-β receptors, a type IV receptor of 60 kDa was identified in pituitary cells,[259] and a type V receptor of 40 kDa was purified from rat liver,[260] but their precise functions are unknown.

The cloning and expression of TGF-β receptors has allowed a better knowledge of their structure and function. The type III human and rat TGF-β receptors are polypeptides of 849 and 853 amino acids, respectively, which contain a single transmembrane domain and a short cytoplasmic region (43 amino acids).[261-263] The type III TGF-β receptor has no obvious signaling motif and shows high amino acid sequence similarity with endoglin, a protein present in endothelial cells, hematopoietic progenitors, and mesangial cells in the kidney.[264] The precise function of the type III TGF-β receptor is unknown.

The type II TGF-β receptor is an 80-kDa polypeptide composed of 565 amino acids.[265] The receptor has a hydrophyllic cysteine-rich extracellular domain that contains three consensus *N*-glycosylation sites, a single hydrophobic transmembrane domain, and a cytoplasmic serine/threonine kinase domain. The TGF-β receptor has been characterized recently as a heteromeric protein kinase complex.[266] The mechanism of action of TGF-β is thus associated, at least in part, with the phosphorylation of proteins on serine and threonine residues.

2. Regulation of TGF-β Receptor Expression

A diversity of endogenous and exogenous factors may influence the expression of TGF receptors in different types of cells, although the mechanisms involved in such regulatory effects remain largely uncharacterized. Changes in the binding of TGF-β to its receptor may be associated with the differentiation of neoplastic cells. Examination of two embryonal carcinoma cell lines (F9 and PC-13) for the presence of TGF-β receptors indicated that these cells bind little, if any, TGF-β and do not respond to TGF-β. Treatment of the cell lines with retinoic acid induced the appearance of irreversibly differentiated cells that exhibited TGF-β receptors. In contrast to the parental cells, TGF-β influenced the growth of the differentiated cells cultured in either serum-free or serum-containing media, inducing a decrease in the growth of these cells.[267]

F. POSTRECEPTOR EVENTS OF TGF-β ACTION

The postreceptor mechanisms of TGF-β action are little known. TGF-β does not elicit tyrosine kinase activity and may not exert mitogenic effects by itself.[270] However, recent evidence indicates that the activated TGF-β receptor possesses kinase activity specific for serine/threonine residues.[266] In mesenchymal cells, such as in AKR-2B mouse embryo cells, TGF-β acts as a potent stimulator of DNA synthesis, but only with a prolonged prereplicative phase, as compared to other growth factors (i.e., EGF, PDGF, and FGF).[269] TGF-β can potentiate the mitogenic activity of other growth factors, in particular that of EGF, and can confer on some normal fibroblasts the ability to grow in soft agar. TGF-β may have effects not only on sensitive cells directly exposed to it but also on neighbor cells, through alterations in gap junction-mediated intercellular communication.[270]

1. Activation of GTP-Binding Protein

TGF-β may exert its mitogenic effect, at least in part, by binding to its surface receptor and inducing activation of a G protein, which in turn would stimulate further biochemical events that initiate cell division.[271] Induction of DNA synthesis is stimulated by TGF-β in AKR-2B mouse cells, and this effect is inhibited in a dose-dependent manner by pertussis toxin. However, a comparative study of the effects of pertussis toxin (which ADP ribosylates G_1) and cholera toxin (which catalyzes the ADP ribosylation of G_s) suggests that the diverse biological effects of TGF-β are mediated through multiple G protein-dependent and G protein-independent transducing pathways.[272]

2. Production of Prostaglandins

TGF-β stimulates PGE_2 production by human lung fibroblast cell lines.[273] This effect is associated with a stimulation of the synthesis of total protein, collagen, and fibronectin. Inhibition of prostaglandin synthesis with indomethacin potentiates the stimulatory effect of TGF-β on the production of these proteins. Prostaglandins may function to down-modulate the effects of TGF-β on collagen and fibronectin production, serving as a negative feedback mechanism to limit the increase in extracellular matrix protein production induced by TGF-β.

3. Phosphoinositide Metabolism and Ca^{2+} Mobilization

Phosphoinositides and calcium may have a role as intracellular messengers of TGF-β action. Treatment of serum-deprived Rat-1 fibroblasts with TGF-β results in stimulation of Ca^{2+} influx via voltage-independent channels and in a marked elevation of cellular inositol trisphosphate levels.[274,275] These effects are blocked by actinomycin D, suggesting that RNA synthesis is required for them to develop. Simultaneous treatment of Rat-1 fibroblasts with EGF and TGF-β produces an additive increase in Ca^{2+} influx, suggesting that EGF and TGF-β may interact with distinct populations of Ca^{2+} channels, or may influence distinct channel-activating mechanisms. Expression of the gene β4, which encodes a distinct type of calcium channel, called ryanodine receptor, is induced by TGF-β in cultured mink lung epithelial cells.[276] However, the precise role of these changes in the physiological effects of TGF-β *in vivo* is unknown. TGF-β is a very potent chemoattractant for human peripheral blood neutrophils, and this action is unrelated to changes in the $[Ca^{2+}]_i$ and GTPase activity.[277] Chemoattraction by TGF-β is inhibited by cycloheximide and actinomycin D, suggesting a mediation by the synthesis of yet unknown proteins.

4. Protein Phosphorylation

The cellular mechanism of action of TGF-β is not directly related to tyrosine phosphorylation but appears to be associated with the phosphorylation of cellular proteins on serine and threonine residues. The type II TGF-receptor is a serine/threonine-specific protein kinase.[265] The cellular substrates of the receptor enzyme remain to be characterized. TGF-β-induced growth arrest of human keratinocytes may involve activation of a serine/threonine-specific protein phosphatase, while activation of tyrosine-specific protein phosphatases could represent an additional mechanism for maintaining cells in a growth-arrested state.[278]

5. Regulation of Transcription

TGF-β is positively or negatively involved in regulating the expression of particular genes in different types of cells. TGF-β and IL-6 have antagonistic effects on the levels of expression of both negatively and positively regulated liver genes associated with acute phase response.[279,280] Transcription of the genes for the EGF receptor and fibronectin is increased by TGF-β in NRK rat kidney cells.[281] TGF-β also induces the expression of actin mRNA in AKR-2B cells, which is followed by increased concentrations of β- and γ-actins in the cytoplasm.[282] Actins are major structural protein components of microfilaments and are involved in cellular morphogenesis, motility, and mitosis.

TGF-β exerts a negative regulatory effect on the expression of the gene coding for stromelysin, a member of matrix-degrading metalloproteinases.[283] Stromelysin is involved in wound repair and remodeling, as well as in embryogenesis and angiogenesis. Induction of stromelysin gene expression by EGF and oncoproteins (v-Src and v-K-Ras) is inhibited by TGF-β.[284] Inhibition of stromelysin gene expression by TGF-β is mediated through binding of a nuclear protein complex containing the c-Fos protein to a 10-bp element which is present in the promoter. This element is conserved in the promoter regions of other TGF-β-inhibited genes. Exposure of the mink lung epithelial cell line CCL64 to TGF-β results in induction of the expression of several genes including those encoding collagen-α type I, fibronectin, plasminogen activator inhibitor 1 (PAI-1), and the monocyte chemotactic cell-activating factor (JE gene).[285] Another gene, TI 1, is rapidly and transiently downregulated by TGF-β and serum in CCL64 cells.[286] The TI 1 gene product is related to a family of transmembrane glycoproteins that are expressed on lymphocytes and tumor cells. Transcriptional expression of the gene encoding osteopontin, a 44-kDa glycoprotein component of the bone extracellular matrix, is regulated by TGF-β.[287] Osteopontin has been detected not only in bone matrix, osteoblasts, osteocytes, and preosteoblast-like fibroblasts, but also in neurons and neurosensory cells. TGF-β suppresses the transcriptional expression of the osteocalcin gene.[288] Osteocalcin is one of the abundant noncollagenonous bone matrix proteins and is produced exclusively by osteoblast. The serum level of osteocalcin is used as an indicator of bone metabolism in patients with metabolic bone diseases.

The mechanisms involved in TGF-β-induced alterations in gene expression may implicate the phosphorylation of specific proteins. TGF-β rapidly induces a phosphorylation of the cyclic adenosine-3′,5′-monophosphate (cAMP)-responsive element-binding (CREB) protein through a mechanism that does not involve the action of cAMP-dependent protein kinase.[289] Transcriptional activation by TGF-β is mediated, at least in part, by the TPA-responsive element, TRE, and CREB phosphorylation induced by TGF-β may result in an increase in TRE binding, which would involve the formation of a heterodimer between CREB and another nuclear protein.

6. Regulation of Proto-Oncogene Expression

The action of TGF-β on its target cells may be associated with changes in the expression of specific proto-oncogenes. Expression of c-*sis*/PDGF-B mRNA is induced by TGF-β in cultured human renal microvascular endothelial cells.[290] TGF-β rapidly induces c-*sis* gene expression in G_0-arrested mouse embryo-derived AKR-2B cells.[291] Expression of PDGF, or a PDGF-like protein, in TGF-β-stimulated cells may result in the induction, with delayed kinetics, of some proto-oncogenes associated with the development of a mitogenic response, in particular c-*fos* and c-*myc*. In human leukemia cell lines (HL-60, U937, Jurkat, and MOLT-4) as well as in the HT-1080 fibrosarcoma cell line, but not in K562 erythroleukemia or A549 lung carcinoma cells, TGF-β predominantly induces the expression of PDGF-A chains.[292] No significant changes in c-*sis*/PDGF-B gene expression were observed in cell lines that express PDGF-A transcripts upon TGF-β stimulation.

Expression of the c-*myc* and c-*fos* proto-oncogenes may be altered by the action of TGF-β on its target cells. Growth of the colon carcinoma cell line MOSER is inhibited by TGF-β and this effect is associated with repression of c-*myc* gene expression.[293,294] TGF-β abrogates both cell growth and c-*myc* expression induced by EGF, insulin, and transferrin in well-differentiated human colon carcinoma cell lines, whereas poorly differentiated human colon carcinoma cells do not respond to the growth factors and the levels of c-*myc* expression remains unaltered.[295] TGF-β inhibits both proliferation and c-*myc* expression in the human hepatoma cell line PLC/PRF/5; in contrast, the Mahlavu hepatoma cell line is resistant to the growth-inhibitory effect of TGF-β and this factor does not suppress c-*myc* expression in these cells.[296] In the MDA-468 human mammary carcinoma cell line, TGF-β has little effect on the expression of c-*myc*, but abrogates the induction of c-*myc* elicited by EGF in these cells.[297] Growth of the tumorigenic, EGF-dependent epithelial mouse cell line BALB/MK is irreversibly arrested by TGF-β, and this arrest is associated with a decrease in c-*myc* expression, which is exerted at a posttranscriptional level, while the expression of c-*fos* is not altered.[298] EGF induction of growth and c-*myc* expression in BALB/MK cells is also inhibited by TGF-β. Increased levels of calcium cause terminal differentiation of BALB/MK cells associated with a rapid down-regulation of c-*fos* expression without change in c-*myc* expression.

Modulation of c-*fos* expression by TGF-β depends on the cell type and may occur at the transcriptional and/or posttranscriptional level. In the rat fibroblast line EL2, TGF-β stimulates DNA synthesis and cell proliferation and this effect is associated with a sustained expression of both c-*fos* mRNA and protein.[299] In contrast, in National Institutes of Health (NIH)/3T3 mouse fibroblasts TGF-β exerts an inhibitory effect on cell growth and induces expression of c-*fos* mRNA but has no detectable effects in c-*fos* protein expression.

Suppression of c-*myc* gene expression may be causally related to the growth inhibitory effects of TGF-β. Treatment of BALB/MK mouse keratinocytes with either antisense c-*myc* oligonucleotides or TGF-β inhibits entry of cells into S phase of the cycle.[300] The block in c-*myc* expression by the action of TGF-β occurs at the level of transcriptional initiation, and the mechanism of this inhibitory effect may involve the synthesis or modification of a protein that would interact with a specific element in the 5′ regulatory region of the c-*myc* gene.

Enhanced transcriptional expression of the proto-oncogene c-*jun* and the related gene *jun*-B is an early genomic response to TGF-β stimulation in sensitive cells.[301] In mink lung CC1 64 cells, the growth inhibitory action of TGF-β is associated with rapid phosphorylation of nuclear proteins and transient expression of the *jun*-B gene, but not the c-*fos* gene.[302] In mouse hepatoma cells (BWTG3 cell line), TGF-β causes a rapid and sustained increase in *jun*-B and *fos*-B mRNA levels, and a subsequent decrease in albumin mRNA.[303] In fresh specimens of human renal cancer cells, as well as in renal cancer cell lines, TGF-β induces *jun*-B but not c-*jun* gene expression.[304] The possible physiological significance of the induction of *jun* genes in relation to the mechanism of action of TGF-β is not clear. Stimulation of c-*jun* and *jun*-B gene expression by TGF-β occurs in A549 human lung adenocarcinoma cells, which are growth inhibited by TGF-β, AKR-2B mouse embryo fibroblasts, which are growth stimulated by TGF-β, and K562 human erythroleukemia cells, which are not affected in their growth by TGF-β.

The possible role of TGF-β on c-*ras* proto-oncogene expression is not understood, but both TGF-β_1 and TGF-β_2 rapidly stimulate GTP binding to Ras protein in TGF-β-sensitive rat intestinal epithelial cells (clonal 4-1 cells).[305] The activation of Ras protein occurs at growth inhibitory concentrations of TGF-β, suggesting that Ras proteins are involved in some cases in the negative regulation of cell proliferation.

The mechanisms of TGF-β-induced proto-oncogene expression are little known. PDGF-induced c-*sis*/PDGF-B mRNA expression is blocked by agents that increase the intracellular levels of cAMP, such as the adrenergic agonists isoproterenol and norepinephrine or the adenylyl cyclase activator forskolin.[290] Expression the of c-*sis*, c-*myc*, and c-*fos* genes in AKR-2B cells is markedly reduced by prior treatment with pertussis toxin, a protein that inhibits receptor/G protein coupled systems.[306] Pertussis toxin does not affect, however, TGF-β-stimulated fibronectin and collagen mRNA accumulation nor has any inhibitory effect on TGF-β-induced neoplastic transformation. Thus, TGF-β-stimulated gene expression may be coupled to multiple pathways which can be distinguished by their sensitivity to pertussis toxin.

7. Regulation of Tumor Suppressor Gene Expression

TGF-β may be involved in the regulation of tumor suppressor gene expression. The activity of the retinoblastoma (RB)-associated protein (RB) during the cell cycle depends on its state of phosphorylation, and the growth inhibitory effect of TGF-β on epithelial cells is accompanied by suppression of RB protein phosphorylation.[307,308] In colon carcinoma cells lines, TGF-β controls RB protein function by two distinct mechanisms, the regulation of its phosphorylation and the control of RB mRNA and protein levels.[309] Expression of the transforming proteins from DNA tumor viruses, e.g., human papillomavirus (HPV)-16 E7, adenovirus type 5 E1A, and simian virus 40 (SV40) large T antigen, can block TGF-β-induced inhibition of c-*myc* gene transcription, but this effect is not observed when the viral proteins are mutated in their RB binding site.[310] Autocrine production of TGF-β may negatively regulate the cycling status of early human hematopoietic progenitors through interaction with the RB product.[311] These observations suggest that the RB protein may play an important role in the growth inhibitory effects exerted by TGF-β on some types of cells. However, the mechanism of these inhibitory effects are complex and the precise role of oncoproteins and tumor suppressor proteins in the multiple physiological effects of TGF-β remains to be elucidated. Accumulation of hypophosphorylated RB protein in cultured human epithelial cells (skin keratinocytes) exposed to TGF-β may be a consequence rather than the cause of the G_1 cell growth arrest induced by TGF-β.[312] There is no evidence of a direct effect of TGF-β on the phosphorylation of RB protein.

The nuclear phosphoprotein p53 has an important role in the negative control of cell proliferation. Inhibition of the proliferation of human breast cancer cells MCF-7 by TGF-β is associated with alterations in the state of phosphorylation and subcellular localization of the tumor suppressor protein p53.[313] The growth inhibitory effects of TGF-β may be exerted, at least in part, through the regulation of tumor suppressor gene expression.

G. BIOLOGICAL EFFECTS OF TGF-β

A wide diversity of biological effects have been attributed to TGF-β.[192] However, many studies on this subject have been performed on *in vitro* systems and the relevance of the findings obtained with such systems for the physiological events occurring *in vivo* should be considered with caution. In one study, administration of TGF-β to intact rats did not result in changes in the metabolite content or enzyme activities in the liver.[314] The reason for this apparent lack of effect of TGF-β on liver metabolism *in vivo* is unknown but there is clear evidence that TGF-β regulates the production of acute phase proteins in human hepatoma cell lines, both directly and by modulating the effect of IL-6.[315] The production of some hormones may be regulated by TGF-β *in vitro* and probably also *in vivo*. TGF-β inhibits aldosterone biosynthesis in cultured bovine zona glomerulosa cells.[316]

The results obtained with targeted disruption of the mouse TGF-β_1 gene indicate the important biological effects of the factor *in vivo*.[317] Mice homozygous for the mutated TGF-β allele do not have gross developmental abnormalities, but 20 d after birth they exhibit an acute wasting syndrome followed by death. The most prominent lesions observed in these animals are tissue necrosis in specific organs and multifocal, mixed inflammatory cell infiltration of numerous organs, particularly heart and stomach. In general, TGF-β deficiency leads to death associated with dysfunction of the immune and inflammatory systems. The lack of developmental defects could be due to the existence of redundant genetic pathways which have evolved overlapping functions.

1. Carbohydrate and Protein Metabolism

TGF-β has important effects on the metabolism of carbohydrates and proteins, stimulating glycolysis and the uptake of glucose and amino acids.[318,319] Glucose uptake and expression of glucose transporter mRNA are both stimulated by TGF-β in quiescent Swiss mouse 3T3 cells.[320] Expression of TGF-β is elevated in chronic complications of diabetes mellitus, for example, in human and experimental diabetic nephropathy.[321]

A high rate of aerobic glycolysis, as measured by the rate of lactate production, is an almost universal characteristic of tumor cells.[322] This phenomenon is observed in Rat-1 cells transformed with a ras oncogene and also, although to a lesser degree, in Rat-1 cells exposed to TGF-β.[323] TGF-β stimulates aerobic glycolysis in Rat-1 cells transfected with a *myc* oncogene, but the stimulated rate is considerably lower than that in *ras* transfected cells. The uptake of methylaminoisobutyrate, a specific substrate of system A amino acid transport, is also accelerated after exposure of Rat-1 cells to TGF-β. Methionine causes an inhibition of glycolysis in Rat-1 cells exposed to TGF-β, either the cells are transfected, or not, with c-*ras* or c-*myc* genes. The inhibition of glycolysis by methionine in Rat-1 cells transfected with the *ras* gene requires that the cells be incubated with methionine for several hours in the presence of serum, but serum can be fully substituted by insulin and IGF-I or IGF-II.[324] In general, the results obtained in these experiments suggest a relationship between Ras proteins and proteins induced by TGF-β.[323] TGF-β exerts some inhibitory effects on protein metabolism. Albumin synthesis in normal human hepatocytes and hepatoma HepG2 cells is inhibited by TGF-β.[325]

2. Blood Cells and Immune Responses

TGF-β has an important role in the regulation of hematopoiesis *in vitro* and probably also *in vivo*.[326] The effects of TGF-β on hematopoietic processes may be either stimulatory or inhibitory and may or may not require the action of accessory growth factors. Both TGF-β_1 and TGF-β_2 can modulate the differentiation of erythroid, granulocyte-macrophage, and multilineage progenitors from a population of normal human bone marrow cells devoid of accessory cells.[327] TGF-β may inhibit the development of erythroid and multilineage colony-forming cells, whereas the growth of cells from the granulocyte-macrophage lineage is enhanced. A synergistic effect of CSF-2 is observed in the stimulation of granulopoiesis by TGF-β.[328] The proliferation of macrophages stimulated by CSF-1 and CSF-2 is enhanced by TGF-β.[329] TGF-β_1 and TGF-β_2 inhibit CSF-2- and IL-3-induced proliferation and colony growth in soft agar of normal human marrow cells and their leukemic counterparts.[330] TGF-β exerts a differential effect on human hematopoiesis, being more suppressive for progenitor cells of earlier, as opposed to later, stages of differentiation. TGF-β is capable of enhancing the suppression of human hematopoiesis mediated by TNF-α and IFN-α and can inhibit the growth of a variety of human and murine myeloid leukemia cells *in vitro*.[331,332] In addition to these effects, TGF-β inhibits the growth and endomitosis of megakaryocytes.[333] Experiments performed *in vivo* show that TGF-β_1 can simultaneously augment and suppress distinct cell lineages in peripheral and central hematopoietic compartments in a reversible and dose-dependent fashion. Administration of repeated subcutaneous injections of TGF-β to intact mice results in induction of granulocytopoiesis and increased levels of lymphoid cells in the peripheral blood, whereas the levels of platelets and erythrocytes are decreased in the blood of the treated animals.[334]

TGF-β is a mediator of immunoregulation.[335,336] It can exert important actions on T and B lymphocytes.[211,337] TGF-β is a costimulator of immunoglobulin A (IgA) production,[338] and selectively induces $F_c\gamma$ RIII on human monocytes.[339] TGF-β is a potent chemotactic agent for human peripheral blood monocytes at concentrations ranging from 0.004 to 0.4 p*M*.[340] This level of activity makes TGF-β the most potent known chemotactic substance for monocytes. Following recruitment of monocytes, TGF-β at higher concentrations can activate monocytes to elaborate mitogenic monokines. TGF-β is a potent immunosuppressive agent in a variety of immunological systems and acts as an inhibitor of IFN-γ production by peripheral blood cells.[341] Suppression of immune response by TGF-β may be due, at least in part, to inhibition of the endogenous production of IL-2 and IL-6.[342] Activity of IL-2-induced LAK cells against human tumor cell lines is inhibited by TGF-β in a dose-dependent manner.[343] This inhibition is reversed by the addition of TNF-α at the initiation of culture. TGF-β is stored in blood platelets as a poorly active complex of high molecular weight which may be dissociated and activated in appropriate *in vivo* microenvironments.[344]

3. Blood Vessels

TGF-β has various effects on several components of blood vessels. Bifunctional effects are produced by TGF-β on migration of rat aortic smooth muscle cells in culture.[345] Low density lipoprotein (LDL) receptor-mediated cholesterol metabolism in vascular smooth muscle cells is upregulated by TGF-β.[346] TGF-β displays potent angiogenic effects both *in vitro* and *in vivo*. Application of TGF-β to the chicken chorioalantoic membrane results in an angiogenic response.[347] TGF-β can modulate extracellular matrix organization and cell-cell junctional complex formation during angiogenic processes.[348] The angiogenic properties of TGF-β may be important in wound healing and tissue repair as well as in tumorigenic processes. TGF-β is involved in preserving the integrity of blood vessels, in particular the endothelial functions. Myocardial ischemia causes heart injury that is characterized by an increase in the levels of circulating TNF-α, local production of superoxide anions, loss of coronary vasodilatation in response to agents that release endothelial cell relaxation factor and to cardiac tissue damage (heart infarct). Ischemic injury can be mimicked by TNF-α, but administration of TGF-β before or immediately after this injury reduces the amount of superoxide anions in the coronary circulation, maintenance of endothelial-dependent coronary relaxation, and reduced injury mediated by exogenous TNF-α.[349] Thus, TGF-β may be capable of preventing severe cardiac injury.

4. Connective Tissue and Extracellular Matrix

TGF-β has important effects on connective, vascular, and bone tissue. It rapidly induces fibrosis and angiogenesis *in vivo* and stimulates collagen formation *in vitro*.[350] TGF-β is a potent chemoattractant *in vitro* for human dermal fibroblasts.[351] When injected subcutaneously into newborn mice, TGF-β causes a rapid increase in connective tissue formation. TGF-β has an important role in tissue repair by transiently attracting fibroblasts into a wound and by stimulating the synthesis of matrix molecules such as fibronectin and collagen and the incorporation of these compounds into the extracellular matrix.[283,352-355] The effects of TGF-β on tissue repair are complemented by other growth factors such as PDGF, which is a potent chemoattractant for wound macrophages and fibroblasts and may stimulate these cells to express endogenous growth factors, including TGF-β.[356] The action of TGF-β on human and rat fibroblasts is exerted, at least in part, at a pretranslational level by stimulating the accumulation of fibronectin and type I collagen mRNAs.[357-359] Stimulation of collagen synthesis induced in fibroblasts by TGF-β may not be associated with cellular proliferation.[360]

Fibronectin is an adhesive glycoprotein which is expressed at high levels during embryogenesis and tissue remodeling. Assembly of fibronectin into pericellular matrix involves a receptor complex. Exposure of human fetal lung fibroblasts to TGF-β *in vitro* results in the synthesis of fibronectin and the fibronectin receptor complex.[361] These changes occur at the level of transcription and are blocked by actinomycin D. Glucocorticoids and TNF-β are involved in the regulation of fibronectin expression by cultured human fibroblasts, but while dexamethasone does not change the ratio between total and specific isoforms of fibronectin, TGF-β increases the expression of fibronectin isoforms that contain a specific sequence, the ED-A sequence.[362] The binding of ^{125}I-fibronectin to the surface of cultured human fibroblasts is stimulated by TGF-β in a dose-dependent manner.[363]

Fibronectin receptor expression in fibroblastic and epithelial cells from human, rat, and mouse origin may be regulated by TGFs. The fibronectin receptor is the prototype of a family of membrane glycoproteins called integrins.[364,365] The integrins are receptors that mediate cell adhesion to extracellular matrix and basement membranes, to other cells, and to plasma proteins. The cytoplasmic domains of integrins interact with cytoskeleton components including talin and fibulin. These interactions are likely to mediate the transmission of signals from the extracellular matrix to the interior of the cell. The integrins have a heterodimeric structure and are composed of two subunits (α and β). Both integrin subunits have a large extracellular domain, a transmembrane domain, and a short cytoplasmic domain. The α subunit contains several calmodulin-type divalent cation binding sites. The β subunits contain an extracellular cysteine-rich domain of unknown function and an intracellular sequence with potential tyrosine phosphorylation sites. The essential structure recognized by the integrins in fibronectin is the sequence Arg-Gly-Asp (RGD). The balance of individual integrins is modulated by TGF-β and the levels of expression of integrin subunits is regulated by TGF-β_1 at the mRNA level.[366] TGF-β_1 switches the pattern of integrins expressed in MG-63 human osteosarcoma cells and causes a selective loss of cell adhesion to laminin.[367]

Degradation of extracellular matrix proteins is intimately associated with the invasive and metastatic properties of tumor cells. This degradation is mediated by specific proteases, especially by plasmin, a proteolytic enzyme capable of degrading most glycoprotein components of the matrix. Plasmin is formed by conversion of the proenzyme, plasminogen, to plasmin by action of plasminogen activator. This process is counterbalanced by the action of plasminogen activator inhibitors. Two types of plasminogen activators have been identified in mammalian tissues, the tissue-type plasminogen activator and the urokinase-type plasminogen activator. Several distinct types of plasminogen activator inhibitors (PAIs) are present in mammalian tissues, thus establishing a complex regulatory system for the degradation of extracellular matrix proteins.[368] Expression of both the plasminogen activators and the PAIs is regulated by TGF-β.[369-372] Since the plasminogen activators are the major mediators of pericellular proteolysis, which may be required for anchorage-independent growth *in vitro* and tumor invasion *in vivo,* it may be concluded that TGF-β is significantly involved in modulating both the synthesis and degradation of extracellular matrix components.

Structural changes in the extracellular matrix may have profound effects on cell proliferation and differentiation, and oncogenic transformation is frequently associated with alterations in the expression of matrix components. TGF-β-induced changes in these components may be responsible for some behavioral alterations observed in transformed cells, including anchorage-independent growth *in vitro.* Induction of anchorage-independent growth of normal fibroblasts by TGF-β is mimicked by fibronectin and is specifically blocked by inhibitors of fibronectin binding to its cell surface receptor.[358] TGF-β increases the expression of cell adhesion receptors on the surface of target cells and this response correlates with an enhanced ability of cells to adhere to fibronectin substrates.[373] The effects of TGF-β on adhesion receptors located on the cell surface, which act as a link between the extracellular matrix and the cytoskeleton, may result in modification of cell adhesion and morphology, which may eventually affect cell migration, proliferation, and differentiation.

Human fibroblasts exposed to TGF-β *in vitro* may exhibit secretion of plasminogen activators.[374] The plasminogen activators of either the urokinase-type or the tissue-type are serine proteases involved in the conversion of the abundant extracellular proenzyme plasminogen into the active protease plasmin, which has a broad trypsin-like activity. Since increased proteolytic activity is often associated with the growth of malignant cells, secretion of TGFs may lead to increased amounts of this activity and may thus facilitate the growth and spread of tumors. The primary effect of TGF-β on regulation of plasminogen activator occurs at the level of transcription and may depend on a short-lived negatively regulated specific protein.[375]

5. Bone and Cartilage Tissue

TGF-β exerts important effects on the structure and function of bone tissue. TGFs are involved in early stages of bone development as well as in bone repair and remodelating after trauma.[376] Experimental evidence shows that TGF-β regulates replication and differentiation of mesenchymal precursor cells, chondrocytes, osteoblasts, and osteoclasts.[377,378] Expression of TGF-β in bone tissue is regulated by hormones, in particular by steroid hormones. Ovariectomy in rats reduces deposition of TGF-β in bone, and diminished skeletal TGF-β may play a role in the pathogenesis of bone loss, fractures, and microfractures that occur in estrogen-deficient states.[379]

Cartilage development (chondrogenesis) proceeds through distinct stages, beginning with the condensation of pluripotent mesenchymal cells, followed by their proliferation and differentiation into resting chondrocytes. The role of growth factors in the *in vivo* differentiation of cartilage tissue is poorly understood, but TGF-β is believed to be implicated in chondrogenesis as well as in cartilage tissue repair and turnover. Studies performed *in vitro* showed that chick embryo sternal chondrocytes undergo a process of dedifferentiation in the presence of TGF-β.[380] TGF-β modulated cells lose the typical round shape of differentiated chondrocytes in monolayer culture and produce fewer matrix macromolecules which include collagen I, rather than cartilage collagens II, IX, X, and XI. These effects are reversible and are abolished upon addition to the conditioned medium of a monoclonal antibody against recombinant human TGF-β_2.

TGF-β enhances PTH stimulation of adenylyl cyclase in clonal rat osteosarcoma cells.[381] TGFs and other growth factors, including EGF and PDGF, stimulate bone resorption in neonatal mouse calvariae in organ culture via a prostaglandin-mediated mechanism.[382] The effect depends on the stimulation of arachidonic acid synthesis and is inhibited by indomethacin. Both TGF-β_1 and TGF-β_2 stimulate prostaglandin production, which increases bone resorption in cultures of neonatal mouse calvariae. In comparison, both TGF-β_1 and TGF-β_2 do not stimulate, but rather inhibit, bone resorption in fetal rat long bone cultures, which is due to inhibition of osteoclast precursor proliferation.[383]

In addition to TGFs, fetal rat calvarial cells in culture release a factor, bone-derived growth factor (BDGF), which stimulates bone fibroblast growth but has no transforming activity for NRK cells.[384,385] Calvariae incubated with calcium-mobilizing osteotropic hormones such as parathormone, calcitriol, and

IL-1 have a concentration-dependent increase in TGF-β activity.[386] In contrast, calcitonin-induced inhibition of bone resorption correlates with a decrease in bone TGF-β. Alkaline phosphatase activity increases in osteoblastic rat osteosarcoma cells incubated in the presence of TGF-β.[387] This stimulation is blocked by actinomycin D and cycloheximide, suggesting that it is associated with *de novo* RNA and protein synthesis. Differentially enhanced expression of TGF-β and c-*fos* mRNAs occurs in the growth plates of developing human long bones.[388] It may be concluded that bone represents an important target tissue for TGF-β and that the activity of TGF-β in bone tissue is regulated by hormones, growth factors, and oncoproteins. TGFs and other growth factors such as EGF and PDGF, produced by tumor cells, may contribute to stimulate bone resorption and to produce hypercalcemia, which is frequently observed in patients with different types of cancer.[389-391] Parathyroid hormone (PTH) is a hormone with strong hypercalcemic properties, and a tumor-secreted peptide corresponding to the amino-terminal region of the PTH molecule may be involved in hypercalcemia that occurs in some cancer patients.[392] Growth factors such as EGF and TGF-α may inhibit the response of osteoblasts to PTH, which may also help to explain the decreased bone formation observed in patients with solid tumors secreting excess growth factors.[393] In general, PTH tends to oppose the effects of TGF-β on the processes of osteoblast proliferation and differentiation.[394]

6. Uterus and Mammary Gland

TGF-β may have an important role, in concert with other hormones and growth factors, in the process of embryonic implantation and early development in the uterus.[395] Implantation involves complex interaction between embryonic and uterine cells. In the mouse, TGF-β may regulate some of the interactions that occur between the epithelium and surrounding uterine stroma during the implantation period and between the primary and secondary decidual zones during the postimplantation period. Available evidence suggests that TGF-β may participate in various events of blastocyst implantation, decidualization, placentation, and embryogenesis.

Locally produced TGF-β may have an important role in the function of the mammary gland by limiting the accumulation of milk proteins during pregnancy. All types of TGF-β are able to regulate the synthesis and secretion of caseins by mouse mammary gland explants cultured in the presence of lactation-inducing hormones.[396] The caseins are a family of acid-precipitable milk phosphoproteins whose biological function is to provide supersaturating concentrations of calcium, phosphates, and essential amino acids to the neonate.

H. EFFECTS OF TGF-β ON CELL PROLIFERATION AND DIFFERENTIATION

TGF-β is an important regulator of cellular proliferation and differentiation.[269,397] The action of TGF-β on cell growth is bifunctional, i.e., TGF-β may act as either a positive or a negative regulator of cell division.[398] In general, TGF-β stimulates the proliferation of mesenchyme-derived cells such as NRK cells and AKR-2B cells, but inhibits the proliferation of epithelial and endothelial cells. Conditions such as the structure of the extracellular matrix may alter the inhibitory effect of TGF-β on the proliferation of normal epithelial cells.[399] Whether cells are stimulated or inhibited to proliferate by TGF-β may also depend on their state of differentiation. TGF-β is, for example, a potent inhibitor of early stages of myelopoiesis, but the proliferation of more mature cells from the myeloid lineage, already irreversibly committed, can be enhanced.[400] Clonal subpopulations of the NRK-49F cell line, which is of normal rat kidney fibroblast origin, can exhibit opposite responses to TGF-β, indicating that NRK-49F cells are a mixture of subpopulations differing in their TGF-β response.[401]

Stimulation of cell proliferation by TGF-β may be indirect and may involve the operation of autocrine loops. TGF-β stimulates DNA synthesis in human skin fibroblasts after a prolonged lag period, as compared to other growth factors, and the mechanism of this induction appears to be dependent on the synthesis and secretion of PDGF-related proteins.[402] Antibodies specific to PDGF block TGF-induced DNA synthesis of human fibroblasts. TGF-β induces proliferation of human connective tissue cells (cultured smooth muscle cells) at low concentrations by stimulating autocrine PDGF-AA secretion, which at higher concentrations of TGF-β is decreased by downregulation of PDGF receptor α subunits and perhaps also by direct growth inhibition.[403] These results suggest that TGF-β may act on some types of cells as a mitogen through an autocrine mechanism involving PDGF-like proteins.

The effects of TGF-β on the control of cell proliferation are predominantly negative. In particular, TGF-β is a potent inhibitor of the proliferation of normal epithelial cells. In the rat intestinal epithelium, expression of TGF-β is mainly localized to the villus tip, i.e., the region of the crypt villus unit that is characterized by the cessation of cell proliferation and the expression of differentiated functions related to food digestion and absorption.[404] In contrast, TGF-β expression is very low in the proliferative zone (crypt) of the epithelium of the small intestine where differentiated function is not found. TGF-β may have a role in the maintenance of a highly differentiated and low proliferative state in the adult liver. A chalone with potent growth inhibitory properties for hepatocellular proliferation was shown to be represented by TGF-β when purified to homogeneity.[405] TGF-β can also have an important role in the negative control of hematopoietic cell proliferation.[406] The suppressive effects of TGF-β on primitive hematopoietic progenitors are abrogated by HGFs such as IL-6 and CSF-3.[407] TGF-β functions as a potent inhibitor of vascular endothelial cell proliferation, and this effect is correlated with the number of EGF receptors expressed on the cell surface.[408] The proliferation of not only normal cells but also of many different types of tumor cells is inhibited, either directly or indirectly, by TGF-β. Proliferation of the prolactin-secreting GH_4 rat pituitary tumor cell line is inhibited by TGF-α and TGF-β.[409] Anchorage-independent growth of both estrogen receptor-negative and -positive human breast cancer cell lines is equipotently inhibited by TGF-β_1 and TGF-β_2.[410] However, some tumor cell lines are resistant to the growth-inhibitory action of TGF-β, even those that possess TGF-β receptors.

The effects of hormones and growth factors on cell proliferation may be mediated by TGF-β. Estrogen inhibits the growth of UMR106 cells, a clonal cell line derived from rat osteosarcoma, and this effect is probably mediated by an autocrine or paracrine mechanism involving the endogenous production and secretion of TGF-β.[411] Monoclonal antibodies to TGF-β block the growth inhibitory effects of estrogen on UMR106 cells. Interaction between TGF-β, acting as a growth inhibitor, and other growth factors, which may stimulate tumor cell growth, may be most important for the regulation of tumor growth. Proliferation of KB human epidermoid carcinoma cells is not affected by TGF-β, but TGF-β augments the inhibition of KB cell growth induced by EGF.[412] On the other hand, TGF-β alone has striking effects on the morphology of KB cells, inducing actin stress fibers and producing flattening of these cells.

TGF-β is secreted in a latent form and must be activated before it binds to cell surface receptors. Whereas the heat-activated latent form of TGF-β was found to inhibit the growth of several human leukemia cell lines (HL-60, U-937, and KG-1), the untreated latent form of TGF-β had only a marginal effect on these cells.[413] In contrast, the latent form of TGF-β was found to be as potent as the heat-activated factor for inhibiting the growth of a human erythroleukemia cell line (HEL cells), suggesting that HEL cells may be capable of activating the latent form of TGF-β. Since HEL cells produce TGF-β themselves, TGF-β may act as an autocrine negative growth factor on these cells. Inhibition of carcinoma and melanoma cell growth by TGF-β *in vitro* is highly dependent on the presence of polyunsaturated fatty acids.[414] The precise nature of the synergistic interaction between TGF-β and polyunsaturated fatty acids is unknown.

Viral infection of cells may alter their response to TGF-β. While the proliferation of normal human B lymphocytes is inhibited by TGF-β, the growth of EBV-infected normal B cells as well as EBV-positive Burkitt's lymphoma cell lines is slightly stimulated by TGF-β.[415]

1. TGF-β and Cell Proliferation

Although TGF-β was first considered as a growth stimulator, in many systems it acts as an inhibitor of cell growth. The growth inhibitory properties of TGF-β have been documented mainly *in vitro,* and the growth of both normal and malignant cells can be inhibited by the addition of TGF-β to the culture medium. The sensitivity of cells to the growth inhibitory properties of TGF-β depends on the type of cell and the culture conditions.[416] In certain conditions the growth of oncogene-transformed cells can be inhibited by TGF-β. For example, the proliferation of K-MuSV-transformed adherent rat cells NRK-49F is slowed down by porcine TGF-β, and this effect is reversed by EGF.[417] The TGF-β-induced inhibition of cell growth is not associated with a cytotoxic action of TGF-β but is due to a lengthening of the cell population doubling time. However, the inhibitory effects of TGF-β in some types of cultured cells, e.g., normal rat epithelial cells, are irreversible.[398] In certain types of cells, TGF-induced inhibition of cell proliferation is associated with the appearance of some characteristics of differentiation, as observed in an intestinal epithelial cell line.[418] However, there may be a discrepancy between TGF-β effects on growth inhibition and induction of differentiation. In human keratinocytes, for example, TGF-β appears to induce a direct growth inhibitory effect not coupled to induction of differentiation.[419]

The growth inhibitory properties of TGF-β have also been observed *in vivo*. Powerful inhibition of mammary growth and morphogenesis is observed when slow-release plastic pellets containing TGF-β are implanted in developing mouse mammary glands.[420] The inhibitory effect of TGF-β on the mouse mammary gland is fully reversible and is not associated with pathologic changes of the developing epithelial tissue, suggesting that TGF-β may be an important physiologic growth inhibitor. The potent growth inhibitory activity TGF-β on hepatocytes *in vitro* suggests a potential role for maintaining these cells in a quiescent state *in vivo*. TGF-β produced by nonparenchymal cells in the liver may act as an important paracrine regulatory mechanism that is activated during liver regeneration, probably being capable of preventing uncontrolled hepatocyte growth during this process.[421] Norepinephrine, acting through the α_1-adrenergic receptor, may block the inhibitory effect of TGF-β on hepatocyte proliferation and may thus allow these cells to escape from the negative control exerted by TGF-β during liver regeneration.[422]

Hematopoietic cell proliferation may be inhibited by TGF-β both *in vitro* and *in vivo*. Proliferation of clonogenic AML blast precursors are reversibly inhibited by TGF-β without induction of differentiation.[406] Administration of a single dose of TGF-β inhibited the baseline and IL-3-driven proliferation of mouse marrow cells.[423] This inhibition was relatively selective for the earlier multipotential granulocyte, erythroid, megakaryocyte, and macrophage CFU progenitors since they are completely inhibited while the more differentiated CFU assayed in culture were inhibited by about 50%. Both TGF-β_1 and TGF-β_2 are potent inhibitors of the growth of early human hematopoietic progenitor cells, either normal or leukemic.[330] TGF-β_1 and TGF-β_2 are also equally potent inhibitors of IL-1-mediated thymocyte proliferation in spite of the fact that they have different tertiary structures, since antibodies raised against TGF-β_1 only neutralize the biological action of TGF-β_1 but do not affect the biological action of TGF-β_2.[424] TGF-β inhibits CSF-1-dependent macrophage precursor proliferation *in vitro* in a dose-dependent manner, suggesting that it may play a role in the negative regulation of macrophage production *in vivo*.[425]

The mechanisms involved in the inhibitory action of TGF-β on cellular proliferation are little understood. In T lymphocytes TGF-β blocks cell proliferation but not the early signaling events.[426] The antiproliferative effect of TGF-β may occur at a cellular level distal to the receptors for growth-activating factors.[427] TGF-β inhibits, in a reversible manner, EGF-induced stimulation of DNA synthesis in the murine keratinocyte (MK) cell line, and this inhibition is not overcome by insulin.[428] The modulatory action of TGF-β on EGF-induced stimulation of DNA synthesis in MK cells appears to take place at a point distal to EGF binding. EGF-stimulated proliferation of cultured rat aortic smooth muscle cells is also inhibited by TGF-β.[429] An increase in collagen secretion may play an important role in the inhibition of NRK cell growth in serum-free monolayer culture by TGF-β.[430] Human TGF-β inhibits, in a dose-dependent manner, the TNF-α-induced proliferation of diploid human fibroblasts (WI-38 cells) through a mechanism involving *de novo* protein synthesis.[431] Inhibition of the proliferation of mink lung epithelial cells by TGF-β occurs when TGF-β is added throughout the prereplicative G_1 phase of the cycle and involves the regulation of cdc2 kinase activity at the G_1/S transition.[432] Associated with these TGF-β-induced growth arrests, there is a decrease in the phosphorylation and kinase activity of histone H1. The negative regulation of G_1/S transition by TGF-β in mammalian cells may involve inhibition of cyclin E-dependent kinase.[433]

Both TGF-β_1 and TGF-β_2 can display growth inhibitory actions on several cellular systems, but in certain types of cells only one of these two types of TGFs exhibits potent growth inhibitory effects. TGF-β_2 is much less potent than TGF-β_1 in inhibiting DNA synthesis in cultured bovine aortic endothelial cells.[434] TGF-β_1, but not TGF-β_2, is a potent inhibitor of murine hematopoietic progenitor cell proliferation *in vitro*.[435] However, there may be species-specific differences in the actions of TGF-β_1 and TGF-β_2: both forms of TGF-β are equally potent as inhibitors of early human hematopoietic progenitors.[330] An inhibitor of megakaryocyte colony formation, stored in the α granules of human platelets, is identical with TGF-β.[436] TGF-β_1 can inhibit the growth of differentiated-arrested, growth factor-dependent, and growth factor-independent leukemic myelomonocytic cells, but is not capable of inhibiting the growth of leukemias blocked later in their lineage.[432]

Tumor cells may produce TGF-β and the factor may exert an autocrine effect with stimulatory or inhibitory effects on tumor cell growth. Estrogen receptor-negative, estrogen-independent human breast cancer cells have receptors for, are inhibited by, and secrete TGF-β activity, suggesting that this polypeptide is an autocrine inhibitor for human breast cancer.[437] In certain instances, when the tumor cells producing TGF-β are themselves susceptible to growth inhibition by TGF-β, TGF-α may also be produced and override the effect of TGF-β, resulting in a positive overall autocrine effect.[37] The growth of nontransformed murine keratinocytes (BALB/MK cells) may depend on an autocrine mechanism involving the opposing effects of endogenously produced TGF-α which acts as a growth stimulator, and TGF-β which acts as a growth inhibitor.[438] Moreover, TGF-α expression in BALB/MK cells may be modulated by EGF or TGF-α itself. It is thus clear that very complex interactions between the different types of TGFs, as well as between TGFs and other growth factors, may play a central role in regulating the growth of tumor cell populations.

Chemically induced inhibition of cell growth and induction of differentiation in transformed cells may be associated with the production of TGF-β. Treatment of the transformed mouse embryo fibroblast cell line AKR-MCA with 1% *N,N*-dimethylformamide (DMF) results in the restoration of a nontransformed phenotype and inhibition of cell growth. The serum-free conditioned medium of these cells contains a factor capable of inhibiting the anchorage-independent growth of the human colon carcinoma cell line MOSER.[439] Fractionation of the crude conditioned medium indicated the presence of a 20-kDa inhibitory protein fraction which was represented by TGF-β. Although human colon carcinoma cells do not usually respond to the inhibitory effects of TGF-β, the MOSER line, which shows a relatively high degree of sensitivity to other growth factors, is responsive to TGF-β.[440] TGF-β inhibits the proliferation of MOSER cells both in monolayer culture and soft agarose, and the cells exposed to TGF-β show differentiation-like morphological changes. There is evidence that MOSER cells specifically secrete TGF-β_2 and that TGF-β_2 acts as an autocrine inhibitory factor for these cells.[441] However, spontaneously arising subclones of MOSER cells may be specifically resistant to the growth inhibitory action of TGF-β.[442] The mechanisms involved in this resistance are probably heterogeneous at the cellular level and may be associated with reduced binding affinity of the TGF-β receptor or with other alterations. Interestingly, some TGF-β resistant MOSER cell subclones express the c-*myc* proto-oncogene at higher levels than do the parental cells in the absence of TGF-β. The identification of spontaneously resistant clones in a TGF-β-sensitive parental cell line such as MOSER cells suggests that human tumors consist of a heterogenous population of cells that may exhibit a mosaic of responses to TGF-β and other endogenous growth-stimulating and growth-inhibiting factors. In general, well-differentiated colon carcinoma cell lines respond better to the growth-inhibitory effects of TGF-β than do the poorly differentiated cell lines.[443] Expression of the c-*myc* and TGF-α genes is rapidly altered by treatment of the well-differentiated colon carcinoma cells with TGF-β in association with reduction in the mitogenic responses to both nutrients alone and nutrients plus exogenous stimulatory factors (EGF, insulin, and transferrin). In contrast, TGF-β has no effect on c-*myc* and TGF-α mRNA expression in poorly differentiated colon carcinoma cells.

The presence of receptors for TGF-β in tumor cell lines does not necessarily indicate that they will respond to the growth inhibitory effect of TGF-β. The human hepatoma cell lines PLC/PRF/5 and Mahlavu both possess receptors for TGF-β, but only the PLC/PRF/5 line is growth-inhibited by TGF-β, indicating that the Mahlavu line may have some abnormality in the postreceptor signal transduction of TGF-β. However, in the presence of phorbol ester (TPA) TGF-β inhibits the growth of Mahlavu cells in a dose-dependent manner.[444] Interestingly, the effect of TPA on inducing cell growth inhibition by TGF-β is not cancelled by the addition of protein kinase C inhibitors such as H7 and staurosporin, suggesting that the effect is not mediated by protein kinase C. Several lines of evidence indicate that some biological actions of phorbol esters are mediated by mechanisms not involving the activation of protein kinase C.

The mechanism of resistance to the growth-inhibitory effects of TGF-β have been studied in a murine keratinocyte cell line (KCR cells derived from v-K-*ras*-transformed BALB/MK keratinocytes) which is resistant to TGF-β.[445] A cDNA probe that corresponded to an mRNA that was downregulated in TGF-β-sensitive cells, but remained unregulated in the resistant KCR cells, was demonstrated to correspond to murine disulfide isomerase, a multifunctional protein involved in posttranslational modifications of secreted proteins. The exact role of disulfide isomerase in resistance to TGF-β is unknown but overexpression of this enzyme could result in aberrant posttranslational modification of proteins involved in the TGF-β signal transduction pathway.

TGF-β inhibits the anchorage-independent growth of many human tumor cell lines at concentrations in the same range as those that enhance the anchorage-independent growth of rodent cell lines.[231] TGF-β inhibits the clonal growth of normal human epithelial cells in a defined medium in a dose-dependent fashion.[446] The same factor inhibits the proliferation and regeneration of endothelial cells.[447,448] The growth of A431 human epidermoid carcinoma cells in monolayer culture is stimulated by TGF-β, but the stimulatory effect of EGF on the growth of the same cells in soft agar is antagonized by TGF-β.[449] TGF-β causes a reversible, cell cycle-dependent growth arrest of normal human prokeratinocytes when added to cultures growing logarithmically. In contrast, the clonal growth of a human squamous cell carcinoma (SCC-25) cultured under similar conditions is not affected by the addition of TGF-β, which may be attributed to the absence of TGF-β receptors in the noninhibited carcinoma cells.[449] TGF-β has an inhibitory effect on the soft agar growth response of NRK cells to PDGF.[450] TGF-β also inhibits the growth of AKR-MCA cells in monolayer culture.

Differential effects of TGF-β on cellular proliferation are observed in normal and malignant rat liver epithelial cells in culture.[398] The growth of early-passage propagable adult rat liver epithelial cells is inhibited by TGF-β in a reversible manner, and this inhibition occurs at two specific points of the cell cycle: the G_1/S border and the G_0 or early G_1 phase.[451] TGF-β acts on rat hepatocytes in primary culture as an inhibitor of DNA synthesis induced by either insulin plus EGF or HGF.[452] Rat platelets contain two types of growth inhibitor for adult rat hepatocytes, and one of them is identical with TGF-β. Platelets may contain both a growth-stimulating factor for hepatocytes (HGF) and one or more growth-inhibitory factors, one of which is TGF-β.

The effect of TGF-β on the proliferation of normal and malignant cells may depend on environmental conditions, including the presence of other growth factors. Growth of cultured androgen-dependent normal rat prostate epithelial cells is androgen-independent but requires the presence of both EGF and acidic FGF; the growth of these cells is inhibited by TGF-β. In contrast, epithelial cells derived from the transplantable Dunning R3327H rat prostate tumor exhibits a density-dependent alteration in response to EGF and acidic FGF. The tumor cells responds to either EGF or acidic FGF at low density. Whereas EGF-stimulated tumor epithelial cell growth is severely inhibited by TGF-β, this inhibition is largely attenuated when the tumor cell growth is supported by acidic FGF.[453] These results suggest that prostate tumor cells may escape the inhibitory effects of TGF-β if acidic FGF is present at adequate levels in the local tumor environment.

2. TGF-β and Cell Differentiation

TGF-β is able to modulate not only the processes related to cell proliferation but also those associated with cell differentiation. The effects of TGF-β on differentiation are variable, however. Whereas in some cellular systems TGF-β acts as a suppressor of differentiation, in other systems it stimulates the expression of differentiated functions. TGF-β acts as a potent inducer of the differentiation of rat liver epithelial cells (hepatocytes) *in vivo* and *in vitro*.[454] TGF-β and basic FGF are crucially involved in the induction of mesoderm during vertebrate development.[455] In contrast, TGF-β acts as a potent inhibitor of the differentiation of B lymphocytes, blocking the transition of pre-B cells to mature, functional B cells.[456] The inhibitory effect of TGF-β on B-cell differentiation appears to be exerted at the transcriptional level. TGF-β inhibits the morphologic differentiation of cultured BALB/c 3T3 T proadipocytes and the expression in these cells of differentiation-specific genes such as lipoprotein lipase (LPL) and glycerol-3-phosphate dehydrogenase (GPD) when added prior to the onset of differentiation, but once morphologic differentiation begins, TGF-β is ineffective in blocking differentiation.[457]

The process of terminal differentiation of muscle cells (myogenic differentiation) is very complex and may depend on interactions between stimulatory growth factors such as FGFs and IGFs and the inhibitory action of TGF-β.[458] Myogenic differentiation involves withdrawal of proliferating myoblasts out of the cell cycle, their subsequent fusion to form multinucleated myotubes, and the coordinate induction of a number of muscle-specific gene products, including creatine kinase and acetylcholine receptors. TGF-β can inhibit myoblast fusion and can prevent expression of the muscle-specific gene products.[459] TGF-β is a very potent inhibitor of differentiation of Yaffe L6 strain of rat myoblasts and is identical to an inhibitor secreted by Buffalo rat liver cells.[460] Inhibition of terminal differentiation of mammalian and avian myoblasts by TGF-β is not restricted to established myoblast cell lines but is also seen in primary cultures of embryo myoblasts.[461] The differentiation inhibitory effects of TGF-β are not associated with cell proliferation and are exerted at the level of muscle-specific mRNA accumulation. TGF-β causes transient induction of c-*fos* and c-*myc* mRNA in BC_3H1 cells.[462] However, persistent c-*myc* gene expression is not required for growth factor-mediated inhibition of myogenic differentiation. The products of mutated, oncogenic forms of c-*ras* genes, but not the products of normal c-*ras* genes, interfere with the differentiation of myogenic cell lines.[463] The effects of activated proto-oncogenes on cellular differentiation may be mediated by the production of TGF-β and/or other similar factors. In certain mitogen-rich environments, TGF-β may act not as an inhibitor but as an inducer of myogenic differentiation, with permanent withdrawal of myoblasts from the cell cycle.[464]

The differentiation-inhibiting effects of TGF-β on muscle tissue are not limited to myoblasts but may also include skeletal muscle satellite cells, a population of myogenic stem cells that are located adjacent to muscle fibers and that can be activated during injury, which may lead to the formation of new multinucleated fibers.[465] Proliferation of skeletal muscle satellite cells is stimulated by IGF-I, IGF-II, and FGF, and this stimulation may be counterbalanced by the inhibitory action of TGF-β. An interplay between stimulatory and inhibitory factors would have a crucial role in the regulation of muscle tissue regeneration.

Unscheduled proliferation of vascular smooth muscle cells is an important component of the atherogenesis. The smooth muscle cells grow *in vitro* with a characteristic pattern of "hill-and-valley" morphology and this pattern is altered by TGF-β in a manner which depends on the plating cell density.[466] Moreover, TGF-β may alter the synthesis of secreted proteins in vascular smooth muscle cells. TGF-β appears to have an important role in directing the organizational, proliferative, and biosynthetic behavior of vascular smooth muscle cells during development, during angiogenesis, and after injury.

TGF-β inhibits the proliferation and induces the differentiation of cultured normal bronchial or tracheal epithelial cells of either human or rabbit origin by a process which is apparently independent of the intracellular concentration of cAMP.[467,468] In contrast, TGF-β does not inhibit DNA synthesis of human lung carcinoma cells even though the cells possess comparable numbers of TGF-β receptors with similar affinities for the factor. Some bronchial carcinoma cell lines may require 100 times more TGF-β to inhibit cell proliferation than normal cells do, although the growth of one carcinoma cell line was not inhibited but was stimulated by TGF-β.

TGF-β, which is abundant in bone and promotes the differentiation of muscle-derived mesenchymal cells into cartilage, inhibits the proliferation and suppresses the differentiation of MC3TE1 cells, a clonal nontransformed murine bone cell line which differentiates in culture.[469] On the other hand, in the clonal osteoblastic sarcoma cell line ROS 17/2.8, TGF-β induces alkaline phosphatase activity and increases the rate of collagen synthesis per cell.[470] TGF-β may be an important regulator of local bone remodeling.

The differentiation-inducing effects of TGF-β on cultured cells may not be necessarily associated with inhibition of cellular proliferation. TGF-β induces hemoglobin synthesis in the human erythroleukemia (HEL) cell line without causing a negative effect of cellular proliferation.[471] Binding of TGF-β to its receptor on differentiating mouse 3T3-L1 preadipocytes results in inhibition of adipogenic differentiation without changing the time-dependent program of DNA synthesizing activity.[472]

TGF-β, alone or in combination with other growth factors, is capable of inducing the differentiation of some tumor cell lines. TGF-β induces, in a dose-dependent manner, the differentiation of U-937 human monocytic leukemia cells into cells with macrophage-like characteristics.[473] When combined with TNF-α, TGF-β induces various markers of U-937 cell maturation, except phagocytic activity. Other human myelogenous leukemic cell lines responded in different ways to the differentiation-inducing effects of TGF-β, but the reason for these differences is not understood. The physiological significance of the results obtained with TGF-β-induced differentiation of tumor cell lines is unclear. Induction of differentiation in cultured neoplastic cells by certain agents may be associated with production of TGF-β. Differentiation of human embryonal carcinoma cells induced by treatment with retinoic acid is accompanied by a marked increase in the levels of TGF-β mRNA as well as by changes in expression in TGF-β receptors, suggesting the operation of autocrine mechanisms involving TGF-β.[474,475] TPA-induced megakaryocyte differentiation of the CML cell line K562 is associated with a severalfold increase in the synthesis of TGF-β mRNA and protein.[476] TGF-β can be detected in the medium of TPA-differentiated K562 cells. On the other hand, the action of some differentiation inducers may be counteracted by TGF-β. Differentiation of mouse myeloid leukemia M1 cells induced by dexamethasone can be inhibited by TGF-β in a dose-dependent fashion.[477]

3. TGF-β-Induced Apoptosis

Programmed cell death (apoptosis) is significantly involved in the growth of both normal and tumor tissues, and TGF-β may have a role in the mechanism of apoptotic processes.[478-480] Apoptosis observed in hepatocyte cultures may be associated with the action of TGF-β.[481,482] Apoptotic hepatocytes show immunostaining for epitopes TGF-β, and addition of TGF-β to hepatocyte cultures can induce apoptosis. Expression of TGF-β in the rat ventral prostate occurs during castration-induced apoptosis.[483,484] TGF-β induces apoptosis in culture uterine endometrial cells.[485] Apoptosis induced by TGF-β in cultured human gastric carcinoma cells requires protein synthesis and is associated with Ca^{2+}- and Mg^{2+}-dependent activation of endonuclease present within the cell nucleus, followed by DNA fragmentation.[486]

I. MODULATION OF ENDOCRINE GLANDS AND HORMONE ACTIONS BY TGF-β

TGF-β may exert important regulatory effects on the action of hormones in their respective target cells. These effects may be either stimulatory or inhibitory, depending on the type of cell and the predominant physiological conditions.

1. Gonadal Function

TGF-β is involved in the control of gonadal functions at both the pituitary and the local level. TGF-β is a potent activator of FSH secretion by pituitary cells in culture.[487] TGF-β-like activity has been identified in conditioned medium obtained from immature porcine Sertoli cell-enriched cultures, and the secretion of this activity into the medium is decreased to low levels by FSH.[488] The production of lactate in cultured porcine Sertoli cells is regulated by TGF-β,[489] which regulates calcium metabolism in these cells.[490] TGF-β acts on Leydig cells in the testis as a potent inhibitor of gonadotropin and cAMP-induced steroidogenesis.[491]

The TGFs may have an important role in ovarian physiology. Both TGF-α and TGF-β are differentially involved in the regulation of growth and steroidogenesis during antral follicle development.[492] Ovarian theca cells produce TGF-β, and this factor acts as a potent growth inhibitor for granulosa cells in the rat.[493] In addition, TGF-β may exert a negative control on ovarian androgen synthesis.[494] Normal ovarian epithelium is growth-inhibited by TGF-β, which may function as an autocrine growth factor for normal ovarian epithelial cells.[495] Interestingly, there are striking homologies in the amino acid sequences of TGF-β and the ovarian follicular fluid protein inhibin, an inhibitor of pituitary FSH secretion.[496] However, the functional significance of this homology, if any, is not clear. TGF-β is a potent *in vitro* stimulator of oocyte maturation in the rat,[497] but actions of inhibin on oocyte maturation have not been demonstrated. Moreover, in contrast to the stimulatory actions of TGF-β in cultured rat granulosa cells, this growth factor is an inhibitor of growth and differentiated function in cultured porcine granulosa cells.[498] The action of FSH on the maturation of ovarian granulosa cells is modulated by several growth factors, including EGF, PDGF, insulin, IGF-I, and TGF-β.[499] This maturation is reflected in cAMP production, steroidogenesis, and LH receptor formation, and TGF-β has a dual effect on this process. TGF-β enhances the stimulatory actions of low levels of FSH but selectively inhibits the development of granulosa cells and the induction of luteinizing hormone (LH) receptors by higher levels of FSH.[500] TGF-β amplifies the induction of aromatase activity by FSH in granulosa cells.[487] The bifunctional modulating effects of TGF-β on cAMP formation and LH receptor expression in granulosa cells could influence the ability of these cells to respond to LH during ovarian follicular maturation and ovulation. However, TGF-β may enhance FSH-stimulated LH receptor induction and steroidogenesis in rat granulosa cells by mechanisms that do not further increase cellular cAMP accumulation.[501] TGF-β enhances the induction of EGF receptors by FSH and regulates the inhibitory action of EGF during granulosa cell differentiation.[502] Consequently, granulosa cells are less able to produce cAMP and cAMP-induced LH receptors in the presence of EGF and TGF-β. The role of TGF-β and other growth factors in ovarian granulosa cell differentiation induced by FSH *in vivo* is unknown.

2. Adrenal Function

Adrenocortical cells represent a highly differentiated cell system in which a specific enzymatic cascade leads to the synthesis of major corticosteroid hormones such as cortisol, and this pathway is acutely activated by adrenocorticotropin (ACTH) and angiotensin II. TGF-β acts as a potent modulator of this system *in vitro,* inducing a decrease in basal as well as ACTH- and angiotensin II-stimulated steroidogenic activity.[503] A major target of TGF-β in adrenocortical cells may be the 17 α-hydroxylase activity, one of the key enzymes in the adrenocortical steroidogenic pathway. TGF-β suppresses basal as well as ACTH-stimulated steroid hormone formation by bovine adrenocortical cells in culture.[504] This effect is dose-dependent, is not accompanied by any change in adrenocortical cell growth, and occurs distal to the formation of cAMP but proximal to the formation of cholesterol. The physiological significance of this effect is not understood but there is evidence of the existence of corticostatic factors that act locally to modulate adrenocortical functions and TGF-β may represent one of these factors.

3. Thyroid Function

The action of TGF-β on thyroid cells has been examined using the FRTL-5 rat thyroid follicular cell line.[505] At sufficient doses of TGF-β the TSH-stimulated growth of FRTL-5 cells is reduced, and this reduction is accompanied by a decrease in the number of TSH receptors and in the activity of adenylyl cyclase. In contrast to these effects on thyroid cell proliferation, TGF-β increases both thyroid-stimulating hormone (TSH)-dependent and -independent iodine uptake by FRTL-5 cells at concentrations equivalent to those required for growth inhibition. TGF-β may represent an autocrine mechanism that controls the growth response to TSH in thyroid follicular cells, while allowing the continuance of differentiated function.

4. Insulin Secretion

Studies with isolated rat pancreatic islets indicate that TGF-β may act as a potent stimulator of insulin secretion.[506] TGF-β_1 and TGF-β_2 are equally potent in the stimulation of pancreatic cells. Activin A, another member of the family of polypeptides structurally related to TGF-β, has a similar effect. TGF-β may act as a local regulator of insulin secretion.

J. INTERACTIONS BETWEEN TGF-β AND OTHER GROWTH FACTORS

Complex interactions with other types of growth is crucial for the expression of the different biological actions of TGF-β. In certain systems, for example, in the murine fibroblast cell line AKR-2B, TGF-β acts as a competence factor, inducing a state of enhanced sensitivity to other growth factors (insulin, IGFs, or EGF).[507] Moreover, the action of TGF-β in G_0-arrested AKR-2B cells is associated with production of mitogenic factors such as PDGF, which may act in an autocrine fashion.[291] In normal human mammary epithelial cells, TGF-β induces PDGF mRNA and secretion of PDGF while inhibiting the growth of the cells through a paracrine mechanism which may involve an interaction between the mammary epithelial cells and fibroblasts.[508] In the nontumorigenic epithelial cell line BALB/MK, TGF-β abrogates the mitogenic effects of EGF.[298] In the tumorigenic NRK-PT14 cell line, derived from rat kidney, TGF-β only stimulates growth when EGF is present in the medium and when EGF is added before TGF-β.[509] Autocrine stimulation by TGF-β is required for EGF-induced anchorage-independent growth of NRK-PT14 cells.

The transforming properties of TGF-β may partially depend on interactions with other growth factors. Cooperation between TGF-β and EGF or TGF-α is required for the induction of anchorage-independent growth in nontransformed cells such as the NRK-49F cell line. A transformed derivative of NRK cells, the cell line NRK-PT14, requires EGF for anchorage-independent growth but has lost the additional requirement for TGF-β.[510] Stimulation of rodent fibroblasts by IGFs is required for transformation induced by TGF-β *in vitro*.[511] In general, full mitogenic stimulation of rodent fibroblasts by an appropriated set of growth factors is a requisite for the induction of a transformed phenotype by TGF-β. The sensitivity of the MCF-7 human breast cancer cell line to TGF-β is correlated with the cellular response to estradiol: estrogen receptor-positive MCF-7 cells are killed by TGF-β, whereas variants of these cells containing defective estrogen receptors are highly resistant to TGF-β.[512] The mechanism of TGF-β resistance in the variant cells is unknown but the repertoire of secreted proteins is markedly altered in the MCF-7 cell variants.

1. Modulation of Heterologous Growth Factor Receptors

Some of the multiple biological effects of TGF-β may be mediated by modulation of the receptors of other growth factors. TGF-β modulates the expression of high-affinity receptors for both EGF and TGF-α on the cell surface.[127,513] In the MDA-468 human breast carcinoma cell line, TGF-β exerts a growth inhibitory effect due to its ability to regulate EGF receptor expression.[514] Increased expression of EGF receptors in MDA-468 cells may lead to growth inhibition. An early effect of TGF-β in NRK cells is the induction of a rapid increase in the number of membrane receptors for EGF, and some delayed effects of TGF-β may be indirect consequences of its ability to regulate EGF receptors and thereby amplify EGF-induced cellular responses.[513] However, in another study using the same cells it was found that exposure to biologically active concentrations of TGF-β induced not an increase, but a rapid decrease in the binding of EGF and TGF-α.[515] It is possible that TGF-β has a dual action on EGF and TGF-α receptors according to the physiological state of the cell, with an inhibitory effect predominating in situations in which cells evolve from a growth-arrested state to a mitogenically active one. In NRK cells, treatment with TGF-β results in biphasic effects on EGF binding. Initially, EGF binding is inhibited by TGF-β as a result of transient decrease in EGF receptor affinity, but thereafter the binding is stimulated as a result of a persistent effect in EGF receptor number.[516]

2. Induction of DNA Synthesis

TGF-β can act either synergistically or antagonistically with other growth factors, including EGF and PDGF, in the regulation of DNA synthesis and cell proliferation. TGF-β is a potent stimulator of DNA synthesis in AKR-2B mouse embryo cells with a prolonged lag phase relative to the stimulation obtained with EGF, PDGF, or FGF.[269] In AKR-2B cells, TGF-β inhibits the early peak of DNA synthesis produced by EGF and insulin. TGF-β is a potent inhibitor of EGF-induced DNA synthesis in both rat hepatocytes and AKR-2B mouse fibroblasts.[507,517] TGF-β and EGF may have opposite and selective effects on the production of secreted proteins by murine 3T3 cells and human fibroblasts through a mechanism that does not affect the overall rate of protein synthesis.[518] The opposite effects of TGF-α, as a stimulator, and TGF-β, as an inhibitor of DNA synthesis, may have implications for the normal development of physiologic processes such as liver regeneration.[519] TGF-β exerts a marked inhibitory effect on DNA synthesis in both normal and neoplastic human B lymphocytes stimulated to proliferate with anti-Igs and brain-derived growth factor (BCGF).[520] Treatment of lymphocytes with TGF-β leads to arrest of the cells in G_1 phase, prior to transferrin receptor expression.

3. Expression of Cell Surface Antigens

The levels of expression of cell surface antigens may be modulated by TGF-β. IFN-γ increases the expression of HLA-DR antigens in a diversity of human and murine cells, including monocytes, fibroblasts, and endothelial cells as well as melanoma cells, and these levels are reduced when the cells are simultaneously treated with TGF-β_1.[521] The possible role of the modulation of cellular antigens by TGF-β in tumorigenic processes is unknown.

4. Expression of a Transformed Phenotype

TGF-β alone is capable of inducing, in a reversible manner, the anchorage-independent growth of mesenchymal cells *in vitro*.[13] However, particular types of interaction with other growth factors or with oncoproteins may result in greater stimulation of colony formation in soft agar. For example, transfection of C3H/10T1/2 mouse fibroblasts with a c-*myc* gene results in a 20-fold or greater stimulation of colony formation in soft agar.[522] Rat 3T3 fibroblasts transfected with a c-*myc* gene can be induced to grow and form colonies in soft agar by treatment either with EGF alone or the combination of PDGF and TGF-β.[397] In this system, TGF-β can function as either an inhibitor or an enhancer of anchorage-independent growth, depending on the particular set of growth factors operating on the cell together with TGF-β. Human fibroblasts may be induced to anchorage-independent growth by a combined treatment with TGF-β and EGF.[523]

Induction of anchorage-independent growth by different sets of growth factors involves different cellular pathways which can be distinguished by their sensitivity to retinoic acid.[524] Colony formation induced by the combined action of PDGF and TGF-β is 100-fold more sensitive to inhibition by retinoic acid than is colony formation induced by treatment of c-*myc*-transfected cells with EGF. Moreover, retinoic acid is inhibitory for colony growth wherever TGF-β is present, regardless of whether the effects of TGF-β are stimulatory as occurs in the presence of PDGF, or inhibitory as found in the presence of EGF.[524] TGF-β and retinoic acid modulate phenotypic transformation of NRK cells induced by EGF and PDGF, and there are both parallels and differences in the activities of TGF-β and retinoic acid in this cellular system.[525] Specific combinations of growth factors and modulating agents may be required for the phenotypic transformation of NRK cells and other types of cells.

K. ROLE OF TGF-β IN NEOPLASTIC PROCESSES

In spite of numerous clinical and experimental studies, the role of TGF-β in neoplastic processes remains enigmatic. However, there is little doubt that TGF-β is involved, in conjunction with other growth factors, in regulating the growth of tumor cells and that different tumors respond in different ways to TGF-β. The factor acts as a growth-stimulatory agent for some types of tumor cells, whereas for other types of tumor cells it may exert a growth inhibitory effect. TGF-β functions as a potent stimulator of the anchorage-independent growth of both poorly and highly metastatic murine melanoma cell clones, but this stimulatory action may require the cooperation of other growth factors.[526] In contrast, treatment of *ras*-transformed hepatocytes with TGF-β may cause partial suppression of the transformed phenotype, inhibiting anchorage-independent growth of these cells.[527] Repression of autocrine TGF-β_1 and TGF-β_2 in quiescent CBS colon carcinoma cells leads to relaxation of the requirements to exit the quiescent state and allows enhanced expression of tumorigenic potential.[528]

The production of TGF-β by malignant cells shows large variation. Constitutive overexpression of TGF-β mRNA is seen in carcinomas of the mouse skin during experiments of multistage carcinogenesis, but not in benign premalignant lesions, suggesting that an excessive production of TGF-β may be associated with malignant progression of the lesions.[529] The responsiveness of different types of tumor cells to TGF-β also exhibits high variability. Transformed rat tracheal epithelial cells (RTE cells) in culture secrete far less TGF-β-like activity than the primary cells and many, but not all, of the transformed RTE cells are hyporesponsive or unresponsive to the growth inhibitory effects of TGF-β.[530] In spite of marked differences in TGF-β secretion between the normal and transformed cells, their levels of TGF-β mRNA expression may be similar, suggesting alterations in the posttranscriptional regulation of TGF-β expression in the transformed cells.

TGF-β may be involved in autocrine or paracrine loops that may have an important role in the regulation of tumor growth. A large proportion of human small-cell lung cancer (SCLC) cell lines produce TGF-β and express TGF-β receptors, suggesting the function of an autocrine mechanism in their growth.[531] Adult hepatocytes are one of the epithelial cell types in which no TGF-β transcripts are expressed, but they possess high-affinity receptors for TGF-β, and TGF-β secreted by nonparenchymal liver cells may have a role during chemical hepatocarcinogenesis through a paracrine type of mechanism.[532] Elevated levels of TGF-β have been found in the plasma of patients with hepatocellular carcinoma.[533] Low-grade astrocytomas, anaplastic astrocytomas, and glioblastomas ubiquitously produce *in vitro* TGF-β mRNA, which may act as a positive or negative autocrine factor for these tumors.[534] Human myelogenous leukemia cells may secrete TGF-β and may stimulate fibroblasts to produce the same growth factor, which in turn promotes leukemic cell growth, thus creating a paracrine loop that favors the growth of the tumor cells.[535] There is evidence that TGF-β may be involved in paracrine and autocrine regulation of the growth of normal and pathologic mammary gland.[536] Altered expression of TGF-β isotypes at the protein level in mammary epithelia does not appear to be a major feature of most human breast lesions, including breast cancer, raising the possibility that altered cellular response to TGF-β already present in the gland may play a role in the development of breast disease.

Progression of tumors to hormone independence may be associated with the operation of autocrine or paracrine mechanisms involving the action of TGF-β. When deprived of steroid in the long term, T-47D human breast cancer cells lose estrogen sensitivity of cell growth, and this change is accompanied by an increase in TGF-β expression and the acquisition of higher sensitivity to TGF-β.[537]

The level of TGF-β receptor expression as well as the rate of reexpression of the receptor on the cell surface may have a crucial regulatory role on the functional activity of TGF-β. There is a correlation between the greater sensitivity of oncogene-transformed murine myeloid cell lines and a more rapid downmodulation and reappearance of surface TGF-β receptors.[538] However, factors in addition to receptor expression may have an important role in determining the sensitivity of tumor cells to the growth inhibitory action of TGF-β. Human colon carcinoma cell lines may possess receptors for TGB-β_1 and TGF-β_2 and exhibit a variety of differentiation changes in response to TGF-β.[539] The growth-inhibitory effects of TGF-β on these lines are associated with the induction of synthesis of certain extracellular matrix glycoproteins, including laminin and fibronectin. Expression of the nucleolar protein B23, also called numatrin or nucleophosmin, which may represent a positive regulator of cell proliferation, is downregulated by TGF-β in the colon carcinoma cell lines. There is a correlation between the degree of differentiation of these cell lines and their sensitivity to the antiproliferative and differentiation-promoting effects of TGF-β.[540] Well-differentiated cell lines are usually sensitive to TGF-β, whereas poorly differentiated cell lines may not exhibit a significant response to TGF-β.

Resistance to TGF-β-induced inhibition of cell growth accompanies neoplastic progression of rat tracheal epithelial cells.[541] The effects of TGF-β on human oral squamous cell carcinoma cell lines indicate that, although these cells are equally sensitive to EGF, their responses to TGF-β are variable.[542] Whereas in some cell lines a dose-dependent inhibition of DNA synthesis and cell growth is observed upon treatment with TGF-β, in other cell lines the growth inhibitory effects of TGF-β are marginal. In some of these cell lines treatment with TGF-β results in a more than 300-fold increase in fibronectin secretion into the medium, but in other lines TGF-β does not induce this effect. Scatchard analysis of the binding of TGF-β to these cell lines indicates that all of them have similar binding properties, exhibiting two classes of binding sites for TGF-β. The existence of reduced levels of TGF-β receptors in malignant tumors such as gastric carcinomas may result in diminished growth inhibitory action of the factor.[543] In general, the sensitivity of different tumor cells to the growth inhibitory effects of TGF-β depends on both receptor and postreceptor signal transduction pathways, suggesting that the tumor cells may have defects at different points in the control of cell cycle-associated biochemical processes.[544]

Application of the tumor promoter phorbol ester TPA to mouse skin results in the induction of very high levels of TGF-β mRNA in the suprabasal differentiating epidermal cells.[545] Using an *in vitro* two stage BALB/c 3T3 cell transformation assay with 3-methylcholanthrene (MCA) as an initiator agent, it has been shown that TGF-β fulfills the properties of a tumor promoter.[546] TGF-β increases the number of transformed foci in MCA-initiated cells in a concentration-dependent manner, and the obtained transformed foci produce tumors when injected into nude mice. The mechanism of the tumor-promoting action of TGF-β is not understood, and the possible role of TGF-β in natural carcinogenic processes occurring *in vivo* is unknown. There is evidence that TGF-β may induce the production of growth factors with transforming capability. Mouse 3T3 fibroblasts in culture produce a leukemic cell growth-promoting factor (LGF), which is represented by a 18-kDa glycoprotein, and production of this protein is increased several times by the addition of low concentrations of TGF-β to the medium.[547] The precise nature of LGF is unknown but it would be related to interleukin 11 (IL-11).

Primary human tumors, including renal cell carcinomas and liver carcinomas, express TGF-β mRNA and protein.[548] Three types of TGF-β (TGF-β_1, TGF-β_2, and TGF-β_3) are expressed by human carcinoid tumors of the gut, which typically produce serotonin and often produce a number of biologically active neuropeptides.[549] Human carcinoid tumors are frequently associated with a desmoplastic reaction, with fibrosis, and endothelial cell proliferation, which may be due to the action of serotonin, TGFs, and other growth factors, including PDGF-B and basic FGF. Human B-cell CLL cells produce and release TGF-β.[550,551] Both normal B cells and CLL cells contain discrete storage sites of TGF-β in their cytoplasm. However, while TGF-β inhibits DNA synthesis in normal B-cells, this effect is reduced in CLL cells. Other types of human tumors produce one or more forms of TGF-β but, in general, the role of this expression in tumor growth remains enigmatic.

TGF-β may act as either a positive or negative factor for tumor cell growth. The invasive capacity of rat tumor cell lines (e.g., AH 130 ascites hepatoma cells and AT-3 prostatic carcinoma cells) can be potentiated by pretreating the cells with TGF-β.[552,553] TGF-β stimulates the growth of MC3T3-E1 osteoblasts established from mouse calvaria, suggesting that TGF-β may have an important role in the production of bone metastases in human prostatic cancer. In other tumor cell systems, TGF-β may act rather as a negative regulator of tumor cell growth and invasion. The MCF-7 human breast cancer cells secrete TGF-β, and this factor may act in these cells as an estrogen-regulated growth inhibitor with autocrine and paracrine functions.[554] Studies with transfection of an expression vector encoding an antisense RNA for TGF-β indicate that TGF-β acts as a negative growth regulator of human colon carcinoma cells *in vitro* and *in vivo*.[555] The ability of HT1000 human fibrosarcoma cells to invade a reconstituted basement membrane is inhibited by TGF-β, and this effect is mediated by an increased expression of a tissue inhibitor of metalloprotease.[556] Growth inhibition of the PC-3 androgen-refractory human prostate cancer cell line by cAMP is selectively associated with production of TGF-β_2.[557]

Tumor cells may be resistant to the growth-inhibitory action of TGF-β. For example, the KC tumor cell line, derived from K-MuSV-induced transformation of mouse keratinocytes, loses its response to the growth-inhibitory effect of TGF-β when maintained for extended periods in the presence of TGF-β.[558] The TGF-β-resistant variants, called KCR cells, have altered TGF-β receptors. Human B-cell CLL cells may express TGF-β *in vivo,* and escape of these cells from the negative regulation exerted by TGF-β may give a growth advantage to tumor cells in the hematopoietic tissues.[559]

Expression of viral genes with tumorigenic potential may be regulated by TGF-β. Induction of tumors in chickens by Rous sarcoma virus (RSV) may crucially require the local production of TGF-β.[560] Transcription of mouse mammary tumor virus (MMTV) DNA in cultured mouse mammary cells is repressed by TGF-β.[561] TGF-β is overproduced by adult T-cell leukemia (ATL) cells,[562] and the p40^x (Tax) protein of the HTLV-I virus, which is associated with ATL, augments TGF-β gene expression by transactivating the TGF-β promoter.[563] The excessive production of TGF-β by ATL cells could have a role in the pathogenesis of hypercalcemia and the immunosuppression frequently observed in ATL. The human papillomaviruses HPV-16 and HPV-18 are frequently associated with neoplasia of the uterine cervix and both the growth and virus expression of HPV-transformed cells may be responsive to the growth inhibitory action of TGF-β.[564,565] It is possible that TGF-β mediates regression of early, preneoplastic HPV-associated lesions of the human cervix, and that loss of responsiveness to TGF-β results in the progression of a number of these lesions to cervical cancer.

The growth response of different types of tumor cells to TGF-β shows great variability, which may also vary with time for a given type of cell, according to the genotypic and phenotypic evolution of the cells *in vitro* or *in vivo.* In some cases, the tumor cells, including oncogene-transformed cells, may become partially or totally resistant to the growth inhibitory action of TGF-β. Mammalian epithelial cells transfected with a *ras* oncogene may lose their growth arrest response to TGF-β as well as the regulatory action of the cell cycle-associated protein on cell division.[566] Rat liver epithelial cell lines (RLE cells) may become resistant to the growth inhibitory effects of TGF-β following transformation by the v-*raf* oncogene.[567] The more aberrant oncogene-transformed RLE cell lines are particularly resistant to the growth inhibitory action of TGF-β.

TGF-β acts as a potent growth inhibitor for many types of tumor cells, but it may act as a growth stimulator for other types of tumor cells. Production of TGF-β is markedly elevated in rat prostate cancer cells (Dunning R3327 sublines), and the growth, viability, and aggressiveness of the malignant prostate cells *in vivo* is stimulated by the endogenously produced growth factor.[568]

Activation of proto-oncogenes, or inactivation of tumor suppressor genes may cause resistance to TGF-β. The growth inhibitory effects of TGF-β on rat intestinal epithelial cells IEC-18 are attenuated by the expression of a mutated human c-H-*ras* proto-oncogene, which induces morphological transformation and anchorage-independent growth in these cells.[569] A similar loss of TGF-β-induced growth inhibition is observed in multistage carcinogenesis of thyroid follicular cells, and this effect is accompanied by inactivation of p53 tumor suppressor protein expression.[570] The p53 protein may mediate or be required for the growth inhibitory signal normally induced by TGF-β in epithelial cells.

The sensitivity of tumor cells to TGF-β may be related in some way to their metastatic potential. While TGF-β inhibits DNA synthesis in 10T1/2 cells and in a nonmetastatic tumor, the growth of 10T1/2 cells transformed by transfection with a T24 c-H-*ras* gene, which confers a high metastatic ability, is markedly stimulated by TGF-β.[571] Interestingly, EGF abrogates TGF-β-induced inhibition of the parental 10T1/2 cells, but has no effect on the TGF-β response of metastatic cells. The mechanisms responsible for the conversion of an inhibitory to a stimulatory signal by TGF-β are unknown.

IV. TGF-LIKE GROWTH FACTORS

TGF-β_1 is the prototype of a superfamily of polypeptides involved in the control of physiologic processes associated with cellular proliferation and differentiation. Members of the TGF-related superfamily are present in a wide variety of species including insects, amphibia, and birds. In mammals, the best characterized of these activities are represented by TGF-β_1, TGF-β_2, TGF-β_3, and TGF-β_5, but other related polypeptides have been identified in mammalian and nonmammalian species. The TGF-β superfamily of proteins has been classified into five main groups.[572] The superfamily includes the following agents: (1) the Müllerian inhibiting substance (MIS), which causes the regression of the Müllerian duct during male sexual differentiation; (2) the inhibins and activins, which can regulate pituitary FSH secretion; (3) the bone morphogenetic proteins (BMPs), which induce cartilage and bone formation; (4) the product of the VG-1 gene of *Xenopus,* whose expression is restricted to the vegetal pole of the egg, and (5) the product of the decapentaplegic gene of *Drosophila, dpp,* which influences pattern formation within the developing embryo.[572] The 60A gene of *Drosophila,* which is related to *dpp,* encodes a 455-amino acid protein with the potential to be secreted and processed to yield a subunit of about 120 amino acids capable of dimerization, as is typical of the other TGF-β family members.[573] The *Drosophila* 60A gene is expressed at high levels at several different times during the development of the insect.

A. THE BSC-1 GROWTH INHIBITOR

A growth inhibitor purified from African green monkey kidney cells (BSC-1 cells) is a polypeptide with TGF-β activity.[574] It competes with TGF-β for binding to the same cellular receptors. The BSC-1 cell growth inhibitor (BSC-1 factor) inhibits DNA synthesis and cell proliferation in certain types of cells in culture, including BSC-1 cells. The BSC-1 inhibitor is a protein with an apparent molecular weight of 25 kDa, is effective at physiological concentrations, and its inhibitory effects are reversible. TGF-β derived from human platelets and the inhibitor derived from BSC-1 cells are closely related molecules.

The BSC-1 inhibitor inhibits the growth of concanavalin A-stimulated mouse thymocytes and this inhibitory effect can be reversed by the addition of IL-2.[575] In certain cases, the BSC-1 factor may behave not as a growth inhibitor but as a growth stimulator. The BSC-1 factor synergizes with growth factors and hormones such as bombesin, vasopressin, and PDGF in stimulating DNA synthesis in confluent monolayers of 3T3 mouse cells.[576] The complete amino acid sequence of the BSC-1 inhibitor, deduced from the nucleotide sequence of the respective cDNA, is identical to that of TGF-β_2.[577] Thus, the BSC-1 growth inhibitor and TGF-β_2 appear to be identical molecules. However, the BSC-1 inhibitor would contain, in addition to TGF-β_2, about 10% of TGF-β_1.[578]

B. TGF-β_2

A growth inhibitory polypeptide purified from porcine blood as well as from the human prostatic adenocarcinoma cell line PC-3 exhibited significant homology to TGF-β and was named type 2 TGF-β (TGF-β_2).[203,204] This form of TGF-β is a homodimer of two 12-kDa polypeptide chains. A heterodimer containing one TGF-β_1 chain and one TGF-β_2 chain was named TGF-$\beta_{1,2}$. A 280-kDa form of the TGF-β receptor displays high affinity for both TGF-β_1 and TGF-β_2, whereas receptor forms of 65 and 85 kDa have higher affinity for TGF-β_1 than for TGF-β_2. The existence of different forms of TGF-β exhibiting differential interactions with a family of TGF-β receptors may provide a greater flexibility for the regulation of cell growth and differentiation in different tissues.

DNA sequence analysis of a TGF-β_2 clone derived from the human prostatic adenocarcinoma cell line PC-3 indicated that TGF-β_2 is synthesized as a large precursor polypeptide of 442 amino acids, the carboxyl terminus of which is cleaved to yield the mature 112-amino acid TGF-β_2 monomer. The complete amino acid sequence of TGF-β_2 has been determined.[237] Comparison of the amino acid sequences of TGF-β_2 and the classical TGF-β molecule, TGF-β_1, revealed a high degree of homology suggesting that TGF-β_1 and TGF-β_2 have evolved from a common ancestor. Overall sequence homology between the TGF-β_1 and TGF-β_2 molecules is 71% throughout the mature form of the precursor, but there is only 31% homology between the amino-terminal precursor regions of TGF-β_1 and TGF-β_2.[579]

Human TGF-β_2 has been quantitated on the basis of one of its properties, the inhibition of the proliferation of the mink lung epithelial cell line CCl-64. The factor can also be detected and quantitated by means of specific antisera developed in immunized turkeys.[580] The use of such antisera indicates that whereas TGF-β_1 and TGF-β_2 are secreted by a variety of cultured cells, some cells secrete predominantly either TGF-β_1 or TGF-β_2, and other cells produce both factors in nearly equal amounts. The expression of each of the two forms of TGF-β may be independently regulated. Examination of normal and benign hyperplastic prostates indicated that although both TGF-β_1 and TGF-β_2 were expressed in all of the samples, TGF-β_2 was expressed at increased levels in benign prostate hyperplasia as compared to those of the normal organ.[581]

The specific functions of TGF-β_2 are little known. During mouse embryogenesis, TGF-β_2 mRNA shows widespread epithelial expression, and the localization of TGF-β mRNA in neuronal tissue may be correlated with differentiation.[572] A potent immunosuppressive molecule released by non-T small lymphocytic suppressor cells in murine allopregnancy, the decidual suppressor factor (DSF), is closely related to TGF-β_2.[582] A murine amniotic fluid immunosuppressive factor may be identical to TGF-β_2.[583] These results suggest that TGF-β_2 and DSF may represent species of TGF-β that accumulate in a physiologically active form in amniotic fluid. High levels of TGF-β_2 mRNA accumulate in the pregnant uterus of the mouse, and hormones and growth factors involved in pregnancy may directly regulate the production of this mRNA. The precise mechanism by which TGF-β_2 can suppress immune responses in pregnancy is unknown.

Little is known about the possible role of TGF-β_2 in neoplastic processes but TGF-β_2 can inhibit IL-2-induced LAK cell activity against human tumor cell lines.[343] TGF-β_2 could have some role in the development of prostatic carcinoma and other tumors. The RB tumor supressor protein activates expression of the human TGF-β_2 gene through the transcription factor ATF-2.[584] The three types of TGF (TGF-β_1, TGF-β_2, and TGF-β_3) are differentially expressed in glioblastoma cells, astrocytes, and microglia.[585]

C. TGF-β_3

Overlapping cDNA clones isolated from human umbilical cord tissue contain a sequence coding for a distinct member of the TGF-β gene family, TGF-β_3.[205] The TGF-β_3 gene is located on human chromosome 14, at region 14q23-q24.[586] The promoter of this gene has almost no structural or functional similarity to the TGF-β_1 gene, despite high conservation of amino acid sequence between both types of TGF-β, suggesting a differential mechanism of regulation of their expression.[587] A cAMP-responsive element (CRE) regulates both the basal and cAMP-induced activity of the TGF-β_3 gene promoter. The product of the TGF-β_3 gene is synthesized as a precursor of 412 amino acids. The 112 carboxyl-terminal residues of this precursor share 80% sequence identity with TGF-β_1 and TGF-β_2. The high conservation of the TGF-β_3 protein in the human and the homologous product of the chicken (111/112 amino acids of the carboxyl-terminal portion) indicate the fundamental role of the TGF-β_3 protein in normal cell physiology. The function of the mature TGF-β_3 protein can be predicted to be multiple. TGF-β_3 is a potent regulator of functions associated with bone formation, enhancing DNA and collagen synthesis, and decreasing alkaline phosphatase activity in osteoblast-enriched cultures from fetal rat bone.[588]

D. TGF-E

TGF-e is a growth factor present in normal and neoplastic tissues of mostly epithelial origin, plasma, and platelets, and it has been purified to homogeneity from bovine kidney.[589] TGF-e is an acid- and heat-stable polypeptide of 22 to 25 kDa which requires disulfide bonds for maximal activity. It stimulates both anchorage-dependent and -independent growth of epithelial and fibroblastic cells of nonneoplastic origin.[590] TGF-e functions as a progression factor for both AKR-2B and BALB/c 3T3 cells. It is a potent mitogen for normal human epidermal keratinocytes and may play a role in epidermal growth and regeneration. Both TGF-β and TGF-e are present in the human pituitary.[591]

E. GDF-1

A cDNA clone derived from a library prepared from mouse embryos encoded a new member, termed GDF-1, of the TGF-β family.[592] The nucleotide sequence of GDF-1 predicts a protein of 357 amino acids with a mol wt of 38,600. GDF-1 is 52% homologous to VG-1, a factor localized in the vegetative pole of *Xenopus* eggs. Genomic Southern analysis showed that GDF-1 is highly conserved across species. Most likely, GDF-1 is an extracellular factor involved in the mediation and/or regulation of cell differentiation events during embryonic development.

F. CDGF

A TGF-β-like mitogenic activity for fibroblastic cells found in the serum-free medium of growth-arrested primary cultures of chicken embryo fibroblasts (CEF cells) was designated CEF-derived growth factor (CDGF).[593,594] The CDGF is a 32-kDa heterodimeric protein consisting of two disulfide-linked subunits of 15 and 17 kDa. CDGF acts synergistically with certain growth factors and may play a role in the cellular homeostasis of fibroblastic cells by autocrine and/or paracrine action. The precise relationships between CDGF and other members of the TGF-β family of proteins remain to be determined.

G. TUMOR-DERIVED TGF-β-LIKE FACTORS

Factors structurally and/or functionally related to TGF-β may be produced by certain types of malignant cells *in vitro* and perhaps also *in vivo*. The human glioblastoma cell line 308 secretes a 12.5-kDa peptide, termed glioblastoma-derived T-cell suppressor factor or glioblastoma-derived growth inhibitor (GDGI), which has suppressive effects on IL-2-dependent T-cell growth.[595] The amino acid sequence of GDGI, deduced from cDNA clones, revealed that it is 110-amino acids long and that it shares 71% homology with TGF-β_1.[596] GDGI is identical to TGF-β_2 purified from human glioblastoma cells and porcine platelets. GDGI/TGF-β_2 can suppress the IL-2-induced generation of LAK cells from both normal peripheral blood lymphocytes and peripheral blood lymphocytes obtained from brain tumor patients.[597] Human glioblastomas could exert an inhibitory influence on the generation of an immune response *in vivo* through the production of TGF-β_2.

The fibrosarcoma cell line 8387 produces two polypeptides of 16 and 12 kDa, respectively, that inhibit the growth in soft agar of A549 lung adenocarcinoma cells and decrease the secretion of plasminogen activator by human fibroblasts.[598] Antibodies to TGF-β efficiently inhibited the effects of the 16-kDa factor but not those of the 12-kDa factor in cell culture assays. The results suggest that 8387 fibrosarcoma cells produce two major growth inhibitors, one of which is closely related to TGF-β.

The RSV-transformed rat cerebral microvascular endothelial cell line RCE-T1 produces and secretes a growth factor that exhibits biological properties similar to those of TGF-β.[599] The RCE-T1-derived TGF-β-like growth factor is secreted in an active form and its secretion is elevated in late passage cultures comprised of rapidly growing, less differentiated cells exhibiting reduced sensitivity to the growth-inhibitory effects of exogenous TGF-β_1.

V. OTHER TRANSFORMING GROWTH FACTORS

In addition to TGF-α and TGF-β, other types of TGFs may exist. TGF-like mitogens apparently different from TGF-α and TGF-β have been detected in the human anterior pituitary gland.[600] These TGFs are proteins of around 15 kDa which may contain disulfide bonds since their activities are markedly reduced by treatment with dithiothreitol. The normal physiological roles of pituitary TGFs are unknown.

Multiple types of TGFs may be produced by transformed cells. A factor, ND-TGF, was obtained from a mouse neuroblastoma cell line, but is not a specific neuronal growth factor.[601] ND-TGF is a strong mitogen for different cell lines, it does not compete with EGF for receptor binding as does TGF-α, and, unlike TGF-β it has strong mitogenic activity. Whereas EGF alone is unable to induce any degree of anchorage-independent growth in NRK cells, ND-TGF is able to induce progressively growing colonies of these cells in soft agar, even without the addition of EGF.[602]

A TGF produced by the human adrenocortical cell line SW13 as well as by human colon carcinoma tissues, designated human epithelial transforming factor (h-TGFe), does not bind heparin and appears to be represented by a protein of 59 kDa.[603] h-TGFe stimulates anchorage-independent growth of SW13 cells. A similar activity is present in human milk as well as in conditioned medium from epithelial cell lines.

Another TGF-like substance was isolated and purified from the ASV-transformed rat cell line 77N1.[22,23] This factor is represented by a 12-kDa protein that does not compete with EGF binding to EGF membrane receptors and is capable of inducing DNA synthesis in growth-arrested BALB/3T3 cells and promoting anchorage-independent growth of nontransformed BALB/3T3 cells in soft agar. TGFs with apparently unique biological activities were isolated from the conditioned medium of ASV-transformed chicken and hamster cells, human rhabdomyosarcoma cells A637, and salivary gland epithelial cells CSG 211 chemically transformed *in vitro*.[16,28,58,604] Increased EGF-like activity detected in some primary human ovarian carcinomas may correspond to TGF-β related growth factors.[605] Analyses of acid extracts of urine from normal human donors and cancer patients revealed the presence of five EGF-related and two non-EGF-related TGFs, the latter having soft agar colony-stimulating activity only in the presence of added EGF.[606] Two of the five EGF-related TGFs were elevated in the urine of cancer patients, which could be due to changes in normal TGF metabolism. A TGF activity of high molecular weight purified from the urine of patients with malignant astrocytoma appears to be similar, if not identical, to high molecular weight EGF which appears to be present in only trace amounts in normal urine and is elevated in the urine of some cancer patients.[607]

Vaccinia virus-infected cells release an acid-stable mitogen that competes with EGF for binding to receptors and shares biological properties similar to those of both EGF and TGFs.[608] The factor is immunologically unrelated to EGF and TGFs and would consist of a much larger protein.

Expression of a gene, *cripto,* that encodes a 188-amino acid protein which shows structural homology to EGF, TGF-α, and amphirregulin, results in neoplastic transformation of NIH/3T3 mouse fibroblasts and NOG-8 mouse mammary epithelial cells.[609,610] High levels of *cripto* mRNA are present in primary human gastric and colorectal carcinomas.[611] The normal function of the *cripto*-encoded protein is unknown.

REFERENCES

1. **Todaro, G.J., De Larco, J.E., and Cohen, S.,** Transformation by murine and feline sarcoma viruses specifically blocks binding of epidermal growth factor to cells, *Nature,* 264, 26, 1976.
2. **De Larco, J.E. and Todaro, G.J.,** Growth factors from murine sarcoma virus-transformed cells, *Proc. Natl. Acad. Sci. U.S.A.,* 75, 4001, 1978.
3. **Anzano, M.A., Roberts, A.B., Smith, J.M., Sporn, M.B., and De Larco, J.E.,** Sarcoma growth factor from conditioned medium of virally transformed cells is composed of both type α and type β transforming growth factors, *Proc. Natl. Acad. Sci. U.S.A.,* 80, 6264, 1983.
4. **Bascom, C.C., Sipes, N.J., Coffey, R.J., and Moses, H.L.,** Regulation of epithelial cell proliferation by transforming growth factors, *J. Cell. Biochem.,* 39, 25, 1989.
5. **Sporn, M.B., Roberts, A.B., Shull, J.H., Smith, J.M., Ward, J.M., Sodek, J.,** Polypeptide transforming growth factors isolated from bovine sources and used for wound healing *in vivo, Science,* 219, 1329, 1983.
6. **Fausto, N. and Mead, J.E.,** Regulation of liver growth: proto-oncogenes and transforming growth factors, *Lab. Invest.,* 60, 4, 1989.
7. **Sporn, M.B. and Todaro, G.J.,** Autocrine secretion and malignant transformation of cells, *N. Engl. J. Med.,* 303, 878, 1980.
8. **Todaro, G.J., De Larco, J.E., Fryling, C., Johnson, P.A., and Sporn, M.B.,** Transforming growth factors (TGFs): properties and possible mechanisms of action, *J. Supramol. Struct.,* 15, 287, 1981.
9. **Todaro, G.J., De Larco, J.E., and Fryling, C.M.,** Sarcoma growth factor and other transforming peptides produced by human cells: interactions with membrane receptors, *Fed. Proc.,* 41, 2996, 1982.
10. **Roberts, A.B., Frolik, C.A., Anzano, M.A., and Sporn, M.B.,** Transforming growth factors from neoplastic and nonneoplastic cells, *Fed. Proc.,* 42, 2621, 1983.
11. **Lawrence, D.A.,** Transforming growth factors — an overview, *Biol. Cell,* 53, 93, 1985.
12. **Roberts, A.B. and Sporn, M.B.,** Transforming growth factors, *Cancer Surv.,* 4, 683, 1985.
13. **Lyons, R.M. and Moses, H.L.,** Transforming growth factors an the regulation of cell proliferation, *Eur. J. Biochem.,* 187, 467, 1990.
14. **Morita, H., Noda, K., Umeda, M., and Ono, T.,** Activities of transforming growth factors on cell lines and their modification by other growth factors, *Jpn. J. Cancer Res.,* 75, 403, 1984.

15. **Roberts, A.B., Lamb, L.C., Newton, D.L., Sporn, M.B., De Larco, J.E., and Todaro, G.J.,** Transforming growth factors: isolation of polypeptides from virally and chemically transformed cells by acid/ethanol extraction, *Proc. Natl. Acad. Sci. U.S.A.,* 77, 3494, 1980.
16. **Moses, H.L., Branum, E.L., Proper, J.A., and Robinson, R.A.,** Transforming growth factor production by chemically transformed cells, *Cancer Res.,* 41, 2842, 1981.
17. **Kryceve-Martinerie, C., Lawrence, D.A., Crochet, J., Julien, P. and Vigier, P.,** Cells transformed by Rous sarcoma virus release transforming growth factors, *J. Cell. Physiol.,* 113, 365, 1982.
18. **Roberts, A.B., Anzano, M.A., Lamb, L.C., Smith, J.M., Frolik, C.A., Marquardt, M., Todaro, G.J., and Sporn, M.B.,** Isolation from the murine sarcoma cells of novel transforming growth factors potentiated by EGF, *Nature,* 295, 417, 1982.
19. **Marquardt, H., Hunkapiller, M.W., Hood, L.E., Twardzik, D.R., De Larco, J.E., Stephenson, J.R., and Todaro, G.J.,** Transforming growth factors produced by retrovirus-transformed rodent fibroblasts and human melanoma cells: amino acid sequence homology with epidermal growth factor, *Proc. Natl. Acad. Sci. U.S.A.,* 80, 4684, 1983.
20. **Chua, C.C., Geiman, D., and Ladda, R.L.,** Transforming growth factors released from Kirsten sarcoma virus transformed cells do not compete for epidermal growth factor membrane receptors, *J. Cell. Physiol.,* 117, 116, 1983.
21. **Massagué, J.,** Epidermal growth factor-like transforming growth factor. I. Isolation, chemical characterization, and potentiation by other transforming growth factors from feline sarcoma virus-transformed cells, *J. Biol. Chem.,* 258, 13606, 1983.
22. **Hirai, R., Yamaoka, K., and Mitsui, H.,** Isolation and partial purification of a new class of transforming growth factors from an avian sarcoma virus-transformed rat cell line, *Cancer Res.,* 43, 5742, 1983.
23. **Yamaoka, K., Hirai, R., Tsugita, A., and Mitsui, H.,** The purification of an acid- and heat-labile transforming growth factor from an avian sarcoma virus-transformed rat cell line, *J. Cell. Physiol.,* 119, 307, 1984.
24. **Massagué, J.,** Type β transforming growth factor from feline sarcoma virus-transformed rat cells: isolation and biological properties, *J. Biol. Chem.,* 259, 9756, 1984.
25. **Pircher, R., Lawrence, D.A., and Julien, P.,** Latent β-transforming growth factor in nontransformed and Kirsten sarcoma virus-transformed normal rat kidney cells, clone 49F, *Cancer Res.,* 44, 5538, 1984.
26. **Anzano, M.A., Roberts, A.B., De Larco, J.E., Wakefield, L.M., Assoian, R.K., Roche, N.S., Smith, J.M., Lazarus, J.E., and Sporn, M.B.,** Increased secretion of transforming growth factor accompanies viral transformation of cells, *Mol. Cell. Biol.,* 5, 242, 1985.
27. **Kryceve-Martinerie, C., Lawrence, D.A., Crochet, J., Jullien, P., and Vigier, P.,** Further study of β-TGFs released by virally transformed and non-transformed cells, *Int. J. Cancer,* 35, 553, 1985.
28. **Wigley, C.B., Trejdosiewicz, L.K., Southgate, J., Coventry, R., and Ozanne, B.,** Growth factor production during multistage transformation of epithelium *in vitro.* I. Partial purification and characterisation of the factor(s) from a fully transformed epithelial cell line, *J. Cell. Physiol.,* 125, 156, 1985.
29. **De Larco, J.E., Pigott, D.A., and Lazarus, J.A.,** Ectopic peptides released by a human melanoma cell line that modulate the transformed phenotype, *Proc. Natl. Acad. Sci. U.S.A.,* 82, 5015, 1985.
30. **Richmond, A., Lawson, D.H., Lawson, D.W., and Chawla, R.K.,** Characterization of autostimulatory and transforming growth factors from human melanoma cells, *Cancer Res.,* 45, 6390, 1985.
31. **Coffey, R.J., Jr., Shipley, G.D., and Moses, H.L.,** Production of transforming growth factors by human colon cancer cell lines, *Cancer Res.,* 46, 1164, 1986.
32. **van Zoelen, E.J.J., van Rooijen, M.A., van Oostwaard, T.M.J., and de Laat, S.W.,** Production of transforming growth factors by simian sarcoma virus-transformed cells, *Cancer Res.,* 47, 1582, 1987.
33. **Siegfried, J.M.,** Detection of human lung epithelial cell growth factors produced by a lung carcinoma cell line: use in culture of primary solid lung tumors, *Cancer Res.,* 47, 2903, 1987.
34. **Peres, R., Betsholtz, Westermakr, B., and Heldin, C.-W.,** Frequent expression of growth factors for mesenchymal cells in human mammary carcinoma cell lines, *Cancer Res.,* 47, 3425, 1987.
35. **Coffey, R.J., Jr., Goustin, A.S., Soderquist, A.M., Shipley, G.D., Wolfshohl, J., Carpenter, G., and Moses, H.L.,** Transforming growth factor α and β expression in human colon cancer lines: implications for an autocrine model, *Cancer Res.,* 47, 4590, 1987.
36. **Yamada, Y. and Serrero, G.,** Characterization of transforming growth factors produced by the insulin-independent teratoma-derived cell line 1246-3A, *J. Cell. Physiol.,* 140, 254, 1989.

37. **Liu, C., Tsao, M.-S., and Grisham, J.W.,** Transforming growth factors produced by normal and neoplastically transformed rat liver epithelial cells in culture, *Cancer Res.,* 48, 850, 1988.
38. **Myrdal, S.E. and Auersperg, N.,** An agent or agents produced by virus-transformed cells cause unregulated ruffling in untransformed cells, *J. Cell Biol.,* 102, 1224, 1986.
39. **Myrdal, S.E., Twardzik, D.R., and Auersperg, N.,** Cell-mediated co-action of transforming growth factors: incubation of type β with normal rat kidney cells produces a soluble activity that prolongs the ruffling response to type α, *J. Cell Biol.,* 102, 1230, 1986.
40. **Noda, M., Selinger, Z., Scolnick, E.M., and Bassin, R.H.,** Flat revertants isolated from Kirsten sarcoma virus-transformed cells are resistant to the action of specific oncogenes, *Proc. Natl. Acad. Sci. U.S.A.,* 80, 5602, 1983.
41. **Salomon, D.S., Zwiebel, J.A., Noda, M., and Bassin, R.H.,** Flat revertants derived from Kirsten murine sarcoma virus-transformed cells produce transforming growth factors, *J. Cell. Physiol.,* 121, 22, 1984.
42. **Nickell, K.A., Halper, J., and Moses, H.L.,** Transforming growth factors in solid human malignant neoplasms, *Cancer Res.,* 43, 1966, 1983.
43. **Nakamura, H., Komatsu, K., Akedo, H., Hosokawa, M., Shibata, H., and Masaoka, T.,** Human leukemic cells contain transforming growth factor, *Cancer Lett.,* 21, 133, 1983.
44. **Salomon, D.S., Zwiebel, J.A., Bano, M., Losonczy, I., Fehnel, P., and Kidwell, W.R.,** Presence of transforming growth factors in human breast cancer cells, *Cancer Res.,* 44, 4069, 1984.
45. **Hamburger, A.W., White, C.P., and Dunn, F.E.,** Production of transforming growth factors by primary human tumour cells, *Br. J. Cancer,* 51, 9, 1985.
46. **Derynck, R., Goeddel, D.V., Ullrich, A., Gutterman, J.U., Williams, R.D., Bringman, T.S., and Berger, W.H.,** Synthesis of messenger RNAs for transforming growth factors α and β and the epidermal growth factor receptor by human tumors, *Cancer Res.,* 47, 707, 1987.
47. **Wigley, C.B., Paraskeva, C., and Coventry, R.,** Elevated production of growth factor by human premalignant colon adenomas and a derived epithelial cell line, *Br. J. Cancer,* 54, 799, 1986.
48. **Sairenji, M., Suzuki, K., Murakami, K., Motohashi, H., Okamoto, T., and Umeda, M.,** Transforming growth factor activity in pleural and peritoneal effusions from cancer and non-cancer patients, *Jpn. J. Cancer Res.,* 78, 814, 1987.
49. **Roberts, A.B., Anzano, M.A., Lamb, L.C., Smith, J.M., and Sporn, M.B.,** New class of transforming growth factors potentiated by epidermal growth factor: isolation from non-neoplastic tissues, *Proc. Natl. Acad. Sci. U.S.A.,* 78, 5339, 1981.
50. **Stromberg, K., Pigott, D.A., Ranchalis, J.E., and Twardzik, D.R.,** Human term placenta contains transforming growth factors, *Biochem. Biophys. Res. Commun.,* 106, 354, 1982.
51. **Frolik, C.A., Dart, L.L., Meyers, C.A., Smith, D.M., and Sporn, M.B.,** Purification and initial characterization of a type β transforming growth factor from human placenta, *Proc. Natl. Acad. Sci. U.S.A.,* 80, 3676, 1983.
52. **Roberts, A.B., Anzano, M.A., Meyers, C.A., Wideman, J., Blacher, R., Pan, Y.-C.E., Stein, S., Lehrman, S.R., Smith, J.M., Lamb, L.C., and Sporn, M.B.,** Purification and properties of a type β transforming growth factor from bovine kidney, *Biochemistry,* 22, 5692, 1983.
53. **Noda, K., Umeda, M., and Ono, T.,** Transforming growth factor activity in human colostrum, *Jpn. J. Cancer Res.,* 75, 109, 1984.
54. **Lawrence, D.A., Pircher, R., Kryceve-Martinerie, C., and Jullien, P.,** Normal embryo fibroblasts release transforming growth factors in a latent form, *J. Cell. Physiol.,* 121, 184, 1984.
55. **Zwiebel, J.A., Bano, M., Nexo, E., Salomon, D.S., and Kidwell, W.R.,** Partial purification of transforming growth factors from human milk, *Cancer Res.,* 46, 933, 1986.
56. **Hill, D.J., Strain, A.J., and Milner, R.D.G.,** Presence of transforming growth factor β-like activity in multiple fetal rat tissues, *Cell Biol. Int. Rep.,* 10, 915, 1986.
57. **Nishimura, R., Okumura, H., Noda, K., Yasumitsu, H., and Umeda, M.,** High level of β-type transforming growth factor activity in human urine obtained from cancer patients, *Jpn. J. Cancer Res.,* 77, 560, 1986.
58. **Stromberg, K. and Hudgins, W.R.,** Urinary transforming growth factors in neoplasia: separation of ^{125}I-labeled transforming growth factor-α from epidermal growth factor in human urine, *Cancer Res.,* 46, 6004, 1986.
59. **Hanauske, A.-R., Arteaga, C.L., Clark, G.M., Buchok, J., Marshall, M., Hazarika, P., Pardue, R.L., and Von Hoff, D.D.,** Determination of transforming growth factor activity in effusions from cancer patients, *Cancer,* 61, 1832, 1988.

60. **Yeh, J. and Yeh, Y.-C.,** Transforming growth factor-α and human cancer, *Biomed. Pharmacother.,* 43, 651, 1989.
61. **Salomon, D.S., Kim, N., Saeki, T., and Ciardiello, F.,** Transforming growth factor-α: an oncodevelopmental growth factor, *Cancer Cells,* 2, 389, 1990.
62. **Ignotz, R.A., Kelly, B., Davis, R.J., and Massagué, J.,** Biologically active precursor for transforming growth factor type α, released by retrovirally transformed cells, *Proc. Natl. Acad. Sci. U.S.A.,* 83, 6307, 1986.
63. **Lee, D.C., Rose, T.M., Webb, N.R., and Todaro, G.J.,** Cloning and sequence analysis of a cDNA for rat transforming growth factor-α, *Nature,* 313, 489, 1985.
64. **Yeh, Y.-C., Guh, J-Y., Yeh, J., and Yeh, W-H.,** Transforming growth factor type α in normal human adult saliva, *Mol. Cell. Endocrinol.,* 67, 247, 1989.
65. **Coffey, R.J., Jr., Derynck, R., Wilcox, J.N., Bringman, T.S., Goustin, A.S., Moses, H.L., and Pittelkow, M.R.,** Production and auto-induction of transforming growth factor-α in human keratinocytes, *Nature,* 328, 817, 1987.
66. **Ono, M., Okamura, K., Nakayama, Y., Tomita, I.M., Sato, Y., Komatsu, Y., and Kuwano, M.,** Induction of human microvascular endothelial tubular morphogenesis by human keratinocytes: involvement of transforming growth factor-α, *Biochem. Biophys. Res. Commun.,* 189, 601, 1992.
67. **Ju, W.D., Schiller, J.T., Kazempour, M.K., and Lowy, D.R.,** TGFα enhances locomotion of cultured human keratinocytes, *J. Invest. Dermatol.,* 100, 628, 1993.
68. **Elder, J.T., Fisher, G.J., Lindquist, P.B., Bennet, G.L., Pittelkow, M.R., Coffey, R.J., Jr., Ellingsworth, L., Derynck, R., and Voorhees, J.J.,** Overexpression of transforming growth factor α in psoriatic epidermis, *Science,* 243, 811, 1989.
69. **Mann, G.B., Fowler, K.J., Gabriel, A., Nice, E.C., Williams, R.L., and Dunn, A.R.,** Mice with a null mutation of the TGFα gene have abnormal skin architecture, wavy hair, and curly whiskers and often develop corneal inflammation, *Cell,* 73, 249, 1993.
70. **Luetteke, N.C., Qiu, T.H., Peiffer, R.L., Oliver, P., Smithies, O., and Lee, D.C.,** TGFα deficiency results in hair follicle and eye abnormalities in targeted and waved-1 mice, *Cell,* 73, 263, 1993.
71. **Madtes, D.K., Raines, E.W., Sakariassen, K.S., Assoian, R.K., Sporn, M.B., Bell, G.I., and Ross, R.,** Induction of transforming growth factor-α in activated human alveolar macrophages, *Cell,* 53, 285, 1988.
72. **Wong, D.T.W., Weller, P.F., Galli, S.J., Elovic, A., Rand, T.H., Gallagher, G.T., Chiang, T., Chou, M.Y., Matopssian, K., McBride, J., and Todd, R.,** Human eosinophils express transforming growth factor α, *J. Exp. Med.,* 172, 673, 1990.
73. **Cartlidge, S.A. and Elder, J.B.,** Transforming growth factor α and epidermal growth factor levels in normal human gastrointestinal mucosa, *Br. J. Cancer,* 60, 657, 1989.
74. **Rhodes, J.A., Tam, J.P., Finke, U., Saunders, M., Bernanke, J., Silen, W., and Murphy, R.A.,** Transforming growth factor α inhibits secretion of gastrin acid, *Proc. Natl. Acad. Sci. U.S.A.,* 83, 3844, 1986.
75. **Evarts, R.P., Nakatsukasa, H., Marsden, E.R., Hu, Z., and Thorgeirsson, S.S.,** Expression of transforming growth factor-α in regenerating liver and during hepatic differentiation, *Mol. Carcinogenesis,* 5, 25, 1992.
76. **Hoffmann, B. and Paul, D.,** Basic fibroblast growth factor and transforming growth factor-α are hepatotrophic mitogens *in vitro, J. Cell. Physiol.,* 142, 149, 1990.
77. **Gomella, L.G., Sargent, E.R., Wade, T.P., Anglard, P., Linehan, W.M., and Kasid, A.,** Expression of transforming growth factor α in normal human adult kidney and enhanced expression of transforming growth factors α and β1 in renal cell carcinoma, *Cancer Res.,* 49, 6972, 1989.
78. **Derynck, R.,** Transforming growth factor-α: structure and biological activities, *J. Cell. Biochem.,* 32, 293, 1986.
79. **Hanauske, A.R., Buchok, J., Scheithauer, W., and Von Hoff, D.D.,** Human colon cancer cell lines secrete α TGF-like activity, *Br. J. Cancer,* 55, 57, 1987.
80. **Zhang, M., Woo, D.D.L., and Howard, B.D.,** Transforming growth factor α and a PC12-derived growth factor induce neurites in PC12 cells and enhance the survival of embryonic brain neurons, *Cell Regul.,* 1, 511, 1990.
81. **Massagué, J.,** Transforming growth factor-α — A model for membrane-anchored growth factors, *J. Biol. Chem.,* 265, 21393, 1990.

82. **Pain, B., Woods, C.M., Saez, J., Flickinger, T., Raines, M., Peyrol, S., Moscovici, C., Moscovici, M.G., Kung, H.-J., Jurdic, P., Lazarides, E., and Samarut, J.,** EGF-R as a hemopoieitic growth factor receptor: the c-*erb*B product is present in chicken erythrocytic progenitors and controls their self-renewal, *Cell,* 65, 37, 1991.
83. **Twardzik, D.R.,** Differential expression of transforming growth factor-α during prenatal development of the mouse, *Cancer Res.,* 45, 5413, 1985.
84. **Lee, D., Rochford, R., Todaro, G.J., and Villareal, L.P.,** Developmental expression of rat transforming growth factor-α mRNA, *Mol. Cell. Biol.,* 5, 3644, 1985.
85. **Gruppuso, P.A.,** Expression of hepatic transforming growth factor receptors during late gestation in the fetal rat, *Endocrinology,* 125, 3037, 1989.
86. **Han, V.K.M., Hunter, E.S., III, Pratt, R.M., Zendegui, J.G., and Lee, D.C.,** Expression of rat transforming growth factor α mRNA during development occurs predominantly in the maternal decidua, *Mol. Cell. Biol.,* 7, 2335, 1987.
87. **Smith, J.M., Sporn, M.B., Roberts, A.B., Derynck, R., Winkler, M.E., and Gregory, H.,** Human transforming growth factor-α causes precocious eyelid opening in newborn mice, *Nature,* 315, 515, 1985.
88. **Skinner, M.K. and Coffey, R.J., Jr.,** Regulation of ovarian cell growth through the local production of transforming growth factor-α by theca cells, *Endocrinology,* 123, 2632, 1988.
89. **Adashi, E.Y., Resnick, C.E., and Twardzik, D.R.,** Transforming growth factor-α attenuates the acquisition of aromatase activity by cultured rat granulosa cells, *J. Cell. Biochem.,* 33, 1, 1987.
90. **Skinner, M.K., Takacs, K., and Coffey, R.J.,** Transforming growth factor-α gene expression and action in the seminiferous tubule: peritubular cell-Sertoli cell interactions, *Endocrinology,* 124, 845, 1989.
91. **Mullaney, B.P. and Skinner, M.K.,** Transforming growth factor-α and epidermal growth factor receptor gene expression and action during pubertal development of the seminiferous tubule, *Mol. Endocrinol.,* 6, 2103, 1992.
92. **Nakhal, A.M. and Tam, J.P.,** Transforming growth factor is a potent stimulator of testicular ornithine decarboxylase in immature mouse, *Biochem. Biophys. Res. Commun.,* 132, 1180, 1985.
93. **Vonderhaar, B.K.,** Local effects of EGF, α-TGF, and EGF-like growth factors on lobuloalveolar development of the mouse mammary gland *in vivo, J. Cell. Physiol.,* 132, 581, 1987.
94. **Takahashi, N., MacDonald, B.R., Hon, J., Winkler, M.E., Derynck, R., Mundy, G.R., and Roodman, G.D.,** Recombinant human transforming growth factor-α stimulates the formation of osteoclast-like cells in long-term human marrow cultures, *J. Clin. Invest.,* 78, 894, 1986.
95. **Gan, B.S., MacCannell, K.L., and Hollenberg, M.D.,** Epidermal growth factor-urogastrone causes vasodilatation in the anesthetized dog, *J. Clin. Invest.,* 80, 199, 1987.
96. **Gan, B.S., Hollenberg, M.D., MacCannell, K.L., Lederis, K., Winkler, M.E., and Derynck, R.,** Distinct vascular actions of epidermal growth factor-urogastrone and transforming growth factor-α, *J. Pharmacol. Exp. Ther.,* 242, 331, 1987.
97. **Schreiber, A.B., Winkler, M.E., and Derynck, R.,** Transforming growth factor-α: a more potent angiogenic mediator than epidermal growth factor, *Science,* 232, 1250, 1986.
98. **Derynck, R., Roberts, A.B., Winkler, M.E., Chen, E.Y., and Goeddel, D.V.,** Human transforming growth factor-α: precursor structure and expression in E. coli, *Cell,* 38, 287, 1984.
99. **Saeki, T., Cristiano, A., Lynch, M.J., Brattain, M., Kim, N., Normanno, N., Kenney, N., Ciardiello, F., and Salomon, D.S.,** Regulation by estrogen through the 5′-flanking region of the transforming growth factor α gene, *Mol. Endocrinol.,* 5, 1955, 1991.
100. **Brissenden, J.E., Derynck, R., and Francke, U.,** Mapping of transforming growth factor α gene on human chromosome 2 close to the breakpoint of the Burkitt's lymphoma t(2;8) variant translocation, *Cancer Res.,* 45, 5593, 1985.
101. **Blasband, A.J., Rogers, K.T., Chen, X., Azizkhan, J.C., and Lee, D.C.,** Characterization of the rat transforming growth factor α gene and identification of promoter sequences, *Mol. Cell. Biol.,* 10, 2111, 1990.
102. **Gentry, L.E., Twardzik, D.R., Lim, G.J., Ranchalis, J.E., and Lee, D.C.,** Expression and characterization of transforming growth factor α precursor protein in transfected mammalian cells, *Mol. Cell. Biol.,* 7, 1585, 1987.
103. **Marquardt, H. and Todaro, G.J.,** Human transforming growth factor: production by a melanoma cell line, purification, and characterization, *J. Biol. Chem.,* 257, 5220, 1982.

104. **Winkler, M.E., Bringman, T., and Marks, B.J.,** The purification of fully active recombinant transforming growth factor α produced in *Escherichia coli, J. Biol. Chem.,* 261, 13838, 1986.

105. **Lazar, E., Vicenzi, E., Van Obberghen-Schilling, E., Wolff, B., Dalton, S., Watanabe, S., and Sporn, M.B.,** Transforming growth factor α: an aromatic side chain at position 38 is essential for biological activity, *Mol. Cell. Biol.,* 9, 860, 1989.

106. **Yang, S-G., Winkler, M.E., and Hollenberg, M.D.,** Contribution of the C-terminal dipeptide of transforming growth factor-α to its activity: biochemical and pharmacologic profiles, *Eur. J. Pharmacol.,* 188, 289, 1990.

107. **Tam, J.P., Sheikh, M.A., Solomon, D.S., and Ossowski, L.,** Efficient synthesis of human type α transforming growth factor: its physical and biological characterization, *Proc. Natl. Acad. Sci. U.S.A.,* 83, 8082, 1986.

108. **Linsley, P.S., Hargreaves, W.R., Twardzik, D.R., and Todaro, G.J.,** Detection of larger polypeptides structurally and functionally related to type I transforming growth factor, *Proc. Natl. Acad. Sci. U.S.A.,* 82, 356, 1985.

109. **Bringman, T.S., Lindquist, P.B., and Derynck, R.,** Different transforming growth factor-α species are derived from a glycosylated and palmitoylated transmembrane precursor, *Cell,* 48, 429, 1987.

110. **Wong, S.T., Winchell, L.F., McCune, B.K., Earp, H.S., Teixidó, J., Massagué, J., Herman, B., and Lee, D.C.,** The TGF-α precursor expressed on the cell surface binds to the EGF receptor on adjacent cells, leading to signal transduction, *Cell,* 56, 495, 1989.

111. **Marquardt, H., Hunkapiller, M.W., Hood, L.E., and Todaro, G.J.,** Rat transforming growth factor type I: structure and relation to epidermal growth factor, *Science,* 223, 1079, 1984.

112. **Tam, J.P., Marquardt, H., Rosberger, D.F., Wong, T.W., and Todaro, G.J.,** Synthesis of biologically active rat transforming growth factor I, *Nature,* 309, 376, 1984.

113. **Nestor, J.J., Newman, S.R., DeLustro, B., Todaro, G.J., and Schreiber, A.B.,** A synthetic fragment of rat transforming growth factor α with receptor binding and antigenic properties, *Biochem. Biophys. Res. Commun.,* 129, 226, 1985.

114. **Darlak, K., Franklin, G., Woost, P., Sonnenfeld, E., Twardzik, D., Spatola, A., and Schultz, G.,** Assessment of biological activity of synthetic fragments of transforming growth factor-α, *J. Cell. Biochem.,* 36, 341, 1988.

115. **Lazar, E., Watanabe, S., Dalton, S., and Sporn, M.B.,** Transforming growth factor α: mutation of aspartic acid 47 and leucine 48 results in different biological activities, *Mol. Cell. Biol.,* 8, 1247, 1988.

116. **DeFeo-Jones, D., Tai, J.Y., Wegrzyn, R.J., Vuocolo, G.A., Baker, A.E., Payne, L.S., Garsky, V.M., Oliff, A., and Riemen, M.W.,** Structure-function analysis of synthetic and recombinant derivatives of transforming growth factor α, *Mol. Cell. Biol.,* 8, 2999, 1988.

117. **Wu, D., Wang, L., Chi, Y., Sato, G.H., and Sato, J.D.,** Human epidermal growth factor receptor residue covalently cross-linked to epidermal growth factor, *Proc. Natl. Acad. Sci. U.S.A.,* 87, 3151, 1990.

118. **Carpenter, G., Stoscheck, C.M., Preston, Y.A., and De Larco, J.E.,** Antibodies to the epidermal growth factor receptor block the biological activities of sarcoma growth factor, *Proc. Natl. Acad. Sci. U.S.A.,* 80, 5627, 1983.

119. **Massagué, J.,** Epidermal growth factor-like transforming growth factor. II. Interaction with epidermal growth factor receptor in human placental membranes and A431 cells, *J. Biol. Chem.,* 258, 13614, 1983.

120. **Fernandez-Pol, J.A., Klos, D.J., and Hamilton, P.D.,** Modulation of transforming growth factor α-dependent expression of epidermal growth factor receptor gene by transforming growth factor β, triiodothyronine, and retinoic acid, *J. Cell. Biochem.,* 41, 159, 1989.

121. **Korc, M. and Finman, J.E.,** Attenuated processing of epidermal growth factor in the face of marked degradation of transforming growth factor-α, *J. Biol. Chem.,* 264, 14990, 1989.

122. **Ebner, R. and Derynck, R.,** Epidermal growth factor and transforming growth factor-α: differential intracellular routing and processing of ligand-receptor complexes, *Cell Regul.,* 2, 599, 1991.

123. **Decker, S.J.,** Epidermal growth factor and transforming growth factor-α induce differential processing of the epidermal growth factor receptor, *Biochem. Biophys. Res. Commun.,* 166, 615, 1990.

124. **Winkler, M.E., O'Connor, L., Winget, M., and Fendly, B.,** Epidermal growth factor and transforming growth factor α bind differently to the epidermal growth factor receptor, *Biochemistry,* 28, 6373, 1989.

125. **Lax, I., Johnson, A., Howk, R., Sap, J., Bellot, F., Winkler, M., Ullrich, A., Vennstrom, B., Schlessinger, J., and Givol, D.,** Chicken epidermal growth factor (EGF) receptor: cDNA cloning, expression in mouse cells, and differential binding of EGF and transforming growth factor α, *Mol. Cell. Biol.,* 8, 1970, 1988.
126. **Reynolds, F.H., Jr., Todaro, G.J., Fryling, C., and Stephenson, J.R.,** Human transforming growth factors induce tyrosine phosphorylation of EGF receptors, *Nature,* 292, 259, 1981.
127. **Frolik, C.A., Wakefield, L.M., Smith, D.M., and Sporn, M.B.,** Characterization of a membrane receptor for transforming growth factor-β in normal rat kidney fibroblasts, *J. Biol. Chem.,* 259, 10995, 1984.
128. **Tam, J.P.,** Physiological effects of transforming growth factor in the newborn mice, *Science,* 229, 673, 1985.
129. **Pike, L.J., Marquardt, H., Todaro, G.J., Gallis, B., Casnellie, J.E., Bornstein, P., and Krebs, E.G.,** Transforming growth factor and epidermal growth factor stimulate the phosphorylation of a synthetic, tyrosine-containing peptide in a similar manner, *J. Biol. Chem.,* 257, 14628, 1982.
130. **Murthy, U., Rieman, D.J., and Rodeck, U.,** Inhibition of TGF-α-induced second messengers by anti-EGF receptor antibody-425, *Biochem. Biophys. Res. Commun.,* 172, 471, 1990.
131. **Hanauske, A.-R., Buchok, J.B., Pardue, R.L., Muggia, V.A., and Von Hoff, D.D.,** Assessment of biological activity of linear α-transforming growth factor and monospecific antibodies to α-transforming growth factor and linear α-transforming growth factor, *Cancer Res.,* 46, 5567, 1986.
132. **Cutry, A.R., Kinniburgh, A.J., Twardzik, D.R., and Wenner, C.E.,** Transforming growth factor α (TGF α) induction of c-*fos* and c-*myc* expression in C3H 10T1/2 cells, *Biochem. Biophys. Res. Commun.,* 152, 216, 1988.
133. **Brenner, D.A., Koch, K.S., and Leffert, H.L.,** Transforming growth factor-α stimulates proto-oncogene c-*jun* expression and a mitogenic program in primary cultures of adult rat hepatocytes, *DNA,* 8, 279, 1989.
134. **Imanishi, K., Yamaguchi, K., Suzuki, M., Honda, S., Yanaihara, N., and Abe, K.,** Production of transforming growth factor-α in human tumour cell lines, *Br. J. Cancer,* 59, 761, 1989.
135. **Moody, T.W., Lee, M., Kris, R.M., Bellot, F., Bepler, G., Oie, H., and Gazdar, A.F.,** Lung carcinoid cell lines have bombesin-like peptides and EGF receptors, *J. Cell. Biochem.,* 43, 139, 1990.
136. **Reiss, M., Stash, E.B., Vellucci, V.F., and Zhou, Z.,** Activation of the autocrine transforming growth factor α pathway in human squamous carcinoma cells, *Cancer Res.,* 51, 6254, 1991.
137. **Mulder, K.M.,** Differential regulation of c-*myc* and transforming growth factor-α messenger RNA expression in poorly differentiated and well-differentiated colon carcinoma cells during the establishment of a quiescent state, *Cancer Res.,* 51, 2256, 1991.
138. **Ziober, B.L., Willson, J.K.V., Hymphrey, L.E., Childress-Fields, K., and Brattain, M.G.,** Autocrine transforming growth factor-α is associated with progression of transformed properties in human colon cancer cells, *J. Biol. Chem.,* 268, 691, 1993.
139. **Imamoto, A., Beltrán, L.M., and DiGiovanni, J.,** Evidence for autocrine/paracrine growth stimulation by transforming growth factor-α during the process of skin tumor promotion, *Mol. Carcinogenesis,* 4, 52, 1991.
140. **Pittelkow, M.R., Lindquist, P.B., Abraham, R.T., Graves-Deal, R., Derynck, R., and Coffey, R.J., Jr.,** Induction of transforming growth factor-α expression in human keratinocytes by phorbol esters, *J. Biol. Chem.,* 264, 5164, 1989.
141. **Raymond, V.W., Lee, D.C., Grisham, J.W., and Earp, H.S.,** Regulation of transforming growth factor α messenger RNA expression in a chemically transformed rat hepatic cell line by phorbol ester and hormones, *Cancer Res.,* 49, 3608, 1989.
142. **Ohmura, E., Okada, M., Onoda, N., Kamiya, Y., Murakami, H., Tsushima, T., and Shizume, K.,** Insulin-like growth factor I and transforming growth factor α as autocrine growth factors in human pancreatic cancer cell growth, *Cancer Res.,* 50, 103, 1990.
143. **Chung, C.K. and Antoniades, H.N.,** Expression of c-*sis*/ platelet-derived growth factor B, insulin-like growth factor I, and transforming growth factor α messenger RNAs and their respective receptor messenger RNAs in primary human gastric carcinomas: *in vivo* studies with *in situ* hybridization and immunocytochemistry, *Cancer Res.,* 52, 3453, 1992.
144. **Tanaka, S., Imanishi, K., Yoshihara, Haruma, K., Sumii, K., Kajiyama, G., and Akamatsu, S.,** Immunoreactive transforming growth factor α is commonly present in colorectal neoplasia, *Am. J. Pathol.,* 139, 123, 1991.

145. **Mydlo, J.H., Michaeli, J., Cordon-Cardo, C., Goldenberg, A.S., Heston, W.D.W., and Fair, W.R.,** Expression of transforming growth factor α and epidermal growth factor receptor messenger RNA in neoplastic and nonneoplastic human kidney tissue, *Cancer Res.,* 49, 3407, 1989.
146. **Tateishi, M., Ishida, T., Mitsudomi, T., Sugimachi, K.,** Prognostic implication of transforming growth factor α in adenocarcinoma of the lung — an immunohistochemical study, *Br. J. Cancer,* 63, 130, 1991.
147. **Aasland, R., Akslen, L.A., Varhaug, J.E., and Lillehaug, J.R.,** Co-expression of the genes encoding transforming growth factor-α and its receptor in papillary carcinomas of the thyroid, *Int. J. Cancer,* 46, 382, 1990.
148. **Yang, Y., Chisholm, G.D., and Habib, F.K.,** Epidermal growth factor and transforming growth factor α concentrations in BPH and cancer of the prostate: their relationships with tissue androgen levels, *Br. J. Cancer,* 67, 152, 1993.
149. **Kommoss, F., Wintzer, H.O., von Kleist, S., Kohler, M., Walker, R., Langton, B., van Tran, K., Pfleiderer, A., and Bauknecht, T.,** *In situ* distribution of transforming growth factor α in normal human tissues and in malignant tumours of the ovary, *J. Pathol.,* 162, 223, 1990.
150. **Morishige, K., Kurachi, H., Amemiya, K., Fujita, Y., Yamamoto, T., Miyake, A., and Tanizawa, O.,** Evidence for the involvement of transforming growth factor α and epidermal growth factor receptor autocrine growth mechanism in primary human ovarian cancers *in vitro, Cancer Res.,* 51, 5322, 1991.
151. **Bauknecht, T., Angel, P., Kohler, M., Kommoss, F., Birmelin, G., Pfleiderer, A., and Wagner, E.,** Gene structure and expression analysis of the epidermal growth factor receptor, transforming growth factor-α, *myc, jun,* and metallothionein in human ovarian carcinomas. Classification of malignant phenotypes, *Cancer,* 71, 419, 1993.
152. **Driman, D.K., Kobrin, M.S., Kudlow, J.E., and Asa, S.L.,** Transforming growth factor-α in normal and neoplastic human endocrine tissues, *Hum. Pathol.,* 23, 1360, 1992.
153. **Zhou, L. and Leung, B.S.,** Growth regulation of ovarian cancer cells by epidermal growth factor and transforming growth factors α and β1, *Biochim. Biophys. Acta,* 1180, 130, 1992.
154. **Markowitz, S.D., Molkentin, K., Gerbic, C., Jackson, J., Stellato, T., and Willson, J.K.V.,** Growth stimulation by coexpression of transforming growth factor-α and epidermal growth factor-receptor in normal and adenomatous human colon epithelium, *J. Clin. Invest.,* 86, 356, 1990.
155. **Stromberg, K., Hudgins, W.R., and Orth, D.N.,** Urinary TGFs in neoplasia: immunoreactive TGF-α in the urine of patients with disseminated breast carcinoma, *Biochem. Biophys. Res. Commun.,* 144, 1059, 1987.
156. **Ellis, D.L., Kafka, S.P., Chow, J.C., Nanney, L.B., Inman, W.H., McCadden, M.E., and King, L.E., Jr.,** Melanoma, growth factors, acanthosis nigricans, the sign of Leser-Trélat, and multiple acrochordons. A possible role for α-transforming growth factor in cutaneous paraneoplastic syndromes, *N. Engl. J. Med.,* 317, 1582, 1987.
157. **Liu, C., Woo, A., and Tsao, M.-S.,** Expression of transforming growth factor-α in primary human colon and lung carcinomas, *Br. J. Cancer,* 62, 425, 1990.
158. **Hanauske, A.-R., Hanauske, U., Buchok, J., and Von Hoff, D.D.,** Recombinant human transforming growth factor-α stimulates *in vitro* colony formation of fresh human tumor specimens, *Int. J. Cell Cloning,* 6, 221, 1988.
159. **Wong, D.T.W., Gallagher, G.T., Gertz, R., Chang, A.L.C., and Shklar, G.,** Transforming growth factor α in chemically transformed hamster oral keratinocytes, *Cancer Res.,* 48, 3130, 1988.
160. **Imanishi, K., Yamaguchi, K., Kuranami, M., Kyo, E., Hozumi, T., and Abe, K.,** Inhibition of growth of human lung adenocarcinoma cell lines by anti-transforming growth factor-α monoclonal antibody, *J. Natl. Cancer Inst.,* 81, 220, 1989.
161. **Hofer, D.R., Sherwood, E.R., Bromberg, W.D., Mendelsohn, J., Lee, C., and Kozlowski, J.M.,** Autonomous growth of androgen-independent human prostatic carcinoma cells: role of transforming growth factor α, *Cancer Res.,* 51, 2780, 1991.
162. **Stromberg, K., Collins, T.J., Gordon, A.W., Jackson, C.L., and Johnson, G.R.,** Transforming growth factor-α acts as an autocrine growth factor in ovarian carcinoma cell lines, *Cancer Res.,* 52, 341, 1992.
163. **Walker, C., Everitt, J., Freed, J.J., Knudson, A.G., Jr., and Whiteley, L.O.,** Altered expression of transforming growth factor-α in hereditary rat renal cell carcinoma, *Cancer Res.,* 51, 2973, 1991.

164. **Finzi, E., Kilkenny, A., Strickland, J.E., Balaschak, M., Bringman, T., Derynck, R., Aaronson, S., and Yuspa, S.H.,** TGF α stimulates growth of skin papillomas by autocrine and paracrine mechanisms but does not cause neoplastic progression, *Mol. Carcinogen.,* 1, 7, 1988.
165. **Singletary, S.E., Williams, N.N., Rodeck, U., Larry, L., Tucker, S., Spitzer, G., and Herlyn, M.,** Transforming growth factor-α secretion by epidermal growth factor-dependent human tumor cell lines, *Anticancer Res.,* 10, 1501, 1990.
166. **Rosenthal, A., Lindquist, P.B., Bringman, T.S., Goeddel, D.V., and Derynck, R.,** Expression in rat fibroblasts of a human transforming growth factor-α cDNA results in transformation, *Cell,* 46, 301, 1986.
167. **Watanabe, S., Lazar, E., and Sporn, M.B.,** Transformation of normal rat kidney (NRK) cells by an infectious retrovirus carrying a synthetic rat type α transforming growth factor gene, *Proc. Natl. Acad. Sci. U.S.A.,* 84, 1258, 1987.
168. **Shankar, V., Ciardiello, F., Kim, N., Derynck, R., Liscia, D.S., Merlo, G., Langton, B.C., Sheer, D., Callahan, R., Bassin, R.H., Lippman, M.E., Hynes, N., and Salomon, D.S.,** Transformation of an established mouse mammary epithelial cell line following transfection with a human transforming growth factor α cDNA, *Mol. Carcinogen.,* 2, 1, 1989.
169. **Hynes, N.E., Taverna, D., Harwerth, I.M., Ciardiello, F., Salomon, D.S., Yamamoto, T., and Groner, B.,** Epidermal growth factor receptor, but not c-*erbB-2,* activation prevents lactogenic hormone induction of the β-casein gene in mouse mammary epithelial cells, *Mol. Cell. Biol.,* 10, 4027, 1990.
170. **McGeady, M.L., Kerby, S., Shankar, V., Ciardiello, F., Salomon, D., and Seidman, M.,** Infection with a TGF-α retroviral vector transforms normal mouse mammary epithelial cells but not normal rat fibroblasts, *Oncogene,* 4, 1375, 1989.
171. **Blasband, A.J., Gilligan, D.M., Winchell, L.F., Wong, S.T., Luetteke, N.C., Rogers, K.T., and Lee, D.C.,** Expression of the TGF α integral membrane precursor induces transformation of NRK cells, *Oncogene,* 5, 1213, 1990.
172. **Brondyk, W.H., Kujoth, G.C., and Fahl, W.E.,** Transforming growth factor-α expression produces only morphological transformants of diploid human fibroblasts, *Cancer Res.,* 53, 2162, 1993.
173. **Gabrilovic, J., Moens, G., Thiery, J.P., and Jouanneau, J.,** Expression of transfected transforming growth factor α induces a motile fibroblast-like phenotype with extracellular matrix-degrading potential in a rat bladder carcinoma cell line, *Cell Regul.,* 1, 1003, 1990.
174. **Yoshida, K., Kyo, E., Tsuda, T., Tsujino, T., Ito, M., Niimoto, M., and Tahara, E.,** EGF and TGF-α, the ligands of hyperproduced EGFR in human esophageal carcinoma cells, act as autocrine growth factors, *Int. J. Cancer,* 45, 131, 1990.
175. **Smith, J.J., Derynck, R., and Korc, M.,** Production of transforming growth factor α in human pancreatic cancer cells: evidence for a superagonist autocrine cycle, *Proc. Natl. Acad. Sci. U.S.A.,* 84, 7567, 1987.
176. **Dickson, R.B., McManaway, M.E., and Lippman, M.E.,** Estrogen-induced factors of breast cancer cells partially replace estrogen to promote tumor growth, *Science,* 232, 1540, 1986.
177. **Bates, S.E., Davidson, N.E., Valverius, E.M., Freter, C.E., Dickson, R.B., Tam, J.P., Kudlow, J.E., Lippman, M.E., and Salomon, D.S.,** Expression of transforming growth factor α and its messenger ribonucleic acid in human breast cancer: its regulation by estrogen and its possible functional significance, *Mol. Endocrinol.,* 2, 543, 1988.
178. **Arteaga, C.L., Coronado, E., and Osborne, C.K.,** Blockade of the epidermal growth factor receptor inhibits transforming growth factor α-induced but not estrogen-induced growth of hormone-dependent human breast cancer, *Mol. Endocrinol.,* 2, 1064, 1988.
179. **Liu, S.C., Sanfilippo, B., Perroteau, I., Derynck, R., Salomon, D.S., and Kidwell, W.R.,** Expression of transforming growth factor α (TGF α) in differentiated rat mammary tumors: estrogen induction of TGF α production, *Mol. Endocrinol.,* 1, 683, 1987.
180. **Murphy, L.C. and Dotzlaw, H.,** Regulation of transforming growth factor α and transforming growth factor β messenger ribonucleic acid abundance in T-47D, human breast cancer cells, *Mol. Endocrinol.,* 3, 611, 1989.
181. **Sandgren, E.P., Luetteke, N.C., Palmiter, R.D., Brinster, R.L., and Lee, D.C.,** Overexpression of TGF-α in transgenic mice: induction of epithelial hyperplasia, pancreatic metaplasia, and carcinoma of the breast, *Cell,* 61, 1121, 1990.

182. **Jhappan, C., Stahle, C., Harkins, R.N., Fausto, N., Smith, G.H., and Merlino, G.T.,** TGF α overexpression in transgenic mice induces liver neoplasia and abnormal development of the mammary gland and pancreas, *Cell,* 61, 1137, 1990.
183. **Matsui, Y., Halter, S.A., Holt, J.T., Hogan, B.L.M., and Coffey, R.J.,** Development of mammary hyperplasia and neoplasia in MMTV-TGF α transgenic mice, *Cell,* 61, 1147, 1990.
184. **Sandgren, E.P., Luetteke, N.C., Qiu, T.H., Palmitter, R.D., Brinster, R.L., and Lee, D.C.,** Transforming growth factor α dramatically enhances oncogene-induced carcinogenesis in transgenic mouse pancreas and liver, *Mol. Cell. Biol.,* 13, 320, 1993.
185. **Ciardiello, F., Hynes, N., Kim, N., Valverius, E.M., Lippman, M.E., and Salomon, D.S.,** Transformation of mouse mammary epithelial cells with the Ha-*ras* but not with the *neu* oncogene results in a gene dosage-dependent increase in transforming growth factor-α production, *FEBS Lett.,* 250, 474, 1989.
186. **Perucho, M. and Massagué, J.,** Reversible induction of transforming growth factor-α by human *ras* oncogenes, *J. Tumor Marker Oncol.,* 1, 81, 1986.
187. **Buick, R.N., Filmus, J., and Quaroni, A.,** Activated H-*ras* transforms rat intestinal epithelial cells with expression of α-TGF, *Exp. Cell Res.,* 170, 300, 1987.
188. **Ciardiello, F., Kim, N., Hynes, N., Jaggi, R., Redmont, S., Liscia, D.S., Sanfilippo, B., Merlo, G., Callahan, R., Kidwell, W.R., and Salomon, D.S.,** *Mol. Endocrinol.,* 2, 1202, 1988.
189. **Yamamoto, T., Hattori, T., and Tahara, E.,** Interaction between transforming growth factor-α and c-Ha-*ras* p21 in progression of human gastric carcinoma, *Pathol. Res. Pract.,* 183, 663, 1988.
190. **Murakami, H., Sanderson, N.D., Nagy, P., Marino, P.A., Merlino, G., and Thorgeirsson, S.S.,** Transgenic mouse model for synergistic effects of nuclear oncogenes and growth factors in tumorigenesis: interaction of c-*myc* and transforming growth factor α in hepatic oncogenesis, *Cancer Res.,* 53, 1719, 1993.
191. **Vassar, R., Hutton, M.E., and Fuchs, E.,** Transgenic overexpression of transforming growth factor α bypasses the need for c-Ha-*ras* mutations in mouse skin tumorigenesis, *Mol. Cell. Biol.,* 12, 4643, 1992.
192. **Sporn, M.B., Roberts, A.B., Wakefield, L.M., and Assoian, R.K.,** Transforming growth factor-β: biological function and chemical structure, *Science,* 233, 532, 1986.
193. **Moses, H.L., Coffey, R.J., Jr., Leof, E.B., Lyons, R.M., and Keski-Oja, J.,** Transforming growth factor β regulation of cell proliferation, *J. Cell. Physiol.,* Suppl. 5, 1, 1987.
194. **Roberts, A.B. and Sporn, M.B.,** Transforming growth factor-β, *Adv. Cancer Res.,* 51, 107, 1988.
195. **Rodland, K.D., Muldoon, L.L., and Magun, B.E.,** Cellular mechanisms of TGF-β action, *J. Invest. Dermatol.,* 94, 33S, 1990.
196. **Barnard, J.A., Lyons, R.M., and Moses, H.L.,** The cell biology of transforming growth factor β, *Biochim. Biophys. Acta,* 1032, 79, 1990.
197. **Sporn, M.B. and Roberts, A.B.,** TGF-β — problems and prospects, *Cell Regul.,* 1, 875, 1990.
198. **Massagué, J., Cheifetz, S., Laiho, M., Ralph, D.A., Weis, F.M.B., and Zentella, A.,** Transforming growth factor β, *Cancer Surv.,* 12, 81, 1992.
199. **Sporn, M.B. and Roberts, A.B.,** Transforming growth factor β — Recent progress and new challenges, *J. Cell Biol.,* 119, 1017, 1992.
200. **Laiho, M. and Keski-Oja, J.,** Transforming growth factors-β as regulators of cellular growth and phenotype, *Crit. Rev. Oncogenesis,* 3, 1, 1992.
201. **Border, W.A., Okuka, S., Languino, L.R., Sporn, M.B., and Ruoslahti, E.,** Suppression of experimental glomerulonephritis by antiserum against transforming growth factor β1, *Nature,* 346, 371, 1990.
202. **Massagué, J.,** The TGF-β family of growth and differentiation factors, *Cell,* 49, 437, 1987.
203. **Massagué, J.,** The transforming growth factor-β family, *Annu. Rev. Cell Biol.,* 6, 597, 1990.
204. **Burt, D.W. and Paton, I.R.,** Evolutionary origins of the transforming growth factor-β gene family, *DNA Cell Biol.,* 11, 497, 1992.
205. **Jakowlew, S.B., Dillard, P.J., Sporn, M.B., and Roberts, A.B.,** Complementary deoxyribonucleic acid cloning of a messenger ribonucleic acid encoding transforming growth factor β4 from chick embryo chondrocytes, *Mol. Endocrinol.,* 2, 1186, 1988.
206. **Bascom, C.C., Wolfshohl, J.R., Coffey, R.J., Jr., Madisen, L, Webb, N.R., Purchio, A.R., Derynck, R., and Moses, H.L.,** Complex regulation of transforming growth factor β1, β2, and β3 mRNA expression in mouse fibroblasts and keratinocytes by transforming growth factors β1 and β2, *Mol. Cell. Biol.,* 9, 5508, 1989.

207. **Millan, F.A., Denhez, F., Kondaiah, P., and Akhurst, R.J.,** Embryonic gene expression patterns of TGF β1, β2, and β3 suggest different developmental functions *in vivo, Development,* 111, 131, 1991.
208. **Assoian, R.K., Komoriya, A., Meyers, C.A., Miller, D.M., and Sporn, M.B.,** Transforming growth factor-β in human platelets: identification of a major storage site, purification and characterization, *J. Biol. Chem.,* 258, 7155, 1983.
209. **Keski-Oja, J., Leof, E.B., Lyons, R.M., Coffey, R.J., Jr., and Moses, H.L.,** Transforming growth factors and control of neoplastic growth, *J. Cell. Biochem.,* 33, 95, 1987.
210. **Grotendorst, G.R., Smale, G., and Pencev, D.,** Production of transforming growth factor β by human peripheral blood monocytes and neutrophils, *J. Cell. Physiol.,* 140, 396, 1989.
211. **Kehrl, J.H., Wakefield, L.M., Roberts, A.B., Jakowlew, S., Alvarez-Mon, M., Derynck, R., Sporn, M.B., and Fauci, A.S.,** Production of transforming growth factor β by human T lymphocytes, and its potential role in the regulation of cell growth, *J. Exp. Med.,* 163, 1037, 1986.
212. **Fava, R.A., Casey, T.T., Wilcox, J., Pelton, R.W., Moses, H.L., and Nanney, L.B.,** Synthesis of transforming growth factor-β1 by megakaryocytes and its localization to megakaryocyte and platelet α-granules, *Blood,* 76, 1946, 1990.
213. **Gerwin, B.I., Lechner, J.F., Reddel, R.R., Roberts, A.B., Robbins, K.C., Gabrielson, E.W., and Harris, C.C.,** Comparison of production of transforming growth factor-β and platelet-derived growth factor by normal human mesothelial cells and mesothelioma cell lines, *Cancer Res.,* 47, 6180, 1987.
214. **Nakamura, T.,** Growth control of mature hepatocytes by growth factor and growth inhibitor from platelets, *Gunma Symp. Endocrinol. (Tokyo),* 25, 179, 1988.
215. **Kanzaki, T., Olofsson, A., Morén, A., Wernstedt, C., Hellman, U., Miyazono, K., Claesson-Welsh, L., and Heldin, C.-H.,** TGF-β1 binding protein: a component of the large latent complex of TGF-β1 with multiple repeat sequences, *Cell,* 61, 1051, 1990.
216. **Lyons, R.M., Keski-Oja, J., and Moses, H.L.,** Proteolytic activation of latent transforming growth factor-β from fibroblast-conditioned medium, *J. Cell Biol.,* 106, 1659, 1988.
217. **Antonelli-Orlidge, A., Saunders, K.B., Smith, S.R., and D'Amore, P.A.,** An activated form of transforming growth factor β is produced by cocultures of endothelial cells and pericytes, *Proc. Natl. Acad. Sci. U.S.A.,* 86, 4544, 1989.
218. **Stromberg, K. and Twardzik, D.R.,** A β-type transforming growth factor, present in conditioned cell culture medium independent of cell transformation, may derive from serum, *J. Cell. Biochem.,* 27, 443, 1985.
219. **Kothari, V., Advani, S.H., and Rao, S.G.A.,** Growth factors in chronic myelogenous leukemia, *Cancer Lett.,* 32, 285, 1986.
220. **Nakamura, T., Kitazawa, T., and Ichihara, A.,** Partial purification and characterization of masking protein for β-type transforming growth factor from rat platelets, *Biochem. Biophys. Res. Commun.,* 141, 176, 1986.
221. **Tsuji, T., Okada, F., Yamaguchi, K., and Nakamura, T.,** Molecular cloning of the large subunit of transforming growth factor type β masking protein and expression of the mRNA in various rat tissues, *Proc. Natl. Acad. Sci. U.S.A.,* 87, 8835, 1990.
222. **O'Connor-McCourt, M.D. and Wakefield, L.M.,** Latent transforming growth factor-β in serum. A specific complex with α_2-macroglobulin, *J. Biol. Chem.,* 262, 14090, 1987.
223. **Huang, S.S., O'Grady, P., and Huang, J.S.,** Human transforming growth factor β-α_2-macroglobulin complex is a latent form of transforming growth factor β, *J. Biol. Chem.,* 263, 1535, 1988.
224. **Coffey, R.J., Jr., Kost, L.J., Lyons, R.M., Moses, H.L., and LaRusso, N.F.,** Hepatic processing of transforming growth factor β in the rat. Uptake, metabolism, and biliary excretion, *J. Clin. Invest.,* 80, 750, 1987.
225. **Ranganathan, G., Lyons, R., Jiang, N.-S., and Moses, H.,** Transforming growth factor type β in normal human urine, *Biochem. Biophys. Res. Commun.,* 148, 1503, 1987.
226. **Fujii, D., Brissenden, J.E., Derynck, R., and Francke, U.,** Transforming growth factor β gene maps to human chromosome 19 long arm and to mouse chromosome 7, *Somatic Cell Mol. Genet.,* 12, 281, 1986.
227. **Derynck, R., Jarrett, J.A., Chen, E.Y., Eaton, D.H., Bell, J.R., Assoian, R.K., Roberts, A.B., Sporn, M.B., and Goeddel, D.V.,** Human transforming growth factor-β complementary DNA sequence and expression in normal and transformed cells, *Nature,* 316, 701, 1985.
228. **Kim, S.-J., Jeang, K.-T., Glick, A.B., Sporn, M.B., and Roberts, A.B.,** Promoter sequences of the human transforming growth factor-β1 gene responsive to transforming growth factor-β1 autoinduction, *J. Biol. Chem.,* 264, 7041, 1989.

229. **Geiser, A.G., Kim, S.-J., Roberts, A.B., and Sporn, M.B.,** Characterization of the mouse transforming growth factor-β1 promoter and activation by the Ha-*ras* oncogene, *Mol. Cell. Biol.,* 11, 84, 1991.
230. **Kim, S.-J., Lee, H.-D., Robbins, P.D., Busam, K., Sporn, M.B., and Roberts, A.M.,** Regulation of transforming growth factor β1 gene expression by the product of the retinoblastoma-susceptibility gene, *Proc. Natl. Acad. Sci. U.S.A.,* 88, 3052, 1991.
231. **Fernandez-Pol, J.A., Klos, D.J., and Grant, G.A.,** Purification and biological properties of type β transforming growth factor from mouse transformed cells, *Cancer Res.,* 46, 5153, 1986.
232. **Gentry, L.E., Lioubin, M.N., Purchio, A.F., and Marquardt, H.,** Molecular events in the processing of recombinant type 1 pre-pro-transforming growth factor β to the mature polypeptide, *Mol. Cell. Biol.,* 8, 4162, 1988.
233. **Brunner, A.M., Gentry, L.E., Cooper, J.A., and Purchio, A.F.,** Recombinant type 1 transforming growth factor β precursor produced in Chinese hamster ovary cells is glycosylated and phosphorylated, *Mol. Cell. Biol.,* 8, 2229, 1988.
234. **Sha, X., Brunner, A.M., Purchio, A.F., and Gentry, L.E.,** Transforming growth factor β1: importance of glycosylation and acidic proteases for processing and secretion, *Mol. Endocrinol.,* 3, 1090, 1989.
235. **Derynck, R., Jarrett, J.A., Chen, E.Y., and Goeddel, D.V.,** The murine transforming growth factor-β precursor, *J. Biol. Chem.,* 261, 4377, 1986.
236. **Okada, F., Yamaguchi, K., Ichihara, A., and Nakamura, T.,** One of two subunits of masking protein in latent TGF-β is a part of pro-TGF-β, *FEBS Lett.,* 242, 240, 1989.
237. **Marquardt, H., Lioubin, M.N., and Ikeda, T.,** Complete amino acid sequence of human transforming growth factor type β2, *J. Biol. Chem.,* 262, 12127, 1987.
238. **Mason, A.J., Hayflick, J.S., Ling, N., Esch, F., Ueno, N., Ying, S.-H., Guillemin, R., Niall, H., and Seeburg, P.H.,** Complementary DNA sequences of ovarian follicular fluid inhibin show precursor structure homology with transforming growth factor-β, *Nature,* 318, 659, 1985.
239. **Ying, S.-Y., Becker, A., Ling, N., Ueno, N., and Guillemin, R.,** Inhibin and β type transforming growth factor (TGF β) have opposite modulating effects on the follicle stimulating hormone (FSH)-induced aromatase activity of cultured rat granulosa cells, *Biochem. Biophys. Res. Commun.,* 136, 969, 1986.
240. **Seyedin, S.M., Thomas, T.C., Thompson, A.Y., Rosen, D.M., and Piez, K.A.,** Purification and characterization of two cartilage-inducing factors from bovine demineralized bone, *Proc. Natl. Acad. Sci. U.S.A.,* 82, 2267, 1985.
241. **Seyedin, S.M., Thompson, A.Y., Bentz, H., Rosen, D.M., McPherson, M., Conti, A., Siegel, N.R., Galluppi, G.R., and Piez, K.A.,** Cartilage-inducing growth factor-A: apparent identity to transforming growth factor-β, *J. Biol. Chem.,* 261, 5693, 1986.
242. **Ellingsworth, L.R., Brennan, J.E., Fok, K., Rosen, D.M., Bentz, H., Piez, K.A., and Seyedin, S.M.,** Antibodies to the N-terminal portion of cartilage-inducing factor A and transforming growth factor β: immunochemistry localization and association with differentiating cells, *J. Biol. Chem.,* 261, 12362, 1986.
243. **Seyedin, S.M., Segarini, P.R., Rosen, D.M., Thompson, A.Y., Bentz, H., and Graycar, J.,** Cartilage-inducing factor-B is a unique protein structurally and functionally related to transforming growth factor-β, *J. Biol. Chem.,* 262, 1946, 1987.
244. **Ranchalis, J.E., Gentry, L., Ogawa, Y., Seyedin, S.M., McPherson, J., Purchio, A., and Twardzik, D.R.,** Bone-derived and recombinant transforming growth factor β are potent inhibitors of tumor cell growth, *Biochem. Biophys. Res. Commun.,* 148, 783, 1987.
245. **Segarini, P.R., Roberts, A.B., Rosen, D.M., and Seyedin, S.M.,** Membrane binding characteristics of two forms of transforming growth factor-β, *J. Biol. Chem.,* 262, 14655, 1987.
246. **Padgett, R.W., St. Johnston, R.D., and Gelbart, W.M.,** A transcript from a *Drosophila* pattern gene predicts a protein homologous to the transforming growth factor-β family, *Nature,* 325, 81, 1987.
247. **Panganiban, G.E.F., Rashka, K.E., Neitzel, M.D., and Hoffmann, F.M.,** Biochemical characterization of the *Drosophila dpp* protein, a member of the transforming growth factor β family of growth factors, *Mol. Cell. Biol.,* 10, 2669, 1990.
248. **Tucker, R.F., Branum, E.L., Shipley, G.D., Ryan, R.J., and Moses, H.L.,** Specific binding to cultured cells of ^{125}I-labeled type β transforming growth factor from human platelets, *Proc. Natl. Acad. Sci. U.S.A.,* 81, 6757, 1984.

249. **Coffey, R.J., Jr., Leof, E.B., Shipley, G.D., and Moses, H.L.,** Suramin inhibition of growth factor receptor binding and mitogenicity in AKR-2B cells, *J. Cell. Physiol.,* 132, 143, 1987.
250. **Massagué, J. and Kelly, B.,** Internalization of transforming growth factor-β and its receptor in BALB/c 3T3 fibroblasts, *J. Cell. Physiol.,* 128, 216, 1986.
251. **Massagué, J.,** Subunit structure of a high-affinity receptor for type-β transforming growth factor: evidence for a disulfide-linked glycosylated receptor complex, *J. Biol. Chem.,* 260, 7059, 1985.
252. **Massagué, J., Cheifetz, S., Ignotz, R.A., and Boyd, F.T.,** Multiple type-β transforming growth factors and their receptors, *J. Cell. Physiol.,* Suppl. 5, 43, 1987.
253. **Massagué, J.,** Receptors for the TGF-β family, *Cell,* 69, 1067, 1992.
254. **Cheifetz, S., Like, B., and Massagué, J.,** Cellular distribution of type I and type II receptors for transforming growth factor-β, *J. Biol. Chem.,* 261, 9972, 1986.
255. **Cheifetz, S., Andres, J.L., and Massagué, J.,** The transforming growth factor-β receptor type III is a membrane proteoglycan. Domain structure of the receptor, *J. Biol. Chem.,* 263, 16984, 1988.
256. **Segarini, P.R. and Seyedin, S.M.,** The high molecular weight receptor to transforming growth factor-β contains glycosaminoglycan chains, *J. Biol. Chem.,* 263, 8366, 1988.
257. **Segarini, P.R., Rosen, D.M., and Seyedin, S.M.,** Binding of transforming growth factor-β to cell surface proteins varies with cell type, *Mol. Endocrinol.,* 3, 261, 1989.
258. **Ellingsworth, L., Nakayama, D., Dasch, J., Segarini, P., Carrillo, P., and Waegell, W.,** Transforming growth factor β1 (TGF-β1) receptor expression on resting and mitogen-activated T cells, *J. Cell. Biochem.,* 39, 489, 1989.
259. **Cheifetz, S., Ling, N., Guillemin, R., and Massagué, J.,** A surface component on GH3 pituitary cells that recognizes transforming growth factor-β, activin, and inhibin, *J. Biol. Chem.,* 263, 17225, 1988.
260. **O'Grady, P., Kuo, M.-D., Baldassare, J.J., Huang, S.S., and Huang, J.S.,** Purification of a new type high molecular weight receptor (type V receptor) of transforming growth factor β (TGFβ), *J. Biol. Chem.,* 266, 8583, 1991.
261. **López-Casillas, F., Cheifetz, S., Doody, J., Andres, J.L., Lane, W.S., and Massagué, J.,** Structure and expression of the membrane proteoglycan betaglycan, a component of the TGF-β receptor system, *Cell,* 67, 785, 1991.
262. **Wang, X.-F., Lin, H.Y., Ng-Eaton, E., Downward, J., Lodish, H.F., and Weinberg, R.A.,** Expression cloning and characterization of the TGF-β type III receptor, *Cell,* 67, 797, 1991.
263. **Morén, A., Ichijo, H., and Miyazono, K.,** Molecular cloning and characterization of the human and porcine transforming growth factor-β type III receptors, *Biochem. Biophys. Res. Commun.,* 189, 356, 1992.
264. **Gougos, A. and Letarte, M.,** Primary structure of endoglin, an RGD-containing glycoprotein of human endothelial cells, *J. Biol. Chem.,* 265, 8361, 1990.
265. **Lin, H.Y., Wang, X.-F., Ng-Eaton, E., Weinberg, R.A., and Lodish, H.F.,** Expression cloning of the TGF-β type II receptor, a functional transmembrane serine/threonine kinase, *Cell,* 68, 775, 1992.
266. **Wrana, J.L., Attisano, L., Carcamo, J., Zentella, A., Doody, J., Laiho, M., Wang, X.F., and Massagué, J.,** TGF-β signals through a heteromeric protein kinase receptor complex, *Cell,* 71, 1003, 1992.
267. **Rizzino, A.,** Appearance of high affinity receptors for type β transforming growth factor during differentiation of murine embryonal carcinoma cells, *Cancer Res.,* 47, 4386, 1987.
268. **Libby, J., Martinez, R., and Weber, M.J.,** Tyrosine phosphorylation in cells treated with transforming growth factor-β, *J. Cell. Physiol.,* 129, 159, 1986.
269. **Shipley, G.D., Tucker, R.F., and Moses, H.L.,** Type β transforming growth factor/growth inhibitor stimulates entry of monolayer cultures of AKR-2B cells into S phase after a prolonged prereplicative interval, *Proc. Natl. Acad. Sci. U.S.A.,* 82, 4147, 1985.
270. **Madhukar, B.V., Oh, S.Y., Chang, C.C., Wade, M., and Trosko, J.E.,** Altered regulation of intercellular communication by epidermal growth factor, transforming growth factor-β and peptide hormones in normal human keratinocytes, *Carcinogenesis,* 10, 13, 1989.
271. **Murthy, U.S., Anzano, M.A., Stadel, J.M., and Greig, R.,** Coupling of TGF-β-induced mitogenesis to G-protein activation in AKR-2B cells, *Biochem. Biophys. Res. Commun.,* 152, 1228, 1988.
272. **Howe, P.H., Bascom, C.C., Cunningham, M.R., and Leof, E.B.,** Regulation of transforming growth factor β1 action by multiple transducing pathways: evidence for both G protein-dependent and -independent signaling, *Cancer Res.,* 49, 6024, 1989.

273. **Diaz, A., Varga, J., and Jimenez, S.A.,** Transforming growth factor-β stimulation of lung fibroblast prostaglandin E_2 production, *J. Biol. Chem.,* 264, 11554, 1989.
274. **Muldoon, L.L., Rodland, K.D., and Magun, B.E.,** Transforming growth factor β modulates epidermal growth factor-induced phosphoinositide metabolism and intracellular calcium levels, *J. Biol. Chem.,* 263, 5030, 1988.
275. **Muldoon, L.L., Rodland, K.D., and Magun, B.E.,** Transforming growth factor β and epidermal growth factor alter calcium influx and phosphoinositol turnover in Rat-1 fibroblasts, *J. Biol. Chem.,* 263, 18834, 1988.
276. **Giannini, G., Clementi, E., Ceci, R., Marziali, G., and Sorrentino, V.,** Expression of ryanodine receptor-Ca^{2+} channel that is regulated by TGF-β, *Science,* 257, 91, 1992.
277. **Reibman, J., Meixler, S., Lee, T.C., Gold, L.I., Cronstein, B.N., Haines, K.A., Kolasinski, S.L., and Weissmann, G.,** Transforming growth factor β1, a potent chemoattractant for human neutrophils, bypasses classic signal-transduction pathways, *Proc. Natl. Acad. Sci. U.S.A.,* 88, 6805, 1991.
278. **Gruppuso, P.A., Mikumo, R., Brautigan, D.L., and Braun, L.,** Growth arrest induced by transforming growth factor β1 is accompanied by protein phosphatase activation in human keratinocytes, *J. Biol. Chem.,* 266, 3444, 1991.
279. **Hassan, J.H., Chelucci, C., Peschle, C., and Sorrentino, V.,** Transforming growth factor β (TGF-β) inhibits expression of fibrinogen and factor VII in a hepatoma cell line, *Thromb. Haemost.,* 67, 478, 1992.
280. **Morrone, G., Poli, V., Hassan, J.H., and Sorrentino, V.,** Effect of TGF β on liver genes expression. Antagonistic effect of TGF β on IL-6-stimulated genes in Hep 3B cells, *FEBS Lett.,* 301, 1, 1992.
281. **Thompson, K.L., Assoian, R., and Rosner, M.R.,** Transforming growth factor-β increases transcription of the genes encoding the epidermal growth factor receptor and fibronectin in normal rat kidney fibroblasts, *J. Biol. Chem.,* 263, 19519, 1988.
282. **Leof, E.B., Proper, J.A., Getz, M.J., and Moses, H.L.,** Transforming growth factor type β regulation of actin mRNA, *J. Cell. Physiol.,* 127, 83, 1986.
283. **Matrisian, L.M., Ganser, G.L., Kerr, L.D., Pelton, R.W., and Wood, L.D.,** Negative regulation of gene expression by TGF-β, *Mol. Reprod. Develop.,* 32, 111, 1992.
284. **Kerr, L.D., Olashaw, N.E., and Matrisian, L.M.,** Transforming growth factor β1 and cAMP inhibit transcription of epidermal growth factor- and oncogene-induced transin RNA, *J. Biol. Chem.,* 263, 16999, 1988.
285. **Kerr, L.D., Miller, D.B., and Matrisian, L.M.,** TGF-β1 inhibition of transin/stromelysin gene expression is mediated through a Fos binding sequence, *Cell,* 61, 267, 1990.
286. **Kallin, B., de Martin, R., Etzold, T., Sorrentino, V., and Philipson, L.,** Cloning of a growth arrest-specific and transforming growth factor β-regulated gene, TI 1, from an epithelial cell line, *Mol. Cell. Biol.,* 11, 5338, 1991.
287. **Noda, M., Yoon, K., Prince, C.W., Butler, W.T., and Rodan, G.A.,** Transcriptional regulation of osteopontin production in rat osteosarcoma cells by type β transforming growth factor, *J. Biol. Chem.,* 263, 13916, 1988.
288. **Noda, M.,** Transcriptional regulation of osteocalcin production by transforming growth factor-β in rat osteoblast-like cells, *Endocrinology,* 124, 612, 1989.
289. **Kramer, I., Koornneef, I., de Laat, S.W., and van den Eijnden-van Raaij, A.J.M.,** TGF-β1 induces phosphorylation of the cyclic AMP responsive element binding protein in ML-CC164 cells, *EMBO J.,* 10, 1083, 1991.
290. **Daniel, T.O., Gibbs, V.C., Milfay, D.F., and Williams, L.T.,** Agents that increase cAMP accumulation block endothelial c-*sis* induction by thrombin and transforming growth factor-β, *J. Biol. Chem.,* 262, 11893, 1987.
291. **Leof, E.B., Proper, J.A., Goustin, A.S., Shipley, G.D., DiCorleto, P.E., and Moses, H.L.,** Induction of c-*sis* mRNA and activity similar to platelet-derived growth factor by transforming growth factor β: a proposed model for indirect mitogenesis involving autocrine activity, *Proc. Natl. Acad. Sci. U.S.A.,* 83, 2453, 1986.
292. **Mäkelä, T.P., Alitalo, R., Paulsson, Y., Westermark, B., Heldin, C.-H., and Alitalo, K.,** Regulation of platelet-derived growth factor gene expression by transforming growth factor β and phorbol ester in human leukemia cell lines, *Mol. Cell. Biol.,* 7, 3656, 1987.
293. **Mulder, K.M., Levine, A.E., Hernandez, X., McKnight, M.K., Brattain, D.E., and Brattain, M.G.,** Modulation of c-*myc* by transforming growth factor-β in human colon carcinoma cells, *Biochem. Biophys. Res. Commun.,* 150, 711, 1988.

294. **Mulder, K.M. and Brattain, M.G.,** Alterations in c-*myc* expression in relation to maturational status of human colon carcinoma cells, *Int. J. Cancer,* 42, 64, 1988.
295. **Mulder, K.M., Humphrey, L.E., Choi, H.G., Childress-Fields, K.E., and Brattain, M.G.,** Evidence for c-*myc* in the signaling pathway for TGF-β in well-differentiated human colon carcinoma cells, *J. Cell. Physiol.,* 145, 501, 1990.
296. **Ito, N., Kawata, S., Tamura, S., Takaishi, K., Saitoh, R., and Tarui, S.,** Modulation of c-*myc* expression by transforming growth factor β1 in human hepatoma cell lines, *Jpn. J. Cancer Res.,* 81, 216, 1990.
297. **Fernandez-Pol, J.A., Talkad, V.D., Klos, D.J., and Hamilton, P.D.,** Suppression of the EGF-dependent induction of c-*myc* proto-oncogene expression by transforming growth factor β in a human breast carcinoma cell line, *Biochem. Biophys. Res. Commun.,* 144, 1197, 1987.
298. **Coffey, R.J., Jr., Bascom, C.C., Sipes, N.J., Graves-Deal, R., Weissman, B.E., and Moses, H.L.,** Selective inhibition of growth-related gene expression in murine keratinocytes by transforming growth factor β, *Mol. Cell. Biol.,* 8, 3088, 1988.
299. **Liboi, E., Di Francesco, P., Gallinari, P., Testa, U., Rossi, G.B., and Peschle, C.,** TGF β induces a sustained c-*fos* expression associated with stimulation or inhibition of cell growth in EL2 or NIH 3T3 fibroblasts, *Biochem. Biophys. Res. Commun.,* 151, 298, 1988.
300. **Pietenpol, J.A., Holt, J.T., Stein, R.W., and Moses, H.L.,** Transforming growth factor β1 suppression of c-*myc* gene transcription: role in inhibition of keratinocyte proliferation, *Proc. Natl. Acad. Sci. U.S.A.,* 87, 3758, 1990.
301. **Pertovaara, L., Sistonen, L., Bos, T.J., Vogt, P.K., Keski-Oja, J., and Alitalo, K.,** Enhanced *jun* gene expression is an early genomic response to transforming growth factor β stimulation, *Mol. Cell. Biol.,* 9, 1255, 1989.
302. **Kramer, I., Koorneeff, I., de Vries, C., de Groot, R.P., de Laat, S.W., van de Eijnden-van Raaij, A.J.M., and Kruijer, W.,** Phosphorylation of nuclear protein is an early event in TGF β1 action, *Biochem. Biophys. Res. Commun.,* 175, 816, 1991.
303. Beauchamp, R.D., Sheng, H.-M., Ishizuka, J., Townsend, C.M., Jr., and Thompson, J.C., Transforming growth factor (TGF)-β stimulates hepatic *jun*-B and *fos*-B proto-oncogenes and decreases albumin mRNA, *Ann. Surg.,* 216, 300, 1992.
304. **Koo, A.S., Chiu, R., Soong, J., Dekernion, J.B., and Belldegrun, A.,** The expression of c-*jun* and *jun*B mRNA in renal cell cancer and *in vitro* regulation by transforming growth factor β1 and tumor necrosis factor α1, *J. Urol.,* 148, 1314, 1992.
305. **Mulder, K.M. and Morris, S.L.,** Activation of $p21^{ras}$ by transforming growth factor β in epithelial cells, *J. Biol. Chem.,* 267, 5029, 1992.
306. **Howe, P.H., Cunningham, M.R., and Leof, E.B.,** Distinct pathways regulate transforming growth factor β1-stimulated proto-oncogene and extracellular matrix gene expression, *J. Cell. Physiol.,* 142, 39, 1990.
307. **Laiho, M., DeCaprio, J.A., Ludlow, J.W., Livingston, D.M., and Massagué, J.,** Growth inhibition by TGF-β linked to suppression of retinoblastoma protein phosphorylation, *Cell,* 62, 175, 1990.
308. **Whitson, R.H., Jr. and Itakura, K.,** TGF-β_1 inhibits DNA synthesis and phosphorylation of the retinoblastoma gene product in a rat liver epithelial cell line, *J. Cell. Biochem.,* 48, 305, 1992.
309. **Yan, Z., Hsu, S., Winawer, S., and Friedman, E.,** Transforming growth factor β1 (TGF-β1) inhibits retinoblastoma gene expression but not pRB phosphorylation in TGF-β1-growth stimulated colon carcinoma cells, *Oncogene,* 7, 801, 1992.
310. **Pietenpol, J.A., Stein, R.W., Moran, E., Yaciuk, P., Schlegel, R., Lyons, R.M., Pittelkow, M.R., Münger, K., Howley, P.M., and Moses, H.L.,** TGF-β1 inhibition of c-*myc* transcription and growth in keratinocytes is abrogated by viral transforming proteins with pRB binding domains, *Cell,* 61, 777, 1990.
311. **Hatzfeld, J., Li, M.-L., Brown, E.L., Sookdeo, H., Levesque, J.-P., O'Toole, T., Gurney, C., Clark, S.C., and Hatzfeld, A.,** Release of early human hematopoietic progenitors from quiescence by antisense transforming growth factor β1 or Rb oligonucleotides, *J. Exp. Med.,* 174, 925, 1991.
312. **Münger, K., Pietenpol, J.A., Pittelkow, M.R., Holt, J.T., and Moses, H.L.,** Transforming growth factor β_1 regulation of c-*myc* expression, pRB phosphorylation, and cell cycle progression in keratinocytes, *Cell Growth Differ.,* 3, 291, 1992.
313. **Suzuki, K., Ono, T., and Takahashi, K.,** Inhibition of DNA synthesis by TGF-β1 coincides with inhibition of phosphorylation and cytoplasmic translocation of p53 protein, *Biochem. Biophys. Res. Commun.,* 183, 1175, 1992.

314. **Reed, B.Y., King, M.T., Gitomer, W.L., and Veech, R.L.,** Early metabolic effects of platelet-derived growth factor and transforming growth factor-β in rat liver *in vivo, J. Biol. Chem.,* 262, 8712, 1987.
315. **Mackiewicz, A., Ganapathi, M.K., Schultz, D., Brabenec, A., Weinstein, J., Kelley, M.F., and Kushner, I.,** Transforming growth factor β_1 regulates production of acute phase proteins, *Proc. Natl. Acad. Sci. U.S.A.,* 87, 1581, 1990.
316. **Gupta, P., Franco-Saenz, R., and Mulrow, P.J.,** Transforming growth factor-β1 inhibits aldosterone biosynthesis in cultured bovine zona glomerulosa cells, *Endocrinology,* 132, 1184, 1993.
317. **Shull, M.M., Ormsby, I., Kier, A.B., Pawlowski, S., Diebold, R.J., Yin, M., Allen, R., Sidman, C., Proetzel, G., Calvin, D., Annunziata, N., and Doetschman, T.,** Targeted disruption of the mouse transforming growth factor-β1 gene results in multifocal inflammatory disease, *Nature,* 359, 693, 1992.
318. **Inman, W.H. and Colowick, S.P.,** Stimulation of glucose uptake by transforming growth factor β: evidence for the requirement of epidermal growth factor-receptor activation, *Proc. Natl. Acad. Sci. U.S.A.,* 82, 1346, 1985.
319. **Boerner, P., Resnick, R.J., and Racker, E.,** Stimulation of glycolysis and amino acid uptake in NRK-49F cells by transforming growth factor β and epidermal growth factor, *Proc. Natl Acad. Sci. U.S.A.,* 82, 1350, 1985.
320. **Kitagawa, T., Masumi, A., and Akamatsu, Y.,** Transforming growth factor-β1 simulates glucose uptake and the expression of glucose transporter messenger RNA in quiescent Swiss mouse 3T3 cells, *J. Biol. Chem.,* 266, 18066, 1991.
321. **Yamamoto, T., Nakamura, T., Noble, N.A., Ruoslahti, E., and Border, W.A.,** Expression of transforming growth factor-β is elevated in human and experimental diabetic nephropathy, *Proc. Natl. Acad. Sci. U.S.A.,* 90, 1814, 1993.
322. **Warburg, O.,** On the origin of cancer cells, *Science,* 123, 309, 1956.
323. **Racker, E., Resnick, R.J., and Feldman, R.,** Glycolysis and methylaminoisobutyrate uptake in rat-1 cells transfected with *ras* or *myc* oncogenes, *Proc. Natl. Acad. Sci. U.S.A.,* 82, 3535, 1985.
324. **Resnick, R.J., Feldman, R., Willard, J., and Racker, E.,** Effect of growth factors and methionine on glycolysis and methionine transport in rat fibroblasts and fibroblasts transfected with *myc* and *ras* genes, *Cancer Res.,* 46, 1800, 1986.
325. **Busso, N., Chesne, C., Delers, F., Morel, F., and Guillouzo, A.,** Transforming growth factor-β (TGF-β) inhibits albumin synthesis in normal human hepatocytes and in hepatoma HepG2 cells, *Biochem. Biophys. Res. Commun.,* 171, 647, 1990.
326. **Hooper, W.C.,** The role of transforming growth factor-β in hematopoiesis — A review, *Leukemia Res.,* 15, 179, 1991.
327. **Ottmann, O.G., and Pelus, L.M.,** Differential proliferative effects of transforming growth factor-β on human hematopoietic progenitor cells, *J. Immunol.,* 140, 2661, 1988.
328. **Keller, J.R., Jacobsen, S.E.W., Sill, K.T., Ellingsworth, L.R., and Ruscetti, F.W.,** Stimulation of granulopoiesis by transforming growth factor-β. Synergy with granulocyte-macrophate colony-stimulating factor, *Proc. Natl. Acad. Sci. U.S.A.,* 88, 7190, 1991.
329. **Celada, A. and Maki, R.A.,** Transforming growth factor-β enhances M-CSF and GM-CSF-stimulated proliferation of macrophages, *J. Immunol.,* 148, 1102, 1992.
330. **Sing, G.K., Keller, J.R., Ellingsworth, L.R., and Ruscetti, F.W.,** Transforming growth factor β selectively inhibits normal and leukemic human bone marrow cell growth *in vitro, Blood,* 72, 1504, 1988.
331. **Sing, G.K., Keller, J.R., Ellingsworth, L.R., and Ruscetti, F.W.,** Transforming growth factor-β1 enhances the suppression of human hematopoiesis by tumor necrosis factor-α or recombinant interferon-α, *J. Cell. Biochem.,* 39, 107, 1989.
332. **Keller, J.R., Sing, G.K., Ellingsworth, L.R., and Ruscetti, F.W.,** Transforming growth factor β: possible roles in the regulation of normal and leukemic hematopoietic cell growth, *J. Cell. Biochem.,* 39, 175, 1989.
333. **Kuter, D.J., Gminski, D.M., and Rosenberg, R.D.,** Transforming growth factor-β inhibits megakaryocyte growth and endomitosis, *Blood,* 79, 619, 1992.
334. **Carlino, J.A., Higley, H.R., Creson, J.R., Avis, P.D., Ogawa, Y., and Ellingsworth, L.R.,** Transforming growth factor β1 systemically modulates granuloid, erythroid, lymphoid, and thrombocytic cells in mice, *Exp. Hematol.,* 20, 943, 1992.
335. **Kehrl, J.H.,** Transforming growth factor-β — An important mediator of immunoregulation, *Int. J. Cell Cloning,* 9, 438, 1991.
336. **Sasaki, H., Pollard, R.B., Schmitt, D., and Suzuki, F.,** Transforming growth factor-β in the regulation of the immune response, *Clin. Immunol. Immunother.,* 65, 1, 1992.

337. **Kehrl, J.H., Roberts, A.B., Wakefield, L.M., Jakowlew, S., Sporn, M.B., and Fauci, A.S.,** Transforming growth factor β is an important immunomodulatory protein for human B lymphocytes, *J. Immunol.,* 137, 3855, 1986.
338. **Kim, P.-H. and Kagnoff, M.F.,** Transforming growth factor-β1 is a costimulator for IgA production, *J. Immunol.,* 144, 3411, 1990.
339. **Welch, G.R., Wong, H.L., and Wahl, S.M.,** Selective induction of FcγRIII on human monocytes by transforming growth factor-β, *J. Immunol.,* 144, 3444, 1990.
340. **Wahl, S.M., Hunt, D.A., Wakefield, L.M., McCartney-Francis, N., Wahl, L.M., Roberts, A.B., and Sporn, M.B.,** Transforming growth factor type β induces monocyte chemotaxis and growth factor production, *Proc. Natl. Acad. Sci. U.S.A.,* 84, 5788, 1987.
341. **Espevik, T., Figari, I.S., Shalaby, M.R., Lackides, G.A., Lewis, G.A., Shepard, H.M., and Palladino, M.A., Jr.,** Inhibition of cytokine production by cyclosporin A and transforming growth factor β, *J. Exp. Med.,* 166, 571, 1987.
342. **Espevik, T., Waage, A., Faxvaag, A., and Shalaby, M.R.,** Regulation of interleukin-2 and interleukin-6 production from T-cells: involvement of interleukin-1β and transforming growth factor-β, *Cell. Immunol.,* 126, 47, 1990.
343. **Espevik, T., Figari, I.S., Ranges, G.E., and Palladino, M.A., Jr.,** Transforming growth factor-β_1 (TGF-β_1) and recombinant human tumor necrosis factor-α reciprocally regulate the generation of lymphokine-activated killer cell activity. Comparison between natural porcine platelet-derived TGF-β_1, TGF-β_2, and recombinant human TGF-β_1, *J. Immunol.,* 140, 2312, 1988.
344. **Pircher, R., Jullien, P., and Lawrence, D.A.,** β-Transforming growth factor is stored in human blood platelets as a latent high molecular weight complex, *Biochem. Biophys. Res. Commun.,* 136, 30, 1986.
345. **Koshikawa, T., Morisaki, N., Saito, Y., and Yoshida, S.,** Bifunctional effects of transforming growth factor-β on migration of cultured rat aortic smooth muscle cells, *Biochem. Biophys. Res. Commun.,* 169, 725, 1990.
346. **Nicholson, A.C. and Hajjar, D.P.,** Transforming growth factor-β up-regulates low density lipoprotein receptor-mediated cholesterol metabolism in vascular smooth muscle cells, *J. Biol. Chem.,* 267, 25982, 1992.
347. **Yang, E.Y. and Moses, H.L.,** Transforming growth factor β1-induced changes in cell migration, proliferation, and angiogenesis in the chicken chorioallantoic membrane, *J. Cell Biol.,* 111, 731, 1990.
348. **Merwin, J.R., Anderson, J.M., Kocher, O., Van Itallie, C.M., and Madri, J.A.,** Transforming growth factor β_1 modulates extracellular matrix organization and cell-cell junctional complex formation during *in vitro* angiogenesis, *J. Cell. Physiol.,* 142, 117, 1990.
349. **Lefer, A.M., Tsao, P., Aoki, N., and Palladino, M.A., Jr.,** Mediation of cardioprotection by transforming growth factor-β, *Science,* 249, 61, 1990.
350. **Roberts, A.B., Sporn, M.B., Assoian, R.K., Smith, J.M., Roche, N.S., Wakefield, L.M., Heine, U.I., Liotta, L.A., Falanga, V., Kehrl, J.H., and Fauci, A.S.,** Transforming growth factor type β: rapid induction of fibrosis and angiogenesis *in vivo* and stimulation of collagen formation *in vitro, Proc. Natl. Acad. Sci. U.S.A.,* 83, 4167, 1986.
351. **Postlethwaite, A.E., Keski-Oja, J., Moses, H.L., and Kang, A.H.,** Stimulation of the chemotactic migration of transforming growth factor β, *J. Exp. Med.,* 165, 251, 1987.
352. **Ignotz, R.A. and Massagué, J.,** Transforming growth factor-β stimulates the expression of fibronectin and collagen and their incorporation into the extracellular matrix, *J. Biol. Chem.,* 261, 4337, 1986.
353. **Varga, J. and Jimenez, S.A.,** Stimulation of normal human fibroblast collagen production and processing by transforming growth factor-β, *Biochem. Biophys. Res. Commun.,* 138, 974, 1986.
354. **Müller, G., Behrens, J., Nussbaumer, U., Böhlen, P., and Birchmeier, W.,** Inhibitory action of transforming growth factor β on endothelial cells, *Proc. Natl. Acad. Sci. U.S.A.,* 84, 5600, 1987.
355. **Penttinen, R.P., Kobayashi, S., and Bornstein, P.,** Transforming growth factor β increases mRNA for matrix proteins in the presence and in the absence of changes in mRNA stability, *Proc. Natl. Acad. Sci. U.S.A.,* 85, 1105, 1988.
356. **Pierce, G.F., Mustoe, T.A., Lingelbach, J., Masakowski, V.R., Griffin, G.L., Senior, R.M., and Deuel, T.F.,** Platelet-derived growth factor and transforming growth factor-β enhance tissue repair activities by unique mechanisms, *J. Cell Biol.,* 109, 429, 1989.
357. **Raghow, R., Postlethwaite, A.E., Keski-Oja, J., Moses, H.L., and Kang, A.H.,** Transforming growth factor-β increases steady state levels of type I procollagen and fibronectin messenger RNAs posttranscriptionally in cultured human dermal fibroblasts, *J. Clin. Invest.,* 79, 1285, 1987.

358. **Ignotz, R.A., Endo, T., and Massagué, J.,** Regulation of fibronectin and type I collagen mRNA levels by transforming growth factor-β, *J. Biol. Chem.,* 262, 6443, 1987.
359. **Varga, J., Rosenbloom, J., and Jimenez, S.A.,** Transforming growth factor β (TGF β) causes a persistent increase in steady-state amounts of type I and type III collagen and fibronectin mRNAs in normal human dermal fibroblasts, *Biochem. J.,* 247, 597, 1987.
360. **Fine, A. and Goldstein, R.H.,** The effect of transforming growth factor-β on cell proliferation and collagen formation by lung fibroblasts, *J. Biol. Chem.,* 262, 3897, 1987.
361. **Roberts, C.J., Birkenmeier, T.M., McQuillan, J.J., Akiyama, S.K., Yamada, S.S., Chen, W.-T., Yamada, K.M., and McDonald, J.A.,** Transforming growth factor β stimulates the expression of fibronectin and of both subunits of the human fibronectin receptor by cultured human lung fibroblasts, *J. Biol. Chem.,* 263, 4586, 1988.
362. **Balza, E., Borsi, L., Allemanni, G., and Zardi, L.,** Transforming growth factor β regulates the levels of different fibronectin isoforms in normal human cultured fibroblasts, *FEBS Lett.,* 228, 42, 1988.
363. **Allen-Hoffmann, B.L., Crankshaw, C.L., and Mosher, D.F.,** Transforming growth factor β increases cell surface binding and assembly of exogenous (plasma) fibronectin by normal human fibroblasts, *Mol. Cell. Biol.,* 8, 4234, 1988.
364. **Ruoslahti, E. and Pierschbacher, M.D.,** New perspectives in cell adhesion: RGD and integrins, *Science,* 238, 491, 1987.
365. **Ruoslahti, E. and Giancotti, F.G.,** Integrins and tumor cell dissemination, *Cancer Cells,* 1, 119, 1989.
366. **Heino, J., Ignotz, R.A., Hemler, M.E., Crouse, C., and Massagué, J.,** Regulation of cell adhesion receptors by transforming growth factor-β. Concomitant regulation of integrins that share a common β_1 subunit, *J. Biol. Chem.,* 264, 380, 1989.
367. **Heino, J. and Massagué, J.,** Transforming growth factor-β_1 switches the pattern of integrins expressed in MG-63 human osteosarcoma cells and causes a selective loss of cell adhesion to laminin, *J. Biol. Chem.,* 264, 21806, 1989.
368. **Gerard, R.D., Chien, K.R., and Meidell, R.S.,** Molecular biology of tissue plasminogen activator and endogenous inhibitors, *Mol. Biol. Med.,* 3, 449, 1986.
369. **Laiho, M., Saksela, O., and Keski-Oja, J.,** Transforming growth factor-β induction of type 1 plasminogen activator inhibitor: pericellular deposition and sensitivity to exogenous urokinase, *J. Biol. Chem.,* 262, 17467, 1987.
370. **Keski-Oja, J., Blasi, F., Leof, E.B., and Moses, H.L.,** Regulation of the synthesis and activity of urokinase plasminogen activator in A549 human lung carcinoma cells by transforming growth factor β, *J. Cell Biol.,* 106, 451, 1988.
371. **Keski-Oja, J., Raghow, R., Sawdey, M., Loskutoff, D.J., Postlethwaite, A.E., Kang, A.H., and Moses, H.L.,** Regulation of mRNAs for type-1 plasminogen activator inhibitor, fibronectin, and type I procollagen by transforming growth factor-β: divergent responses in lung fibroblasts and carcinoma cells, *J. Biol. Chem.,* 263, 3111, 1988.
372. **Newman, M.J., Lane, E.A., Iannotti, A.M., Nugent, M.A., Pepinsky, R.B., and Keski-Oja, J.,** Characterization and purification of a secreted plasminogen activator inhibitor (PAI-1) induced by transforming growth factor-β1 in normal rat kidney (NRK) cells: decreased PAI-1 expression in transformed NRK cells, *Endocrinology,* 126, 2936, 1990.
373. **Ignotz, R.A. and Massagué, J.,** Cell adhesion protein receptors as targets for transforming growth factor-β action, *Cell,* 51, 189, 1987.
374. **Laiho, M., Saksela, O., and Keski-Oja, J.,** Transforming growth factor β alters plasminogen activator activity in human skin fibroblasts, *Exp. Cell Res.,* 164, 399, 1986.
375. **Lund, L.R., Riccio, A., Andreasen, P.A., Nielsen, L.S., Kristensen, P., Laiho, M., Saksela, O., Blasi, F., and Dano, K.,** Transforming growth factor-β is a strong and fast acting positive regulator of the level of type-1 plasminogen activator inhibitor mRNA in WI-38 human lung fibroblasts, *EMBO J.,* 6, 1281, 1987.
376. **Centrella, M., Massagué, J., and Canalis, E.,** Human platelet-derived transforming growth factor-β stimulates parameters of bone growth in fetal rat calvariae, *Endocrinology,* 119, 2306, 1986.
377. **Centrella, M., McCarthy, T.L., and Canalis, E.,** Skeletal tissue and transforming growth factor β, *FASEB J.,* 2, 3066, 1988.
378. **Centrella, M., McCarthy, T.L., and Canalis, E.,** Effects of transforming growth factors on bone cells, *Connect. Tissue Res.,* 20, 267, 1989.

379. **Finkelman, R.D., Bell, N.H., Strong, D.D., Demers, L.M., and Baylink, D.J.,** Ovariectomy selectively reduces the concentration of transforming growth factor β in rat bone: implications for estrogen deficiency-associated bone loss, *Proc. Natl. Acad. Sci. U.S.A.,* 89, 12190, 1992.
380. **Tschan, T., Böhme, K., Conscience-Egli, M., Zenke, G., Winterhalter, K.H., and Bruckner, P.,** Autocrine or paracrine transforming growth factor-β modulates the phenotype of chick embryo sternal chondrocytes in serum-free agarose culture, *J. Biol. Chem.,* 268, 5156, 1993.
381. **Gutierrez, G.E., Mundy, G.R., Manning, D.R., Hewlett, E.L., and Katz, M.S.,** Transforming growth factor β enhances parathyroid hormone stimulation of adenylate cyclase in clonal osteoblast-like cells, *J. Cell. Physiol.,* 144, 438, 1990.
382. **Tashjian, A.H., Jr., Voelkel, E.F., Lazzaro, M., Singer, F.R., Roberts, A.B., Derynck, R., Winkler, M.E., and Levine, L.,** Alpha and β human transforming growth factors stimulate prostaglandin production and bone resorption in cultured mouse calvaria, *Proc. Natl. Acad. Sci. U.S.A.,* 82, 4535, 1985.
383. **Pfeilschifter, J., Seyedin, S.M., and Mundy, G.R.,** Transforming growth factor β inhibits bone resorption in fetal rat long bone cultures, *J. Clin. Invest.,* 82, 680, 1988.
384. **Centrella, M. and Canalis, E.,** Transforming and nontransforming growth factors are present in medium conditioned by fetal rat calvariae, *Proc. Natl. Acad. Sci. U.S.A.,* 82, 7335, 1985.
385. **Canalis, E. and Centrella, M.,** Isolation of a nontransforming bone-derived growth factor from medium conditioned by fetal rat calvariae, *Endocrinology,* 118, 2002, 1986.
386. **Pfeilschifter, J. and Mundy, G.R.,** Modulation of type β transforming growth factor activity in bone cultures by osteotropic hormones, *Proc. Natl. Acad. Sci. U.S.A.,* 84, 2024, 1987.
387. **Noda, M. and Rodan, G.A.,** Type β transforming growth factor (TGF β) regulation of alkaline phosphatase expression and other phenotype-related mRNAs in osteoblastic rat osteosarcoma cells, *J. Cell. Physiol.,* 133, 426, 1987.
388. **Sandberg, M., Vuorio, T., Hirvonen, H., Alitalo, K., and Vuorio, E.,** Enhanced expression of TGF-β and c-*fos* mRNAs in the growth plates of developing human long bones, *Development,* 102, 461, 1988.
389. **Ibbotson, K.J., D'Souza, S.M., Ng, K.W., Osborne, C.K., Niall, M., Martin T.J., and Mundy, G.R.,** Tumor-derived growth factor increases bone resorption in a tumor associated with humoral hypercalcemia of malignancy, *Science,* 221, 1292, 1983.
390. **Mundy, G.R., Ibbotson, K.J., and D'Souza, S.M.,** Tumor products and the hypercalcemia of malignancy, *J. Clin. Invest.,* 76, 391, 1985.
391. **Linkhart, T.A., Mohan, S., Jennings, J.C., and Baylink, D.J.,** Copurification of osteolytic and transforming growth factor β activities produced by human lung tumor cells associated with humoral hypercalcemia of malignancy, *Cancer Res.,* 49, 271, 1989.
392. **Horiuchi, N., Caulfield, M.P., Fisher, J.E., Goldman, M.E., McKee, R.L., Reagan, J.E., Levy, J.J., Nutt, R.F., Rodan, S.B., Schofield, T.L., Clemens, T.L., and Rosenblatt, M.,** Similarity of synthetic peptide from human tumor to parathyroid hormone *in vivo* and *in vitro, Science,* 238, 1566, 1987.
393. **Gutierrez, G.E., Mundy, G.R., Derynck, R., Hewlett, E.L., and Katz, M.S.,** Inhibition of parathyroid hormone-responsive adenylate cyclase in clonal osteoblast-like cells by transforming growth factor α and epidermal growth factor, *J. Biol. Chem.,* 262, 15845, 1987.
394. **Centrella, M., McCarthy, T.L., and Canalis, E.,** Parathyroid hormone modulates transforming growth factor β activity and binding in osteoblast-enriched cell cultures from fetal rat parietal bone, *Proc. Natl. Acad. Sci. U.S.A.,* 85, 5889, 1988.
395. **Tamada, H., McMaster, M.T., Flanders, K.C., Andrews, G.K., and Dey, S.K.,** Cell type-specific expression of transforming growth factor-β1 in the mouse uterus during the periimplantation period, *Mol. Endocrinol.,* 4, 965, 1990.
396. **Robinson, S.D., Roberts, A.B., and Daniel, C.W.,** TGFβ suppresses casein synthesis in mouse mammary explants and may play a role in controlling milk levels during pregnancy, *J. Cell Biol.,* 120, 245, 1993.
397. **Roberts, A.B., Anzano, M.A., Wakefield, L.M., Roche, N.S., Stern, D.F., and Sporn, M.B.,** Type β transforming growth factor: a bifunctional regulator of cell growth, *Proc. Natl. Acad. Sci. U.S.A.,* 82, 119, 1985.
398. **McMahon, J.B., Richards, W.L., del Campo, A.A., Song, M-K.H., and Thorgeirsson, S.S.,** Differential effects of transforming growth factor-β on proliferation of normal and malignant rat liver epithelial cells in culture, *Cancer Res.,* 46, 4665, 1986.

399. **Takahashi, K., Suzuki, K., and Ono, T.,** Loss of growth inhibitory activity of TGF-β toward normal human mammary epithelial cells grown within collagen gel matrix, *Biochem. Biophys. Res. Commun.,* 173, 1239, 1990.
400. **Aglietta, M., Stacchini, A., Severino, A., Sanavio, F., Ferrando, M.L., and Piacibello, W.,** Interaction of transforming growth factor-β1 with hemopoietic growth factors in the regulation of human normal and leukemic myelopoiesis, *Exp. Hematol.,* 17, 296, 1989.
401. **Tang, N., Cunningham, K., and Enger, M.D.,** TGF β elicits opposite responses in clonal subpopulations of NRK-49F cells, *Exp. Cell Res.,* 196, 13, 1991.
402. **Soma, Y. and Grotendorst, G.R.,** TGF-β stimulates primary human skin fibroblast DNA synthesis via an autocrine production of PDGF-related peptides, *J. Cell. Physiol.,* 140, 246, 1989.
403. **Battegay, E.J., Raines, E.W., Seifert, R.A., Bowen-Pope, D.F., and Ross, R.,** TGF-β induces bimodal proliferation of connective tissue cells via complex control of an autocrine PDGF loop, *Cell,* 63, 515, 1990.
404. **Barnard, J.A., Beauchamp, R.D., Coffey, R.J., and Moses, H.L.,** Regulation of intestinal epithelial cell growth by transforming growth factor type β, *Proc. Natl. Acad. Sci. U.S.A.,* 86, 1578, 1989.
405. **Strain, A.J.,** Transforming growth factor β and inhibition of hepatocellular proliferation, *Scand. J. Gastroenterol.,* 23(Suppl. 151), 37, 1988.
406. **Tessier, N. and Hoang, T.,** Transforming growth factor β inhibits the proliferation of the blast cells of acute myeloblastic leukemia, *Blood,* 72, 159, 1988.
407. **Kishi, K., Ellingsworth, L.R., and Ogawa, M.,** The suppressive effects of type β transforming growth factor (TGF β) on primitive murine hemopoietic progenitors are abrogated by interleukin-6 and granulocyte colony-stimulating factor, *Leukemia,* 3, 687, 1989.
408. **Takehara, K., LeRoy, E.E., and Grotendorst, G.R.,** TGF-β inhibition of endothelial cell proliferation: alteration of EGF binding and EGF-induced growth-regulatory (competence) gene expression, *Cell,* 49, 415, 1987.
409. **Ramsdell, J.S.,** Transforming growth factor-α and -β are potent and effective inhibitors of GH_4 pituitary tumor cell proliferation, *Endocrinology,* 128, 1981, 1991.
410. **Zugmaier, G., Ennis, B.W., Deschauer, B., Katz, D., Knabbe, C., Wilding, G., Daly, P., Lippman, M.E., and Dickson, R.B.,** Transforming growth factors type β1 and β2 are equipotent growth inhibitors of human breast cancer cell lines, *J. Cell. Physiol.,* 141, 353, 1989.
411. **Gray, T.K., Lipes, B., Linkhart, T., Mohan, S., and Baylink, D.,** Transforming growth factor β mediates the estrogen induced inhibition of UMR106 cell growth, *Connect. Tissue Res.,* 20, 23, 1989.
412. **Koyasu, S., Kadowaki, T., Nishida, E., Tobe, K., Abe, E., Kasuga, M., Sakai, H., and Yahara, I.,** Alteration in growth, cell morphology, and cytoskeletal structures of KB cells induced by epidermal growth factor and transforming growth factor-β, *Exp. Cell Res.,* 176, 107, 1988.
413. **Piao, Y.-F., Ichijo, H., Miyagawa, K., Ohashi, H., Takaku, F., and Miyazono, K.,** Latent form of transforming growth factor-β1 acts as a potent growth inhibitor on a human erythroleukemia cell line, *Biochem. Biophys. Res. Commun.,* 167, 27, 1990.
414. **Newman, M.J.,** Inhibition of carcinoma and melanoma cell growth by type I transforming growth factor β is dependent on the presence of polyunsaturated fatty acids, *Proc. Natl. Acad. Sci. U.S.A.,* 87, 5543, 1990.
415. **Blomhoff, H.K., Smeland, E., Mustafa, A.A., Godal, T., and Ohlsson, R.,** Epstein-Barr virus mediates a switch in responsiveness to transforming growth factor, type β, in cells of the B cell lineage, *Eur. J. Immunol.,* 17, 299, 1987.
416. **Rollins, B.J., O'Connell, T.M., Bennett, G., Burton, L.E., Stiles, C.D., and Rheinwald, J.G.,** Environment-dependent growth inhibition of human epidermal growth keratinocytes by recombinant human transforming growth factor-β, *J. Cell. Physiol.,* 139, 455, 1989.
417. **Jullien, P., Berg, T.M., de Lannoy, C., and Lawrence, D.A.,** Bifunctional activity of transforming growth factor type β on the growth of NRK-49F cells, normal and transformed by Kirsten murine sarcoma virus, *J. Cell. Physiol.,* 136, 175, 1988.
418. **Kurokowa, M., Lynch, K., and Podolsky, D.K.,** Effects of growth factors on an intestinal epithelial cell line: transforming growth factor β inhibits proliferation and stimulates differentiation, *Biochem. Biophys. Res. Commun.,* 142, 775, 1987.
419. **Matsumoto, K., Hashimoto, K., Hashiro, M., Yoshimasa, H., and Yoshikawa, K.,** Modulation of growth and differentiation in normal human keratinocytes by transforming growth factor-β, *J. Cell. Physiol.,* 145, 95, 1990.

420. **Silberstein, G.B. and Daniel, C.W.,** Reversible inhibition of mammary gland growth by transforming growth factor-β, *Science,* 237, 291, 1987.
421. **Braun, L., Mead, J.E., Panzica, M., Mikumo, R., Bell, G.I., and Fausto, N.,** Transforming growth factor β mRNA increases during liver regeneration: a possible paracrine mechanism of growth regulation, *Proc. Natl. Acad. Sci. U.S.A.,* 85, 1539, 1988.
422. **Houck, K.A., Cruise, J.L., and Michalopoulos, G.,** Norepinephrine modulates the growth-inhibitory effect of transforming growth factor-β in primary rat hepatocyte cultures, *J. Cell. Physiol.,* 135, 551, 1988.
423. **Goey, H., Keller, J.R., Back, T., Longo, D.L., Ruscetti, F.W., and Wiltrout, R.H.,** Inhibition of early murine hemopoietic progenitor cell proliferation after *in vivo* locoregional administration of transforming growth factor-β1, *J. Immunol.,* 143, 877, 1989.
424. **Ellingsworth, L.R., Nakayama, D., Segarini, P., Dasch, J., Carrillo, P., and Waegell, W.,** Transforming growth factor-βs are equipotent growth inhibitors of interleukin-1-induced thymocyte proliferation, *Cell. Immunol.,* 114, 41, 1988.
425. **Strassmann, G., Cole, M.D., and Newman, W.,** Regulation of colony-stimulating factor 1-dependent macrophage precursor proliferation by type β transforming growth factor, *J. Immunol.,* 140, 2645, 1988.
426. **Morris, D.R., Kuepfer, C.A., Ellingsworth, L.R., Ogawa, Y., and Rabinovitch, P.S.,** Transforming growth factor-β blocks proliferation but not early mitogenic signaling events in T-lymphocytes, *Exp. Cell Res.,* 185, 529, 1989.
427. **Like, B. and Massagué, J.,** The antiproliferative effect of type β transforming growth factor occurs at a level distal from receptors for growth-activating factors, *J. Biol. Chem.,* 261, 13426, 1986.
428. **Zendegui, J.G., Inman, W.H., and Carpenter, G.,** Modulation of the mitogenic response of an epidermal growth factor-dependent keratinocyte cell line by dexamethasone, insulin, and transforming growth factor-β, *J. Cell. Physiol.,* 136, 257, 1988.
429. **Ouchi, Y., Hirosumi, J., Watanabe, M., Hattori, A., Nakamura, T., and Orimo, H.,** Inhibitory effect of transforming growth factor-β on epidermal growth factor-induced proliferation of cultured rat aortic smooth muscle cells, *Biochem. Biophys. Res. Commun.,* 157, 301, 1988.
430. **Nugent, M.A. and Newman, M.J.,** Inhibition of normal rat kidney cell growth by transforming growth factor-β is mediated by collagen, *J. Biol. Chem.,* 264, 18060, 1989.
431. **Kamijo, R., Takeda, K., Naguno, M. and Konno, K.,** Suppression of TNF-stimulated proliferation of diploid fibroblasts and TNF-α induced cytotoxicity against transformed fibroblasts by TGF-β, *Biochem. Biophys. Res. Commun.,* 158, 155, 1989.
432. **Howe, P.H., Draetta, G., and Leof, E.B.,** Transforming growth factor β1 inhibition of $p34^{cdc2}$ phosphorylation and histone H1 kinase activity is associated with G1/S-phase growth arrest, *Mol. Cell. Biol.,* 11, 1185, 1991.
433. **Koff, A., Ohtsuki, M., Polyak, K., Roberts, J.M., and Massagué, J.,** Negative regulation of G1 in mammalian cells —Inhibition of cyclin E-dependent kinase by TGF-β, *Science,* 260, 536, 1993.
434. **Jennings, J.C., Mohan, S., Linkhart, T.A., Widstrom, R., and Baylink, D.J.,** Comparison of the biological actions of TGF β-1 and TGF β-2: differential activity in endothelial cells, *J. Cell. Physiol.,* 137, 167, 1988.
435. **Keller, J.R., Mantel, C., Sing, G.K., Ellingsworth, L.R., Ruscetti, S.K., and Ruscetti, F.W.,** Transforming growth factor β1 selectively regulates early murine hematopoietic progenitors and inhibits the growth of IL-3-dependent myeloid leukemia cell lines, *J. Exp. Med.,* 168, 737, 1988.
436. **Mitjavila, M.T., Vinci, G., Villeval, J.L., Kieffer, N., Henri, A., Testa, U., Breton-Gorius, J., and Vainchenker, W.,** Human platelet α granules contain a nonspecific inhibitor of megakaryocyte colony formation: its relationship to type β transforming growth factor (TGF-β), *J. Cell. Physiol.,* 134, 93, 1988.
437. **Arteaga, C.L., Tandon, A.K., Von Hoff, D.D., and Osborne, C.K.,** Transforming growth factor β: potential autocrine growth inhibitor of estrogen receptor-negative human breast cancer cells, *Cancer Res.,* 48, 3898, 1988.
438. **Coffey, R.J., Jr., Sipes, N.J., Bascom, C.C., Graves-Deal, R., Pennington, C.Y., Weissman, B.E., and Moses, H.L.,** Growth modulation of mouse keratinocytes by transforming growth factors, *Cancer Res.,* 48, 1596, 1988.
439. **Levine, A.E., Crandall, C.A., and Brattain, M.G.,** Regulation of growth inhibitory activity in transformed mouse embryo fibroblasts, *Exp. Cell Res.,* 171, 357, 1987.
440. **Hoosein, N.M., Brattain, D.E., McKnight, M.K., Levine, A.E., and Brattain, M.G.,** Characterization of the inhibitory effects of transforming growth factor-β on a human colon carcinoma cell line, *Cancer Res.,* 47, 2950, 1987.

441. **Levine, A.E. and Lewis, L.R.,** Transforming growth factor-β2 is an autocrine growth inhibitory factor for the MOSER human colon carcinoma cell line, *Cancer Lett.,* 68, 33, 1993.
442. **Mulder, K.M., Ramey, M.K., Hoosein, N.M., Levine, A.E., Hinshaw, X.H., Brattain, D.E., and Brattain, M.G.,** Characterization of transforming growth factor-β-resistant subclones isolated from a transforming growth factor-β-sensitive human colon carcinoma cell line, *Cancer Res.,* 48, 7120, 1988.
443. **Mulder, K.M., Zhong, Q., Choi, H.G., Humphrey, L.E., and Brattain, M.G.,** Inhibitory effects of transforming growth factor β_1 on mitogenic response, transforming growth factor α, and c-*myc* in quiescent, well-differentiated colon carcinoma cells, *Cancer Res.,* 50, 7581, 1990.
444. **Takaishi, K., Kawata, S., Ito, N., Tamura, S., Shirat, Y., and Tarui, S.,** Effects of phorbol ester on cell growth inhibition by transforming growth factor β1 in human hepatoma cell lines, *Biochem. Biophys. Res. Commun.,* 171, 91, 1990.
445. **Sipes, N.J., Miller, D.A., Bascom, C.C., Winkler, J.K., Matrisian, L.M., and Moses, H.L.,** Altered regulation of protein disulfide isomerase in cells resistant to the growth-inhibitory effects of transforming growth factor β1, *Cell Growth Differ.,* 1, 241, 1990.
446. **Shipley, G.D., Pittelkow, M.R., Wille, J.J., Jr., Scott, R.E., and Moses, H.L.,** Reversible inhibition of normal human prokeratinocyte proliferation by type β transforming growth factor-growth inhibitor in serum-free medium, *Cancer Res.,* 46, 2068, 1986.
447. **Frater-Schröder, M., Müller, G., Birchmeier, W., and Böhlen, P.,** Transforming growth factor-β inhibits endothelial cell proliferation, *Biochem. Biophys. Res. Commun.,* 137, 295, 1986.
448. **Heimark, R.L., Twardzik, D.R., and Schwartz, S.M.,** Inhibition of endothelial regeneration by type-β transforming growth factor from platelets, *Science,* 233, 1078, 1986.
449. **Lee, K., Tanaka, M., Hatanaka, M., and Kuze, F.,** Reciprocal effects of epidermal growth factor and transforming growth factor β on the anchorage-dependent and -independent growth of A431 epidermoid carcinoma cells, *Exp. Cell Res.,* 173, 156, 1987.
450. **Rizzino, A., Ruff, E., and Rizzino, H.,** Induction and modulation of anchorage-independent growth by platelet-derived growth factor, fibroblast growth factor, and transforming growth factor-β, *Cancer Res.,* 46, 2816, 1986.
451. **Lin, P., Liu, C., Tsao, M.-S., and Grisham, J.W.,** Inhibition of proliferation in cultured rat liver epithelial cells at specific cell cycle stages by transforming growth factor-β, *Biochem. Biophys. Res. Commun.,* 143, 26, 1987.
452. **Nakamura, T., Tomita, Y., Hirai, R., Yamaoka, K., Kaji, K., and Ichihara, A.,** Inhibitory effect of transforming growth factor-β on DNA synthesis of adult rat hepatocytes in primary culture, *Biochem. Biophys. Res. Commun.,* 133, 1042, 1985.
453. **McKeehan, W.L. and Adams, P.S.,** Heparin-binding growth factor/prostatropin attenuates inhibition of rat prostate tumor epithelial cell growth by transforming growth factor type β, *In Vitro Cell. Develop. Biol.,* 24, 243, 1988.
454. **Nagy, P., Evarts, R.P., McMahon, J.B., and Thorgeirsson, S.S.,** Role of TGF-β in normal differentiation and oncogenesis in rat liver, *Mol. Carcinogenesis,* 2, 345, 1989.
455. **Kimelman, D. and Kirschner, M.,** Synergistic induction of mesoderm by FGF and TGF-β and the identification of an mRNA coding for FGF in the early Xenopus embryo, *Cell,* 51, 869, 1987.
456. **Lee, G., Ellingsworth, L.R., Gillis, S., Wall, R., and Kinkade, P.W.,** β Transforming growth factors are potential regulators of B lymphopoiesis, *J. Exp. Med.,* 166, 1290, 1987.
457. **Sparks, R.L., Allen, B.J., and Strauss, E.E.,** TGF-β blocks early but not late differentiation-specific gene expression and morphologic differentiation of 3T3 T proadipocytes, *J. Cell. Physiol.,* 150, 568, 1992.
458. **Florini, J.R. and Magri, K.A.,** Effects of growth factors on myogenic differentiation, *Am. J. Physiol.,* 256, C701, 1989.
459. **Olson, E.N., Sternberg, E., Hu, J.S., Spizz, G., and Wilcox, C.,** Regulation of myogenic differentiation by type β transforming growth factor, *J. Cell Biol.,* 103, 1799, 1986.
460. **Florini, J.R., Roberts, A.B., Ewton, D.Z., Falen, S.L., Flanders, K.C., and Sporn, M.B.,** Transforming growth factor-β: a very potent inhibitor of myoblast differentiation, identical to the differentiation inhibitor secreted by Buffalo rat liver cells, *J. Biol. Chem.,* 261, 16509, 1986.
461. **Massagué, J., Cheifetz, S., Endo, T., and Nadal-Ginard, B.,** Type β transforming growth factor is an inhibitor of myogenic differentiation, *Proc. Natl. Acad. Sci. U.S.A.,* 83, 8206, 1986.
462. **Spizz, G., Hu, J.-S., and Olson, E.N.,** Inhibition of myogenic differentiation by fibroblast growth factor of type β transforming growth factor does not require persistent c-*myc* expression, *Dev. Biol.,* 123, 500, 1987.

463. **Olson, E.N., Spizz, G., and Tainsky, M.A.,** The oncogenic forms of N-*ras* or H-*ras* prevent skeletal myoblast differentiation, *Mol. Cell. Biol.,* 7, 2104, 1987.
464. **Zentella, A. and Massagué, J.,** Transforming growth factor β induces myoblast differentiation in the presence of mitogens, *Proc. Natl. Acad. Sci. U.S.A.,* 89, 5176, 1992.
465. **Allen, R.E. and Boxhorn, L.K.,** Inhibition of skeletal muscle satellite cell differentiation by transforming growth factor-β, *J. Cell. Physiol.,* 133, 567, 1987.
466. **Majack, R.A.,** Beta-type transforming growth factor specifies organizational behavior in vascular smooth muscle cell cultures, *J. Cell Biol.,* 105, 465, 1987.
467. **Masui, T., Wakefield, L.M., Lechner, J.F., LaVeck, M.A., Sporn, M.B., and Harris, C.C.,** Type β transforming growth factor is the primary differentiation-inducing serum factor for normal human bronchial epithelial cells, *Proc. Natl. Acad. Sci. U.S.A.,* 83, 2438, 1986.
468. **Jetten, A.M., Shirley, J.E., and Stoner, G.,** Regulation of proliferation and differentiation of respiratory tract epithelial cells by TGF β, *Exp. Cell Res.,* 167, 539, 1986.
469. **Noda, M. and Rodan, G.A.,** Type-β transforming growth factor inhibits proliferation and expression of alkaline phosphatase in murine osteoblast-like cells, *Biochem. Biophys. Res. Commun.,* 140, 56, 1986.
470. **Pfeilschifter, J., D'Souza, S.M., and Mundy, G.R.,** Effect of transforming growth factor-β on osteoblastic osteosarcoma cells, *Endocrinology,* 121, 212, 1987.
471. **Hooper, W.C, Pruckler, J., Jackson, D., and Evatt, B.L.,** Transforming growth factor-β induces hemoglobin synthesis in a human erythroleukemia cell line, *Biochem. Biophys. Res. Commun.,* 165, 145, 1989.
472. **Ignotz, R.A. and Massagué, J.,** Type β transforming growth factor controls the adipogenic differentiation of 3T3 fibroblasts, *Proc. Natl. Acad. Sci. U.S.A.,* 82, 8530, 1985.
473. **Kamijo, R., Takeda, K., Nagumo, M., and Konno, K.,** Effects of combinations of transforming growth factor-β_1 and tumor necrosis factor on induction of differentiation of human myelogenous leukemic cell lines, *J. Immunol.,* 144, 1311, 1990.
474. **Weima, S.M., van Rooijen, M.A., Feijen, A., Mummery, C.L., van Zoelen, E.J.J., de Laat, S.W., and van den Eijnden-van Raaij, A.J.M.,** Transforming growth factor-β and its receptor are differentially regulated in human embryonal carcinoma cells, *Differentiation,* 41, 245, 1989.
475. **Falk, L.A., De Benedetti, F., Lohrey, N., Birchenall-Roberts, M.C., Elligsworth, L.W., Faltynek, C.R., and Ruscetti, F.W.,** Induction of transforming growth factor-β_1 (TGF-β_1) receptor expression and TGF-β_1 protein production in retinoic acid-treated HL-60 cells: possible TGF-β_1-mediated autocrine inhibition, *Blood,* 77, 1248, 1991.
476. **Alitalo, R., Mäkelä, T.P., Koskinen, P., Andersson, L.C., and Alitalo, K.,** Enhanced expression of transforming growth factor β during megakaryoblastic differentiation of K562 leukemia cells, *Blood,* 71, 899, 1988.
477. **Okabe-Kado, J., Honma, Y., Hayashi, M., and Hozumi, M.,** Inhibitory action of transforming growth factor-β on induction of differentiation of myeloid leukemia cells, *Jpn. J. Cancer Res.,* 80, 228, 1989.
478. **Gerschenson, L.E. and Rotello, R.J.,** Apoptosis — A different type of cell death, *FASEB J.,* 6, 2450, 1992.
479. **Wyllie, A.H.,** Apoptosis, *Br. J. Cancer,* 67, 205, 1993.
480. **Schwartzman, R.A. and Cidlowski, J.A.,** Apoptosis — The biochemistry and molecular biology of programmed cell death, *Endocrine Rev.,* 14, 133, 1993.
481. **Oberhammer, F.A., Pavelka, M., Sharma, S., Tiefenbacher, R., Purchio, A.F., Bursch, W., and Schulte-Hermann, R.,** Induction of apoptosis in cultured hepatocytes and in regressing liver by transforming growth factor β1, *Proc. Natl. Acad. Sci. U.S.A.,* 89, 5408, 1992.
482. **Bursch, W., Oberhammer, F., Jirtle, R.L., Askari, M., Sedivy, R., Grasl-Kraupp, B., Purchio, A.F., and Schulte-Hermann, R.,** Transforming growth factor-β1 as a signal for induction of cell death by apoptosis, *Br. J. Cancer,* 67, 531, 1993.
483. **Kyprianou, N. and Isaacs, J.T.,** Expression of transforming growth factor in the rat ventral prostate during castration-induced programmed cell death, *Mol. Endocrinol.,* 3, 1515, 1989.
484. **Martikainen, P., Kyprianou, N., and Isaacs, J.T.,** Effect of transforming growth factor-β_1 on proliferation and death of rat prostatic cells, *Endocrinology,* 127, 2963, 1990.
485. **Rotello, R.J., Lieberman, R.C., Purchio, A.F., and Gerschenson, L.E.,** Coordinated regulation of apoptosis and cell proliferation by transforming growth factor β_1 in cultured uterine epithelial cells, *Proc. Natl. Acad. Sci. U.S.A.,* 88, 3412, 1991.

486. **Yanagihara, K. and Tsumuraya, M.,** Transforming growth factor β1 induces apoptotic cell death in cultured human gastric carcinoma cells, *Cancer Res.,* 52, 4042, 1992.

487. **Ying, S.-Y., Becker, A., Baird, A., Ling, N., Ueno, N., Esch, F., and Guillemin, R.,** Type β transforming growth factor (TGF-β) is a potent stimulator of the basal secretion of follicle stimulating hormone (FSH) in a pituitary monolayer system, *Biochem. Biophys. Res. Commun.,* 135, 950, 1986.

488. **Benahmed, M., Cochet, C., Keramidas, M., Chauvin, M.A., and Morera, A.M.,** Evidence for a FSH dependent secretion of a receptor reactive transforming growth factor β-like material by immature Sertoli cells in primary culture, *Biochem. Biophys. Res. Commun.,* 154, 1222, 1988.

489. **Esposito, G., Keramidas, M., Mauduit, C., Feige, J.J., Morera, A.M., and Benahmed, M.,** Direct regulating effect of transforming growth factor-β1 on lactate production in cultured porcine Sertoli cells, *Endocrinology,* 128, 1441, 1991.

490. **Reichert, L.E., Sporn, M.B., and Santacoloma, T.A.,** Transforming growth factor β1 modulates calcium metabolism in Sertoli cells, *Endocrinology,* 132, 1745, 1993.

491. **Lin, T., Blaisdell, J., and Haskell, J.F.,** Transforming growth factor-β inhibits Leydig cell steroidogenesis in primary culture, *Biochem. Biophys. Res. Commun.,* 146, 387, 1987.

492. **Roberts, A.J. and Skinner, M.K.,** Transforming growth factor-α and factor-β differentially regulate growth and steroidogenesis during antral follicle development, *Endocrinology,* 129, 2041, 1991.

493. **Skinner, M.K., Keski-Oja, J., Osteen, K.G., and Moses, H.L.,** Ovarian thecal cells produce transforming growth factor-β which can regulate granulosa cell growth, *Endocrinology,* 121, 786, 1987.

494. **Hernandez, E.R., Hurwitz, A., Payne, D.W., Dharmarajan, A.M., Purchio, A.F., and Adashi, E.Y.,** Transforming growth factor-β1 inhibits ovarian androgen production: gene expression, cellular localization, mechanisms(s), and site(s) of action, *Endocrinology,* 127, 1804, 1990.

495. **Berchuck, A., Rodriguez, G., Olt, G., Whitaker, R., Boente, M.P., Arrick, B.A., Clarke-Pearson, D.L., and Bast, R.C.,** Regulation of growth of normal ovarian epithelial cell lines by transforming growth factor-β, *Am. J. Obstet. Gynecol.,* 166, 676, 1992.

496. **Mason, A.J., Hayflick, J.S., Ling, N., Esch, F., Ueno, N., Ying, S-Y., Guillemin, R., Niall, H., and Seeburg, P.H.,** Complementary DNA sequences of ovarian follicular fluid inhibin show precursor structure and homology with transforming growth factor-β, *Nature,* 318, 659, 1985.

497. **Feng, P., Catt, K.J., and Knecht, M.,** Transforming growth factor-β stimulates meiotic maturation of the rat oocyte, *Endocrinology,* 122, 181, 1988.

498. **Mondschein, J.S., Canning, S.F., and Hammond, J.M.,** Effects of transforming growth factor-β on the production of immunoreactive insulin-like growth factor I and progesterone and on [^{3}H]thymidine incorporation in porcine granulosa cell cultures, *Endocrinology,* 123, 1970, 1988.

499. **Knecht, M., Feng, P., and Catt, K.J.,** Transforming growth factor-β regulates the expression of luteinizing hormone receptors in ovarian granulosa cells, *Biochem. Biophys. Res. Commun.,* 139, 800, 1986.

500. **Knecht, M., Feng, P., and Catt, K.,** Bifunctional role of transforming growth factor-β during granulosa cell development, *Endocrinology,* 120, 1243, 1987.

501. **Dodson, W.C. and Schomberg, D.W.,** The effect of transforming growth factor-β on follicle-stimulating hormone-induced differentiation of cultured rat granulosa cells, *Endocrinology,* 120, 512, 1987.

502. **Feng, P., Catt, K.J., and Knecht, M.,** Transforming growth factor β regulates the inhibitory actions of epidermal growth factor during granulosa cell differentiation, *J. Biol. Chem.,* 261, 14167, 1986.

503. **Feige, J.J., Cochet, C., and Chambaz, E.M.,** Type β transforming growth factor is a potent modulator of differentiated adrenocortical cell functions, *Biochem. Biophys. Res. Commun.,* 139, 693, 1986.

504. **Hotta, M. and Baird, A.,** Differential effects of transforming growth factor type β on the growth and function of adrenocortical cells *in vitro, Proc. Natl. Acad. Sci. U.S.A.,* 83, 7795, 1986.

505. **Morris, J.C., III, Raganathan, G., Hay, I.D., Nelson, R.E., and Jiang, N-S.,** The effects of transforming growth factor-β on growth and differentiation of the continuous rat thyroid follicular cell line, FRTL-5, *Endocrinology,* 123, 1385, 1988.

506. **Totsuka, Y., Tabuchi, M., Kojima, I., Eto, Y., Shibai, H., and Ogata, E.,** Stimulation of insulin secretion by transforming growth factor-β, *Biochem. Biophys. Res. Commun.,* 158, 1060, 1989.

507. **Goustin, A.S., Nuttall, G.A., Leof, E.B., Ranganathan, G., and Moses, H.L.,** Transforming growth factor type β can act as a potent competence factor for AKR-2B cells, *Exp. Cell Res.,* 172, 293, 1987.

508. **Bronzert, D.A., Bates, S.E., Sheridan, J.P., Lindsey, R., Valverius, E.M., Stampfer, M.R., Lippman, M.E., and Dickson, R.B.,** Transforming growth factor β induces platelet-derived growth factor (PDGF) messenger RNA and PDGF secretion while inhibiting growth in normal human mammary epithelial cells, *Mol. Endocrinol.,* 4, 981, 1990.

509. **Nugent, M.A., Lane, E.A., Keski-Oja, J., Moses, H.L., and Newman, M.J.,** Growth stimulation, altered regulation of epidermal growth factor receptors, and autocrine transformation of spontaneously transformed normal rat kidney cells by transforming growth factor β, *Cancer Res.,* 49, 3884, 1989.
510. **Newman, M.J., Lane, E.A., Nugent, M.A., and Racker, E.,** Induction of anchorage-independent growth by epidermal growth factor and altered sensitivity to type β transforming growth factor in partially transformed rat kidney cells, *Cancer Res.,* 46, 5842, 1986.
511. **Massagué, J., Kelly, B., and Mottola, C.,** Stimulation by insulin-like growth factors is required for cellular transformation by type β transforming growth factor, *J. Biol. Chem.,* 260, 4551, 1985.
512. **Hagino, Y., Mawatari, M., Yoshimura, A., Kohno, K., Kobayashi, M., and Kuwano, M.,** Estrogen inhibits the growth of MCF-7 cell variants to transforming growth factor-β, *Jpn. J. Cancer Res.,* 79, 74, 1988.
513. **Assoian, R.K., Frolik, C.A., Roberts, A.B., Miller, D.M., and Sporn, M.B.,** Transforming growth factor-β controls receptor levels for epidermal growth factor in NRK fibroblasts, *Cell,* 36, 35, 1984.
514. **Fernandez-Pol, J.A., Klos, D.J., Hamilton, P.D., and Talkad, V.D.,** Modulation of epidermal growth factor receptor gene expression by transforming growth factor-β in a human breast carcinoma cell line, *Cancer Res.,* 47, 4260, 1987.
515. **Massagué, J.,** Transforming growth factor-β modulates the high-affinity receptors for epidermal growth factor and transforming growth factor-α, *J. Cell Biol.,* 100, 1508, 1985.
516. **Assoian, R.K.,** Biphasic effects of type β transforming growth factor on epidermal growth factor receptors in NRK fibroblasts: functional consequences for epidermal growth factor-stimulated mitosis, *J. Biol. Chem.,* 260, 9613, 1985.
517. **Carr, B.I., Hayashi, I., Branum, E.L., and Moses, H.L.,** Inhibition of DNA synthesis in rat hepatocytes by platelet-derived type β transforming growth factor, *Cancer Res.,* 46, 2330, 1986.
518. **Chiang, C.-P. and Nilsen-Hamilton, M.,** Opposite and selective effects of epidermal growth factor and human platelet transforming growth factor-β on the production of secreted proteins by murine 3T3 cells and human fibroblasts, *J. Biol. Chem.,* 261, 10478, 1986.
519. **Mead, J.E. and Fausto, N.,** Transforming growth factor α may be a physiological regulator of liver regeneration by means of an autocrine mechanism, *Proc. Natl. Acad. Sci. U.S.A.,* 86, 1558, 1989.
520. **Smeland, E.B., Blomhoff, H.K., Holte, H., Ruud, E., Beiske, K., Funderud, S., Godal, T., and Ohlsson, R.,** Transforming growth factor type β (TGF β) inhibits G_1 to S transition, but not activation of human B lymphocytes, *Exp. Cell Res.,* 171, 213, 1987.
521. **Czarniecki, C.W., Chiu, H.H., Wong, G.H.W., McCabe, S.M., and Palladino, M.A.,** Transforming growth factor-β_1 modulates the expression of class II histocompatibility antigens on human cells, *J. Immunol.,* 140, 4217, 1988.
522. **Leof, E.B., Proper, J.A., and Moses, H.L.,** Modulation of transforming growth factor type β action by activated *ras* and c-*myc*, *Mol. Cell. Biol.,* 7, 2649, 1987.
523. **Palmer, H., Maher, V.M., and McCormick, J.J.,** Platelet-derived growth factor or basic fibroblast growth factor induce anchorage-independent growth of human fibroblasts, *J. Cell. Physiol.,* 137, 588, 1988.
524. **Roberts, A.B., Roche, N.S., and Sporn, M.B.,** Selective inhibition of the anchorage-independent growth of *myc*-transfected fibroblasts by retinoic acid, *Nature,* 315, 237, 1985.
525. **van Zoelen, E.J.J., van Oostwaard, T.M.J., and de Laat, S.W.,** Transforming growth factor-β and retinoic acid modulate phenotypic transformation of normal rat kidney cells induced by epidermal growth factor and platelet-derived growth factor, *J. Biol. Chem.,* 261, 5003, 1986.
526. **Mooradian, D.L., Purchio, A.F., and Furcht, L.T.,** Differential effects of transforming growth factor β1 on the growth of poorly and highly metastatic murine melanoma cells, *Cancer Res.,* 50, 273, 1990.
527. **Serra, R., Verderame, M.F., and Isom, H.C.,** Transforming growth factor β_1 partially suppresses the transformed phenotype of *ras*-transformed hepatocytes, *Cell Growth Differ.,* 3, 693, 1992.
528. **Wu, S.P., Sun, L.-Z., Willson, J.K.V., Humphrey, L., Kerbel, R., and Brattain, M.G.,** Repression of autocrine transforming growth factor β_1 and β_2 in quiescent CBS colon carcinoma cells leads to progression of tumorigenic properties, *Cell Growth Differ.,* 4, 115, 1993.
529. **Krieg, P., Schnapke, R., Fürstenberger, G., Vogt, I., and Marks, F.,** TGF-β1 and skin carcinogenesis: antiproliferative effect *in vitro* and TGF-β1 mRNA expression during epidermal hyperproliferation and multistage tumorigenesis, *Mol. Carcinogenesis,* 4, 129, 1991.
530. **Steigerwalt, R.W., Rundhaug, J.E., and Nettesheim, P.,** Transformed rat tracheal epithelial cells exhibit alterations in transforming growth factor-β secretion and responsiveness, *Mol. Carcinogenesis,* 5, 32, 1992.

531. **Damstrup, L., Rygaard, K., Spang-Thomsen, M., and Poulsen, H.S.,** Expression of transforming growth factor β (TGF β) receptors and expression of TGFβ_1, TGFβ_2 and TGFβ_3 in human small cell lung cancer cell lines, *Br. J. Cancer,* 67, 1015, 1993.
532. **Nakatsukasa, H., Evarts, R.P., Hsia, C., Marsden, E., and Thorgeirsson, S.S.,** Expression of transforming growth factor-β1 during chemical hepatocarcinogenesis in the rat, *Lab. Invest.,* 65, 511, 1991.
533. **Shirai, Y., Kawata, S., Ito, N., Tamura, S., Takaishi, K., Kiso, S., Tsushima, H., and Matsuzawa, Y.,** Elevated levels of plasma TGF-β in patients with hepatocellular carcinoma, *Jpn. J. Cancer Res.,* 83, 876, 1992.
534. **Jennings, M.T., Maciunas, R.J., Carver, R., Bascom, C.C., Juneau, P., Misulis, K., and Moses, H.L.,** TGFβ_1 and TGFβ_2 are potential growth regulators for low-grade and malignant gliomas *in vitro:* evidence in support of an autocrine hypothesis, *Int. J. Cancer,* 49, 129, 1991.
535. **Komatsu, K., Nakamura, H., Shinkai, K., and Akedo, H.,** Secretion of transforming growth factor-β by human myelogenous leukemic cells and its possible role in proliferation of the leukemic cells, *Jpn. J. Cancer Res.,* 80, 928, 1989.
536. **McCune, B.K., Mullin, B.R., Flanders, K.C., Jaffurs, W.J., Mullen, L.T., and Sporn, M.B.,** Localization of transforming growth factor-β isotypes in lesions of the human breast, *Hum. Pathol.,* 23, 13, 1992.
537. **Daly, R.J., King, R.J.B., and Darbre, P.D.,** Interaction of growth factors during progression towards steroid independence in T-47-D human breast cancer cells, *J. Cell. Biochem.,* 43, 199, 1990.
538. **Birchenall-Roberts, M.C., Falk, L.A., Kasper, J., Keller, J., Faltynek, C.R., and Ruscetti, F.W.,** Differential expression of transforming growth factor-β1 (TGF-β1) receptors in murine myeloid cell lines transformed with oncogenes. Correlation with differential growth inhibition by TGF-β1, *J. Biol. Chem.,* 266, 9617, 1991.
539. **Chakrabarty, S., Fan, D., and Varani, J.,** Modulation of differentiation and proliferation in human colon carcinoma cells by transforming growth factor β1 and β2, *Int. J. Cancer,* 46, 493, 1990.
540. **Hoosein, N.M., McKnight, M.K., Levine, A.E., Mulder, K.M., Childress, K.E., Brattain, D.E., and Brattain, M.G.,** Differential sensitivity of subclasses of human colon carcinoma cell lines to the growth inhibitory effects of transforming growth factor-β1, *Exp. Cell Res.,* 181, 442, 1989.
541. **Hubbs, A.F., Hahn, F.F., and Thomassen, D.G.,** Increased resistance to transforming growth factor β accompanies neoplastic progression of rat tracheal epithelial cells, *Carcinogenesis,* 10, 1599, 1989.
542. **Ichijo, H., Momose, F., and Miyazono, K.,** Biological effects and binding properties of transforming growth factor-β on human oral squamous cell carcinoma cells, *Exp. Cell Res.,* 187, 263, 1990.
543. **Ito, M., Yasui, W., Nakayama, H., Yokozaki, H., Ito, H., and Tahara, E.,** Reduced levels of transforming growth factor-β type-I receptor in human gastric carcinomas, *Jpn. J. Cancer Res.,* 83, 86, 1992.
544. **Hébert, C.D. and Birnbaum, L.S.,** Lack of correlation between sensitivity to growth inhibition and receptor number for transforming growth factor β in human squamous carcinoma cell lines, *Cancer Res.,* 49, 3196, 1989.
545. **Akhurst, R.J., Fee, F., and Balmain, A.,** Localized production of TGF-β mRNA in tumour promoter-stimulated mouse epidermis, *Nature,* 331, 363, 1988.
546. **Hamel, E., Katoh, F., Mueller, G., Birchmeier, W., and Yamasaki, H.,** Transforming growth factor β as a potent promoter in two-stage BALB/c 3T3 cell transformation, *Cancer Res.,* 48, 2832, 1988.
547. **Komatsu, K., Nakamura, H., and Akedo, H.,** Transforming growth factor (TGF)-β1 induces leukemic cell-growth-promoting activity in fibroblastic cells, *Cell Biol. Int. Rep.,* 17, 433, 1993.
548. **Ito, N., Kawata, S., Tamura, S., Takaishi, K., Shirai, Y., Kiso, S., Yabuuchi, I., Matsuda, Y., Nishioka, M., and Tarui, S.,** Elevated levels of transforming growth factor-β messenger RNA and its polypeptide in human hepatocellular carcinoma, *Cancer Res.,* 51, 4080, 1991.
549. **Beauchamp, R.D., Coffey, R.J., Jr., Lyons, R.M., Perkett, E.A., Townsend, C.M., Jr., and Moses, H.L.,** Human carcinoid cell production of paracrine growth factors that stimulate fibroblast and endothelial cell growth, *Cancer Res.,* 51, 5253, 1991.
550. **Kremer, J.P., Reisbach, G., Nerl, C., and Dörmer, P.,** B-cell chronic lymphocytic leukemia cells express and release transforming growth factor β, *Br. J. Haematol.,* 80, 480, 1992.
551. **Israels, L.G., Israels, S.J., Begleiter, A., Verburg, L., Schwartz, L., Mowat, M.R.A., and Johnston, J.B.,** Role of transforming growth factor-β in chronic lymphocytic leukemia, *Leukemia Res.,* 17, 81, 1993.
552. **Mukai, M., Shinkai, K., Komatsu, K., and Akedo, H.,** Potentiation of invasive capacity of rat ascites hepatoma cells by transforming growth factor-β, *Jpn. J. Cancer Res.,* 80, 107, 1989.

553. **Matuo, Y., Nishi, N., Takasuka, H., Masuda, Y., Nishikawa, K., Isaacs, J.T., Adams, P.S., McKeehan, W.L., and Sato, G.H.,** Production and significance of TGF-β in AT-3 metastatic cell line established from the Dunning rat prostatic adenocarcinoma, *Biochem. Biophys. Res. Commun.,* 166, 840, 1990.
554. **Knabbe, C., Lippman, M.E., Wakefield, L.M., Flanders, K.C., Kasid, A., Derynck, R., and Dickson, R.B.,** Evidence that transforming growth factor-β is a hormonally regulated negative growth factor in human breast cancer cells, *Cell,* 48, 417, 1987.
555. **Wu, S., Theodorescu, D., Kerbel, R.S., Willson, J.K.V., Mulder, K.M., Humphrey, L.E., and Brattain, M.G.,** TGF-β_1 is an autocrine-negative growth regulator of human colon carcinoma FET cells *in vivo* as revealed by transfection of an antisense expression vector, *J. Cell. Biol.,* 116, 187, 1992.
556. **Kubota, S., Fridman, R., and Yamada, Y.,** Transforming growth factor-β suppresses the invasiveness of human fibrosarcoma cells in vitro by increasing expression of tissue inhibitor metalloprotease, *Biochem. Biophys. Res. Commun.,* 176, 129, 1991.
557. **Bant, Y.-J., Kim, S.-J., Danielpour, D., O'Reilly, M.A., Kim, K.Y., Myers, C.E., and Trepel, J.B.,** Cyclic AMP induces transforming growth factor β2 gene expression and growth arrest in the human androgen-independent prostate carcinoma cell line PC-3, *Proc. Natl. Acad. Sci. U.S.A.,* 89, 3556, 1992.
558. **Sipes, N.J., Lyons, R.M., and Moses, H.L.,** Isolation and characterization of Kirsten murine sarcoma virus-transformed mouse keratinocytes resistant to transforming growth factor β, *Mol. Carcinogenesis,* 3, 12, 1990.
559. **Kremer, J.-P., Reisbach, G., Nerl, C., and Dörmer, P.,** B-cell chronic lymphocytic leukaemia cells express and release transforming growth factor-β, *Br. J. Haematol.,* 80, 480, 1992.
560. **Sieweke, M.H., Thompson, N.L., Sporn, M.B., and Bissell, M.J.,** Mediation of wound-related Rous sarcoma virus tumorigenesis by TGF-β, *Science,* 248, 1656, 1990.
561. **Cato, A.C.B., Mink, S., Nierlich, B., Ponta, H., Schaap, D., Schuuring, E., and Sonnenberg, A.,** Transforming growth factor-β represses transcription of the mouse mammary tumor virus DNA in cultured mouse mammary cells, *Oncogene,* 5, 103, 1990.
562. **Niitsu, Y., Urushizaki, Y., Koshida, Y., Terui, K., Mahara, K., Kohgo, Y., and Urushizaki, I.,** Expression of the TGF-β gene in adult T cell leukemia, *Blood,* 71, 263, 1988.
563. **Kim, S.-J., Kehrl, J.H., Burton, J., Tendler, C.L., Jeang, K.-T., Danielpour, D., Thevenin, C., Kim, K.Y., Sporn, M.B., and Roberts, A.,** Transactivation of the transforming growth factor β1 (TGF-β1) by human T lymphotrophic virus type 1 Tax: a potential mechanism for the increased production of TGF-β1 in adult T cell leukemia, *J. Exp. Med.,* 172, 121, 1990.
564. **Woodworth, C.D., Notario, V., and DiPaolo, J.A.,** Transforming growth factor β 1 and 2 transcriptionally regulate human papillomavirus (HPV) type 16 early gene expression in HPV-immortalized human genital epithelial cells, *J. Virol.,* 64, 4767, 1990.
565. **Braun, L., Dürst, M., Mikumo, R., Crowley, A., and Robinson, M.,** Regulation of growth and gene expression in human papillomavirus-transformed keratinocytes by transforming growth factor-β: implications for the control of papillomavirus infection, *Mol. Carcinogenesis,* 6, 100, 1992.
566. **Longstreet, M., Miller, B., and Howe, P.H.,** Loss of transforming growth factor β_1 (TGF-β_1)-induced growth arrest and p34(cdc2) regulation in *ras*-transfected epithelial cells, *Oncogene,* 7, 1549, 1992.
567. **Huggett, A.C., Hampton, L.L., Ford, C.P., Wirth, P.J., and Thorgeirsson, S.S.,** Altered responsiveness of rat liver epithelial cells to transforming growth factor β_1 following their transformation with v-*raf, Cancer Res.,* 50, 7468, 1990.
568. **Steiner, M.S. and Barrack, E.R.,** Transforming growth factor-β 1 overproduction in prostate cancer: effects on growth *in vivo* and *in vitro, Mol. Endocrinol.,* 6, 15, 1992.
569. **Zhao, J. and Buick, R.N.,** Relationship of levels and kinetics of H-*ras* expression to transformed phenotype and loss of TGF-β1-mediated growth regulation in intestinal epithelial cells, *Exp. Cell Res.,* 204, 82, 1993.
570. **Wyllie, F.S., Dawson, T., Bond, J.A., Goretzki, P., Game, S., Prime, S., and Wynford-Thomas, D.,** Correlated abnormalities of transforming growth factor-β1 response and p53 expression in thyroid epithelial cell transformation, *Mol. Cell. Endocrinol.,* 76, 13, 1991.
571. **Schwarz, L.C., Gingras, M.-C., Goldberg, G., Greenberg, A.H., and Wright, J.A.,** Loss of growth factor dependence and conversion of a transforming growth factor-β_1 inhibition to stimulation in metastatic H-*ras*-transformed murine fibroblasts, *Cancer Res.,* 48, 6999, 1988.
572. **Burt, D.W.,** Evolutionary grouping of the transforming growth factor-β superfamily, *Biochem. Biophys. Res. Commun.,* 184, 590, 1992.

573. **Wharton, K.A., Thomsen, G.H., and Gelbart, W.M.,** *Drosophila* A60 gene, another transforming growth factor β family member, is closely related to human bone morphogenetic proteins, *Proc. Natl. Acad. Sci. U.S.A.,* 88, 9214, 1991.
574. **Tucker, R.F., Shipley, G.D., Moses, H.L., and Holley, R.W.,** Growth inhibitor from BSC-1 cells closely related to platelet type β transforming growth factor, *Science,* 226, 705, 1984.
575. **Ristow, H.-J.,** BSC-1 growth inhibitor/type β transforming growth factor is a strong inhibitor of thymocyte proliferation, *Proc. Natl. Acad. Sci. U.S.A.,* 83, 5531, 1986.
576. **Brown, K.D. and Holley, R.W.,** Insulin-like synergistic stimulation of DNA synthesis in Swiss 3T3 cells by the BSC-1 cell-derived growth inhibitor related to transforming growth factor type β, *Proc. Natl. Acad. Sci. U.S.A.,* 84, 3743, 1987.
577. **Hanks, S.K., Armour, R., Baldwin, J.H., Maldonado, F., Spiess, J., and Holley, R.W.,** Amino acid sequence of the BSC-1 cell growth inhibitor (polyergin) deduced from the nucleotide sequence of the cDNA, *Proc. Natl. Acad. Sci. U.S.A.,* 85, 79, 1988.
578. **McPherson, J.M., Sawamura, S.J., Ogawa, Y., Dineley, K., Carrillo, P., and Piez, K.A.,** The growth inhibitor of African green monkey (BSC-1) cells is transforming growth factors β1 and β2, *Biochemistry,* 28, 3442, 1989.
579. **Madisen, L., Webb, N.R.,Rose, T.M., Marquardt, H., Ikeda, T., Twardzik, D., Seyedin, S., and Purchio, A.F.,** Transforming growth factor-β2: cDNA cloning and sequence analysis, *DNA,* 7, 1, 1988.
580. **Danielpour, D., Dart, D.L., Flanders, K.C., Roberts, A.B., and Sporn, M.B.,** Immunodetection and quantitation of the two forms of transforming growth factor-β (TGF-β1 and TGF-β2) secreted by cells in culture, *J. Cell. Physiol.,* 138, 79, 1989.
581. **Mori, H., Maki, M., Oishi, K., Jaye, M., Igarashi, K., Yoshida, O., and Hatanaka, M.,** Increased expression of genes for basic fibroblast growth factor and transforming growth factor type β2 in human benign prostatic hyperplasia, *Prostate,* 16, 71, 1990.
582. **Clark, D.A., Flanders, K.C., Banwatt, D., Millar-Book, W., Manuel, J., Stedronska-Clark, J., and Rowley, B.,** Murine pregnancy decidua produces a unique immunosuppressive molecule related to transforming growth factor β2, *J. Immunol.,* 144, 3008, 1990.
583. **Altman, D.J., Schneider, S.L., Thompson, D.A., Cheng, H.-L., and Tomasi, T.B.,** A transforming growth factor β2 (TGF-β2)-like immunosuppressive factor in amniotic fluid and localization of TGF-β2 mRNA in the pregnant uterus, *J. Exp. Med.,* 172, 1391, 1990.
584. **Kim, S.J., Wagner, S., Liu, F., O'Reilly, M.A., Robbins, P.D., and Green, M.R.,** Retinoblastoma gene product activates expression of the human TGF-β_2 gene through transcription factor ATF-2, *Nature,* 358, 331, 1992.
585. **Constam, D.B., Philipp, J., Malipiero, U.V., ten Dijke, P., Schachner, M., and Fontana, A.,** Differential expression of transforming growth factor-β1, factor-β2, and factor-β3 by glioblastoma cells, astrocytes, and microglia, *J. Immunol.,* 148, 1404, 1992.
586. **ten Dijke, P., Geurts van Kessel, A.H.M., Foulkes, J.G., and Le Beau, M.M.,** Transforming growth factor type β3 maps to human chromosome 14, region q23-q24, *Oncogene,* 3, 721, 1988.
587. **Lafyatis, R., Lechleider, R., Kim, S.-J., Jakowlew, S., Roberts, A.B., and Sporn, M.B.,** Structural and functional characterization of the transforming growth factor β3 promoter. A cAMP-responsive element regulates basal and induced transcription, *J. Biol. Chem.,* 265, 19128, 1990.
588. **ten Dijke, P., Iwata, K.K., Goddard, C., Pieler, C., Canalis, E., McCarthy, T.L., and Centrella, M.,** Recombinant transforming growth factor type β3: biological activities and receptor-binding properties in isolated bone cells, *Mol. Cell. Biol.,* 10, 4473, 1990.
589. **Parnell, P.G., Wunderlich, J., Carter, B., and Halper, J.,** Purification of transforming growth factor type e, *J. Cell. Biochem.,* 42, 111, 1990.
590. **Brown, C.A. and Halper, J.,** Mitogenic effects of transforming growth factor type e on epithelial and fibroblastic cells — Comparison with other growth factors, *Exp. Cell Res.,* 190, 233, 1990.
591. **Halper, J., Parnell, P.G., Carter, B.J., Ren, P., and Scheithauer, B.W.,** Presence of growth factors in human pituitary, *Lab. Invest.,* 66, 639, 1992.
592. **Lee, S-J.,** Identification of a novel member (GDF-1) of the transforming growth factor-β superfamily, *Mol. Endocrinol.,* 4, 1034, 1990.
593. **Geistlich, A. and Gehring, H.,** Isolation and characterization of a novel type of growth factor derived from serum-free conditioned medium of chicken embryo fibroblasts, *Eur. J. Biochem.,* 207, 147, 1992.
594. **Geistlich, A. and Gehring, H.,** CDGF (chicken embryo fibroblast-derived growth factor) is mitogenically related to TGF-β and modulates PDGF, bFGF, and IGF-I action on sparse NIH/3T3 cells, *Exp. Cell Res.,* 204, 329, 1993.

595. **Wrann, M., Bodmer, S., de Martin, R., Siepl, C., Hofer-Warbinek, R., Frei, K., Hofer, E., and Fontana, A.,** T cell suppressor factor from human glioblastoma cells is a 12.5-kd protein closely related to transforming growth factor-β, *EMBO J.,* 6, 1633, 1987.
596. **de Martin, R., Haendler, B., Hofer-Warbinek, R., Gaugitsch, H., Wrann, M., Schlüsener, H., Seifert, J.M., Bodmer, S., Fontana, A., and Hofer, E.,** Complementary DNA for human glioblastoma-derived T cell suppressor factor, a novel member of the transforming growth factor-β gene family, *EMBO J.,* 6, 3673, 1987.
597. **Kuppner, M.C., Hamou, M.-F., Bodmer, S., Fontana, A., de Tribolet, N.,** The glioblastoma-derived T-cell suppressor factor/transforming growth factor β_2 inhibits the generation of lymphokine-activated killer (LAK) cells, *Int. J. Cancer,* 42, 562, 1988.
598. **Laiho, M.,** Modulation of extracellular proteolytic activity and anchorage-independent growth of cultured cells by sarcoma cell-derived factors: relationship to transforming growth factor-β, *Exp. Cell Res.,* 176, 297, 1988.
599. **Mooradian, D.L. and Diglio, C.A.,** Production of a transforming growth factor-β-like growth factor by RSV-transformed rat cerebral microvascular endothelial cells, *Tumor Biol.,* 12, 171, 1991.
600. **Tsushima, T., Ohba, Y., Emoto, N., Shizume, K., and Imai, Y.,** Identification and initial characterization of transforming growth factor-like mitogen(s) in human anterior pituitary, *Biochem. Biophys. Res. Commun.,* 133, 951, 1985.
601. **van Zoelen, E.J.J., Twardzik, D.R., van Oostwaard, T.M.J., van der Saag, P.T., de Laat, S.W., and Todaro, G.J.,** Neuroblastoma cells produce transforming growth factors during exponential growth in a defined hormone-free medium, *Proc. Natl. Acad. Sci. U.S.A.,* 81, 4085, 1984.
602. **van Zoelen, E.J.J., van Oostwaard, T.M.J., van der Saag, P.T., and de Laat, S.W.,** Phenotypic transformation of normal rat kidney cells in a growth-factor-defined medium: induction by a neuroblastoma-derived transforming growth factor independently of the EGF receptor, *J. Cell. Physiol.,* 123, 151, 1985.
603. **Dunnington, D.J., Scott, R.G., Anzano, M.A., and Greig, R.,** Characterization and partial purification of human epithelial transforming growth factor, *J. Cell. Biochem.,* 44, 229, 1990.
604. **Stromberg, K., Hudgins, W.R., Fryling, C.M., Hazarika, P., Dedman, J.R., Pardue, R.L., Hargreaves, W.R., and Orth, D.N.,** Human A673 cells secrete high molecular weight EGF-receptor binding growth factors that appear to be immunologically unrelated to EGF or TGF-α, *J. Cell. Biochem.,* 32, 247, 1986.
605. **Bauknecht, T., Kiechle, M., Bauer, G., and Siebers, J.W.,** Characterization of growth factors in human ovarian carcinomas, *Cancer Res.,* 46, 2614, 1986.
606. **Kimball, E.S., Bohn, W.H., Cockley, K.D., Warren, T.C., and Sherwin, S.A.,** Distinct high-performance liquid chromatography pattern of transforming growth factor activity in urine of cancer patients as compared with that of normal individuals, *Cancer Res.,* 44, 3613, 1984.
607. **Stromberg, K., Hudgins, W.R., Dorman, L.S., Henderson, L.E., Sowder, R.C., Sherrell, B.J., Mount, C.D., and Orth, D.N.,** Human brain tumor-associated urinary high molecular weight transforming growth factor: a high molecular weight form of epidermal growth factor, *Cancer Res.,* 47, 1190, 1987.
608. **Twardzik, D.R., Brown J.P., Ranchalis, J.E., Todaro, G.J., and Moss, B.,** Vaccinia virus-infected cells release a novel polypeptide functionally related to transforming and epidermal growth factors, *Proc. Natl. Acad. Sci. U.S.A.,* 82, 5300, 1985.
609. **Ciccodicola, A., Dono, R., Obici, S., Simeone, A., Zollo, M., and Persico, M.G.,** Molecular characterization of a gene of the "EGF family" expressed in undifferentiated human NTERA2 teratocarcinoma cells, *EMBO J.,* 8, 1987, 1989.
610. **Ciardiello, F., Dono, R., Kim, N., Persico, M.G, and Salomon, D.S.,** Expression of *cripto,* a novel gene of the epidermal growth factor gene family, leads to *in vitro* transformation of a normal mouse mammary epithelial cell line, *Cancer Res.,* 51, 1050, 1991.
611. **Kumiyasu, H., Yoshida, K., Yokozaki, H., Yasui, W., Ito, H., Toge, T., Ciardiello, F., Persico, M.G., Saeki, T., Salomon, D.S., and Tahara, E.,** Expression of *cripto,* a novel gene of the epidermal growth factor family, in human gastrointestinal carcinomas, *Jpn. J. Cancer Res.,* 82, 969, 1991.

Chapter 9

Regulatory Peptides with Growth Factor-Like Properties

I. INTRODUCTION

The cells of many organs and tissues are maintained *in vivo* in a nonproliferating state, but they can be induced to resume DNA synthesis and cell proliferation by exposure to hormones and growth factors or to small regulatory peptides. Some regulatory peptides were originally found in the nervous system and are known as neuropeptides, and other regulatory peptides may function as postganglionic receptors or neurotransmitters.[1] Such peptides may act not only in processes associated with the rapid transmission of signals in the nervous system but can display potent mitogenic activity, stimulating DNA synthesis and cell proliferation in a variety of tissues, thus acting in a manner similar to that of growth factors. Neuropeptides such as the gastrin-releasing peptide (GRP)/bombesin, bradykinin, vasoactive intestinal peptide (VIP), and endothelin can act as growth factors for cultured cells and probably also for different types of cells *in vivo*. However, the regulatory peptides function rather as comitogens and require the presence of serum, insulin, or other factors for displaying mitogenic activity. Several neuropeptides have important actions in both the central nervous system and the digestive tract and may be considered as brain-gut hormones.[2,3] The mitogenic regulatory peptides frequently act by paracrine or autocrine mechanisms and may have an important role not only in the function and growth of many normal tissues but also in tumorigenesis. The proliferation of aggressive tumors, such as in lung cancer, is stimulated by mitogenic peptides that are produced either by the cancer cells themselves or by cells from neighboring tissues.[4]

II. BOMBESIN AND BOMBESIN-LIKE PEPTIDES

Bombesin and bombesin-like peptides such as the gastrin-releasing peptide (GRP) are neurohormones with potent mitogenic effects on a variety of cell types, including epithelial fibroblastic cells.[5] Bombesin is a 14-amino acid peptide (tetradecapeptide) initially isolated from two species of European frogs of the genus *Bombina*. GRP, the mammalian counterpart of amphibian bombesin, has a carboxyl-terminal heptapeptide which is identical to that of bombesin. The bombesin-related peptides identified in vertebrates have been classified in three subfamilies: bombesin, ranatensin, and litorin. In mammals, the bombesin-related peptide neuromedin-B is an important member of the bombesin family. These peptides have powerful effects on cell proliferation *in vitro* and *in vivo* and are also able to elicit a wide range of physiological responses including smooth-muscle contraction and the release of peptide hormones. High concentrations of bombesin and neuromedin-B are present in the hypothalamus and pituitary gland as well as in the adrenal zona medullaris of various animal species. Bombesin may be an example of an integrative neuropeptide with important influence on the regulation of food intake and other complex physiological processes.[6]

A. THE BOMBESIN RECEPTOR

Murine Swiss 3T3 cells possess a single class of high-affinity receptors for bombesin which mediate the biological responses of these cells to the peptide.[7] Analysis of the nucleotide sequence of a cDNA coding for the bombesin receptor in these cells revealed a single, long, open reading frame encoding a 384-amino acid protein with a predicted molecular weight of 43 kDa.[8] The deduced bombesin receptor protein has seven hydrophobic segments corresponding to seven transmembrane domains which are characteristically found in members of the superfamily of receptors coupled to G proteins. A functional membrane-associated, high-affinity bombesin receptor was isolated from the solubilized membranes of human small-cell lung carcinoma NCI-H345 cells.[9] The mature bombesin receptor in human and murine cells is represented by glycoproteins of 65 to 75 kDa.[10,11] However, available evidence suggests the existence of at least two functionally distinct bombesin receptor subtypes, one preferring bombesin as ligand and the other preferring neuromedin-B. Both receptors belong to the G protein-coupled receptor family but their distribution among different mammalian tissues may show variations. Moreover, the two types of receptors may be linked to different signaling intracellular pathways or mediators in cells such as those of the rat adrenal cortex, which respond to bombesin and neuromedin-B with increased proliferative activity.[12]

B. POSTRECEPTOR MECHANISM OF BOMBESIN ACTION

Bombesin and bombesin-like peptides induce DNA synthesis and mitosis in sensitive cells but the postreceptor mechanisms of their action are little understood. The bombesin receptor does not possess intrinsic protein-tyrosine kinase activity, but bombesin rapidly stimulates tyrosine phosphorylation in 3T3 fibroblasts, inducing the phosphorylation of a 115-kDa membrane protein.[13] In addition to its effect on tyrosine phosphorylation, bombesin acts in murine and human cells through the phosphoinositide signaling pathway, stimulating the breakdown of phosphatidylinositol 4,5-bisphosphate and the production of inositol trisphosphates and 1,2-diacylglycerol and inducing a transient elevation of $[Ca^{2+}]_i$, with mobilization of an intracellular pool of Ca^{2+} and alteration of Ca^{2+} flux rates and the pH_i.[14-17] However, stimulation of Ca^{2+} mobilization in human small-cell lung carcinoma (SCLC) cells by neuropeptides such as bombesin and tachykinin does not result in cell proliferation.[18] Bombesin stimulation of National Institutes of Health (NIH)/3T3 mouse fibroblasts in which the N-*ras* gene is overexpressed stimulates the production of inositol 1,4,5-trisphosphate and leads to $[Ca^{2+}]_i$ elevation.[19] However, the bombesin-stimulated inositol phosphate response in the NIH/3T3 cells transformed by N-*ras* overexpression is only increased when the cells are cultured under subconfluent conditions and is attenuated by the production of autocrine factors by the transformed cells.[20] Bombesin may utilize a Ca^{2+}-sensitive site other than phospholipase C and protein kinase C in bombesin-induced growth stimulation.[21] The hormone stimulates a rapid activation of phospholipase A_2-catalyzed phosphatidylcholine hydrolysis in Swiss 3T3 cells.[22] Further studies are required to characterize the cellular mechanisms involved in the mitogenic effects of bombesin.

C. BOMBESIN-INDUCED ALTERATION OF PROTO-ONCOGENE EXPRESSION

Bombesin may induce alterations in the levels of proto-oncogene expression. The levels of expression of c-*fos* and c-*myc* proto-oncogenes are increased in mouse 3T3 fibroblasts exposed to nanomolar concentrations of bombesin.[23-25] Transcripts of the c-*fos* gene markedly increase 15 min after the addition of bombesin to the culture medium and decrease 30 to 60 min thereafter. The induction of c-*myc* occurs later, about 1 h after bombesin treatment. No changes in c-K-*ras* expression are observed in the bombesin-treated cells, but it has been shown that expression of a mutated human c-H-*ras* gene desensitizes the intracellular Ca^{2+}-mobilizing system to serum growth factors and bombesin by a mechanism involving a decrease in the affinity of the inositol 1,4,5-trisphosphate receptor.[26] In the rat brain, peripheral bombesin can induce c-Fos protein expression.[27]

D. ROLE OF BOMBESIN IN CARCINOGENIC PROCESSES

Bombesin and bombesin-like peptides are growth factors to normal human bronchial epithelial cells and may be involved in the autocrine proliferation of some human lung cancers (SCLCs).[28-31] These tumors produce high levels of bombesin-like peptides and have receptors for bombesin.[32] In addition, SCLCs produce several other hormones, including adrenocorticotropin (ACTH), calcitonin, arginine-vasopressin, and neurotensin. It was assumed that SCLCs originated in endocrine Kultschitzy cells of the lung, but recent evidence suggests their derivation is from a hematopoietic stem cell.[33] The release of bombesin/GRP by cells of SCLCs is stimulated by agents such as secretin and vasoactive intestinal peptide (VIP), which increase the intracellular concentration of cyclic adenosine-3′,5′-monophosphate (cAMP).[34] Moreover, injection of secretin to patients with metastatic SCLCs results in increased levels of plasma bombesin. High-affinity receptors for bombesin and GRP-like peptides are present in SCLC cell lines.[35,36] An autocrine model for the growth of SCLC, involving bombesin as a growth factor, is also supported by the observation that a bombesin receptor-specific antagonist inhibits, in the absence of exogenous bombesin, the growth of a human SCLC cell line (NCI-H345) that constitutively produces GRP.[37] The bombesin receptor antagonist that inhibits the growth of NCI-H345 human SCLC cells also blocks the increase in phosphoinositide turnover and cytoplasmic free Ca^{2+} stimulated by bombesin in these cells.

Human prostatic carcinoma cells express bombesin receptors and, in a dose-dependent manner, bombesin can stimulate the growth rate of these tumor cell lines.[38] The effect of bombesin on prostatic carcinoma cells is specifically inhibited by anti-bombesin antibodies. Bombesin could be an important growth factor for the prostate gland and may be involved in the growth of human prostatic tumors.

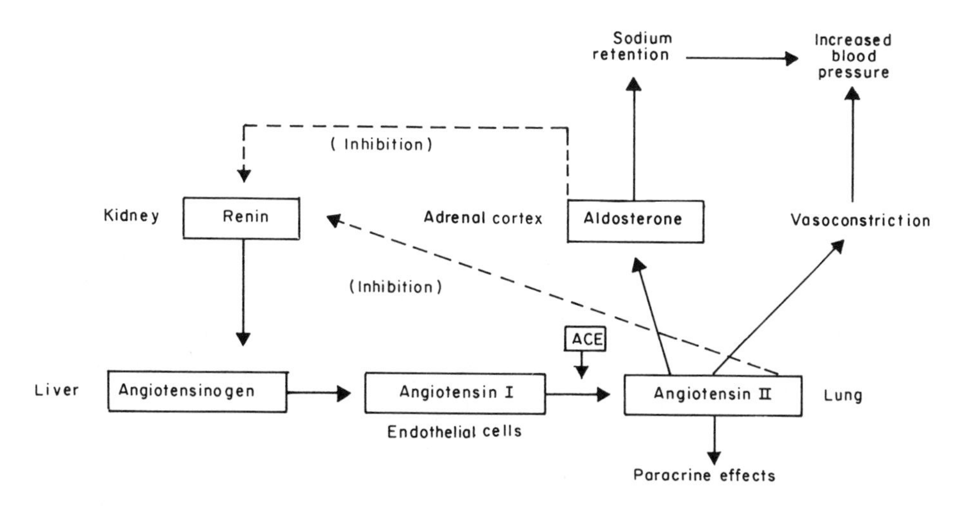

Figure 9.1. The renin-angiotensin system.

III. ANGIOTENSIN

The renin-angiotensin system is involved in the regulation of blood pressure and extracellular fluid homeostasis.[39-43] Human renin is biosynthesized in the kidney from a prorenin precursor of 406 amino acids which is cleaved by proteolytic enzymes. The mature renin is a 41-kDa single-chain glycosylated carboxypeptidase that belongs to the aspartyl proteinase family, which also includes pepsin, chymosin, and lysosomal cathepsins. The human genome contains a single renin gene which resides on chromosome 1. Expression of the renin gene in the kidney is stimulated by cAMP and inhibited by angiotensin II. Renin biosynthesis is modulated by a number of stimuli, including changes in the flux of Na^+ across the distal tubule and the influence of baroreceptors sensitive to changes in the renal perfusion pressure and blood pressure in the right atrium. Other factors involved in the control of renin synthesis are plasma K^+ concentration, catecholamines, vasopressin, and aldosterone (which exerts a negative feedback). In addition to these agents, a multitude of hormones, growth factors, cytokines, and neurotransmitters have been found to modify renin release from the kidney, but the precise physiological significance of these influences remains to be elucidated. Regulation of renin synthesis and secretion by external agents is mediated by changes in cAMP, cGMP, calcium, and ion channels. Renin is released from the juxtaglomerular cells lining the afferent arterioles of the glomerulus. Once renin is released into the blood, it acts on α-globulin (angiotensinogen), which is a tetradecapeptide which is synthesized in the liver and serves as a substrate for renin kinase activity. This reaction results in the synthesis of the decapeptide angiotensin I, which circulates through the lung and is converted into an octapeptide, angiotensin II, by the action of an angiotensin-converting enzyme (ACE). Angiotensin II is the final active product of the renin-angiotensin system (Figure 9.1).

A. BIOLOGICAL PROPERTIES OF ANGIOTENSIN II

Angiotensin II is a powerful vasoconstrictor which has an important role in the pathogenesis of hypertension. It affects blood pressure by increasing vascular resistance through direct effects on blood vessels and by facilitating sympathetic nerve transmission. It also influences water and salt excretion through the stimulation of aldosterone synthesis and secretion by the zona glomerulosa of the adrenal gland. In addition, angiotensin II has a stimulatory effect on the thirst center in the central nervous system. Angiotensin II is also implicated in regulating the growth of vascular smooth muscle (VSM) cells through its interaction with cell surface receptors. However, angiotensin II is not mitogenic for VSM cells and it may cause hypertrophy of these cells, without increasing their number, by stimulating protein synthesis. In contrast, platelet-derived growth factor (PDGF) is mitogenic for VSM cells and can also stimulate protein synthesis in these cells. Both angiotensin II and noradrenaline increase PDGF-BB receptors and potentiate PDGF-BB-stimulated DNA synthesis in VSM cells.[44] Angiotensin II can stimulate glucose transport activity in VSM cells.[45] Actions of angiotensin II in the kidney include vascular effects, with diminished renal blood flow and changes in the glomerular filtration rate, and tubular effects, with increased Na^+/H^+ antiporter activity. In addition to its effects on blood vessels and kidney, angiotensin

II is involved in other functions, including the regulation of ovarian function. Angiotensin II may also have effects in other tissues, including gut, adrenal, liver, vascular smooth muscle, and nervous system. In certain cases it may display growth factor-like activities and may be involved, for example, in the pathogenesis of hypertrophy of the heart.[46]

B. ANGIOTENSIN RECEPTORS

Angiotensin II exerts its cellular effects through binding and activation of specific receptors located on the surface of cells from various organs and tissues, including the vasculature, adrenal cortex, adrenal medulla, brain, liver, and kidney. Two distinct isoforms of angiotensin II receptors, type 1 (ATR-1) and type 2 (ATR-2), were identified in the rat adrenal gland by means of nonpeptide antagonists.[47] ATR-1 and ATR-2 bind angiotensin II with similar avidity and potency but have differential tissue distribution.[48] Ovarian granulosa cells exclusively express an ATR-2 of 79 kDa.[49] Only ATR-1 appears to mediate the classical functions assigned to angiotensin II and the functions of ATR-2 remain to be assigned.

The cloning and expression of cDNAs encoding angiotensin receptors has been reported.[50,51] The rat ATR-1 receptor encodes a protein of 367 amino acids.[52] ATR-1 is a member of the G protein coupled family of receptors with seven transmembrane regions. There are two ATR-1 subtypes, ATR-1A and ATR-1B, which may be related to the differential regulation of their expression rather than to different functional properties of the receptor proteins.[53] However, there is evidence that in rat renal mesangial cells both ATR-1A and ATR-1B mediate the effects of angiotensin II through activation of adenylyl cyclase and intracellular mobilization of Ca^{2+}, whereas the stimulation of protein synthesis induced by angiotensin II in these cells would depend only on the activation of ATR-1A receptors.[54] ATR-1A and ATR-1B mRNAs are differentially expressed in the brain.[55] The human gene encoding ATR-1A has been cloned, expressed, and characterized.[56] The cloning of a rat gene encoding an additional angiotensin II receptor subtype, ATR-3, has been reported.[57]

It has been proposed that the receptor for angiotensin is encoded by the *mas* gene.[58] This putative proto-oncogene was first detected using tumorigenicity assay in nude mice following transfection with DNA isolated from human epidermoid carcinoma.[59] However, there is no evidence that the *mas* gene contributes to tumor formation *in vivo*. The Mas protein possesses seven transmembrane domains and shows a general similarity to the family of G-protein linked receptors. Mas shows a hydrophobic profile similar to that of rhodopsin and an α-subunit of acetylcholine receptor. It is homologous to several neurotransmitter receptors, including the adrenergic and muscarinic receptors, as well as the receptors for dopamine D_2, substance K, and substance P. Transcripts of the *mas* gene have been detected in the rat hippocampus and cerebral cortex.[60] The *mas* gene is only expressed in a small subpopulation of cells in the central nervous system, primarily in hippocampal pyramidal cells and granule cells of the dentate gyrus.[61] The extremely restricted pattern of *mas* gene expression suggests that its product may have a function in determining the morphology and connections of specific cell types in the hippocampus. The Mas protein may lead to increased responsiveness to the angiotensin II signaling system, but Mas is probably not a typical conventional receptor for angiotensin II.[62] The exact function of the Mas protein remains to be elucidated but it is possible that it may represent a specific subclass of angiotensin II receptor expressed in the hippocampus.

Multiple endogenous factors are involved in the regulation of angiotensin II receptor expression and function. A short exposure of bovine adrenal zona glomerulosa cells to ACTH *in vitro* is able to decrease the expression of cell surface angiotensin II receptors and to reduce the production of aldosterone induced by angiotensin II.[63]

C. POSTRECEPTOR MECHANISMS OF ACTION OF ANGIOTENSIN

The postreceptor mechanisms of angiotensin action have been studied in various target systems, including the proximal tubular epithelium of the kidney, where they are linked to decrements in the production of cAMP, activation of the phospholipase A_2, and activation of Ca^2 channels.[64] In adrenal glomerulosa cells, angiotensin stimulates adenylyl cyclase via prostaglandins,[65] and activates the Na^+/H^+ antiporter in the membrane.[66] An increase in $[Ca^{2+}]_1$ is an important component of the mechanism of action of angiotensin II in adrenal glomerulosa cells, and this effect is due to both a stimulation of Ca^{2+} entry across the membrane and to mobilization of Ca^{2+} from intracellular stores.[67] Angiotensin II, as well as ACTH, stimulates aldosterone secretion by adrenal zona glomerulosa cells, and this effect is associated with an increase in $[Ca^{2+}]_1$.[68] Angiotensin II rapidly and transiently induces a biphasic increase of $[Ca^{2+}]_1$ in cells isolated from the rat kidney early proximal tubule.[69] It produces similar increases of the $[Ca^{2+}]_1$ in follicular oocytes from *Xenopus laevis*.[70] These cells express angiotensin II receptors, and, in them, the hormone can increase the rate of maturation induced by progesterone. These results suggest the existence of a calcium-mobilizing signal transferred from follicle cells to the oocyte through gap junctions, which may play a physiological role in oocyte maturation.

The effects of angiotensin II on rat VSM cells are linked to phosphoinositide metabolism and Ca^{2+} mobilization. These effects are independent of Na^+/H^+ exchange or changes in pH_i but are associated with protein kinase C activation and the accumulation of c-*fos* mRNA.[71] A tight coupling exists between the angiotensin II receptor and the formation of inositol trisphosphate in VSM cells.[72] In these cells, the hormone stimulates phosphorylation of nuclear lamins via a protein kinase C-dependent mechanism.[73] Angiotensin elicits a rapid and transient increase in hepatocyte protein kinase C activity.[74] Angiotensin II receptor subtypes could play opposite roles in regulating phosphatidylinositol hydrolysis in slices of rat skin.[75] In rat hepatocytes, ATR-1 mediates the activation of phosphorylase *a* by angiotensin II and is coupled to turnover of phosphoinositides.[76] In VSM cells, angiotensin II stimulates the formation of 1,2-diacylglycerol from phosphatidylcholine.[77] However, the role of phosphoinositide metabolism in the mechanisms of action of angiotensin II is not clear. The hormone can induce enhanced phosphatidylinositol turnover and increases the amount of diacylglycerol in liver cells, but does not exert mitogenic effects in these cells.[78] Mitogenic stimulation with angiotensin of the neuronal cell line NG115-401L-C3, transfected with an angiotensin receptor gene, is not correlated with the levels of phosphoinositide metabolites or with changes in the activity of the enzyme, phosphatidylinositol 3-kinase.[79] Angiotensin binding to ATR-2 in ovarian granulosa cells may not result in changes of phosphoinositide turnover, $[Ca^{2+}]_i$, or cAMP and cGMP levels. ATR-2 is not coupled to G proteins.[48] ATR-1 and ATR-2 may be coupled with distinct signal transduction mechanisms in different types of cells.[80] Astrocyte cultures from neonatal rat brains contain predominantly ATR-1 that are coupled with a stimulation of inositolphospholipid hydrolysis. In contrast, neuron cultures of the same origin contain ATR-2 molecules that are coupled to a reduction in basal cAMP levels, although a smaller population of ATR-1 is present in the neurons.

Protein phosphorylation may have a key role in the mechanism of action of angiotensin II. In rat VSM cells, angiotensin II stimulates activity of pp44 and pp42 microtubule-associated protein (MAP) kinases and tyrosine kinases.[81] Treatment of quiescent VSM cells with angiotensin results in the stimulation of a distinct set of cellular proteins, and this effect is accompanied by the activation of several serine/threonine protein kinases, suggesting that activation of a complex protein kinase cascade is involved in angiogensin II-mediated signal transduction.[82] In PC12 cells and rat adrenal glomerulosa cells, binding of angiotensin II to AT2-R may signal through stimulation of protein tyrosine phosphatase activity by a mechanism implicating inhibition of basal and atrial natriuretic factor (ANF)-stimulated particulate guanylyl cyclase activity.[83]

Angiotensin II is involved in the regulation of gene expression in particular types of cells. Angiotensin II regulates, for example, the expression of the gene coding for tenascin in VSM cells.[84] Tenascin functions as a modulator of cell growth.[85] The role of proto-oncogene products in the physiological effects of angiotensin II is not understood. In adrenal glomerulosa cells, which are a major site of angiotensin II action, the hormone stimulates phosphoinositide hydrolysis and Ca^{2+} mobilization, and this effect is accompanied by the induction of several early response genes, including c-*fos*, c-*jun*, *jun*-B, and *Krox*-24.[86] Expression of c-*fos*, c-*jun*, and *jun*-B mRNA induced by angiotensin II in adrenocortical fasciculata cells is mediated by AT-1 receptors.[87] In isolated rat hepatocytes, both angiotensin II and phorbol ester induce early expression of the c-*fos*, c-*myc*, and c-*mos* proto-oncogenes, as well as increased expression of the β-actin gene.[88] In cultures of VSM cells, exposure to angiotensin II results in increased expression of c-*fos* and c-*jun*.[89] The Fos and Jun proteins form homodimeric or heterodimeric complexes which may act to induce other genes by binding to a consensus DNA sequence, the TRE/AP-1 binding site. Angiotensin regulates gene expression by the TRE via a protein kinase C-dependent pathway.[90] Myocardial infarction produced in rats by ligation of the left coronary artery is followed by reactive cardiac hypertrophy, and this alteration is associated with enhanced expression of angiotensin II receptors and induction of c-*myc* and c-*jun* gene expression in the surviving ventricular myocytes.[91] These observations suggest that the increase in systolic stress produced by sudden coronary artery occlusion may be coupled with the induction of early growth-related genes that participate in the modulation of cardiomyocyte growth. Angiotensin II shows only weak or no mitogenic activity in adult rat kidney glomerular mesangial cells in culture. In this system, angiotensin II induces a rise in $[Ca^{2+}]_i$ and activation of phospholipases C and A_2, but the levels of expression of immediate early genes such as c-*fos*, c-*jun*, and *egr*-1 remain unaltered.[92] In contrast, arginine vasopressin acts as a strong mitogen in this system and induces the expression of immediate early genes.

IV. ENDOTHELIN

Endothelin is a potent vasoconstrictor peptide isolated from the conditioned medium of porcine aortic endothelial cells.[93] It is secreted by endothelial cells and plays a role in the regulation of blood pressure through its action on VSM cells. However, endothelin activity is represented by a family of homologous acidic peptides of 21 amino acids with widespread biological activities.[95-100] At least four distinct isoforms of endothelin have been identified: ET-1, ET-2, ET-3, and VIC (vasoactive intestinal contractor). In addition to vascular functions, endothelin activities involve cardiac, pulmonary, and renal functions as well as mitogenesis and tissue remodeling.

A. PRODUCTION AND BIOLOGICAL PROPERTIES OF ENDOTHELIN

Endothelins are produced by cells other than endothelial cells and has important biological actions on various organs and tissues. Endothelins exhibit properties similar to those of hormones and growth factors. Human macrophages, but not neutrophils or lymphocytes, produce endothelins suggesting an important role for these peptides in the microenvironment of tissue macrophages.[101] Cultured human keratinocytes and neonatal rat cardiac myocytes synthesize and secrete ET-1.[102,103]

Different types of cells exhibit varied specific responses to endothelin. In synergy with other mitogens, endothelin stimulates DNA synthesis in mouse 3T3 fibroblasts.[104,105] Endothelin can also promote growth in VSM cells and renal mesangial cells. ET-1 is synthesized by and released from mesangial cells, suggesting the possibility that it may act as an autocrine factor.[106] Endothelin action on DNA synthesis in human vascular endothelial cells may be mediated by an autocrine and/or paracrine mechanism.[107] ET-1 mediates endothelial cell-dependent proliferation of vascular pericytes.[108] The production of superoxide or monocyte adhesion to endothelial cells is not stimulated by endothelin, which is not a chemoattractant factor for monocytes.[109] Vasopressin-stimulated water permeability in the terminal inner medullary collecting duct is inhibited by endothelin.[110] In the perfused rat liver, endothelin behaves as an agonist, producing sustained vasoconstriction and increased hepatic glycogenolysis and glucose output.[111] Endothelin secreted by keratinocytes displays mitogenic properties for human melanocytes and may be involved in the process of melanization of this type of cells.[112,113] Cerebral rat astrocytes possess abundant receptors of the ETR-B isoform, and ET-1 modulates the differentiation of these cells caused by cAMP analogues in culture and can induce an active proliferation of the astrocytes.[114] ET-1 can suppress the differentiation of rat adipocyte precursors in serum-free culture.[115]

Endothelins are involved in the regulation of hormone secretion. ET-3 stimulates LHRH secretion from LHRH neurons by a mechanism involving PGE_2 production.[116] It also inhibits the secretion of prolactin and stimulates the secretion of gonadotropins and TSH by pituitary cells in culture.[117] ET-1 promotes steroidogenesis and enhances c-*myc* and c-*jun* gene expression in the murine Leydig tumor cell line MA-10.[118] Adrenal zona glomerulosa cells contain high-affinity ET-1 receptors, and addition of ET-1 to these cells may result in stimulation of aldosterone secretion. Incubation of adrenal zona glomerulosa cells with ET-1 results in down-regulation of endothelin receptors and an antimitogenic effect with decreased incorporation of labeled thymidine.[119] Phenylmercuric acetate (PMA) has similar effects to those of ET-1 in zona glomerulosa cells. Endothelin enhances ACTH-stimulated aldosterone release from cultured bovine adrenal cells.[120] Endothelin stimulates testosterone secretion by rat Leydig cells.[121] Parathyroid epithelial cells express both the ET-1 peptide and ET-1 receptors. The levels of ET-1 mRNA and peptide in parathyroid epithelial cells are rapidly regulated by extracellular Ca^{2+}, suggesting that endothelin mediates, at least in part, the effects of calcium on the parathyroid system through an autocrine mechanism.[122]

Endothelin may have hormone-like effects and may also have functions similar to those of neuropeptides. Both ET-1 and ET-3, but not ET-2, are present in the porcine brain as well as in the spinal cord.[123] ET-3 is abundantly present in rat intestine, pituitary gland, and brain, suggesting that it may have a physiological function in the neuroendocrine system.[124] The pituitary may be an important source of endothelin synthesis and secretion as well as an effector organ of endothelin action. Endothelin produces a selective inhibition of prolactin release by pituitary cells *in vitro* and stimulates the secretion of other pituitary hormones such as LH, FSH, and TSH.[125,126] A prolactin release-inhibiting activity present in the neurointermediate lobe of the rat and called vasoactive intestinal contractor (VIC) is identical with ET-2.[127,128] VIC may be involved in the physiological regulation of prolactin secretion.

Secretion of ET-3 by rat pituitary cells in culture is stimulated by IGF-I.[129] Endothelin is synthesized in the posterior pituitary and may be involved in neurosecretory functions.[130] ET-1-like immunoreactivity is present in human cerebrospinal fluid.[131] The ET-3 gene is expressed in the human placenta and the ET-3 peptide is actively secreted into the fetal circulation by placental endothelial cells.[132,133] The plasma levels of endothelin in maternal and fetal blood are related. It is thus clear that endothelin has a wide diversity of physiological effects in different organs and tissues at different stages of development.

B. ENDOTHELIN STRUCTURE AND THE ENDOTHELIN GENES

Endothelin was originally identified as a 21-amino acid peptide of 2492 mol wt and with a primary structure which is different from that of other bioactive peptides of mammalian origin. The cloning and sequencing of cDNAs encoding the endothelin precursor polypeptide, preproendothelin, and the identity of human and porcine endothelin have been reported.[134,135] Cathepsin-D catalyzes the proteolytic processing of the porcine ET-1 precursor.[136] The preproendothelin gene contains on its 5′-flanking region a phorbol ester (TPA)-responsive element. Three distinct endothelin genes (ET-1, ET-2, and ET-3) have been identified in the human genome.[137] An additional member of the family, the vasointestinal constrictor (VIC), is present in the mouse.[138] The ET-1 gene codes for "classical" endothelin. The nucleotide sequences of the three human ET genes are highly conserved within the regions encoding the mature endothelin peptides. The ET-1 gene contains an AP-1 site that is recognized by complexes formed between Fos and Jun proteins, which are involved in the regulation of ET-1 gene expression.[139] Insulin stimulates ET-1 gene expression in endothelial cells.[140]

C. REGULATION OF ENDOTHELIN SYNTHESIS AND SECRETION

Synthesis and secretion of endothelin is regulated by a variety of endogenous and exogenous factors. The different endothelin types may be regulated in differential manners in different organs and tissues *in vivo* as well as in cells cultured *in vitro*. ET-1 stimulates its own synthesis in cultured human endothelial cells.[141] Elevated glucose levels induce enhanced secretion of ET-1 by cultured bovine aortic endothelial cells.[142] Exposure to IL-1 results in increased production of ET-1 by endothelial cells.[143] EGF decreases the production of ET-2 by cultured human renal adenocarcinoma cells.[144] Bombesin and glucocorticoids stimulate human breast cancer cells to produce endothelin, which functions as a paracrine mitogen for breast stromal cells.[145] ANF inhibits the production and secretion of ET-1 from cultured endothelial cells.[146] This effect may have physiological significance, since ANF is a vasodilator and lowers blood pressure, whereas ET-1 is a vasoconstrictor and pressor protein. ET-3 gene transcripts are abundantly expressed in the eye ball of the rat, and the brain, small intestine, kidney, and submandibular gland of the animal contain relatively large amounts of ET-3 mRNA.[147] In contrast, the rat stomach and spleen contain small amounts of ET-3 mRNA. The possible biological significance of the different distribution of endothelin isoforms in different organs and tissues is unknown.

D. ENDOTHELIN RECEPTORS

Specific binding sites for endothelin are present in various fetal and adult organs and tissues of mammals, including lung, kidney, heart, intestine, adrenal gland, eye, and brain. Swiss 3T3 mouse fibroblasts have endothelin receptors, and endothelin potentiates DNA synthesis stimulated in these cells by PDGF, basic FGF, and insulin.[148] Adrenal zona glomerulosa cells have receptors for ET-1, and incubation of the these cells with ET-1 results in stimulation of aldosterone secretion in a dose-dependent manner, although the effect is less potent than that of angiotensin II. An endothelin receptor protein purified to homogeneity from human placenta is a 40-kDa polypeptide.[149]

The effects of endothelins on different types of cells may be related to differential expression of receptors with relative specificity for the ET-1, ET-2, and ET-3. Two distinct types of endothelin receptors have been identified in tissues from different vertebrate species, ETR-A, which is specific for ET-1, and ETR-B, which is nonspecific for the different types of endothelins (ET-1, ET-2, and ET-3).[150] Cloning and expression of ETR-A and ETR-B indicate that these receptors are members of the G-protein coupled family of receptors possessing seven transmembrane domains.[151-153] The endothelin receptor sequences are highly conserved in different mammalian species. Analysis of a cDNA coding for ETR-A, isolated from a human placental cDNA library, showed that the entire ETR-A sequence contains 427 amino acids and has a calculated molecular weight of 48,722 Da.[154] Different tissues exhibit different expression of endothelin receptor subtypes. ETR-A may transduce the effects of the hormone in the anterior pituitary gland.[155] High levels of ETR-A mRNA are expressed in the uterus and testis of Cynomolgus monkeys. Functionally distinct ETR-B receptors are expressed in vascular endothelium and smooth muscle.[156] ETR-A and ETR-B receptors, as well as ET-1 and ET-3 growth factor isopeptides, are expressed by human adrenal cortex, suggesting the potential role of endothelins as local mediators in the adrenal gland.[157] ETR-A and ETR-B may use different mechanisms for signal transduction.[158] mRNAs coding for these two types endothelin receptors are differentially expressed in normal human arteries and advanced atherosclerotic plaques.[159]

In addition to ETR-A and ETR-B, there is evidence of the existence of endothelin receptors with binding properties typical of "super-high" affinity sites in the picomolar range.[160] These super-high affinity sites are expressed in both central nervous system and peripheral tissues and may be related to the vasodilatation property of endothelin, whereas the classic high-affinity sites participate in the vasoconstrictive action of the hormone.

Little is known about the regulation of endothelin receptor expression in different tissues by endogenous and exogenous factors. The level of ETR-B receptor transcripts in ROS17/2 rat osteosarcoma cells is downregulated by endothelin through decreasing mRNA stability.[161] Phorbol esters and angiotensin II downregulate ETR-A expression in human VSM cells by a mechanism involving protein kinase C.[162,163] Exposure of VSM cells to the glucocorticoid dexamethasone results in downregulation of endothelin receptor expression.[164]

E. POSTRECEPTOR MECHANISMS OF ENDOTHELIN ACTION

Binding of endothelin to its receptor in different types of cells can evoke the stimulation of a number of signal transduction pathways, including G protein-coupled activation of phospholipase C as well as activation of protein kinase C, phospholipase A_2, and receptor-operated and voltage-dependent Ca^{2+} channels, which may lead to changes in gene expression and mitogenesis.[165,166] The mitogenic effects of endothelin in rat mesangial cells are associated with activation of phospholipase C and with increased phosphoinositide turnover and alterations in the levels of $[Ca^{2+}]_i$. In VSM cells, the endothelin receptor may be coupled to phospholipase C via a pertussis toxin-insensitive G protein.[167] In these cells, endothelin stimulates 1,2-diacylglycerol accumulation and activates protein kinase C.[168] In rat mesangial cells, ET-1 stimulates phospholipase D activity through a mechanism involving activation of protein kinase C, which results in the formation of the putative second messenger phosphatidic acid.[169] Endothelin is a complete mitogen for cultured Rat-1 fibroblasts, and in these cells it stimulates the hydrolysis of phosphatidylinositol 4,5-bisphosphate and phosphatidylcholine.[170] The effects of endothelin on phospholipid metabolism have been confirmed in isolated rat hearts.[171]

Monovalent and divalent ions may be involved in the mechanism of action of endothelin. In mesangial cells maintained in culture, endothelin activates the exchange of Na^+/H^+, causing cytosolic alkalinization. Endothelin-induced activation of Na^+/H^+ exchange in brain capillary endothelial cells may occur via a high-affinity ET-3 receptor that may not be coupled to phospholipase C.[172] A sustained increase in $[Ca^{2+}]_i$ is an important component of the mechanism of action of endothelin in its target cells. In intact blood vessels, the effects of endothelin include a slow tonic contraction and Ca^{2+} influx. A rapid effect exerted by endothelin on the $[Ca^{2+}]_i$ of VSM cells is due to intracellular mobilization of Ca^{2+}.[173,174] In addition to this mechanism, endothelin may also affect, either directly or indirectly, the function of voltage-operated and receptor-operated membrane Ca^{2+} channels. ET-1 and ET-3 may stimulate calcium mobilization by different mechanisms in VSM cells.[175] In human ciliary muscle cells, endothelin induces a reversible membrane voltage depolarization and a rise in $[Ca^{2+}]_i$.[176] In human myometrial cells, endothelin and oxytocin stimulate the Ca^{2+} second messenger system but via binding to distinct specific receptors.[177] Endothelin, as oxytocin, may be an important endogenous modulator of uterine contractility. An initial transient peak of $[Ca^{2+}]_i$ induced by endothelin in mouse Swiss 3T3 fibroblasts and MC3T3-E1 osteoblastic cells is due to Ca^{2+} release from intracellular stores, and a later more sustained increase of the ion is associated with Ca^{2+} influx across the plasma membrane.[178,179]

Endothelin can activate phospholipase A_2, which results in the release of arachidonic acid.[180] This may be followed by the conversion of arachidonate to prostaglandins and thromboxane, depending on the enzymatic capabilities of the target cells. In the distal lung of the rat, ET-1 stimulates arachidonate 15-lipooxigenase activity and oxygen radical formation.[181] Endothelin stimulates the secretion of steroid hormones (corticosterone and aldosterone) by frog adrenal gland by a mechanism that is mediated by prostaglandins and requires the presence of extracellular calcium.[182]

At least some of the cellular actions of endothelin may depend on the modulation of the adenylyl cyclase system. This system is inhibited by endothelin in brain capillary endothelial cells.[183] In addition to the stimulation of phospholipase C via G_q protein, ETR-A and ETR-B are coupled to adenylyl cyclase via G_s in rat VSM cells and G_i in bovine endothelial cells, respectively.[184] However, it is not known whether different G proteins coupled to endothelin receptor subtypes in the vasculature induce different cellular responses.

Phosphorylation of cellular proteins on serine/threonine and tyrosine residues is stimulated by the action of endothelin. A rapid stimulation of MAP kinase activity occurs in rat mesangial cells and ventricular cardiomyocytes exposed to endothelin.[185,186] The mechanism of endothelin action may involve phosphorylation of distinct cellular protein on tyrosine residues. In mouse 3T3 fibroblasts, endothelin stimulates tyrosine phosphorylation of specific proteins. In rat glomerular mesangial cells, ET-1 stimulates autophosphorylation of the c-Src protein as well as c-Src-catalyzed phosphorylation of a peptide substrate specific for tyrosine kinase activity.[187] ET-1-stimulated phospholipase D activity in A10 VSM cells is dependent on tyrosine kinase.[188] A cross-talk between G protein-coupled endothelin receptors such as the endothelin receptors and nonreceptor tyrosine kinases, such as the c-Src oncoprotein, may have an important role in the regulation of gene transcription and mitogenic signaling by growth factors and regulatory peptides.

Endothelin may influence DNA replication and gene expression in its target cells. These effects may be mediated, at least in part, through modification of topoisomerase activity. The topoisomerases I and II are enzymes that catalyze interconversions among DNA topological isomers and are involved in the regulation of replication and transcription. Cultured mesangial cells contain high-affinity endothelin receptors, and exposure of these cells to endothelin results in a transient increase in topoisomerase I activity.[189] This increase is inhibited by incubation of the cells with pertussis toxin, suggesting the involvement of GTP-binding protein(s) in endothelin-mediated action. Expression of specific genes may be regulated by endothelin in its target cells. The hormone stimulates cardiac α- and β-myosin heavy chain gene expression.[190] ET-1 activates the transcriptional expression of the urokinase receptor gene.[191]

Proto-oncogenes may have a role in the cellular mechanism of endothelin action. In astrocytes and astrocytoma cells, the three types of endothelins stimulate nerve growth factor (NGF) gene expression by a mechanism involving induction of c-*fos* gene expression.[192] In cultured rat VSM cells, endothelin-induced changes in $[Ca^{2+}]_i$ are associated with a transient increase in c-*fos* and c-*myc* mRNA expression, which may be followed by DNA synthesis and cell proliferation.[193] ET-1 stimulates c-*fos* mRNA expression and acts as a modulator of rat FRTL5 thyroid cell proliferation.[194] Expression of c-*fos* and c-*jun* is stimulated and expression of the proenkephalin gene is downregulated by EG-1 in CG glioma cells.[195] In Rat-1 cells, endothelin-induced transcription of genes of the *fos* and *jun* families depends, at least in part, on increases in the $[Ca^{2+}]_i$ stimulated by the hormone.[196] Differential regulation in the expression of c-*fos*, *fra*-1, c-*jun*, and *jun*-B may contribute to the signaling mechanism of endothelin isopeptides.[197] Activation of AP-1 *cis*-elements may contribute to the nuclear signaling by endothelin. The ability of different endothelin isopeptides to increase AP-1 activity correlates with their ability to stimulate cell growth.

F. ROLE OF ENDOTHELIN IN NEOPLASTIC PROCESSES

Human cancer cell lines and primary tumors of epithelial origin may produce relatively large amounts of ET-1 and its precursor molecule, big-ET, and TGF-β markedly stimulates the production of endothelin by some of these cells.[198-202] The cell lines contain a single class of specific binding sites for ET-1, and ET-1 induces increases in the concentration of cytosolic Ca^{2+} and stimulates proliferation of the cells, suggesting its role as an autocrine factor in tumor cell growth. ET-1 may also play a modulatory role in the growth of stromal cells surrounding cancer cells, acting as a paracrine factor in the tumorigenic process.

V. ATRIAL NATRIURETIC FACTOR AND RELATED PEPTIDES

The atrial natriuretic factor/peptide (ANF/ANP), also called atriopeptin, is a hormone released by mammalian heart atrial muscle cells.[203-211] The discovery of ANF represented the first evidence that the heart functions as an endocrine organ. The secreted ANF plays an important role in water and sodium movement and excretion. It has diuretic, natriuretic, and vasodilator properties, inducing a decrease in blood pressure and intravascular volume. In addition to its cardiovascular and renal effects, ANF exerts important actions in various organs and tissues, including liver, adrenal, lung, and intestine, as well as in the lymphatic, reproductive, and endocrine systems.[212] ANF modulates sympathetic nervous activity and may inhibit the release of other hormones, including renin, aldosterone, and ADH. ANF is involved in the negative regulation of ACTH secretion by the pituitary. A corticotropin-release inhibiting hormone (CRIH) produced in the hypothalamus is identical with ANF.[213]

ANF is involved in the modulation of some processes related to cell proliferation and differentiation. Increased numbers of ANF receptors are found in regenerating rat liver.[214] ANF has an antimitogenic effect on mesangial cells, which may be mediated by alterations of intracellular Ca^{2+} dynamics.[215] ANF inhibits, in a dose-dependent manner, the PDGF-stimulated proliferation of rat VSM cells in culture.[216] Immunoreactive ANF is present in the acrosome and the flagellum of spermatids and elongating spermatozoa of mammalian testis, suggesting that it may participate in capacitation, sperm motility, and/or sperm development.[217] An ANF hormonal system exists even in plants, where it could be involved in the movement of water from the roots up through stems to the leaves, perhaps through dilatation of vessels in the xylem.[218]

In addition to ANF, there are other related peptides with natriuretic activity: the brain natriuretic peptide (BNP), isolated from brain and heart, and the C-type natriuretic peptide (CNP), purified from porcine brain.[209] Like ANF, both BNP and CNP can elicit vasorelaxant, natriuretic, and diuretic responses.

A. SYNTHESIS AND STRUCTURE OF ANF AND RELATED PEPTIDES

ANF is synthesized in cardiocytes in the form of a precursor (pro-ANF) and is stored in the atrial granules in the form of heterogeneous pro-ANF molecules.[219] ANF is present in extracardiac localizations such as the brain and the kidney. The amino acid sequence of human and mouse ANF has been deduced from the nucleotide sequence of the respective genes.[220] Human ANF is synthesized in a prepro-ANF form of 151 amino acids which contains a sequence that is identical to the consensus glucocorticoid receptor binding sequence. The circulating form of ANF is a 28-amino acid peptide with a ring structure due to the presence of a central disulfide bridge. Variant molecular forms of ANF have been detected in human urine and kidney, as well as in the testis and brain.

Sequences conserved at the 5′ flanking side of the ANF gene may be involved in the regulation of ANF gene expression by different types of agents. Endothelin may directly stimulate the expression of the ANF gene in rat cardiocytes, enhancing the synthesis and secretion of ANF.[221] Endothelin is a potent stimulus for ANF secretion by superfused rat cardiac atria.[222] Endothelin-stimulated secretion of ANF partially depends on calcium influx through voltage-dependent calcium channels, but is independent of endogenous calcium release from the sarcoplasmic reticulum.

The ANF-related peptides, BNP and CNP, were isolated from porcine brain on the basis of their potent relaxant effects on chick rectum.[223,224] Like ANF, these proteins are synthesized in the form of large precursor polypeptides, and the mature, active hormones have a 17-amino acid loop formed by an intramolecular disulfide bridge. Human BNP and CNP have a similar structure, but the amino- and carboxyl-terminal tails vary in both length and composition in relation to ANF.[225,226] The amino acid sequence of ANP and CNP is highly conserved among species, whereas the primary structure of BNP is variable. The three natriuretic peptides are synthesized at distinct sites and may have different functions. ANP and BNP have been detected in the central nervous system and the adrenal medulla, as well as in the heart, and the plasma levels of both peptides show a marked increase in patients with severe congestive heart failure.[227] In contrast, CNP appears to be localized exclusively in the central nervous system or in cells derived from the neural crest. CNP may function as a neurotransmitter to coordinate central aspects of salt and water balance and blood pressure.[209]

B. NATRIURETIC PEPTIDE RECEPTORS

The ANF receptor is a member of the guanylyl cyclase family of receptors;[228-231] and is a membrane form of guanylyl cyclase.[232] The enzyme is divided by a single transmembrane domain into an amino-terminal, extracellular ANF-binding domain, and a carboxyl-terminal, intracellular catalytic domain which has guanylyl cyclase activity. There are three distinct molecular forms of ANF receptors, termed NPR-A, NPR-B, and NPR-C.[233-236] NPR-A and NPR-B are proteins of 140 kDa, and their intracellular regions can be divided in two parts, one distal domain of approximately 250 amino acids which is involved in the guanylyl cyclase activity and the other, of approximately 280 amino acids, the kinase homology domain (KHD), located proximal to the plasma membrane. The precise role of the KHD is not known, but it could act as a negative regulator of guanylyl cyclase activity, and hormone binding to the extracellular domain of the receptor would release this inhibition.[237] The KHD of NPR-A would be involved in the modulation of NPR-A ligand affinity.[238]

In contrast to NPR-A and NPR-B, NPR-C is a protein of 65 kDa which contains a very short (37 amino acids) cytoplasmic tail that exhibits no homology to the intracellular domains of any other known receptors. The extracellular domain of NPR-C is similar to those of NPR-A and NPR-B, all three domains being composed of about 440 amino acids. The precise physiological role of NPR-C is not understood, but it could function as ANF clearance-storage binding protein, serving to remove large amounts of ANF from the circulation or to store and release ANF slowly. NPR-C could also mediate the action of ANF through messengers other than cGMP. The three molecular forms of natriuretic peptide receptors recognize the different natriuretic peptides in a differential manner. In some types of cells, BNP is much less potent than ANF to stimulate NPR-A guanylyl cyclase activity. Expression of NPR-C, but not NPR-A, in cultured vascular endothelial cells is very sensitive to changes in salt (NaCl) concentration in the medium.[239] cGMP response is greatly exaggerated in spite of a dramatic loss of NPR-C in the NaCl-treated cells. ANF inhibits the production and secretion of endothelin by cultured endothelial cells, and this effect is mediated by NPR-C.[146] This effect may have physiological significance, since ANF is a vasodilator and lowers blood pressure, whereas ET-1 is a vasoconstrictor and -pressor protein. The mechanisms involved in the regulation of the activity of natriuretic peptide receptors are not understood but may include phosphorylation of the receptor protein at specific amino acid residues by protein kinase C or other enzymes.

C. POSTRECEPTOR MECHANISMS OF ACTION OF NATRIURETIC PEPTIDES

The natriuretic peptide receptors NPR-A and NPR-B possess intrinsic guanylyl cyclase activity, and cGMP is a second messenger in the postreceptor mechanisms of action of ANF. One of the earliest events following binding of ANF to its receptor is an increase in cGMP concentrations, indicating that cGMP is a mediator of the physiological effects of the hormone. Regulation of endothelial cell permeability and other events elicited by the action of ANF are mediated by cGMP.[240] All molecular forms of natriuretic peptides, but in particular CNP, increase the cGMP accumulation in primary cultures of pituitary cells as well as in a gonadotrope-derived cell line and GH_3 pituitary cells.[241] Site-directed mutational analysis of a rat adrenal gland-derived cDNA clone encoding membrane guanylyl cyclase revealed that the residue Leu-364 of the enzyme plays a critical role in ligand binding and ANF signal transduction.[242] Endothelin inhibits the ANF-stimulated production of cGMP by activating the protein kinase C in rat aortic muscle cells.[243]

ANF modifies the phosphorylation of distinct cellular proteins and exerts an inhibitory effect on the autophosphorylation of protein kinase C. Activation of cGMP-dependent protein kinase by the ANF receptor and phosphorylation of ductal ion channels in the renal inner-medullary collecting duct may play an important role in the regulation of Na^+ absorption in the kidney.[244] Inhibition of vascular contraction by ANF may be explained, at least in part, by a reduction in the basal cytosolic concentration of Ca^{2+}.

Transgenic mice carrying fusions between the transcriptional regulatory DNA sequences of ANF and those encoding the simian virus 40 (SV40) T antigen develop an hyperplastic growth of the right atrium, while the left atrium remains relatively normal in size.[245] The atrial hyperplasia is accompanied by a progressive increase in both the frequency and severity of abnormalities in the atrial conduction system, which ultimately result in death.

D. ROLE OF ANF IN NEOPLASTIC PROCESSES

The syndrome of inappropriate antidiuretic hormone secretion is observed frequently in patients with a specific type of lung cancer (SCLC). It is characterized by hyponatremia, low serum osmolality, and high urine osmolality. In some of these patients, the tumor cells ectopically produce and secrete high levels of arginine vasopressin (AVP).[246] In other cases, the syndrome is associated with an ectopic production of ANF, as suggested by the presence of ANF mRNA in tumor specimens from patients with SCLC and the syndrome of inappropriate antidiuretic hormone secretion.[247]

VI. VASOACTIVE INTESTINAL PEPTIDE

The vasoactive intestinal peptide (VIP) is a 28-amino acid peptide hormone isolated from porcine intestine.[248] It was so named because of its potent hypotensive effect and vasodilatatory properties. VIP is a member of the glucagon-secretin neuropeptide family, which includes the 38-residue pituitary adenylyl cyclase activating polypeptide (PACAP).[249]

VIP and VIP-immunoreactive substances are expressed in many organs and tissues, including lung, liver, and brain. In the brain, VIP is found at relatively high concentrations in the hippocampus, amygdala, and cerebral cortex. In the lung, VIP containing neurons are found in airway smooth muscle cells, around glands, and in arterial walls, and it acts in the lung as a bronchodilator. In the liver, VIP functions as a potent stimulator of glycogenolysis and gluconeogenesis. VIP modulates the activities of exocrine and endocrine glands. It regulates the proliferation of various types of cells, including smooth muscle cells, sympathetic neuroblasts, and hippocampal cells. VIP modulates the growth of the human keratinocyte cell line HaCaT and early events in migration of human keratinocytes.[250,251]

VIP can act as both a growth factor and a neurotransmitter. It displays growth factor-like functions in whole cultured mouse embryos.[252] VIP is produced in the pituitary and the hypothalamus and has an important role in neuroendocrine regulation. VIP inhibits the steroid hormone-induced luteinizing hormone (LH) surge in ovariectomized rats.[253] In turn, estrogen can stimulate the expression of VIP mRNA in the brain and the pituitary.[254] VIP participates in the modulation of the functions of the immune system. High-affinity VIP receptors are expressed in murine and human peripheral blood lymphocytes. VIP can decrease NK cell activity of human large granular lymphocytes.[255] VIP may be one of the neurotransmitters that induces the increase in blood flow and engorgement of the human vagina during sexual arousal.[256]

A. THE VIP RECEPTOR

Receptors for VIP are expressed in a wide diversity of cell types. The human melanoma cell line IGR39 displays both high- and low-affinity VIP receptors, which are inhibited by fetal calf serum.[257] IGR39 cells, as well as the HT29 cell line derived from human colonic adenocarcinoma, exhibit a high number of VIP receptors, which are downregulated by the polyanionic compound suramin.[258] VIP receptors expressed on murine peripheral blood lymphocytes turn over rapidly.[259] Pituitary lactotrophs express VIP receptors, and one of the most important functions of VIP in the pituitary may consist of the control of prolactin secretion.

The VIP receptor in rat lung membranes and bovine aorta is represented by a single-chain polypeptide of 54 kDa which is physically and functionally associated in form of a ternary complex with the guanine nucleotide-binding stimulatory protein G_s and adenylyl cyclase.[260,261] Cloning of the rat VIP receptor indicated that it is a protein of 459 amino acids containing seven transmembrane domains.[262] The VIP receptor is structurally related to the secretin, calcitonin, and PTH receptors, which may constitute a subfamily of G_s protein-coupled receptors. VIP receptor mRNA is expressed in various rat tissues, including liver, lung, intestine, pituitary, and brain.

B. POSTRECEPTOR MECHANISMS OF VIP ACTION

The postreceptor mechanism of VIP action is exerted in most or all tissues through the stimulatory nucleotide-binding protein (G_s) and the activation of adenylyl cyclase.[263] The modulatory effects of VIP on cell proliferation occur by mechanisms that involve changes in the intracellular cAMP concentration. VIP and forskolin (an agent that favors the accumulation of cAMP) exert synergistic inhibitory effects on the synthesis of DNA and the proliferation of HT29 human colon adenocarcinoma cells.[264] In other types of cells, VIP may act as a mitogen. In mouse 3T3 cells, VIP exhibits mitogenic properties, stimulating DNA synthesis when added in the presence of insulin and modulators of cAMP metabolism.[265] When added to 3T3 cells at mitogenic concentrations, VIP promotes rapid accumulation of cAMP, but does not induce intracellular mobilization of Ca^{2+}, nor does it stimulate protein kinase C activity. However, VIP acts as a growth factor for the human keratinocyte cell line HaCaT by a mechanism that may involve changes in Ca^{2+}/calmodulin and phosphoinositide metabolism.[266] Pituitary adenylyl cyclase activating peptide and VIP induce an increase in the $[Ca^{2+}]_i$ of cultured rat hippocampal neurons.[267] In spite of the fact that the VIP receptor does not possess tyrosine kinase activity, binding of VIP to its receptor on the cell surface stimulates the rapid phosphorylation of the c-Src protein in cell cultures of chick embryonic retinal pigment epithelium.[268,269] The c-Src tyrosine kinase may be an integral part of the VIP signal transduction pathway.

C. ROLE OF VIP IN NEOPLASTIC PROCESSES

Little is known about the possible role of VIP in neoplasia, but there is evidence that VIP may be an autocrine growth factor in human neuroblastomas which coexpress VIP and its receptor.[270] VIP may have a role in the progression of human lung cancer. A VIP antagonist is capable of inhibiting NSCLC growth.[271]

VII. BRADYKININ

Bradykinin is a nonpeptide of the kinin family, a class of extracellular signaling agents that act locally in a manner similar to that of the hormones.[272,273] Recent evidence indicates that bradykinin possesses growth factor-like effects and can act through paracrine mechanisms. It is generated from the higher molecular weight precursor kininogens during tissue damage by the action of proteolytic enzymes that are members of the kallikrein family. When activated by trauma, burns, inflammation, shock, allergy, or other conditions, the kininogens release, in addition to bradykinin, Lys-bradykinin (kallidin) and Met-Lys-bradykinin. These three kinins can contribute to the inflammatory response in acute and chronic diseases. Bradykinin can cause localized pain, inflammation, vasodilatation, edema, smooth muscle spasms, and alterations in vascular permeability. It is involved in the regulation of blood pressure and smooth muscle activity.

A. BRADYKININ RECEPTORS

Receptors for bradykinin have been detected in a diversity of organs and tissues, including uterus, intestine, heart, aorta, kidney, and spinal cord, as well as in endothelial, epithelial, fibroblastic, and neuronal cells maintained in culture *in vitro*. According to their affinity for bradykinin fragments and analogues, the receptors for bradykinin and kallidin have been classified in types B_1 and B_2.[274,275] Most of the physiological effects of bradykinin, including smooth muscle contraction, are associated with the B_2 receptor. The entire nucleotide sequences of the human bradykinin B_2 receptor gene have been determined, and the human gene has been assigned to chromosome 14.[276] The B_2 receptor consists of 364 amino acids and has a molecular weight of 41,442 Da. It shares sequence similarity to other G protein-coupled receptors and contains seven transmembrane segments and several potential sites of glycosylation. Additional subtypes of B_2 bradykinin receptors have been identified in the guinea pig.[277]

B. POSTRECEPTOR MECHANISMS OF BRADYKININ ACTION

Bradykinin binding to its receptor in cells such as rat cerebellar astrocytes induces activation of phospholipase C, phosphoinositide metabolism, and Ca^{2+} mobilization, which results in activation of protein kinase C.[278] In human IMR-90 lung fibroblasts, bradykinin stimulates Ca^{2+} efflux.[279] In human gingival fibroblasts, the ligand-activated B_2 bradykinin receptor induces a transient increase in $[Ca^{2+}]_i$ that is associated with intracellular Ca^{2+} mobilization and is secondary to an initial effect of bradykinin on phosphoinositide metabolism.[280] Bradykinin stimulates arachidonic acid release and PGE_2 biosynthesis in the same cells. In PC12 cells, bradykinin promotes formation of 1,2-diacylglycerol through mechanisms that are either dependent or independent of phospholipase D.[281]

C. ROLE OF BRADYKININ IN NEOPLASTIC PROCESSES

Bradykinin could have a role in cell growth control during tumorigenesis. Transformation of cells cultured *in vitro* by expression of a transforming *ras* oncogene may be associated with increased bradykinin receptor numbers.[282,283] The increased growth factor-stimulated phosphatidylinositol hydrolysis and DNA synthesis observed in the *ras*-transformed cells would occur via alterations in the expression of PDGF and bradykinin receptors. Bradykinin has mitogenic action in certain types of cells. The mitogenic effects of insulin and bradykinin in rat cell lines may be synergistic and are increased by expression of a mutant *ras* oncogene.[284]

REFERENCES

1. **Rozengurt, E.,** Neuropeptides as cellular growth factors: role of multiple signalling pathways, *Eur. J. Clin. Invest.,* 21, 123, 1991.
2. **Cooper, P.E. and Martin, J.B.,** Neuroendocrinology and brain peptides, *Arch. Neurol.,* 8, 551, 1980.
3. **Krieger, D.T.,** Brain peptides: what, where and why?, *Science,* 222, 975, 1983.
4. **Schüller, H.M.,** Receptor-mediated mitogenic signals and lung cancer, *Cancer Cells,* 3, 496, 1991.
5. **Zachary, I. and Rozengurt, E.,** High-affinity receptors for peptides of the bombesin family in Swiss 3T3 cells, *Proc. Natl. Acad. Sci. U.S.A.,* 82, 7616, 1985.
6. **McCoy, J.G. and Avery, D.D.,** Bombesin: potential integrative peptide for feeding and satiety, *Peptides,* 11, 595, 1990.
7. **Brown, K.D., Laurie, M.S., Littlewood, C.J., Blakeley, D.M., and Corps, A.N.,** Characterization of the high-affinity receptors on Swiss 3T3 cells which mediate the binding, internalization and degradation of the mitogenic peptide bombesin, *Biochem. J.,* 252, 227, 1988.
8. **Battey, J.F., Way, J.M., Corjay, M.H., Shapira, H., Kusano, K., Harkins, R., Wu, J.M., Slattery, T., Mann, E., and Feldman, R.I.,** Molecular cloning of the bombesin/gastrin-releasing peptide receptor from Swiss 3T3 cells, *Proc. Natl. Acad. Sci. U.S.A.,* 88, 395, 1991.

9. **Kane, M.A., Aguayo, S.M., Portanova, L.B., Ross, S.E., Holley, M., Kelley, K., and Miller, Y.E.,** Isolation of the bombesin/gastrin-releasing peptide receptor from human small lung carcinoma NCI-H435 cells, *J. Biol. Chem.,* 266, 9486, 1991.
10. **Kris, R.M., Hazan, R., Villines, J., Moody, T.W., and Schlessinger, J.,** Identification of the bombesin receptor on murine and human cells by cross-linking experiments, *J. Biol. Chem.,* 262, 11215, 1987.
11. **Narayan, S., Guo, Y.-S., Townsend, C.M., Jr., and Singh, P.,** Specific binding and growth effects of bombesin-related peptides on mouse colon cancer cells *in vitro, Cancer Res.,* 50, 6772, 1990.
12. **Markowska, A., Nussdorfer, G.G., and Malendowicz, L.K.,** Effects of bombesin and neuromedin-B on the proliferative activity of the rat adrenal cortex, *Histol. Histopathol.,* 8, 359, 1993.
13. **Cirillo, D.M., Gaudino, G., Naldini, L., and Comoglio, P.M.,** Receptor for bombesin with associated tyrosine kinase activity, *Mol. Cell. Biol.,* 6, 4641, 1986.
14. **Takuwa, N., Takuwa, Y., Bollag, W.E., and Rasmussen, H.,** The effects of bombesin on polyphosphoinositide and calcium metabolism in Swiss 3T3 cells, *J. Biol. Chem.,* 262, 182, 1987.
15. **Moody, T.W., Murphy, A., Mahmoud, S., and Fiskum, G.,** Bombesin-like peptides elevate cytosolic calcium in small cell lung cancer cells, *Biochem. Biophys. Res. Commun.,* 147, 189, 1987.
16. **Bierman, A.J., Koenderman, L., Tool, A.J., and de Laat, S.W.,** Epidermal growth factor and bombesin differ strikingly in the induction of early responses in Swiss 3T3 cells, *J. Cell. Physiol.,* 142, 441, 1990.
17. **Patel, K.V. and Schrey, M.P.,** Activation of inositol phospholipid signaling and Ca^{2+} efflux in human breast cancer cells by bombesin, *Cancer Cells,* 50, 235, 1990.
18. **Takuwa, N., Takuwa, Y., Ohue, Y., Mukai, H., Endoh, K., Yamashita, K., Kumada, M., and Munekata, E.,** Stimulation of calcium mobilization but not proliferation by bombesin and tachykinin neuropeptides in human small cell lung cancer cells, *Cancer Res.,* 50, 240, 1990.
19. **Lloyd, A.C., Davies, S.A., Crossley, I., Whitaker, M., Houslay, M.D., Hall, A., Marshall, C.J., and Wakelam, M.J.O.,** Bombesin stimulation of inositol 1,4,5-trisphosphate generation and intracellular calcium release is amplified in a cell line overexpressing the N-*ras* proto-oncogene, *Biochem. J.,* 260, 813, 1989.
20. **Wakelam, M.J.O.,** Inhibition of the amplified bombesin-stimulated inositol phosphate response in N-*ras* transformed cells by high density culturing, *FEBS Lett.,* 228, 182, 1988.
21. **Takuwa, N., Iwamoto, A., Kumada, M., Yamashita, K., and Takuwa, Y.,** Role of Ca^{2+} in bombesin-induced mitogenesis in Swiss 3T3 fibroblasts, *J. Biol. Chem.,* 266, 1403, 1991.
22. **Currie, S., Smith, G.L., Crichton, C.A., Jackson, C.G., Hallam, C., and Wakelam, M.J.O.,** Bombesin stimulates the rapid activation of phospholipase A_2-catalyzed phosphatidylcholine hydrolysis in Swiss 3T3 cells, *J. Biol. Chem.,* 267, 6056, 1992.
23. **Palumbo, A.P., Rossino, P., and Comoglio, P.M.,** Bombesin stimulation of c-*fos* and c-*myc* gene expression in cultures of Swiss 3T3 cells, *Exp. Cell Res.,* 167, 276, 1986.
24. **Letterio, J.J., Coughlin, S.R., and Coughlin, L.T.,** Pertussis toxin-sensitive pathway in the stimulation of c-*myc* expression and DNA synthesis by bombesin, *Science,* 234, 1117, 1986.
25. **Bravo, R., MacDonald-Bravo, H., Müller, R., Hübsch, D., and Almendral, J.M.,** Bombesin induces c-*fos* and c-*myc* expression in quiescent Swiss 3T3 cells. Comparative study with other mitogens, *Exp. Cell Res.,* 170, 103, 1987.
26. **Maly, K., Kiani, A., Oberhuber, H., and Grunicke, H.,** Interference of Ha-*ras* with inositol trisphosphate-mediated Ca^{2+}-release, *FEBS Lett.,* 291, 113, 1991.
27. **Bonaz, B., DeGiorgio, R., and Tache, Y.,** Peripheral bombesin induces c-*fos* protein in the rat brain, *Brain Res.,* 600, 353, 1993.
28. **Cuttitta, F., Carney, D.N., Mulshine, J., Moody, T.W., Fedorko, J., Fischler, A., and Minna, J.D.,** Bombesin-like peptides can function as autocrine growth factors in human small-cell lung cancer, *Nature,* 316, 823, 1985.
29. **Willey, J.C., Lechner, J.F., and Harris, C.C.,** Bombesin and the C-terminal tetradecapeptide of gastrin-releasing peptide are growth factors for normal human bronchial epithelial cells, *Exp. Cell Res.,* 153, 245, 1984.
30. **Alexander, R.W., Upp, J.R., Jr., Poston, G.J., Gupta, V., Townsend, C.M., Jr., and Thompson, J.C.,** Effects of bombesin on growth of human small cell lung carcinoma *in vivo, Cancer Res.,* 48, 1439, 1988.
31. **Moody, T.W., Lee, M., Kris, R.M., Bellot, F., Bepler, G., Oie, H., and Gazdar, A.F.,** Lung carcinoid cell lines have bombesin-like peptides and EGF receptors, *J. Cell. Biochem.,* 43, 139, 1990.

32. **Moody, T.W. and Cuttitta, F.,** Growth factor and peptide receptors in small cell lung cancer, *Life Sci.,* 52, 1161, 1993.
33. **Ruff, M.R. and Pert, C.B.,** Small cell carcinoma of the lung macrophage-specific antigens suggest hematopoietic stem cell origin, *Science,* 225, 1034, 1984.
34. **Korman, L.Y., Carney, D.N., Citron, M.L., and Moody, T.W.,** Secretin/vasoactive intestinal peptide-stimulated secretion of bombesin/gastrin releasing peptide from human small cell carcinoma of the lung, *Cancer Res.,* 46, 1214, 1986.
35. **Moody, T.W., Carney, D.N., Cuttitta, F., Quattrocchi, K., and Minna, J.D.,** High affinity receptors for bombesin/GRP-like peptides on human small cell lung cancer, *Life Sci.,* 37, 105, 1985.
36. **Gaudino, G., Cirillo, D., Naldini, L., Rossino, P., and Comoglio, P.M.,** Activation of the protein-tyrosine kinase associated with the bombesin receptor complex in small cell lung carcinomas, *Proc. Natl. Acad. Sci. U.S.A.,* 85, 2166, 1988.
37. **Trepel, J.B., Moyer, J.D., Cuttitta, F., Frucht, H., Coy, D.H., Natale, R.B., Mushine, J.L., Jensen, R.T., and Sausville, E.A.,** A novel bombesin receptor antagonist inhibits autocrine signals in a small cell lung carcinoma cell line, *Biochem. Biophys. Res. Commun.,* 156, 1383, 1988.
38. **Bologna, M., Festuccia, C., Muzi, P., Biordi, L., and Ciomei, M.,** Bombesin stimulates growth of human prostatic cancer cells *in vitro, Cancer,* 63, 1714, 1989.
39. **Katz, A.M.,** Angiotensin II: hemodynamic regulator or growth factor?, *J. Mol. Cell. Cardiol.,* 22, 739, 1990.
40. **Ichikawa, I. and Harris, R.C.,** Angiotensin actions in the kidney: renewed insight into the old hormone, *Kidney Int.,* 40, 583, 1991.
41. **Johnston, C.I.,** Renin angiotensin system — A dual tissue and hormonal system for cardiovascular control, *J. Hypertension,* 10, S13, 1992.
42. **Eggena, P. and Barrett, J.D.,** Regulation and functional consequences of angiotensinogen gene expression, *J. Hypertension,* 10, 1307, 1992.
43. **Skott, O. and Jensen, B.L.,** Cellular and intrarenal control of renin secretion, *Clin. Sci.,* 84, 1, 1993.
44. **Bobik, A., Grinpukel, S., Little, P.J., Grooms, A., and Jackman, G.,** Angiotensin II and noradrenaline increase PDGF0BB receptors and potentiate PDGF-BB stimulated DNA synthesis in vascular smooth muscle, *Biochem. Biophys. Res. Commun.,* 166, 580, 1990.
45. **Low, B.C., Ross, I.K., and Grigor, M.R.,** Angiotensin II stimulates glucose activity in cultured vascular smooth muscle cells, *J. Biol. Chem.,* 267, 20740, 1992.
46. **Schelling, P., Fischer, H., and Ganten, D.,** Angiotensin and cell growth: a link to cardiovascular hypertrophy?, *J. Hypertension,* 9, 3, 1991.
47. **Chiu, A.T., Herblin, W.F., McCall, D.E., Ardecky, R.J., Carini, D.J., Duncia, J.V., Pease, L.J., Wong, P.C., Wexler, R.R., Johnson, A.L., and Timmermans, P.B.M.W.M.,** Identification of angiotensin II receptor subtypes, *Biochem. Biophys. Res. Commun.,* 165, 196, 1989.
48. **Kitami, Y., Okura, T., Marumoto, K., Wakamiya, R., and Hiwada, K.,** Differential gene expression and regulation of type-1 angiotensin II receptor subtypes in the rat, *Biochem. Biophys. Res. Commun.,* 188, 446, 1992.
49. **Pucell, A.G., Hodges, J.C., Sen, I., Bumpus, F.M., and Husain, A.,** Biochemical properties of the ovarian granulosa cell type 2-angiotensin II receptor, *Endocrinology,* 128, 1947, 1991.
50. **Sasaki, K., Yamano, Y., Bardhan, S., Iwai, N., Murray, J.J., Hasegawa, M., Matsuda, Y., and Inagami, T.,** Cloning and expression of a complementary DNA encoding a bovine adrenal angiotensin-II type-1 receptor, *Nature,* 351, 230, 1991.
51. **Murphy, T.J., Alexander, R.W., Griendling, K.K., Runge, M.S., and Bernstein, K.E.,** Isolation of a cDNA encoding the vascular type-1 angiotensin-II receptor, *Nature,* 351, 233, 1991.
52. **Langford, K., Frenzel, K., Martin, B.M., and Bernstein, K.E.,** The genomic organization of the rat AT_1 angiotensin receptor, *Biochem. Biophys. Res. Commun.,* 183, 1025, 1992.
53. **Kakar, S.S., Sellers, J.C., Devor, D.C., Musgrove, L.C., and Neill, J.D.,** Angiotensin II type-1 receptor subtype cDNAs: differential tissue expression and hormonal regulation, *Biochem. Biophys. Res. Commun.,* 183, 1090, 1992.
54. **Madhun, Z.T., Ernsberger, P., Ke, F.-C., Zhou, J., Hopfer, U., and Douglas, J.G.,** Signal transduction mediated by angiotensin II receptor subtypes expressed in rat renal mesangial cells, *Regul. Peptides,* 44, 149, 1993.
55. **Mauzy, C.A., Hwang, O., Egloff, A.M., Wu, L.H., and Chung, F.Z.,** Cloning, expression and characterization of a gene encoding the human angiotensin II type 1A receptor, *Biochem. Biophys. Res. Commun.,* 186, 277, 1992.

56. **Kakar, S.S., Riel, K.K., and Neill, J.D.,** Differential expression of angiotensin-II receptor subtype messenger RNAs (AT-1A and AT-1B) in the brain, *Biochem. Biophys. Res. Commun.,* 185, 688, 1992.
57. **Sandberg, K., Ji, H., Clark, A.J.L., Shapira, H., and Catt, K.J.,** Cloning and expression of a novel angiotensin II receptor subtype, *J. Biol. Chem.,* 267, 9455, 1992.
58. **Jackson, T.R., Blair, L.A.C., Marshall, J., Goedert, M., and Hanley, M.R.,** The *mas* oncogene encodes an angiotensin receptor, *Nature,* 335, 437, 1988.
59. **Young, D., Waitches, G., Birchmeier, C., Fasano, O., and Wigler, M.,** Isolation and characterization of a new cellular oncogene encoding a protein with multiple potential transmembrane domains, *Cell,* 45, 711, 1986.
60. **Hanley, M.R.,** Proto-oncogenes in the nervous system, *Neuron,* 1, 175, 1988.
61. **Martin, K.A., Grant, S.G.N., and Hockfield, S.,** The *mas* proto-oncogene is developmentally regulated in the rat central nervous system, *Develop. Brain Res.,* 68, 75, 1992.
62. **Ambroz, C., Clark, A.J.L., and Catt, K.J.,** The *mas* oncogene enhances angiotensin-induced $[Ca^{2+}]_i$ responses in cells with pre-existing angiotensin II receptors, *Biochim. Biophys. Acta,* 1133, 107, 1991.
63. **Yoshida, A., Nishikawa, T., Tamura, Y., and Yoshica, S.,** ACTH-induced inhibition of the action of angiotensin II in bovine zona glomerulosa cells — A modulatory effect of cyclic AMP on the angiotensin II receptor, *J. Biol. Chem.,* 266, 4288, 1991.
64. **Douglas, J.G., Romero, M., and Hopofer, U.,** Signaling mechanisms coupled to the angiotensin receptor of proximal tubular epithelium, *Kidney Int.,* 38(Suppl. 30), S43, 1990.
65. **Shima, S. and Fukase, H.,** Angiotensin stimulates adenylate cyclase activity via prostaglandins in the adrenal glomerulosa membrane, *Endocrinol. Jpn.,* 37, 529, 1990.
66. **Conlin, P.R., Kim, S.Y., Williams, G.H., and Canessa, M.L.,** Na^+-H^+ exchanger kinetics in adrenal glomerulosa cells and its activation by angiotensin II, *Endocrinology,* 127, 236, 1990.
67. **Ambroz, C. and Catt, K.J.,** Angiotensin II receptor-mediated calcium influx in bovine adrenal glomerulosa cells, *Endocrinology,* 131, 408, 1992.
68. **Tremblay, E., Payet, M.-D., and Gallo-Payet, N.,** Effects of ACTH and angiotensin II on cytosolic calcium in cultured adrenal glomerulosa cells. Role of cAMP production in the ACTH effect, *Cell Calcium,* 12, 655, 1991.
69. **Jung, K.Y. and Endou, H.,** Biphasic increasing effect of angiotensin-II on intracellular free calcium in isolated rat early proximal tubule, *Biochem. Biophys. Res. Commun.,* 165, 1221, 1989.
70. **Sandberg, K., Bor, M., Ji, H., Markwick, A., Millan, M.A., and Catt, K.J.,** Angiotensin II-induced calcium mobilization in oocytes by signal transfer through gap junctions, *Science,* 249, 298, 1990.
71. **Taubman, M.B., Berk, B.C., Izumo, S., Tsuda, T., Alexander, R.W., and Nadal-Ginard, B.,** Angiotensin II induces c-*fos* mRNA in aortic smooth muscle. Role of Ca^{2+} mobilization and protein kinase C activation, *J. Biol. Chem.,* 264, 526, 1989.
72. **Ullian, M.E. and Linas, S.L.,** Angiotensin II surface receptor coupling to inositol trisphosphate formation in vascular smooth muscle cells, *J. Biol. Chem.,* 265, 195, 1990.
73. **Tsuda, T. and Alexander, R.W.,** Angiotensin II stimulates phosphorylation of nuclear lamins via a protein kinase C-dependent mechanism in cultured vascular smooth muscle cells, *J. Biol. Chem.,* 265, 1165, 1990.
74. **Tang, E.K.Y. and Houslay, M.D.,** Glucagon, vasopressin and angiotensin all elicit a rapid, transient increase in hepatocyte protein kinase C activity, *Biochem. J.,* 283, 341, 1992.
75. **Gyurko, R., Kimura, B., Kurian, P., Crews, F.T., and Phillips, M.I.,** Angiotensin II receptor subtypes play opposite role in regulating phosphatidylinositol hydrolysis in rat skin slices, *Biochem. Biophys. Res. Commun.,* 186, 285, 1992.
76. **García-Sáinz, J. and Macías-Silva, M.,** Angiotensin II stimulates phosphoinosite turnover and phosphorylase through AII-1 receptors in isolated rat hepatocytes, *Biochem. Biophys. Res. Commun.,* 172, 780, 1990.
77. **Kondo, T., Konishi, F., Inui, H., and Inagami, T.,** Diacylglycerol formation from phosphatidylcholine in angiotensin II stimulated vascular smooth muscle cells, *Biochem. Biophys. Res. Commun.,* 187, 1460, 1992.
78. **Dean, N.M. and Boynton, A.L.,** Angiotensin II causes phosphatidylinositol turnover and increases 1,2-diacylglycerol mass but is not mitogenic in rat liver T51B cells, *Biochem. J.,* 269, 347, 1990.
79. **Poyner, D.R., Hawkins, P.T., Benton, H.P., and Hanley, M.R.,** Changes in inositol lipids and phosphates after stimulation of the MAS-transfected NG115-401L-C3 cell line by mitogenic and non-mitogenic stimuli, *Biochem. J.,* 271, 605, 1990.

80. **Sumners, C., Tang, W., Zelezna, B., and Raizada, M.K.**, Angiotensin II receptor subtypes are coupled with distinct signal-transduction mechanisms in neurons and astrocytes from rat brain, *Proc. Natl. Acad. Sci. U.S.A.*, 88, 7567, 1991.
81. **Duff, J.L., Berk, B.C., and Corson, M.A.**, Angiotensin-II stimulates the pp44 and pp42 mitogen-activated protein kinases in cultured rat aortic smooth muscle cells, *Biochem. Biophys. Res. Commun.*, 188, 257, 1992.
82. **Molloy, C.J., Taylor, D.S., and Weber, H.**, Angiotensin II stimulation of rapid protein tyrosine phosphorylation and protein kinase activation in rat aortic smooth muscle cells, *J. Biol. Chem.*, 268, 7338, 1993.
83. **Bottari, S.P., King, I.N., Reichlin, S., Dahstroem, I., Lydon, N., and de Gasparo, M.**, The angiotensin AT_2 receptor stimulates protein tyrosine phosphatase activity and mediates inhibition of particulate guanylate cyclase, *Biochem. Biophys. Res. Commun.*, 183, 206, 1992.
84. **Sharifi, B.G., Lafleur, D.W., Pirola, C.J., Forrester, J.S., and Fagin, J.A.**, Angiotensin II regulates tenascin gene expression in vascular smooth muscle cells, *J. Biol. Chem.*, 267, 23910, 1992.
85. **End, P., Panayotou, G., Entwistle, A., Watefield, M.D., and Chiquet, M.**, Tenascin: a modulator of cell growth, *Eur. J. Biochem.*, 209, 1041, 1992.
86. **Clark, A.J.L., Balla, T., Jones, M.R., and Catt, K.J.**, Stimulation of early gene expression by angiotensin II in bovine adrenal glomerulosa cells: roles of calcium and protein kinase C, *Mol. Endocrinol.*, 6, 1889, 1992.
87. **Viard, I., Jaillard, C., Ouali, R., and Saez, J.M.**, Angiotensin II-induced expression of proto-oncogene (c-*fos*, *jun*-B and c-*jun*) messenger RANA in bovine adrenocortical fasciculata cells (BAC) is mediated by AT-1 receptors, *FEBS Lett.*, 313, 43, 1992.
88. **González-Espinosa, C. and García-Sáinz, J.A.**, Angiotensin II and active phorbol esters induce proto-oncogene expression in isolated rat hepatocytes, *Biochim. Biophys. Acta*, 1136, 309, 1992.
89. **Naftilan, A.J., Gilliland, G.K., Eldridge, C.S., and Kraft, A.S.**, Induction of the proto-oncogene c-*jun* by angiotensin II, *Mol. Cell. Biol.*, 10, 5536, 1990.
90. **Takeuchi, K., Nakamura, N., Cook, N.S., Pratt, R.E., and Dzau, V.J.**, Angiotensin II can regulate gene expression by the AP-1 binding sequence via a protein kinase C-dependent pathway, *Biochem. Biophys. Res. Commun.*, 172, 1189, 1990.
91. **Reiss, K., Capasso, J.M., Huang, H., Meggs, L.G., Li, P., and Anversa, P.**, ANG II receptors, c-*myc*, and c-*jun* in myocytes after myocardial infarction and ventricular failure, *Am. J. Physiol.*, 264, H760, 1993.
92. **Schulze-Lohoff, E., Köhler, M., Fees, H., Reindl, N., and Sterzel, R.B.**, Divergent effects of arginine vasopressin and angiotensin II on proliferation and expression of the immediate early genes c-*fos*, c-*jun* and *Egr*-1 in cultured rat glomerular mesangial cells, *J. Hypertension*, 11, 127, 1993.
93. **Yanagisawa, M., Kurihara, H., Kimura, S., Tomobe, Y., Kobayashi, M., Mitsui, Y., Yazaki, Y., Goto, K., and Masaki, T.**, A novel potent vasoconstrictor peptide produced by vascular endothelial cells, *Nature*, 332, 411, 1988.
95. **Le Monnier de Gouville, A.-C., Lippton, H.L., Cavero, I., Summer, W.R., and Hyman, A.L.**, Endothelin — a new family of endothelium-derived peptides with widespread biological properties, *Life Sci.*, 45, 1499, 1989.
96. **Simonson, M.S. and Dunn, M.J.**, Cellular signaling by peptides of the endothelin gene family, *FASEB J.*, 4, 2989, 1990.
97. **Rubanyi, G.M. and Botelho, L.H.P.**, Endothelins, *FASEB J.*, 5, 2713, 1991.
98. **Luscher, T.F., Oemar, B.S., Boulanger, C.M., and Hahn, A.W.A.**, Molecular and cellular biology of endothelin and its receptors, *J. Hypertension*, 11, 7 and 121, 1993.
99. **Battistini, B., Chailler, P., Dorleans-Juste, P., Briere, N., and Sirois, P.**, Growth regulatory properties of endothelins, *Peptides*, 14, 385, 1993.
100. **Haynes, W.G. and Webb, D.J.**, The endothelin family of peptides — Local hormones with diverse roles in health and disease, *Clin. Sci.*, 84, 485, 1993.
101. **Ehrenreich, H., Anderson, R.W., Fox, C.H., Rieckmann, P., Hoffman, G.S., Travis, W.D., Coligan, J.E., Kehrl, J.H., and Fauci, A.S.**, Endothelins, peptides with potent vasoactive properties, are produced by human macrophages, *J. Exp. Med.*, 172, 1741, 1990.
102. **Yohn, J.J., Morelli, J.G., Walchak, S.J., Rundell, K.B., Norris, D.A., and Zamora, M.R.**, Cultured human keratinocytes synthesize and secrete endothelin-1, *J. Invest. Dermatol.*, 100, 23, 1993.
103. **Suzuki, T., Kumazaki, T., and Mitsui, Y.**, Endothelin-1 is produced and secreted by neonatal rat cardiac myocytes *in vitro*, *Biochem. Biophys. Res. Commun.*, 191, 823, 1993.

104. **Kusuhara, M., Yamaguchi, K., Ohnishi, A., Abe, K., Kimura, S., Oono, H., Hori, S., and Nakamura, Y.,** Endothelin potentiates growth-stimulated DNA synthesis in Swiss 3T3 cells, *Jpn. J. Cancer Res.,* 80, 302, 1989.
105. **Brown, K.D. and Littlewood, C.J.,** Endothelin stimulates DNA synthesis in Swiss 3T3 cells. Synergy with polypeptide growth factors, *Biochem. J.,* 263, 977, 1989.
106. **Sakamoto, H., Sasaki, S., Hirata, Y., Imai, T., Ando, K., Ida, T., Sakurai, T., Yanagisawa, M., Masaki, T., and Marumo, F.,** Production of endothelin-1 by rat cultured mesangial cells, *Biochem. Biophys. Res. Commun.,* 169, 462, 1990.
107. **Takagi, Y., Fukase, M., Takata, S., Yoshimi, H., Tokunaga, O., and Fujita, T.,** Autocrine effect of endothelin on DNA synthesis in human vascular endothelial cells, *Biochem. Biophys. Res. Commun.,* 168, 537, 1990.
108. **Yamagishi, S., Hsu, C.C., Kobayashi, K., and Yamamoto, H.,** Endothelin-1 mediates endothelial cell-dependent proliferation of vascular pericytes, *Biochem. Biophys. Res. Commun.,* 191, 840, 1993.
109. **Bath, P.M.W., Mayston, S.A., and Martin, J.F.,** Endothelin and PDGF do not stimulate peripheral blood monocyte chemotaxis, adhesion to endothelium, and superoxide production, *Exp. Cell Res.,* 187, 339, 1990.
110. **Nadler, S.P., Zimpelmann, J.A., and Hebert, R.L.,** Endothelin inhibits vasopressin-stimulated water permeability in rat terminal inner medullary collecting duct, *J. Clin. Invest.,* 90, 1458, 1992.
111. **Ghandhi, C.R., Stephenson, K., and Olson, M.S.,** Endothelin, a potent peptide agonist in the liver, *J. Biol. Chem.,* 265, 17432, 1990.
112. **Yada, Y., Higuchi, K., and Imokawa, G.,** Effects of endothelins on signal transduction and proliferation in human melanocytes, *J. Biol. Chem.,* 266, 18352, 1991.
113. **Imokawa, G., Yada, Y., and Miyagishi, M.,** Endothelins secreted from human keratinocytes are intrinsic mitogens for human melanocytes, *J. Biol. Chem.,* 267, 24675, 1992.
114. **Hama, H., Sakurai, T., Kasuya, Y., Fukiji, M., Masaki, T., and Goto, K.,** Action of endothelin-1 on rat astrocytes through the ET_B receptor, *Biochem. Biophys. Res. Commun.,* 186, 355, 1992.
115. **Shinohara, O., Murata, Y.I., and Shimizu, M.,** Endothelin-1 suppression of rat adipocyte precursor cell differentiation in serum-free culture, *Endocrinology,* 130, 2031, 1992.
116. **Moretto, M., López, F.J., and Negro-Vilar, A.,** Endothelin-3 stimulates luteinizing hormone-releasing hormone (LHRH) secretion from LHRH neurons by a prostaglandin-dependent mechanism, *Endocrinology,* 132, 789, 1993.
117. **Kanyicska, B., Burris, T.P., and Freeman, M.E.,** Endothelin-3 inhibits prolactin and stimulates LH, FSH and TSH secretion from pituitary cell culture, *Biochem. Biophys. Res. Commun.,* 174, 338, 1991.
118. **Ergul, A., Glassberg, M.K., Majercik, M.H., and Puett, D.,** Endothelin-1 promotes steroidogenesis and stimulates proto-oncogene expression in transformed murine Leydig cells, *Endocrinology,* 132, 598, 1993.
119. **Cozza, E.N. and Gomez-Sanchez, C.E.,** Effects of endothelin-1 on its receptor concentration and thymidine incorporation in calf adrenal zona glomerulosa cells: a comparative study with phorbol esters, *Endocrinology,* 127, 549, 1990.
120. **Rosolowski, L.J. and Campbell, W.B.,** Endothelin enhances adrenocorticotropin-stimulated aldosterone release from cultured bovine adrenal cells, *Endocrinology,* 126, 1860, 1990.
121. **Conte, D., Questino, P., Fillo, S., Nordio, M., Isidori, A., and Romanelli, F.,** Endothelin stimulates testosterone secretion by rat Leydig cells, *J. Endocrinol.,* 136, 121, 1993.
122. **Fujii, Y., Moreira, J.E., Orlando, C., Maggi, M., Aurbach, G.D., Brandi, M.L., and Sakaguchi, K.,** Endothelin as an autocrine factor in the regulation of parathyroid cells, *Proc. Natl. Acad. Sci. U.S.A.,* 88, 4235, 1991.
123. **Shinmi, O., Kimura, S., Sawamura, T., Sugita, Y., Yoshizawa, T., Uchiyama, Y., Tanagisawa, M., Goto, K., Masaki, T., and Kanazawa, I.,** Endothelin-3 is a novel neuropeptide: isolation and sequence determination of endothelin-1 and endothelin-3 in porcine brain, *Biochem. Biophys. Res. Commun.,* 164, 587, 1989.
124. **Matsumoto, H., Suzuki, N., Onda, H., and Fujino, M.,** Abundance of endothelin-3 in rat intestine, pituitary gland and brain, *Biochem. Biophys. Res. Commun.,* 164, 74, 1989.
125. **Samson, W.K., Skala, K.D., Alexander, B.D., and Huang, F.-L.S.,** Pituitary site of action of endothelin: selective inhibition of prolactin release *in vitro, Biochem. Biophys. Res. Commun.,* 169, 737, 1990.
126. **Kanyicska, B., Burris, T.P., and Freeman, M.E.,** Endothelin-3 inhibits prolactin and stimulates LH, FSH and TSH secretion from pituitary cell culture, *Biochem. Biophys. Res. Commun.,* 174, 338, 1991.

127. **Samson, W.K., Skala, K.D., Alexander, B.D., Huang, F.-L.S., and Gomez-Sanchez, C.,** A prolactin release inhibiting activity isolated from neurointermediate lobe extracts is an endothelin-like peptide, *Regul. Peptides,* 39, 103, 1992.
128. **Samson, W.K. and Skala, K.D.,** Comparison of the pituitary effects of the mammalian endothelins: vasoactive intestinal contractor (endothelin-beta, rat endothelin-2) is a potent inhibitor of prolactin secretion, *Endocrinology,* 130, 2964, 1992.
129. **Matsumoto, H., Suzuki, N., Shiota, K., Inoue, K., Tsuda, M., and Fujino, M.,** Insulin-like growth factor-I stimulates endothelin-3 secretion from rat pituitary cells in primary culture, *Biochem. Biophys. Res. Commun.,* 172, 661, 1990.
130. **Yoshizawa, T., Shinmi, O., Giaid, A., Yanagisawa, M., Gibson, S.J., Kimura, S., Uchiyama, Y., Polak, J.M., Masaki, T., and Kanazawa, I.,** Endothelin: a novel peptide in the posterior pituitary system, *Science,* 247, 462, 1990.
131. **Hirata, Y., Matsunaga, T., Ando, K., Furukawa, T., Tsukagoshi, H., and Marumo, F.,** Presence of endothelin-1-like immunoreactivity in human cerebrospinal fluid, *Biochem. Biophys. Res. Commun.,* 166, 1274, 1990.
132. **Onda, H., Ohkubo, S., Ogi, K., Kowaka, T., Kimura, C., Matsumoto, H., Suzuki, N., and Fujino, M.,** One of the endothelin gene family, endothelin-3 gene, is expressed in the placenta, *FEBS Lett.,* 261, 327, 1990.
133. **Nakamura, T., Kasai, K., Konuma, S., Emoto, T., Banba, N., Ishikawa, M., and Shimoda, S.,** Immunoreactive endothelin concentrations in maternal and fetal blood, *Life Sci.,* 46, 1045, 1990.
134. **Itoh, Y., Yanagisawa, M., Ohkubo, S., Kimura, C., Kosaka, T., Inoue, A., Ishida, N., Mitsui, Y., Onda, H., Fujino, M., and Masaki, T.,** Cloning and sequence analysis of cDNA encoding the precursor of a human endothelium-derived vasoconstrictor peptide, endothelin: identity of human and porcine endothelin, *FEBS Lett.,* 231, 440, 1988.
135. **Inoue, A., Yanagisawa, M., Takuwa, Y., Mitsui, Y., Kobayashi, M., and Masaki, T.,** The human preproendothelins-1 gene. Complete nucleotide sequence and regulation of expression, *J. Biol. Chem.,* 264, 14954, 1989.
136. **Takaoka, M., Hukumori, Y., Shiragami, K., Ikegawa, R., Matsumura, Y., and Morimoto, S.,** Proteolytic processing of porcine big endothelin-1 catalyzed by cathepsin-D, *Biochem. Biophys. Res. Commun.,* 173, 1218, 1990.
137. **Inoue, A., Yanagisawa, M., Kimura, S., Kasuya, Y., Miyauchi, T., Goto, K., and Masaki, T.,** The human endothelin family: three structurally and pharmacologically distinct isopeptides predicted by three separate genes, *Proc. Natl. Acad. Sci. U.S.A.,* 86, 2863, 1989.
138. **Saida, K., Mitsui, Y., and Ishida, N.,** A novel peptide, vasoactive intestinal vasoconstrictor, of a new (endothelin) peptide family, *J. Biol. Chem.,* 264, 14613, 1989.
139. **Lee, M.-E., Dhadly, M.S., Temizer, D.H., Clifford, J.A., Yoshizumi, M., and Quertermous, T.,** Regulation of endothelin-1 gene expression by Fos and Jun, *J. Biol. Chem.,* 266, 19034, 1991.
140. **Oliver, F.J., Delarubia, G., Feener, E.P., Lee, M.E., Loeken, M.R., Shiba, T., Quertermous, T., and King, G.L.,** Stimulation of endothelin-1 gene expression by insulin in endothelial cells, *J. Biol. Chem.,* 266, 23251, 1991.
141. **Saijonmää, O., Nyman, T., and Fyhrquist, F.,** Endothelin-1 stimulates its own synthesis in human endothelial cells, *Biochem. Biophys. Res. Commun.,* 188, 286, 1992.
142. **Yamauchi, T., Ohanaka, K., Takayanagi, R., Umeda, F., and Nawata, H.,** Enhanced secretion of endothelin-1 by elevated glucose levels from cultured bovine aortic endothelial cells, *FEBS Lett.,* 267, 16, 1990.
143. **Yoshizumi, M., Kurihara, H., Morita, T., Yamashita, T., Ohnashi, Y., Sugiyama, T., Takaku, F., Yanagisawa, M., Masaki, T., and Yazaki, Y.,** Interleukin-1 increases the production of endothelin-1 by cultured endothelial cells, *Biochem. Biophys. Res. Commun.,* 166, 324, 1990.
144. **Tokito, F., Suzuki, N., Hosoya, M., Matsumoto, H., Ohkubo, S., and Fujino, M.,** Epidermal growth factor (EGF) decreased endothelin-2 (ET-2) production in human renal adenocarcinoma cells, *FEBS Lett.,* 295, 17, 1991.
145. **Schrey, M.P., Patel, K.V., and Tezapsidis, N.,** Bombesin and glucocorticoids stimulate human breast cancer cells to produce endothelin, a paracrine mitogen for breast stromal cells, *Cancer Res.,* 52, 1786, 1992.
146. **Hu, R.M., Levin, E.R., Pdram, A., and Frank, H.J.L.,** Atrial natriuretic peptide inhibits the production and secretion of endothelin from cultured endothelial cells — Mediation through the C-receptor, *J. Biol. Chem.,* 167, 17384, 1992.

147. **Shiba, R., Sakurai, T., Yamada, G., Morimoto, H., Saito, A., Masaki, T., and Goto, K.,** Cloning and expression of rat preproendothelin-3 cRNA, *Biochem. Biophys. Res. Commun.,* 186, 588, 1992.
148. **Cozza, E.N., Gomez-Sanchez, C.E., Foecking, M.F., and Chiou, S.,** Endothelin binding to cultured calf adrenal zona glomerulosa cells and stimulation of aldosterone secretion, *J. Clin. Invest.,* 84, 1032, 1989.
149. **Wada, K., Tabuchi, H., Ohba, R., Satoh, M., Tachibana, Y., Akiyama, N., Hiraoika, O., Asakura, A., Miyamoto, C., and Furuichi, Y.,** Purification of an endothelin receptor from human placenta, *Biochem. Biophys. Res. Commun.,* 167, 251, 1990.
150. **Williams, D.L., Jr., Jones, K.L., Colton, C.D., and Nutt, R.F.,** Identification of high affinity endothelin-1 receptor subtypes in human tissues, *Biochem. Biophys. Res. Commun.,* 180, 475, 1991.
151. **Sakurai, T., Yanagisawa, M., Takuwa, Y., Miyazaki, H., Kimura, S., Goto, K., and Masaki, T.,** Cloning of a cDNA encoding a non-isopeptide-selective subtype of the endothelin receptor, *Nature,* 348, 732, 1990.
152. **Hosoda, K., Nakao, K., Hiroshirai, S., Suga, S., Ogawa, Y., Mukoyama, M., Shirakami, G., Saito, Y., Nakanishi, S., and Imura, H.,** Cloning and expression of human endothelin-1 receptor cDNA, *FEBS Lett.,* 287, 23, 1991.
153. **Elshourbagy, N.A., Korman, D.R., Wu, H.-L., Sylvester, D.R., Lee, J.A., Nuthalaganti, P., Bergsma, D.J., Kumar, C.S., and Nambi, P.,** Molecular characterization and regulation of the human endothelin receptors, *J. Biol. Chem.,* 268, 3873, 1993.
154. **Adachi, M., Yang, Y.-Y., Furuichi, Y., and Miyamoto, C.,** Cloning and characterization of cDNA encoding human A-type endothelin receptor, *Biochem. Biophys. Res. Commun.,* 180, 1265, 1991.
155. **Samson, W.K.,** The endothelin-A receptor subtype transduces the effects of the endothelins in the anterior pituitary gland, *Biochem. Biophys. Res. Commun.,* 187, 590, 1992.
156. **Shetty, S.S., Okada, T., Webb, R.L., Delgrande, D., and Lappe, R.W.,** Functionally distinct endothelin-B receptors in vascular endothelium and smooth muscle, *Biochem. Biophys. Res. Commun.,* 191, 459, 1993.
157. **Imai, T., Hirata, Y., Eguchi, S., Kanno, K., Ohta, K., Emori, T., Sakamoto, A., Yanagisawa, M., Masaki, T., and Marumo, F.,** Concomitant expression of receptor subtype and isopeptide of endothelin by human adrenal gland, *Biochem. Biophys. Res. Commun.,* 182, 1115, 1992.
158. **Ohnishi-Suzaki, A., Yamaguchi, K., Kusuhara, M., Adachi, I., Abe, K., and Kimura, S.,** Comparison of biological activities of endothelin-1, -2, and -3 in murine and human fibroblast cell lines, *Biochem. Biophys. Res. Commun.,* 166, 608, 1990.
159. **Winkles, J.A., Alberts, G.F., Brogi, E., and Libby, P.,** Endothelin-1 and endothelin receptor mRNA expression in normal and atherosclerotic human arteries, *Biochem. Biophys. Res. Commun.,* 191, 1081, 1993.
160. **Sokolovsky, M., Ambar, I., and Galron, R.,** A novel subtype of endothelin receptors, *J. Biol. Chem.,* 267, 20551, 1992.
161. **Sakurai, T., Morimoto, H., Kasuya, Y., Takuwa, Y., Nakauchi, H., Masaki, T., and Goto, K.,** Level of ETB receptor messenger RNA is down-regulated by endothelin through decreasing the intracellular stability of messenger RNA molecules, *Biochem. Biophys. Res. Commun.,* 186, 342, 1992.
162. **Roubert, P., Gillard, V., Plas, P., Guillon, J.-M., Chabrier, P.E., and Braquet, P.,** Angiotensin II and phorbol-esters potently down-regulate endothelin (ET-1) binding sites in vascular smooth muscle cells, *Biochem. Biophys. Res. Commun.,* 164, 809, 1989.
163. **Resink, T.J., Scott-Burden, T., Weber, E., and Bühler, F.R.,** Phorbol ester promotes a sustained down-regulation of endothelin receptors and cellular responses to endothelin in human vascular smooth muscle cells, *Biochem. Biophys. Res. Commun.,* 166, 1213, 1990.
164. **Nambi, P., Pullen, M., Wu, H.L., Nuthulaganti, P., Elshourbagy, N., and Kumar, C.,** Dexamethasone down-regulates the expression of endothelin receptors in vascular smooth muscle cells, *J. Biol. Chem.,* 267, 19555, 1992.
165. **Simonson, M.S., Wann, S., Mené, P., Dubyak, G.R., Kester, M., Nakazato, Y., Sedor, J.R., and Dunn, M.J.,** Endothelin stimulates phospholipase C, Na^+/H^+ exchange, c-*fos* expression, and mitogenesis in rat mesangial cells, *J. Clin. Invest.,* 83, 708, 1989.
166. **Muldoon, L.L., Rodland, K.D., Forsythe, M.L., and Magun, B.E.,** Stimulation of phosphatidylinositol hydrolysis, diacylglycerol release, and gene expression in response to endothelin, a potent new agonist for fibroblasts and smooth muscle cells, *J. Biol. Chem.,* 264, 8529, 1989.

167. **Takuwa, Y., Kasuya, Y., Takuwa, N., Kudo, M., Yanagisawa, M., Goto, K., Masaki, T., and Yamashita, K.,** Endothelin receptor is coupled to phospholipase-C via a pertussis toxin insensitive guanine nucleotide binding regulatory protein in vascular smooth muscle cells, *J. Clin. Invest.,* 85, 653, 1990.
168. **Griendling, K.K., Tsuda, T., and Alexander, R.W.,** Endothelin stimulates diacylglycerol accumulation and activates protein kinase C in cultured vascular smooth muscle cells, *J. Biol. Chem.,* 264, 8237, 1989.
169. **Kester, M., Simonson, M.S., McDermott, R.G., Baldi, E., and Dunn, M.J.,** Endothelin stimulates phosphatidic acid formation in cultured rat mesangial cells: role of a protein kinase C-regulated phospholipase D, *J. Cell. Physiol.,* 150, 578, 1992.
170. **MacNulty, E.E., Plevin, R., and Wakelam, M.J.O.,** Stimulation of the hydrolysis of phosphatidylinositol 4,5-bisphosphate and phosphatidylcholine by endothelin, a complete mitogen for Rat-1 fibroblasts, *Biochem. J.,* 272, 761, 1990.
171. **Prasad, M.R.,** Endothelin stimulates degradation of phospholipids in isolated rat hearts, *Biochem. Biophys. Res. Commun.,* 174, 952, 1991.
172. **Vigne, P., Ladoux, A., and Frelin, C.,** Endothelins activate Na^+/H^+ exchange in brain capillary endothelial cells via a high affinity endothelin-3 receptors that is not coupled to phospholipase-C, *J. Biol. Chem.,* 266, 5925, 1991.
173. **Meyer-Lehnert, H., Wanning, C., Predel, H.-G., Bäcker, A., Stelkens, H., and Kramer, H.J.,** Effects of endothelin on sodium transport mechanisms: potential role in cellular Ca^{2+} mobilization, *Biochem. Biophys. Res. Commun.,* 163, 458, 1989.
174. **Bialecki, R.A., Izzo, N.J., Jr., and Colucci, W.S.,** Endothelin-1 increases intracellular calcium mobilization but not calcium uptake in rabbit vascular smooth muscle cells, *Biochem. Biophys. Res. Commun.,* 164, 474, 1989.
175. **Little, P.J., Neylon, C.B., Trachuk, V.A. and Bobik, A.,** Endothelin-1 and endothelin-3 stimulate calcium mobilization by different mechanisms in vascular smooth muscle, *Biochem. Biophys. Res. Commun.,* 183, 694, 1992.
176. **Korbmacher, C., Helbig, H., Haller, H., Erickson-Lamy, K.A., and Wiederholt, M.,** Endothelin depolarizes membrane voltage and increases intracellular calcium concentration in human ciliary muscle cells, *Biochem. Biophys. Res. Commun.,* 164, 1031, 1989.
177. **Maher, E., Bardequez, A., Gardner, J.P., Goldsmith, L., Weiss, G., Mascarina, M., and Aviv, A.,** Endothelin- and oxytocin-induced calcium signaling in cultured human myometrial cells, *J. Clin. Invest.,* 87, 1251, 1991.
178. **Ohnishi, A., Yamaguchi, K., Kusuhara, M., Abe, K., and Kimura, S.,** Mobilization of intracellular calcium by endothelin in Swiss 3T3 cells, *Biochem. Biophys. Res. Commun.,* 161, 489, 1989.
179. **Takuwa, Y., Ohue, Y., Takuwa, N., and Yamashita, K.,** Endothelin-1 activates phospholipase C and mobilize Ca^{2+} from extra- and intracellular pools in osteoblastic cells, *Am. J. Physiol.,* 257, E803, 1989.
180. **Resink, T.J., Scott-Burden, T., and Buhler, F.R.,** Activation of phospholipase A_2 by endothelin in cultured vascular smooth muscle cells, *Biochem. Biophys. Res. Commun.,* 158, 279, 1989.
181. **Nagase, T., Fukuchi, Y., Jo, C., Teramoto, S., Uejima, Y., Ishida, K., Shimizu, T., and Orimo, H.,** Endothelin-1 stimulates arachidonate 15-lipoxigenase activity and oxygen radical formation in the rat distal lung, *Biochem. Biophys. Res. Commun.,* 168, 485, 1990.
182. **Delarue, C., Delton, I., Fiorini, F., Homo-Delarche, F., Fasolo, A., Braquet, P., and Vaudry, H.,** Endothelin stimulates steroid secretion by frog adrenal gland *in vitro:* evidence for the involvement of prostaglandins and extracellular calcium in the mechanism of action of endothelin, *Endocrinology,* 127, 2001, 1990.
183. **Ladoux, A. and Frelin, C.,** Endothelins inhibit adenylate cyclase in brain capillary endothelial cells, *Biochem. Biophys. Res. Commun.,* 180, 169, 1991.
184. **Eguchi, S., Hirata, Y., Imai, T., and Marumo, F.,** Endothelin receptor subtypes are coupled to adenylate cyclase via different guanyl nucleotide-binding proteins in vasculature, *Endocrinology,* 132, 524, 1993.
185. **Wang, Y.Z., Simonson, M.S., Pouysségur, J., and Dunn, M.J.,** Endothelin rapidly stimulates mitogen-activated protein kinase activity in rat mesangial cells, *Biochem. J.,* 287, 589, 1992.
186. **Bogoyevitch, M.A., Glennon, P.E., and Sugden, P.H.,** Endothelin-1, phorbol esters and phenylepinephrine stimulate MAP kinase activities in ventricular cardiomyocytes, *FEBS Lett.,* 317, 271, 1993.

187. **Simonson, M.S. and Herman, W.H.,** Protein kinase C and protein tyrosine kinase activity cotribute to mitogenic signaling by endothelin-1. Cross-talk between G protein-coupled receptors and pp60^{c-src}, *J. Biol. Chem.,* 268, 9347, 1993.
188. **Wilkes, L.C., Patel, V., Purkiss, J.R., and Boarder, M.R.,** Endothelin-1 stimulated phospholipase D in A10 vascular smooth muscle derived cells is dependent on tyrosine kinase —Evidence for involvement in stimulation of mitogenesis, *FEBS Lett.,* 322, 147, 1993.
189. **Nambi, P., Wu, H.-L., Woessner, R.D., and Mattern, M.R.,** Inhibition of endothelin-mediated topoisomerase I activation by pertussis toxin, *FEBS Lett.,* 276, 17, 1990.
190. **Wang, D.-L., Chen, J.-J., Shin, N.-L., Kao, Y.-C., Hsu, K.-H., Huang, W.-Y., and Liew, C.-C.,** Endothelin stimulates cardiac α-myosin and β-myosin heavy chain gene expression, *Biochem. Biophys. Res. Commun.,* 183, 1260, 1992.
191. **He, C.J., Nguyen, G., Li, X.M., Peraldi, M.N., Adida, C., Rondeau, E., and Sraer, J.D.,** Transcriptional activation of the urokinase receptor gene by endothelin 1, *Biochem. Biophys. Res. Commun.,* 186, 1631, 1992.
192. **Ladenheim, R.G., Lacroix, I., Foignant-Chaverot, N., Strosberg, A.D., and Couraud, P.O.,** Endothelins stimulate c-*fos* and nerve growth factor expression in astrocytes and astrocytoma, *J. Neurochem.,* 60, 260, 1993.
193. **Komuro, I., Kurihara, H., Sugiyama, T., Yoshizumi, M., Takaku, F., and Yazaki, Y.,** Endothelin stimulates c-*fos* and c-*myc* expression and proliferation of vascular smooth muscle cells, *FEBS Lett.,* 238, 249, 1988.
194. **Miyakawa, M., Tsuchima, T., Isozaki, O., Demura, H., Shizume, K., and Arai, M.,** Endothelin-1 stimulates c-*fos* mRNA expression and acts as a modulator on cell proliferation of rat FRTL5 thyroid cells, *Biochem. Biophys. Res. Commun.,* 184, 231, 1992.
195. **Yin, J., Lee, J.A., and Howels, R.D.,** Stimulation of c-*fos* and c-*jun* gene expression and down-regulation of proenkephalin gene expression in CG glioma cells by endothelin 1, *Mol. Brain Res.,* 14, 213, 1992.
196. **Pribnow, D., Muldoon, L.L., Fajardo, M., Theodor, L., Chen, L.Y.S., and Magun, B.E.,** Endothelin induces transcription of *fos/jun* family genes — A prominent role for calcium ion, *Mol. Endocrinol.,* 6, 1003, 1992.
197. **Simonson, M.S., Jones, J.M., and Dunn, M.J.,** Differential regulation of *fos* and *jun* gene expression and AP-1 *cis*-element activity by endothelin isopeptides. Possible implications for mitogenic signaling by endothelin, *J. Biol. Chem.,* 267, 8643, 1992.
198. **Suzuki, N., Matsumoto, H., Kitada, C., Kimura, S., and Fujino, M.,** Production of endothelin-1 and big-endothelin-1 by human tumor cells with epithelial-like morphology, *J. Biochem.,* 106, 736, 1989.
199. **Kusuhara, M., Yamaguchi, K., Nagasaki, K., Hayashi, C., Suzaki, A., Hori, S., Handa, S., Nakamura, Y., and Abe, K.,** Production of endothelin in human cancer cell lines, *Cancer Res.,* 50, 3257, 1990.
200. **Shichiri, M., Hirata, Y., Nakayima, T., Ando, K., Imai, T., Yanagisawa, M., Masaki, T., and Marumo, F.,** Endothelin-1 is an autocrine/paracrine growth factor for human cancer cell lines, *J. Clin. Invest.,* 87, 1867, 1991.
201. **Pekonen, F., Saijonmaa, O., Nyman, T., and Fyhrquist, F.,** Human endometrial adenocarcinoma cells express endothelin-1, *Mol. Cell. Endocrinol.,* 84, 203, 1992.
202. **Ishibashi, M., Fujita, M., Nagai, K., Kako, M., Furue, H., Haku, E., Osamura, Y., and Yamaji, T.,** Production and secretion of endothelin by hepatocellular carcinoma, *J. Clin. Endocrinol. Metab.,* 76, 378, 1993.
203. **Flynn, T.G. and Davies, P.L.,** The biochemistry and molecular biology of atrial natriuretic factor, *Biochem. J.,* 232, 313, 1985.
204. **de Bold, A.J.,** Atrial natriuretic factor: a hormone produced by the heart, *Science,* 230, 767, 1985.
205. **Inagami, T.,** Atrial natriuretic factor, *J. Biol. Chem.,* 264, 3043, 1989.
206. **Sagnella, G.A. and MacGregor, G.A.,** Atrial natriuretic peptides, *Q. J. Med.,* 77, 1001, 1990.
207. **Rosenzweig, A. and Seidman, C.E.,** Atrial natriuretic factor and related peptide hormones, *Annu. Rev. Biochem.,* 60, 229, 1991.
208. **Samson, W.K.,** Natriuretic peptides — A family of hormones, *Trends Endocrinol. Metab.,* 3, 86, 1992.
209. **Koller, K.J. and Goeddel, D.V.,** Molecular biology of the natriuretic peptides and their receptors, *Circulation,* 86, 1081, 1992.
210. **Lang, C.C., Choy, A.-M., and Struthers, A.D.,** Atrial and brain natriuretic peptides: a dual natriuretic peptide system potentially involved in circulatory homeostasis, *Clin. Sci.,* 83, 519, 1992.

211. **Christensen, G.,** Release of atrial natriuretic factor, *Scand. J. Clin. Lab. Invest.,* 53, 91, 1993.
212. **Vollmar, A.M.,** Atrial natriuretic peptide in peripheral organs other than the heart, *Klin. Wochenschr.,* 68, 699, 1990.
213. **Antoni, F.A., Hunter, E.F.M., Lowry, P.J., Noble, J.M., and Seckl, J.R.,** Atriopeptin: an endogenous corticotropin-release inhibiting hormone, *Endocrinology,* 130, 1753, 1992.
214. **Nair, B.G., Steinke, L., Yu, Y.M., Rashed, H.M., Seyer, J.M., and Patel, T.B.,** Increase in the number of atrial natriuretic hormone receptors in regenerating rat liver, *J. Biol. Chem.,* 266, 267, 1991.
215. **Johnson, A., Lermioglu, F., Garg, U.C., Morgan-Boyd, R., and Hassid, A.,** A novel biological effect of atrial natriuretic hormone: inhibition of mesangial cell mitogenesis, *Biochem. Biophys. Res. Commun.,* 152, 893, 1988.
216. **Abell, T.J., Richards, A.M., Ikram, H., Espiner, E.A., and Yandle, T.,** Atrial natriuretic factor inhibits proliferation of vascular smooth muscle cells stimulated by platelet-derived growth factor, *Biochem. Biophys. Res. Commun.,* 160, 1392, 1989.
217. **Pandey, K.N. and Orgebin-Crist, M.-C.,** Atrial natriuretic factor in mammalian testis: immunological detection in spermatozoa, *Biochem. Biophys. Res. Commun.,* 180, 437, 1991.
218. **Vesely, D.L. and Giordano, A.T.,** Atrial natriuretic peptide hormonal system in plants, *Biochem. Biophys. Res. Commun.,* 179, 695, 1991.
219. **Thibault, G., Lazure, C., Chrétien, M., and Cantin, M.,** Molecular heterogeneity of pro-atrial natriuretic factor, *J. Biol. Chem.,* 264, 18796, 1989.
220. **Seidman, C.E., Bloch, K.D., Klein, K.A., Smith, J.A., and Seidman, J.G.,** Nucleotide sequences of the human and mouse atrial natriuretic factor genes, *Science,* 226, 1206, 1984.
221. **Fukuda, Y., Hirata, Y., Taketani, S., Kojima, T., Oikawa, S., Nakazato, H., and Kobayashi, Y.,** Endothelin stimulates accumulations of cellular atrial natriuretic peptide and its messenger RNA in rat cardiocytes, *Biochem. Biophys. Res. Commun.,* 164, 1431, 1989.
222. **Schiebinger, R.J. and Gomez-Sanchez, C.E.,** Endothelin: a potent stimulus of atrial natriuretic peptide secretion by superfused rat atria and its dependency on calcium, *Endocrinology,* 127, 119, 1990.
223. **Sudoh, T., Kangawa, K., Minamino, N., and Matsuo, H.,** A new natriuretic peptide in porcine brain, *Nature,* 332, 78, 1988.
224. **Sudoh, T., Minamino, N., Kangawa, K., and Matsuo, H.,** C-type natriuretic peptide (CNP): a new member of natriuretic peptide family identified in porcine brain, *Biochem. Biophys. Res. Commun.,* 168, 863, 1990.
225. **Kambayashi, Y., Nakao, K., Mukoyama, M., Saito, Y., Ogawa, Y., Shiono, S., Inouye, K., Yoshida, N., and Imura, H.,** Isolation and sequence determination of human brain natriuretic peptide in human atrium, *FEBS Lett.,* 259, 341, 1990.
226. **Tawaragi, Y., Fuchimara, K., Tanaka, S., Minamino, N., Kangawa, K., and Matsuo, H.,** Gene precursor structures of human C-type natriuretic peptide, *Biochem. Biophys. Res. Commun.,* 175, 645, 1991.
227. **Mukoyama, M., Nakao, K., Hosoda, K., Suga, S., Saito, Y., Ogawa, Y., Shirakami, G., Jougasaki, M., Obata, K., Yasue, H., Kambayashi, Y., Inoye, K., and Imura, H.,** Brain natriuretic peptide as a novel cardiac hormone in humans: evidence for an exquisite dual natriuretic peptide system, ANP and BNP, *J. Clin. Invest.,* 87, 1402, 1991.
228. **Chinkers, M. and Garbers, D.L.,** Signal transduction by guanylyl cyclases, *Annu. Rev. Biochem.,* 60, 553, 1991.
229. **Wong, S.K.F. and Garbers, D.L.,** Receptor guanylyl cyclases, *J. Clin. Invest.,* 90, 299, 1992.
230. **Garbers, D.L.,** Guanylyl cyclase receptors and their endocrine, paracrine, and autocrine ligands, *Cell,* 71, 1, 1992.
231. **Nakao, K., Ogawa, Y., Suga, S., and Imura, H.,** Molecular biology and biochemistry of the natriuretic peptide system. 2. Natriuretic peptide receptors, *J. Hypertension,* 10, 1111, 1992.
232. **Chinkers, M., Garbers, D.L., Chang, M.-S., Lowe, D.G., Chin, H., Goeddel, D.V., and Schulz, S.,** A membrane form of guanylate cyclase is an atrial natriuretic peptide receptor, *Nature,* 338, 78, 1989.
233. **Fuller, F., Porter, J.G., Arfsten, A.E., Miller, J., Schilling, J.W., Scarborough, R.M., Lewicki, J.A., and Shenk, D.B.,** Atrial natriuretic peptide clearance receptor, *J. Biol. Chem.,* 263, 9395, 1988.
234. **Chang, M.S., Lowe, D.G., Lewis, M., Hellmiss, R., Chen, E., and Goeddel, D.V.,** Differential activation by atrial and brain natriuretic peptides of two different receptor guanylate cyclases, *Nature,* 341, 68, 1989.

235. **Watt, V.M. and Yip, C.C.,** HeLa cells contain the atrial natriuretic peptide receptor with guanylate cyclase activity, *Biochem. Biophys. Res. Commun.,* 164, 671, 1989.
236. **Lowe, D.G., Camerato, T.R., and Goeddel, D.V.,** cDNA sequence of the human atrial natriuretic peptide clearance receptor, *Nucleic Acids Res.,* 18, 3412, 1990.
237. **Chinkers, M. and Garbers, D.L.,** The protein kinase domain of the ANP receptor is required for signalling, *Science,* 245, 1392, 1989.
238. **Jewett, J.R.S., Koller, K.J., Goeddel, D.V., and Lowe, D.G.,** Hormonal induction of low affinity receptor guanylyl cyclase, *EMBO J.,* 12, 769, 1993.
239. **Katafuchi, T., Mizuno, T., Hagiwara, H., Itakura, M., Ito, T., and Hirose, S.,** Modulation by NaCl of atrial natriuretic peptide receptor levels and cyclic GMP responsiveness to atrial natriuretic peptide of cultured vascular endothelial cells, *J. Biol.,* 267, 7624, 1992.
240. **Lofton, C.E., Newman, W.H., and Currie, M.G.,** Atrial natriuretic peptide regulation of endothelial cell permeability is mediated by cGMP, *Biochem. Biophy. Res. Commun.,* 172, 793, 1990.
241. **McArdle, C.A., Poch, A., and Käppler, K.,** Cyclic guanosine monophosphate production in the pituitary: stimulation by C-type natriuretic peptide and inhibition by gonadotropin-releasing hormone in αT3-1 cells, *Endocrinology,* 132, 2065, 1993.
242. **Duda, T., Garaczniak, R.M., and Sharma, R.K.,** Site-directed mutational analysis of a membrane guanylate cyclase cDNA reveals the atrial natriuretic factor signaling site, *Proc. Natl. Acad. Sci. U.S.A.,* 88, 7882, 1991.
243. **Jaiswal, R.K.,** Endothelin inhibits the atrial natriuretic factor stimulated cGMP production by activating the protein kinase-C in rat aortic smooth muscle cells, *Biochem. Biophys. Res. Commun.,* 182, 395, 1992.
244. **Light, D.B., Corbin, J.D., and Stanton, B.A.,** Dual ion-channel regulation by cyclic GMP and cyclic GMP-dependent protein kinase, *Nature,* 344, 336, 1990.
245. **Field, L.J.,** Atrial natriuretic factor-SV40 T antigen transgenes produce tumors and cardiac arrhythmias in mice, *Science,* 239, 1029, 1988.
246. **Utiger, R.D.,** Inappropriate antidiuresis and carcinoma of the lung: detection of arginine vasopressin in tumor extracts by immunoassay, *J. Clin. Endocrinol. Metab.,* 26, 970, 1966.
247. **Bliss, D.P., Jr., Battey, J.F., Linnoila, R.I., Birrer, M.J., Gazdar, A.F., and Johnson, B.E.,** Expression of the atrial natriuretic factor gene in small cell lung cancer tumors and tumor cell lines, *J. Natl. Cancer Inst.,* 82, 305, 1990.
248. **Said, S.I.,** Vasoactive intestinal polypeptide: current status, *Peptides,* 5, 143, 1984.
249. **Miyata, A., Arimura, A., Dahl, R.R., Minamino, N., Uehara, A., Jiang, L., Culler, M.D., and Coy, D.H.,** Isolation of a novel 38 residue hypothalamic polypeptide which stimulates adenylate cyclase in pituitary cells, *Biochem. Biophys. Res. Commun.,* 184, 567, 1989.
250. **Wollina, U., Bonnekoh, B., and Mahrle, G.,** Vasoactive intestinal peptide (VIP) modulates the growth fraction of epithelial skin cells, *Int. J. Oncol.,* 1, 17, 1992.
251. **Wollina, U. and Knopf, B.,** Vasoactive intestinal peptide (VIP) modulates early events of migration in human keratinocytes, *Int. J. Oncol.,* 2, 229, 1993.
252. **Gressens, P., Hill, J.M., Gozes, I., Fridkin, M., and Brenneman, D.E.,** Growth factor function of vasoactive intestinal peptide in whole cultured mouse embryos, *Nature,* 362, 155, 1993.
253. **Weick, R.F. and Stobie, K.M.,** Vasoactive intestinal peptide inhibits the steroid-induced LH surge in the ovariectomized rat, *J. Endocrinol.,* 133, 433, 1992.
254. **Kasper, S., Popescu, R.A., Torsello, A., Vrontakis, M.E., Ikejiani, C., and Friesen, H.G.,** Tissue-specific regulation of vasoactive intestinal peptide messenger ribonucleic acid levels by estrogen in the rat, *Endocrinology,* 130, 1796, 1992.
255. **Sirianni, M.C., Annibale, B., Tagliaferri, F., Fais, S., De Luca, S., Pallone, F., Delle Fave, G., and Aiuti, F.,** Modulation of human natural killer activity by vasoactive intestinal peptide (VIP) family. VIP, glucagon and GHRF specifically inhibit NK activity, *Regul. Peptides,* 38, 79, 1992.
256. **Levin, R.J.,** VIP, vagina, clitoral and periurethral glans — An update on human female genital arousal, *Exp. Clin. Endocrinol.,* 98, 61, 1991.
257. **Bellan, C., Fabre, C., Secchi, J., Marvaldi, J., Pichon, J., and Luis, J.,** Modulation of the expression of the VIP receptor by serum factors on the human melanoma cell line IGF39, *Exp. Cell Res.,* 200, 34, 1992.
258. **Bellan, C., Pic, P., Marvaldi, J., Fantini, J., and Pichon, J.,** Suramin inhibits vasoactive intestinal peptide (VIP) binding and VIP-induced cAMP accumulation into two human cancerous cell lines, *Sec. Messenger Phosphoprot.,* 13, 163, 1991.

259. **Ottaway, C.A.,** Receptors for vasoactive intestinal peptide on murine lymphocytes turn over rapidly, *J. Neuroimmunol.,* 38, 241, 1992.
260. **Kermode, J.C., DeLuca, A.W., Zilberman, A., Valliere, J., and Shreeve, S.M.,** Evidence for the formation of a functional complex between vasoactive intestinal peptide, its receptor, and G_s in lung membranes, *J. Biol. Chem.,* 267, 3382, 1992.
261. **Shreeve, S.M., DeLuca, A.W., Diehl, N.L., and Kermode, J.C.,** Molecular properties of the vasoactive intestinal peptide receptor in aorta and other tissues, *Peptides,* 13, 919, 1992.
262. **Ishihara, T., Shigemoto, R., Mori, K., Takahashi, K., and Nagata, S.,** Functional expression and tissue distribution of a novel receptor for vasoactive intestinal polypeptide, *Neuron,* 8, 811, 1992.
263. **Segura, J.J., Guerrero, J.M., Goberna, R., and Calvo, J.R.,** Stimulatory effect of vasoactive intestinal peptide (VIP) on cycle AMP production in rat peritoneal macrophages, *Regul. Peptides,* 37, 195, 1992.
264. **Gamet, L., Murat, J.-C., Remaury, A., Remesy, C., Valet, P., Paris, H., and Denis-Pouxviel, C.,** Vasoactive intestinal peptide and forskolin regulate proliferation of the HT29 human colon adenocarcinoma cell line, *J. Cell. Physiol.,* 150, 501, 1992.
265. **Zurier, R.B., Kozma, M., Sinnett-Smith, J., and Rozengurt, E.,** Vasoactive intestinal peptide synergistically stimulates DNA synthesis in mouse 3T3 cells: role of cAMP, Ca^{2+} and protein kinase C, *Exp. Cell Res.,* 176, 155, 1988.
266. **Wollina, U., Bonnekoh, B., Klinger, R., Wetzker, R., and Mahrle, G.,** Vasoactive intestinal peptide (VIP) acting as a growth factor for human keratinocytes, *Neuroendocrinol. Lett.,* 14, 21, 1992.
267. **Tatsuno, I., Yada, T., Vigh, S., Hidaka, H., and Arimura, A.,** Pituitary adenylate cyclase activating polypeptide and vasoactive intestinal peptide increase cytosolic free calcium concentration in cultured rat hippocampal neurons, *Endocrinology,* 131, 73, 1992.
268. **Koh, S.-W.M.,** Signal transduction through the vasoactive intestinal peptide receptor stimulates phosphorylation of the tyrosine kinase $pp60^{c\text{-}src}$, *Biochem. Biophys. Res. Commun.,* 174, 452, 1991.
269. **Koh, S.-W.M.,** The $pp60^{c\text{-}src}$ in retinal pigment epithelium and its modulation by vasoactive intestinal peptide, *Cell Biol. Int. Rep.,* 16, 1003, 1992.
270. **D'Orisio, M.S., Fleshman, D.J., Qualman, S.J., and D'Orisio, T.M.,** Vasoactive intestinal peptide: autocrine growth factor in neuroblastoma, *Regul. Peptides,* 37, 213, 1992.
271. **Moody, T.W., Zia, F., Draoui, M., Brenneman, D.E., Fridkin, M., Davidson, A., and Gozes, I.,** A vasoactive intestinal peptide antagonist inhibits non-small cell lung cancer growth, *Proc. Natl. Acad. Sci. U.S.A.,* 90, 4345, 1993.
272. **Rocha e Silva, M.,** *Kinin Hormones,* Charles C Thomas, Springfield, IL, 1970.
273. **Regoli, D. and Barabé, J.,** Pharmacology of bradykinin and related kinins, *Pharmacol. Rev.,* 31, 1, 1980.
274. **Regoli, D., Rhaleb, N.E., Dion, S., and Drapeau, G.,** New selective bradykinin receptor antagonists and bradykinin B_2 receptor characterization, *Trends Pharmacol. Sci.,* 11, 156, 1990.
275. **Hall, J.M.,** Bradykinin receptors — Pharmacological properties and biological roles, *Pharmacol. Ther.,* 56, 131, 1992.
276. **Powell, S.J., Slynn, G., Thomas, C., Hopkins, B., Briggs, I., and Graham, A.,** Human bradykinin B2 receptor: nucleotide sequence analysis and assignment to chromosome 14, *Genomics,* 15, 435, 1993.
277. **Seguin, L., Widdowson, P.S., and Giesen-Crouse, E.,** Existence of three subtypes of bradykinin B_2 receptors in guinea pig, *J. Neurochem.,* 59, 2125, 1992.
278. **Lin, W.-W. and De-Maw, C.,** Regulation of bradykinin-induced phosphoinositide turnover in cultured cerebellar astrocytes: possible role of protein kinase C, *Neurochem. Int.,* 21, 573, 1992.
279. **Sawutz, D.G., Faunce, D.M.,** Characterization of bradykinin B_2 receptors on human IMR-90 lung fibroblasts: stimulation of $^{45}Ca^{2+}$ efflux by D-Phe^7 substituted bradykinin analogues, *Eur. J. Pharmacol.,* 277, 309, 1992.
280. **Lerner, U.H., Brunius, G., Andurén, I., Berggren, P.-O., Juntti-Berggren, L., and Modéer, T.,** Bradykinin induces a B2 receptor-mediated calcium signal linked to prostanoid formation in human gingival fibroblasts *in vitro, Agents Actions,* 37, 44, 1992.
281. **Horwitz, J. and Ricanati, S.,** Bradykinin and phorbol dibutyrate activate phospholipase D in PC12 cells by different mechanisms, *J. Neurochem.,* 59, 1474, 1992.
282. **Parries, G., Hoebel, R., and Racker, E.,** Opposing effects of a *ras* oncogene on growth factor-stimulated phosphoinositide hydrolysis: desensitization to platelet-derived growth factor and enhanced sensitivity to bradykinin, *Proc. Natl. Acad. Sci. U.S.A.,* 84, 2648, 1987.

283. **Downward, J., de Gunzburg, J., Riehl, R., and Weinberg, R.A.,** p21^{ras}-induced responsiveness of phosphatidylinositol turnover to bradykinin is a receptor number effect, *Proc. Natl. Acad. Sci. U.S.A.*, 85, 5774, 1988.
284. **Roberts, R.A. and Gullick, W.J.,** Bradykinin receptor number and sensitivity to ligand stimulation of mitogenesis is increased by expression of a mutant *ras* oncogene, *J. Cell Sci.*, 94, 527, 1989.

INDEX

A

B

C

D

E

F

G

H

I

K

L

M

N

O

P

R

S

T

U

V

W